Fundamental Physics

This book presents the two foundational pillars of modern physics, quantum mechanics and general relativity, in a coherent form aimed primarily at upper-level undergraduate students. It provides a unified picture of the underlying field of mathematical physics and the fruitful interactions between different methods and applications. To help students navigate both topics, the book takes a systematic, pedagogical and uniform approach to the underlying theories and applications. This includes detailed discussions of applications, such as in atomic and particle physics, measurements of space and time, black holes and cosmology.

Fundamental Physics: A Guide to Relativity and Quantum Theory is essential for upper-level undergraduate students looking to develop a cohesive understanding of the ways in which quantum mechanics and general relativity are central not only to the challenges and approaches to quantum theories of gravity but also to gravitational implications on sensitive quantum effects. Worked examples and homework problems are available throughout.

Key Features:

- This class-tested textbook combines two fundamental topics in modern physics: quantum mechanics and general relativity.
- Detailed conceptual discussions of physics insights are combined with careful mathematical constructions.
- Worked examples are available throughout.
- Chapter-end homework problems are also included.

Martin Bojowald is a Professor at The Pennsylvania State University, USA. His current research focuses on quantum gravity and the physical phenomena it implies, possibly leading to crucial tests of its theoretical ingredients.

Textbook Series in Physical Sciences

This textbook series offers pedagogical resources for the physical sciences. It publishes high-quality, high-impact texts to improve understanding of fundamental and cutting edge topics, as well as to facilitate instruction. The authors are encouraged to incorporate numerous problems and worked examples, as well as making available solutions manuals for undergraduate and graduate level course adoptions. The format makes these texts useful as professional self-study and refresher guides as well. Subject areas covered in this series include condensed matter physics, quantum sciences, atomic, molecular, and plasma physics, energy science, nanoscience, spectroscopy, mathematical physics, geophysics, environmental physics and so on, in terms of both theory and experiment.

Understanding Nanomaterials
Malkiat S. Johal

Concise Optics: Concepts, Examples, and Problems
Ajawad I. Haija, M. Z. Numan, W. Larry Freeman

A Mathematica Primer for Physicists
Jim Napolitano

Understanding Nanomaterials, Second Edition
Malkiat S. Johal, Lewis E. Johnson

Physics for Technology, Second Edition: With Applications in Industrial Control Electronics
Daniel H. Nichols

Time-Resolved Spectroscopy: An Experimental Perspective
Thomas Weinacht; Brett J. Pearson

No-Frills Physics: A Concise Study Guide for Algebra-Based Physics
Matthew D. McCluskey

Quantum Principles and Particles, Second Edition
Walter Wilcox

Electronic Conduction: Classical and Quantum Theory to Nanoelectronic Devices
John P. Xanthakis

Physics: A Focused Introduction
Colin Campbell

Quantum Many-Body Physics: A Path Integral Approach
Luca Salasnich

Semiconductors and Nanostructures: An Introduction
Rui Vilão

Fundamental Physics: A Guide to Relativity and Quantum Theory
Martin Bojowald

For more information about this series, please visit:
www.crcpress.com/Textbook-Series-in-Physical-Sciences/book-series/TPHYSCI

Fundamental Physics

A Guide to Relativity and Quantum Theory

Martin Bojowald

CRC Press
Taylor & Francis Group
Boca Raton London New York

CRC Press is an imprint of the
Taylor & Francis Group, an **informa** business

Designed cover image: Shutterstock

First edition published 2027
by CRC Press
2385 NW Executive Center Drive, Suite 320, Boca Raton, FL 33431

and by CRC Press
4 Park Square, Milton Park, Abingdon, Oxon, OX14 4RN

CRC Press is an imprint of Taylor & Francis Group, LLC

ISBN: 978-1-032-96234-4 (hbk)
ISBN: 978-1-032-96235-1 (pbk)
ISBN: 978-1-003-58870-2 (ebk)

DOI: 10.1201/9781003588702

Typeset in Nimbus Roman
by KnowledgeWorks Global Ltd.

Contents

Preface

Quantum mechanics and relativity theory are the main pillars of modern physics. In contrast to classical physics, their understanding is challenging not only because of the required mathematical methods but also owing to several counter-intuitive physical features. While quantum mechanics and relativity are fundamental and therefore apply on all energy or length scales, their effects become relevant beyond sensitive measurements mainly on extreme scales that we are not used to from day-to-day experience. Analyzing these phenomena therefore requires strict mathematical methods as well as careful conceptual thinking.

The original and by now well-understood implications of quantum mechanics and relativity theory were seen in opposite regimes, such as small atomic distances for quantum effects compared with vast astronomical or cosmological scales for general relativity. Also their basic mathematical formulations are sufficiently independent of each other to warrant separate introductions in unrelated courses. On closer examination, however, quantum mechanics and relativity theory share several broad mathematical features such as geometrical ideas, a prominent role of tensor products or tensor calculus, as well as the importance and implementation of symmetry. Conceptually, it is useful to recognize a kinship in how both quantum mechanics and relativity theory start with idealized notions that turn out to be inadequate for actual physical situations, but can be amended in suitable ways in order to be broadly applicable. In quantum mechanics, the main example is the transition from pure states and wave functions of closed systems to mixed states and density operators of open systems. In relativity, there is an important step from global inertial frames to local inertial frames, even within special relativity before its generalization to curved space-time.

A large number of fascinating phenomena exist in which both quantum and relativistic effects are relevant. Some of the outstanding puzzles of fundamental physics can be found in the sought-after combination of quantum physics and relativity in an elusive theory of quantum gravity. Even future quantum applications may be expected to require knowledge of general relativity because of the ever-increasing sensitivity of measurements and control mechanisms that, in some cases, are subject to gravitational time dilation. More generally, gravitational tidal forces in curved space-time may affect the quantum state of, say, a qubit. A quantum computer on a satellite may have to be operated with suitable adjustments, compared with one on Earth.

The first two parts of this book serve as independent introductions to quantum mechanics and relativity theory, respectively. Nevertheless, several cross-references show some of the mathematical connections between these two major frameworks. Except for more-advanced material toward the ends, each part corresponds to a one-semester introductory course. These two parts emerged from years of teaching quantum mechanics based on the first part of *Introduction to Quantum Mechanics* by David J. Griffiths, and teaching special and general relativity based on about the first half of *Gravity: An Introduction to Einstein's General Relativity* by James B. Hartle. I am indebted to these authors, and to Deirdre Shoemaker who had established the course Special and General Relativity at The Pennsylvania State University before I took over. Some of the ingredients are likely still recognizable, although much material has been changed, added, or rearranged during almost 20 years of teaching and in the course of writing this book.

Examples of major new contributions that can be found here are a discussion of the quantum measurement problem based on positive-operator valued measures, modern concepts from quantum information theory, and an extended presentation of various geometrical features of both quantum mechanics as well as special and general relativity. Also, the second part contains an introduction to the Hamiltonian formulation of general relativity because of its importance for quantum gravity and quantum reference frames, and the role of constraints is foreshadowed in new examples for quantum mechanics in the first part. While the prominent role of time in relativity is usually contrasted by a neglectful treatment in quantum mechanics, the quantum nature of time is developed here in more detail at the end of the first part and revisited in the third.

Part three of this book combines results from both preceding parts in various ways, ranging from new ingredients of proposed lab experiments to foundational questions of quantum gravity. The nature of these topics means that this part is largely open-ended, focusing on an emphasis of open problems and challenges where more research is required.

Section I

Quantum Physics

1 Basic Properties

The modern theory of quantum mechanics describes the dynamics of a variety of systems as well as possible measurement outcomes. While quantum theory is fundamental and therefore applies to all systems, including macroscopic ones, its prominent effects are not directly visible and have to be revealed by dedicated sensitive measurements. It is often necessary to keep track of possible influences from unintended interactions with the system's surroundings. In contrast to classical physics, which would relate such interactions to small and often negligible exchanges of energy, quantum physics is highly susceptible to such influences because they can change the information potentially gained from a measurement, or they may adversely affect a controlled process such as a quantum computation.

An analysis that keeps track of all possible interactions requires much care as well as advanced mathematical methods. Nevertheless, many important properties of quantum physics can be derived by ignoring such interactions in a first approximation, considering the simplified case of isolated systems. One of the basic concepts of quantum mechanics applied to isolated systems is the wave function, the main topic of this chapter. It has several counterintuitive features because it describes properties that we do not experience in day-to-day interactions. The necessity of introducing this concept had been slowly revealed by analyzing the limitations of classical physics in new experiments that became possible by the early 20th century.

1.1 CLASSICAL PHYSICS AND ITS LIMITATIONS

Classical physics has two distinct basic concepts, given by particles and waves. Particles are often thought of as point-like and lacking significant extensions, while waves are always spread out. Mathematically, the main difference is the application of ordinary differential equations for the former and partial differential equations for the latter.

1.1.1 PARTICLES

A particle is characterized by its mass $m > 0$ and moves along a trajectory $\vec{x}(t)$ that specifies the spatial position $\vec{x} = (x, y, z)$ as a function of time t. If we know the trajectory, we can compute the velocity $\vec{v}(t) = \mathrm{d}\vec{x}/\mathrm{d}t = \dot{\vec{x}}(t)$, the momentum $\vec{p}(t) = m\dot{\vec{x}}(t)$ and the acceleration $\vec{a}(t) = \ddot{\vec{x}}(t)$. Newton's second law then tells us that the force acting on the particle in order to make it move along the trajectory $\vec{x}(t)$ is given by $\vec{F}(t) = m\vec{a}(t)$.

The inverse problem is much harder, in which we begin with a given force law $\vec{F}(\vec{x}, t)$, usually derived within a non-mechanical branch of physics such as electrodynamics, and attempt to find the trajectory of a particle with mass m by solving the second-order differential equation $m\ddot{\vec{x}}(t) = \vec{F}(\vec{x}, t)$. If there is friction or a magnetic field, the force may also depend on the velocity of the particle. In general, if we know the initial position $\vec{x}(0)$ and velocity $\vec{v}(0)$, the second-order equation has a unique solution $\vec{x}(t)$ with these initial values, but it may be hard to find this function in closed form.

As a simple example, a constant force $\vec{F} = F\vec{n}$ with magnitude $F = |\vec{F}|$ pointing in the direction of a unit vector $\vec{n}$, implies the equation $m\ddot{\vec{x}}(t) = F\vec{n}$. Integrating once, we obtain

$$\dot{\vec{x}}(t) = \frac{F}{m} t\vec{n} + \vec{c}_1 \tag{1.1}$$

with a constant vector $\vec{c}_1$. A second integration results in

$$\vec{x}(t) = \frac{F}{2m} t^2\vec{n} + t\vec{c}_1 + \vec{c}_2 \tag{1.2}$$

DOI: 10.1201/9781003588702-2

with another constant vector $\vec{c}_2$. Evaluating both equations at $t = 0$, we identify the constant vectors with initial values according to $\vec{c}_1 = \dot{\vec{x}}(0)$ and $\vec{c}_2 = \vec{x}(0)$.

Solving the differential equation implied by Newton's second law is more difficult if the force depends on the position. However, the first part of the solution procedure, integrating once to obtain $\dot{\vec{x}}(t)$ from $\ddot{\vec{x}}(t)$, can be done more easily by using **energy conservation**, provided that the force is conservative. A force is called conservative if it does not explicitly depend on time and is the negative gradient of a potential $V(\vec{x})$,

$$\vec{F}(\vec{x}) = -\text{grad}\, V(\vec{x}) = -\begin{pmatrix} \partial V/\partial x \\ \partial V/\partial y \\ \partial V/\partial z \end{pmatrix} . \tag{1.3}$$

Any trajectory of a particle subject to a conservative force with potential $V(\vec{x})$ is such that the energy

$$E = \frac{|\vec{p}|^2}{2m} + V(\vec{x}) \tag{1.4}$$

is time-independent, and thus conserved, along any solution of the equation of motion $m\ddot{\vec{x}} = \vec{F}(\vec{x}) = -\text{grad}\, V(\vec{x})$.

Mathematically, E is called a **first integral** of the equation of motion because it tells us something about solutions and depends only on first-order time derivatives of $\vec{x}$ through $\vec{p}$. Unlike Newton's law, the expression does not contain any second-order derivatives. We will see several times later on that first integrals or conserved quantities such as the energy are strictly related to the presence of **symmetry**, given here by the time independence or time-translation invariance of the potential.

Energy conservation in a conservative potential can be used to find qualitative properties of the motion without solving differential equations. This procedure is called the method of **potential landscapes** because it makes use of an intuitive understanding of motion down mountain slopes under the gravitational force. As visualized in Figure 1.1, we draw the potential as a function $V(x)$

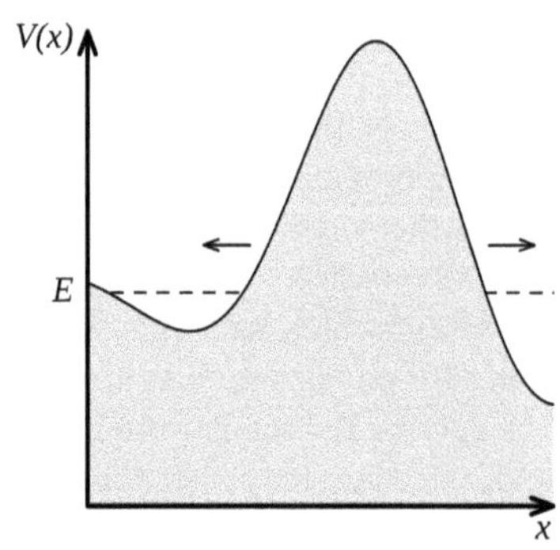

Figure 1.1 Potential landscape for a conservative force with potential $V(x)$. Since the kinetic energy is always positive, the conserved energy E cannot be smaller than the potential at any point where a particle is moving. For a given energy, the x-axis is therefore split into allowed regions (dashed lines) where a particle with this energy can move, and forbidden regions where it cannot be found. If an allowed region is bounded, the particle oscillates between a minimum and a maximum x given by the endpoints of the allowed region. These endpoints, called turning points, can be computed by evaluating the energy condition with zero momenta. If an allowed region is unbounded, a particle follows scattering motion in the potential. The force, defined as the negative gradient of the potential and shown here by the example of two arrows, points toward the direction of descent of the potential at any point where it is evaluated.

in a diagram, together with the energy E of a solution. (It is easier to draw such a diagram for a single spatial coordinate x, rather than all three dimensions in $\vec{x}$.) Since $|\vec{p}|^2/(2m) \geq 0$, we always have $E \geq V(\vec{x})$ at any $\vec{x}$ where the particle may be moving. If we determine all $\vec{x}$ where this inequality is fulfilled, we obtain the **allowed regions** in the potential for the given energy. All other values of $\vec{x}$ form the **forbidden regions**. Allowed and forbidden regions give us the following characteristic properties of motion in the potential:

1. A particle always stays in a fixed allowed region and cannot jump to another allowed region because this would require more energy than it has available.
2. An allowed region may be bounded, in which case the particle is oscillating between the two endpoints, or unbounded, in which case the particle moves in from far away, is reflected or scattered by the force of the potential, and then moves back to where it came from.
3. The endpoints of allowed regions are called **turning points** because the direction in which the particle is moving changes when it reaches them. Since the energy equals the potential at the turning points, their positions can be computed by evaluating the condition $E = V(\vec{x})$.

For this procedure, the potential is required to be time independent because it would be impossible to keep track of motion through a changing potential landscape within a single diagram.

1.1.2 WAVES

A wave is described by a function $f(\vec{x},t)$ in which the components of $\vec{x}$ together with t are four independent coordinates. The physical meaning of f depends on the kind of wave, for instance, the air pressure for a sound wave or a component of the electric or magnetic field for an electromagnetic wave such as light. Because f depends independently on $\vec{x}$ and t, it has independent partial derivatives $\partial f/\partial t$, $\partial f/\partial x$, $\partial f/\partial y$ and $\partial f/\partial z$. In this part, we assume that space (x,y,z) and time t are simply given by coordinates assigned to points and moments. The following discussion of relativity will reveal the mathematical structure and geometry behind these basic concepts.

The equation of motion for a wave, or a **wave equation**, is given by a relationship that the partial derivatives, as well as second-order or higher versions, always obey. A common example that is realized, for instance, for light in vacuum and with negligible gravity is the second-order partial differential equation

$$\frac{\partial^2 f}{\partial x^2} + \frac{\partial^2 f}{\partial y^2} + \frac{\partial^2 f}{\partial z^2} - \frac{1}{v^2}\frac{\partial^2 f}{\partial t^2} = 0 \tag{1.5}$$

with a constant v. A family of solutions is given by

$$f(x,y,z,t) = a\cos(k(x-vt)) + b\sin(k(x-vt)) \tag{1.6}$$

with more constants, a, b and k. Unlike the constant v, the new constants a, b and k are not universal properties of the wave phenomena described by (1.5), but rather depend on features of an individual solution such as the energy supplied when the wave was started.

The specific solutions (1.6) are called **plane waves** because the function f takes constant values on the y-z planes at any fixed time t. Since the partial differential equation (1.5) is linear in f, the sum of two or more solutions is also a solution. We can therefore generalize our family of solutions to

$$f(x,t) = \sum_k (a(k)\cos(k(x-vt)) + b(k)\sin(k(x-vt))) \tag{1.7}$$

where we sum over a finite number of real numbers k, and the coefficients $a(k)$ and $b(k)$ may depend on these values. Since the number of values of k we sum over is not limited, the general solution has two free functions, $a(k)$ and $b(k)$, which are a feature of partial differential equations: Instead of finitely many initial values for an ordinary differential equation, we need infinitely many numbers to specify the initial wave $f(\vec{x},0)$ and its time derivative at $t=0$. Depending on the type of wave equation, there may also be free choices in the values assumed at a spatial boundary.

The solution (1.6) evaluated at a fixed time t is sinusoidal. If we consider a local minimum or maximum, its position changes if we look at the same function sometime later. Since a local maximum or minimum of a linear combination of $\sin(\phi)$ and $\cos(\phi)$ is given by a specific value of the **phase** $\phi=\phi_0$ on which the functions depend, we can express the motion of local minima and maxima by the equation $k(x-vt)=\phi_0$, using the argument in (1.6). This equation implies a relationship between x and t that can be solved for

$$x(t)=\phi_0/k+vt\,. \tag{1.8}$$

The result shows that local minima and maxima of the wave move at speed v in the x-direction. The constant v in (1.5) therefore determines the speed of the wave, and it equals the speed of light for electromagnetic waves in vacuum. For a general wave, v is called the **phase velocity** because it is derived from the phase ϕ.

The meaning of the parameter k can be seen by evaluating the solution (1.6) at a fixed time t, such as $t=0$. The phase is then $\phi=kx$, and since the sine and cosine functions are periodic with period 2π in ϕ, the wave is periodic with period $2\pi/k$ in x. The spatial period of a wave is known as the wave length, $\lambda=2\pi/k$. Therefore, the parameter k, called the **wave number**, equals $k=2\pi/\lambda$. Similarly, the period in time can be determined by evaluating (1.6) at a fixed x, such as $x=0$. The phase is then $\phi=kvt$, and since it still appears in sine and cosine functions with period 2π, the wave has the period $T=2\pi/(kv)$ in time. The frequency ω of the wave, or of any oscillation, is defined through $T=2\pi/\omega$. Therefore,

$$\omega=kv \tag{1.9}$$

relates the three wave properties, the frequency, wave length or number, and phase velocity. The relationship (1.9) is valid for any wave that solves equation (1.5). It is called the **dispersion relation** of the wave because the velocity dependence of the frequency determines how a wave disperses as it evolves. For electromagnetic waves, for instance, the wave equation and the dispersion relation would change if the wave enters a non-vacuum region. Different frequency contributions may then travel at different speeds, which explains several well-known optical phenomena.

The remaining parameters a and b in (1.6) are the amplitudes, and the intensity of the plane wave is given by a^2+b^2. For a classical wave, the intensity determines how much energy the wave transports.

1.1.3 PARTICLES AND WAVES

The mathematical properties of particles and waves are rather different from each other. Particles and waves also behave in distinguishable ways in experiments. In classical physics, a general characteristic is that particles cast sharp shadows because they move along narrow trajectories, while waves are spread out, can move around corners, and interfere with each other when they overlap. For light, we are familiar with sharp shadows. There is therefore a particle description of light in terms of light rays. Interference also exists for light, but it is hard to detect because the wave length of visible light is very small. It was first observed for light in the **double-slit experiment** by Thomas Young in 1803. The fundamental classical description of light is therefore as a wave, or an electromagnetic wave as explained by Maxwell's equations.

In 1887, Heinrich Hertz discovered the **photoelectric effect** when he noticed that electrons were emitted from a metal surface by shining light on it. Electrons are electrically charged particles that are bound to atomic nuclei of the opposite charge. In a metal, some of the atomic electrons can easily move between atoms and in this way make it possible for the metal to conduct electricity. A qualitative potential landscape can be used to describe this situation as shown in Figure 1.2. The metal nuclei, aligned in a crystal structure, form a periodic electric potential in which the electrons

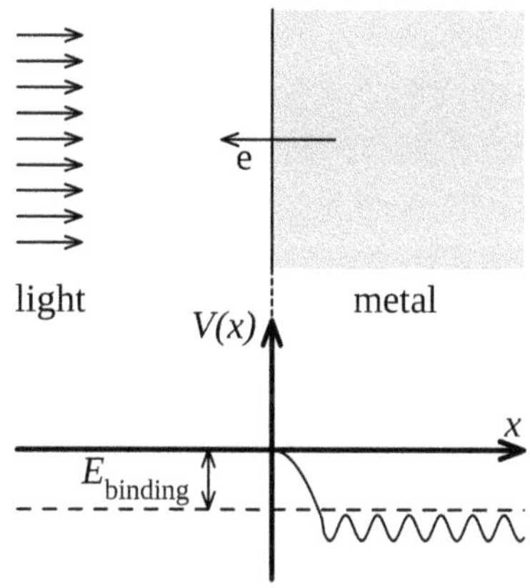

Figure 1.2 A model of the photoelectric effect. Electrons are bound to the metal by a potential derived from the electric force between nuclei and electrons. Electrons can move freely through vacuum surrounding the metal in an experiment, but the energy required to enter the vacuum region is greater than the energy of metal electrons because the vacuum does not contain positive charges. If the energy difference is supplied by light that reaches the metal surface, metal electrons can be emitted into the vacuum region. An electric field in the vacuum region (not shown here) can be used to accelerate the freed electrons away from the surface to a wire in which they build a measurable electric current.

move. Electrons that participate in electric conductance have an energy that is greater than the local maxima of the periodic potential, but less than the potential on the boundary of the metal and in a surrounding vacuum region. These electrons are bound to the metal but can move freely within it. If light reaches the metal, it may be absorbed by some of the electrons close to the surface. If the absorbed energy is greater than the potential difference between metal and vacuum, called the binding energy, some of the electrons can move from the metal to the vacuum, building a weak electric current through the vacuum region that can be measured. Hertz did not directly measure the current but demonstrated the presence of free electrons after shining light on a metal surface by inducing sparks in a narrow gap between two electrodes.

Later experiments initiated by Joseph John Thomson (who also confirmed the existence of the electron as a negatively charged particle) made it possible to measure properties of the emitted electrons in detail. In particular, the strength of the electric current is proportional to the rate of electrons emitted by light. Moreover, by letting these electrons move against a potential difference, it was possible to measure their initial kinetic energy by increasing the potential until electrons were unable to make it across the vacuum region.

These measurement results were intended to test a classical model of the photoelectric effect based on the electromagnetic theory that describes how light can transmit energy to an electron. In the classical model, light is a wave whose amplitude is given by the electric and magnetic fields. These fields exert a force on an electrically charged particle such as the electron. They accelerate the electron and thereby supply it with additional energy, $E_{\rm light}$. If this additional energy is greater than the binding energy $E_{\rm binding}$, the electron can enter the vacuum region if it moves toward the metal surface. The remaining energy, $E_{\rm e} = E_{\rm light} - E_{\rm binding}$, can be measured as the kinetic energy of the electron that allows it to move against a potential difference, provided the latter is small enough. This measured energy, according to the classical picture, should increase if the electric and magnetic fields of the light wave are bigger, and therefore as the amplitude or intensity of the wave are increased.

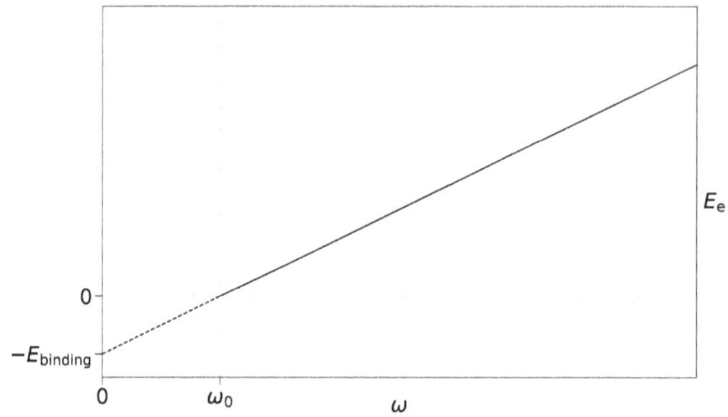

Figure 1.3 Sketch of experimental results for the kinetic energy E_e of electrons emitted in the photoelectric effect, shown as a function of the frequency ω of light.

The experiments did not confirm this expectation:

1. The energy of individual electrons did not depend on the intensity of light. Instead, more electrons were emitted as the intensity was increased, corresponding to a larger electric current. However, all these electrons were stopped by the same potential difference as required at a lower intensity. The kinetic energy of emitted electrons is therefore independent of the intensity of the wave, or of the electromagnetic acceleration experienced by the electrons.
2. Instead, the kinetic energy depends on the frequency of light used in the experiment, as indicated in Figure 1.3. The dependence is linear, in such a way that electrons are emitted only above a certain threshold frequency ω_0. An extension of the linear plot intersects the energy axis at the negative binding energy, $-E_{\text{binding}}$.

These observations were explained by a new model proposed by Albert Einstein in 1905. In this model, light used in the photoelectric effect does not behave like a wave but instead has particle properties of "light quanta." If a light particle, or a **photon** as it was named later, has the energy $E_{\text{photon}} = \hbar\omega$ in a light ray with frequency ω, where $\hbar$ is a constant to be determined by experiments, the energy of a free electron after emission in the photoelectric effect is given by a simple difference,

$$E_e = E_{\text{photon}} - E_{\text{binding}} = \hbar\omega - E_{\text{binding}}\,. \tag{1.10}$$

This linear equation for the electron energy as a function of the frequency explains all features seen in the plot of Figure 1.3. Moreover, if the intensity of light is increased, it contains more photons, and therefore more electrons can be emitted by absorbing photons, also in agreement with observations. This non-classical model provides a quantized picture of light: Light of a given frequency ω cannot be broken up into portions with energies less than $\hbar\omega$, in contrast to the classical understanding of a wave where the energy can be decreased continuously by changing the intensity.

Sometime later, in 1925 in experiments by Clinton Davisson and Lester Germer, wave properties were observed in an electron beam by detecting interference effects. The same relationship, $E = \hbar\omega$, describes the dependence of the electron energy on the frequency consistent with the interference pattern, with the same value of the constant $\hbar$. Moreover, the momentum was related to the wave number by $p = \hbar k$. These observations demonstrate the universality of $\hbar$ as a fundamental constant, called **Planck's constant**, for Max Planck had introduced it earlier in thermodynamical considerations. They show that microscopic phenomena of physics combine particle and wave properties. Whether we notice particle or wave properties depends on how we perform an experiment, or on the measurement process, but not on the type of physical object.

The mathematical description therefore requires a formulation that combines properties of particles and waves. Waves are described by functions that depend on more independent variables than the trajectories of particles. They give us more freedom in a mathematical formulation, and therefore, it is convenient to combine particle and wave properties by formulating particles as waves.

However, there is no universal wave equation that we could use for this purpose. We should refer to some of the experimental results, in particular the relationships $E = \hbar\omega$ and $p = \hbar k$, and see if they can determine properties of a suitable wave equation for particles. The right-hand sides of these two equations contain wave properties, while the left-hand sides also appear in important particle equations, such as

$$E = \frac{p^2}{2m} \tag{1.11}$$

for a free particle with mass m. If we replace E and p with $\hbar\omega$ and $\hbar k$, respectively, we obtain

$$\omega = \frac{\hbar k^2}{2m}. \tag{1.12}$$

Just as (1.9), this equation has the form of a dispersion relation that tells us how the frequency and wave number of a wave are related, but its mathematical form is different for particles compared with electromagnetic waves. A particle therefore corresponds to a different wave phenomenon and is described by a different partial differential equation. Our first task is to find this new equation.

We can turn dispersion relations into mathematical operations such as derivatives if we use sample wave functions given by plane waves, $f(x,t) = a\cos(kx - \omega t)$. According to the chain rule, taking derivatives by x or t directly corresponds to multiplication by functions of k and ω, as in

$$\frac{\partial f}{\partial t} = a\omega\sin(kx - \omega t) \quad \text{and} \quad \frac{\partial^2 f}{\partial^2 t} = -a\omega^2\cos(kx - \omega t) = -\omega^2 f(x,t) \tag{1.13}$$

as well as

$$\frac{\partial f}{\partial x} = -ak\sin(kx - \omega t) \quad \text{and} \quad \frac{\partial^2 f}{\partial^2 x} = -ak^2\cos(kx - \omega t) = -k^2 f(x,t). \tag{1.14}$$

We have to go to the second derivatives in order to reproduce the original trigonometric function, which also gives us a negative sign. The equations therefore suggest that we may identify the energy $E = \hbar\omega$ with $\hbar$ times the "square root of the negative second derivative by t," and the momentum $p = \hbar k$ with the "square root of the negative second derivative by x."

Taking square roots of derivatives themselves (as opposed to square roots of the functions obtained by taking derivatives) is possible in the sense of mathematical operations. The square root of a second derivative, as an operation, is a first-order derivative because in the inverse procedure, the combination of two first-order derivatives implies a second-order derivative. This result is also suggested by the usual square notation of second-order derivatives. In the present case, we have to include the square root of negative one, which equals the imaginary unit i of complex analysis. We arrive at the relationships

$$E = \pm i\hbar\frac{\partial}{\partial t} \quad \text{and} \quad p = \pm i\hbar\frac{\partial}{\partial x} \tag{1.15}$$

where i is such that $i^2 = -1$. The two $\pm$ choices are undetermined at this stage.

These relationships open up a large number of mathematical questions, which we will discuss in due course. For instance, we should look for a suitable mathematical interpretation that makes it possible to equate numbers (E and p) with mathematical operations ($\partial/\partial t$ and $\partial/\partial x$). For now, if we directly insert (1.15) in the energy equation

$$E = \frac{p^2}{2m} + V(x) \tag{1.16}$$

of a particle with mass m moving in a potential $V(x)$, and apply the resulting derivative operators to a **wave function** which we call $\Psi(x,t)$, we obtain the partial differential equation

$$i\hbar\frac{\partial\Psi}{\partial t} = -\frac{\hbar^2}{2m}\frac{\partial^2\Psi}{\partial x^2} + V(x)\Psi. \tag{1.17}$$

As we will see later, the plus choice should be made for E in (1.15), as used in this wave equation. The choice for p does not matter at this point, but it will turn out to be the negative one.

For a free particle, this equation has plane-wave solutions that produce the expected dispersion relation (1.12). In quantum mechanics, it is postulated as the correct equation not only for the free particle but for any (non-relativistic) particle subject to a conservative force. It is called the **Schrödinger equation**, was introduced by Erwin Schrödinger in 1925, and has since been tested numerous times by comparisons of its predictions with experiments. In particular, it explains atomic spectra and various other concepts of microscopic and particle physics.

Before we can discuss physics implications, we have to understand a number of mathematical and conceptual questions:

1. Comparisons with experiments require predictions derived from solutions of the Schrödinger equation. It is a linear partial differential equation, which is not quite as complicated as a non-linear one. Nevertheless, the fact that it has a non-constant coefficient $V(x)$ if the particle is not free presents some mathematical challenges. We will learn how to solve the Schrödinger equation in several special cases of physical interest.
2. We know that a solution Ψ of Schrödinger's equation describes a wave that represents a particle, but this basic understanding does not provide enough information to relate mathematical solutions to potential observations in experiments. We will have to find a suitable physical property corresponding to Ψ and understand its meaning. Unlike for classical waves, the meaning of Ψ is not related to a tangible physical phenomenon such as air pressure in a sound wave or the electric and magnetic fields of an electromagnetic wave. After all, the conceptual understanding of the physical object described by the wave function is still as a point particle, which classically has a position but not an independent position-dependent property. This question about the meaning of Ψ will lead us to probability theory as another mathematical ingredient.
3. Schrödinger's equations contain i in one of its coefficients. Generic solutions $\Psi(x,t)$ are therefore complex rather than real numbers. This property again shows that the physical meaning of Ψ is tricky because physical observables (shown on a measuring device) are always real numbers. We will begin the discussion of mathematical background with properties of complex numbers.

1.1.4 COMPLEX NUMBERS

Complex numbers allow us to take a square root of a negative number in the following way:

1. We introduce a symbol i such that $i^2 = -1$. This symbol does not represent a real number, but is an example of a complex number. It is called the imaginary unit.
2. We can now solve the equation $x^2 = -a$, where $a > 0$ by $x = \pm i\sqrt{a}$. The two solutions are complex numbers, or, more specifically, imaginary numbers because they are proportional to i. A general complex number is the sum of a real and an imaginary number. We write a general complex number as

$$z = a + bi \tag{1.18}$$

where a and b are real numbers. The real number $a = \mathrm{Re}(z)$ is called the real part of z and the real number $b = \mathrm{Im}(z)$, the imaginary part. Notice that the imaginary part of a complex

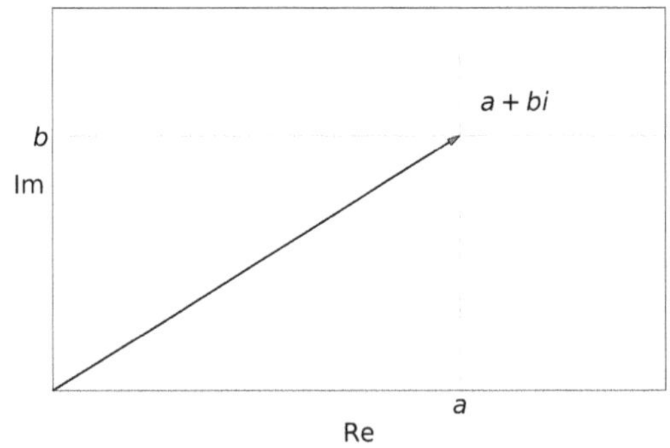

Figure 1.4 Visualization of a complex number $a+bi$ in the 2-dimensional plane with real and imaginary parts as independent directions.

number is not an imaginary number, but a real number. This is just how the imaginary part is commonly defined.
A real number is also a complex number, one with a vanishing imaginary part. A complex number has two independent real components and can therefore be visualized as a vector in the 2-dimensional plane; see Figure 1.4.

3. For solutions of general quadratic or higher-degree polynomials, or for other purposes, we may need to add or multiply complex numbers. We do so by observing the usual rules of commutativity, associativity, and distributivity, as well as $i^2 = -1$ whenever this square shows up in a product. As equations, the general rules are

   Commutativity: $z_1 + z_2 = z_2 + z_1$ and $z_1 z_2 = z_2 z_1$

   Associativity: $z_1 + (z_2 + z_3) = (z_1 + z_2) + z_3$ and $z_1(z_2 z_3) = (z_1 z_2) z_3$

   Distributivity: $z_1(z_2 + z_3) = z_1 z_2 + z_1 z_3$

   for any complex numbers z_1, z_2 and z_3.
4. Example: The complex number

$$z = (1+i+2+i)i = (3+2i)i = 3i+2i^2 = -2+3i \tag{1.19}$$

   has real part $\text{Re}(z) = -2$ and imaginary part $\text{Im}(z) = 3$.

Every complex number that results from a calculation can be brought to the form $z = a+bi$ with real a and b. It is usually a good idea to derive this form because it makes it easier to use z in follow-up calculations. For instance, combining the preceding definitions and rules, we obtain explicit equations for the mathematical operations defined for complex numbers, given by addition

$$(a+bi)+(c+di) = (a+c)+(b+d)i \tag{1.20}$$

and multiplication

$$(a+bi)(c+di) = ac+bci+adi+bdi^2 = (ac-bd)+(bc+ad)i\,. \tag{1.21}$$

Sometimes, we have to be careful and make sure that we use only the allowed operations. An example of an operation that is not allowed is given by an application of standard rules for the real square root, which has not appeared in our definitions for complex numbers. We may be tempted to interpret the valid equation $i^2 = -1$ as $i = \sqrt{-1}$, but this step is sometimes problematic for complex numbers. (Note also that the square root is defined only up to a sign choice.) For instance, we will use the expression $1/i$ a few times, which should have a unique answer as the inverse of multiplication by i. We can attempt this calculation in two ways,

$$\frac{1}{i}=\frac{i}{i^2}=\frac{i}{-1}=-i \tag{1.22}$$

or

$$\frac{1}{i}=\frac{1}{\sqrt{-1}}\text{“}=\text{”}\sqrt{\frac{1}{-1}}=\sqrt{-1}=i\,, \tag{1.23}$$

with different results. In the first version, we only use the allowed rules of multiplication (applied to numerator and denominator at the same time) as well as $i^2=-1$. In the second line, we are trying to apply rules for square roots that are valid for real and positive numbers, but have not been defined for complex numbers. (See the second “equality” in this line.) This result, which is the opposite of the first one, should therefore be discarded, and we conclude that

$$\frac{1}{i}=-i\,. \tag{1.24}$$

There are additional operations for complex numbers, some of which can be derived from multiplication and others that generalize useful concepts from real numbers:

Norm: For a complex number $z=a+bi$, we define the norm as

$$|z|=|a+bi|=\sqrt{a^2+b^2}\geq 0\,, \tag{1.25}$$

which is always a non-negative real number. The norm can be visualized as the length of the vector in Figure 1.4 according to Euclidean geometry.

Complex conjugation: For a complex number $z=a+bi$, we define the complex conjugate as another complex number, obtained by flipping the sign of the imaginary part:

$$z^*=(a+bi)^*=a-bi\,. \tag{1.26}$$

It follows that the product of a complex number and its conjugate equals the square of its norm:

$$(a+bi)(a+bi)^*=(a+bi)(a-bi)=a^2+abi-abi-b^2i^2=a^2+b^2=|a+bi|^2\,. \tag{1.27}$$

The square of the norm is not the same as the square of the complex number, unless $b=0$:

$$(a+bi)^2=a^2-b^2+2abi\neq|a+bi|^2\,. \tag{1.28}$$

It is important to remember this inequality because we are not used to it from real numbers.

Using complex conjugation, the real and imaginary parts of a complex number z can be expressed as

$$\mathrm{Re}(z)=\frac{1}{2}(z+z^*)\quad\text{and}\quad\mathrm{Im}(z)=\frac{1}{2i}(z-z^*)\,. \tag{1.29}$$

Inverse number: The norm gives us a useful method to describe some information about a complex number by a real number. Since multiplication of a complex number by its complex conjugate produces the real norm squared, we can apply this operation in such a way that the imaginary unit is moved to a different place. For instance, unless $a=b=0$, we can use these operations in order to compute the inverse of the complex number $a+bi$ under multiplication:

$$(a+bi)^{-1}=\frac{1}{a+bi}=\frac{a-bi}{(a+bi)(a-bi)}=\frac{a}{a^2+b^2}-\frac{b}{a^2+b^2}i\,. \tag{1.30}$$

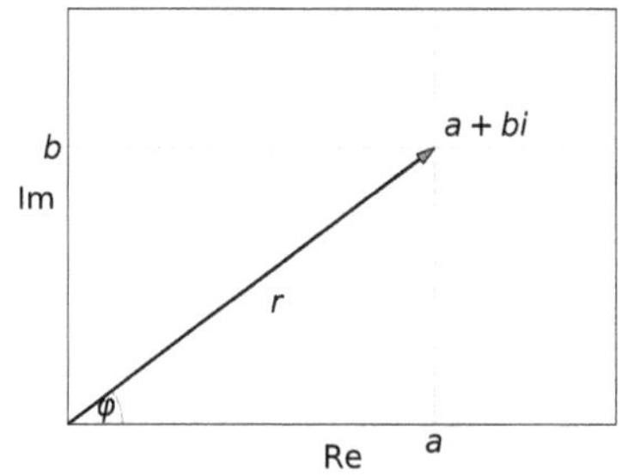

Figure 1.5 Polar coordinates on the plane provide an alternative parameterization of complex numbers as $z = a + bi = r\cos(\varphi) + r\sin(\varphi)i$.

Division: The inverse (1.30) allows us to define division by a complex number as multiplication by its inverse.

Complex conjugation is a new operation for complex numbers. It can be combined with addition, multiplication, and with itself, such that the following rules are satisfied:

$$(z_1 + z_2)^* = z_1^* + z_2^* \quad , \quad (z_1 z_2)^* = z_1^* z_2^* \quad , \quad (z^*)^* = z \tag{1.31}$$

for any complex numbers z_1, z_2, and z. These equalities can be checked by direct calculations, writing $z = a + bi$.

The real and imaginary parts of a complex number behave like coordinates on a 2-dimensional plane used to visualize the number as in Figure 1.4. As usual, we are free to apply coordinate transformations that may be non-linear. A convenient transformation is often given by the transition from Cartesian coordinates used in Figure 1.4 to polar coordinates, as shown in Figure 1.5. In equations, the transformation takes the well-known form

$$r = \sqrt{a^2 + b^2} \quad \text{and} \quad \varphi = \arctan(b/a) \tag{1.32}$$

with the inverse

$$a = r\cos(\varphi) \quad , \quad b = r\sin(\varphi) . \tag{1.33}$$

We can therefore write a complex number as

$$z = a + bi = r\cos(\varphi) + r\sin(\varphi)i \tag{1.34}$$

where $r = |z|$ is called the **amplitude** of z and φ the **phase**, thanks to a resemblance of (1.34) with a plane wave.

The result of this transformation can be written more compactly as

$$z = re^{i\varphi} \quad \text{or} \quad z = r\exp(i\varphi) \tag{1.35}$$

where the exponential function of an imaginary number is defined as

$$e^{i\varphi} = \exp(i\varphi) = \cos(\varphi) + \sin(\varphi)i . \tag{1.36}$$

Indeed, if we ignore the fact that $i\varphi$ is not a real number, the coefficients of the Taylor expansion of $\exp(i\varphi)$ are the same as the coefficients of the Taylor expansion of $\cos(\varphi) + i\sin(\varphi)$.

The exponential function and sine and cosine have the useful property that standard rules of differentiation and integration can be applied to them even if they are evaluated for complex numbers. In particular, we have

$$\frac{\mathrm{d}e^z}{\mathrm{d}z} = e^z \tag{1.37}$$

as well as

$$\frac{\mathrm{d}\cos(z)}{\mathrm{d}z} = -\sin(z) \quad \text{and} \quad \frac{\mathrm{d}\sin(z)}{\mathrm{d}z} = \cos(z)\,. \tag{1.38}$$

These operations are consistent with (1.36):

$$\frac{\mathrm{d}e^{i\varphi}}{\mathrm{d}\varphi} = ie^{i\varphi} = i(\cos(\varphi)+\sin(\varphi)i) = -\sin(\varphi)+\cos(\varphi)i = \frac{\mathrm{d}}{\mathrm{d}\varphi}(\cos(\varphi)+\sin(\varphi)i)\,. \tag{1.39}$$

We use the standard rules for exponentials, in particular

$$e^{z_1+z_2} = e^{z_1}e^{z_2}\,, \tag{1.40}$$

and complex conjugation can be moved to the argument of the exponential function:

$$(e^z)^* = e^{z^*}\,. \tag{1.41}$$

For an imaginary number, this equation follows directly from (1.36) if we use $\cos(-\varphi) = \cos(\varphi)$ and $\sin(-\varphi) = -\sin(\varphi)$. In this way, we show that (1.35) is consistent with the definition of the norm and its appearance in the amplitude $r = |z|$:

$$|z|^2 = zz^* = re^{i\varphi}re^{-i\varphi} = r^2e^{i\varphi-i\varphi} = r^2\,. \tag{1.42}$$

Applications of complex exponentials can provide useful shortcuts for the (re)derivation of trigonometric identities, as in

$$\begin{aligned} e^{2i\varphi} &= \cos(2\varphi)+\sin(2\varphi)i \\ =(e^{i\varphi})^2 &= (\cos(\varphi)+\sin(\varphi)i)^2 = \cos(\varphi)^2-\sin(\varphi)^2+2\sin(\varphi)\cos(\varphi)i \end{aligned} \tag{1.43}$$

Comparing real and imaginary parts in the first and last lines shows that

$$\cos(2\varphi) = \cos(\varphi)^2-\sin(\varphi)^2 \qquad \text{and} \qquad \sin(2\varphi) = 2\sin(\varphi)\cos(\varphi)\,. \tag{1.44}$$

1.2 STATISTICAL PROPERTIES

We now have enough knowledge about complex numbers to return to the wave function $\Psi(x,t)$ of quantum mechanics, which as a solution of the Schrödinger equation must in general take complex values. Since measurement results are always real numbers, we can first try to find a physical meaning for the squared norm of $\Psi(x,t)$, given by $|\Psi(x,t)|^2$. This function is defined by the squared norm of the complex number $\Psi(x,t)$ at any fixed x and t. The squared norm $|\Psi(x,t)|^2$ is a real function, but we need an independent definition that tells us what this function should mean and how it could be measured in an experiment. Such a definition is not obvious. It has been found and established only after comparing calculations with many measurements in atomic physics. It turned out that the proposed meaning of $|\Psi(x,t)|^2$ compatible with measurement results introduces a conceptually new ingredient in fundamental physics, given by unavoidable statistical behavior.

Recall that the wave function $\Psi(x,t)$ is a wave from a mathematical viewpoint and therefore spread out and extended, but it is supposed to describe a point particle. If the physical object we are dealing with is a point particle, a single observation should yield a sharp result for the position x at a given time t. This expectation is fully compatible with a wave only if the latter is very sharp and drops off quickly if we move away from the measured x, but this is not the generic behavior of a wave or of a solution of Schrödinger's equation.

In order to reconcile the competing expectations for point particles and waves, we interpret $|\Psi(x,t)|^2$ not as a complete description of a particle at time t, but as a **probability density** that describes statistical properties for the outcomes of a large number of position measurements at time

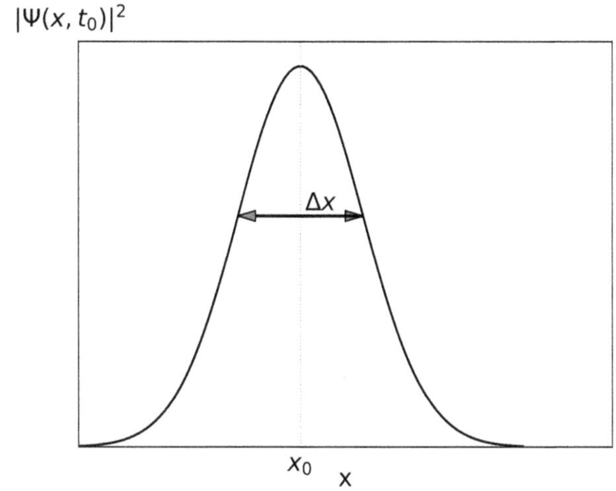

Figure 1.6 Example of a wave function and its statistical properties. The probability density at a given time t_0 is centered around the mean value x_0 with standard deviation Δx.

t. The term "probability" shows that the concept is fundamentally statistical, and the term "density" means that we should multiply $|\Psi(x,t)|^2$ by a length (in one spatial dimension) or a volume (in three spatial dimensions) in order to obtain the probability for position measurements to be within this length or volume. This interpretation is called the **Born rule**, named after Max Born who proposed it in 1926.

In a slightly more detailed version, this rule states that $|\Psi(x,t)|^2\mathrm{d}x$ is the probability for the point particle to be measured somewhere between x and $x+\mathrm{d}x$ at time t, where $\mathrm{d}x$ is a small increment of x. If the interval we are interested in is large rather than infinitesimal, we have to integrate in order to find the probability in a given range of x:

$$P_t([x_1,x_2]) = \int_{x_1}^{x_2} |\Psi(x,t)|^2\mathrm{d}x. \tag{1.45}$$

The subscript t indicates that the probability is valid at some fixed time t, but t does not take part in the integration or in probabilistic statements. The question of how to represent time measurements in quantum mechanics is tricky, as we will see. At this point, t is a parameter that labels evolving instances of statistical information $|\Psi(x,t)|^2$ but, unlike x, is itself not subject to statistical measurements.

The Born rule implies that quantum mechanics has a fundamental statistical behavior that cannot be reduced to measurement uncertainties but is part of the mathematical description. This conclusion is inevitable if we assume that the wave function contains all possible information about the system given by a point particle. There have been attempts to construct "hidden-variable theories" that include additional information, but it is hard to make them consistent with measurement results.

In quantum mechanics, in general, we cannot predict the outcome of a single measurement. We can only predict probabilities for certain values to occur, or statistical quantities such as means and standard deviations; see Figure 1.6. Trajectories $x(t)$ of point particles exist only as approximations when the wave function is very narrow at any given time. In practice, therefore, for a trajectory to be a good approximation, the fundamental standard deviation from the mean predicted by the wave function has to be less than observational imprecision.

1.2.1 PROBABILITY

In quantum mechanics, properties of an isolated point particle are described by a wave function. The propagating degree of freedom, subject to a partial differential equation, is not matter or energy, as in classical waves, but probability. We will therefore use methods from probability theory in our physical evaluations.

1.2.1.1 Probability Distribution

Probability theory determines the likelihood of finding specific outcomes in a set of possibilities. We first assume that there are finitely many objects (as they are commonly called in probability theory), whose number is given by a positive integer N, with properties that are described by assigning each object to a class c_j. The set of classes is labeled by another number, j, which is often but not always an integer. The assignment of an object to a class must always be possible (the classes then provide a complete description of the properties) and unique (no object can be placed in two or more classes). Classical probability theory requires such a choice of a complete, non-overlapping set of classes.

In quantum mechanics, we may consider N measurements of the position of a particle, such as an electron. There is then only one physical object, the electron, but probability theory would consider each individual measurement result as an "object." If the position in a single measurement can be determined within a certain precision, the classes c_j may be chosen as non-overlapping intervals that cover the entire x-axis, each one with a length given by the measurement precision. In experiments, the classes c_j are also referred to as bins. For instance, if the number j is an integer, we may define the bin c_j as the x-interval between $j\varepsilon$ and $(j+1)\varepsilon$, where ε is the precision. More precisely, we write $c_j = \{x \in \mathbb{R}: j\varepsilon \leq x < (j+1)\varepsilon\}$ in this example, making sure that the intervals do not overlap even at the boundaries, and cover the full range of x.

It is possible that some individual measurements yield the same result c_j. After N measurements, we therefore expect a certain non-negative integer number $N(j)$ of measurement results in each class c_j. Since we know that we have made N measurements and require that the classes are complete, we have

$$\sum_j N(j) = N\,. \tag{1.46}$$

If the summation range is not mentioned explicitly in an equation, it refers to the complete range required for an inclusion of all possible measurement outcomes or bins c_j. The **probability** for an object (that is, a measurement result) to be in a class c_j is defined as

$$P(j) = \frac{N(j)}{N}\,. \tag{1.47}$$

Using (1.46) and the fact that both N and $N(j)$ are non-negative, we conclude that

$$0 \leq P(j) \leq 1 \tag{1.48}$$

for each j. A value of $P(j) = 0$ means that the outcome given by class c_j is impossible in the given set, and $P(j) = 1$ means that the outcome given by class c_j is certain. In the latter case, all other $N(j)$ are zero.

The probability of finding an object in a subset of all the classes is given by the sum of the corresponding probabilities. For instance, the probability for an object to be in class c_{j_1} or in class c_{j_2} equals

$$P(j_1, j_2) = \frac{N(j_1) + N(j_2)}{N} = P(j_1) + P(j_2)\,. \tag{1.49}$$

For this statement, it is important that the classes do not overlap, such that it is impossible for an object to be in more than one class. The probability of finding an object somewhere, in any one of the possible classes, equals one because we know that the object exists. (In quantum mechanics, we know that we just made a measurement, and therefore should have an outcome.) Therefore,

$$\sum_j P(j) = 1 \tag{1.50}$$

if we sum over all the classes. Using (1.47), we can check directly that this equation is correct:

$$\sum_j P(j) = \frac{1}{N}\sum_j N(j) = \frac{1}{N}N = 1\,. \tag{1.51}$$

More generally, we call any set of non-negative numbers $P(j)$ that satisfies (1.50) a **probability distribution**. The condition (1.50), called the **normalization condition**, makes sure that the set of classes c_j is complete, such that we are guaranteed to find each object in one of the classes.

1.2.1.2 Expectation Values

Given a probability distribution $P(j)$, we define the **average** or **expectation value** of j as

$$\langle j\rangle = \sum_j jP(j)\,. \tag{1.52}$$

The angular brackets are common notation in quantum mechanics and simply indicate that we are computing the average of the class number j. These brackets do not mean that the left-hand side is a function of j, which it is not because we are summing over j on the right, making it impossible to evaluate the expression at some fixed value of j. As another example, if we use a set of colors to describe possible outcomes, we would write something like $\langle\text{color}\rangle$ which would be less likely to be confused with a function. However, we prefer numbers j to describe outcomes because we can then do calculations with $\langle j\rangle$ and use it in other expressions.

We may also be interested in the average not of j directly but of some function of j, given by $f(j)$. We define the average or expectation value of $f(j)$ as

$$\langle f(j)\rangle = \sum_j f(j)P(j)\,. \tag{1.53}$$

Unlike $P(j)$, the function $f(j)$ is not restricted to be positive or to obey the normalization condition (1.50). In general, $\langle f(j)\rangle \neq f(\langle j\rangle)$. The positions of different parentheses therefore matter. The former is the expectation value of the function f applied to each outcome, while the latter is the function f directly applied to the expectation value of all outcomes. As a detailed equation,

$$\langle f(j)\rangle = \sum_j f(j)P(j) \neq f\left(\sum_j jP(j)\right) = f(\langle j\rangle) \tag{1.54}$$

with an inequality in the middle unless f is a linear function. As an example, the concept of statistical variance, to be introduced soon, is based on the inequality $\langle j^2\rangle \neq \langle j\rangle^2$. It is easy to confirm this inequality by using the example of a symmetric probability distribution over the integers, such that $P(-j) = P(j)$, and assuming $P(0) < 1$. In this case, $\langle j\rangle^2 = 0$ while $\langle j^2\rangle$ is the non-zero sum of non-negative numbers with at least one strictly positive contribution thanks to $P(0) < 1$.

If we already computed $\langle j\rangle$ for a probability distribution $P(j)$, we may be interested in the average of the difference $\Delta j = j - \langle j\rangle$, which could be interpreted as a measure of how much the distribution is spread out around $\langle j\rangle$. However, this choice is not useful because we can compute

$$\langle \Delta j\rangle = \langle j - \langle j\rangle\rangle = \sum_j (j - \langle j\rangle)P(j) \tag{1.55}$$

for any $P(j)$. We first expand the sum as $\sum_j jP(j) - \sum_j \langle j \rangle P(j)$. In the second term, remembering that $\langle j \rangle$ is no longer a function of j because we are summing over j in its definition, (1.52), we can pull this factor out and continue with

$$\langle \Delta j \rangle = \sum_j jP(j) - \langle j \rangle \sum_j P(j) = \langle j \rangle - \langle j \rangle \cdot 1 = 0. \tag{1.56}$$

Here, we again used the definition (1.52) in the first term, and the normalization condition (1.50) in the second term. The resulting difference is zero for any probability distribution $P(j)$.

1.2.1.3 Variance and Standard Deviation

In many cases, it is however of interest to determine how much a distribution is spread out around the average. Our preceding attempt failed because positive deviations from $\langle j \rangle$ to the right cancel out completely with negative deviations to the left. We can easily correct this problem by averaging

$$(\Delta j)^2 = (j - \langle j \rangle)^2 = j^2 - 2\langle j \rangle j + \langle j \rangle^2. \tag{1.57}$$

For this function of j, we have the expectation value

$$\langle (\Delta j)^2 \rangle = \langle j^2 - 2\langle j \rangle j + \langle j \rangle^2 \rangle = \langle j^2 \rangle - 2\langle j \rangle \langle j \rangle + \langle j \rangle^2 \tag{1.58}$$

where, as before, we used the fact that $\langle j \rangle$ does not depend on j (therefore, $\langle \langle j \rangle j \rangle = \langle j \rangle \langle j \rangle$) and the normalization condition (1.50) (therefore, $\langle \langle j \rangle^2 \rangle = \langle j \rangle^2$). The result can be simplified to

$$\langle (\Delta j)^2 \rangle = \langle j^2 \rangle - \langle j \rangle^2. \tag{1.59}$$

Notice the different positions of squares and parentheses in the two terms of this expression, which imply that the result is in general non-zero. In probability theory, $\langle (\Delta j)^2 \rangle$ is called the **variance** of the distribution, and it is more commonly expressed through the **standard deviation**

$$\sigma_j = \sqrt{\langle (\Delta j)^2 \rangle} = \sqrt{\langle j^2 \rangle - \langle j \rangle^2}. \tag{1.60}$$

If the physical property described by j has units, such as the position x of a particle, the expectation value $\langle j \rangle$ and the standard deviation σ_j have the same units and can easily be compared. The calculation in (1.59) shows that the standard deviation is always real, even though it is defined as the square root of a difference of two numbers.

As an example, consider the probability distribution defined by $P(9) = P(10) = 1/2$ and $P(j) = 0$ for all other j. We choose the finite range of integers j from one to ten as our bins. It is easy to compute the average $\langle j \rangle = 9.5$ of this distribution. The standard deviation requires slightly more work and follows from

$$\sigma_j = \sqrt{\frac{1}{2}(9^2 + 10^2) - 9.5^2} = \sqrt{\frac{1}{2}(9^2 + 10^2) - \left(\frac{9+10}{2}\right)^2} = \sqrt{\frac{(9-10)^2}{4}} = \frac{1}{2} \tag{1.61}$$

Looking at Figure 1.7, this result is not unexpected because there are two bins of size 1 right next to each other, such that their centers are displaced from the average value by a distance of $1/2$.

In quantum mechanics, the average is usually referred to as the expectation value. The suggested meaning is that this value gives us the most likely outcome of a measurement, which we therefore expect to happen. However, the name may sometimes be confusing because in a bimodal or multimodal distribution (one with two or more peaks) it may happen that the expectation value is never realized as a single outcome. A simple example is shown in Figure 1.8, a slight modification of Figure 1.7. In this distribution, with $P(8) = P(10) = 1/2$, we compute $\langle j \rangle = 9$, but we have $P(9) = 0$ which implies that the outcome 9 is never obtained: $P(\langle j \rangle) = 0$. The expectation value therefore does not refer to the outcome of a single measurement or a single choice of an object, but only to the average value obtained in multiple measurements. The standard deviation of the probability distribution shown in Figure 1.8 equals $\sigma = 1$.

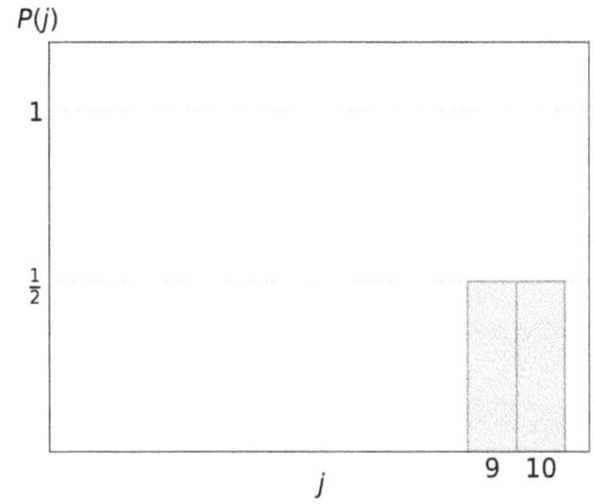

Figure 1.7 Example of a probability distribution with $P(9) = P(10) = 1/2$ and $P(j) = 0$ for all other j.

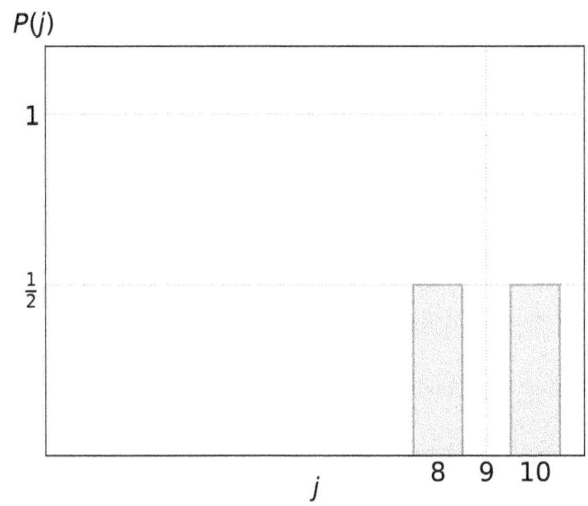

Figure 1.8 Example of a probability distribution with $P(8) = P(10) = 1/2$ and $P(j) = 0$ for all other j.

1.2.1.4 Probability Density

Measurements are always made with a certain precision, which implies that their results can always be collected in discrete bins. Nevertheless, it is often useful to work with an idealized situation in which the size of the bins shrinks to zero. The mathematical description is then independent of experimental features such as a detailed estimate of the precision. In the limit in which the bin size is taken to zero we obtain a continuous distribution of possible outcomes or classes c_j, each of which has an infinitesimal size.

If we only reduce the bin sizes, increasing their number in order to ensure completeness, it becomes less and less likely that a given bin contains an outcome or an object. Most probabilities $N(j)/N$ in the distribution will then be zero, unless we collect more objects or more measurement results. It is then convenient to take a limit of a potentially infinite number of outcomes, turning the probability distribution into a continuous function $\rho(x)$ of a continuous variable x, as shown in Figure 1.9. Since we never make an infinite number of measurements, it is not meaningful to interpret a single value of $\rho(x)$ as a probability. Instead, we use this function as a **probability density**: We are free to choose any range $x \in [a,b]$, such that the probability of finding x in the range

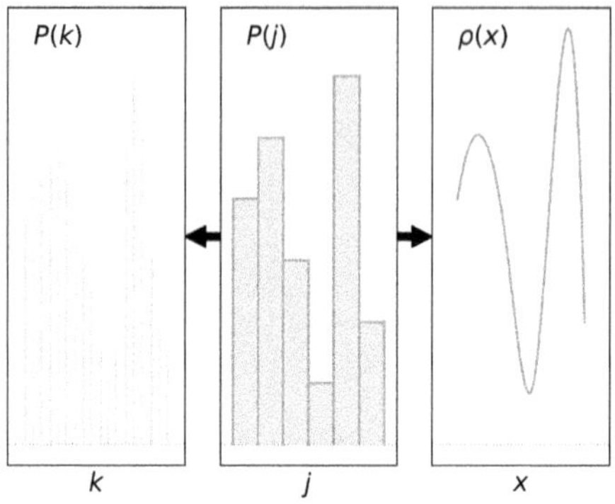

Figure 1.9 Two limits of a probability distribution (center) for vanishing bin sizes. On the left, we obtain a distribution that is still discrete but has finitely many very narrow peaks. On the right, the number of bins grows to infinity as their sizes shrink. The result is a continuous function, or a probability density, depending on a real number x.

$a \leq x \leq b$ is given by

$$P_{ab} = \int_a^b \rho(x) \mathrm{d}x \,. \tag{1.62}$$

A probability density therefore allows us to work with flexible bin sizes and positions, depending on the questions we are trying to answer. If we know that there is an object, it is certain that we find it at some x. Thus, we obtain the **normalization condition**

$$\int_{-\infty}^{\infty} \rho(x) \mathrm{d}x = 1 \,. \tag{1.63}$$

If x takes values in a finite range, (1.63) can still be used assuming that $\rho(x)$ vanishes outside of this range.

The integral in (1.62) is motivated by the additive behavior of probabilities: Before we reach the limit of infinitesimal bin size, the probability of finding an object in a range of bins c_j, where $j_1 \leq j \leq j_2$ is given by

$$P(j_1, j_2) = \sum_{j=j_1}^{j_2} P(j) \,. \tag{1.64}$$

In the limit of infinitesimal sizes, the sum is turned into an integral. A continuous probability distribution is called a probability density because only the integral (1.62) but not an individual value of $\rho(x)$ can be interpreted as a probability. Correspondingly, the function $\rho(x)$ is not subject to the condition that it always be less than or equal to one for all x. Only non-negativity $\rho(x) \geq 0$ and normalization are required properties.

The expectation value of x or a function $f(x)$ in a probability density $\rho(x)$ is defined as

$$\langle f(x) \rangle = \int_{-\infty}^{\infty} f(x) \rho(x) \mathrm{d}x \,. \tag{1.65}$$

The standard deviation looks the same as before,

$$\sigma_x = \sqrt{\langle x^2 \rangle - \langle x \rangle^2} \,, \tag{1.66}$$

where the two expectation values are computed using the corresponding functions in (1.65).

As an example, consider the **Gaussian** probability density

$$\rho(x) = A\exp(-\lambda(x-a)^2) \tag{1.67}$$

with three constants A, λ and a. We will need Gaussian integrals in order to derive the normalization condition as well as expectation values. Using substitution or integration by parts, the required integrals can be reduced to the basic Gaussian integral

$$\int_{-\infty}^{\infty} e^{-z^2}\mathrm{d}z = \sqrt{\pi}. \tag{1.68}$$

This value (or half this value if we start integrating at zero rather than $-\infty$) can be obtained from standard integrations if we first square the integral, interpret the square as an integral on the 2-dimensional plane, and transform to polar coordinates:

$$\begin{aligned}\left(\int_0^{\infty} e^{-z^2}\mathrm{d}z\right)^2 &= \int_0^{\infty}\int_0^{\infty} e^{-(x^2+y^2)}\mathrm{d}x\mathrm{d}y = \int_0^{\infty}\int_0^{\pi/2} e^{-r^2} r\mathrm{d}r\mathrm{d}\varphi \\ &= \frac{\pi}{2}\left(-\frac{1}{2}e^{-r^2}\right)\Big|_0^{\infty} = \frac{\pi}{4}.\end{aligned} \tag{1.69}$$

Here, we used $r^2 = x^2 + y^2$ and $\mathrm{d}x\mathrm{d}y = r\mathrm{d}r\mathrm{d}\varphi$, and integrate over the first quadrant where φ ranges from 0 to $\pi/2$. Taking a square root results in

$$\int_0^{\infty} e^{-z^2}\mathrm{d}z = \frac{\sqrt{\pi}}{2}. \tag{1.70}$$

Reflection symmetry of the integrand then implies (1.68).

We can now proceed with using (1.67) as a probability density. We first make sure that it is normalized:

$$1 = \int_{-\infty}^{\infty} \rho(x)\mathrm{d}x = A\int_{-\infty}^{\infty} e^{-\lambda x^2}\mathrm{d}x = A\sqrt{\frac{\pi}{\lambda}} \tag{1.71}$$

where we obtained the final expression from (1.68) after substituting $z = \sqrt{\lambda}x$. The normalization condition therefore requires

$$A = \sqrt{\frac{\lambda}{\pi}} \tag{1.72}$$

while λ and a remain free at this point.

The expectation value of x in (1.67) is given by

$$\langle x\rangle = \sqrt{\frac{\lambda}{\pi}}\int_{-\infty}^{\infty} xe^{-\lambda(x-a)^2}\mathrm{d}x = \sqrt{\frac{\lambda}{\pi}}\int_{-\infty}^{\infty}(y+a)e^{-\lambda y^2}\mathrm{d}y = a \tag{1.73}$$

where we substituted $y = x-a$ and used the fact that

$$\int_{-\infty}^{\infty} ye^{-\lambda y^2}\mathrm{d}y = 0 \tag{1.74}$$

because we are integrating an odd function over a symmetric range. The constant a in (1.67) is therefore equal to the expectation value of x in this probability density. This result is not surprising because the Gaussian function (1.67) is symmetric around $x = a$, in which case the expected value of x is at the center of the distribution.

The variance requires additional integrations. We first compute

$$\begin{aligned}\langle x^2\rangle &= \sqrt{\frac{\lambda}{\pi}}\int_{-\infty}^{\infty} x^2e^{-\lambda(x-a)^2}\mathrm{d}x = \sqrt{\frac{\lambda}{\pi}}\int_{-\infty}^{\infty}(y+a)^2e^{-\lambda y^2}\mathrm{d}y \\ &= \sqrt{\frac{\lambda}{\pi}}\int_{-\infty}^{\infty}(y^2+2ay+a^2)e^{-\lambda y^2}\mathrm{d}y\end{aligned} \tag{1.75}$$

using again the substitution $y = x - a$. Linearity allows us to split this integral into three. The last one can be obtained from the normalization condition, the middle one vanishes because we are integrating an odd function over a symmetric range, and the first one can be derived using integration by parts in

$$\int_{-\infty}^{\infty} e^{-\lambda y^2} \mathrm{d}y = -\int_{-\infty}^{\infty} y(-2\lambda y e^{-\lambda y^2}) \mathrm{d}y = 2\lambda \int_{-\infty}^{\infty} y^2 e^{-\lambda y^2} \mathrm{d}y. \tag{1.76}$$

This calculation follows the general rule

$$\int u'v \mathrm{d}y = uv - \int uv' \mathrm{d}y \tag{1.77}$$

for $u = y$ and $v = e^{-\lambda y^2}$. Since the integral on the left-hand side of (1.76) equals $\sqrt{\pi/\lambda}$, we obtain

$$\int_{-\infty}^{\infty} y^2 e^{-\lambda y^2} \mathrm{d}y = \frac{1}{2\lambda}\sqrt{\frac{\pi}{\lambda}} \tag{1.78}$$

from the right-hand side. Alternatively, we can compute this integral by taking a derivative of $\int_{-\infty}^{\infty} e^{-\lambda y^2} \mathrm{d}y = \sqrt{\pi/\lambda}$ by λ.

With this integral, we obtain

$$\sigma_x = \sqrt{\langle x^2 \rangle - \langle x \rangle^2} = \frac{1}{\sqrt{2\lambda}} \tag{1.79}$$

for the standard deviation of the probability density (1.67). These results allow us to write a normalized Gaussian probability density with expectation value $\langle x \rangle$ and standard deviation σ_x as

$$\rho_{\langle x \rangle, \sigma_x}(x) = \frac{1}{\sqrt{2\pi}\,\sigma_x} \exp\left(-\frac{(x - \langle x \rangle)^2}{2\sigma_x^2}\right). \tag{1.80}$$

The maximum of $\rho_{\langle x \rangle, \sigma_x}(x)$ is at $x = \langle x \rangle$, and if we move a distance of σ_x to the left or the right, we obtain

$$\rho_{\langle x \rangle, \sigma_x}(\langle x \rangle \pm \sigma_x) = \frac{1}{\sqrt{2\pi}\sigma_x} e^{-1/2} \approx 0.6 \rho_{\langle x \rangle, \sigma_x}(\langle x \rangle). \tag{1.81}$$

At $x = \langle x \rangle \pm \sigma_x$, the density is about three fifth of its maximum value at $x = \langle x \rangle$; see Figure 1.10. The standard deviation is therefore a good measure of how much the density is spread out on the x-axis.

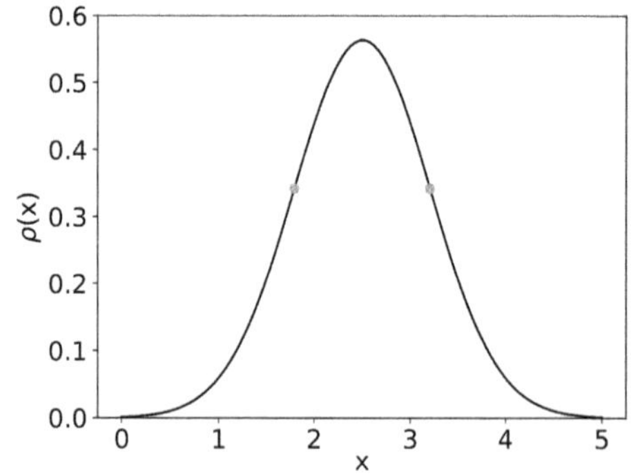

Figure 1.10 Gaussian probability density with $\lambda = 1$ and $a = 2.5$, or $\sigma_x = 1/\sqrt{2} \approx 0.7$ and $\langle x \rangle = 2.5$. The circles show the values at $x = \langle x \rangle \pm \sigma_x$ computed in (1.81).

1.2.2 NORMALIZATION OF THE WAVE FUNCTION

As defined in Eq. (1.45), the wave function $\Psi(x,t)$ determines a probability density $\rho_t(x) = |\Psi(x,t)|^2$ for measurements of the position x of a particle at some time t. The probability of finding the particle somewhere in a range $a \leq x \leq b$ is therefore given by

$$P_{a,b} = \int_a^b |\Psi(x,t)|^2 \mathrm{d}x \,. \tag{1.82}$$

For this statement to be meaningful, $|\Psi(x,t)|^2$ must obey the normalization condition of probability densities, such that the probability of finding the particle anywhere on the x-axis equals one:

$$\int_{-\infty}^{\infty} |\Psi(x,t)|^2 \mathrm{d}x = 1 \quad \text{for all } t \,. \tag{1.83}$$

If this condition were violated, we would only have a fraction of a particle if the integral is between zero and one, or more than one particle if the integral is greater than one.

For the integral (1.83) to be able to equal one, it must first of all be finite. This property is not guaranteed for a generic wave function because we are integrating over an infinite range. A necessary condition for finiteness is that the wave function approach zero in the limit of large $|x|$:

$$\Psi(x,t) \to 0 \quad \text{if } x \to \pm\infty, \quad \text{for all } t \,. \tag{1.84}$$

This condition is also useful if we integrate by parts because it often implies that boundary terms vanish.

If $\int_{-\infty}^{\infty} |\Psi(x,t_0)|^2 \mathrm{d}x = N$ for a fixed t_0 is finite but not equal to one, we can always redefine the wave function by dividing it by $\sqrt{N}$. (The integral, and therefore N, cannot be negative, and it is non-zero for any continuous $\Psi(x,t_0)$ that is not identically zero.) The normalization condition is then satisfied for $\tilde{\Psi}(x,t) = \Psi(x,t)/\sqrt{N}$. Moreover, since the Schrödinger equation (1.17), given by

$$i\hbar \frac{\partial \Psi}{\partial t} = -\frac{\hbar^2}{2m} \frac{\partial^2 \Psi}{\partial x^2} + V(x)\Psi \,, \tag{1.85}$$

is linear in Ψ, it is satisfied by $\tilde{\Psi}(x,t)$ if it is satisfied by $\Psi(x,t)$, provided that N is time-independent.

The normalization integral is in fact time independent, as another consequence of the Schrödinger equation. This demonstration requires a longer calculation and leads to a new concept, the probability current. Time independence is a conservation law. In the present case, if the normalization integral is time-independent, the statistical interpretation of the wave function is conserved and can be maintained in time. More physically, we can interpret this result as a conservation of particle numbers because a value of one in (1.83) means that we have a single particle, which does not disappear or split up into multiple particles. Actual particles can sometimes split up into multiple ones in processes such as pair creation, but these events cannot be described by quantum mechanics and instead require quantum field theory. Quantum field theory works with wave functions of a different form than $\Psi(x,t)$, and therefore we need not be concerned here about changing particle numbers.

A more familiar **conservation law** is that of electric charge. If we define the charge $Q_{a,b}$ in the region $a \leq x \leq b$ through a charge density $\rho(x,t)$, we have an equation

$$Q_{a,b}(t) = \int_a^b \rho(x,t) \mathrm{d}x \tag{1.86}$$

very similar to the probabilities defined by the wave function (but without a normalization or non-negativity condition). If a and b are finite, the charge contained between these points may change if charged particles move through the boundaries. The time derivative of $Q_{a,b}(t)$ may therefore be non-zero. However, it must be related to the sum of charges moving in through the boundaries minus the sum of charges moving out.

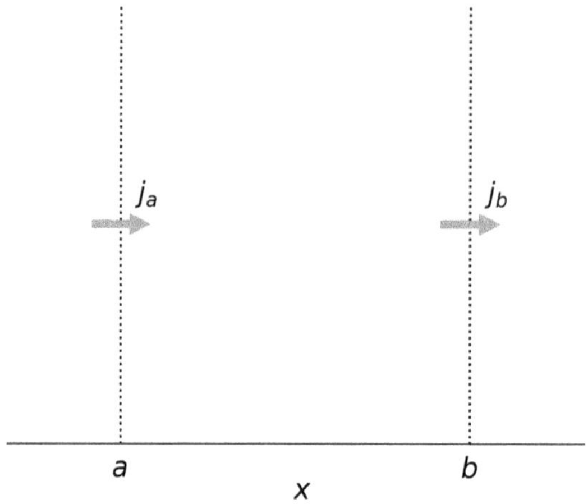

Figure 1.11 Electric currents through boundaries of the region $a \le x \le b$. If positive values indicate currents pointing to the right, charges move into the region on the left if $j_a > 0$ and move out on the right if $j_b > 0$.

Moving charges are described by currents. The charge is conserved, taking into account currents through the boundaries, if it obeys the equation

$$\frac{\mathrm{d}Q_{a,b}(t)}{\mathrm{d}t} = j_a(t) - j_b(t) \tag{1.87}$$

where we use sign conventions as shown in Figure 1.11. There is a positive current if charges are moving toward greater x, and a negative current if charges are moving toward smaller x. This equation is implied upon spatial integration if the charge density $\rho(x,t)$ and the current $j(x,t)$ obey the differential equation

$$\frac{\partial \rho}{\partial t} = -\frac{\partial j}{\partial x}, \tag{1.88}$$

called the **continuity equation.**

For probabilities obtained from the wave function, we know the analog of the charge Q and its density, given by P and $|\Psi|^2$. If they are subject to a conservation law, we should be able to derive a corresponding **probability current** such that an equation of the form (1.88) is satisfied. In order to derive the probability current, we begin by testing the time dependence of the normalization integral by taking a time derivative:

$$\frac{\mathrm{d}}{\mathrm{d}t}\int_{-\infty}^{\infty} |\Psi(x,t)|^2 \mathrm{d}x = \int_{-\infty}^{\infty} \frac{\partial}{\partial t}\left(\Psi(x,t)^*\Psi(x,t)\right)\mathrm{d}x = \int_{-\infty}^{\infty}\left(\frac{\partial \Psi^*}{\partial t}\Psi + \Psi^*\frac{\partial \Psi}{\partial t}\right)\mathrm{d}x. \tag{1.89}$$

We are denoting the derivative by d if it is applied to a function that depends only on one variable, in this case t because x is integrated out on the left. We are using a partial derivative ∂ when it acts on a function of multiple independent variables, here t and x in the integrand.

The time derivatives in the last integral are not arbitrary but determined by the Schrödinger equation. Dividing by $i\hbar$ (recalling $1/i = -i$) and taking a complex conjugate, we obtain

$$\frac{\partial \Psi}{\partial t} = \frac{i\hbar}{2m}\frac{\partial^2 \Psi}{\partial x^2} - \frac{i}{\hbar}V(x)\Psi \quad \text{and} \quad \frac{\partial \Psi^*}{\partial t} = -\frac{i\hbar}{2m}\frac{\partial^2 \Psi^*}{\partial x^2} + \frac{i}{\hbar}V(x)\Psi^*. \tag{1.90}$$

Therefore, the integrand of (1.89) reads

$$\frac{\partial}{\partial t}|\Psi(x,t)|^2 = \left(-\frac{i\hbar}{2m}\frac{\partial^2 \Psi^*}{\partial x^2} + \frac{i}{\hbar}V(x)\Psi^*\right)\Psi + \Psi^*\left(\frac{i\hbar}{2m}\frac{\partial^2 \Psi}{\partial x^2} - \frac{i}{\hbar}V(x)\Psi\right)$$

$$= -\frac{i\hbar}{2m}\left(\frac{\partial^2\Psi^*}{\partial x^2}\Psi - \Psi^*\frac{\partial^2\Psi}{\partial x^2}\right). \tag{1.91}$$

The left-hand side is equal to $\partial\rho/\partial t$, just as needed for a continuity equation (1.88). If there is a corresponding probability current, the right-hand side must be a derivative by x, which is not what it looks like right now. However, if we add and subtract the term $(\partial\Psi^*/\partial x)(\partial\Psi/\partial x)$, an intervention that does not change the validity of the equation, we have

$$\frac{\partial}{\partial t}|\Psi(x,t)|^2 = -\frac{i\hbar}{2m}\left(\frac{\partial^2\Psi^*}{\partial x^2}\Psi - \Psi^*\frac{\partial^2\Psi}{\partial x^2} + \frac{\partial\Psi^*}{\partial x}\frac{\partial\Psi}{\partial x} - \frac{\partial\Psi^*}{\partial x}\frac{\partial\Psi}{\partial x}\right) = -\frac{\partial}{\partial x}j(x,t) \tag{1.92}$$

with the probability current

$$j(x,t) = \frac{i\hbar}{2m}\left(\frac{\partial\Psi^*}{\partial x}\Psi - \Psi^*\frac{\partial\Psi}{\partial x}\right). \tag{1.93}$$

The existence of a probability current demonstrates that the normalization integral is time-independent:

$$\frac{\mathrm{d}}{\mathrm{d}t}\int_{-\infty}^{\infty}|\Psi(x,t)|^2\mathrm{d}x = -\int_{-\infty}^{\infty}\frac{\partial j(x,t)}{\partial x}\mathrm{d}x = -\left. j(x,t)\right|_{x=-\infty}^{x=\infty} = 0 \tag{1.94}$$

where we used the condition that Ψ and Ψ^* in j must approach zero at $x \to \pm\infty$. It is therefore sufficient to normalize the wave function at one time. If it solves the Schrödinger equation, it will remain normalized at all times.

1.3 PHYSICAL PROPERTIES

A probability density $\rho(x)$ depending on a real number x can be used to describe the statistics of measurements of the position x of a particle. In classical physics, the use of a probability density rather than a single sharp value x might be required in order to take into account unavoidable experimental imprecision of a position measurement, for instance because the particle might be moved slightly by pressure fluctuations in the surrounding air, or if it is charged and in vacuum, might be deflected slightly by stray magnetic fields caused by the surrounding electronic devices. Individual measurement results then lie in a range around the expectation value with a size given by the standard deviation.

1.3.1 POSITION MEASUREMENTS

In quantum mechanics, a probability density $\rho_t(x) = |\Psi(x,t)|^2$ is used because the position is fundamentally determined by a wave function, which is spread out and does not imply a single sharp value x for a measurement. Since the wave function usually depends on time, the probability density and therefore the position expectation value

$$\langle x\rangle(t) = \int_{-\infty}^{\infty} x\rho_t(x)\mathrm{d}x = \int_{-\infty}^{\infty} x|\Psi(x,t)|^2\mathrm{d}x \tag{1.95}$$

also depend on time.

The statistical interpretation requires us to make multiple measurements because a single one would not tell us how much the distribution is spread out. In order to do multiple measurements in a way suitable for the underlying statistics, we should set up the same experiment several times in an identical way that corresponds to the statistics of $\Psi(x,t_0)$ at some initial t_0, and let each of these instances run for the same amount of time, Δt, before we record the position measurement. We do not have to start all of them at exactly the same time according to a laboratory clock, which may be difficult in practice. But all instances should start by preparing the same $\Psi(x,t_0)$, and the same time

Δt should lapse between the initial setup and the measurement. In this way, we obtain independent measurements that can be used to determine the statistics at a given time $t = t_0 + \Delta t$. The average of the outcomes gives us the expectation value $\langle x \rangle(t)$ at time t, and the standard deviation gives as $\sigma_x(t)$ at the same time.

Since the statistical interpretation (valid at constant t) and the time dependence are independent features of a moving particle in quantum mechanics, we should be careful not to confuse them. For instance, it would not be correct to use a single particle and perform position measurements at different times. In this way, we would also obtain multiple measurements, but their statistics are not described by our probability density $\rho_t(x)$, which determines expectation values such as (1.95) by integrating over x at a fixed time t. Physically, the procedure of using multiple measurements done at different times does not describe the correct position expectation value, even if the particle is slow-moving, because every measurement interacts with the particle (for instance by light being scattered off the particle in a microscope) and changes the statistics. The measurement at the second time would therefore observe a different statistics than the measurement done at the first time, such that there is not a single probability density for the outcomes of this procedure.

As the notation $\langle x \rangle(t)$ indicates, we first compute the expectation value, indicated by brackets $\langle \cdot \rangle$ around x, by integrating over x at a given time t, which in practice requires multiple measurement outcomes taken at this time. If we are interested in how the particle is moving, on average, we can repeat this procedure at later times, using multiple measurements of x at each time value that we are interested in. The results $\{\langle x \rangle(t_1), \langle x \rangle(t_2), \ldots\}$ can then be interpolated to obtain a curve $\langle x \rangle(t)$ for the time dependence of the expectation value.

Performing a single position measurement in quantum mechanics has the following effect on the system. After the measurement is complete, the position of the particle becomes certain because we have just measured it, but only within a certain precision range of the measurement procedure. A measurement therefore changes the wave function drastically because it reduces its initial spread to be within its precision range. It implies a sharper density that is non-zero only in a certain interval around the measured value, x_0.

The transition from a spread-out wave function to a more narrow peak in a measurement is referred to as the "collapse" of the wave function, but it is only the support of the wave function on the x-axis that collapses. (The support of a function is the set of all points where it is non-zero.) The wave function is still there and remains normalized because it retains its statistical meaning. The peak value therefore increases if the support gets narrowed down. There is no physical collapse, just our information about the position collapses to a sharper value because we have just measured it.

The precise wave function after a measurement depends on details of how the experiment is performed. This difficulty can be addressed in two ways:

1. Many observable features of quantum mechanics are related to discrete quantities such that individual measurement results are restricted to a discrete subset of the real line, such as some integers. This property holds irrespective of wave functions describing the system and is realized before any measurement is made. If the measurement precision is smaller than the spacing between the allowed discrete values, any outcome amounts to a sharp value without any spread. Mathematically, we can use a probability distribution rather than a density. The support of the collapsed wave function is then a single point and we do not need any additional information to determine how it is spread out.
2. A measurement of this form is called **repeatable** because a second measurement performed right after the first one (leaving no time for dynamics or interactions to change the system) yields the same value: the first measurement causes the initial wave function to collapse to a certain probability distribution. Even in the discrete case, repeatability may be an idealization because measurement procedures interact with the system in order to extract information. A skillfully designed measurement minimizes such interactions, but they cannot always be completely removed. A complete description of measurements in quantum mechanics

therefore requires a generalization of the collapse statement to a mathematical operation that can take into account non-repeatability. Such descriptions exist, and they can also be used for a better formulation of measurements of continuous quantities, such as the position. We will need to develop more formalism before we can discuss a suitable mathematical formulation in Section 5.3.2.

There is another implication of the latter complicating factor in real experiments, given by the fact that they function by letting the measured particle interact with other physical degrees of freedom such as light in a microscope. In a complete description, not only the particle but also light and other relevant parts of the measurement apparatus should be described by a wave function combined with $\Psi(x,t)$ for the position. A measurement then introduces **correlations** between the degrees of freedom because a position x_0 is related to a corresponding display on the apparatus. The quantum description of a real experiment therefore requires wave functions that depend on a large number of independent variables, and are subject to partial differential equations with many independent derivatives. Understanding the measurement process in this way is still an open and very active research question.

1.3.2 MOMENTUM

If no active measurement is performed and time simply passes, the position expectation value should change because the wave function evolves according to the Schrödinger equation. We can compute this change using the latter equation. The first part of this calculation follows directly from an integration of x times the expressions in (1.92)–(1.93):

$$\frac{\mathrm{d}\langle x\rangle}{\mathrm{d}t} = \int_{-\infty}^{\infty} x\frac{\partial}{\partial t}(\Psi^*\Psi)\,\mathrm{d}x = -\frac{i\hbar}{2m}\int_{-\infty}^{\infty} x\frac{\partial}{\partial x}\left(\frac{\partial\Psi^*}{\partial x}\Psi - \Psi^*\frac{\partial\Psi}{\partial x}\right)\mathrm{d}x. \tag{1.96}$$

The last result is simplified if we integrate by parts,

$$\frac{\mathrm{d}\langle x\rangle}{\mathrm{d}t} = \frac{i\hbar}{2m}\int_{-\infty}^{\infty}\left(\frac{\partial\Psi^*}{\partial x}\Psi - \Psi^*\frac{\partial\Psi}{\partial x}\right)\mathrm{d}x, \tag{1.97}$$

where the boundary term vanishes because a normalizable Ψ or Ψ^* vanishes at infinity. Integrating by parts again gives

$$\frac{\mathrm{d}\langle x\rangle}{\mathrm{d}t} = -\frac{i\hbar}{m}\int_{-\infty}^{\infty}\Psi^*\frac{\partial\Psi}{\partial x}\mathrm{d}x. \tag{1.98}$$

Since we have taken a time derivative of the position expectation value, we can define it as the velocity expectation value,

$$\langle v\rangle = \frac{\mathrm{d}\langle x\rangle}{\mathrm{d}t} \tag{1.99}$$

or, since (1.98) explicitly depends on the mass, introduce the **momentum expectation value**

$$\langle p\rangle = m\langle v\rangle = \int_{-\infty}^{\infty}\Psi^*\left(\frac{\hbar}{i}\frac{\partial}{\partial x}\right)\Psi\mathrm{d}x. \tag{1.100}$$

This definition, unlike $\langle v\rangle$, does not explicitly depend on m. In this sense, the momentum expectation value is more universal, or more fundamental, than the velocity expectation value because the same expression can be used for all particles.

We started with a probability density for position values, and derived an expectation value for measurements of the momentum, even though we do not (yet) have a probability density for momentum measurements. The momentum is independent of the position because different particles can have different momentum values at the same position, and these values, unlike the force, are not determined by the potential. We were able to derive this independent information because the wave function depends on time. The Schrödinger equation, used in the derivation, then provides additional mathematical properties that uniquely determine the expression for $\langle p\rangle$. The complex-valuedness of the wave function also plays a role: The position expectation value

$$\langle x\rangle = \int_{-\infty}^{\infty} x|\Psi|^2 \mathrm{d}x = \int_{-\infty}^{\infty} \Psi^* x \Psi \mathrm{d}x \tag{1.101}$$

depends only on the norm of the wave function. Since a complex number $\Psi(x,t)$ is equivalent to two real numbers, there is an additional degree of freedom that independently determines the momentum expectation value. The derivative in (1.100) extracts this degree of freedom.

We can now define additional statistical properties, such as the momentum variance, by squaring the p-expression in the momentum expectation value (1.100). Unlike x in (1.101), the corresponding expression in the momentum is the partial derivative $(\hbar/i)\partial/\partial x$, Squaring it means taking a second-order derivative. The momentum variance is therefore defined as

$$\sigma_p^2 = \int_{-\infty}^{\infty} \Psi^* \left(-\hbar^2 \frac{\partial^2 \Psi}{\partial x^2}\right) \mathrm{d}x - \left(\int_{-\infty}^{\infty} \Psi^* \frac{\hbar}{i}\frac{\partial \Psi}{\partial x} \mathrm{d}x\right)^2 . \tag{1.102}$$

In contrast to σ_x, it is not easy to estimate the value of σ_p by looking at the graph of a wave function $\Psi(x,t)$. As an example, $\sigma_p = 0$ for a constant function with vanishing derivatives, but as a wave function this choice is not permissible because it is not normalizable. Nevertheless, we may have a wave function that is close to a constant or a plane wave (which has constant norm) in a wide range of x and then falls off to zero at plus and minus infinity. Thus, a wave function with small σ_p looks very different from a wave function with small σ_x.

Individual position measurements spread less around the expectation value $\langle x\rangle$ if the wave function is sharply peaked, such as a Gaussian with small σ_x. However, this is not the case for momentum measurements. Since momentum expectation values are obtained from a derivative operation, we need a nearly constant derivative for a small momentum variance. A sharply peaked Gaussian does not obey this condition. Instead, we have small momentum variance if the wave function is close to a plane wave, $\Psi(x,t_0) \propto e^{ikx}$ at some time t_0, with a constant k. Applying the derivative in the momentum operation then produces a factor of $\hbar k = 2\pi\hbar/\lambda$ with the wave length λ. A momentum measurement in quantum mechanics is a measurement of the wave length of the wave function. In order to have a sharp outcome, we therefore need a stretch of a plane wave that is sufficiently long, containing several wave lengths such that we can unambiguously read off the wave function. A plane wave several wave lengths long, however, has a large position variance.

If we have small position variance with large momentum variance, the wave function is narrow at an initial time but has contributions that move at different momenta. We expect the initially narrow peak to spread out quickly, and the position will be much less certain at later times. If we are interested in staying close to a sharply defined position as the wave function evolves, we need small position variance together with small momentum variance. We have to find a compromise between the spread in position and the spread in wave length. Such a state is called a wave packet and can be pictured as in Figure 1.12. With additional methods, we will later demonstrate that the product of the standard deviations of position and momentum always have to obey the inequality

$$\sigma_x \sigma_p \geq \frac{\hbar}{2} \tag{1.103}$$

called Heisenberg's **uncertainty relation**. Therefore, σ_x can be small only at the price of σ_p being large, and vice versa. It is impossible to have $\sigma_x = 0$ and $\sigma_p = 0$, which would correspond to the classical behavior of a sharp instant on a particle trajectory. The wave function implies fundamental uncertainty expressed by non-zero standard deviations.

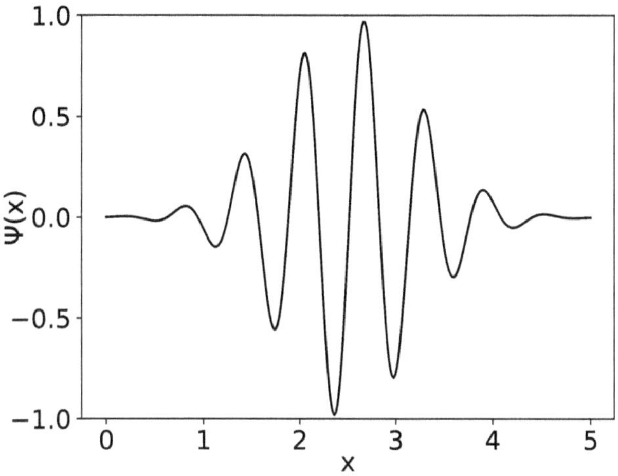

Figure 1.12 Example of a (non-normalized) wave packet $\Psi(x)$, given by $\sin(10x)e^{-(x-2.5)^2}$.

1.3.3 OPERATORS

The form of the momentum and position expectation values, (1.100) and (1.101), suggests a useful notation. The two expressions differ only by the mathematical operation inserted between Ψ^* and Ψ in the integral, given by multiplication with x in (1.101) and the x-derivative times $\hbar/i$ in (1.100). Mathematically, these operations are called **operators** and denoted with a hat to indicate that they are not real or complex numbers:

$$\hat{x} = x \quad \text{and} \quad \hat{p} = \frac{\hbar}{i}\frac{\partial}{\partial x}. \tag{1.104}$$

The first operator, in this notation, is not the value x of a specific particle, but rather the mathematical operation that multiplies a function $\Psi(x,t)$ by x, producing a new function $x\Psi(x,t)$.

Using the two basic operators (1.104), we can turn any function $Q(x,p)$ of x and p into an operator by defining

$$\widehat{Q}(x,p) = Q(\hat{x},\hat{p}). \tag{1.105}$$

In general we have to be careful because multiplying the derivative of a function with x is not the same as taking the derivative of the function multiplied by x: $xf' \neq (xf)'$ if $f \neq 0$. Products of x and p therefore do not have a unique operator version, but there are interesting functions that do not contain such a product and therefore are assigned unique operators. If we have an operator, we can define the corresponding expectation value by following the pattern seen in (1.101) and (1.100):

$$\langle \widehat{Q}(x,p) \rangle = \int_{-\infty}^{\infty} \Psi^* Q\left(x, \frac{\hbar}{i}\frac{\partial}{\partial x}\right) \Psi \mathrm{d}x \tag{1.106}$$

and the standard deviation

$$\sigma_Q = \sqrt{\langle \widehat{Q}^2 \rangle - \langle \widehat{Q} \rangle^2}. \tag{1.107}$$

From now on, we place hatted symbols in expectation values in order to indicate the underlying operator.

Useful examples are the kinetic and potential energy, classically given by $T(p) = p^2/(2m)$ and $V(x)$, respectively. Since they depend on either x or p, they correspond to uniquely defined operators in quantum mechanics,

$$\hat{T} = -\frac{\hbar^2}{2m}\frac{\partial^2}{\partial x^2} \quad \text{and} \quad \hat{V} = V(x), \tag{1.108}$$

the latter as a multiplication operator just like x itself. The kinetic and potential energy therefore have expectation values

$$\langle \hat{T} \rangle = -\frac{\hbar^2}{2m}\int_{-\infty}^{\infty} \Psi^* \frac{\partial^2 \Psi}{\partial x^2}\mathrm{d}x \tag{1.109}$$

and

$$\langle \hat{V} \rangle = \int_{-\infty}^{\infty} V(x)|\Psi(x)|^2 \mathrm{d}x. \tag{1.110}$$

Their sum is the energy of the particle, with operator

$$\hat{E} = \hat{T} + \hat{V} = -\frac{\hbar^2}{2m}\frac{\partial^2}{\partial x^2} + V(x) \tag{1.111}$$

and expectation value

$$\langle \hat{E} \rangle = \int_{-\infty}^{\infty} \Psi^* \left(-\frac{\hbar^2}{2m}\frac{\partial^2}{\partial x^2} + V(x)\right)\Psi \mathrm{d}x. \tag{1.112}$$

Acting on a wave function that solves Schrödinger's equation, this operator produces the same result as

$$\hat{E} = i\hbar\frac{\partial}{\partial t} \tag{1.113}$$

with expectation value

$$\langle \hat{E} \rangle = i\hbar \int_{-\infty}^{\infty} \Psi^* \frac{\partial \Psi}{\partial t}\mathrm{d}x. \tag{1.114}$$

The operators $\hat{p}$ and $\hat{E}$ provide a meaningful mathematical explanation of the heuristic relations (1.15), and they specify the previously undetermined signs. Note, however, that the dependence of Ψ on t is conceptually different from the dependence on x because it does not participate in the statistical role of the probability density. Correspondingly, the formal operator $\hat{E}$ has to be distinguished from the operator $\hat{p}$. We will revisit the challenging nature of time in quantum mechanics throughout this book.

1.3.4 SEMICLASSICAL DYNAMICS

Further dynamical properties are revealed by taking time derivatives of $\langle \hat{x} \rangle(t)$ beyond first order. At second order, we have

$$m\frac{\mathrm{d}^2\langle \hat{x} \rangle}{\mathrm{d}t^2} = \frac{\mathrm{d}\langle \hat{p} \rangle}{\mathrm{d}t} = \frac{\hbar}{i}\int_{-\infty}^{\infty}\frac{\partial}{\partial t}\left(\Psi^*\frac{\partial \Psi}{\partial x}\right)\mathrm{d}x = \frac{\hbar}{i}\int_{-\infty}^{\infty}\left(\frac{\partial \Psi^*}{\partial t}\frac{\partial \Psi}{\partial x} + \Psi^*\frac{\partial^2 \Psi}{\partial x \partial t}\right)\mathrm{d}x \tag{1.115}$$

following our previous strategy. Inserting the Schrödinger equation turns this expression into

$$m\frac{\mathrm{d}^2\langle \hat{x} \rangle}{\mathrm{d}t^2} = \frac{\hbar}{i}\int_{-\infty}^{\infty}\left(\left(-\frac{i\hbar}{2m}\frac{\partial^2 \Psi^*}{\partial x^2} + \frac{i}{\hbar}V(x)\Psi^*\right)\frac{\partial \Psi}{\partial x} + \Psi^*\frac{\partial}{\partial x}\left(\frac{i\hbar}{2m}\frac{\partial^2 \Psi}{\partial x^2} - \frac{i}{\hbar}V(x)\Psi\right)\right)\mathrm{d}x \tag{1.116}$$

in which the derivative terms cancel out upon integration by parts. We are left with

$$m\frac{\mathrm{d}^2\langle \hat{x} \rangle}{\mathrm{d}t^2} = \int_{-\infty}^{\infty}\left(V(x)\Psi^*\frac{\partial \Psi}{\partial x} - \Psi^*\frac{\partial}{\partial x}(V(x)\Psi)\right)\mathrm{d}x = -\int_{-\infty}^{\infty}\Psi^*\frac{\mathrm{d}V}{\mathrm{d}x}\Psi\mathrm{d}x = -\left\langle \widehat{\frac{\mathrm{d}V}{\mathrm{d}x}} \right\rangle. \tag{1.117}$$

The final result, which is usually referred to as **Ehrenfest's theorem**, is again of the form of an expectation value. The equation closely resembles Newton's second law of classical mechanics, $\mathrm{d}p/\mathrm{d}t = -\mathrm{d}V/\mathrm{d}x$ for a conservative potential. The classical and quantum equations would be identical if $\langle V'(\hat{x}) \rangle$ in (1.117) were the same as $V'(\langle \hat{x} \rangle)$ because we could then just substitute x in the classical equation with $\langle\hat{x}\rangle$. However, these two expressions with differently placed brackets are in general unequal, as a single example demonstrates: If we have a cubic potential, $V(x) = \frac{1}{3}x^3$, the difference of the two expressions,

$$\langle V'(\hat{x})\rangle - V'(\langle\hat{x}\rangle) = \langle\hat{x}^2\rangle - \langle\hat{x}\rangle^2 = \sigma_x^2 \tag{1.118}$$

is equal to the position variance of the wave function, which in most cases is non-zero. Therefore, quantum evolution of expectation values is different from classical evolution.

For some potentials, $\langle V'(\hat{x})\rangle$ is the same as $V'(\langle\hat{x}\rangle)$ for any wave function. This is the case if $V(x)$ is quadratic in x and of the form $V(x) = a + bx + cx^2$ with constant a, b and c. If only a is non-zero, we have a free particle without any forces, the b-term contributes a constant force, and the c-term a harmonic-oscillator force. The derivative of the potential is then linear in x, such that the variance seen in (1.118) does not appear.

For a general potential, we can use a Taylor expansion

$$V(x) = \sum_{k=0}^{\infty} \frac{1}{k!} V^{(k)}(0) x^k \tag{1.119}$$

around $x = 0$, where $V^{(k)}$ is the k-th order derivative of $V(x)$. The expectation value of $V'(x)$ then takes the form

$$\langle V'(\hat{x})\rangle = \left\langle \sum_{k=1}^{\infty} \frac{1}{(k-1)!} V^{(k)}(0)\hat{x}^{k-1} \right\rangle = \sum_{n=0}^{\infty} \frac{1}{n!} V^{(n+1)}(0)\langle\hat{x}^n\rangle \tag{1.120}$$

after substituting $n = k-1$, while V' evaluated at $\langle\hat{x}\rangle$ equals

$$V'(\langle\hat{x}\rangle) = \sum_{k=1}^{\infty} \frac{1}{(k-1)!} V^{(k)}(0)\langle\hat{x}\rangle^{k-1} = \sum_{n=0}^{\infty} \frac{1}{n!} V^{(n+1)}(0)\langle\hat{x}\rangle^n . \tag{1.121}$$

Comparing coefficients, we need $\langle\hat{x}^n\rangle = \langle\hat{x}\rangle^n$ for the expressions to be equal, but for a generic wave function this is true only if $n \leq 1$ or $k \leq 2$, as the integral formula for expectation values shows:

$$\int_{-\infty}^{\infty} x^n |\Psi(x,t)|^2 \mathrm{d}x \neq \left(\int_{-\infty}^{\infty} x |\Psi(x,t)|^2 \mathrm{d}x \right)^n \tag{1.122}$$

unless $n = 1$ or $n = 0$.

If the probability density $|\Psi(x,t)|^2$ has a sharp peak around $x = \langle\hat{x}\rangle$ at a given t, σ_x and the higher-order versions $\langle\hat{x}^k\rangle - \langle\hat{x}\rangle^k$ (called **moments** of the wave function) are small for all non-zero k. The wave function then approximately follows the classical trajectory because all terms in the Taylor expansion

$$\langle V'(\hat{x})\rangle - V'(\langle\hat{x}\rangle) = \sum_{n=2}^{\infty} \frac{1}{n!} V^{(n+1)}(0) \left(\langle\hat{x}^n\rangle - \langle\hat{x}\rangle^n\right) \tag{1.123}$$

are small. Quantum corrections to classical evolution can be estimated to be as large as $\langle V'(\hat{x})\rangle - V'(\langle\hat{x}\rangle)$ and parameterized by the moments of an evolving wave function. If some of these corrections, for instance the first few terms of the Taylor expansion, are included in the equations of motion, one obtains a **semiclassical approximation** of quantum mechanics.

If we include the first term in the expansion (1.123) in (1.117) as an equation of motion, we obtain

$$m \frac{\mathrm{d}^2\langle\hat{x}\rangle}{\mathrm{d}t^2} = -V'(\langle\hat{x}\rangle) + \frac{1}{2} V'''(0)\sigma_x^2 . \tag{1.124}$$

This equation cannot be solved directly for $\langle\hat{x}\rangle$ unless $V'''(0)=0$ because σ_x is, in general, a time-dependent function just like $\langle\hat{x}\rangle$. We need another differential equation that determines $\mathrm{d}\sigma_x/\mathrm{d}t$ or its second-order derivative. Proceeding as in (1.96), we have

$$\begin{aligned}\frac{\mathrm{d}\langle\hat{x}^2\rangle}{\mathrm{d}t} &= \int_{-\infty}^{\infty} x^2\frac{\partial}{\partial t}(\Psi^*\Psi)\,\mathrm{d}x = -\frac{i\hbar}{2m}\int_{-\infty}^{\infty} x^2\frac{\partial}{\partial x}\left(\frac{\partial\Psi^*}{\partial x}\Psi-\Psi^*\frac{\partial\Psi}{\partial x}\right)\mathrm{d}x \\ &= \frac{i\hbar}{m}\int_{-\infty}^{\infty} x\left(\frac{\partial\Psi^*}{\partial x}\Psi-\Psi^*\frac{\partial\Psi}{\partial x}\right)\mathrm{d}x = \frac{1}{m}\int_{-\infty}^{\infty}\Psi^*\left(x\frac{\hbar}{i}\frac{\partial\Psi}{\partial x}+\frac{\hbar}{i}\frac{\partial(x\Psi)}{\partial x}\right)\mathrm{d}x\,.\end{aligned} \tag{1.125}$$

In the last step, we have moved all operators such that they act on Ψ rather than Ψ^*. We can write this equation in the short form

$$\frac{\mathrm{d}\langle\hat{x}^2\rangle}{\mathrm{d}t} = \frac{1}{m}\langle\hat{x}\hat{p}+\hat{p}\hat{x}\rangle\,, \tag{1.126}$$

an example of an expression that contains the non-commuting operators $\hat{x}$ and $\hat{p}$ in a **symmetric ordering**. The time derivative of the variance $\sigma_x^2=\langle\hat{x}^2\rangle-\langle\hat{x}\rangle^2$ then equals

$$\frac{\mathrm{d}\sigma_x^2}{\mathrm{d}t} = \frac{1}{m}\langle\hat{x}\hat{p}+\hat{p}\hat{x}\rangle - 2\langle\hat{x}\rangle\frac{\langle\hat{p}\rangle}{m} = \frac{2}{m}C_{xp}\,. \tag{1.127}$$

The last step introduces a new parameter, the **covariance**

$$C_{xp} = \frac{1}{2}\langle\hat{x}\hat{p}+\hat{p}\hat{x}\rangle - \langle\hat{x}\rangle\langle\hat{p}\rangle\,, \tag{1.128}$$

as another independent statistical quantity. A more familiar statistical concept is given by **correlations**, defined as

$$c_{xp} = \frac{C_{xp}}{\sigma_x\sigma_p}\,. \tag{1.129}$$

When correlations are non-zero, measurements of x and p are not fully independent. For a correlated wave function, for instance, we may be more likely to find a right-moving particle at positive x, and a left-moving particle at negative x.

Continuing to derive a complete set of differential equations, we now have to determine how the covariance changes in time. We compute

$$\begin{aligned}\frac{\mathrm{d}\langle\hat{x}\hat{p}\rangle}{\mathrm{d}t} &= \frac{\hbar}{i}\int_{-\infty}^{\infty} x\frac{\partial}{\partial t}\left(\Psi^*\frac{\partial\Psi}{\partial x}\right)\mathrm{d}x \\ &= \frac{\hbar}{i}\int_{-\infty}^{\infty} x\left(\left(-\frac{i\hbar}{2m}\frac{\partial^2\Psi^*}{\partial x^2}+\frac{i}{\hbar}V(x)\Psi^*\right)\frac{\partial\Psi}{\partial x}+\Psi^*\frac{\partial}{\partial x}\left(\frac{i\hbar}{2m}\frac{\partial^2\Psi}{\partial x^2}-\frac{i}{\hbar}V(x)\Psi\right)\right)\mathrm{d}x\end{aligned} \tag{1.130}$$

The additional factor of x prevents some of the cancellations seen previously. Moving all operations to Ψ rather than Ψ^*, we obtain

$$\frac{\mathrm{d}\langle\hat{x}\hat{p}\rangle}{\mathrm{d}t} = -\frac{\hbar^2}{m}\int_{-\infty}^{\infty}\Psi^*\frac{\partial^2\Psi}{\partial x^2}\mathrm{d}x - \int_{-\infty}^{\infty}\Psi^* xV'(x)\Psi\mathrm{d}x = \frac{1}{m}\langle\hat{p}^2\rangle - \langle\hat{x}V'(\hat{x})\rangle\,. \tag{1.131}$$

The time derivative of $\langle\hat{p}\hat{x}\rangle$ turns out to be the same, such that

$$\frac{\mathrm{d}C_{xp}}{\mathrm{d}t} = \frac{1}{m}(\langle\hat{p}^2\rangle-\langle\hat{p}\rangle^2) - \langle\hat{x}V'(\hat{x})\rangle + \langle\hat{x}\rangle\langle V'(\hat{x})\rangle\,. \tag{1.132}$$

Using the Taylor expansion of $V'(x)\approx V'(0)+V''(0)x$, the right-hand side depends on the two variances,

$$\frac{\mathrm{d}C_{xp}}{\mathrm{d}t} \approx \frac{\sigma_p^2}{m} - V''(0)\sigma_x^2\,. \tag{1.133}$$

As the last equation of motion, we have

$$\frac{d\sigma_p^2}{dt} \approx -2V''(0)C_{xp} \tag{1.134}$$

after similar derivations.

For a free particle or a particle subject to a constant force, such that $V''(0) = 0$, the momentum variance σ_p^2 is constant and does not change in time. Using this result in (1.133) then implies a linear $C_{xp}(t) = \sigma_p^2 t/m + C_{xp}(0)$ with an initial value $C_{xp}(0)$ at time zero. The position variance is quadratic according to (1.127),

$$\sigma_x^2(t) = \frac{\sigma_p^2 t^2}{m^2} + \frac{2}{m}C_{xp}(0)t + \sigma_x(0)^2 . \tag{1.135}$$

This function has a minimum at

$$t_{\min} = -\frac{mC_{xp}(0)}{\sigma_p^2} \tag{1.136}$$

with the value

$$\sigma_x^2(t_{\min}) = \frac{\sigma_x(0)^2\sigma_p^2 - C_{xp}(0)^2}{\sigma_p^2} . \tag{1.137}$$

Looking back at the uncertainty relation (1.103) applied at time $t_{\min}$, we see that $\sigma_x^2(t_{\min})\sigma_p^2 \geq \hbar^2/4$ and therefore

$$\sigma_x^2\sigma_p^2 - C_{xp}^2 \geq \frac{\hbar^2}{4} \tag{1.138}$$

from (1.137). This inequality holds at any time since our choice of the initial time, which we called $t = 0$, was generic because we did not restrict the initial values of σ_x, σ_p and C_{xp}. Thus, $\sigma_x\sigma_p \geq \sqrt{\frac{1}{4}\hbar^2 + C_{xp}^2}$ gives as a greater lower bound than what is implied by (1.103), whenever $C_{xp} \neq 0$. (We did not include the C_{xp}-term when we applied the uncertainty relation at time $t_{\min}$, but our solutions demonstrate that $C_{xp}(t_{\min})$ vanishes.)

Taking into account additional statistical information, given here by the covariance or correlations between x and p, often allows us to improve results that were initially obtained only from more intuitive concepts such as the variances. When we derive the uncertainty relation, we will see that (1.138) is indeed a sharpened version of (1.103) and is always correct, not just in this example.

EXERCISES

1. Use the classical equations of motion to demonstrate that the energy

$$E = \frac{p^2}{2m} + V(x)$$

of a particle of mass m moving in a potential $V(x)$ is conserved in time.
2. Derive the classically allowed and forbidden regions for the potential $V(x) = 1 - x^2 + x^4$ if the energy is equal to

   (i) $E_1 = 7/8$,
   (ii) $E_2 = 3/2$.

3. Use the chain rule for multiple variables in order to transform the 1-dimensional wave equation

$$\frac{\partial^2 f}{\partial x^2} - \frac{1}{v^2}\frac{\partial^2 f}{\partial t^2} = 0 \tag{1.139}$$

to new independent coordinates $u_\pm = x \pm vt$. Confirm that the new equation takes the form

$$\frac{\partial^2 f}{\partial u_+ \partial u_-} = 0.$$

Use this equation to show that the most general solution of (1.139) is given by

$$f(x,t) = f_1(x+vt) + f_2(x-vt)$$

with arbitrary functions f_1 and f_2 that each depend on a single argument.

4. What are the real and imaginary parts of $\frac{1}{2+i}$ and $\frac{1+2i}{3+4i}$?
5. Compute $|e^{a+ib}|^2$ for real numbers a and b.
6. Prove $\cos(a+b) = \cos a \cos b - \sin a \sin b$ and $\sin(a+b) = \sin a \cos b + \cos a \sin b$, using $e^{i(a+b)} = \cos(a+b) + i\sin(a+b)$ with a and b real.
7. The probability density for measurements of a continuous variable x is given by a normalized non-negative function $\rho(x)$.

   (i) Given an integer j, what is the probability $P(j)$ of finding x between $j-\frac{1}{2}$ and $j+\frac{1}{2}$?
   (ii) Show that the numbers $P(j)$, defined for all integers j, form a probability distribution.
   (iii) We can now define two expectation values, $\langle j\rangle$ using $P(j)$ and $\langle x\rangle$ using $\rho(x)$. Do they always take the same value? Why/why not?
   (iv) Assume that

   $$\rho(x) = c\cdot\begin{cases} x+1 & \text{if } -1\leq x\leq 1 \\ -x+3 & \text{if } 1<x\leq 3 \\ 0 & \text{otherwise}\end{cases}$$

   with a constant $c>0$. Compute the value of c, plot $\rho(x)$ and derive the distribution $P(j)$ associated with $\rho(x)$ as defined above.
   What are the values of $\langle j\rangle$, $\langle x\rangle$, σ_j and σ_x?

8. Consider a wave function $\Psi_\alpha(x,t_0) = (m\omega/\pi\hbar)^{1/4}\exp(E(x))$ with the exponent

$$E(x) = -\frac{m\omega}{2\hbar}x^2 - \sqrt{\frac{2m\omega}{\hbar}}\,xe^{-i\alpha} - \frac{1}{2}(1+e^{-2i\alpha})$$

for some real number α. The constants m and ω may be interpreted as mass and frequency, respectively, of a harmonic oscillator.

   (i) Show that the real part of $E(x)$ is equal to

   $$R(x) = -\frac{m\omega}{2\hbar}\left(x+\sqrt{\frac{2\hbar}{m\omega}}\cos(\alpha)\right)^2.$$

   (ii) Compute the probability density determined by $\Psi_\alpha(x,t_0)$.
   (iii) Derive $\langle x\rangle$ and σ_x by comparing the probability density of $\Psi_\alpha(x,t_0)$ with a Gaussian (1.80).

9. Take a wave function $\Psi(x,t) = (\lambda/\pi)^{1/4}\exp(-\frac{1}{2}\lambda x^2)$ with $\lambda>0$ (real).
Compute σ_x, σ_p, and the product $\sigma_x\sigma_p$.

10. Compute $\langle \hat{x} \rangle$, $\langle \hat{p} \rangle$, σ_x, σ_p and C_{xp} for a complex Gaussian

$$\Psi(x,t_0) = A\exp(-(a+bi)x^2 + (c+di)x + x_0)$$

where A, a, b, c, d and x_0 are real numbers.
Evaluate the uncertainty relations (1.103) and (1.138).

11. Compute the probability current for a plane wave $\Psi(x,t) = A\exp(i(kx - \omega t))$ and compare your result with the particle velocity.
12. Derive (1.134).
13. Define the wave function

$$\Psi_z(x,t) = \left(\frac{m\omega}{\pi\hbar}\right)^{\frac{1}{4}} \exp\left(-\frac{1}{2}\left(i\omega t + |z|^2 - z^2 e^{-2i\omega t} + \frac{m\omega}{\hbar}x^2 - 2i\sqrt{\frac{2m\omega}{\hbar}}zxe^{-i\omega t}\right)\right)$$

for an arbitrary complex number z and a real constant ω.

(i) Show that $\Psi_z(x,t)$ is a solution of the Schrödinger equation with potential $V(x) = \frac{1}{2}m\omega^2x^2$ (the harmonic oscillator).
(ii) Compute the probability density $\rho_z(x,t) = |\Psi_z(x,t)|^2$.
(iii) Derive the time-dependent expectation value $\langle \hat{x} \rangle(t)$ and the variance $\sigma_x(t)$ by comparing the probability density from part (ii) with that of a standard Gaussian.

14. Solve the coupled equations (1.127), (1.133) and (1.134) for a harmonic oscillator, $V(x) = \frac{1}{2}m\omega^2x^2$. Compare with your results for the preceding exercise.

2 Examples

The Schrödinger equation

$$i\hbar\frac{\partial\Psi}{\partial t}=-\frac{\hbar^2}{2m}\frac{\partial^2\Psi}{\partial x^2}+V(x)\Psi \tag{2.1}$$

with a conservative potential $V(x)$ determines the time dependence of wave functions $\Psi(x,t)$ in one spatial dimension. If we can solve this equation, we obtain complete statistical information about the system, at all times t. Compared with evolving expectation values and variances or correlations, the wave function $\Psi(x,t)$ describes infinitely many statistical parameters, which one may label by the infinitely many values $\Psi(x,t_0)$ at any given time t_0. Alternatively, complete statistical information requires knowledge of all the infinitely many moments that appeared in (1.123), as well as their analogs using p and combinations of x with p as in (1.128). In terms of $\Psi(x,t)$, the infinite number of independent parameters that characterize a solution means that we have to solve a partial differential equation rather than coupled ordinary differential equations.

The Schrödinger equation is a linear partial differential equation, but the non-constant coefficient $V(x)$ (unless there is no force) still makes it hard to find solutions. Fortunately, the assumption that $V(x)$ does not depend on t for a conservative force allows us to apply the method of separating variables. This method initially leads to a set of non-evolving wave functions called stationary states, akin to standing waves. These functions do not tell us how particles move and change their statistics, as in the example of (1.135), but they have two major implications. First, they allow us to derive discrete energy values of bound states, and they can be used as a basis to construct fully evolving time-dependent wave functions.

2.1 SEPARATION OF VARIABLES

Separating variables, we look for specific solutions $\Psi(x,t)$ of the form $\Psi(x,t)=\psi(x)f(t)$ with two functions, ψ and f, that each depend only on one of the two independent variables, x and t, respectively. Functions of this form are not the most general solutions of the Schrödinger equation, but they do provide important information. Moreover, we will see that the general solution can be constructed from solutions of separable form.

2.1.1 STATIONARY STATES

Given $\Psi(x,t)=\psi(x)f(t)$, we compute the partial derivatives required for the terms of Schrödinger's equation:

$$\frac{\partial\Psi}{\partial t}=\psi\frac{\mathrm{d}f}{\mathrm{d}t}\quad\text{and}\quad\frac{\partial^2\Psi}{\partial x^2}=\frac{\mathrm{d}^2\psi}{\mathrm{d}x^2}f\,. \tag{2.2}$$

The Schrödinger equation then implies

$$i\hbar\psi\frac{\mathrm{d}f}{\mathrm{d}t}=-\frac{\hbar^2}{2m}f\frac{\mathrm{d}^2\psi}{\mathrm{d}x^2}+V(x)\psi f \tag{2.3}$$

or, upon division by ψf (where this product is non-zero),

$$i\hbar\frac{1}{f(t)}\frac{\mathrm{d}f(t)}{\mathrm{d}t}=-\frac{\hbar^2}{2m}\frac{1}{\psi(x)}\frac{\mathrm{d}^2\psi}{\mathrm{d}x^2}+V(x)\,. \tag{2.4}$$

The left-hand side of this equation does not depend on x, and the right-hand side does not depend on t. Here, it is important that we assumed a conservative force. If the two sides are equal, they can depend on neither x nor t, and therefore both equal the same constant, which we call E.

DOI: 10.1201/9781003588702-3

The original partial differential equation, applied to separable solutions, then implies two independent ordinary differential equations,

$$E = i\hbar \frac{1}{f_E(t)} \frac{\mathrm{d}f_E(t)}{\mathrm{d}t} = i\hbar \frac{\mathrm{d}\ln f_E(t)}{\mathrm{d}t} \tag{2.5}$$

and

$$-\frac{\hbar^2}{2m}\frac{\mathrm{d}^2\psi_E(x)}{\mathrm{d}x^2} + V(x)\psi_E(x) = E\psi_E(x)\,. \tag{2.6}$$

In both cases, the subscript E on f and ψ indicates that the solutions depend on E because this constant is a new coefficient in these equations. The first equation, (2.5), is independent of the potential and is solved by

$$f_E(t) = f_0 \exp(-iEt/\hbar) \tag{2.7}$$

with an integration constant f_0 that we may set equal to one because it can be combined with ψ in the product ψf.

The second equation, (2.6) is called the **time-independent Schrödinger equation**. It depends on the potential and cannot be solved in general. It is an ordinary differential equation (if we consider only one spatial dimension, as we do for now) and has no complex coefficients. Solutions may therefore be real, or if there is a complex solution, its real and imaginary parts provide two independent real solutions: Since the equation is linear in ψ_E, if ψ_E is a solution, the complex conjugate ψ_E^* is also a solution because the differential equation has real coefficients. By linearity, the real part $\frac{1}{2}(\psi_E + \psi_E^*)$ and the imaginary part $-\frac{1}{2}i(\psi_E - \psi_E^*)$ are then solutions. Instead of looking for complex solutions, it is therefore sufficient to look for a complete set of real solutions.

The product solution

$$\Psi_E(x,t) = \psi_E(x)\exp(-iEt/\hbar) \tag{2.8}$$

is called a **stationary state** because it depends on time only through the phase. Therefore, the probability density $|\Psi_E(x,t)|^2 = |\psi_E(x)|^2$ is time-independent, and so are all expectation values

$$\langle\widehat{Q}(x,p)\rangle = \int \psi_E(x)^* Q(x, -i\hbar\mathrm{d}/\mathrm{d}x)\psi_E(x)\mathrm{d}x\,. \tag{2.9}$$

The general term "state" denotes the way a specific system has been set up for an experiment. Classically, the state is determined by the initial values of position and momentum. In quantum mechanics, the state is determined by the initial wave function. "State" and "wave function" are often used synonymously in basic quantum mechanics, although we will see later that the concept of a state is more general and includes mathematical objects that are not wave functions.

2.1.2 ENERGY

The classical energy as a function of x and p is given by the Hamilton function,

$$H(x,p) = \frac{p^2}{2m} + V(x)\,. \tag{2.10}$$

In quantum mechanics it is replaced by the operator

$$\hat{E} = \hat{H} = \frac{\hat{p}^2}{2m} + V(\hat{x}) = -\frac{\hbar^2}{2m}\frac{\partial^2}{\partial x^2} + V(x)\,. \tag{2.11}$$

Using this operator, the time-independent Schrödinger equation takes the compact form

$$\hat{H}\psi_E = E\psi_E\,, \tag{2.12}$$

or

$$\hat{H}\Psi_E = i\hbar\frac{\partial\Psi_E}{\partial t} = E\Psi_E \tag{2.13}$$

for the full stationary state.

The energy expectation value therefore equals

$$\langle\hat{H}\rangle = \int_{-\infty}^{\infty}\Psi_E^*\hat{H}\Psi_E\mathrm{d}x = \int_{-\infty}^{\infty}\Psi_E^*E\Psi_E\mathrm{d}x = E\int_{-\infty}^{\infty}\Psi_E^*\Psi_E\mathrm{d}x = E\,, \tag{2.14}$$

using normalization of Ψ_E. Similarly, the expectation value of $\hat{H}^2$ equals

$$\langle\hat{H}^2\rangle = \int_{-\infty}^{\infty}\Psi_E^*\hat{H}(\hat{H}\Psi_E)\mathrm{d}x = E\int_{-\infty}^{\infty}\Psi_E^*\hat{H}\Psi_E\mathrm{d}x = E^2\,. \tag{2.15}$$

The energy variance

$$\sigma_H^2 = \langle\hat{H}^2\rangle - \langle\hat{H}\rangle^2 = 0 \tag{2.16}$$

vanishes in a stationary state, and therefore the energy has a sharp value, $\langle\hat{H}\rangle = E$.

If we measure the energy in several copies of a system prepared in the same stationary state, each individual measurement result equals E. An important physical application is spectroscopy of atoms or molecules, which results in sharp frequencies for electromagnetic waves emitted or absorbed by an atom. The spectral lines are sharp because, via $E = \hbar\omega$, their frequencies correspond to energy differences of sharp energy values of stationary states. A single line is visible because it is made up of multiple photons emitted by different atoms in the same initial state. Since the energy is sharp in a stationary state, a sharp spectral line is implied.

In precise measurements, it can be seen that spectral lines have a non-zero width. The lines are therefore not completely sharp, which is explained by the fact that interactions between the atom and the electromagnetic field, which are required for atomic transitions and the emission or absorption of photons, destabilize stationary states. Actual energy states are not exactly stationary, but if they are sufficiently stable, they can be approximated well by stationary states and their sharp energy values.

2.1.3 TIME-DEPENDENT STATES

Using linearity, we can add multiples of different stationary states and obtain a new solution of the time-dependent Schrödinger equation, a **superposition**. If the contributions have different energies, energy measurements in the resulting state are no longer sharp; they rather form a statistical ensemble with the given energy values used in the superposition as possible outcomes. The resulting state is no longer stationary because different phase factors of the form $\exp(-iE_jt/\hbar)$ with more than one E_j cannot be factored out completely, leaving some time dependence in expectation values.

Adding multiples of stationary states therefore allows us to construct more general states. It is possible to show that stationary states form a basis as used in linear algebra, which means that every solution of the time-dependent Schrödinger equation can be written as a sum or series of stationary states:

$$\Psi(x,t) = \sum_n c_n\psi_{E_n}(x)\exp(-iE_nt/\hbar) \tag{2.17}$$

for a suitable collection of energy values E_n and (complex) constants c_n. Sometimes we may have to extend the sum to a series of infinitely many terms, and sometimes we may have to replace the summation with an integral. Possible values of E_n and corresponding wave functions ψ_{E_n} are determined by the time-independent Schrödinger equation.

We will now demonstrate these and other properties in specific examples. Solving the time-independent Schrödinger equation for a given potential $V(x)$ is usually challenging, and therefore there are only a few systems in which one can obtain complete analytical solutions. The following list of tractable examples is standard and highlights crucial properties of quantum-mechanical systems, as well as useful mathematical methods.

2.2 INFINITE SQUARE WELL

The first example is a model of a particle or an atom in a trap with sharp confining forces, in one spatial dimension. We define the potential as

$$V(x) = \begin{cases} 0 & \text{if } 0 \le x \le a \\ \infty & \text{otherwise} \end{cases} \tag{2.18}$$

with a constant $a > 0$. For positive energy, the particle moves freely in its allowed region between $x = 0$ and $x = a$. It turns around immediately when it hits a turning point at $x = 0$ or $x = a$. Because of the infinitely high walls, the particle is never found at $x < 0$ or $x > a$.

2.2.1 ENERGY

We solve the time-independent Schrödinger equation (2.6), first in the forbidden regions. The second-order derivative $\mathrm{d}^2\psi_E/\mathrm{d}x^2$ of a well-defined wave function evaluated at a fixed x is some finite number, while $V(x)\psi_E(x)$ is infinite in a forbidden region unless $\psi_E(x) = 0$. The product must be finite if it is consistent with a finite energy E in the time-independent Schrödinger equation. This conclusion is sufficient to solve the Schrödinger equation in the forbidden regions, where we have $\psi_E(x) = 0$ if $x < 0$ or $x > a$. When we move into the allowed region, we should make sure that the wave function is continuous, such that it can be differentiable. After all, we need the second derivative for the wave equation, so the first derivative must be available too. This conclusion gives us the boundary conditions $\psi_E(0) = 0$ and $\psi_E(a) = 0$.

Since the potential vanishes in the allowed region, the time-independent Schrödinger equation for $0 < x < a$ takes the form

$$-\frac{\hbar^2}{2m}\frac{\mathrm{d}^2\psi_E}{\mathrm{d}x^2} = E\psi_E \tag{2.19}$$

or, more compactly,

$$\frac{\mathrm{d}^2\psi}{\mathrm{d}x^2} = -k^2\psi \tag{2.20}$$

with a constant

$$k = \frac{\sqrt{2mE}}{\hbar}. \tag{2.21}$$

Solutions ψ still depend on E or k, but the subscript is omitted now for the sake of convenience.

Equation (2.20) is a linear, second-order, ordinary differential equation with constant coefficients, familiar from the harmonic oscillator in classical mechanics. It has the following solutions:

1. If $E < 0$ (which is impossible classically), $k = i\kappa$ is imaginary, with a real number $\kappa = \sqrt{-2mE}/\hbar \neq 0$. Since $-k^2 > 0$ in (2.20) in this case, we have the general solution

$$\psi(x) = Ce^{\kappa x} + De^{-\kappa x} \tag{2.22}$$

   with two constants, C and D. The boundary conditions require $\psi(0) = C + D = 0$, and therefore $D = -C$, as well as $\psi(a) = C(e^{\kappa a} - e^{-\kappa a}) = 0$. Since C cannot be zero for a normalizable wave function, this equation has no non-zero real solution for κ because it implies $e^{\kappa a} = e^{-\kappa a}$, or $e^{2\kappa a} = 1$, and therefore $\kappa = 0$ if $a \neq 0$. However, we assumed $\kappa \neq 0$ because we started this discussion with negative E. This contradiction rules out negative energy values for the infinite square well in quantum mechanics.
2. If $E = 0$, $k = 0$ is real. The differential equation then tells us that the second derivative of the wave function vanishes, and therefore the wave function is linear. However, the boundary conditions at $x = 0$ and $x = a$ only allow a constant zero as the compatible linear solution, which is not normalizable. Therefore, E has to be strictly positive.

3. If E is positive, k is real. Equation (2.20) is then solved by

$$\psi(x) = A\sin(kx) + B\cos(kx) \tag{2.23}$$

with two constants, A and B. The boundary conditions require that $\psi(0) = B = 0$ and $\psi(a) = A\sin(ka) = 0$. Since A and B cannot both be zero for a normalizable ψ, we obtain the condition $\sin(ka) = 0$ which has infinitely many solutions,

$$k_n = \frac{n\pi}{a} \tag{2.24}$$

labeled by an integer n. We can restrict n to positive integers, $n = 1, 2, \ldots$, because $n = 0$ would imply $\psi(x) = B = 0$, which is not normalizable, and negative n would just multiply the wave function by -1, which does not change any expectation values.

The boundary conditions therefore imply a restriction on possible values of E, which must be in the discrete set

$$E_n = \frac{\hbar^2 k_n^2}{2m} = \frac{n^2\pi^2\hbar^2}{2ma^2} \quad , \quad n = 1, 2, 3, \ldots \tag{2.25}$$

In contrast to the classical case, where any positive value of E (or zero) is possible, the energy of a particle in an infinite square well in quantum mechanics is discrete and non-zero.

2.2.2 STATES

So far, the constant A in a stationary state has not been determined. Its value, or rather the value of A_n for a given k_n, is given by normalization of the function $\psi_n(x) = A_n \sin(k_n x)$ in the range $0 \leq x \leq a$. We obtain

$$\begin{aligned} 1 &= \int_{-\infty}^{\infty} |\psi_n(x)|^2 \mathrm{d}x = \int_0^a |A_n|^2 \sin(k_n x)^2 \mathrm{d}x = |A_n|^2 \int_0^a \sin(n\pi x/a)^2 \mathrm{d}x \\ &= |A_n|^2 \frac{a}{n\pi} \int_0^{n\pi} \sin(y)^2 \mathrm{d}y = |A_n|^2 \frac{a}{\pi} \int_0^{\pi} \sin(y)^2 \mathrm{d}y = \frac{1}{2} a |A_n|^2 . \end{aligned} \tag{2.26}$$

Note the restriction of integration to a finite range $0 \leq x \leq a$, taking into account the fact that $\psi_n(x)$ vanishes if $x < 0$ or $x > a$. In the second line we substituted $y = n\pi x/a$, used the period of π for $\sin(y)^2$, and then evaluated $\int_0^\pi \sin(y)^2 \mathrm{d}y = \frac{1}{2}\pi$. This integral can be obtained from integrating by parts $\int \cos(y)^2 \mathrm{d}y$ in

$$\begin{aligned} \pi &= \int_0^{\pi} \mathrm{d}y = \int_0^{\pi} (\sin(y)^2 + \cos(y)^2) \mathrm{d}y \\ &= \int_0^{\pi} \sin(y)^2 \mathrm{d}y + (\sin(y)\cos(y))|_0^{\pi} + \int_0^{\pi} \sin(y)^2 \mathrm{d}y = 2\int_0^{\pi} \sin(y)^2 \mathrm{d}y . \end{aligned} \tag{2.27}$$

Therefore, $|A_n|^2 = 2/a$. We can always choose normalization factors to be positive real numbers because the sign or a constant phase $e^{i\phi}$ would not change any expectation values. Here, we write

$$A_n = \sqrt{\frac{2}{a}} . \tag{2.28}$$

We conclude that the set of all stationary states in the infinite square well is given by

$$\psi_n(x) = \sqrt{\frac{2}{a}} \sin(n\pi x/a) \quad \text{for } n = 1, 2, \ldots \tag{2.29}$$

This function has units of $1/\sqrt{\text{length}}$, which is correct because it squares to a (probability) density in one dimension.

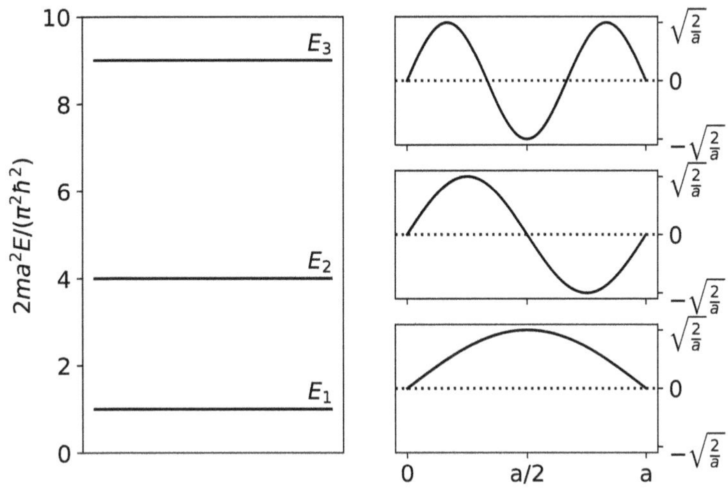

Figure 2.1 The ground state and the first two excited states of the infinite square well.

As already noted, the energy values (2.25) are always positive. Unlike in the classical case, the energy cannot be zero (a particle at rest in the well) because a non-zero, continuous wave function always needs some oscillations, which imply non-zero momentum. The smallest possible energy,

$$E_1 = \frac{\pi^2\hbar^2}{2ma^2}, \tag{2.30}$$

is the **ground-state** energy, with ground-state wave function $\psi_1(x)$. The stationary states ψ_n with $n \geq 2$ are called **excited states.**

2.2.3 MATHEMATICAL PROPERTIES

Stationary states of the infinite square well have several useful properties, some of which can be seen in Figure 2.1. We will show later that the same observations can be made about more general systems.

Listed by increasing values of the positive integer n, the stationary states are alternately even and odd with respect to the center of the well at $x = a/2$. As n increases, the wave function oscillates more, as determined by the number of zeros: $\psi_n(x)$ has $n-1$ zeros in the interior of the well. This property is always realized for a potential with reflection symmetry around a center, here at $x = a/2$.

2.2.3.1 Orthonormality

Stationary states at different energy levels are **orthogonal** to each other in the sense that

$$\int_{-\infty}^{\infty} \psi_m(x)^* \psi_n(x)\mathrm{d}x = 0 \quad \text{if } m \neq n. \tag{2.31}$$

If we interpret the normalization integral $\int_{-\infty}^{\infty} |\psi_n(x)|^2 \mathrm{d}x$ as some kind of squared vector length since we are summing squares of the values of $\psi_n(x)$, the generalization (2.31) of the integral to two different wave functions plays the role of a dot product or of an inner product used to define orthogonality.

Equation (2.31) follows immediately if m is even and n is odd or vice versa because we are then integrating an odd function over a symmetric range (centered at $x = a/2$). It is correct also if both n

and m are even, or if both are odd, as follows by using trigonometric identities in the integration

$$\begin{aligned}\int_{-\infty}^{\infty}\psi_m(x)^*\psi_n(x)\mathrm{d}x &= \frac{2}{a}\int_0^a \sin(m\pi x/a)\sin(n\pi x/a)\mathrm{d}x \\ &= \frac{1}{a}\int_0^a (\cos((m-n)\pi x/a)-\cos((m+n)\pi x/a))\,\mathrm{d}x \\ &= \frac{1}{\pi}\left(\frac{\sin((m-n)\pi)}{m-n}-\frac{\sin((m+n)\pi)}{m+n}\right)=0 \end{aligned} \tag{2.32}$$

for all $m \neq n$.

It is convenient to express orthogonality and normalization through a single equation,

$$\int_{-\infty}^{\infty}\psi_m(x)^*\psi_n(x)\mathrm{d}x = \delta_{mn} = \begin{cases} 0 & \text{if } m\neq n \\ 1 & \text{if } m=m \end{cases} \tag{2.33}$$

called **orthonormality**. The symbol δ_{mn} defined in this equation is the Kronecker delta. It represents the components of an identity matrix, in this case in an infinite-dimensional space because the values of m and n can be any positive integers.

The stationary states of the infinite square well are **complete** in the sense that any (sufficiently well-behaved) function $f(x)$ defined in the range $0\leq x\leq a$ with boundary values $f(0)=0=f(a)$ can be expressed as a sum or an infinite series of the stationary states with suitable constant coefficients c_n:

$$f(x)=\sum_{n=1}^{\infty}c_n\psi_n(x)=\sqrt{\frac{2}{a}}\sum_{n=1}^{\infty}c_n\sin(n\pi x/a)\,. \tag{2.34}$$

This expression is the Fourier series of $f(x)$, and the completeness property follows from Fourier analysis.

2.2.3.2 General solutions and time dependence

Orthonormality and completeness are important because they allow us to extend our separable solutions to the most general solution of the time-dependent Schrödinger equation for the infinite square well. Given a function $f(x)$ in the range $0\leq x\leq a$ with the required boundary conditions, we use orthonormality in order to derive the coefficients c_n in (2.34):

$$\int_{-\infty}^{\infty}\psi_m(x)^*f(x)\mathrm{d}x=\sum_{n=1}^{\infty}c_n\int_{-\infty}^{\infty}\psi_m(x)^*\psi_n(x)\mathrm{d}x=\sum_{n=1}^{\infty}c_n\delta_{mn}=c_m\,. \tag{2.35}$$

In the third step we used the orthonormality condition (2.33). In the last step, the infinite series has only one non-zero term because $c_n\delta_{mn}$ is non-zero only if $m=n$, and this non-zero term equals c_m. Without orthogonality, deriving the c_n would require solving an infinite set of linear equations, obtained for instance by evaluating (2.34) at a countably infinite number of points between $x=0$ and $x=a$. This procedure provides one equation for each c_n, but the system of equations would be coupled and hard or impossible to solve directly.

An example for the choice of $f(x)$ is a suitable initial wave function $\Psi(x,0)$ in the well, at time $t=0$. In general, this initial wave function need not be a stationary state, but we can write it as a linear combination of stationary states if we compute the corresponding c_n as in (2.35),

$$c_n=\int_{-\infty}^{\infty}\psi_n(x)^*\Psi(x,0)\mathrm{d}x\,, \tag{2.36}$$

such that

$$\Psi(x,0)=\sum_{n=1}^{\infty}c_n\psi_n(x)\,. \tag{2.37}$$

We know the time dependence of a stationary state that equals $\Psi_n(x,0) = \psi_n(x)$ at time $t = 0$ because it is of the separable form

$$\Psi_n(x,t) = \psi_n(x)\exp(-iE_nt/\hbar)\,. \tag{2.38}$$

Since the Schrödinger equation is linear, the time dependence of a generic initial state $\Psi(x,0)$, expanded as a series of stationary states in (2.37), is obtained by replacing $\psi_n(x)$ with $\Psi_n(x,t)$ at time t:

$$\Psi(x,t) = \sum_{n=1}^{\infty} c_n\psi_n(x)\exp(-iE_nt/\hbar) \tag{2.39}$$

with c_n given in (2.36).

If there are at least two contributions with different n, the state is no longer stationary because we cannot factor out all time dependent terms. Expectation values then also depend on time. In this way, using separable wave functions to simplify the original partial differential equation gives us access to all solutions, thanks to linearity which allows us to combine different stationary states. For the infinite square well, the most general solution is given by

$$\Psi(x,t) = \sqrt{\frac{2}{a}}\sum_{n=1}^{\infty} c_n\sin(n\pi x/a)\exp(-in^2\pi^2\hbar t/(2ma^2)) \tag{2.40}$$

with coefficients c_n such that $\sum_{n=1}^{\infty}|c_n|^2 = 1$ for normalization. If we compute the c_n from (2.36), the normalization condition of the c_n is automatically satisfied because each $\psi_n(x)$ is normalized.

2.2.4 STATISTICAL PROPERTIES

We now have a large number of wave functions, given by stationary states, for which we can compute expectation values. The position expectation value in a stationary state ψ_n, which equals the position expectation value of the corresponding $\Psi_n(x,t)$ at all times t, is given by

$$\begin{aligned}\langle\hat{x}\rangle_n &= \int_{-\infty}^{\infty}\psi_n(x)^*x\psi_n(x)\mathrm{d}x = \frac{2}{a}\int_0^a x\sin(n\pi x/a)^2\mathrm{d}x\\ &= \frac{2}{a}\int_{-a/2}^{a/2}(x+a/2)\sin(n\pi x/a+n\pi/2)^2\mathrm{d}x\\ &= \frac{2}{a}\int_{-a/2}^{a/2}x\sin(n\pi x/a+n\pi/2)^2\mathrm{d}x+\int_{-a/2}^{a/2}\sin(n\pi x/a+n\pi/2)^2\mathrm{d}x\,.\end{aligned} \tag{2.41}$$

In the second line, we shifted the integration variable by a constant $-a/2$, using substitution, in order to apply symmetries of the integrand in the next step. The first integral in the third line is then of an odd function over a symmetric range, and thus equals zero. (The function $\sin(n\pi x/a+n\pi/2)^2$ equals $\sin(n\pi x/a)^2$ if n is even, and $\cos(n\pi x/a)^2$ if n is odd. Both functions are even in x, and multiplication by x turns them into odd functions.) The remaining integral equals $a/2$, as we saw when we normalized stationary states. Therefore, for all n, the position expectation values

$$\langle\hat{x}\rangle_n = \frac{1}{2}a \tag{2.42}$$

take the same value for all n, right in the middle of the well. For a stationary state, this result is obtained for all t: $\langle\hat{x}\rangle_n(t) = a/2$.

The momentum expectation values equal

$$\langle\hat{p}\rangle_n = \frac{\hbar}{i}\int_{-\infty}^{\infty}\psi_n(x)^*\psi_n'(x)\mathrm{d}x = \frac{\hbar}{2i}\int_{-\infty}^{\infty}\frac{\mathrm{d}\psi_n(x)^2}{\mathrm{d}x}\mathrm{d}x = \frac{\hbar}{2i}\psi_n(x)^2|_{-\infty}^{\infty} = 0 \tag{2.43}$$

for all n. Here, we used the fact that the stationary states are real, $\psi_n(x)^* = \psi_n(x)$, and obey the boundary conditions $\psi_n(0) = 0 = \psi_n(a)$ or, more generally, $\lim_{x\to\pm\infty}\psi(x) = 0$. These properties do not refer to the specific functional form of our stationary states found here, and are therefore correct for stationary states in any potential. An alternative derivation uses the general equation (1.99) and thus

$$\langle\hat{p}\rangle = m\langle\hat{v}\rangle = m\frac{\mathrm{d}\langle\hat{x}\rangle}{\mathrm{d}t}\,. \tag{2.44}$$

If $\langle\hat{x}\rangle(t)$ is constant as a function of t, as found here, the momentum expectation value vanishes.

A single measurements of the momentum may nevertheless result in a non-zero value because the stationary states have non-vanishing momentum variance: The momentum has quantum fluctuations around the value zero. We can obtain the momentum variance by observing that the squared momentum of a particle in the infinite square well equals $2m$ times its energy, which has only the kinetic contribution. Since we already computed the energy values of stationary states and found that they are non-zero, the momentum variances must be non-zero too. Specifically, they equal

$$\sigma_{p,n}^2 = \langle\hat{p}^2\rangle_n = 2mE_n = \left(\frac{n\pi\hbar}{a}\right)^2 \tag{2.45}$$

because energy and the squared momentum are in this case related by $\hat{H}\psi_n = (2m)^{-1}\hat{p}^2\psi_n = E_n\psi_n$.

The expectation value of position squared requires a longer calculation by direct integration, which results in

$$\langle\hat{x}^2\rangle_n = a^2\left(\frac{1}{3} - \frac{1}{2n^2\pi^2}\right)\,. \tag{2.46}$$

We have the standard deviations

$$\sigma_{x,n} = \frac{1}{2}a\sqrt{\frac{1}{3} - \frac{2}{n^2\pi^2}} \quad \text{and} \quad \sigma_{p,n} = \frac{n\pi\hbar}{a} \tag{2.47}$$

with products

$$\sigma_{x,n}\sigma_{p,n} = \frac{1}{2}\hbar\sqrt{\frac{1}{3}n^2\pi^2 - 2} \geq \frac{1}{2}\hbar\sqrt{\frac{\pi^2}{3} - 2} \approx 1.14\cdot\frac{\hbar}{2} > \frac{\hbar}{2}\,. \tag{2.48}$$

In the infinite square well, there is no stationary state, labeled by $n \geq 1$, that takes the minimal value allowed by the uncertainty relation (1.103).

2.3 HARMONIC OSCILLATOR

The infinite square well has a bounded allowed region and therefore has oscillating solutions, with the position or its expectation value bouncing back and forth unless the state is stationary. Boundary conditions imposed at the sharp walls imply a clear mathematical reason for the discreteness of possible energy values of bound states in quantum mechanics. However, they also imply an infinite force, which can never be realized in an experiment. A more general description of oscillations requires a smooth position-dependent potential $V(x)$ that is finite everywhere except perhaps in the limits of $x \to \pm\infty$.

Given a potential $V(x)$, a stable equilibrium point is defined as a point $x = x_0$ at which the force, $F(x_0) = -V'(x_0)$, vanishes, such that there is a rightward pointing force if the particle is displaced slightly to the left of x_0, and a leftward pointing force if the particle is displaced slightly to the right of x_0. A particle moving at a sufficiently small speed close to x_0 therefore experiences forces that tend to restore it to the position x_0, but if it has some kinetic energy, it cannot stay exactly at x_0. If the kinetic energy at x_0 is sufficiently small, the particle oscillates around the point x_0.

The two conditions that define a stable equilibrium point can be expressed by two equations,

$$F(x_0) = -V'(x_0) = 0 \quad \text{(equilibrium)} \tag{2.49}$$

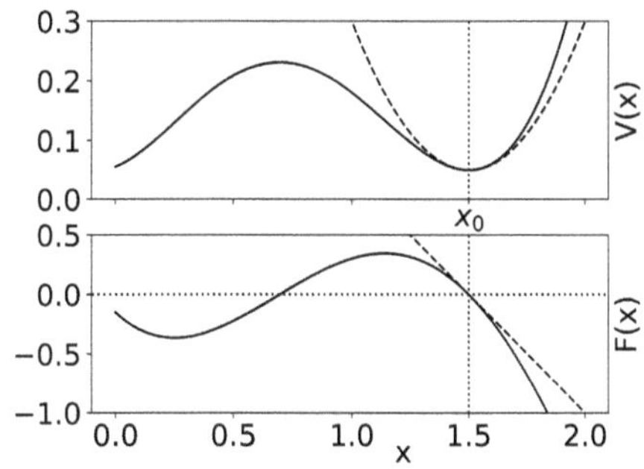

Figure 2.2 Potential landscape (top) of a potential with a stable equilibrium point at $x = x_0$, together with the force function (bottom). At the equilibrium point, the force vanishes. If the equilibrium point is stable, the force is positive on the left (pointing to the right) and negative on the right of x_0 (pointing to the left). The force therefore tends to restore the particle to the position $x = x_0$. For small oscillation amplitudes, the potential around x_0 can be approximated by a parabola (dashed) derived from a second-order Taylor expansion. The corresponding force approximation is linear around x_0, given by the tangent to the force function.

and

$$F'(x_0) = -V''(x) < 0 \quad \text{(stability)} \tag{2.50}$$

They are visualized in Figure 2.2. If the oscillation amplitude, given by the maximum of $|x(t) - x_0|$ along the particle trajectory $x(t)$, is small, we can replace the potential by a second-order Taylor expansion,

$$V(x) = V(x_0) + V'(x_0)(x - x_0) + \frac{1}{2}V''(x_0)(x - x_0)^2 + \cdots \tag{2.51}$$

The first term, $V(x_0)$, is constant and does not contribute to the force. The second term, $V'(x_0)(x - x_0)$, is zero if the Taylor expansion is applied at an equilibrium point. The third term, $\frac{1}{2}V''(x_0)(x - x_0)^2$, is non-zero for $x \neq x_0$ and has a positive coefficient if the equilibrium point is stable.

The second-order Taylor expansion is equivalent to a harmonic potential

$$V(x) \approx \frac{1}{2}V''(x_0)(x - x_0)^2 = \frac{1}{2}K(x - x_0)^2 \tag{2.52}$$

with a positive coefficient $K = V''(x_0)$. The force

$$F(x) \approx -K(x - x_0) \tag{2.53}$$

is a linear function of x. Newton's second law therefore implies the linear differential equation

$$\frac{\mathrm{d}^2x}{\mathrm{d}t^2} = -\frac{K}{m}(x - x_0). \tag{2.54}$$

There are standard procedures for solving linear ordinary differential equations with constant coefficients, using linear combinations of trigonometric or exponential functions. Here, the general solution is

$$x(t) = x_0 + A\sin(\omega t) + B\cos(\omega t) \tag{2.55}$$

with two integration constants, A and B, and the oscillation frequency

$$\omega = \sqrt{\frac{K}{m}}. \tag{2.56}$$

The integration constants are related to initial values by

$$A = \frac{\dot{x}(0)}{\omega} \quad \text{and} \quad B = x(0) - x_0 . \tag{2.57}$$

They determine the amplitude $\sqrt{A^2+B^2}$ of oscillations and the phase $\phi_0 = -\arctan(B/A)$ such that $x(t) = x_0$ when $\omega t = \phi_0$. These statements can be derived by applying a trigonometric identity, writing the general solution as

$$x(t) = x_0 + \text{sgn}(A)\sqrt{A^2+B^2}\,\sin(\omega t + \arctan(B/A)) . \tag{2.58}$$

2.3.1 ANALYTIC SOLUTIONS

An analysis of the harmonic oscillator in quantum mechanics requires us to solve the time-independent Schrödinger equation

$$\hat{H}\psi = -\frac{\hbar^2}{2m}\frac{\mathrm{d}^2\psi}{\mathrm{d}x^2} + \frac{1}{2}m\omega^2 x^2\psi = E\psi . \tag{2.59}$$

We simplify the constant coefficients by substituting $\xi = x\sqrt{m\omega/\hbar}$, which implies the new differential equation

$$\frac{\mathrm{d}^2\psi}{\mathrm{d}\xi^2} = (\xi^2 - G)\psi \tag{2.60}$$

with a single constant,

$$G = \frac{2E}{\hbar\omega} . \tag{2.61}$$

In general, second-order differential equations with non-constant coefficients constitute a difficult problem. It is often possible to determine some properties of possible solutions by inserting a power series for the unknown function, and rewriting the differential equation as a set of relationships between the series coefficients. If we are lucky, we can even find a complete set of solutions, which turns out to be the case for the harmonic oscillator and a few other standard examples of quantum mechanics.

If we are looking for solutions in the form of a power series, we first check that a finite-degree polynomial does not work (even if we set aside the normalizability requirement). For instance, the attempt $\psi(\xi) = a + b\xi + c\xi^2$ would imply $\psi'' = 2c$ for the left-hand side of the differential equation, and $(\xi^2 - G)\psi = -Ga - Gb\xi + (a - Gc)\xi^2 + b\xi^3 + c\xi^4$. Comparing coefficients of the two polynomials that are supposed to be equal if $\psi'' = (\xi^2 - G)\psi$, we obtain $2c = -Ga$, $0 = -Gb$, $0 = a - Gc$, $0 = b$, and $0 = c$. The last three equations imply that $\psi(\xi) = 0$, which is not allowed for a probability density $|\psi(\xi)|^2$. Similarly, any finite-degree polynomial results in a vanishing solution because the degrees of the left-hand side and the right-hand side of the differential equation do not match.

For valid solutions, we have to look for a power series $\psi(\xi) = \sum_{j=0}^{\infty} c_j\xi^j$ with infinitely many terms. We evade the previous problem because both sides of the differential equation with such a $\psi(\xi)$ have the same (infinite) degrees. This observation is the starting point of the analytic or power-series method, which consists of two steps:

Step 1: We find an approximate solution valid for large ξ. Such a solution will help us to implement the condition of normalizability. Within an approximation, large values of ξ allow us to ignore some terms in the differential equation, making it easier to find analytical solutions. The solutions should respect the boundary condition $\lim_{\xi\to\pm\infty}\psi(\xi) = 0$ in order to have a normalizable wave function.

Since G is a constant in the differential equation, ξ^2 can easily be much greater than G if $|\xi|$ is large. In the approximation of large $|\xi|$, we can therefore ignore G in (2.60), such that

$$\frac{d^2\psi}{d\xi^2} \approx \xi^2\psi\,. \tag{2.62}$$

Even without the G-term, it is not easy to solve this equation. If ξ^2 on the right were a constant ξ_0^2, we would have simple solutions $\psi(\xi) = A\exp(\pm\xi_0\xi)$ with a constant A. For non-constant ξ, this observation suggests to use a function of the form $\psi(\xi) \approx A\exp(-\frac{1}{2}\xi^2)$. Replacing the constant ξ_0 with ξ leads to a quadratic exponent, and the additional factor of $\frac{1}{2}$ is useful because we are taking derivatives of this exponent if we apply the chain rule. We only consider a negative exponent in order to have normalizability.

We now check that this function is indeed an approximate solution. Its first-order derivative equals $d\psi/d\xi = -A\xi\exp(-\frac{1}{2}\xi^2)$ and the second-order derivative is

$$\frac{d^2\psi}{d\xi^2} = A(\xi^2-1)\exp(-\tfrac{1}{2}\xi^2) \approx \xi^2\psi \tag{2.63}$$

again for large $|\xi|$, where we can ignore -1 compared with ξ^2.

Completing the first step, our normalizable approximate solution is therefore

$$\psi(\xi) \approx A\exp(-\tfrac{1}{2}\xi^2)\,. \tag{2.64}$$

We do not normalize this function yet because the normalization integral requires an exact solution for all ξ, but we respect our boundary condition $\lim_{\xi\to\pm\infty}\psi(\xi) = 0$.

Step 2: Using the approximate solution, we now try to find exact solutions of the form

$$\psi(\xi) = h(\xi)\exp(-\tfrac{1}{2}\xi^2) \tag{2.65}$$

with a polynomial $h(\xi)$. A polynomial would not be normalizable over an infinite x-range, but it is normalizable if it is multiplied by the exponential factor. (This is why we first performed Step 1.) The exponential factor also implies infinitely many terms in a Taylor expansion, which means that $h(\xi)$ may be polynomial even though there are no polynomial solutions $\psi(\xi)$. We are therefore using infinitely many but explicitly known terms in the power series of the exponential function, and introduce a finite (but so far unlimited) number of free coefficients by the polynomial factor of $h(\xi)$. The remainder of the method consists in evaluating conditions on the coefficients of $h(\xi)$.

We first compute derivatives of the product form (2.65):

$$\frac{d\psi}{d\xi} = \left(\frac{dh}{d\xi} - \xi h\right)\exp(-\tfrac{1}{2}\xi^2) \tag{2.66}$$

$$\frac{d^2\psi}{d\xi^2} = \left(\frac{d^2h}{d\xi^2} - 2\xi\frac{dh}{d\xi} + (\xi^2-1)h\right)\exp(-\tfrac{1}{2}\xi^2)\,. \tag{2.67}$$

The differential equation (2.60) therefore requires that h obey the equation

$$\frac{d^2h}{d\xi^2} - 2\xi\frac{dh}{d\xi} + (G-1)h = 0\,. \tag{2.68}$$

This equation looks more complicated than (2.60), but it is more convenient because it has polynomial solutions: h and $\xi dh/d\xi$ have the same polynomial degree if h is polynomial, while $\mathrm{d}^2h/\mathrm{d}\xi^2$ has lower degree. In contrast to the original version of the differential equation, it is now possible for all terms to cancel out on the left-hand side of (2.68) because there are two terms of the highest polynomial degree.

In order to solve for polynomials $h(\xi)$ of so-far undetermined degree, we write them in the form of a general power series,

$$h(\xi) = a_0 + a_1\xi + a_2\xi^2 + a_3\xi^3 + \cdots = \sum_{j=0}^{\infty} a_j\xi^j \tag{2.69}$$

and compute derivatives:

$$\frac{\mathrm{d}h}{\mathrm{d}\xi} = a_1 + 2a_2\xi + 3a_3\xi^2 \cdots = \sum_{j=1}^{\infty} ja_j\xi^{j-1} = \sum_{j=0}^{\infty} ja_j\xi^{j-1} \tag{2.70}$$

and thus

$$\xi\frac{\mathrm{d}h}{\mathrm{d}\xi} = \sum_{j=0}^{\infty} ja_j\xi^j \tag{2.71}$$

as needed for the differential equation. At the end of (2.70), we brought back the $j = 0$-term because it vanishes anyway, given the factor of j. This will make it easier to combine all three terms in (2.68).

The second-order derivative equals

$$\begin{aligned}\frac{\mathrm{d}^2h}{\mathrm{d}\xi^2} &= 2a_2 + 6a_3\xi + \cdots = \sum_{j=2}^{\infty} j(j-1)a_j\xi^{j-2} \\ &= \sum_{k=0}^{\infty}(k+1)(k+2)a_{k+2}\xi^k \end{aligned}\tag{2.72}$$

after shifting the summation label by two, substituting $k = j-2$. In the first line, we are able to start summing at $j = 2$ because the coefficient $j(j-1)$ vanishes for $j = 0$ and $j = 1$. All three terms needed for (2.68) — h itself, (2.71) and (2.72) — now have the same power-series form with a factor of ξ raised to the order of each term (j or k).

We combine all three contributions and, factoring out ξ^j, obtain the equation

$$\sum_{j=0}^{\infty}\left((j+1)(j+2)a_{j+2} - 2ja_j + (G-1)a_j\right)\xi^j = 0\,. \tag{2.73}$$

A power series vanishes if and only if all coefficients vanish:

$$(j+1)(j+2)a_{j+2} - 2ja_j + (G-1)a_j = 0 \tag{2.74}$$

for all $j = 0, 1, \ldots$ The coefficients a_j are therefore subject to the recursion formula

$$a_{j+2} = \frac{2j+1-G}{(j+1)(j+2)}a_j \tag{2.75}$$

if $h(\xi) = \sum_{j=0}^{\infty} a_j\xi^j$ is a solution, for a given G.

The 2-step recursion formula, relating a_j to a_{j+2}, determines the solution up to two free constants, which is what we expect for a solution of a second-order differential equation. If we choose a_0 and a_1, the recursion formula can be solved for a_2 using a_0, for a_3 using a_1 and so on. All a_j are uniquely determined in terms of a_0 and a_1.

We are interested in normalizable solutions $\psi(\xi) = h(\xi)\exp(-\frac{1}{2}\xi^2)$. If $h(\xi)$ is a polynomial, the product is normalizable because the exponential function approaches zero rapidly at infinity. However, for generic values of a_0 and a_1, a recursion formula is expected to lead to infinitely many non-zero a_j. The function $h(\xi)$ is then a power series, and it is not guaranteed that its product with $\exp(-\frac{1}{2}\xi^2)$ is normalizable.

If $h(\xi)$ is a non-polynomial power series, we analyze normalizability by solving our recursion relation for asymptotically large j because the corresponding powers of ξ are relevant for potential violations of this condition: Any finite-j coefficient may be changed by adding a polynomial to $h(\xi)$ without affecting normalizability properties of $\psi(\xi)$.

Again, we use approximate solutions, this time by ignoring constants compared with large j in (2.75). The approximate recursion formula valid for large j is

$$a_{j+2} \approx \frac{2}{j} a_j . \tag{2.76}$$

Since we are interested in large j, dividing by j, which has a general range starting at zero, is allowed. We solve and use this equation in several substeps of Step 2:

Substep 2.1: For approximate solutions, it is sufficient to consider only even j because there is no major difference between j and $j+1$ if j is large. Then, substituting $k = j/2$ (an integer) and $b_k = a_{2k}$, we eliminate the factor of two in (2.76) and obtain the new approximate recursion formula

$$b_{k+1} \approx \frac{1}{k} b_k \tag{2.77}$$

with a step size of one in k, rather than two in j. An approximate solution is therefore determined by a single constant, b_1. Solving the equation recursively, we first divide b_1 by one to obtain b_2, then divide b_2 by two to obtain b_3, then divide b_3 by three and so on. The product of all these coefficients, up to some finite k, produces the factorial $k!$. Therefore, $b_k \approx b_1/k!$, or

$$a_j \approx \frac{c}{(j/2)!} \tag{2.78}$$

with a constant c, if we transform back to a_j. In this result, we skip odd j such that $(j/2)!$ is defined. The large-j contributions of the odd terms in $h(\xi)$ equal ξ times the large-j contributions of the even terms, and they do not change normalizability features of the resulting $\psi(\xi)$ to be derived now.

If a_j has the approximate behavior (2.78), the corresponding power series $h(\xi)$ has the large-j contribution

$$h(\xi) \approx c \sum_j \frac{\xi^j}{(j/2)!} = c \sum_k \frac{\xi^{2k}}{k!} = c\exp(\xi) \tag{2.79}$$

where we substituted the same $k = j/2$ as used in the recursion formula. Since we only know the large-j behavior of a_j, $h(\xi)$ may not exactly equal $c\exp(\xi)$, but if we include smaller-j coefficients, we can always write $h(\xi)$ as $c\exp(\xi)$ with a constant c, plus a finite-degree polynomial in ξ. For large $|\xi|$, the exponential function with a positive exponent is much larger than any polynomial.

Inserting this $h(\xi)$ in $\psi(\xi) = h(\xi)\exp(-\frac{1}{2}\xi^2)$ results in $\psi(x) \approx c\exp(+\frac{1}{2}\xi^2)$, which is not normalizable and therefore inadmissible as a wave function. The only way to avoid this conclusion is to work with power-series solutions $h(\xi)$ for which the recursion relation does not reach large j because all a_j vanish for sufficiently large j. The power series can then be reduced to a polynomial.

Substep 2.2: We obtain a normalizable solution $\psi(\xi)$ only if the recursion formula for a_j stops at a finite j, such that there is a final $a_n \neq 0$ while all a_j with $j > n$ are zero. The function $h(\xi)$ is then a polynomial of degree n.

Applying the recursion formula to $j = n$ implies

$$a_{n+2} = \frac{2n+1-G}{(n+1)(n+2)} a_n \tag{2.80}$$

where, by assumption, $a_n \neq 0$ and $a_{n+2} = 0$. These two conditions can be valid only if the numerator of the fraction vanishes:

$$G_n = 2n+1 \tag{2.81}$$

with a new subscript n that labels different admissible values of the constant G in (2.60). For each G_n, there is a different polynomial $h_n(\xi)$ and a corresponding normalizable solution $\psi_n(\xi)$.

This formal result about normalizability has strong implications because G depends on the energy value of the time-independent Schrödinger equation. Using (2.61), we can thus determine the allowed energies of stationary states,

$$E_n = \frac{1}{2}\hbar\omega G_n = (n+\tfrac{1}{2})\hbar\omega . \tag{2.82}$$

As in the infinite square well, the energies of stationary states are discrete, but here they are evenly spaced because E_n is linear in n.

Substep 2.3: The condition (2.81) implies that $a_{n+2} = 0$ for any solution with the given G_n. It then follows from the recursion formula that all $a_{n+2k} = 0$ with even increments for $k = 1, 2, \ldots$, but we need all a_j with $j > n$ to vanish. Since we already fixed G_n for a given n, the coefficients a_{n+2k-1} with odd increments of n can vanish only if all the a_j with odd j vanish if n is even, and all the a_j with even j must vanish if n is odd. These conditions apply even for coefficients with $j < n$.

The restrictions on coefficients implied by normalizability are compatible with the 2-step recursion formula, which provides two independent sets of conditions for a_n with even and odd n, respectively. Therefore, if n is even, we should choose $a_1 = 0$ (while $a_0 \neq 0$), and for odd n, we should choose $a_0 = 0$ (while $a_1 \neq 0$). Only one free constant then remains for each state, a_0 for even n and a_1 for odd n, which can be determined by normalization.

To summarize, we derived the following properties of stationary states from our multi-step solution procedure:

1. All stationary states of the harmonic oscillator are of the form

$$\psi_n(\xi) = h_n(\xi)\exp(-\tfrac{1}{2}\xi^2) \tag{2.83}$$

with a polynomial $h_n(\xi)$ of degree n. The corresponding energies are $E_n = (n+1/2)\hbar\omega$. The ground state with energy $E_0 = \hbar\omega$ is obtained for $n = 0$, and the energy of excited states increases with the degree n.

2. If n is odd, we have to choose $a_0 = 0$ for $h_n(\xi)$ to be of finite degree. The recursion formula then implies that all $a_{2k} = 0$, and $\psi_n(\xi)$ is an odd function. If n is even, we have to choose $a_1 = 0$ for $h_n(\xi)$ to be of finite degree. The recursion formula then implies that all $a_{2k+1} = 0$, and $\psi_n(\xi)$ is an even function. Since $\exp(-\frac{1}{2}\xi^2)$ is an even function, we obtain the same symmetry properties as in the infinite square well.

As this procedure shows, the normalizability condition first implies that $h_n(\xi)$ has to be a polynomial of finite degree, which then fixes G_n in (2.81) through the recursion formula. Mathematically, therefore, normalizability implies the discreteness of energies.

We have obtained the same qualitative features as seen in the infinite square well: The stationary states are alternately even and odd, and the number of zeros, which depends on the polynomial degree of $h_n(\xi)$, is determined by the integer n. It turns out that they also share the orthonormality and completeness properties. Completeness is related to the fact that we included polynomials of all degrees. In general, this property is hard to prove in infinite-dimensional function spaces because it is not sufficient to count elements in a subset of linearly independent states and conclude that they form a basis if their number agrees with the dimension. In infinite dimensions, if we remove a finite number of states from a basis, we still have infinitely many states and their number formally agrees with the dimension. But they would no longer be complete because some basis states have been removed and cannot be obtained as linear combinations of the remaining states. Orthonormality can be derived more easily by using an alternative method of solving the time-independent Schrödinger equation for the harmonic oscillator, introduced next.

2.3.2 ALGEBRAIC METHOD

We return to the differential equation

$$-\frac{\hbar^2}{2m}\frac{\mathrm{d}^2\psi}{\mathrm{d}x^2}+\frac{1}{2}m\omega^2x^2\psi = E\psi\,. \tag{2.84}$$

The classical energy of the harmonic oscillator is quadratic in x and p and therefore can be factored into two linear terms,

$$E=\frac{p^2}{2m}+\frac{1}{2}m\omega^2x^2=\frac{1}{2m}(ip+m\omega x)(-ip+m\omega x)\,. \tag{2.85}$$

In quantum mechanics, linear expressions in x and p imply linear differential equations, which can more easily be solved than second-order equations. We therefore try to make use of the two operators obtained by inserting quantum expressions for the two linear factors in E,

$$\hat{a}_\pm=\frac{1}{\sqrt{2m\hbar\omega}}(\mp i\hat{p}+m\omega\hat{x})=\frac{1}{\sqrt{2m\hbar\omega}}\left(\mp\hbar\frac{\mathrm{d}}{\mathrm{d}x}+m\omega x\right)\,. \tag{2.86}$$

The coefficients are chosen such that $\hat{a}_+$ and $\hat{a}_-$ have no units.

2.3.2.1 Factorization

The classical energy equals $\hbar\omega$ times the product of the classical versions of $\hat{a}_\pm$,

$$E=\hbar\omega a_+a_- \tag{2.87}$$

Products of operators, such as $\hat{a}_+\hat{a}_-$, may depend on the ordering, and in this case we will see that $\hat{a}_+\hat{a}_-\neq\hat{a}_-\hat{a}_+$. We should therefore first determine which version of the product, if any, equals the quantum energy or the Hamilton operator $\hat{H}$.

The product of two operators is defined by how they act on a wave function, starting with the rightmost operator that is written next to the wave function:

$$\begin{aligned}\hat{a}_+\hat{a}_-\psi(x) &= \frac{1}{2m\hbar\omega}\left(-\hbar\frac{\mathrm{d}}{\mathrm{d}x}+m\omega x\right)\left(\hbar\frac{\mathrm{d}}{\mathrm{d}x}+m\omega x\right)\psi(x)\\ &= \frac{1}{2m\hbar\omega}\left(-\hbar\frac{\mathrm{d}}{\mathrm{d}x}+m\omega x\right)\left(\hbar\psi'(x)+m\omega\psi(x)\right)\end{aligned}$$

$$= \frac{1}{2m\hbar\omega}\left(-\hbar^2\psi''(x) - \hbar m\omega\frac{\mathrm{d}}{\mathrm{d}x}(x\psi(x)) + \hbar m\omega x\psi'(x) + m^2\omega^2x^2\psi(x)\right)$$

$$= \frac{1}{2m\hbar\omega}\left(\left(\frac{\hbar}{i}\frac{\mathrm{d}}{\mathrm{d}x}\right)^2 + (m\omega x)^2 - \hbar m\omega\right)\psi(x). \tag{2.88}$$

Since the wave function $\psi(x)$, at this point, is a generic differentiable function used to show how the operators act, we cannot compute its derivatives. However, we can rewrite them in terms of momentum operators, as indicated in the last line. The first two terms of the last line are then proportional to the Hamilton operator, but there is also an extra term, $-\hbar m\omega$. Therefore,

$$\hat{a}_+\hat{a}_-\psi(x) = \left(\frac{\hat{H}}{\hbar\omega} - \frac{1}{2}\right)\psi(x). \tag{2.89}$$

The product of $\hat{a}_+$ and $\hat{a}_-$ is proportional to the Hamilton operator up to a constant shift. Since the preceding equations were valid for generic wave functions, we may remove them from our expressions and simply write operator equations,

$$\hat{a}_+\hat{a}_- = \frac{\hat{H}}{\hbar\omega} - \frac{1}{2}. \tag{2.90}$$

This equation will be crucial in the following solution procedure. Its availability is restricted to the harmonic oscillator (and some other cases that may be related by intricate mathematical operations). If the potential had not been quadratic, the factorization would not have led to linear expressions, and the corresponding shift would not have been constant. The algebraic method for solving the system relies on the constant shift, and is therefore not applicable to most other potentials.

If we follow the same derivation in reversed order of $\hat{a}_+$ and $\hat{a}_-$, we obtain a different result,

$$\hat{a}_-\hat{a}_+ = \frac{\hat{H}}{\hbar\omega} + \frac{1}{2}. \tag{2.91}$$

Therefore, the product of $\hat{a}_+$ and $\hat{a}_-$ is not commutative. The failure to commute is expressed by the **commutator**

$$\hat{a}_-\hat{a}_+ - \hat{a}_+\hat{a}_- = 1. \tag{2.92}$$

The left-hand side is commonly abbreviated as

$$[\hat{a}_-, \hat{a}_+] = \hat{a}_-\hat{a}_+ - \hat{a}_+\hat{a}_- \tag{2.93}$$

and mathematically corresponds to the composition rule of a Lie algebra, replacing the product in standard algebra and arithmetic. The fact that the commutator equals a constant, one, reveals a mathematical structure underlying the Hamilton operator of the harmonic oscillator that makes it possible to apply the algebraic method. We will see additional related features in Section 3.2.4.

2.3.2.2 Stationary states

For stationary states, we have $\hat{H}\psi = E\psi$ with an energy constant E that is a number, not an operator. Therefore,

$$\hat{a}_+\hat{a}_-\psi = \left(\frac{E}{\hbar\omega} - \frac{1}{2}\right)\psi \tag{2.94}$$

and

$$\hat{a}_-\hat{a}_+\psi = \left(\frac{E}{\hbar\omega} + \frac{1}{2}\right)\psi. \tag{2.95}$$

produce constant factors on stationary states. These mathematical results have several useful properties that help us to rederive the stationary states of the harmonic oscillator and their energy values. They also give us new methods for computations of additional state properties.

We immediately find one stationary state if we solve the linear first-order differential equation $\hat{a}_- \psi_0 = 0$. Equation (2.90) then implies that

$$\hat{H}\psi_0 = \hbar\omega(\hat{a}_+\hat{a}_- + 1/2)\psi_0 = \frac{1}{2}\hbar\omega\psi_0 \,. \tag{2.96}$$

Therefore, ψ_0 such that $\hat{a}_- \psi_0$ is a solution of the time-independent Schrödinger equation with energy $E_0 = \frac{1}{2}\hbar\omega$.

Assuming that we know a stationary state ψ with energy value E, such that $\hat{H}\psi = E\psi$, we can apply $\hat{a}_+$ or $\hat{a}_-$ to ψ and obtain new stationary states with energy values different from E. To see this, given ψ, define a new state $\psi_+ = \hat{a}_+\psi$. We compute $\hat{H}\psi_+$ by using $\hat{H} = \hbar\omega(\hat{a}_+\hat{a}_- + 1/2)$:

$$\begin{aligned} \hat{H}\psi_+ &= \hbar\omega(\hat{a}_+\hat{a}_- + 1/2)\hat{a}_+\psi = \hbar\omega(\hat{a}_+\hat{a}_-\hat{a}_+ + \tfrac{1}{2}\hat{a}_+)\psi \\ &= \hbar\omega\hat{a}_+(\hat{a}_-\hat{a}_+ + \tfrac{1}{2})\psi = \hbar\omega\hat{a}_+(\hat{a}_-\hat{a}_+ - \tfrac{1}{2})\psi + \hbar\omega\hat{a}_+\psi \,. \end{aligned} \tag{2.97}$$

In the second step, we multiplied by $\hat{a}_+$ from the right, maintaining the original order of factors. In the third step, we factor out $\hat{a}_+$ to the left. Since there is a factor of $\hat{a}_+$ in the original $\hat{a}_+\hat{a}_-$, we do not have to commute $\hat{a}_+$ with $\hat{a}_-$. In the final step, we rearranged constants in anticipation of using the $\hat{a}_-\hat{a}_+$ version of the energy relationship, (2.95).

If we insert $\hat{a}_-\hat{a}_+ - 1/2 = \hat{H}/(\hbar\omega)$, the first term of the last result in (2.97) is proportional to the original energy E of ψ, and the second term implies a shift:

$$\hat{H}\psi_+ = \hat{a}_+(E\psi + \hbar\omega\psi) = (E + \hbar\omega)\psi_+ \,. \tag{2.98}$$

Therefore, if ψ is a stationary state with energy E, $\hat{a}_+\psi$ is another stationary state with energy $E + \hbar\omega$. Similar derivations show that $\hat{a}_-$ lowers the energy:

$$\hat{H}(\hat{a}_-\psi) = (E - \hbar\omega)(\hat{a}_-\psi) \,. \tag{2.99}$$

If ψ is a stationary state with energy E, $\hat{a}_-\psi$ is another stationary state with energy $E - \hbar\omega$.

The operators $\hat{a}_+$ and $\hat{a}_-$ are called **ladder operators** because, starting with an initial stationary state such as ψ_0, they can be applied multiple times in order to construct a ladder of stationary states with equally spaced energy levels, separated by $\hbar\omega$. The operator $\hat{a}_+$ is the raising operator, and the operator $\hat{a}_-$ is the lowering operator. Because the potential is bounded from below, there should be a smallest possible energy value, the ground-state energy E_0. The ground state ψ_0 must be such that $\hat{a}_-\psi_0 = 0$ in order to make sure that there is no state with smaller energy, $E_0 - \hbar\omega$: If $\hat{a}_-\psi_0 = 0$, $\hat{a}_-^n\psi_0 = 0$ and the ladder ends at the bottom rung given by ψ_0 with energy E_0.

The condition $\hat{a}_-\psi_0 = 0$ can be evaluated explicitly because it is equivalent to a first-order differential equation for ψ_0:

$$\hat{a}_-\psi_0 = \frac{1}{\sqrt{2m\hbar\omega}}(\hbar\psi_0' + m\omega x\psi_0) = 0 \tag{2.100}$$

implies

$$\frac{\psi_0'}{\psi_0} = \frac{\mathrm{d}\ln\psi_0}{\mathrm{d}x} = -\frac{m\omega}{\hbar}x \tag{2.101}$$

and is solved by

$$\ln\psi_0 = -\frac{1}{2}\frac{m\omega}{\hbar}x^2 + c \tag{2.102}$$

with a constant c, or

$$\psi_0(x) = A_0 e^{-m\omega x^2/(2\hbar)} \tag{2.103}$$

with $A_0 = e^c$ to be determined by normalization. Since this is a unique solution, up to the constant factor, there is only one ladder of states and energies. The ground-state energy follows from

$$E_0\psi_0 = \hat{H}\psi_0 = \hbar\omega(\hat{a}_+\hat{a}_- + \tfrac{1}{2})\psi_0 = \frac{1}{2}\hbar\omega\psi_0 \tag{2.104}$$

using $\hat{a}_-\psi_0 = 0$. Therefore,

$$E_0 = \frac{1}{2}\hbar\omega\,. \tag{2.105}$$

We now have the ground state $\psi_0(x)$ with the wave function (2.103) and energy E_0. Applying the raising operator multiple times, the set of all stationary states is given by

$$\psi_n(x) = A_n\hat{a}_+^n e^{-m\omega x^2/(2\hbar)} \tag{2.106}$$

with energies

$$E_n = (n+1/2)\hbar\omega \tag{2.107}$$

for non-negative integers, $n = 0,1,2,\ldots$. The constants A_n are required for normalization, which is not necessarily preserved by the operator $\hat{a}_+$.

The energies (2.107) show an interesting relationship with the photon picture that explains the photoelectric effect: If we have an electrically charged particle in a trap that can be modeled by a harmonic oscillator, it interacts with an electromagnetic wave that hits the trap. Classically, we expect maximal absorption if the oscillator is in resonance with the electromagnetic wave, such that they have the same frequency $\omega_{\text{light}} = \omega$. In quantum mechanics, this means that a single photon in the electromagnetic wave has an energy $E_{\text{photon}} = \hbar\omega_{\text{light}} = \hbar\omega$. Such a photon can be completely absorbed by the oscillator because it provides just the right amount of energy to jump from an energy level E_n to the next level, E_{n+1}. Suitable interactions between an electromagnetic wave and an electron can be expressed in terms of the ladder operators.

As two examples of excited states, we consider the first and second ones. The first excited state is obtained by applying $\hat{a}_+$ to $\psi_0(x)$, up to normalization. Using (2.86), we compute the resulting wave function as

$$\hat{a}_+\psi_0 = \frac{A_0}{\sqrt{2m\hbar\omega}}\left(-\hbar\frac{\mathrm{d}}{\mathrm{d}x} + m\omega x\right)e^{-m\omega x^2/(2\hbar)} = A_0\sqrt{\frac{2m\omega}{\hbar}}\,xe^{-m\omega x^2/(2\hbar)}\,. \tag{2.108}$$

A second application of the raising operator implies

$$\hat{a}_+^2\psi_0 = \frac{A_0}{\hbar}\left(-\hbar\frac{\mathrm{d}}{\mathrm{d}x} + m\omega x\right)\left(xe^{-m\omega x^2/(2\hbar)}\right) = A_0\left(2\frac{m\omega}{\hbar}x^2 - 1\right)e^{-m\omega x^2/(2\hbar)}\,. \tag{2.109}$$

We see the same pattern of symmetries as in the infinite square well: The ground state and the second excited state are even functions of x, while the first excited state is odd; see Figure 2.3. In general, the n-th excited state is a polynomial of degree n times the exponential function, maintaining the relationship between symmetries of the wave function and n being even or odd. These features agree with what we obtained from the analytic method, but are derived here in rather different ways.

2.3.2.3 Normalization

The stationary states are still to be normalized. There would be some tricky integrations if we used the specific expressions of potentially high-degree polynomials times the exponential function.

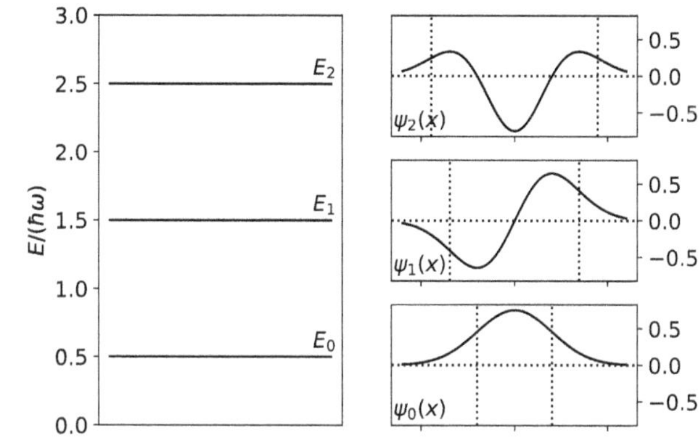

Figure 2.3 Energy values and wave functions of the ground state and the first two excited states of the harmonic oscillator. Vertical lines on the right indicate classical turning points at the corresponding energies. The wave functions therefore oscillate in the classically allowed regions, but they also imply non-zero probabilities in the classically forbidden regions. The asymptotic exponential behavior of stationary states means that the probability of finding a position far from a classically allowed region decreases rapidly with the distance.

Fortunately, this step can be completed rather quickly if we again use properties of the ladder operators. Starting with an explicit integration for the ground state, we obtain

$$1 = \int_{-\infty}^{\infty} |\psi_0(x)|^2 \mathrm{d}x = |A_0|^2 \int_{-\infty}^{\infty} e^{-m\omega x^2/\hbar} \mathrm{d}x = \sqrt{\frac{\hbar}{m\omega}} |A_0|^2 \int_{-\infty}^{\infty} e^{-y^2} \mathrm{d}y = \sqrt{\frac{\pi\hbar}{m\omega}} |A_0^2| \qquad (2.110)$$

using the same methods as in the example (1.67) of a Gaussian probability density. Choosing a real and positive A_0, we obtain

$$A_0 = \left(\frac{m\omega}{\pi\hbar}\right)^{1/4}. \qquad (2.111)$$

We use the method of recursion for normalization of all the excited states. We will apply the integral formula

$$\int_{-\infty}^{\infty} (\hat{a}_+ f(x))^* g(x) \mathrm{d}x = \int_{-\infty}^{\infty} f(x)^* (\hat{a}_- g(x)) \mathrm{d}x \qquad (2.112)$$

which is valid for any pair of normalizable functions, $f(x)$ and $g(x)$, and can be derived directly from the definition (2.86) of $\hat{a}_\pm$ and integration by parts:

$$\begin{aligned} \int_{-\infty}^{\infty} (\hat{a}_+ f(x))^* g(x) \mathrm{d}x &= \int_{-\infty}^{\infty} \frac{1}{\sqrt{2m\hbar\omega}} \left(-\hbar \frac{\mathrm{d}f(x)^*}{\mathrm{d}x} + m\omega x f(x)^*\right) g(x) \mathrm{d}x \\ &= \frac{1}{\sqrt{2m\hbar\omega}} \left(\hbar \int_{-\infty}^{\infty} f(x)^* \frac{\mathrm{d}g(x)}{\mathrm{d}x} + m\omega \int_{-\infty}^{\infty} f(x)^* x g(x) \mathrm{d}x\right) \\ &= \int_{-\infty}^{\infty} f(x)^* (\hat{a}_- g(x)) \mathrm{d}x. \end{aligned} \qquad (2.113)$$

When we integrate by parts in the second line, boundary terms do not contribute because $f(x)$ and $g(x)$ have to approach zero at $x \to \pm\infty$ in order to be normalizable. The integral formula is an operator version of complex conjugation, applied to $\hat{a}_\pm$.

Applied to the normalization integral of $\psi_{n+1}(x) = C_{n+1} \hat{a}_+ \psi_n(x)$, we obtain

$$\int_{-\infty}^{\infty} |\hat{a}_+ \psi_n(x)|^2 \mathrm{d}x = \int_{-\infty}^{\infty} (\hat{a}_+ \psi_n(x))^* (\hat{a}_+ \psi_n(x)) \mathrm{d}x = \int_{-\infty}^{\infty} \psi_n(x)^* (\hat{a}_- \hat{a}_+ \psi_n(x)) \mathrm{d}x. \qquad (2.114)$$

Equation (2.95) allows us to evaluate the product $\hat{a}_-\hat{a}_+$ when it acts on a stationary state, without the need to compute explicit derivatives:

$$\begin{aligned}\frac{1}{C_{n+1}^2}\int_{-\infty}^{\infty}|\psi_{n+1}(x)|^2\mathrm{d}x &= \int_{-\infty}^{\infty}|\hat{a}_+\psi_n(x)|^2\mathrm{d}x = \int_{-\infty}^{\infty}\psi_n(x)^*(\hat{a}_-\hat{a}_+\psi_n(x))\mathrm{d}x \\ &= \int_{-\infty}^{\infty}\psi_n(x)^*\left(\frac{\hat{H}}{\hbar\omega}+\frac{1}{2}\right)\psi_n(x)\mathrm{d}x \\ &= \left(\frac{E_n}{\hbar\omega}+\frac{1}{2}\right)\int_{-\infty}^{\infty}|\psi_n(x)|^2\mathrm{d}x = (n+1)\int_{-\infty}^{\infty}|\psi_n(x)|^2\mathrm{d}x\,.\end{aligned} \tag{2.115}$$

We are now ready to apply the argument of recursion: If we assume that $\psi_n(x)$ has already been normalized, the preceding result tells us that

$$\frac{1}{C_{n+1}^2}\int_{-\infty}^{\infty}|\psi_{n+1}(x)|^2\mathrm{d}x = n+1\,. \tag{2.116}$$

The next state, $\psi_{n+1}(x)$, is then normalized provided $C_n = 1/\sqrt{n+1}$. The result,

$$\psi_{n+1}(x) = \frac{\hat{a}_+\psi_n(x)}{\sqrt{n+1}}\,, \tag{2.117}$$

allows us to normalize all excited states by recursion, deriving conditions for A_n in $\psi_n(x) = A_n\hat{a}_+^n e^{-m\omega x^2/(2\hbar)}$ from (2.106). The ground state has already been normalized,

$$\psi_0(x) = A_0 e^{-m\omega x^2/(2\hbar)} \quad \text{with } A_0 = \left(\frac{m\omega}{\pi\hbar}\right)^{1/4}. \tag{2.118}$$

Given A_0, we obtain two versions of the first excited state,

$$\psi_1(x) = A_1\hat{a}_+ e^{-m\omega x^2/(2\hbar)} = A_1\frac{\hat{a}_+\psi_0(x)}{A_0} \tag{2.119}$$

from (2.106), and

$$\psi_1(x) = \frac{\hat{a}_+\psi_0(x)}{\sqrt{0+1}} \tag{2.120}$$

from (2.117), both containing $\hat{a}_+\psi_0(x)$. Comparing coefficients implies $A_1 = A_0$. For the next state, we have

$$\psi_2(x) = A_2\hat{a}_+^2 e^{-m\omega x^2/(2\hbar)} = A_2\frac{\hat{a}_+\psi_1(x)}{A_1} \tag{2.121}$$

and

$$\psi_2(x) = \frac{\hat{a}_+\psi_1(x)}{\sqrt{1+1}} \tag{2.122}$$

such that $A_2 = A_1/\sqrt{2} = A_0/\sqrt{2}$.

For general n,

$$\psi_n(x) = A_n(\hat{a}_+)^n e^{-m\omega x^2/(2\hbar)} = A_n\frac{\hat{a}_+\psi_{n-1}(x)}{A_{n-1}} \tag{2.123}$$

and

$$\psi_n(x) = \frac{\hat{a}_+\psi_{n-1}(x)}{\sqrt{n}} \tag{2.124}$$

imply $A_n = A_{n-1}/\sqrt{n} = \cdots = A_0/\sqrt{n!}$. The normalized stationary states are therefore

$$\psi_n(x) = \left(\frac{m\omega}{\pi\hbar}\right)^{1/4}\frac{1}{\sqrt{n!}}\hat{a}_+^n e^{-m\omega x^2/(2\hbar)}\,. \tag{2.125}$$

The algebraic method relies on the existence of ladder operators linear in $\hat{x}$ and $\hat{p}$, which is implied by a quadratic potential. The method is not applicable in general to higher-degree polynomials or non-polynomial potentials. The analytic or power-series method is more general, in which we try to find solutions of the time-independent Schrödinger equation in the form of a power series in x, times a suitable factor that makes it normalizable. For the harmonic oscillator, both methods are useful because one or the other may be more tractable for a specific question.

2.3.3 TIME-DEPENDENT EXAMPLES

Each stationary state has the simple time dependence

$$\Psi_n(x,t) = \psi_n(x)e^{-\frac{1}{2}i\omega t}\exp(-in\omega t). \tag{2.126}$$

The first phase factor, $e^{-\frac{1}{2}i\omega t}$, is independent of n and therefore can be factored out of any superposition of stationary states,

$$\Psi(x,t) = e^{-\frac{1}{2}i\omega t}\sum_{n=0}^{\infty} c_n\psi_n(x)\exp(-in\omega t). \tag{2.127}$$

The linear n-dependence in the remaining exponent then makes it easy to construct specific time-dependent behaviors.

A quantum clock can be modeled by choosing $c_n = 1/\sqrt{N}$ for $n = 1,\ldots,N$, while $c_n = 0$ for $n > N$ and for $n = 0$. At time $t = 0$, we have

$$\Psi(x,0) = \frac{1}{\sqrt{N}}\sum_{n=1}^{N}\psi_n(x). \tag{2.128}$$

At a later time, $t_1 = 2\pi/(\omega N)$, the state is

$$\Psi(x,t_1) = \frac{1}{\sqrt{N}}\sum_{n=1}^{N}\psi_n(x)\exp(-2\pi in/N). \tag{2.129}$$

There is an orthogonality relation between these two states because

$$\int_{-\infty}^{\infty}\Psi(x,0)^*\Psi(x,t_1)\mathrm{d}x = \frac{1}{N}\sum_{n=1}^{N}\exp(-2\pi i/N)^n = \frac{1}{N}\frac{\exp(-2\pi i)-1}{1-\exp(2\pi i/N)} = 0 \tag{2.130}$$

using the summation formula

$$\sum_{n=1}^{N} z^N = \frac{z^N-1}{1-z^{-1}} \tag{2.131}$$

that can be proved by induction. This pattern repeats at every integer multiple of t_1:

$$\begin{aligned}\int_{-\infty}^{\infty}\Psi(x,jt_1)^*\Psi(x,kt_1)\mathrm{d}x &= \frac{1}{N}\sum_{n=1}^{N}\exp(-2\pi i(k-j)/N)^n \\ &= \frac{1}{N}\frac{\exp(-2\pi(k-j)i)-1}{1-\exp(2\pi i(k-j)/N)} = 0\end{aligned} \tag{2.132}$$

are orthogonal for any j and k such that $0 < k-j < N$. Starting at $\Psi(x,0)$, the quantum clock moves through $N-1$ orthogonal states at equal time increments, until it returns to the initial state at time $t = 2\pi/\omega$.

Another example is a quantum gate constructed from the harmonic oscillator. Quantum gates are dynamical processes that implement basic logic operations for a quantum computer. An important example is the controlled NOT (or CNOT) gate, which flips the value of a bit provided a control bit is set to one: The control-computing pairs (n_1, n_2) of bits are transformed according to $(0,0) \mapsto (0,0)$, $(0,1) \mapsto (0,1)$, $(1,0) \mapsto (1,1)$ and $(1,1) \mapsto (1,0)$. The harmonic oscillator can be used to model such a bit flip because its half-cycle evolution (waiting half a period $\delta t = \pi/\omega$) changes the sign of odd states ψ_{2j+1} while it preserves even states ψ_j. Since the bit pairs $(0,0)$ and $(0,1)$ do not change, we can represent them by even states, such as $(0,0) = \psi_0$ and $(0,1) = \psi_2$. The remaining two bit pairs should be flipped, but we cannot use the same odd state with opposite signs (such as ψ_1 and $-\psi_1$) since the statistical information in the state does not depend on its overall sign. Instead, we use $(1,0) = \frac{1}{\sqrt{2}}(\psi_0 + \psi_1)$ and $(1,1) = \frac{1}{\sqrt{2}}(\psi_0 - \psi_1)$. Half-cycle evolution then performs a CNOT gate on any given initial bit pair.[1]

2.3.4 PROPERTIES OF STATES AND EXPECTATION VALUES

Using the analytic method, we can compute specific stationary states with small $n \geq 2$ by directly applying the recursion formula. The ground state ($n = 0$) and first excited state ($n = 1$) are determined directly by the free constants a_0 and a_1 and therefore do not require recursion steps.

For $n = 0$, we obtain a constant $h_0(\xi) = a_0^{(0)}$. We use the superscript on $a_0^{(n)}$ because the zeroth-order coefficient of the polynomial $h_n(\xi)$ may depend on n, and similarly for the remaining coefficients $a_j^{(n)}$. The wave function then equals $\psi_0(\xi) = a_0^{(0)} \exp(-\frac{1}{2}\xi^2)$ and is Gaussian, with energy value $E_0 = \frac{1}{2}\hbar\omega$. For $n = 1$, an odd number, we have to choose $a_0^{(1)} = 0$. The coefficient $a_1^{(1)}$ is then free. We obtain the linear polynomial $h_1(\xi) = a_1^{(1)}\xi$ and the wave function $\psi_1(\xi) = a_1^{(1)}\xi \exp(-\frac{1}{2}\xi^2)$ with energy $E_1 = \frac{3}{2}\hbar\omega$.

If $n = 2$, an even number, we have to choose $a_1^{(2)} = 0$. The coefficient $a_0^{(2)}$ is free, and, using $G_2 = 2n + 1 = 5$, we derive

$$a_2^{(2)} = \frac{G_2 - 1}{2} a_0^{(2)} = -2a_0^{(2)} \tag{2.133}$$

from the recursion formula. We obtain the polynomial $h_2(\xi) = a_0^{(2)}(1 - 2\xi^2)$ and the wave function $\psi_2(\xi) = a_0^{(2)}(1 - 2\xi^2)\exp(-\frac{1}{2}\xi^2)$ with energy $E_2 = \frac{5}{2}\hbar\omega$. After relating ξ to x, these results agree with the functions (2.108) and (2.109) obtained by applying the raising operator to the ground state.

2.3.4.1 Hermite polynomials

It would be tedious to apply the recursion formula or ladder operators in order to derive stationary states with large n. Fortunately, in the 19th century mathematicians studied many classes of polynomials with coefficients subject to second-order recursion formulas, and one example fits our equation (2.75). This class is defined as the **Hermite polynomials** $H_n(\xi)$, named after the French mathematician Charles Hermite. In addition to the recursion formula for their coefficients, these polynomials also obey the recursion formula

$$H_{n+1}(\xi) = 2\xi H_n(\xi) - 2nH_{n-1}(\xi) \tag{2.134}$$

directly for their ξ-dependence. Since the Hermite polynomials are nowadays used in many applications, they are available in common computer algebra programs and in several programming languages.

The original mathematical definition did not consider normalization as needed in quantum mechanics, which in any case cannot be applied directly to a polynomial without the factor of $\exp(-\frac{1}{2}\xi^2)$. Instead, unique polynomial coefficients were defined by requiring that the highest-degree coefficient, given by $H_{n,n}$ in a polynomial expansion $H_n(\xi) = \sum_{j=0}^{n} H_{n,j}\xi^j$, equals $H_{n,n} = 2^n$.

[1] M. A. Nielsen and I. L. Chuang: Quantum Computation and Quantum Information, Cambridge University Press, 2021.

This condition, unlike quantum-mechanics normalization, is preserved by the recursion formula (2.134). The low-degree Hermite polynomials are therefore $H_0(\xi) = 1$ and $H_1(\xi) = 2\xi$, from which all other ones are defined by applying (2.134).

Mathematicians also studied integrals of products of Hermite polynomials, from which quantum-mechanics normalization of the corresponding $\psi_n(\xi)$ can be derived. The result is

$$\psi_n(\xi) = \left(\frac{m\omega}{\pi\hbar}\right)^{1/4} \frac{1}{\sqrt{2^n n!}} H_n(\xi) \exp(-\tfrac{1}{2}\xi^2). \tag{2.135}$$

There is a factor of $(m\omega/\pi\hbar)^{1/4}$ because we normalize ψ_n by integrating $|\psi_n(\xi)|^2$ over x, and we have $\sqrt{m\omega/(\pi\hbar)}\,\mathrm{d}x = \mathrm{d}\xi$.

Examples of Hermite polynomials and stationary states are shown in Figures 2.4 and 2.5. For large $|\xi|$ or $|x|$, the wave functions and their probability densities approach zero, as required for normalizability, but they are not exactly zero except for a finite number of points, given by the zeros of the polynomial $h(\xi)$. Even though the energy is fixed and finite in a stationary state and the potential keeps growing for large $|x|$, there is a non-vanishing likelihood of finding the particle at any value of x, even at places where the harmonic potential is greater than a given energy value. This result is our first taste of quantum tunneling.

2.3.4.2 Expectation values

The stationary states can be used to determine expectation values of $\hat{x}$ and $\hat{p}$. For generic n, this calculation would be tedious if we use Hermite polynomials, but the ladder operators provide a simple method. It is easy to compute the expectation values of $\hat{a}_+$ and $\hat{a}_-$ in any stationary state $\psi_n(x)$ because of orthogonality:

$$\langle \hat{a}_\pm \rangle = \int_{-\infty}^{\infty} \psi_n(x)^* \hat{a}_\pm \psi_n(x) \mathrm{d}x \propto \int_{-\infty}^{\infty} \psi_n(x)^* \psi_{n\pm 1}(x) \mathrm{d}x = 0. \tag{2.136}$$

This result immediately implies

$$\langle \hat{x} \rangle_n = 0 \quad \text{and} \quad \langle \hat{p} \rangle_n = 0 \tag{2.137}$$

because $\hat{x}$ and $\hat{p}$ are linearly related to $\hat{a}_\pm$ by

$$\hat{x} = \sqrt{\frac{\hbar}{2m\omega}} (\hat{a}_+ + \hat{a}_-) \quad \text{and} \quad \hat{p} = \sqrt{\frac{m\hbar\omega}{2}} i(\hat{a}_+ - \hat{a}_-). \tag{2.138}$$

These two equations are the inverse of definition (2.86).

Standard deviations of x and p in a stationary state cannot be zero, as a consequence of the uncertainty relation. Mathematically, non-zero contributions are obtained because squaring $\hat{x}$ and $\hat{p}$ in (2.138) implies the cross-terms $\hat{a}_+\hat{a}_-$ and $\hat{a}_-\hat{a}_+$, which have non-zero expectation values. In particular, we have

$$\langle \hat{x}^2 \rangle_n = \frac{\hbar}{2m\omega} \langle (\hat{a}_+ + \hat{a}_-)^2 \rangle_n = \frac{\hbar}{2m\omega} \langle \hat{a}_+^2 + \hat{a}_+\hat{a}_- + \hat{a}_-\hat{a}_+ + \hat{a}_-^2 \rangle_n \tag{2.139}$$

In the last expression, we have to pay attention to the ordering because $\hat{a}_+$ and $\hat{a}_-$ do not commute. The ordering of each individual term is determined by the original ordering of all operators in the product $(\hat{a}_+ + \hat{a}_-)^2 = (\hat{a}_+ + \hat{a}_-)(\hat{a}_+ + \hat{a}_-)$, which implies two different cross-terms, $\hat{a}_+\hat{a}_-$ and $\hat{a}_-\hat{a}_+$.

Using (2.90) and (2.91) as well as $\langle \hat{a}_+^2 \rangle_n = 0 = \langle \hat{a}_-^2 \rangle_n$ from orthogonality, we simplify the expectation value of $\hat{x}^2$ to

$$\langle \hat{x}^2 \rangle_n = \frac{\hbar}{2m\omega} \cdot \frac{2E_n}{\hbar\omega} = \frac{\hbar}{m\omega}(n + 1/2) \tag{2.140}$$

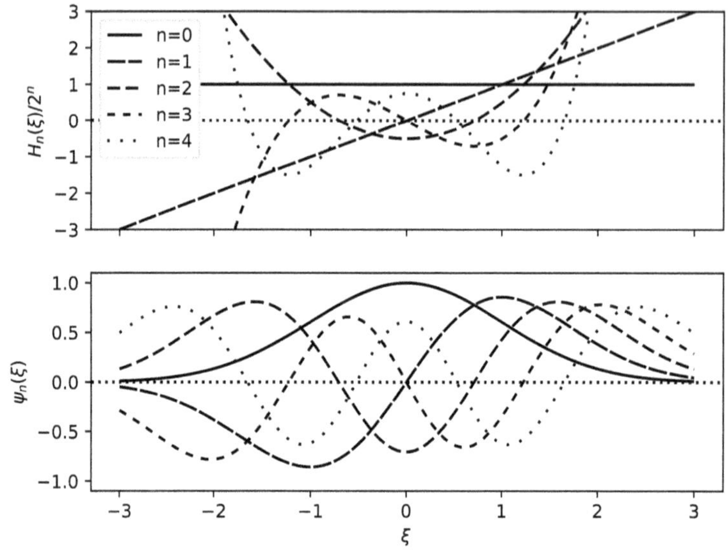

Figure 2.4 The first five Hermite polynomials (divided by 2^n) and stationary states of the harmonic oscillator.

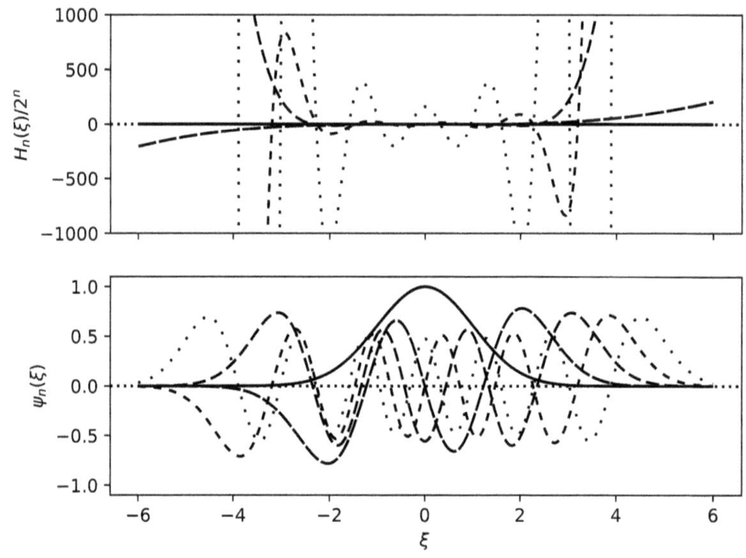

Figure 2.5 The five Hermite polynomials $H_{3j}(x)$ for $0 \leq j \leq 4$ (divided by 2^{3j}) and stationary states of the harmonic oscillator. The polynomials $H_n(x)$ increase rapidly for larger n, while normalization with the exponential factor of the stationary states keeps their amplitude nearly constant in a finite range.

where $\langle \hat{H} \rangle_n = E_n = (n + 1/2)\hbar\omega$. The position variance therefore equals

$$\sigma_{x,n}^2 = \frac{\hbar}{m\omega}(n + 1/2) . \tag{2.141}$$

The expectation value of $\hat{p}^2$ can be computed in a similar way or, more easily, by relating it to the kinetic energy. The potential energy is non-zero, but for a quadratic potential it is proportional to the expectation value of $\hat{x}^2$ that we just computed. Therefore,

$$\langle \hat{H} \rangle_n = \frac{\langle \hat{p}^2 \rangle_n}{2m} + \frac{1}{2} m\omega^2 \langle \hat{x}^2 \rangle_n = E_n = \hbar\omega(n+1/2) \tag{2.142}$$

can be solved for

$$\langle \hat{p}^2 \rangle_n = 2mE_n - m^2\omega^2 \langle \hat{x}^2 \rangle_n = mE_n\,, \tag{2.143}$$

which then determines the momentum variance

$$\sigma_{p,n}^2 = m\hbar\omega(n+1/2)\,. \tag{2.144}$$

The product

$$\sigma_{x,n}\sigma_{p,n} = \hbar(n+1/2) \geq \frac{\hbar}{2} \tag{2.145}$$

obeys the uncertainty relation and saturates it only for $n=0$, the ground state.

2.4 FREE PARTICLE

If there is no force acting on a particle, the potential $V(x)=0$ is zero (or constant) and the time-independent Schrödinger equation

$$-\frac{\hbar^2}{2m}\frac{\mathrm{d}^2\psi}{\mathrm{d}x^2} = E\psi \tag{2.146}$$

has constant coefficients. For positive energy E, this equation is solved by

$$\psi(x) = A'\sin(kx) + B'\cos(kx) = A\exp(ikx) + B\exp(-ikx) \tag{2.147}$$

where

$$k = \frac{\sqrt{2mE}}{\hbar} \tag{2.148}$$

is a positive real number. The coefficient pairs (A',B') and (A,B) present two equivalent versions of the general solution. We will mainly work with the exponential form because it is more convenient.

2.4.1 STATIONARY STATES

We encountered the same solutions in the infinite square well, but in that case they were valid only in a finite range $0 \leq x \leq a$. A free particle is not constrained to the well or any finite region, and therefore the plane-wave form (2.147) extends to all values of x. Physically, this property is consistent with the fact that a free particle can be found at any value of x if we just wait long enough, assuming it moves in the right direction. There should then be a non-zero probability density for any x at a suitable time. A stationary state must realize this non-zero value at any time.

However, if this is the case, $\psi(x)$ does not have a well-defined limit for $x \to \pm\infty$. These observations have implications that require a new treatment of wave functions. In previous examples, we have seen that boundary conditions on wave functions imply discrete energy values. Since we do not have strict boundary conditions for the free particle, there should be no restrictions on possible energy values E, as long as they are positive. We should not expect discrete energies for the free particle. This result is not inconsistent with observations because we do spectroscopy with discrete spectral lines for bound-state systems such as electrons in an atom or a molecule, but not for free particles.

Since the plane waves (2.147) are not normalizable, the time-independent Schrödinger equation has no solutions with a suitable statistical interpretation. We arrived at this time-independent equation after making a mathematical assumption, that of separability for special solutions $\Psi(x,t) = \psi(x)f(t)$ of the time-dependent equation. This assumption is not valid for physically allowed solutions of the free particle. Nevertheless, it may still be useful mathematically as an intermediate step of constructing more general solutions.

If physically allowed solutions are not separable, the energy cannot be sharp and must instead have some non-zero standard deviation σ_E. For a free particle, normalizability is in fact closely related to unsharp energies because (i) the energy is purely kinetic and directly related to the momentum $p=\sqrt{2mE}$, and (ii) the momentum variance is restricted by the uncertainty relation. Sharp energies would imply sharp momenta, but this outcome is consistent with the uncertainty relation only if position measurements have infinite standard deviation, as they do in non-normalizable states such as a plane wave. Mathematically, all these statements are compatible with one another. We only have to determine how much physical meaning we can attach to stationary states of the free particle.

We conclude that position measurements of the free particle can only be described by a time-dependent, non-stationary state. Mathematically, we are allowed to construct such states as superpositions of the non-normalizable plane-wave solutions (2.147), but the superpositions must be normalizable.

2.4.2 WAVE PACKETS

The time-dependent version of a plane wave with energy $E(k)=\hbar^2k^2/(2m)$ is given by

$$\Psi_k(x,t)=A\exp(i(kx-E(k)t/\hbar))=A\exp(i(kx-\hbar k^2t/(2m))) \tag{2.149}$$

where we combined the two independent solutions in (2.147) by allowing k to be any real number, including negative ones. A common distinction between the signs is reflected in the names "right-moving plane waves" for solutions with $k>0$, and correspondingly "left-moving plane waves" for solutions with $k<0$. These names are motivated by the observation that a wave feature at constant phase ϕ_0, such as a local maximum of the real part of $\Psi_k(x,t)$, moves according to $x-\hbar kt/(2m)=\phi_0/k$ if we display a movie of the time-dependent real part of (2.149). Solving this equation for x as a function of time, we obtain $x(t)=\phi_0/k+v_{\rm phase}t$ with the **phase velocity**

$$v_{\rm phase}=\frac{\hbar k}{2m}\,. \tag{2.150}$$

The plane-wave solution of the time-independent Schrödinger equation corresponds to the dispersion relation

$$\omega(k)=\frac{E(k)}{\hbar}=\frac{\hbar k^2}{2m}\,. \tag{2.151}$$

With this function, the phase velocity can be defined as

$$v_{\rm phase}=\frac{\omega(k)}{k}\,. \tag{2.152}$$

The sign of the velocity is equal to the sign of k, which determines the direction in which the wave is moving. The phase velocity equals half the classical velocity $v=p/m$ if we use $p=\hbar k$. A factor of $1/2$ compared with the classical velocity of a free particle might seem like an unacceptable contradiction with basic observations. However, the phase velocity does not describe the changing measured position of a free particle. If we perform measurements compatible with the statistical content of the wave function, we may set up a large number of experiments starting with the same initial state, and then measure the position for each of them at a different time. We do not repeat position measurements at different times on the same system in order to avoid influences of measurements on the system. We may prepare a state that resembles a plane wave in a large but finite region, and falls off to zero outside of this region in order to be normalizable.

If we then line up the outcomes as a function of the measurement time, they are unlikely to accumulate near a single local maximum of our approximate plane wave. Instead, after measuring one position, it is more likely to obtain the next value at a distant part of the wave because its probability density is constant in the finite region. For the same reason, the lined-up measured positions do not move in a specific direction but jump back and forth at random. Therefore, the phase velocity should not be used in a comparison with the classical velocity. In spite of the non-zero value of $v_{\rm phase}$, and in spite of common terminology such as "left-moving" and "right-moving," a plane wave is not moving at all because it is stationary as a solution of the time-independent Schrödinger equation. If we are interested in understanding the motion of a free particle, we should construct a normalizable solution of the time-dependent Schrödinger equation. Normalizability prevents such a solution from being strictly constant, implying specific features of the wave that can be followed in order to determine how it is moving. If the position variance is sufficiently small, we obtain a near-classical trajectory for the expectation value $\langle\hat{x}\rangle(t)$.

Heuristically, normalizable solutions must be superpositions of plane waves such that their individual contributions cancel out at large $|x|$. It is possible to implement this condition by using the mathematical method of Fourier transformation. For discrete energies, we have seen how the general solution of the time-dependent Schrödinger equation can be written as a superposition by summing up all stationary states for the allowed energy values. For the free particle, the continuum of all positive energy values is possible. The sum is therefore replaced by an integral:

$$\Psi(x,t) = \frac{1}{\sqrt{2\pi}} \int_{-\infty}^{\infty} \phi(k) \exp(i(kx - \hbar k^2 t/(2m)))\mathrm{d}k\,. \tag{2.153}$$

The factor of $1/\sqrt{2\pi}$ is used by convention in this context, the function $\phi(k)$ replaces the previous coefficients c_n in (2.39), and the exponential function is the product of a time-independent stationary state $\psi_k(x) = \exp(ikx)$ with the corresponding time-dependent factor, $f_k(t) = \exp(-iE(k)t/\hbar)$.

If $\phi(k)$ is non-zero for at least two values of k, we have a wave packet in which k does not have a sharp value. Since energy and momentum are directly related to k, they do not have sharp values either. It is therefore possible to respect the uncertainty relation at finite σ_x. Moreover, it follows from Fourier analysis that a finite value of $\int_{-\infty}^{\infty}|\phi(k)|^2\mathrm{d}k$ implies normalizability of $\Psi(x,t)$ for all t. If we start with a normalizable wave function $\Psi(x,0)$ at time zero, the corresponding $\phi(k)$ such that

$$\Psi(x,0) = \frac{1}{\sqrt{2\pi}} \int_{-\infty}^{\infty} \phi(k) \exp(ikx)\mathrm{d}k \tag{2.154}$$

is determined by inverse Fourier transformation:

$$\phi(k) = \frac{1}{\sqrt{2\pi}} \int_{-\infty}^{\infty} \Psi(x,0) \exp(-ikx)\mathrm{d}x\,. \tag{2.155}$$

This expression then implies the full time dependence of the wave function according to (2.153).

2.4.3 GAUSSIAN WAVE PACKET

A tractable example is given by a Gaussian initial state,

$$\Psi(x,0) = \left(\frac{\lambda}{\pi}\right)^{1/4} e^{-\frac{1}{2}\lambda x^2} \tag{2.156}$$

with some real $\lambda > 0$. We compute

$$\phi(k) = \frac{1}{\sqrt{2\pi}} \left(\frac{\lambda}{\pi}\right)^{1/4} \int_{-\infty}^{\infty} e^{-\frac{1}{2}\lambda x^2 - ikx}\mathrm{d}x \tag{2.157}$$

by transforming the integrand to a single Gaussian without a linear term in the exponent. Such a transformation is always possible thanks to the method of completing the square, followed by substitution: We first write

$$-\frac{1}{2}\lambda x^2 - ikx = a(x+b)^2 + c \tag{2.158}$$

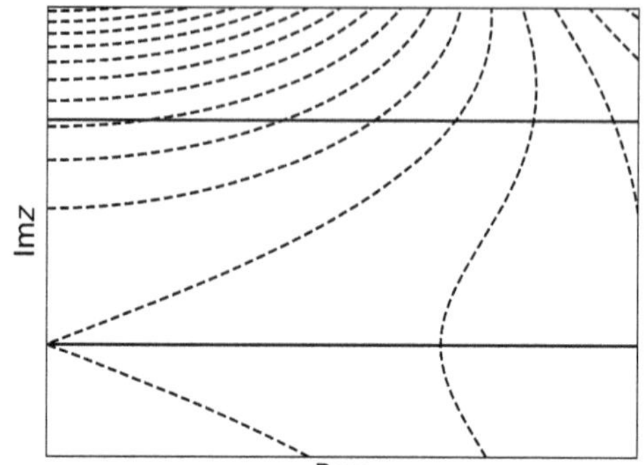

Figure 2.6 Contour plot of the real part of the complex function $f(z) = \exp(-\frac{1}{2}z^2)$. Moving an integration path from one of the solid horizontal lines to the other one implies a different integrand. In this case, however, complex analysis shows that the integrals over the two complete lines, both from $-\infty$ to ∞, are the same.

with suitable constants a, b and c that can be determined by expanding the latter expression to $ax^2 + 2abx + ab^2 + c$ and then comparing coefficients:

$$a = -\frac{1}{2}\lambda \quad , \quad b = -\frac{ik}{2a} = \frac{ik}{\lambda} \quad , \quad c = -ab^2 = -\frac{k^2}{2\lambda} . \tag{2.159}$$

Therefore,

$$\begin{aligned} \phi(k) &= \frac{1}{\sqrt{2\pi}}\left(\frac{\lambda}{\pi}\right)^{1/4} \int_{-\infty}^{\infty} e^{-\frac{1}{2}\lambda(x+ik/\lambda)^2 - \frac{1}{2}k^2/\lambda} \mathrm{d}x \\ &= \frac{1}{\sqrt{2\pi}}\left(\frac{\lambda}{\pi}\right)^{1/4} e^{-\frac{1}{2}k^2/\lambda} \int_{-\infty+ik/\lambda}^{\infty+ik/\lambda} e^{-\frac{1}{2}\lambda z^2} \mathrm{d}x \end{aligned} \tag{2.160}$$

where we moved out the x-independent $e^{-\frac{1}{2}k^2/\lambda}$ and substituted $z = x + ik/\lambda$.

The latter step is mathematically tricky because we substitute a complex number for a real integration variable. We are not just transforming coordinates on the axis or a fixed 1-dimensional line. In addition, we are shifting the integration curve on the 2-dimensional complex plane. As illustrated in Figure 2.6, such a substitution is not always guaranteed to keep the integral unmodified because the integrand changes if we evaluate a complex function on a different curve. For the function considered here, however, methods of complex analysis show that the substitution does not change the value of the integral. Therefore,

$$\phi(k) = (\lambda\pi)^{-1/4} \exp(-k^2/(2\lambda)) \tag{2.161}$$

using our standard Gaussian integral.

The Fourier transformation of a Gaussian $\Psi(x,0)$ is therefore another Gaussian $\phi(k)$. This new function can be interpreted as a wave function that describes measurements of k or of $p = \hbar k$. It determines a probability density

$$|\phi(k)|^2 = \frac{1}{\sqrt{\lambda\pi}} e^{-k^2/\lambda} \tag{2.162}$$

for these measurements. Writing this function in standard Gaussian form,

$$|\phi(k)|^2 = \frac{1}{\sqrt{2\pi}\,\sigma_k} e^{-k^2/(2\sigma_k^2)} , \tag{2.163}$$

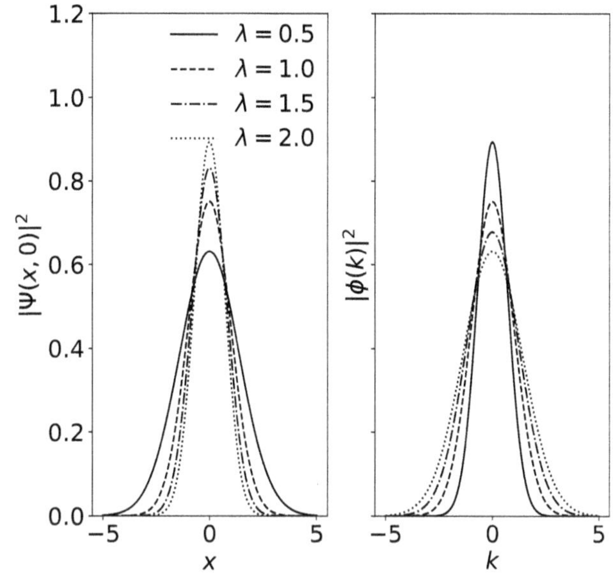

Figure 2.7 Gaussian probability densities $(\lambda/\pi)^{1/4}\exp(-\frac{1}{2}x^2/\lambda)$ (left) and their Fourier transformations (right, using the same line styles for the four values of λ). The standard deviations σ_x and σ_k are inversely proportional to each other.

allows us to read off the standard deviation

$$\sigma_k = \sqrt{\frac{\lambda}{2}} \tag{2.164}$$

for k-measurements, which is different from the standard deviation

$$\sigma_x = \frac{1}{\sqrt{2\lambda}} \tag{2.165}$$

implied by the initial wave function for position measurements. In particular, large λ imply a small position variance but a large k-variance; see Figure 2.7. Multiplying σ_k by $\hbar$ to obtain σ_p, we see that the uncertainty relation is always saturated by real-valued Gaussians as used here:

$$\sigma_x\sigma_p = \hbar\sigma_x\sigma_k = \frac{\hbar}{2}. \tag{2.166}$$

Given the coefficient function $\phi(k)$ for a Gaussian initial wave function $\Psi(x,0)$, we compute the time-dependent solution

$$\Psi(x,t) = \frac{1}{\sqrt{2\pi}(\lambda\pi)^{1/4}}\int_{-\infty}^{\infty}\exp\left(-\frac{k^2}{2\lambda}+ikx-i\hbar\frac{k^2t}{2m}\right)\mathrm{d}k \tag{2.167}$$

by completing the square,

$$\Psi(x,t) = \frac{1}{\sqrt{2\pi}(\lambda\pi)^{1/4}}\int_{-\infty}^{\infty}\exp\left(-\tfrac{1}{2}\left(\frac{1}{\lambda}+\frac{i\hbar t}{m}\right)\left(k-\frac{ix}{1/\lambda+i\hbar t/m}\right)^2-\tfrac{1}{2}\frac{x^2}{1/\lambda+i\hbar t/m}\right)\mathrm{d}k, \tag{2.168}$$

and shifting the integration path in the complex plane:

$$\begin{aligned}\Psi(x,t) &= \frac{1}{\sqrt{2\pi}(\lambda\pi)^{1/4}}\exp\left(-\tfrac{1}{2}\frac{x^2}{1/\lambda+i\hbar t/m}\right)\int_{-\infty-ix/(1/\lambda+i\hbar t/m)}^{\infty-ix/(1/\lambda+i\hbar t/m)}\exp\left(-\left(\frac{1}{\lambda}+\frac{i\hbar t}{m}\right)\frac{K^2}{2}\right)\mathrm{d}K\\ &= \frac{1}{(\lambda\pi)^{1/4}}\frac{1}{\sqrt{1/\lambda+i\hbar t/m}}\exp\left(-\tfrac{1}{2}\frac{\lambda x^2}{1+i\lambda\hbar t/m}\right). \end{aligned} \tag{2.169}$$

Here, we complete the square for the exponent as a polynomial in k because this is the integration variable.

Writing

$$\frac{1}{1+i\lambda\hbar t/m} = \frac{1-i\lambda\hbar t/m}{1+\lambda^2\hbar^2t^2/m^2} \tag{2.170}$$

in the exponent and

$$\left|\frac{1}{\sqrt{z}}\right|^2 = \frac{1}{\sqrt{z^*}}\frac{1}{\sqrt{z}} = \frac{1}{\sqrt{|z|^2}} \tag{2.171}$$

in the prefactor, we obtain the probability density

$$|\Psi(x,t)|^2 = \sqrt{\frac{\lambda}{\pi}}\frac{1}{\sqrt{1+\lambda^2\hbar^2t^2/m^2}}\exp\left(-\frac{\lambda x^2}{1+\lambda^2\hbar^2t^2/m^2}\right). \tag{2.172}$$

This probability density is Gaussian at all times with a time-dependent standard deviation

$$\sigma_x(t) = \frac{1}{\sqrt{2\lambda}}\sqrt{1+\frac{\lambda^2\hbar^2t^2}{m^2}} \tag{2.173}$$

in agreement with our older result (1.135). This function is unbounded at large $|t|$. The momentum variance at time t is still determined by the k-dependent coefficient function $\phi(k)$, which is time independent. Therefore, the momentum variance $\sigma_p = \hbar\sigma_k = \sqrt{\lambda/2}\,\hbar$ implies the product

$$\sigma_x(t)\sigma_p = \frac{\hbar}{2}\sqrt{1+\frac{\lambda^2\hbar^2t^2}{m^2}} \tag{2.174}$$

which is at the minimum of the uncertainty relation only at time $t = 0$. This time value is distinguished because we assumed the wave function to be Gaussian with real exponent at $t = 0$, corresponding to vanishing covariance $C_{xp} = 0$ in the context of (1.135).

2.4.4 MOVING PARTICLE

The wave function of the previous example corresponds to a particle at rest because $\langle\hat{p}\rangle(t) = 0$ follows from the even k-distribution $\phi(k)$. (The state is not stationary, and therefore $\langle\hat{p}\rangle = 0$ is not automatically implied.) For a general time dependence that allows for moving particles, we should choose $\phi(k)$ such that $\langle\hat{k}\rangle = \int_{-\infty}^{\infty} k|\phi(k)|^2\mathrm{d}k \neq 0$. The momentum expectation value $\langle\hat{p}\rangle = \hbar\langle\hat{k}\rangle$ is then non-zero.

If we assume that $\phi(k)$ is sharply peaked around some value $\langle\hat{k}\rangle = k_0$, the specific form of the function does not matter much and we can derive general results about the motion of a free particle. In an approximation, values of k around k_0 are then the most significant ones in the integral

$$\Psi(x,t) = \frac{1}{\sqrt{2\pi}}\int_{-\infty}^{\infty}\phi(k)\exp(i(kx-\omega(k)t))\mathrm{d}k \tag{2.175}$$

with the frequency function

$$\omega(k) = \frac{E(k)}{\hbar} = \frac{\hbar k^2}{2m} \tag{2.176}$$

that determines the dispersion relation of our quantum waves. The dispersion relation of the free particle implies a quadratic relationship between ω and k, but around k_0 we can replace this behavior with a linear function by Taylor expansion:

$$\omega(k) = \omega(k_0) + \left.\frac{d\omega}{dk}\right|_{k=k_0}(k-k_0) + \cdots \approx \omega_0 + v_{\mathrm{group}}(k-k_0) \tag{2.177}$$

where $\omega_0 = \omega(k_0)$, and the **group velocity** is defined as

$$v_{\mathrm{group}}(k_0) = \left.\frac{d\omega}{dk}\right|_{k=k_0} . \tag{2.178}$$

The relationship between the group velocity and motion of the free particle can be seen by completing the derivation of $\Psi(x,t)$ after a Taylor expansion of $\omega(k)$ around k_0:

$$\begin{aligned}\Psi(x,t) &\approx \frac{1}{\sqrt{2\pi}}\int_{-\infty}^{\infty}\phi(k)\exp(i(kx-\omega_0 t - v_{\mathrm{group}}(k-k_0)t))dk \\ &= \frac{1}{\sqrt{2\pi}}\exp(-i(\omega_0 t - v_{\mathrm{group}}k_0 t))\int_{-\infty}^{\infty}\phi(k)\exp(ik(x-v_{\mathrm{group}}t))dk \\ &= \exp(-i(\omega_0 - v_{\mathrm{group}}k_0)t)\Psi(x-v_{\mathrm{group}}t,0)\end{aligned} \tag{2.179}$$

using (2.154) in the last step. The wave function at time t is therefore directly related to the wave function at time $t=0$ by a phase factor and a time-dependent shift in x. For the probability density, this implies

$$|\Psi(x,t)|^2 = |\Psi(x-v_{\mathrm{group}}t,0)|^2 . \tag{2.180}$$

This equation describes a wave packet moving at a speed given by the group velocity. Evaluating the definition (2.178) with the dispersion relation (2.176), we obtain

$$v_{\mathrm{group}} = \frac{d\omega}{dk} = \frac{\hbar k}{m} = v_{\mathrm{classical}} . \tag{2.181}$$

Unlike the phase velocity (2.150), the group velocity agrees with the classical velocity. Indeed, we now have a moving wave packet, and if it is sufficiently narrow, statistical position measurements can reveal its motion.

In our explicit Gaussian example, we saw that the wave packet spreads out in time with a standard deviation (2.173). Equation (2.180) does not include a spreading effect because the probability density is simply translated in x as t changes. We do not see the spreading effect in this calculation because we assumed that $\phi(k)$ is sharply peaked around some k_0. Therefore, σ_k is small, which corresponds to a small λ if the state is Gaussian. The corresponding time dependence of (2.173) is then weak because t has a small coefficient.

2.5 SCATTERING

If a particle encounters a region in which the potential is non-zero, it may be deflected or be subject to different kinds of scattering. Scattering effects are important probes in particle and condensed-matter physics, and they reveal new quantum phenomena.

There are two qualitatively distinct classes of states, depending on the potential $V(x)$ and the energy level E: If $V(x) \to \infty$ for $x \to \pm\infty$, all allowed regions for a given energy E are finite and bounded by two turning points. A particle moving in such a potential is always trapped or restricted

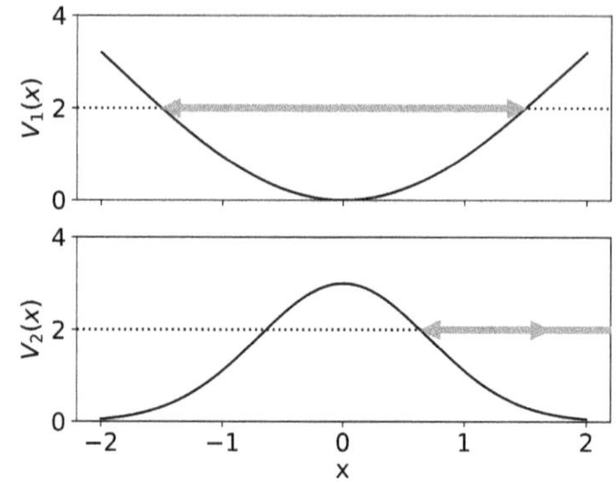

Figure 2.8 A potential $V_1(x)$ with a bound state at energy $E = 2$ and a potential $V_2(x)$ with a scattering state at $E = 2$. In a bound state, the position moves back and forth between two turning points of a finite allowed region. In a scattering state, the position moves in from far away, encounters a single turning point of an unbounded allowed region, and then moves back out.

to a finite region. Corresponding quantum states are bound states, as in our two examples of the infinite square well and the harmonic oscillator. If, however, $V(x)$ has a finite limit for $x \to \infty$ or $x \to -\infty$, a particle may move out to infinity if it has an energy E greater than the finite limit of the potential. This endeavor might take an infinite amount of time, but at any finite time there would be no general upper limit to x or $-x$ determined just by the potential. A corresponding quantum state is called a **scattering state** because it describes how a particle moves in from far away, encounters the potential, and gets scattered or reflected back to infinity. (The name "scattering" derives from the situation in three spatial dimensions, where particles at different energies can get scattered in multiple directions.)

Figure 2.8 shows an example of a classical bound state and a scattering state, related to simple features of the potential landscape. In quantum mechanics, bound states and scattering states are distinguished from each other by their mathematical properties:

1. A **bound state** in quantum mechanics is described by a normalizable wave function for a stationary state with energy E. The allowed energy values form a discrete set after imposing the normalizability condition. Discrete energies are restricted to a range implied by the condition $E < \min \lim_{x\to\pm\infty} V(x)$, where "min" indicates the smaller of the two values obtained for $x \to \pm\infty$.
2. A **scattering state** in quantum mechanics can be stationary with a sharp energy E only if it is non-normalizable. The energy range is continuous because the normalizability condition cannot be imposed on stationary states. Normalizable scattering states are wave packets obtained from superpositions of different energy contributions.

So far, we have seen two examples of bound-state problems, the infinite square well and the harmonic oscillator, with $\lim_{x\to\pm\infty} V(x) = \infty$ in both cases, and one example of a scattering problem (according to our formal definition), given by the free particle. In general, there may be bound states and scattering states for different energies in the same potential. For physical applications, we usually consider bound states for particles confined to the finite region of an atom or an electromagnetic trap, and scattering states for particles in an accelerator beam or electrons in an electric circuit.

To be specific, we usually assume that $\lim_{x\to\pm\infty} V(x) = 0$ for scattering states. If there are bound states, which would require the potential to be negative in some regions, they have negative energies, $E < 0$. The negative part of the energy range is therefore discrete. Scattering states have positive energies $E > 0$ in a continuous range.

2.5.1 DELTA-FUNCTION POTENTIAL

The delta-function potential is a mathematical construction that constitutes the simplest case of a system in which non-trivial scattering takes place at a point-like barrier or well, resembling an upward or downward spike. In the downward case, it is an example of a system that has a bound state and scattering states.

2.5.1.1 The delta function

The delta function $\delta(x)$ is defined by two conditions: $\delta(x) = 0$ if $x \neq 0$ and $\int_{-\infty}^{\infty} \delta(x) \mathrm{d}x = 1$. This definition does not result in a function in the strict mathematical sense because its value at $x = 0$ cannot be finite if its integral is to be non-zero. The correct mathematical term is a **distribution** (not to be confused with a probability distribution) or a generalized function, which means that it can be integrated as specified in the second part of its definition. The integral condition can be interpreted as quantifying the degree of infinity the delta function has at $x = 0$. In particular, it implies that $\delta(x) \neq c\delta(x)$ with a constant c, unless $c = 1$. Without the integral condition, the delta function would not be uniquely defined because the first condition would be true for $c\delta(x)$ with any finite c. Since we integrate when we solve differential equations, it is possible to find unique solutions of some differential equations with delta-function coefficients.

One can visualize the delta function as a limit of a sharp peak. For instance, we may model the delta function as a limit of a family of Gaussians,

$$\delta_\varepsilon(x) = \frac{1}{\sqrt{\pi}\,\varepsilon} \exp(-x^2/\varepsilon^2) \tag{2.182}$$

for $\varepsilon \to 0$, illustrated in Figure 2.9. The integral condition is fulfilled for any non-zero ε because the Gaussians are normalized. Moreover,

$$\lim_{\varepsilon \to 0} \delta_\varepsilon(x) = 0 \tag{2.183}$$

for any $x \neq 0$, while

$$\lim_{\varepsilon \to 0} \delta_\varepsilon(0) = \lim_{\varepsilon \to 0} \frac{1}{\sqrt{\pi}\varepsilon} = \infty. \tag{2.184}$$

Therefore,

$$\lim_{\varepsilon \to 0} \delta_\varepsilon(x) = \delta(x) \tag{2.185}$$

because the limit obeys the two conditions that define the delta function. Unlike the delta function in the limit, the model by a family of functions $\delta_\varepsilon(x)$ is not unique. For instance, we could use piecewise constant functions instead of Gaussians.

There are useful equations for the delta function that follow from its definition, independently of a model such as $\delta_\varepsilon(0)$. In particular, we have

$$f(x)\delta(x-a) = f(a)\delta(x-a) \tag{2.186}$$

for any function $f(x)$ and a real constant a. The equation is easily seen to be correct for $x \neq a$, because both sides then equal zero. If $x = a$, the equation also holds because $f(x) = f(a)$ in this case.

Equation (2.186) greatly simplifies integrations of functions multiplied by the delta function: We have

$$\int_{-\infty}^{\infty} f(x)\delta(x-a)\mathrm{d}x = f(a) \int_{-\infty}^{\infty} \delta(x-a)\mathrm{d}x = f(a) \tag{2.187}$$

if we first use (2.186), then pull out the x-independent $f(a)$, and use the second condition for the delta function after substituting $x - a = y$. We can therefore integrate a function times the delta

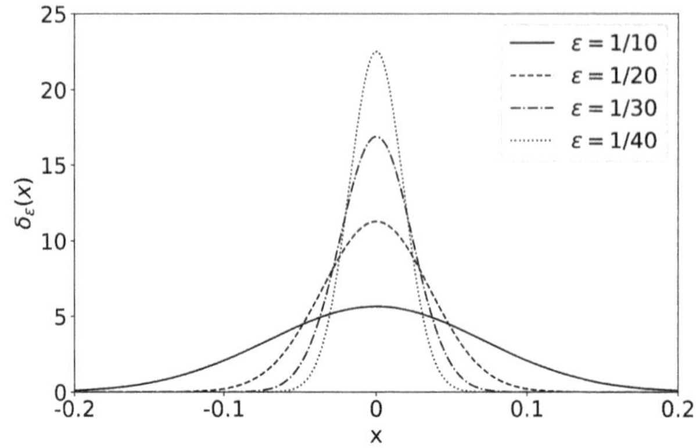

Figure 2.9 The delta function as a limit of Gaussians (2.182). The maximum increases as $\delta_\varepsilon(0) = \pi^{-1/2}\varepsilon^{-1}$, and the width decreases according to $\sigma_{x,\varepsilon} = \varepsilon/\sqrt{2}$.

function even if we do not know the function explicitly. Similarly, if we integrate over a finite range, we just have to check whether the infinity of the delta function is in the range:

$$\int_b^c f(x)\delta(x-a)\mathrm{d}x = \begin{cases} f(a) & \text{if } b < a < c \\ 0 & \text{if } a < b \text{ or } a > c \end{cases}. \tag{2.188}$$

If $a = b$ or $a = c$, we have to use a more careful limit in which the infinity of the delta function approaches the boundary.

2.5.1.2 Bound states

It is possible to use the delta function as a potential with a very sharp peak. We first consider the case of a negative peak (or a sharp well) with potential

$$V(x) = -\alpha\delta(x) \tag{2.189}$$

where $\alpha > 0$, and later extend our results to the case of negative α. The time-independent Schrödinger equation is given by

$$-\frac{\hbar^2}{2m}\frac{\mathrm{d}^2\psi}{\mathrm{d}x^2} - \alpha\delta(x)\psi = E\psi. \tag{2.190}$$

We expect bound states for $E < 0$ because this implies $E < \lim_{x\to\pm\infty} V(x) = 0$. There is a tiny allowed region where $E > V(x) = -\infty$ at a single point, $x = 0$.

We already know the solutions to this equation for $x \neq 0$, where $V(x) = 0$. For these x, we have the differential equation

$$\psi'' = -\frac{2mE}{\hbar^2}\psi = \kappa^2\psi \tag{2.191}$$

with a real

$$\kappa = \frac{\sqrt{-2mE}}{\hbar} > 0 \tag{2.192}$$

for $E < 0$. Given a positive coefficient in the differential equation, we have exponential solutions

$$\psi(x) = A\exp(-\kappa x) + B\exp(\kappa x). \tag{2.193}$$

Since these solutions are valid in two disjoint regions, $x < 0$ and $x > 0$, we have two versions of them:

$$\psi(x) = \begin{cases} A_-\exp(-\kappa x) + B_-\exp(\kappa x) & \text{if } x < 0 \\ A_+\exp(-\kappa x) + B_+\exp(\kappa x) & \text{if } x > 0 \end{cases} \tag{2.194}$$

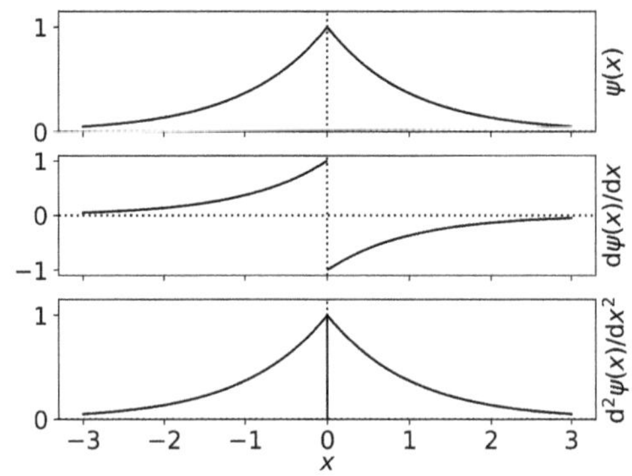

Figure 2.10 Ground state in the delta-function potential. The first derivative is discontinuous at $x=0$, and the second derivative has a negative delta-function peak at $x=0$: For $\kappa=1$ in this plot, we have $\psi(x)=\exp(-|x|)$, $\psi'(x)=-\mathrm{sgn}(x)\exp(-|x|)$ and $\psi''(x)=\exp(-|x|)(1-2\delta(x))$. These results are obtained using the chain and product rules as well as $\mathrm{d}|x|/\mathrm{d}x=\mathrm{sgn}(x)$ and $\mathrm{d}\,\mathrm{sgn}(x)/\mathrm{d}x=2\delta(x)$. The latter result follows from integration by parts in $\int_{-\infty}^{\infty} f(x)\mathrm{sgn}'(x)\mathrm{d}x=\int_{-\infty}^{0} f'(x)\mathrm{d}x-\int_{0}^{\infty} f'(x)\mathrm{d}x=2f(0)=2\int_{-\infty}^{\infty} f(x)\delta(x)\mathrm{d}x$ for any normalizable function $f(x)$.

with four independent constants, $A_\pm$ and $B_\pm$. Normalizability requires $\lim_{x\to\pm\infty}\psi(x)=0$, which is the case if we choose integration constants such that $A_-=0$ and $B_+=0$. We are left with two independent constants in

$$\psi(x)=\begin{cases} B\exp(\kappa x) & \text{if } x<0 \\ A\exp(-\kappa x) & \text{if } x>0 \end{cases}. \tag{2.195}$$

Since $\psi(x)$ must be continuous at $x=0$ for the second derivative to exist, we have $A=B$.

Unlike $\psi(x)$, the derivative $\mathrm{d}\psi/\mathrm{d}x$ is not required to be continuous at $x=0$ because the differential equation (2.190) implies an infinite $\mathrm{d}^2\psi/\mathrm{d}x^2$ at $x=0$ as a result of the delta function, which means that $\mathrm{d}\psi/\mathrm{d}x$ has a discontinuity there. The size of the jump of $\mathrm{d}\psi/\mathrm{d}x$ at $x=0$ depends on "how infinite" the other infinite term in the equation is, given by $-\alpha\delta(x)\psi(x)$. The normalization condition of the delta function implies that it has a specific degree of infinity at $x=0$, which must be matched by the infinity of $\mathrm{d}^2\psi/\mathrm{d}x^2$, or by the size of the discontinuity of $\mathrm{d}\psi/\mathrm{d}x$. As an example, Figure 2.10 shows the ground state in the delta-function potential and its first and second derivatives as an example of the required discontinuities.

Since the normalization condition of the delta function is an integral, we can relate it to the discontinuity by integrating the differential equation (2.190) over a range of x around $x=0$: Choosing some small $\varepsilon>0$, we write

$$-\frac{\hbar^2}{2m}\int_{-\varepsilon}^{\varepsilon}\frac{\mathrm{d}^2\psi}{\mathrm{d}x^2}\mathrm{d}x+\int_{-\varepsilon}^{\varepsilon}V(x)\psi(x)\mathrm{d}x=E\int_{-\varepsilon}^{\varepsilon}\psi(x)\mathrm{d}x \tag{2.196}$$

If $V(x)$ is finite, the equation simply tells us that $\mathrm{d}\psi/\mathrm{d}x$ must be continuous because the integrals evaluate to

$$\int_{-\varepsilon}^{\varepsilon}\frac{\mathrm{d}^2\psi}{\mathrm{d}x^2}\mathrm{d}x=\frac{\mathrm{d}\psi}{\mathrm{d}x}(\varepsilon)-\frac{\mathrm{d}\psi}{\mathrm{d}x}(-\varepsilon) \tag{2.197}$$

while

$$\int_{-\varepsilon}^{\varepsilon}V(x)\psi(x)\mathrm{d}x\approx 2\varepsilon V(0)\psi(0) \tag{2.198}$$

and

$$\int_{-\varepsilon}^{\varepsilon}\psi(x)\mathrm{d}x\approx 2\varepsilon\psi(0). \tag{2.199}$$

In the limit of $\varepsilon \to 0$, the last two integrals vanish, and

$$\lim_{\varepsilon\to 0}\left(\frac{\mathrm{d}\psi}{\mathrm{d}x}(\varepsilon)-\frac{\mathrm{d}\psi}{\mathrm{d}x}(-\varepsilon)\right)=\Delta(\mathrm{d}\psi/\mathrm{d}x)|_{x=0} \tag{2.200}$$

defines the discontinuity of $\mathrm{d}\psi/\mathrm{d}x$ at $x=0$. If all other terms in the integrated differential equation vanish, the discontinuity vanishes.

If we have a delta-function potential, the integral

$$\int_{-\varepsilon}^{\varepsilon} V(x)\psi(x)\mathrm{d}x=-\alpha\int_{-\varepsilon}^{\varepsilon}\delta(x)\psi(x)\mathrm{d}x=-\alpha\psi(0) \tag{2.201}$$

is determined by (2.188). It is independent of ε and does not vanish for $\varepsilon\to 0$ unless $\psi(0)=0$, which is not the case in (2.195). Integrating the differential equation therefore determines a non-zero discontinuity

$$\Delta(\mathrm{d}\psi/\mathrm{d}x)|_{x=0}=-\frac{2m\alpha}{\hbar^2}\psi(0) \tag{2.202}$$

for the wave function at $x=0$.

Using our solution (2.195) with $A=B$, we have

$$\psi'(x)=\begin{cases}-B\kappa e^{-\kappa x} & \text{if } x>0\\ B\kappa e^{\kappa x} & \text{if } x<0\end{cases}. \tag{2.203}$$

Therefore,

$$\Delta(\mathrm{d}\psi/\mathrm{d}x)|_{x=0}=-2B\kappa\,. \tag{2.204}$$

Since $\psi(0)=B$, equation (2.202) implies that $\kappa=m\alpha/\hbar^2$ for $\alpha>0$, while there is no solution if $\alpha<0$ because κ is defined as a positive number. The definition (2.192) of κ then determines the energy

$$E=-\frac{\hbar^2\kappa^2}{2m}=-\frac{m\alpha^2}{2\hbar^2}\,. \tag{2.205}$$

There is a unique bound state, which is the ground state with energy (2.205) if $\alpha>0$, shown in Figure 2.10. There are no bound states if $\alpha<0$.

The ground-state wave function is given by (2.195) with $A=B$, up to normalization. The normalization integral

$$1=\int_{-\infty}^{\infty}|\psi(x)|^2\mathrm{d}x=2|B|^2\int_0^{\infty}e^{-2\kappa x}\mathrm{d}x=\frac{|B|^2}{\kappa} \tag{2.206}$$

implies $B=\sqrt{\kappa}=\sqrt{m\alpha}/\hbar$, and the ground-state wave function is

$$\psi_0(x)=\frac{\sqrt{m\alpha}}{\hbar}\exp(-m\alpha|x|/\hbar^2)\,. \tag{2.207}$$

2.5.1.3 Scattering states

We have scattering states for positive energies $E>0$. If $x\neq 0$, we write the differential equation as

$$\psi''=-\frac{2mE}{\hbar^2}\psi=-k^2\psi \tag{2.208}$$

where

$$k=\frac{\sqrt{2mE}}{\hbar}\,. \tag{2.209}$$

As seen before, this equation has the general solution

$$\psi(x)=Ae^{ikx}+Be^{-ikx} \tag{2.210}$$

given by plane waves, which are not normalizable.

Allowed states that can be interpreted as probability densities are superpositions of plane waves, or normalizable wave packets with unsharp energy. However, since the construction of a specific wave packet is not unique and may be complicated, we work with plane waves and view them as approximations of wide wave packets that contain many wave lengths but eventually approach zero at $x \to \pm\infty$. Physically, this interpretation corresponds to a stream of particles in a scattering experiment that runs at constant intensity for a long time but has to be turned on and, eventually, off. The stream of particles could be electrons in an electric current or bunched particles in an accelerator.

Since we do not impose boundary conditions on non-normalizable states, we expect additional free parameters in their mathematical solutions. (Bound states, by contrast, are uniquely determined by their energy, at least in one dimension.) The physical interpretation of the additional parameters is related to the fact that we can change the energy or frequency of a particle stream, as well as the number of particles in the incoming stream or its intensity. Therefore, the amplitude should not be fixed by normalization, just as it happens for scattering states.

For the delta-function potential, we have two disjoint regions in which there are plane-wave solutions, given by $x < 0$ and $x > 0$, respectively. We begin with independent plane waves in the two regions,

$$\psi(x) = \begin{cases} Ae^{ikx} + B^{-ikx} & \text{if } x < 0 \\ Fe^{ikx} + Ge^{-ikx} & \text{if } x > 0 \end{cases}, \tag{2.211}$$

and impose continuity of $\psi(x)$ at $x = 0$:

$$A + B = F + G. \tag{2.212}$$

The first derivative

$$\frac{\mathrm{d}\psi}{\mathrm{d}x} = \begin{cases} ik(Ae^{ikx} - Be^{-ikx}) & \text{if } x < 0 \\ ik(Fe^{ikx} - Ge^{-ikx}) & \text{if } x > 0 \end{cases} \tag{2.213}$$

should have the required discontinuity at $x = 0$ as given by (2.202). For scattering states, this condition evaluates to

$$\Delta(\mathrm{d}\psi/\mathrm{d}x)|_{x=0} = ik(F - G - A + B) = -\frac{2m\alpha}{\hbar}\psi(0) = -\frac{2m\alpha}{\hbar}(A + B) \tag{2.214}$$

and therefore

$$F - G = A(1 + 2i\beta) - B(1 - 2i\beta) \tag{2.215}$$

with

$$\beta = \frac{m\alpha}{\hbar^2 k}. \tag{2.216}$$

We derived two linear equations, (2.212) and (2.215), for four constants, A, B, F and G. We have to decide which two constants we should consider to be independent, such that the other two constants depend on their values. The decision should reflect the physical meaning we assign to the constants, which in turn depends on the interpretation of plane waves with different signs of k as right-moving and left-moving ones. If we introduce the time dependence of stationary states, we have the functions

$$\Psi(x,t) = \psi(x)e^{-iEt/\hbar} = \begin{cases} Ae^{i(kx-\omega t)} + Be^{-i(kx+\omega t)} & \text{if } x < 0 \\ Fe^{i(kx-\omega t)} + Ge^{-i(kx+\omega t)} & \text{if } x > 0 \end{cases} \tag{2.217}$$

where $\omega = E/\hbar$. In this form, we recognize A and F as the amplitudes of right-moving plane waves in the two regions, and B and G as the amplitudes of left-moving plane waves; see Figure 2.11. As before, the terms "right-moving" and "left-moving" refer to the phase velocity, which does not tell us how the wave function is moving as we have seen in our discussion of the free particle. If the

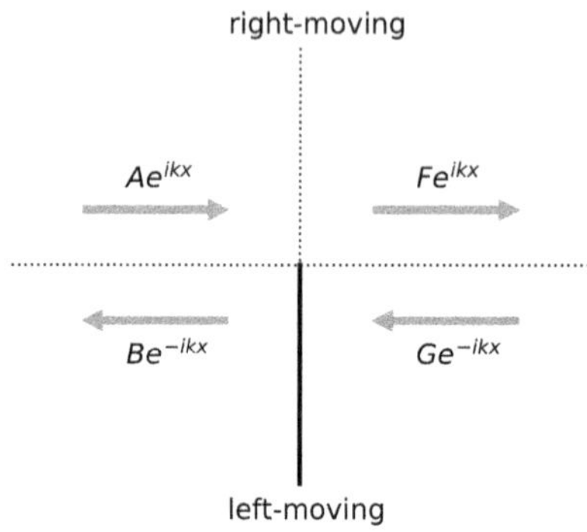

Figure 2.11 Interpretation of the amplitudes of a plane wave (2.217) as incoming and outgoing waves getting scattered at the delta-function potential.

plane waves used here represent long-lasting streams of particles, they are nearly stationary and do not appear to move at all. However, individual particles in each stream are moving. This motion represents local features of the plane wave, which are described well by the phase velocity, or at least by its sign.

2.5.1.4 Scattering coefficients

In a scattering experiment, A and G determine the average numbers of incoming particles from the left and from the right, respectively. These numbers can be changed by tuning the intensity of the particle beams, and should therefore remain free parameters. The constants B and F correspond to the numbers of outgoing particles to the left and to the right, respectively. These numbers should be determined by the numbers of incoming particles, related to A and G, combined with physical properties of the scattering potential implemented here by the (dis)continuity conditions.

We should therefore seek solutions of our two linear equations of the form $B(A,G)$ and $F(A,G)$. To simplify the equations, we can often assume that there are incoming particles only from one side, setting $G=0$. There is then a single free parameter, A, that determines the amplitude of the incident wave, and outgoing waves have two parts: the reflected wave with amplitude B and the transmitted wave with amplitude F. If $G\neq 0$, the distinction between the two outgoing waves is less clear because B then contains particles reflected out of the A-wave and transmitted through from the G-wave, and similarly for F.

The number of reflected particles depends not only on the potential but also on the number of incoming particles. It is more interesting to consider the relative probability of reflection, called the **reflection coefficient**

$$R=\frac{|B|^2}{|A|^2}. \tag{2.218}$$

Similarly, we define the relative probability of transmission, or the **transmission coefficient**

$$T=\frac{|F|^2}{|A|^2}. \tag{2.219}$$

For the delta-function potential, the two equations (2.212) and (2.215) with $G=0$ can directly be solved for

$$B=\frac{i\beta}{1-i\beta}A \quad \text{and} \quad F=\frac{1}{1-i\beta}A \tag{2.220}$$

such that

$$R=\frac{\beta^2}{1+\beta^2} \quad \text{and} \quad T=\frac{1}{1+\beta^2}. \tag{2.221}$$

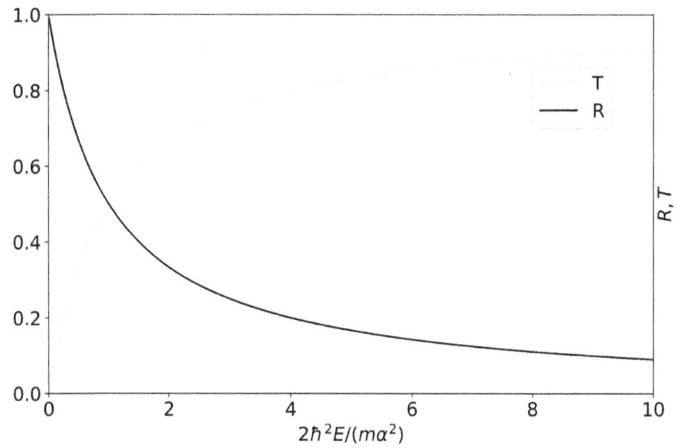

Figure 2.12 The reflection and tunneling coefficients of the delta-function potential.

We have $R+T=1$ for all β, which implies that all particles are either reflected or transmitted. No particles get stuck in the potential or are absorbed. The dependence on the energy follows from (2.216) and (2.209) (see Figure 2.12):

$$R=\frac{1}{1+2\hbar^2E/(m\alpha^2)} \quad \text{and} \quad T=\frac{1}{1+m\alpha^2/(2\hbar^2E)}\,, \tag{2.222}$$

The dependence is continuous because all positive E are allowed for scattering states. At larger energies, more particles are transmitted. But transmission is never complete at any finite energy even though the delta-function peak is negative. In contrast to the classical expectation, some particles are always reflected.

We obtain the same R and T for the repulsive delta-function potential with $\alpha<0$. We can use all the previous scattering solutions, which depend only on α^2 and are insensitive to the sign. (However, the bound state disappears if $\alpha<0$.) There is therefore transmission even though the infinite potential at the delta peak $-\alpha\delta(x)$ is greater than any finite energy if $\alpha<0$. This is an example of the tunneling effect, but in this model it is obtained only for an infinitely thin potential barrier.

2.5.2 FINITE SQUARE WELL

The delta-function potential suggests that it is possible for quantum-mechanical wave functions to tunnel through a potential barrier even if it is higher than their available energy. However, the barrier in this example is infinitely thin, given by a single point $x=0$. If tunneling is a real physical effect, it should also happen for a potential barrier with non-zero width. The barrier should then have a finite height too because an infinite potential in an extended range requires a vanishing wave function, as seen in the example of the infinite square well.

We assume that such a potential is piecewise constant, given by

$$V(x)=\begin{cases} -V_0 & \text{if } -b<x<b \\ 0 & \text{if } |x|\geq b \end{cases} \tag{2.223}$$

for a positive constant $b>0$ and a potential value V_0 that may be positive or negative. The width of the potential equals $a=2b$. It will be convenient to work with a potential that is reflection symmetric around $x=0$. For the infinite square well, by contrast, it was easier to implement one of the boundary conditions at $x=0$, the left side of the well in that case.

2.5.2.1 Bound states

Solving the time-independent Schrödinger equation

$$-\frac{\hbar^2}{2m}\frac{d^2\psi(x)}{dx^2}+V(x)\psi(x)=E\psi(x) \tag{2.224}$$

with this potential, we first look for bound states. For a given $V_0>0$, bound states exist for energies in the range $-V_0<E<0$. A given energy E then implies a differential equation with constant coefficients in each of the three ranges $x<-b$, $-b<x<b$ and $x>b$.

If $x<-b$ or $x>b$, we write the differential equation as $\psi''=\kappa^2\psi$ with

$$\kappa=\frac{\sqrt{-2mE}}{\hbar}>0\,. \tag{2.225}$$

The general solution in each of the two ranges is $\psi(x)=Ae^{-\kappa x}+Be^{\kappa x}$, and the combined solution is normalizable if we choose integration constants such that

$$\psi(x)=\left\{\begin{array}{ll} Be^{\kappa x} & \text{if } x<-b \\ Fe^{-\kappa x} & \text{if } x>b \end{array}\right. . \tag{2.226}$$

If $|x|<b$, the differential equation is of the form $\psi''=-\ell^2\psi$ with a wave number

$$\ell=\frac{\sqrt{2m(E+V_0)}}{\hbar}>0\,. \tag{2.227}$$

The general solution in this finite range is $\psi(x)=C\sin(\ell x)+D\cos(\ell x)$ without restrictions from normalizability.

It is possible to reduce the number of free constants for bound states because we have seen in examples that they are even or odd if the potential is reflection symmetric. Using this property in order to eliminate some of the free constants simplifies the solution procedure.

We first show that the stated symmetry property applies to any reflection-symmetric potential $V(x)$ that obeys the condition $V(-x)=V(x)$. Given this property, if some function $\psi(x)$ solves (2.224), the function $\psi(-x)$ solves the same equation with $V(-x)$ instead of $V(x)$ if we just substitute $-x$ for x everywhere in the equation. If the potential is reflection symmetric, this is the same as the original differential equation, and therefore $\psi(x)$ and $\psi(-x)$ are solutions of the same equation. Since the differential equation is linear, the linear combinations $\psi_+(x)=\frac{1}{2}(\psi(x)+\psi(-x))$ and $\psi_-(x)=\frac{1}{2}(\psi(x)-\psi(-x))$ are also solutions. By definition, $\psi_+(x)$ is an even function and $\psi_-(x)$ is an odd function.

For bound states, it is therefore always possible to assume that solutions are even or odd. Scattering states might not obey this symmetry because they are usually subject to non-symmetric boundary conditions if there are different numbers of particles coming in from the left and from the right, respectively.

For even or odd solutions of bound states, we need to check continuity only at one point, $x=b$, which is then implied also at $x=-b$ by symmetry. First looking at even solutions, we include only the cosine in the well (setting $C=0$) such that

$$\psi(x)=\left\{\begin{array}{ll} Fe^{-\kappa x} & \text{if } x>b \\ D\cos(\ell x) & \text{if } 0<x<b \\ \psi(-x) & \text{if } x<0 \end{array}\right. . \tag{2.228}$$

Continuity of ψ and ψ' at $x=b$ implies

$$Fe^{-\kappa b}=D\cos(\ell b) \quad \text{and} \quad -\kappa Fe^{-\kappa b}=-\ell D\sin(\ell b)\,. \tag{2.229}$$

Dividing (minus) the second equation by the first equation eliminates the constants F and D and gives us a condition

$$\kappa = \ell \tan(\ell b). \tag{2.230}$$

for the bound-state energies.

For odd solutions, we use only the sine function in the well (setting $D = 0$) and obtain

$$\psi(x) = \begin{cases} Fe^{-\kappa x} & \text{if } x > b \\ C\sin(\ell x) & 0 < x < b \\ -\psi(-x) & \text{if } x < 0 \end{cases}. \tag{2.231}$$

Continuity of ψ and ψ' at $x = b$ implies

$$Fe^{-\kappa b} = C\sin(\ell b) \quad \text{and} \quad -\kappa F e^{-\kappa b} = \ell C \cos(\ell b). \tag{2.232}$$

Dividing (minus) the second equation by the first equation gives

$$\kappa = -\ell \cot(\ell b) = -\frac{\ell}{\tan(\ell b)} = \ell \tan(\ell b - \pi/2). \tag{2.233}$$

Equations (2.230) and (2.233) are transcendental equations for the energy E, which appears via the E-dependence of κ in (2.225) and of ℓ in (2.227). Introducing the variables

$$z = \ell b \quad \text{and} \quad z_0 = \frac{b}{\hbar}\sqrt{2mV_0} \tag{2.234}$$

we have $\kappa b = b\sqrt{2mV_0/\hbar^2 - \ell^2} = \sqrt{z_0^2 - z^2}$ (using $\kappa^2 + \ell^2 = 2mV_0/\hbar^2$) and write the energy equation in the form

$$\sqrt{\frac{z_0^2}{z^2} - 1} = \frac{\kappa b}{\ell b} = \begin{cases} \tan z & \text{for even states} \\ \tan(z - \pi/2) & \text{for odd states.} \end{cases} \tag{2.235}$$

The shift in the argument of tan allows us to consider a single equation as a combination of (2.230) and (2.233).

This equation is to be solved for z in terms of z_0, which implies a solution for E in terms of the potential V_0. Exact solutions are hard to find, but general properties can be seen by plotting the relevant functions on the two sides of equation (2.235), as shown in Figure 2.13. We observe: (i) There are a finite number of bound states for any given V_0 because the curves have a finite number of intersection points: a finite number of real solutions to (2.235) exist in the range $0 < z < z_0$. (ii) The spacing between zeros of $\tan z$ and $\tan(z - \pi/2)$, respectively, is $\pi/2$ on the z-axis, while $\sqrt{z_0^2/z^2 - 1}$ intersects the z-axis at z_0. Since all intersections of the curves happen between zero and z_0, their number is given by the smallest integer greater than $2z_0/\pi$. The number of bound states thus increases with increasing z_0 (increasing depth V_0 of the well or increasing width $a = 2b$). (iii) The ground state with the smallest energy E is always even. (iv) Excited states have alternately even and odd wave functions ψ.

For the allowed energies solving (2.235), we have the wave functions

$$\psi_n(x) = \begin{cases} Fe^{-\kappa_n x} & \text{if } x > b \\ F\dfrac{e^{-\kappa_n b}}{\cos(\ell_n b)}\cos(\ell_n x) & \text{if } 0 < x < b \\ \psi_n(-x) & \text{if } x < 0 \end{cases} \tag{2.236}$$

for even n and

$$\psi_n(x) = \begin{cases} Fe^{-\kappa_n x} & \text{if } x > b \\ F\dfrac{e^{-\kappa_n b}}{\sin(\ell_n b)}\sin(\ell_n x) & \text{if } 0 < x < b \\ -\psi_n(-x) & \text{if } x < 0 \end{cases} \tag{2.237}$$

for odd n, with F determined by normalization.

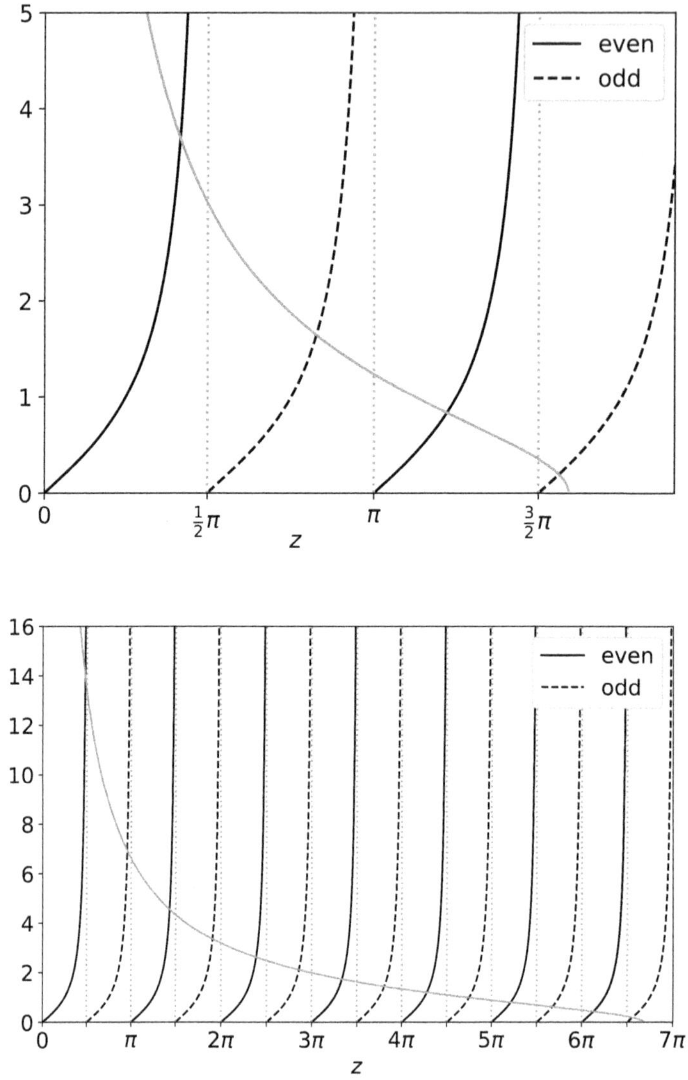

Figure 2.13 Graphical solution of the energy equation (2.235) by intersection points of the curves $\sqrt{z_0^2/z^2-1}$ (which is monotonically decreasing) and $\tan z$ (for even states) or $\tan(z-\pi/2)$ (for odd states). The number of intersection points increases with z_0, which is the point where $\sqrt{z_0^2/z^2-1}$ intersects the z-axis. The values for z_0 are $z_0 = 5$ (top) and $z_0 = 21$ (bottom)

There are two limiting cases of wells with characteristic properties: (i) If z_0 is large, $\sqrt{z_0^2/z^2-1}$ is large for small z. Intersection points are obtained close to $\frac{1}{2}n\pi \approx z_n = \ell_n b = b\sqrt{2m(E_n+V_0)}/\hbar$. The resulting energies are

$$E_n \approx -V_0 + \frac{n^2\pi^2\hbar^2}{8mb^2} = -V_0 + \frac{n^2\pi^2\hbar^2}{2ma^2} \quad \text{for } n = 1,2,\ldots, \tag{2.238}$$

valid as an approximation up to some maximum n such that z_n is much smaller than z_0. The shifted energies $E_n + V_0$ correspond to the values obtained for the infinite square well. (ii) If $z_0 < \pi/2$, there is only one bound state, the ground state. (And one bound state always exists if $V_0 > 0$.) This limiting case can be compared with the delta-function potential modeled as the $b \to 0$ limit of a finite square well with $V_0 = \alpha/(2b) = \alpha/a$: The integral $\int_{-\infty}^{\infty} V(x)\mathrm{d}x = -\alpha$ implies that $V(x) = -\alpha\delta(x)$ in the limit. With these values, $z_0 = \sqrt{m\alpha b}/\hbar \to 0$ for $b \to 0$, and the delta-function potential indeed corresponds to the small-z_0 limit.

2.5.2.2 Scattering states

Scattering states in the finite square well are obtained for $E > 0$. These solutions are non-normalizable and describe scattering of particles that move like free particles far away from the well but are scattered off the well. In one spatial dimension, particles are either transmitted or reflected.

Introducing $k = \sqrt{2mE}/\hbar$ and using $\ell = \sqrt{2m(E+V_0)}/\hbar$ as in (2.227), the general solution now takes the form

$$\psi(x) = \begin{cases} Ae^{ikx} + Be^{-ikx} & \text{if } x < -b \\ C\sin(\ell x) + D\cos(\ell x) & \text{if } -b < x < b \\ Fe^{ikx} + Ge^{-ikx} & \text{if } x > b \end{cases} \tag{2.239}$$

where a term e^{ikx} represents a right-moving plane wave and e^{-ikx} a left-moving one. Recall the usual caveat that the direction refers to the phase velocity and does not indicate time dependence of expectation values in stationary states.

We now choose $G = 0$ to allow incoming particles only from the left. This form of the states, without a contribution from e^{-ikx} for $x > b$ describes waves that come in from the left (a right-moving Ae^{ikx} at $x < -b$), encounter the well and are partially reflected (a left-moving Be^{-ikx} at $x < -b$) or transmitted (a right-moving Fe^{-ikx} at $x > b$). There is no left-moving plane wave for $x > b$, and thus no incoming contribution from the right. This choice makes the solution non-symmetric under reflection of x.

At $x = \pm b$, we impose boundary conditions implementing continuity of ψ and ψ':

ψ at $x = -b$: $Ae^{-ikb} + Be^{ikb} = -C\sin(\ell b) + D\cos(\ell b)$

ψ at $x = b$: $C\sin(\ell b) + D\cos(\ell b) = Fe^{ikb}$

ψ' at $x = -b$: $ik(Ae^{-ikb} - Be^{ikb}) = \ell(C\cos(\ell b) + D\sin(\ell b))$

ψ' at $x = b$: $\ell(C\cos(\ell b) - D\sin(\ell b)) = ikFe^{ikb}$.

Multiplying the second equation by $\sin(\ell b)$ and the fourth one by $\ell^{-1}\cos(\ell b)$, and adding the results, we have

$$C = \left(\sin(\ell b) + i\frac{k}{\ell}\cos(\ell b)\right)e^{ikb}F\,. \tag{2.240}$$

By an analogous procedure, we obtain

$$D = \left(\cos(\ell b) - i\frac{k}{\ell}\sin(\ell b)\right)e^{ikb}F\,. \tag{2.241}$$

Eliminating C and D using these two equations, we have

$$\begin{aligned} Ae^{-ikb} + Be^{ikb} &= \left(\cos(\ell b)^2 - \sin(\ell b)^2 - 2i\frac{k}{\ell}\sin(\ell b)\cos(\ell b)\right)e^{ikb}F \\ &= \left(\cos(2\ell b) - i\frac{k}{\ell}\sin(2\ell b)\right)e^{ikb}F \end{aligned} \tag{2.242}$$

and

$$\begin{aligned} Ae^{-ikb} - Be^{ikb} &= \frac{\ell}{ik}\left(2\sin(\ell b)\cos(\ell b) + i\frac{k}{\ell}\left(\cos(\ell b)^2 - \sin(\ell b)^2\right)\right)e^{ikb}F \\ &= \left(\cos(2\ell b) - i\frac{\ell}{k}\sin(2\ell b)\right)e^{ikb}F \end{aligned} \tag{2.243}$$

from the first and third boundary conditions, respectively. Adding and subtracting these equations provides the expressions

$$A = \left(\cos(2\ell b) - \frac{i}{2}\left(\frac{k}{\ell} + \frac{\ell}{k}\right)\sin(2\ell b)\right)e^{2ikb}F \tag{2.244}$$

$$B = \frac{i}{2}\sin(2\ell b)\left(\frac{\ell}{k} - \frac{k}{\ell}\right)F\,. \tag{2.245}$$

Parameters describing the scattering process are the transmission coefficient

$$\begin{aligned} T &= \left|\frac{F}{A}\right|^2 = \frac{1}{\cos(2\ell b)^2 + \dfrac{(k^2+\ell^2)^2}{4k^2\ell^2}\sin(2\ell b)^2} = \frac{1}{1 + \dfrac{(k^2-\ell^2)^2}{4k^2\ell^2}\sin(2\ell b)^2} \\ &= \frac{1}{1 + \dfrac{V_0^2}{4E(E+V_0)}\sin(2b\sqrt{2m(E+V_0)}/\hbar)^2} \end{aligned} \tag{2.246}$$

and the reflection coefficient

$$\begin{aligned} R &= \left|\frac{B}{A}\right|^2 = \frac{(\ell^2-k^2)^2}{4k^2\ell^2}\,\frac{\sin(2\ell b)^2}{\cos(2\ell b)^2 + \dfrac{(k^2+\ell^2)^2}{4k^2\ell^2}\sin(2\ell b)^2} \\ &= \frac{1}{1 + \dfrac{4k^2\ell^2}{(k^2-\ell^2)^2}\sin(2\ell b)^2} = 1 - T\,. \end{aligned} \tag{2.247}$$

The two functions $T(E)$ and $R(E)$ are shown in Figure 2.14. It is interesting to note that we have perfect transmission, $T = 1$, if the energy $E = E_n$ is such that

$$\frac{2b}{\hbar}\sqrt{2m(E_n+V_0)} = \frac{a}{\hbar}\sqrt{2m(E_n+V_0)} = n\pi\,. \tag{2.248}$$

The solutions E_n happen to agree with the discrete bound-state energies (2.238) in the limit of a wide and deep well.

Figure 2.15 shows an example of a scattering state by plotting the real and imaginary parts as well as the norm $|\psi|$ of the wave function. The potential has a positive V_0, such that $-V_0 < 0$ and there is a finite square well. The energy of the state is $E = V_0/2$. Since we have assumed that there are no incoming particles from the right, $G = 0$, the wave function for $x > b$ is a single plane wave Fe^{ikx} with constant norm. On the left of the well, however, there is a superposition of two plane waves, Ae^{ikx} for the incoming particles and Be^{-ikx} for the reflected wave. The wave function is a sum of these two plane waves for $x < -b$, which shows a non-constant norm with interference phenomena. The wave length λ is smaller in the well because there is more kinetic energy, which implies a larger momentum and a greater $k = 2\pi/\lambda$.

So far, we have interpreted the potential as a square well, using $V_0 > 0$. We obtain similar solutions if we change the sign of V_0, such that $-V_0 = W > 0$. The potential then describes a finite

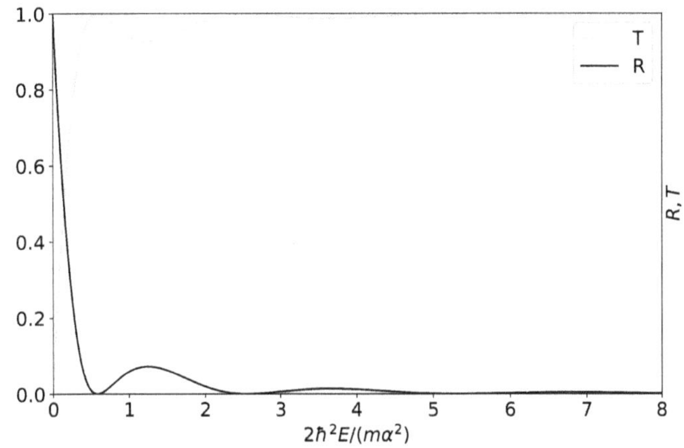

Figure 2.14 Transmission and reflection coefficients for a finite square well, depending on the energy of a plane wave.

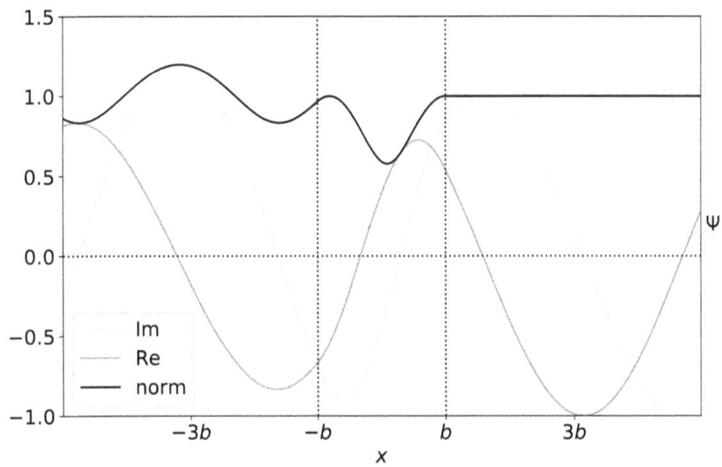

Figure 2.15 Imaginary part, real part, and norm of a solution with energy $E = V_0/2$ in a finite square well of depth $-V_0 < 0$ between $x = -b$ and $x = b$. Amplitudes of the solution are chosen such that $F = 1$ and $G = 0$. The norm to the right of the well is thus constant, while the left shows interference between the incoming wave and the reflected wave. The wave length changes in the well: It is smaller because the particles gain kinetic energy.

square barrier with height W. There are no bound states anymore, but we still have scattering states for all positive energies. We obtain the transmission coefficient $T(E)$ by replacing V_0 in (2.246) with $-W$. The square root $\sqrt{2m(E-W)}$ is then imaginary for $E < W$, which turns the sine function into the hyperbolic sine function sinh: We have

$$\sin(z) = \frac{1}{2i}(e^{iz} - e^{-iz}) \tag{2.249}$$

and therefore

$$\sin(ix) = \frac{1}{2i}(e^{-x} - e^{x}) = i\sinh(x) \tag{2.250}$$

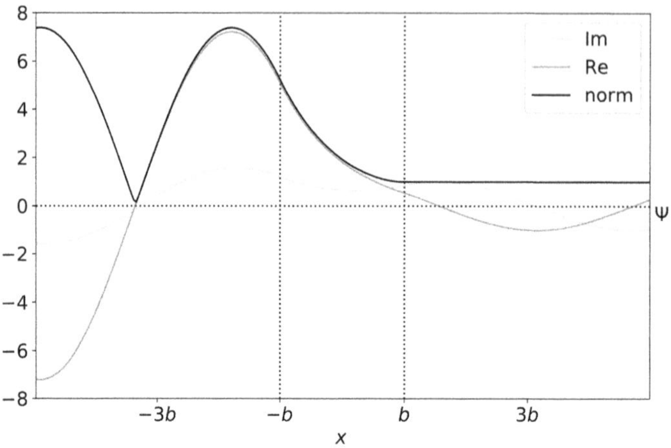

Figure 2.16 Real part, imaginary part and norm of a solution with energy $E = W/2$ for a finite square barrier of height W between the x-values $-b$ and b. Amplitudes of the solution are chosen such that $F = 1$ and $G = 0$. The norm to the right of the well is thus constant, while the left shows interference between the incoming wave and the reflected wave. In the barrier, which a classical particle of energy $W/2 < W$ could not enter, the norm decreases exponentially but remains non-zero. The particle can thus be found in the barrier, and tunneling leads to a non-vanishing transmitted wave to the right.

where $\sinh(x) = \frac{1}{2}(e^x - e^{-x})$ defines the hyperbolic sine. Depending on the energy, the transmission coefficient of a finite square barrier with width $a = 2b$ can now be written as

$$T(E) = \frac{1}{1 + \frac{W^2}{4E(E-W)} \sin(a\sqrt{2m(E-W)}/\hbar)^2} \tag{2.251}$$

if $E > W$, while

$$T(E) = \frac{1}{1 + \frac{W^2}{4E(W-E)} \sinh(a\sqrt{2m(W-E)}/\hbar)^2} \tag{2.252}$$

if $0 < E < W$ and

$$T(E) = \frac{1}{1 + \frac{ma^2W}{2\hbar^2}} \tag{2.253}$$

in the limiting case of $E = W$.

The transmission coefficient is non-zero if the energy is below the barrier, $E < W$, demonstrating tunneling even for a non-zero width of the barrier. Since the sinh function is a linear combination of two exponential functions, the transmission coefficient decreases exponentially as the width a of the barrier or the energy difference $W - E$ increase. An example of a tunneling wave function is shown in Figure 2.16 with an exponentially decreasing norm in the barrier. The norm does not decrease all the way to zero within a finite range of the barrier, ending up with a small non-zero value at the other side that provides the amplitude F of the transmitted wave. Because the transmitted wave has a small amplitude, the reflected wave has an amplitude almost as large as the incoming wave. There are therefore points on the left with nearly complete destructive interference.

Tunneling has several important applications but also some negative implications:

1. Tunneling implies that there are always small leakage currents in electric circuits, making it impossible to construct a perfect insulator.

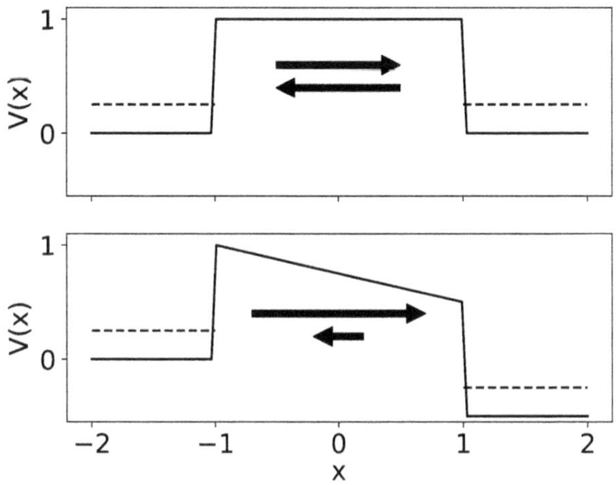

Figure 2.17 Potential barrier between two conducting wires. If the potential difference is zero, electrons tunneling from the left and from the right encounter the same barrier height (top). The net tunneling current is therefore zero. If there is a potential difference across the barrier (bottom), electrons from one side experience a higher barrier than electrons from the other side, which implies a non-zero net tunneling current.

2. A voltage difference across a barrier implies tunneling currents that can be used in different ways. There is no tunneling current through a barrier between two electric wires without a voltage difference because it is then equally likely for electrons to tunnel from the left to the right as from the right to the left. This situation corresponds to a scattering state with $A = G$ in a reflection-symmetric potential. A voltage difference implies a linear but non-constant potential across the barrier, which means that the two sides no longer encounter the same barrier height; see Figure 2.17. There is now a tunneling current because it is more likely for electrons to tunnel from the higher potential to the lower potential than back.
3. The exponential dependence of $T(E)$ on the height of the potential for a given width allows sensitive control mechanisms on the tunneling current. This effect is used in tunneling transistors.
4. The scanning tunneling microscope is an example that makes use of the exponential dependence of the tunneling current on the width a of the barrier. Here, the barrier is a vacuum region surrounding a conducting surface. If a thin conducting probe (a needle) is placed near the surface and connected to it by an electric circuit with a voltage difference, there is a small tunneling current. A scanning mechanism moves the needle through a lattice of points (x,y) parallel to the surface and adjusts the distance z between the needle tip and the surface such that the same tunneling current is measured at all positions. Keeping track of the vertical movements, the mechanism results in a measurement of the function $z(x,y)$, which is related to the surface profile.

2.5.2.3 Piecewise constant potentials

The procedure used here for the finite square well and barrier illustrates the behavior for a general potential $V(x)$: For a given energy E, we obtain oscillating solutions $\psi(x)$ in regions where $E > V(x)$ because we have $\psi'' = -(2m/\hbar^2)(E - V(x))\psi$ with $-(2m/\hbar^2)(E - V(x)) < 0$. Around a fixed x_0 with $V(x_0) < E$, we may use a Taylor expansion in

$$\psi'' = -\frac{2m}{\hbar^2}(E - V(x_0) - V'(x_0)(x - x_0) - \cdots)\psi \approx -\frac{2m}{\hbar^2}(E - V(x_0))\psi. \tag{2.254}$$

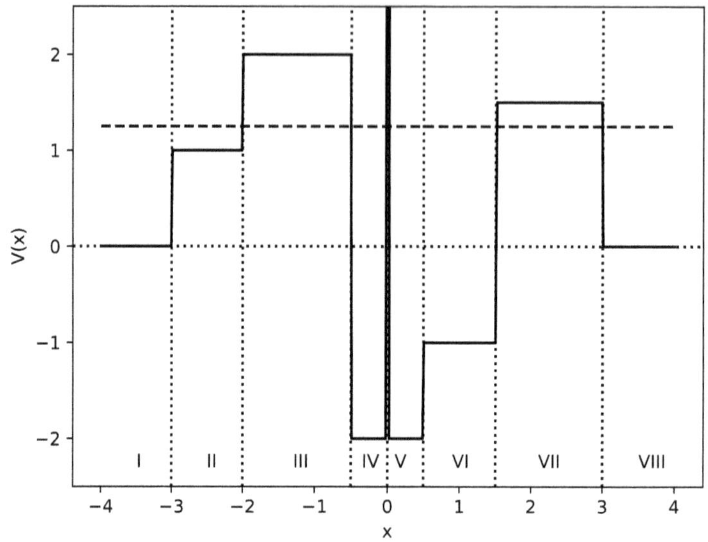

Figure 2.18 Example of a piecewise constant potential with a delta function at $x = 0$. Each discontinuity, including the delta peak, splits the x-axis into disjoint regions (I through VIII) in which the wave function for a given energy (dashed line) can be solved with two free integration constants. Continuity conditions for ψ and ψ' at all the discontinuities of the potential then reduce the constants to two free parameters, or only one if there are no incoming particles from the right. In this example, we have $2 \cdot 8 = 16$ free constants in our initial wave function, on which we have to impose $2 \cdot 7 = 14$ boundary conditions. The remaining free constants can be used as amplitudes of waves incoming from the left and from the right.

If $V(x)$ varies slowly compared with $\psi(x)$, we have

$$V'(x_0) \ll -\frac{\hbar^2}{2m(x-x_0)}\frac{\psi''}{\psi} \approx \frac{E-V(x_0)}{x-x_0} \tag{2.255}$$

and the approximation is justified. We then have a differential equation with nearly constant coefficients and plane-wave approximate solutions $\psi(x) \approx Ae^{ikx} + Be^{-ikx}$ with $k = \sqrt{2m(E-V(x_0))}/\hbar$.

The wave function is exponential in regions where $E < V(x)$. Approximating $\psi'' \approx (2m/\hbar^2)(V(x_0)-E)\psi$ again by a Taylor expansion, the approximate solution is $\psi(x) \approx Ae^{\kappa x} + Be^{-\kappa x}$ with $\kappa = \sqrt{2m(V(x_0)-E)}/\hbar$. We may have a special case if $V(x) = E$ is strictly constant in some region, which implies $\psi'' = 0$ with linear instead of exponential solutions.

Complete solutions of scattering states can be found for piecewise constant potentials that may also include delta functions, as in the example shown in Figure 2.18. At each discontinuity of the potential, counting also delta-function peaks as discontinuities, we introduce an internal boundary on the x-axis. In each remaining continuous region, we solve the wave function by exponential functions ($E < V$), plane waves ($E > V$) or linear functions ($E = V$), in each case with two free integration constants for the general solution of a second-order differential equation. Each internal boundary implies two linear equations for the continuity of ψ and (dis)continuity of ψ'. Since these are linear equations for the integration constants, they can always be solved.

If there are N points of discontinuity, the x-axis is split into $N+1$ separate regions, leading to $2(N+1)$ free integration constants. If we subtract $2N$ for the number of (dis)continuity conditions, we have two remaining parameters, the previous A and G, which can be reduced to a single free constant if there are no particles coming in from the rightmost region. All other constants are then determined by the incoming amplitude, which allows us to compute the reflection and transmission coefficients.

The scattering problem for piecewise linear potentials is always completely solvable. The bound-state problem, however, in general results in intractable energy conditions given by transcendental equations. Nevertheless, properties of discrete energies can be studied using the graphical method. It is then possible to explore general features of quantum systems.

2.5.3 SCATTERING MATRIX

It is possible to derive general properties of 1-dimensional scattering for any potential that has a finite region in which it is non-zero, even if the non-zero part is not specified. We call the region in which the potential is non-zero Region II. It is then bordered to the left (smaller x) by Region I and to the right (larger x) by Region III. Since the potential vanishes in Regions I and III, the Schrödinger equation is solved there by plane waves,

$$\psi(x) = Ae^{ikx} + Be^{-ikx} \tag{2.256}$$

in Region I and

$$\psi(x) = Fe^{ikx} + Ge^{-ikx} \tag{2.257}$$

in Region III. The wave number equals $k = \sqrt{2mE}/\hbar > 0$ with a positive energy E for scattering states.

We now assume that $G \neq 0$, considering the general case of particles coming in from the left and from the right. In addition to A, B, F and G, the general solution of the Schrödinger equation in Region II (which we do not have to know explicitly in the present context) provides two additional integration constants. We therefore have six free constants, constrained by four continuity conditions for ψ and ψ' at the two boundary points of Region II. Imposing these conditions leaves us with two free constants, which we can choose to be A and G for the incoming amplitudes.

Since the continuity conditions are linear, the dependent constants B and F (as well as the two constants from Region II) are linearly related to the independent ones, A and G: We have

$$B = S_{11}A + S_{12}G \quad \text{and} \quad F = S_{21}A + S_{22}G \tag{2.258}$$

with complex coefficients S_{ij}. These coefficients define the **scattering matrix**

$$S = \begin{pmatrix} S_{11} & S_{12} \\ S_{21} & S_{22} \end{pmatrix} \tag{2.259}$$

with components that depend on the energy of the scattering state.

By definition, the scattering matrix relates the incoming amplitudes to the outgoing amplitudes,

$$\begin{pmatrix} B \\ F \end{pmatrix} = S \begin{pmatrix} A \\ G \end{pmatrix}. \tag{2.260}$$

If $G = 0$, the norms squared of some of the matrix components equal our previous scattering coefficients,

$$|S_{11}|^2 = \frac{|B|^2}{|A|^2} = R \tag{2.261}$$

for the reflection coefficient, and

$$|S_{21}|^2 = \frac{|F|^2}{|A|^2} = T \tag{2.262}$$

for the transmission coefficient. The other components, S_{12} and S_{22} are undefined in this case because they are multiplied by zero. For $G \neq 0$, using the scattering matrix instead of two scattering coefficients allows us to keep track of plane waves that may have been partially transmitted or partially reflected from two incoming particle streams.

2.5.3.1 Conservation law

In all specific scattering examples, we have observed that $R+T=1$ is always fulfilled, such that no particles get stuck in the potential. The components of the scattering matrix therefore cannot be completely arbitrary, but rather have to respect relationships such as

$$|S_{11}|^2+|S_{21}|^2=R+T=1\,. \tag{2.263}$$

Before discussing matrix properties, we first confirm the equation $R+T=1$ for a general scattering potential in cases in which $G=0$.

Since we do not know the potential in Region II, let alone solutions of the Schrödinger equation in this region, statements such as $R+T=1$ that require an extension of plane waves across Region II can only be derived from general properties. The most general properties in physics are usually expressed as conservation laws. In addition to energy conservation for a conservative potential, which is implemented here by working with a constant E and a time-independent $V(x)$, we have seen one further example: The normalization of the wave function and therefore its probability content, or more physically the number of particles, are conserved. Mathematically, this result is expressed by equation (1.94) in which the probability current (1.93) appears.

If we apply the definition of the probability current to a stationary state $\psi(x)$, we obtain a time-independent function

$$j(x)=\frac{i\hbar}{2m}\left(\frac{\partial\psi^*}{\partial x}\psi-\psi^*\frac{\partial\psi}{\partial x}\right). \tag{2.264}$$

This current vanishes for a bound state, which can always be chosen real such that $\psi^*=\psi$, but it may be non-zero for a scattering state with complex plane waves. However, the spatial derivative of $j(x)$ vanishes even for scattering states:

$$\frac{\mathrm{d}j}{\mathrm{d}x}=\frac{i\hbar}{2m}\left(\psi\frac{\mathrm{d}^2\psi^*}{\mathrm{d}x^2}-\psi^*\frac{\mathrm{d}^2\psi}{\mathrm{d}x^2}\right)=\frac{i}{\hbar}\left(-(E-V(x))\psi\psi^*+(E-V(x))\psi^*\psi\right)=0 \tag{2.265}$$

upon using the time-independent Schrödinger equation, which takes the same form for ψ and ψ^* because it has real coefficients. The derivative $\mathrm{d}j/\mathrm{d}x$ vanishes for any potential $V(x)$. Therefore, $j(x)=j_0$ is a constant even through Region II. The derivation of this result only used the time-independent Schrödinger equation in this region but no explicit solutions.

We now evaluate the probability current for our known plane-wave solutions in Regions I and III, which must take the same value because $j(x)$ is the same constant j_0 on the entire x-axis. In Region I, we have

$$\begin{aligned} j_0 &= \frac{i\hbar}{2m}\left(ik(-A^*e^{-ikx}+B^*e^{ikx})(Ae^{ikx}+Be^{-ikx})-(A^*e^{-ikx}+B^*e^{ikx})ik(Ae^{ikx}-Be^{-ikx})\right)\\ &= \frac{\hbar k}{m}(|A|^2-|B|^2)\,. \end{aligned} \tag{2.266}$$

If $G=0$ and $\psi(x)=Fe^{ikx}$,

$$j_0=\frac{\hbar k}{m}|F|^2 \tag{2.267}$$

in Region III. Equating (2.266) with (2.267) directly implies

$$R+T=1 \tag{2.268}$$

for any potential in Region II.

If $G\neq 0$, we obtain

$$j_0=\frac{\hbar k}{m}(|A|^2-|B|^2) \tag{2.269}$$

in Region I, as before, and

$$j_0 = \frac{\hbar k}{m}(|F|^2 - |G|^2) \tag{2.270}$$

in Region III. Equating these two expressions and rearranging terms, we obtain

$$|B|^2 + |F|^2 = |A|^2 + |G|^2 \tag{2.271}$$

for any potential in Region II. The scattering process is therefore always such that the norm squared (or the square of the length) of the vector of incoming amplitudes, given by $|A|^2 + |G|^2$, equals the norm squared of the vector of outgoing amplitudes, $|B|^2 + |F|^2$. Using the scattering matrix as defined in (2.260), we have

$$\left\| \begin{pmatrix} B \\ F \end{pmatrix} \right\|^2 = \left\| S \begin{pmatrix} A \\ G \end{pmatrix} \right\|^2 = \left\| \begin{pmatrix} A \\ G \end{pmatrix} \right\|^2 \tag{2.272}$$

where the first equality follows from the definition of S, and the second equality follows from the conserved current. We use double bars to indicate the extension of the norm squared to vectors with complex components by summing their squared norms:

$$\left\| \begin{pmatrix} z_1 \\ z_2 \end{pmatrix} \right\|^2 = |z_1|^2 + |z_2|^2 . \tag{2.273}$$

We may always choose A and G to be real, but B and F are in general complex as seen in our examples.

2.5.3.2 Unitarity

We are free to choose the incoming amplitudes. The result (2.272) therefore shows that the scattering matrix preserves the norm of any vector it multiplies. If the vector components were always real, this would mean that the scattering matrix simply rotates or reflects the vector. It would be an orthogonal matrix: Its column or row vectors would form orthogonal sets.

With complex components, this behavior is expressed by the more general property that the scattering matrix is unitary. As we will see later, a unitary 2×2 matrix such as S has components S_{ij} that obey the conditions

$$|S_{11}|^2 + |S_{21}|^2 = 1 \tag{2.274}$$

and

$$|S_{12}|^2 + |S_{22}|^2 = 1 \tag{2.275}$$

as well as

$$S_{11}^* S_{12} + S_{21}^* S_{22} = 0 . \tag{2.276}$$

The first equation is a generalization of (2.263) to scattering states with $G \neq 0$, while the second equation generalizes a version of (2.263) that can be obtained by setting $A = 0$ and using non-zero G. The third equation is an additional consistency condition that is obeyed by any scattering process. Since the original scattering matrix has four complex components, which are restricted by two real equations and a complex one, it contains information described by four independent real numbers. This count doubles the number of parameters, R and T, seen in one-sided scattering.

It turns out that the scattering matrix contains information not only about scattering states but also about bound states. For bound states, we have $\psi(x) = Be^{\kappa x}$ in Region I and $\psi(x) = Fe^{-\kappa x}$ in Region III with $\kappa = \sqrt{-2mE}/\hbar > 0$ for a negative energy $E < 0$. Using $k = i\kappa$, we obtain the same relationships between the free constants of general solutions as before if we substitute $k = i\kappa$ in the scattering matrix. We still have $B = S_{11}A + S_{12}G$ and $F = S_{21}A + S_{22}G$, but after the substitution we need $A = 0$ and $G = 0$ for normalizable states. We can have non-zero B and F only if there are values of κ (or values of E) such that

$$S_{ij}(k \to i\kappa) = \infty \tag{2.277}$$

for at least some of the components of the scattering matrix. If this is the case for some κ, there is a non-vanishing normalizable state with the energy

$$E = -\frac{\hbar^2 \kappa^2}{2m}. \tag{2.278}$$

We can see this outcome explicitly for the delta-function potential where we derived $S_{11} = B/A$ and $S_{21} = F/A$ for $G = 0$. The result, given in (2.220), is proportional to $1/(1 - i\beta)$ in both cases, where $\beta = m\alpha/(\hbar^2 k)$. If we substitute $k \to i\kappa$, this expression turns into $1/(1 - m\alpha/(\hbar^2\kappa))$ which diverges for $\kappa = m\alpha/\hbar^2$. This value of κ implies precisely the bound-state energy found in (2.205).

The appearance of bound-state energies in scattering coefficients has an important application in elementary particle physics. Experiments at high-energy accelerators measure the scattering coefficients of many interactions produced in collisions of the original particles. These interactions are usually described by scattering states, but sometimes it happens that a new metastable particle is produced in the interactions, for instance the Higgs particle, when the particle beam has sufficient energy to produce it. Since such a particle is stable for sometime, it can be approximated by a bound state. Because an exact bound state would imply a pole of the scattering matrix, the scattering coefficients get very large around the energy required to produce the new particle, called a resonance in particle physics. The scattering coefficients remain finite because a metastable particle is described only approximately by a bound state, but they imply a local maximum of the energy dependence of the scattering coefficients that can be used to measure the mass of the new particle, using the relativistic equation $E = mc^2$.

EXERCISES

1. Assume that a particle in the infinite square well is in one of the stationary states, $\psi_n(x)$.
   (i) Derive the probability of finding the particle between $x = 0$ and $x = ra$, where $0 \leq r \leq 1$.
   (ii) Discuss the two limiting cases $rn \gg 1$ and $rn \ll 1$.
2. A particle with mass m in an infinite square well of size a has the initial wave function $\Psi(x,0) = A(3\psi_1(x) - \psi_3(x))$ at time $t = 0$, where $\psi_1(x)$ and $\psi_3(x)$ are the stationary states with corresponding energy values for $n = 1$ and $n = 3$, respectively.
   (i) Determine the value of A. What is the time-dependent solution $\Psi(x,t)$ of the Schrödinger equation which has the initial values given by $\Psi(x,0)$? Compute $\langle\hat{x}\rangle(t)$ in this state.
   (ii) Repeat for an initial wave function $\Psi(x,0) = A(3\psi_1(x) - \psi_2(x))$.
3. Using the ladder operators $\hat{a}_+$ and $\hat{a}_-$, compute the expectation values $\langle\hat{x}\rangle_n$, $\langle\hat{p}\rangle_n$ and the standard deviations $\sigma_{x,n}$, $\sigma_{p,n}$ in a stationary state ψ_n of the harmonic oscillator.
   What is the smallest value of the product $\sigma_{x,n}\sigma_{p,n}$?
4. A harmonic oscillator has the initial wave function $\Phi_\alpha(x,0) = c(\psi_0(x) - e^{-i\alpha}\psi_1(x))$ in terms of the ground state $\psi_0(x)$ and the first excited state $\psi_1(x)$, for real numbers α and c.
   What is the corresponding time-dependent wave function $\Phi_\alpha(x,t)$?
   Show that $\Phi_\alpha(x,t) = e^{-\frac{1}{2}i\omega t}\Phi_{\alpha+\omega t}(x,0)$.
5. For a given complex number z, consider a wave function $\phi_z(x)$ which obeys the condition $\hat{a}_-\phi_z = iz\phi_z$, using the lowering operator $\hat{a}_-$ of the harmonic oscillator. (These wave functions are not stationary states.)
   (i) Compute the expectation values $\langle\hat{x}\rangle_z$, $\langle\hat{p}\rangle_z$ and the standard deviations $\sigma_{x,z}$, $\sigma_{p,z}$ in such a state.
   (ii) Check the uncertainty relation.

6. (i) Show that $\phi_z = c\sum_{n=0}^{\infty} \frac{1}{\sqrt{n!}}(iz)^n \psi_n$ fulfills the defining relation of ϕ_z given in the preceding problem, where ψ_n are the stationary states of the harmonic oscillator.
   (ii) Compute the constant c from part (i).
   (iii) The Hermite polynomials, which appear in the stationary states of the harmonic oscillator, fulfill the relation $e^{-s^2+2sy} = \sum_{n=0}^{\infty} \frac{1}{n!} s^n H_n(y)$ for arbitrary complex numbers s and y. Use this to derive

$$\phi_z(x) = \left(\frac{m\omega}{\pi\hbar}\right)^{\frac{1}{4}} \exp\left(-\frac{1}{2}\left(|z|^2 - z^2 + \frac{m\omega}{\hbar}x^2 - 2i\sqrt{\frac{2m\omega}{\hbar}}\, zx\right)\right).$$

   (iv) If $\Phi_z(x,t)$ is a solution of the time-dependent Schrödinger equation with initial values $\Phi_z(x,0) = \phi_z(x)$, show that $\Phi_z(x,t) = e^{-i\omega t/2}\phi_{ze^{-i\omega t}}(x)$.

7. Consider a free particle with mass m which at $t = 0$ is described by the wave function $\Psi(x,0) = (\lambda/\pi)^{\frac{1}{4}} \exp(-\frac{1}{2}\lambda x^2 + iKx)$ with real constants $\lambda > 0$ and $K \neq 0$.
   (i) Compute $\langle\hat{x}\rangle$ and $\langle\hat{p}\rangle$ at time $t = 0$.
   (ii) Show that the Fourier transform $\phi(k) = (2\pi)^{-\frac{1}{2}} \int_{-\infty}^{\infty} \Psi(x,0)e^{-ikx}\mathrm{d}x$ of the initial wave function is $\phi(k) = (\lambda\pi)^{-\frac{1}{4}} e^{-(k-K)^2/2\lambda}$.
   (iii) Using $\phi(k)$, derive $\Psi(x,t)$ for arbitrary t. Show that its probability density $|\Psi(x,t)|^2$ is a wave packet in which $\langle\hat{x}\rangle(t)$ moves along the classical trajectory while it spreads out according to $\sigma_x(t) = (2\lambda)^{-\frac{1}{2}}\sqrt{1+\lambda^2\hbar^2 t^2/m^2}$.

8. Normalize the wave function

$$\Psi(x,0) = \begin{cases} A & \text{if } -b \leq x \leq b \\ 0 & \text{otherwise} \end{cases}$$

with a positive constant b. Compute its Fourier transform $\phi(k)$. What is the value of the variance σ_k, obtained by using $|\phi(k)|^2$ as a probability density for k?

9. A particle with mass m is bound in the potential $V(x) = V_1(x) + V_2(x)$ with

$$V_1(x) = \begin{cases} 0 & \text{if } |x| > b \\ W & \text{if } |x| \leq b \end{cases}$$

and

$$V_2(x) = -\alpha\delta(x)$$

where W, b and α are positive constants.
   (i) Derive (but don't solve) the equation for the ground-state energy.
   (ii) Compute the transmission and reflection coefficients for the energy $E = W$, equal to the barrier height.

10. A deuteron nucleus consists of a proton and a neutron. As a model of the binding force between the two particles, treat the proton as moving in a finite square well potential. Using experimental results, we can estimate the radius as $b = 2 \cdot 10^{-15}$m, and the potential well as $V_0 = 36 \cdot 10^6$eV.
    (i) How many bound states do you expect for this system?
    (ii) Find an approximate value for the ground-state energy E_0 by plotting both sides of the equation $\sqrt{z_0^2/z^2 - 1} = \tan z$, where z_0 depends on the above values of b and V_0. Compare your result with the value obtained from the measured binding energy, $E_b = -E_0 = 2.2 \cdot 10^6$eV.

    [The proton has mass $m = 9.4 \cdot 10^8 \mathrm{eV}/c^2$ and $\hbar c = 1.975 \cdot 10^{-7}$eVm, with the speed of light c. An electron Volt, $1\mathrm{eV} = 1.602177 \cdot 10^{-19}$J, is a convenient unit of energy in particle physics.]
11. A potential is defined by $V(x) = V_1(x) + V_2(x)$ with

$$V_1(x) = \begin{cases} 0 & \text{if } |x| > b \\ W & \text{if } |x| \leq b \end{cases}$$

    and

$$V_2(x) = -\alpha\delta(x+b) - \alpha\delta(x-b)$$

    where W, α and b are positive constants.
    (i) Derive an equation for the ground-state energy in this system.
    (ii) Is there a ground state for any choice of (W, α, b)? Justify your answer by using graphical methods.
    (iii) Now assume that a particle moving in this potential has exactly the energy $E = W$. Compute the transmission and reflection coefficients for this energy.
12. For two solutions, ψ_1 and ψ_2, of the time-independent Schrödinger equation with energies E_1 and E_2, respectively, the **Wronskian** is defined as the function

$$W(x) = \psi_1(x)\psi_2'(x) - \psi_1'(x)\psi_2(x).$$

    (i) Write the x-derivative $W'(x)$ in terms of E_1, E_2, $\psi_1(x)$ and $\psi_2(x)$.
    (ii) Show that $W(x)$ is zero if $E_1 = E_2$. Conclude that ψ_1 and ψ_2 are therefore proportional to each other.

    This result shows that there is a unique normalized wave function for each allowed energy value of stationary states $\psi(x)$. That is, there is no **energy degeneracy** in 1-dimensional quantum-mechanical systems, a property that does not hold in two or three dimensions.

3 Quantum Mechanics on a Hilbert Space

The Schrödinger equation is a linear partial differential equation for the wave function. Therefore, any linear combination of two solutions is also a solution. The possibility of forming linear combinations is a well-known property of vectors. Moreover, we normalize solutions for the statistical interpretation, and we have seen several examples of orthogonality relations between bound and other states. These concepts are reminiscent of the length of a vector and the angle between two vectors or the dot product. Scattering is described by a matrix, another concept from linear algebra.

Providing a more general formulation of these properties will help us to develop new methods that make it possible to derive important general results. These include the example of the scattering matrix and its unitary property, as well as a general derivation of the orthogonality of stationary states with different energies in any conservative potential. Perhaps more surprisingly, the uncertainty relation is another example of properties that follow from linear algebra.

A major advantage of taking a more abstract mathematical view is that it allows us to think about objects such as a quantum state or an operator as something that is independent of its realization, or its representation, as a function or as a certain derivative operation. This "something" encapsulates the physical meaning of the system state, or of the observable whose measurement is described by an operator.

The distinction between a physical object and its representation is similar to the difference between a direction and vector components. The basic understanding of a direction is defined as one object pointing toward another object. In our day-to-day experience, we only have to raise our arm and turn a specific way in order to show the direction. It is not necessary to determine and communicate vector components, although we sometimes do so in a more advanced way of providing directions if we use degrees of North, East, South, or West. Mathematically, we specify a direction as a vector and usually express it by determining components in a given coordinate system. The intrinsic meaning of a direction is a statement about the relative locations of two objects. The mathematical description makes directions computable by relating the object's positions to a coordinate system, such that the direction from one to the other can be expressed relative to this system.

For computations, we usually choose a basis or coordinate axes and assign components to a vector, but these numbers depend on our basis choice. Similarly, when we wrote a state at time t_0 as $\Psi(x,t_0)$, we implicitly made a basis choice. The only difference with standard vectors is that $\Psi(x,t_0)$ has infinitely many components. In quantum mechanics, we have to apply linear algebra to infinite-dimensional vector spaces for a general treatment, but interesting finite-dimensional applications exist as well.

3.1 LINEAR ALGEBRA

There are many calculations in quantum mechanics that can be related to familiar concepts of linear algebra. However, they usually require an extension to infinite dimensions or function spaces. The requirements of quantum mechanics have also led to a special kind of notation for vectors that is convenient in several different ways but, at first sight, may not look like what is familiar from linear algebra. In this section, we first review linear algebra in the new notation of quantum mechanics

DOI: 10.1201/9781003588702-4

and then proceed to a generalization to infinite dimensions. One of the advantages of the quantum-mechanics notation of linear algebra is that it makes it easier to distinguish between an abstract vector (formalizing the concept of a direction) and the collection of its components in a given basis. The number of vector components in an infinite-dimensional vector space is infinite, but there is nothing infinite about the concept of a direction. The new notation will therefore facilitate the required transition to infinite dimensions. We will also have to generalize the concept of vector spaces to complex numbers since they appear as values of wave functions.

3.1.1 VECTORS

A vector formalizes the concept of pointing in a specific direction in some space in which different objects are located. Mathematically, vectors are distinguished by the operations that can be performed on them, given by addition and scalar multiplication. A (complex) **vector space** is a set of elements, which we will denote by $|\alpha\rangle$, $|\beta\rangle$ and so on, that obey the following conditions:

1. For any $|\alpha\rangle$ and $|\beta\rangle$ in the vector space, there is another vector given by $|\alpha\rangle + |\beta\rangle$. This operation is commutative: $|\alpha\rangle + |\beta\rangle = |\beta\rangle + |\alpha\rangle$.
2. For any vector $|\alpha\rangle$ and a complex number a (called a **scalar** in order to distinguish it from a vector), there is another vector given by $a|\alpha\rangle$. This operation is associative: $(a_1a_2)|\alpha\rangle = a_1(a_2|\alpha\rangle)$.
3. Addition and scalar multiplication are distributive: $a(|\alpha\rangle + |\beta\rangle) = a|\alpha\rangle + a|\beta\rangle$ and $(a+b)|\alpha\rangle = a|\alpha\rangle + b|\alpha\rangle$.
4. There is a null vector $|0\rangle$ such that $|\alpha\rangle + |0\rangle = |\alpha\rangle$ for all $|\alpha\rangle$.
5. For each $|\alpha\rangle$, there is an inverse $|-\alpha\rangle$ such that $|\alpha\rangle + |-\alpha\rangle = |0\rangle$.

Several implications follow, such as $|0\rangle = 0|\alpha\rangle$ for any $|\alpha\rangle$ because $|\alpha\rangle + 0|\alpha\rangle = (1+0)|\alpha\rangle = |\alpha\rangle$, and $|-\alpha\rangle = (-1)|\alpha\rangle = -|\alpha\rangle$ because $|\alpha\rangle + (-1)|\alpha\rangle = (1-1)|\alpha\rangle = 0|\alpha\rangle = |0\rangle$.

One can easily confirm that all the conditions for a vector space are satisfied if we identify the different vectors $|\alpha\rangle$ with the (non-normalized) solutions of the Schrödinger equation for a given potential. If we consider only normalized solutions, however, we do not obtain a vector space because $|0\rangle$ is not normalizable, and normalization, in general, is not preserved by vector addition or scalar multiplication.

Every vector space has a **basis**, given by a set of vectors $|e_1\rangle$, $|e_2\rangle$, ... such that any vector $|\alpha\rangle$ in the vector space has a unique decomposition

$$|\alpha\rangle = \alpha_1|e_1\rangle + \alpha_2|e_2\rangle + \cdots \tag{3.1}$$

with complex numbers α_i, $i = 1,2,\ldots$. The α_i are called the **components** of the vector $|\alpha\rangle$ in the given basis. If the number n of basis vectors is finite, it is the same for any basis of the given vector space. The number n is then called the **dimension** of the vector space. An infinite-dimensional vector space is a vector space that does not have a finite basis.

The vector components α_i depend on the vector $|\alpha\rangle$ for a given basis but also on the basis for a given $|\alpha\rangle$. It is therefore incorrect to identify a vector $|\alpha\rangle$ with the list of components $(\alpha_1, \alpha_2, \ldots)$ because the former is independent of the basis choice, while the latter depends on the basis. For instance, the direction $|\alpha\rangle$ from us to a certain place is a vector that we can provide by pointing in this direction without specifying a basis. If we say that the place is to the West, we are using a convention or a basis to specify components. (The component in the South-North direction is zero, and the component in the West-East direction is negative one.)

Instead of identifying $|\alpha\rangle$ with its list of components in a basis, we say that the components **represent** the vector $|\alpha\rangle$ in the given basis. A **representation** gives us the following list of correspondences between abstract vector operations, introduced in the definition of a vector space, and standard vector operations based on components:

$$|\alpha\rangle \leftrightarrow (\alpha_1, \alpha_2, \ldots) \tag{3.2}$$
$$|\alpha\rangle + |\beta\rangle \leftrightarrow (\alpha_1 + \beta_1, \alpha_2 + \beta_2, \ldots) \tag{3.3}$$
$$c|\alpha\rangle \leftrightarrow (c\alpha_1, c\alpha_2, \ldots) \tag{3.4}$$
$$|0\rangle \leftrightarrow (0, 0, \ldots) \tag{3.5}$$
$$|-\alpha\rangle \leftrightarrow (-\alpha_1, -\alpha_2, \ldots). \tag{3.6}$$

Objects on the left-hand side of this correspondence are abstract symbols while components on the right are numbers that depend on the basis choice.

The distinction between abstract vectors and basis-dependent components is useful in a discussion of function spaces. As an example, we may consider the set of all quadratic functions of x, given by $a + bx + cx^2$ with complex numbers a, b and c. A simple basis choice is given by $|e_1\rangle = 1$, $|e_2\rangle = x$, and $|e_3\rangle = x^2$. Quadratic functions therefore form a 3-dimensional vector space. In the given basis, we have

$$a + bx + cx^2 = a|e_1\rangle + b|e_2\rangle + c|e_3\rangle \tag{3.7}$$

and the function is represented by the components (a, b, c) in this basis. The first three Hermite polynomials constitute a different basis choice: $|H_0\rangle = 1$, $|H_1\rangle = 2x$, and $|H_2\rangle = 4x^2 - 2$. In this basis, we have

$$a + bx + cx^2 = (a + c/2)|H_0\rangle + \frac{1}{2}b|H_1\rangle + \frac{1}{4}c|H_2\rangle. \tag{3.8}$$

The quadratic function is now represented by a different list of components, $(a + c/2, b/2, c/4)$. But it is the same quadratic function and therefore its role as an abstract vector that could be added to other quadratic functions has not changed.

If we consider all polynomials $a + bx + cx^2 + \cdots$ of unspecified degree, there is no finite-dimensional basis because the functions $1, x, x^2, \ldots, x^N$ for any finite integer N are linearly independent, but they do not allow us to express a polynomial of degree $N + 1$ or higher as a linear combination. Polynomials, or polynomials times a Gaussian function as seen in the bound states of the harmonic oscillator, therefore form an infinite-dimensional vector space.

In general, a vector space does not have a product of vectors that would yield another vector. (The cross product for vectors in a 3-dimensional vector space is a special case that does not generalize in this form to arbitrary dimensions.) Nevertheless, it is possible to multiply vectors by simply juxtaposing them as in $|\alpha\rangle|\beta\rangle$. We interpret this new object after expressing it in a basis of the vector space:

$$\begin{aligned} |\alpha\rangle|\beta\rangle &= (\alpha_1|e_1\rangle + \alpha_2|e_2\rangle + \cdots)(\beta_1|e_1\rangle + \beta_2|e_2\rangle + \cdots) \\ &= \alpha_1\beta_1|e_1\rangle|e_1\rangle + \alpha_1\beta_2|e_1\rangle|e_2\rangle + \cdots \\ &\quad + \alpha_2\beta_1|e_2\rangle|e_1\rangle + \alpha_2\beta_2|e_2\rangle|e_2\rangle + \cdots \\ &= \sum_{j=1}^{n}\sum_{k=1}^{n} \alpha_j\beta_k|e_j\rangle|e_k\rangle. \end{aligned} \tag{3.9}$$

The new components $\alpha_j\beta_k$ have two indices and resemble a matrix rather than a vector. This matrix-valued product of vectors is called the **tensor product**, also written as $|\alpha\rangle \otimes |\beta\rangle$. Matrices can be added and multiplied by a scalar, and together with the zero matrix and the negative matrix we obtain a new vector space. The tensor product can be applied more than once, gaining another component index each time a vector is multiplied in. This product therefore leads to linear objects with multi-index components.

The tensor product (3.9) can be applied to all pairs of vectors in vector spaces with dimensions m and n, respectively, which are not required to be the same. The result is a new vector space, defining the tensor product of the original two spaces, with basis $\{|e_j\rangle \otimes |f_k\rangle : j = 1,\ldots,m; k = 1,\ldots,n\}$ if $\{|e_j\rangle : j = 1,\ldots,m\}$ and $\{|f_k\rangle : k = 1,\ldots,n\}$ are bases of the first and second vector space, respectively. The tensor-product space therefore has dimension mn.

In component form, a tensor is more general than products of components $\alpha_j\beta_k$ of two vectors. Starting with generic $|\alpha\rangle$ and $|\beta\rangle$, the latter has $m+n$ independent components, given by the original α_j and β_k, while a tensor has mn independent components. The tensor product therefore significantly enlarges the original vector spaces by including additional parameters that cannot be expressed as products of vector components. The initial two vector spaces may themselves have been obtained as tensor products of smaller vector spaces. In quantum mechanics, the tensor product allows us to combine independent information from different parts of a system, such as the constituents of an atom.

3.1.2 INNER PRODUCTS

We use the dot product in order to compute the length of a real vector or the angle between two real vectors, according to Euclidean geometry. The common expression for the dot product of two vectors with components α_j and β_k, respectively, is $\sum_{j=1}^{n} \alpha_j\beta_j$. However, this expression depends on the basis choice through the components. We also need generalizations to complex components and to infinite dimensions. As with vector spaces, these questions can more easily be answered if we first look at the abstract definition of an inner product, which generalizes the dot product.

An **inner product** on a complex vector space assigns a complex number $\langle\alpha|\beta\rangle$ to any pair of vectors, $|\alpha\rangle$ and $|\beta\rangle$, such that

1. $\langle\beta|\alpha\rangle = \langle\alpha|\beta\rangle^*$,
2. $\langle\alpha|\alpha\rangle \geq 0$ for all $|\alpha\rangle$ such that $\langle\alpha|\alpha\rangle = 0$ if and only if $|\alpha\rangle = |0\rangle$, and
3. if $|\beta\rangle = b_1|\beta_1\rangle + b_2|\beta_2\rangle$, then $\langle\alpha|\beta\rangle = b_1\langle\alpha|\beta_1\rangle + b_2\langle\alpha|\beta_2\rangle$.

There is an analog of the third condition if the vector $|\alpha\rangle = a_1|\alpha_1\rangle + a_2|\alpha_2\rangle$ on the left of an inner product is a linear combination. In this case, we derive

$$\langle\alpha|\beta\rangle = \langle\beta|\alpha\rangle^* = (a_1\langle\beta|\alpha_1\rangle + a_2\langle\beta|\alpha_2\rangle)^* = a_1^*\langle\alpha_1|\beta\rangle + a_2^*\langle\alpha_2|\beta\rangle \tag{3.10}$$

using conditions 1, 3 and again 1. The reflection of the brackets of $\langle\alpha|$ for the first vector in an inner product serves as a reminder that we have to use complex conjugation if we try to factor out a scalar on this side, such as a_1^* and a_2^* in the preceding equation. According to condition 3, there is no complex conjugation if we factor out a scalar on the right side of an inner product. The inner product is not completely linear in both entries or bilinear, but rather sesquilinear (Latin for "one and a half linear"): It is linear on its right entry, and incompletely linear on its left entry.

Condition 2 allows us to define the norm (or length) of a vector as the real number

$$||\alpha|| = \sqrt{\langle\alpha|\alpha\rangle}\,. \tag{3.11}$$

We say that a vector $|\alpha\rangle$ is normalized if it has norm one, $||\alpha|| = 1$. Two vectors $|\alpha\rangle$ and $|\beta\rangle$ are orthogonal if $\langle\alpha|\beta\rangle = 0$. An orthonormal set of vectors $|\alpha_i\rangle$ has the inner products

$$\langle\alpha_i|\alpha_j\rangle = \delta_{ij} \tag{3.12}$$

for all i and j for which vectors in the set are defined.

If a basis $|e_i\rangle$ is orthonormal, it is possible to compute vector components by projection, relating them to inner products with the basis vectors. If we assume two generic basis decompositions $|\alpha\rangle = \sum_{j=1}^{n} \alpha_j|e_j\rangle$ and $|\beta\rangle = \sum_{k=1}^{n} \beta_k|e_k\rangle$ in the same basis, we have

$$\langle e_k|\alpha\rangle = \sum_{j=1}^{n} \alpha_j\langle e_k|e_j\rangle = \sum_{j=1}^{n} \alpha_j\delta_{kj} = \alpha_k \tag{3.13}$$

and similarly $\beta_k = \langle e_k|\beta\rangle$. Moreover, the inner product of $|\alpha\rangle$ and $|\beta\rangle$ can be computed through the components by

$$\langle\alpha|\beta\rangle = \sum_{k=1}^{n} \beta_k \langle\alpha|e_k\rangle = \sum_{k=1}^{n} \beta_k \left(\sum_{j=1}^{n} \alpha_j^* \langle e_j|e_k\rangle \right) = \sum_{k=1}^{n}\sum_{j=1}^{n} \beta_k \alpha_j^* \delta_{jk} = \sum_{j=1}^{n} \alpha_j^* \beta_j . \quad (3.14)$$

This result generalizes the standard expression of the dot product to complex numbers, but it also shows that it is valid only if the basis is orthonormal. The inner product of two vectors has a different expression in terms of components if the basis is not orthonormal.

The condition $\langle\alpha|\alpha\rangle \geq 0$ for any vector $|\alpha\rangle$ implies an important result, called the **Cauchy–Schwarz inequality**:

$$|\langle\alpha|\beta\rangle|^2 \leq \langle\alpha|\alpha\rangle\langle\beta|\beta\rangle = ||\alpha||^2||\beta||^2 \quad (3.15)$$

for any pair of vectors, $|\alpha\rangle$ and $|\beta\rangle$. In order to derive this inequality, we first define a new vector $|\gamma\rangle$ by

$$|\gamma\rangle = |\beta\rangle - \frac{\langle\alpha|\beta\rangle}{\langle\alpha|\alpha\rangle}|\alpha\rangle \quad (3.16)$$

provided that $|\alpha\rangle \neq |0\rangle$. (If $|\alpha\rangle = |0\rangle$, we have zero on both sides of the Cauchy–Schwarz inequality, which holds in this case.) This new vector $|\gamma\rangle$ is orthogonal to α because

$$\langle\alpha|\gamma\rangle = \langle\alpha|\beta\rangle - \frac{\langle\alpha|\beta\rangle}{\langle\alpha|\alpha\rangle}\langle\alpha|\alpha\rangle = 0 . \quad (3.17)$$

Therefore,

$$\langle\gamma|\gamma\rangle = \langle\beta|\gamma\rangle = \langle\beta|\beta\rangle - \frac{\langle\alpha|\beta\rangle}{\langle\alpha|\alpha\rangle}\langle\beta|\alpha\rangle = \langle\beta|\beta\rangle - \frac{|\langle\alpha|\beta\rangle|^2}{\langle\alpha|\alpha\rangle} \quad (3.18)$$

is closely related to the terms in the Cauchy–Schwarz inequality. In fact, the inequality follows directly from the condition $\langle\gamma|\gamma\rangle \geq 0$.

The Cauchy–Schwarz inequality has two important applications. First, it allows us to define angles in Euclidean geometry because we can always write an inner product of two vectors as the product of the two norms times a factor less than or equal to one:

$$|\langle\alpha|\beta\rangle| = ||\alpha|| \cdot ||\beta|| \cos(\varphi) . \quad (3.19)$$

The angle φ in the cosine is well-defined thanks to the Cauchy–Schwarz inequality because the cosine cannot be greater than one. As a second application, we will see in Section 3.2.3 that the Cauchy–Schwarz inequality in quantum mechanics implies the uncertainty relation (or, more precisely, different versions of uncertainty relations).

The brackets used to denote vectors and inner products follow Dirac notation. Reflecting the symbols, as in replacing $|\alpha\rangle$ with $\langle\alpha|$ for the first entry of an inner product, represents complex conjugation, which has to be used if we factor out scalars on the left. Mathematically, the reflection can also be interpreted as turning a vector $|\alpha\rangle$ into a function on the vector space: Just as we place a function symbol f and a variable x next to each other in the evaluation $f(x)$, we can say that placing $\langle\alpha|$ next to $|\beta\rangle$ evaluates the function $\langle\alpha|$ at the vector $|\beta\rangle$ with a result given by the complex number $\langle\alpha|\beta\rangle$. According to the conditions on inner products, $\langle\alpha|$ is always a linear function of vectors. A linear function of vectors is called a **dual vector**.

A related interpretation of closed parentheses in terms of mathematical properties can be used to double-check equations. In the inner product $\langle\alpha|\beta\rangle$, the brackets are closed, and the result is a single complex number. If the brackets remain open, as in $|\beta\rangle$ or $\langle\alpha|$, we are dealing with multiple components of a vector or a linear function, rather than a single complex number. The vertical bar in this notation is used to indicate the end of a specific expression, which is necessary in particular in the case of open brackets.

3.1.3 LINEAR TRANSFORMATIONS

If we map vectors to other vectors, we distinguish this operation from our notation for vectors and complex numbers by using a hat, as in $|\alpha'\rangle = \hat{T}|\alpha\rangle$, where $\hat{T}$ denotes the mapping or the transformation. In linear algebra and quantum mechanics, we are mainly interested in **linear transformations** that are compatible with the vector operations of addition and scalar multiplication: A transformation $\hat{T}$ is linear if

$$\hat{T}(a|\alpha\rangle + b|\beta\rangle) = a(\hat{T}|\alpha\rangle) + b(\hat{T}|\beta\rangle) \tag{3.20}$$

for all vectors $|\alpha\rangle$ and $|\beta\rangle$ in the vector space, and complex numbers a and b.

A simple example is the transformation $\hat{L}|\alpha\rangle = 5|\alpha\rangle$. As a formal demonstration that this transformation is linear, we calculate

$$\hat{L}(a|\alpha\rangle + b|\beta\rangle) = 5(a|\alpha\rangle + b|\beta\rangle) \tag{3.21}$$

by applying the definition to the full vector $a|\alpha\rangle + b|\beta\rangle$ without using the fact that it is a linear combination. We can now use distributivity of scalar multiplication and vector addition to conclude that

$$5(a|\alpha\rangle + b|\beta\rangle) = 5a|\alpha\rangle + 5b|\beta\rangle \tag{3.22}$$

which is the same as $a\hat{L}|\alpha\rangle + b\hat{L}|\beta\rangle$. The condition for a linear transformation is therefore satisfied.

An example of a transformation that is not linear is given by $\hat{N}|\alpha\rangle = ||\alpha|| \cdot |\alpha\rangle$. This transformation is a small variation of the normalization map transforming a vector $|\alpha\rangle$ to $||\alpha||^{-1/2}|\alpha\rangle$, which is also non-linear but would not be defined on the null vector. We show that $\hat{N}$ is non-linear by finding a simple counter-example to the condition of linearity. For instance, if we apply $\hat{N}$ to $2|\alpha\rangle$, we obtain $||2\alpha|| \cdot 2|\alpha\rangle = 4||\alpha|| \cdot |\alpha\rangle$. This is not the same as $2\hat{N}|\alpha\rangle = 2||\alpha|| \cdot |\alpha\rangle$.

In quantum mechanics, the normalization map is not a linear transformation, but all operators that we have previously denoted by a hat, such as $\hat{x}$ and $\hat{p}$, are linear transformations of wave functions. We often use "operator" as synonymous with "linear transformation."

3.1.3.1 Basis representation

A linear transformation $\hat{T}$ is completely determined by its action on the basis vectors of a vector space. The components T_{ij} of $\hat{T}$ in a given basis of an n-dimensional vector space are defined by

$$\hat{T}|e_j\rangle = \sum_{i=1}^{n} T_{ij}|e_i\rangle \tag{3.23}$$

for $j = 1, 2, \ldots, n$. In order to derive the components T_{ij}, we apply the transformation to each vector in the basis and then write the result as a linear combination of the basis vectors. The n^2 components T_{ij} determine the result of the transformation acting on any vector, using linearity:

$$\hat{T}|\alpha\rangle = \sum_{j=1}^{n} \alpha_j \hat{T}|e_j\rangle = \sum_{j=1}^{n}\sum_{i=1}^{n} \alpha_j T_{ij}|e_i\rangle = \sum_{i=1}^{n}\left(\sum_{j=1}^{n} T_{ij}\alpha_j\right)|e_i\rangle = \sum_{i=1}^{n} \alpha'_i|e_i\rangle, \tag{3.24}$$

reading off the components

$$\alpha'_i = \sum_{j=1}^{n} T_{ij}\alpha_j \tag{3.25}$$

of the new vector $|\alpha'\rangle = \hat{T}|\alpha\rangle$ in the last step.

The components T_{ij} of a linear transformation can be arranged in an $n \times n$-matrix. Because the components and the matrix depend on the basis, they are not the same as the linear transformation, but they represent it just as the vector components $(\alpha_1, \alpha_2, \ldots)$ represent a vector $|\alpha\rangle$ in a given basis. We write this correspondence as

$$\hat{T} \leftrightarrow \begin{pmatrix} T_{11} & T_{12} & \cdots & T_{1n} \\ T_{21} & T_{22} & \cdots & T_{2n} \\ \vdots & \vdots & \ddots & \vdots \\ T_{n1} & T_{n2} & \cdots & T_{nn} \end{pmatrix} = T \tag{3.26}$$

If we drop the hat, we refer to the matrix of a linear transformation in a given basis. Using matrix notation, the action of a linear transformation on a vector, represented by components in the same basis, is given by matrix multiplication: Equation (3.25) is equivalent to

$$\begin{pmatrix} \alpha_1' \\ \alpha_2' \\ \vdots \\ \alpha_n' \end{pmatrix} = \begin{pmatrix} T_{11} & T_{12} & \cdots & T_{1n} \\ T_{21} & T_{22} & \cdots & T_{2n} \\ \vdots & \vdots & \ddots & \vdots \\ T_{n1} & T_{n2} & \cdots & T_{nn} \end{pmatrix} \begin{pmatrix} \alpha_1 \\ \alpha_2 \\ \vdots \\ \alpha_n \end{pmatrix}. \tag{3.27}$$

Standard matrix multiplication requires a representation of vectors by columns.

Matrix components are useful for calculations with linear transformations. They are defined by (3.23) and in general require solving a set of linear equations. This task simplifies if the basis is orthonormal, in which case we can derive a matrix element by a single inner product,

$$T_{ij} = \langle e_i|\hat{T}|e_j\rangle. \tag{3.28}$$

3.1.3.2 Combining matrices by operations

Calculations with linear transformations can be done by using different operations. We can add and multiply linear transformations according to

$$(\hat{S}+\hat{T})|\alpha\rangle = \hat{S}|\alpha\rangle + \hat{T}|\alpha\rangle \quad \leftrightarrow \quad S+T \tag{3.29}$$

$$(\hat{S}\hat{T})|\alpha\rangle = \hat{S}(\hat{T}|\alpha\rangle) \quad \leftrightarrow \quad ST \tag{3.30}$$

using the correspondence between abstract linear transformations on the left and matrix representations in a given basis on the right. In particular, the product of two linear transformations is defined by applying them one after the other.

The result is represented by the matrix product of S and T, which is not commutative: We have

$$(ST)_{ik} = \sum_{j=1}^{n} S_{ij}T_{jk} \tag{3.31}$$

while

$$(TS)_{ik} = \sum_{j=1}^{n} T_{ij}S_{jk} = \sum_{j=1}^{n} S_{jk}T_{ij}. \tag{3.32}$$

Since indices appear in different positions in the two results, they cannot be brought to the same form in general. In the abstract version, we determine the ordering $(\hat{S}\hat{T})|\alpha\rangle = \hat{S}(\hat{T}|\alpha\rangle)$ by first applying the linear transformation $\hat{T}$ that is next to the vector $|\alpha\rangle$ on which the product acts. The failure of the product of two linear transformations to commute is quantified by the **commutator**, written as

$$[\hat{S},\hat{T}] = \hat{S}\hat{T} - \hat{T}\hat{S}. \tag{3.33}$$

There is a second product operation for linear transformations that, unlike the matrix product, acts on a new vector space given by the tensor product of the vector spaces on which the original linear transformations act: The **tensor product** of two linear transformations $\hat{T}$ and $\hat{S}$ is given by

$$(\hat{T}\otimes\hat{S})(|\alpha\rangle\otimes|\beta\rangle) = (\hat{T}|\alpha\rangle)\otimes(\hat{S}|\beta\rangle). \tag{3.34}$$

The two vector spaces need not be identical, such that it is possible to iterate the construction by viewing the original vector spaces as obtained from tensor products of smaller vector spaces.

If we are interested in the components of the tensor product $\hat{T}\otimes\hat{S}$, we have to decide how we list the tensor-product basis vectors $|e_j\rangle\otimes|f_k\rangle$, as in

$$(|e_1\rangle\otimes|f_1\rangle, |e_1\rangle\otimes|f_2\rangle, \ldots, |e_2\rangle\otimes|f_1\rangle, |e_2\rangle\otimes|f_2\rangle, \ldots)\,. \tag{3.35}$$

This choice is not unique, but any other form that includes all the basis vectors would simply amount to a basis change. If $\hat{T}$ has the components T_{ij} in the basis $|e_j\rangle$ and $\hat{S}$ has the components S_{kl} in the basis $|f_k\rangle$, the tensor product $\hat{T}\otimes\hat{S}$ has the component given by the matrix

$$\begin{pmatrix} T_{11}S & T_{12}S & \cdots \\ T_{21}S & T_{22}S & \cdots \\ \vdots & \vdots & \ddots \end{pmatrix} \tag{3.36}$$

in which each entry is the component T_{ij} multiplying the full matrix S. The result is therefore an $mn\times mn$-matrix if T is an $m\times m$-matrix and S is an $n\times n$-matrix.

We confirm the product form by computing the inner products

$$\begin{aligned} \langle e_i|\otimes\langle f_k|(\hat{T}\otimes\hat{S})(|e_j\rangle\otimes|f_l\rangle) &= \langle e_i|\otimes\langle f_k|\left(\hat{T}|e_j\rangle\otimes\hat{S}|f_l\rangle\right) \\ &= \langle e_i|\hat{T}|e_j\rangle\langle f_k|\hat{S}|f_l\rangle = T_{ij}S_{kl}\,. \end{aligned} \tag{3.37}$$

Therefore, if we arrange the pair of indices as determined by the basis list (3.35), we first multiply T_{11} by all the components of S, then move on to T_{12} times all the components of S in the same row, just as it is expressed by the matrix (3.36). As a simple application, it follows that the tensor product of two identity matrices in m and n dimensions, respectively, is the identity matrix in mn dimensions.

The tensor product of linear transformations cannot be commutative because, in general, the two vector spaces on which the individual transformations act need not be the same. If they are not the same, the reverse order of the tensor product is not defined. But even if we have two linear operations, $\hat{T}$ and $\hat{S}$, acting on the same vector space, $\hat{T}\otimes\hat{S}\neq\hat{S}\otimes\hat{T}$ in general. As an example, consider a 2-dimensional vector space and a matrix $T=\begin{pmatrix} 0 & 1 \\ 1 & 0 \end{pmatrix}$ in a given basis, together with the identity matrix S. Using (3.36), we compute

$$T\otimes S=\begin{pmatrix} 0 & S \\ S & 0 \end{pmatrix}=\begin{pmatrix} 0 & 0 & 1 & 0 \\ 0 & 0 & 0 & 1 \\ 1 & 0 & 0 & 0 \\ 0 & 1 & 0 & 0 \end{pmatrix} \tag{3.38}$$

and

$$S\otimes T=\begin{pmatrix} T & 0 \\ 0 & T \end{pmatrix}=\begin{pmatrix} 0 & 1 & 0 & 0 \\ 1 & 0 & 0 & 0 \\ 0 & 0 & 0 & 1 \\ 0 & 0 & 1 & 0 \end{pmatrix}\,. \tag{3.39}$$

We have a non-zero tensor-product commutator

$$T\otimes S-S\otimes T=\begin{pmatrix} 0 & -1 & 1 & 0 \\ -1 & 0 & 0 & 1 \\ 1 & 0 & 0 & -1 \\ 0 & 1 & -1 & 0 \end{pmatrix}\,. \tag{3.40}$$

Even though one of the matrices is the identity matrix and therefore commutes with any other matrix if we compute a matrix product, it does not necessarily commute with another matrix in a tensor product: In general, keeping the first vector in $|\alpha\rangle\otimes|\beta\rangle$ unchanged, is not the same as keeping the second vector unchanged.

3.1.3.3 Operations on a single matrix

A well-known operation that maps a matrix T to another matrix, T^T, is given by **transposition**. It turns out that there is no abstract version of this operation, which therefore cannot be defined for linear transformations without using a basis. For a matrix T, the transpose T^T is obtained by reflecting the components along the diagonal,

$$(T^T)_{ij} = T_{ji}\,. \tag{3.41}$$

Transposition can also be applied to a vector represented by components in a basis, in which case it does not change the components but turns a column vector into a row vector,

$$\begin{pmatrix} \alpha_1 \\ \alpha_2 \\ \vdots \\ \alpha_n \end{pmatrix}^T = (\alpha_1\ \alpha_2\ \cdots\ \alpha_n)\,, \tag{3.42}$$

and vice versa.

A square $n \times n$-matrix is called **symmetric** if it equals its transpose, $T = T^T$ or $T_{ij} = T_{ji}$, and **antisymmetric** if it equals its negative transpose, $T = -T^T$ or $T_{ij} = -T_{ji}$. We can also define the **complex conjugate** of a matrix T^* by taking complex conjugates of all its components, $(T^*)_{ij} = T^*_{ij}$. A matrix is called **real** if it equals its complex conjugate, $T^* = T$ or $T^*_{ij} = T_{ij}$, and **imaginary** if it equals its negative complex conjugate, $T^* = -T$ or $T^*_{ij} = -T_{ij}$.

The combination of transposition and complex conjugation is an important operation. We will see later that this combination can be defined for any linear transformation independently of the basis choice (as long as there is an inner product), unlike transposition and complex conjugation by themselves. The combination is called **Hermitian conjugation**, and the resulting matrix

$$T^\dagger = (T^T)^* = (T^*)^T \tag{3.43}$$

is called the **Hermitian conjugate** or the **adjoint** of T. In components, we have

$$(T^\dagger)_{ij} = T^*_{ji}\,. \tag{3.44}$$

As an example of a 2×2-matrix, we have

$$\begin{pmatrix} 1 & i \\ 2+i & 5i \end{pmatrix}^\dagger = \begin{pmatrix} 1 & 2-i \\ -i & -5i \end{pmatrix}. \tag{3.45}$$

A square matrix is called **Hermitian** or **self-adjoint** if it agrees with its Hermitian conjugate, $T = T^\dagger$ or $T_{ij} = T^*_{ji}$, and **skew-Hermitian** or **anti-Hermitian** if it agrees with its negative Hermitian conjugate, $T = -T^\dagger$ or $T_{ij} = -T^*_{ji}$. The Hermitian conjugate can also be applied to vectors, such as

$$\begin{pmatrix} \alpha_1 \\ \alpha_2 \\ \vdots \\ \alpha_n \end{pmatrix}^\dagger = (\alpha_1^*\ \alpha_2^*\ \cdots\ \alpha_n^*)\,. \tag{3.46}$$

In this way, the inner product of two vectors in an orthonormal basis can be written as the matrix product of the Hermitian conjugate of the first vector with the second vector:

$$\langle\alpha|\beta\rangle = \begin{pmatrix} \alpha_1 \\ \alpha_2 \\ \vdots \\ \alpha_n \end{pmatrix}^\dagger \begin{pmatrix} \beta_1 \\ \beta_2 \\ \vdots \\ \beta_n \end{pmatrix} = (\alpha_1^*\ \alpha_2^*\ \cdots\ \alpha_n^*) \begin{pmatrix} \beta_1 \\ \beta_2 \\ \vdots \\ \beta_n \end{pmatrix} = \alpha_1^*\beta_1 + \alpha_2^*\beta_2 + \cdots + \alpha_n^*\beta_n\,. \tag{3.47}$$

We have the following identities for repeated applications of transposition, the complex conjugate, and Hermitian conjugation:

$$(T^T)^T = T \quad , \quad (T^*)^* = T \quad , \quad (T^\dagger)^\dagger = T \tag{3.48}$$

and

$$(ST)^T = T^T S^T \quad , \quad (ST)^\dagger = T^\dagger S^\dagger \,. \tag{3.49}$$

We can apply transposition and Hermitian conjugation to each factor in a product of matrices, but have to reverse their order.

3.1.3.4 Inverse

The **inverse** $\hat{T}^{-1}$ of a linear transformation $\hat{T}$ is defined by the two equalities in

$$\hat{T}^{-1}\hat{T} = \hat{\mathbb{I}} = \hat{T}\hat{T}^{-1} \tag{3.50}$$

where $\hat{\mathbb{I}}$ is the identity transformation, $\hat{\mathbb{I}}|\alpha\rangle = |\alpha\rangle$ for all $|\alpha\rangle$ in the vector space. In a finite-dimensional vector space, it is possible to show that the first equality in (3.50) implies the second one (or vice versa). Only one of them has to be checked when we compute an inverse. However, in an infinite-dimensional vector space it is possible that only one of the two equalities in (3.50) holds for a given $\hat{T}$, in which case $\hat{T}$ is not considered invertible.

As an example, consider the ladder operators of the harmonic oscillator, which act by $\hat{a}_+\psi_n = \sqrt{n+1}\;\psi_{n+1}$ and $\hat{a}_-\psi_n = \sqrt{n}\;\psi_{n-1}$ on stationary states ψ_n, which form a basis of this infinite-dimensional vector space. Therefore,

$$\hat{a}_-\hat{a}_+\psi_n = (n+1)\psi_n = \left(\frac{\hat{H}}{\hbar\omega} + \frac{1}{2}\right)\psi_n \,. \tag{3.51}$$

If we define the operator

$$\hat{O} = \left(\frac{\hat{H}}{\hbar\omega} + \frac{1}{2}\right)^{-1} \hat{a}_- \tag{3.52}$$

where the inverse operator used here is given by

$$\left(\frac{\hat{H}}{\hbar\omega} + \frac{1}{2}\right)^{-1} \psi_n = \frac{1}{n+1}\psi_n \,, \tag{3.53}$$

we have $\hat{O}\hat{a}_+ = \hat{\mathbb{I}}$. According to the first equality in (3.50), $\hat{O}$ therefore looks like the inverse of $\hat{a}_+$. However, the second equality in (3.50) is not satisfied because

$$\hat{a}_+\hat{O}\psi_0 = \hat{a}_+\left(\frac{\hat{H}}{\hbar\omega} + \frac{1}{2}\right)^{-1} \hat{a}_-\psi_0 = 0 \neq \hat{\mathbb{I}}\psi_0 \tag{3.54}$$

when we act on the ground state.

In a finite-dimensional vector space, it is sufficient to evaluate only one of the two equalities in (3.50) if we compute the inverse. For an explicit computation we use the matrix T that represents $\hat{T}$ in a given basis. The matrix equation

$$T^{-1}T = \mathbb{I} \tag{3.55}$$

(now using the identity matrix $\mathbb{I}$) has n^2 components if n is the dimension of the vector space. The inverse relationship therefore provides n^2 linear equations for the n^2 unknown components of T^{-1}. Such a system of linear equations may or may not have a solution, and therefore the inverse matrix is not guaranteed to exist for a given T, even in finite dimensions. The theory of linear equations demonstrates that a solution of (3.55) for the components $(T^{-1})_{ij}$ exists (and is then unique) if and only if the **determinant** $\det T$ of the coefficient matrix T in (3.55) is non-zero.

For instance, if $n = 1$, a matrix is a single complex number z and its inverse z^{-1} exists if and only if $z = \det(z) \neq 0$. If $n = 2$, there is a closed-form expression

$$\det\begin{pmatrix} T_{11} & T_{12} \\ T_{21} & T_{22} \end{pmatrix} = T_{11}T_{22} - T_{12}T_{21} \tag{3.56}$$

for the determinant of a 2×2-matrix, and its inverse in the case of a non-zero determinant is given by

$$\begin{pmatrix} T_{11} & T_{12} \\ T_{21} & T_{22} \end{pmatrix}^{-1} = \frac{1}{\det(T)}\begin{pmatrix} T_{22} & -T_{12} \\ -T_{21} & T_{11} \end{pmatrix}. \tag{3.57}$$

This matrix is well-defined only if $\det(T) \neq 0$.

The determinant of a 3×3-matrix equals

$$\begin{aligned} &\det\begin{pmatrix} T_{11} & T_{12} & T_{13} \\ T_{21} & T_{22} & T_{23} \\ T_{31} & T_{32} & T_{33} \end{pmatrix} \\ &= T_{11}T_{22}T_{33} + T_{12}T_{23}T_{31} + T_{13}T_{21}T_{32} - T_{31}T_{22}T_{13} - T_{32}T_{23}T_{11} - T_{33}T_{21}T_{12}\,. \end{aligned} \tag{3.58}$$

For general n, there is an iterative procedure that starts with

$$\begin{aligned} &\det\begin{pmatrix} T_{11} & T_{12} & T_{13} & \cdots \\ T_{21} & T_{22} & T_{23} & \cdots \\ T_{31} & T_{32} & T_{33} & \cdots \\ \vdots & \vdots & \vdots & \ddots \end{pmatrix} \\ &= T_{11}\det\begin{pmatrix} T_{22} & T_{23} & \cdots \\ T_{32} & T_{33} & \cdots \\ \vdots & \vdots & \ddots \end{pmatrix} - T_{12}\det\begin{pmatrix} T_{21} & T_{23} & \cdots \\ T_{31} & T_{33} & \cdots \\ \vdots & \vdots & \ddots \end{pmatrix} \pm \cdots \end{aligned} \tag{3.59}$$

If we use the **minors** of T, defined for fixed i and j as the determinant $M(i,j)$ of the sub-matrix of T obtained by removing row i and column j, we can write the general determinant as

$$\det(T) = \sum_{i=1}^{n} (-1)^{i+j} T_{ij} M(i,j)\,. \tag{3.60}$$

Since $M(i,j)$ is defined as a determinant of an $(n-1)\times(n-1)$ matrix, (3.60) implies an iterative method for computing determinants.

Determinants can be used to obtain components of an inverse matrix: The inverse T^{-1} is given by

$$T^{-1} = \frac{C^T}{\det(T)} \tag{3.61}$$

with the **co-matrix** C of T defined by the components $C_{ij} = (-1)^{i+j}M(i,j)$. The components of the inverse matrix therefore equal

$$(T^{-1})_{ij} = (-1)^{i+j}\frac{M(j,i)}{\det(T)}\,. \tag{3.62}$$

Note the flipped positions of j and i in $M(j,i)$.

We prove this equation by writing it as $\det(T)(T^{-1})_{ij} = C_{ji}$ and confirming that we obtain the correct expressions for all components if we multiple by T_{jk} and sum over j:

$$\det(T)\delta_{ik} = \det(T)\sum_{j=1}^{n}(T^{-1})_{ij}T_{jk} = \sum_{j=1}^{n} C_{ji}T_{jk} = \sum_{j=1}^{n}(-1)^{i+j}T_{jk}M(j,i)\,. \tag{3.63}$$

If $i = k$, we have $\det(T)$ on the left, and the right-hand side reproduces the iterative determinant formula (3.60). If $j \neq k$, the right-hand side can be interpreted, again using (3.60), as the determinant of a matrix obtained from T by copying row k into row i. Such a matrix with two identical rows has zero determinant, in agreement with the vanishing δ_{ik} on the left.

The inverse of a matrix product equals the product of the inverses, but in reverse order:

$$(ST)^{-1} = T^{-1}S^{-1}. \tag{3.64}$$

The derivation

$$(ST)^{-1}(ST) = (T^{-1}S^{-1})(ST) = T^{-1}(S^{-1}S)T = T^{-1}T = \mathbb{I} \tag{3.65}$$

uses only associativity of matrix multiplication and can also be applied to the product of two invertible linear transformations in a basis-independent way.

A matrix U is called **unitary** if its inverse $U^{-1} = U^\dagger$ equals its Hermitian conjugate. Therefore,

$$U^\dagger U = \mathbb{I} \tag{3.66}$$

for a unitary matrix. A unitary matrix has a simple inverse because it is easier to compute the Hermitian conjugate which does not require solving a set of linear equations.

3.1.3.5 Basis changes

It is often useful to consider transformations of vector or matrix components by changing the basis, which might simplify calculations or help us to identify basis-independent quantities. If we change or transform the basis, there are two sets of vectors, $|e_j\rangle$ and $|f_i\rangle$, such that a single vector $|\alpha\rangle$ can be expressed in two ways,

$$|\alpha\rangle = \sum_{i=1}^{n} \alpha_i^{(f)}|f_i\rangle = \sum_{j=1}^{n} \alpha_j^{(e)}|e_j\rangle \tag{3.67}$$

with two sets of components, $\alpha_i^{(f)}$ and $\alpha_j^{(e)}$. Each basis vector of one basis can also be expanded in the other basis, as in

$$|e_j\rangle = \sum_{i=1}^{n} S_{ij}|f_i\rangle \tag{3.68}$$

for the n vectors of the e-basis. The coefficients S_{ij} can be arranged in a matrix S, but now they do not belong to a linear transformation of vector components in a fixed basis. If we transform the basis, the direction implied by a given vector remains the same, but we express it with respect to new reference directions.

For instance, if we have a standard basis

$$|e_1\rangle = \begin{pmatrix} 1 \\ 0 \end{pmatrix} \quad , \quad |e_2\rangle = \begin{pmatrix} 0 \\ 1 \end{pmatrix} \tag{3.69}$$

and a second basis

$$|f_1\rangle = \begin{pmatrix} 1 \\ 2 \end{pmatrix} \quad , \quad |f_2\rangle = \begin{pmatrix} 3 \\ 4 \end{pmatrix}, \tag{3.70}$$

we write

$$\begin{pmatrix} 1 \\ 0 \end{pmatrix} = \begin{pmatrix} 3 \\ 4 \end{pmatrix} - 2\begin{pmatrix} 1 \\ 2 \end{pmatrix} \quad , \quad \begin{pmatrix} 0 \\ 1 \end{pmatrix} = \frac{3}{2}\begin{pmatrix} 1 \\ 2 \end{pmatrix} - \frac{1}{2}\begin{pmatrix} 3 \\ 4 \end{pmatrix} \tag{3.71}$$

or

$$|e_1\rangle = -2|f_1\rangle + |f_2\rangle \quad , \quad |e_2\rangle = \frac{3}{2}|f_1\rangle - \frac{1}{2}|f_2\rangle. \tag{3.72}$$

The basis-change matrix equals

$$S = \begin{pmatrix} -2 & 3/2 \\ 1 & -1/2 \end{pmatrix} \tag{3.73}$$

because in (3.68) we defined the e-basis index j as the second index of S_{ij}, which labels different columns.

The basis-change matrix determines how we transform vector components from one basis to another one. Using our basis expansions for some vector $|\alpha\rangle$, we have

$$|\alpha\rangle = \sum_{j=1}^{n} \alpha_j^{(e)}|e_j\rangle = \sum_{j=1}^{n} \alpha_j^{(e)} \sum_{i=1}^{n} S_{ij}|f_i\rangle = \sum_{i=1}^{n}\left(\sum_{j=1}^{n} \alpha_j^{(e)} S_{ij}\right)|f_i\rangle = \sum_{i=1}^{n} \alpha_i^{(f)}|f_i\rangle . \tag{3.74}$$

Comparing coefficients in the f-basis using the last two terms, we obtain

$$\alpha_i^{(f)} = \sum_{j=1}^{n} S_{ij}\alpha_j^{(e)} \tag{3.75}$$

or

$$\alpha^{(f)} = S\alpha^{(e)} \tag{3.76}$$

for the column vectors $\alpha^{(f)}$ and $\alpha^{(e)}$ in matrix notation.

Continuing with the previous example and defining the vector $|\alpha\rangle = 5|e_1\rangle + 6|e_2\rangle$ in the e-basis, it has components

$$\alpha^{(e)} = \begin{pmatrix} 5 \\ 6 \end{pmatrix}_e . \tag{3.77}$$

The transformation matrix S from (3.73) determines the components

$$\alpha^{(f)} = S\alpha^{(e)} = \begin{pmatrix} -2 & 3/2 \\ 1 & -1/2 \end{pmatrix}\begin{pmatrix} 5 \\ 6 \end{pmatrix}_e = \begin{pmatrix} -1 \\ 2 \end{pmatrix}_f \tag{3.78}$$

in the f-basis, using subscripts on column vectors to indicate which basis they belong to. With the computed coefficients of $\alpha^{(f)}$, $|\alpha\rangle = -|f_1\rangle + 2|f_2\rangle$ represents the same vector $|\alpha\rangle$ we started with because

$$-\begin{pmatrix} 1 \\ 2 \end{pmatrix} + 2\begin{pmatrix} 3 \\ 4 \end{pmatrix} = \begin{pmatrix} 5 \\ 6 \end{pmatrix} \tag{3.79}$$

in the standard basis.

Since vector components change if we transform to a new basis, the matrix elements of a linear transformation should also change. A linear transformation $\hat{T}$ represented by the matrix $T^{(e)}$ in the e-basis transforms column vectors according to

$$(\alpha')^{(e)} = T^{(e)}\alpha^{(e)} \tag{3.80}$$

in this basis. The transformation of vector components from the e-basis to the f-basis is given by the matrix S,

$$\alpha^{(f)} = S\alpha^{(e)} . \tag{3.81}$$

These transformations can be combined to obtain

$$(\alpha')^{(f)} = S(\alpha')^{(e)} = ST^{(e)}\alpha^{(e)} = ST^{(e)}S^{-1}\alpha^{(f)} \tag{3.82}$$

using invertibility of S, which is always realized because we can transform back and forth between basis choices. Comparing the last equation with $(\alpha')^{(f)} = T^{(f)}\alpha^{(f)}$ for the transformation $\hat{T}$ represented in the f-basis, we obtain

$$T^{(f)} = ST^{(e)}S^{-1} . \tag{3.83}$$

Two matrices T_1 and T_2 related to each other by $T_2 = ST_1S^{-1}$ with an invertible matrix S are called **similar**. Multiplying a matrix from the left by S and from the right by S^{-1} is called a **similarity transformation** of the matrix. The result (3.83) shows that the matrix of a linear transformation on a vector space changes by a similarity transformation if we change the basis. The matrices are not identical, but they are similar and represent the same linear transformation.

3.1.3.6 Invariants

Equation (3.83) makes it possible to look for basis-independent properties of a matrix, or **invariants**, by finding expressions that do not change if we apply a similarity transformation. Some examples are:

The determinant of a matrix is invariant because it is multiplicative, such that $\det(AB) = \det(A)\det(B)$ for any two matrices A and B. Therefore, if $T_2 = ST_1S^{-1}$ is a similarity transformation,

$$\det(T_2) = \det(S)\det(T_1)\det(S^{-1}) = \det(S)\det(T_1)\frac{1}{\det(S)} = \det(T_1)\,. \tag{3.84}$$

The trace of a matrix T is defined by summing the diagonal components,

$$\mathrm{tr}(T) = \sum_{i=1}^{n} T_{ii}\,. \tag{3.85}$$

It is invariant under similarity transformations because $\mathrm{tr}(AB) = \mathrm{tr}(BA)$, which can be derived directly from the definition in terms of diagonal matrix elements. Therefore,

$$\mathrm{tr}(T_2) = \mathrm{tr}(ST_1S^{-1}) = \mathrm{tr}((ST_1)S^{-1}) = \mathrm{tr}(S^{-1}(ST_1)) = \mathrm{tr}(T_1)\,. \tag{3.86}$$

The eigenvalues of a matrix T are basis-independent, defined as solutions of the equation $\det(T - \lambda\mathbb{I}) = 0$. For a 2×2-matrix,

$$\begin{aligned}\det\begin{pmatrix} T_{11}-\lambda & T_{12} \\ T_{21} & T_{22}-\lambda \end{pmatrix} &= (T_{11}-\lambda)(T_{22}-\lambda) - T_{12}T_{21} \\ &= \lambda^2 - (T_{11}+T_{22})\lambda + T_{11}T_{22} - T_{12}T_{21}\end{aligned} \tag{3.87}$$

with the trace $\mathrm{tr}(T) = T_{11}+T_{22}$ and the determinant $\det(T) = T_{11}T_{22} - T_{12}T_{21}$ as coefficients. The eigenvalues

$$\lambda_\pm = \frac{1}{2}\left(\mathrm{tr}(T) \pm \sqrt{\mathrm{tr}(T)^2 - 4\det(T)}\right) \tag{3.88}$$

are functions of the previous two invariant quantities. If a matrix has more than two independent eigenvalues, there are invariant quantities other than the trace and the determinant.

3.1.4 EIGENVECTORS AND EIGENVALUES

A linear transformation $\hat{T}$ takes a simple form if it is possible to choose basis vectors $|e\rangle_i$ such that

$$\hat{T}|e\rangle_i = \lambda_i|e\rangle_i \tag{3.89}$$

with complex numbers λ_i that may depend on i. It is then easy to compute the determinant of such a transformation in finite dimensions because it is given by the product of all λ_i, or the inverse transformation by inverting each number λ_i (provided it is non-zero).

If

$$\hat{T}|\alpha\rangle = \lambda_\alpha|\alpha\rangle \tag{3.90}$$

holds for a single vector $|\alpha\rangle \neq |0\rangle$, we say that $|\alpha\rangle$ is an **eigenvector** of $\hat{T}$ with **eigenvalue** λ_α. We exclude the null vector from the set of possible eigenvectors because it always obeys (3.90) for a linear transformation but does not determine a unique λ_α. However, the eigenvalue λ_α may well be zero if $\hat{T}|\alpha\rangle = |0\rangle$ for some $|\alpha\rangle \neq |0\rangle$.

3.1.4.1 Characteristic equation

In general, we have to find a basis in which the linear transformation takes the form (3.90), assuming that this is possible: There is no guarantee that the set of all eigenvectors of a linear transformation spans the full vector space. We compute the eigenvalues and eigenvectors of a finite-dimensional linear transformation by choosing an initial basis and converting (3.90) into equations with coefficients derived from its matrix. The eigenvalue equation then reads

$$T\alpha = \lambda_\alpha \alpha \tag{3.91}$$

for the matrix T and a column vector α, or

$$(T - \lambda_\alpha \mathbb{I})\alpha = 0 \tag{3.92}$$

with the unit matrix $\mathbb{I}$. If we fix a possible value for λ_α, this condition provides n linear equations for the n components of α. There are two cases:

1. If the n linear equations have a unique solution for the n components α_i, this solution must be the null vector because it always obeys the eigenvector equation. However, the null vector is not considered an eigenvector.
2. For a valid eigenvector, the set of n linear equations cannot have a unique solution. There are then solutions for $|\alpha\rangle \neq |0\rangle$, and in this case λ_α is an eigenvalue. The set $(T - \lambda\mathbb{I})\alpha = 0$ of linear equations for α (with a fixed λ) has non-unique solutions if and only if the coefficient matrix $T - \lambda\mathbb{I}$ is non-invertible, which means that its determinant vanishes:

$$0 = \det(T - \lambda\mathbb{I}) = \det\begin{pmatrix} T_{11}-\lambda & T_{12} & \cdots & T_{1n} \\ T_{21} & T_{22}-\lambda & \cdots & T_{2n} \\ \vdots & \vdots & \ddots & \vdots \\ T_{n1} & T_{n2} & \cdots & T_{nn}-\lambda \end{pmatrix}. \tag{3.93}$$

Evaluating the determinant results in an n-th order polynomial equation for λ, the **characteristic equation** of T. The equation and its coefficients multiplying powers of λ in the polynomial are basis-independent because

$$\det(STS^{-1} - \lambda\mathbb{I}) = \det(S(T-\lambda\mathbb{I})S^{-1}) = \det(S)\det(T-\lambda\mathbb{I})\det(S)^{-1} = \det(T-\lambda\mathbb{I}) \tag{3.94}$$

if we change T to STS^{-1} by a similarity transformation. In this way, the polynomial gives us n basis-independent quantities defined by its coefficients, or alternatively by its solutions λ which are non-linear functions of the coefficients. (A polynomial of degree n has $n+1$ coefficients, but the characteristic equation implies that the coefficient of the highest order multiplying λ^n equals $(-1)^n$ and is independent of the matrix T.) We conclude that the eigenvalues of a linear transformation are always basis-independent.

The eigenvalues give us useful basis-independent information about a matrix. An n-th degree polynomial always has n solutions (not all of which may be real), but some of them might take the same value. The eigenvalues are determined by factoring the characteristic equation as

$$\det(T - \lambda\mathbb{I}) = (-1)^n(\lambda-\lambda_1)(\lambda-\lambda_2)\cdots(\lambda-\lambda_n) \tag{3.95}$$

into n factors. If the same solution $\lambda = \bar{\lambda}$ appears in several factors, the number of such factors is called the **multiplicity** of the eigenvalue $\bar{\lambda}$. If we add the multiplicities of all distinct solutions of the characteristic equation, the sum always equals n because there are n factors in (3.95). For instance, the identity matrix $T = \mathbb{I}$ has only one distinct eigenvalue, $\lambda = 1$, but it has multiplicity n. Counting multiplicities, there are n eigenvalues. If there is at least one eigenvalue with multiplicity greater than one, the set of eigenvalues is called **degenerate**.

Multiplicity is important because an eigenvalue with multiplicity greater than one may have more than one linearly independent eigenvector. Once we know the eigenvalues λ, we know that the eigenvector equation $(T-\lambda_\alpha\mathbb{I})\alpha=0$ has non-null solutions α for each λ_α in the set of solutions λ of the characteristic equation. These non-null solutions α are not unique because any scalar multiple of α is a solution if α is a solution. If λ_α has multiplicity greater than one, there may be different linearly independent α solving the eigenvector equation with this value (but this is not guaranteed).

3.1.4.2 Eigenvector basis

Solving $(T-\lambda_\alpha\mathbb{I})\alpha=0$ for all solutions λ_α of the characteristic equation, including multiplicity, we obtain n eigenvectors. In general, the set of these eigenvectors is not guaranteed to be linearly independent. But in important special cases, such as a Hermitian matrix T, the normalized eigenvectors are linearly independent. Because there are then n linearly independent eigenvectors in an n-dimensional vector space, they can be used as a basis $|f_i\rangle$, $i=1,\dots,n$, to which we may transform from the original basis $|e_j\rangle$ in which the matrix elements had been defined or computed.

The linear transformation $\hat{T}$ has a simple action in the eigenvector basis, given by

$$\hat{T}|f_i\rangle=\lambda_i|f_i\rangle \tag{3.96}$$

The matrix elements T_{ij} in this basis are defined by $\hat{T}|f_i\rangle=\sum_{j=1}^n T_{ij}|f_j\rangle$, and since the latter equals $\lambda_i|f_i\rangle$, we have $T_{ij}=\lambda_i\delta_{ij}$. In the eigenvector basis, the matrix

$$T=\begin{pmatrix}\lambda_1 & 0 & \cdots & 0\\ 0 & \lambda_2 & \cdots & 0\\ \vdots & \vdots & \ddots & \vdots\\ 0 & 0 & \cdots & \lambda_n\end{pmatrix} \tag{3.97}$$

is diagonal. If a matrix T has eigenvectors that form a basis, it is called **diagonalizable**.

If an eigenvector basis exists for a given T, it simplifies calculations. For instance, the determinant is simply $\det(T)=\lambda_1\lambda_2\cdots\lambda_n$. It is also easy to take a diagonal matrix to some integer power,

$$T^k=\begin{pmatrix}\lambda_1^k & 0 & \cdots & 0\\ 0 & \lambda_2^k & \cdots & 0\\ \vdots & \vdots & \ddots & \vdots\\ 0 & 0 & \cdots & \lambda_n^k\end{pmatrix} \tag{3.98}$$

and to solve differential equations such as a finite-dimensional version

$$i\hbar\frac{\mathrm{d}\Psi}{\mathrm{d}t}=T\Psi \tag{3.99}$$

of the Schrödinger equation with a time-dependent vector $\Psi(t)$ and a time-independent matrix T. If T is diagonalizable, a state that is initially an eigenvector, $\Psi(0)=\alpha$, evolves according to $\Psi(t)=\exp(-i\lambda_\alpha/\hbar)\Psi(0)$ and therefore remains an eigenvector. A general vector evolves as a superposition of these time-dependent eigenvectors. (If $T(t)$ is time-dependent, it is not guaranteed that $\Psi(t)$ remains an eigenvector if it is initially an eigenvector of $T(0)$. In this case, it is more difficult to solve the differential equation.)

If a matrix T is diagonalizable but, in a given basis, not in diagonal form, the eigenvectors allow us to compute the similarity transformation from T to $D=STS^{-1}$ that diagonalizes T. If we have computed all the eigenvectors $\alpha^{(j)}$, $j=1,\dots,n$, we define the similarity transformation by specifying the components $(S^{-1})_{ij}=(\alpha^{(j)})_i$ of the inverse of S, using the i-th component of the eigenvector $\alpha^{(j)}$ with eigenvalue λ_j. As a matrix, we have

$$S^{-1} = \begin{pmatrix} \alpha_1^{(1)} & \alpha_1^{(2)} & \cdots & \alpha_1^{(n)} \\ \alpha_2^{(1)} & \alpha_2^{(2)} & \cdots & \alpha_2^{(n)} \\ \vdots & \vdots & \ddots & \vdots \\ \alpha_n^{(1)} & \alpha_n^{(2)} & \cdots & \alpha_n^{(n)} \end{pmatrix} = (\alpha^{(1)}\ \alpha^{(2)}\ \cdots\ \alpha^{(n)}) \tag{3.100}$$

where we use the i-th eigenvector $\alpha^{(i)}$ as the i-th column of the matrix. We can confirm that this is the correct basis-change matrix by computing STS^{-1} using matrix multiplication of T times the individual column of S^{-1} we just defined:

$$\begin{aligned} STS^{-1} &= ST(\alpha^{(1)}\ \alpha^{(2)}\ \cdots\ \alpha^{(n)}) = S(T\alpha^{(1)}\ T\alpha^{(2)}\ \cdots\ T\alpha^{(n)}) \\ &= S(\lambda_1\alpha^{(1)}\ \lambda_2\alpha^{(2)}\ \cdots\ \lambda_n\alpha^{(n)}) = (\lambda_1 S\alpha^{(1)}\ \lambda_2 S\alpha^{(2)}\ \cdots\ \lambda_n S\alpha^{(n)}) \\ &= \begin{pmatrix} \lambda_1 & 0 & \cdots & 0 \\ 0 & \lambda_2 & \cdots & 0 \\ \vdots & \vdots & \ddots & \vdots \\ 0 & 0 & \cdots & \lambda_n \end{pmatrix}. \end{aligned} \tag{3.101}$$

In the last step, we used $(S\alpha^{(i)})_j = \delta_{ij}$ for the j-th component of the vector $S\alpha^{(i)}$, which follows from

$$SS^{-1} = S(\alpha^{(1)}\ \alpha^{(2)}\ \cdots\ \alpha^{(n)}) = (S\alpha^{(1)}\ S\alpha^{(2)}\ \cdots\ S\alpha^{(n)}) = \mathbb{I}. \tag{3.102}$$

3.1.4.3 Examples

1. If $n = 2$, the characteristic equation of a matrix T is given by $\lambda^2 - \lambda\,\mathrm{tr}(T) + \det(T) = 0$ with solutions

$$\lambda_\pm = \frac{1}{2}\left(\mathrm{tr}(T) \pm \sqrt{\mathrm{tr}(T)^2 - 4\det(T)}\right) = \frac{1}{2}\left(T_{11} + T_{22} \pm \sqrt{(T_{11} - T_{22})^2 + 4T_{12}T_{21}}\right). \tag{3.103}$$

   The two eigenvalues are distinct if $\det T \neq \frac{1}{4}\mathrm{tr}(T)^2$.

2. The matrix

$$T = \begin{pmatrix} 1 & 1 \\ 0 & 1 \end{pmatrix} \tag{3.104}$$

   has only one distinct eigenvalue $\lambda_\pm = 1$ (with multiplicity two) because $\det T = 1$ and $\mathrm{tr}(T) = 2$. The eigenvector equation

$$(T - \lambda\mathbb{I})\alpha = \begin{pmatrix} 0 & 1 \\ 0 & 0 \end{pmatrix}\begin{pmatrix} \alpha_1 \\ \alpha_2 \end{pmatrix} = \begin{pmatrix} \alpha_2 \\ 0 \end{pmatrix} = \begin{pmatrix} 0 \\ 0 \end{pmatrix} \tag{3.105}$$

   implies $\alpha_2 = 0$, and therefore all eigenvectors have the form $\alpha = (c, 0)^T$ with a non-zero constant c. There is only one linearly independent eigenvector, $\alpha = (1, 0)^T$ if we normalize it, and the matrix is not diagonalizable.

3. If a 2×2-matrix T is Hermitian, $T = T^\dagger = (T^T)^*$, its components are restricted by the conditions that T_{11} and T_{22} are real, and $T_{12} = T_{21}^*$. The general solution (3.103) for eigenvalues then implies

$$\lambda_\pm = \frac{1}{2}\left(T_{11} + T_{22} \pm \sqrt{(T_{11} - T_{22})^2 + 4|T_{12}|^2}\right) \tag{3.106}$$

   and therefore both eigenvalues are real. The two eigenvalues are distinct, $\lambda_+ \neq \lambda_-$, if the square root is non-zero, which is the case for $T_{11} \neq T_{22}$ or $T_{12} \neq 0$. In these cases, the matrix is diagonalizable. The two eigenvalues are equal, $\lambda_+ = \lambda_-$, if $T_{11} = T_{22}$ and $T_{12} = 0$, but in this case the matrix is already diagonal and proportional to the identity matrix, $T = \lambda\mathbb{I}$. A Hermitian 2×2-matrix (and, as it turns out, a Hermitian matrix in any dimension) is always diagonalizable.

4. As a specific case of a Hermitian 2×2-matrix, we choose $T_{11} = T_{22} = a$ and $T_{12} = T_{21} = b \neq 0$ with real a and b. The eigenvalues are

$$\lambda_\pm = \frac{1}{2}\left(2a \pm \sqrt{4b^2}\right) = a \pm b \tag{3.107}$$

or

$$\lambda_+ = a+b \quad , \quad \lambda_- = a-b\,. \tag{3.108}$$

The eigenvector equations

$$(T - \lambda_\pm\mathbb{I})\alpha^{(\pm)} = \begin{pmatrix} \mp b & b \\ b & \mp b \end{pmatrix}\alpha^{(\pm)} = \begin{pmatrix} 0 \\ 0 \end{pmatrix} \tag{3.109}$$

imply the vector equation

$$\begin{pmatrix} \mp b & b \\ b & \mp b \end{pmatrix}\begin{pmatrix} \alpha_1^{(\pm)} \\ \alpha_2^{(\pm)} \end{pmatrix} = \begin{pmatrix} \mp b\alpha_1^{(\pm)} + b\alpha_2^{(\pm)} \\ b\alpha_1^{(\pm)} \mp b\alpha_2^{(\pm)} \end{pmatrix} = \begin{pmatrix} 0 \\ 0 \end{pmatrix} \tag{3.110}$$

for $\alpha^{(\pm)}$ which is solved by $\alpha_1^{(\pm)} = \pm\alpha_2^{(\pm)}$. Therefore,

$$\alpha^{(\pm)} = c_\pm \begin{pmatrix} 1 \\ \pm 1 \end{pmatrix} \tag{3.111}$$

with non-zero constants $c_\pm$.

We have two linearly independent eigenvectors and therefore an eigenvector basis. For a Hermitian matrix, it is convenient to normalize eigenvectors, such that

$$\alpha^{(\pm)} = \frac{1}{\sqrt{2}}\begin{pmatrix} 1 \\ \pm 1 \end{pmatrix} \tag{3.112}$$

if we have an inner product in which the standard basis vectors $(1,0)^T$ and $(0,1)^T$ are orthonormal. The basis change from the standard basis to the eigenvector basis is then accomplished by a similarity transformation with

$$S^{-1} = \begin{pmatrix} 1/\sqrt{2} & 1/\sqrt{2} \\ 1/\sqrt{2} & -1/\sqrt{2} \end{pmatrix} \tag{3.113}$$

using the eigenvectors as columns of this matrix.

In the case of a Hermitian matrix T, we will see that the normalized eigenvectors form an orthonormal basis. Therefore, the columns of S^{-1} are orthonormal, which implies that S is unitary, $S^\dagger = S^{-1}$. Since S^{-1} as derived in the present example is real and symmetric, it equals its own Hermitian conjugate, and we obtain $S = S^{-1}$. An explicit calculation indeed results in

$$STS^{-1} = \frac{1}{2}\begin{pmatrix} 1 & 1 \\ 1 & -1 \end{pmatrix}\begin{pmatrix} a & b \\ b & a \end{pmatrix}\begin{pmatrix} 1 & 1 \\ 1 & -1 \end{pmatrix} = \begin{pmatrix} a+b & 0 \\ 0 & a-b \end{pmatrix}. \tag{3.114}$$

3.1.5 HERMITIAN TRANSFORMATIONS

The Hermitian conjugate (or adjoint) of a matrix T is defined by combining transposition and complex conjugation: $T^\dagger = (T^T)^*$. Unlike transposition and complex conjugation by themselves, Hermitian conjugation has a basis-independent definition that can be used also in infinite-dimensional vector spaces. However, Hermitian conjugation requires an inner product, and it depends on this choice.

Given an abstract linear transformation $\hat{T}$ on a vector space that may be infinite-dimensional, on which an inner product $\langle\alpha|\beta\rangle$ has been introduced, the **Hermitian conjugate** $\hat{T}^\dagger$ is defined as the linear transformation that obeys the condition

$$\hat{T}^\dagger|\alpha\rangle = |\gamma\rangle \quad \text{if} \quad \langle\gamma|\beta\rangle = \langle\alpha|\hat{T}|\beta\rangle \quad \text{for all} \quad |\beta\rangle\,. \tag{3.115}$$

This is a somewhat convoluted definition that does not offer a direct pathway to computing the Hermitian conjugate. Moreover, the Hermitian conjugate depends on how the inner product is defined. Both properties are different from what we are used to from the matrix version.

In spite of the indirect wording, the general definition implies that $\hat{T}^\dagger$ is uniquely defined for a given $\hat{T}$ and a fixed inner product: If we have two different vectors $|\gamma\rangle$ and $|\gamma'\rangle$ that obey the required condition for all $|\beta\rangle$, we have $\langle\alpha|\hat{T}|\beta\rangle = \langle\gamma|\beta\rangle = \langle\gamma'|\beta\rangle$ and therefore $\langle\gamma-\gamma'|\beta\rangle$ for all $|\beta\rangle$. Given the conditions on an inner product, this is possible only if $|\gamma-\gamma'\rangle = 0$ or $|\gamma\rangle = |\gamma'\rangle$. The two vectors turn out to be the same, in contrast to our assumption which is therefore ruled out.

Writing $\langle\gamma| = \langle\hat{T}^\dagger\alpha|$ as a reflection of (3.115), turning $|\gamma\rangle$ into a dual vector, the definition of the Hermitian conjugate directly implies that

$$\langle\hat{T}^\dagger\alpha|\beta\rangle = \langle\alpha|\hat{T}|\beta\rangle \tag{3.116}$$

for all $|\alpha\rangle$ and $|\beta\rangle$. This operation therefore allows us to move a linear transformation from the right-hand side of an inner product to a potentially different transformation on the left-hand side of an inner product.

The definition also implies that

$$(\hat{S}\hat{T})^\dagger = \hat{T}^\dagger\hat{S}^\dagger\,, \tag{3.117}$$

reversing the ordering just as we have seen it for matrices. For abstract linear transformations, this equation follows from

$$\langle\alpha|\hat{S}\hat{T}|\beta\rangle = \langle\hat{S}^\dagger\alpha|\hat{T}|\beta\rangle = \langle\hat{T}^\dagger\hat{S}^\dagger\alpha|\beta\rangle = \langle(\hat{S}\hat{T})^\dagger\alpha|\beta\rangle\,. \tag{3.118}$$

In the middle equality, we treat $\hat{S}^\dagger|\alpha\rangle$ or the dual version $\langle\hat{S}^\dagger\alpha|$ as a new vector to which we then apply $\hat{T}^\dagger$. Since we first acted with $\hat{S}^\dagger$ on $|\alpha\rangle$ in the previous step, we obtain the reverse ordering.

A simple example of a linear transformation is given by scalar multiplication, $\hat{T}|\beta\rangle = c|\beta\rangle$ with a complex number c. Using

$$\langle\alpha|\hat{T}|\beta\rangle = c\langle\alpha|\beta\rangle = \langle\gamma|\beta\rangle \tag{3.119}$$

if $|\gamma\rangle = c^*|\alpha\rangle$, the Hermitian conjugate of $\hat{T}$ is given by $\hat{T}^\dagger|\alpha\rangle = c^*|\alpha\rangle$. If the vector space has finite dimension, a $\hat{T}$ of this form is represented by $c\mathbb{I}$ with the identity matrix $\mathbb{I}$.

The abstract definition of the Hermitian conjugate is directly related to the definition in terms of matrix elements if we use an orthonormal basis. In an orthonormal basis, the inner product of two vectors is given by their components according to $\langle\alpha|\beta\rangle = \alpha^\dagger\beta$, where $\alpha^\dagger$ is a row vector with complex conjugates of the components of α; see equation (3.47). We compute

$$\langle\alpha|\hat{T}|\beta\rangle = \alpha^\dagger T\beta = (T^\dagger\alpha)^\dagger\beta = \langle\hat{T}^\dagger\alpha|\beta\rangle \tag{3.120}$$

where $(TS)^\dagger = S^\dagger T^\dagger$ is valid also if S is a vector rather than a matrix. Therefore, if $\hat{T}$ is represented by the matrix T in an orthonormal basis, $\hat{T}^\dagger$ is represented by the matrix $T^\dagger$.

We call $\hat{T}$ a **Hermitian transformation** if it is linear and $\hat{T}^\dagger = \hat{T}$. Thus, $\langle\hat{T}\alpha|\beta\rangle = \langle\alpha|\hat{T}|\beta\rangle$ for all $|\alpha\rangle$ and $|\beta\rangle$ if $\hat{T}$ is Hermitian. Hermitian transformations are important because they have the following general properties in finite and infinite dimensions:

3.1.5.1 Real eigenvalues

All eigenvalues of a Hermitian transformation are real: If $|\alpha\rangle$ is an eigenvector of a Hermitian $\hat{T}$ with eigenvalue λ, we have

$$\langle\alpha|\hat{T}|\alpha\rangle = \langle\alpha|\lambda|\alpha\rangle = \lambda\langle\alpha|\alpha\rangle \tag{3.121}$$

using the eigenvector equation, and

$$\langle\alpha|\hat{T}|\alpha\rangle = \langle\hat{T}\alpha|\alpha\rangle = \langle\lambda\alpha|\alpha\rangle = \lambda^*\langle\alpha|\alpha\rangle \tag{3.122}$$

using the Hermitian property and again the eigenvector equation. Combining both equations and using $\langle\alpha|\alpha\rangle \neq 0$ because an eigenvector cannot be the null vector, we obtain $\lambda = \lambda^*$.

3.1.5.2 Orthogonal eigenvectors

For a Hermitian transformation, eigenvectors with distinct eigenvalues are orthogonal: If $|\alpha\rangle$ is an eigenvector with eigenvalue λ, and $|\beta\rangle$ is an eigenvector with eigenvalue μ such that $\lambda \neq \mu$, we have

$$\mu\langle\alpha|\beta\rangle = \langle\alpha|\hat{T}|\beta\rangle = \langle\hat{T}\alpha|\beta\rangle = \lambda^*\langle\alpha|\beta\rangle = \lambda\langle\alpha|\beta\rangle \tag{3.123}$$

using the first property, which guarantees real eigenvalues of a Hermitian transformation, in the last step. If $\lambda \neq \mu$, this equation is consistent only if $\langle\alpha|\beta\rangle = 0$.

3.1.5.3 Completeness

Every Hermitian transformation has an eigenvector basis and is therefore diagonalizable, but this is harder to prove.

3.2 FUNCTION SPACES

Infinite-dimensional vector spaces usually appear as function spaces, such as the wave functions of quantum mechanics. Since functions of a continuous variable require infinitely many values to be specified, they form an infinite-dimensional space. They do not represent directions in an intuitive sense, but they allow the same mathematical operations that we used to define a vector space in Section 3.1.1.

3.2.1 FUNCTIONS AS VECTORS

Addition and scalar multiplication are defined for functions, and they obey the conditions of commutativity, associativity and distributivity. There is a null vector given by a vanishing function, and an additive inverse for each function. Therefore, functions form a vector space. We use the notation $|f\rangle$ for a function viewed as an abstract vector which we can subject to mathematical vector operations. The vector is represented by the values of the function, such as $f(x)$, which we will identify as vector components in a specific basis. The interpretation as a vector allows us to perform a basis change for the representation of a function, and sometimes simplify calculations by doing so.

There is an inner product on a function space defined by

$$\langle f|g\rangle = \int_{-\infty}^{\infty} f(x)^* g(x) \mathrm{d}x \tag{3.124}$$

if we represent the abstract functions $|f\rangle$ and $|g\rangle$ by $f(x)$ and $g(x)$, respectively. This expression obeys all the conditions imposed on an inner product in Section 3.1.2. Since $f(x)$ are vector components in a basis, the form of the right-hand side of (3.124) and the integrand are basis-dependent. This inner product on a function space is a continuous analog of the expression

$$\langle\alpha|\beta\rangle = \sum_{i=1}^{n} \alpha_i^* \beta_i \tag{3.125}$$

in a finite-dimensional vector space, for which we have seen that it holds only if the basis is orthonormal. As we will see soon, the basis used to write the inner product (3.124) is also orthonormal.

More generally, a finite-dimensional inner product would be

$$\langle\alpha|\beta\rangle = \sum_{j=1}^{n}\sum_{i=1}^{n} K_{ij}\alpha_i^*\beta_j \tag{3.126}$$

if the basis $|e_i\rangle$ is not orthonormal but rather such that $\langle e_i|e_j\rangle = K_{ij}$. The new matrix K_{ij} is Hermitian, $K_{ij} = \langle e_i|e_j\rangle = \langle e_j|e_i\rangle^* = K_{ji}^*$, and obeys $K_{ii} > 0$. The continuous version of the inner product in a non-orthonormal basis looks like

$$\langle f|g\rangle = \int_{-\infty}^{\infty}\int_{-\infty}^{\infty} K(x,y)f(x)^*g(y)\mathrm{d}x\mathrm{d}y \tag{3.127}$$

with a so-called integral kernel $K(x,y)$. Conditions on the inner product require $K(x,y) = K(y,x)^*$ as well as $K(x,x) > 0$, but this leaves room for choices other than $K(x,y) = \delta(x-y)$ for an orthonormal basis. For the sake of convenience, we usually work with an orthonormal basis.

The inner product (3.124) on a function space is defined by an integral over an infinite range, which is not guaranteed to be finite for all functions. We therefore have to restrict the space to those functions for which the inner product gives us a finite value. This condition can be expressed more simply by requiring that all functions to be used in the function space have finite norm $||f|| = \sqrt{\langle f|f\rangle}$.

It might seem that the inner product $\langle f|g\rangle$, defined for two functions, is more general and less unique than the norm squared $||f||^2 = \langle f|f\rangle$ of a single function. However, the conditions on an inner product imply that it can be derived uniquely if we are able to compute the norms squared for all functions (as well as their sums and scalar multiples) in the function space: From

$$||f+g||^2 = \langle f+g|f+g\rangle = ||f||^2 + ||g||^2 + \langle f|g\rangle + \langle g|f\rangle = ||f||^2 + ||g||^2 + 2\mathrm{Re}(\langle f|g\rangle) \tag{3.128}$$

we can compute the real part of $\langle f|g\rangle = \langle g|f\rangle^*$ if we know the three norms squared $||f||^2$, $||g||^2$ and $||f+g||^2$. Similarly, from

$$||f+ig||^2 = \langle f+ig|f+ig\rangle = ||f||^2 + ||g||^2 + i\langle f|g\rangle - i\langle g|f\rangle = ||f||^2 + ||g||^2 - 2\mathrm{Im}(\langle f|g\rangle) \tag{3.129}$$

we can compute the imaginary part, and therefore the inner product is completely determined by the norms. In particular, all inner products in a function space are finite if all admissible functions have finite norm squared,

$$||f||^2 = \int_{-\infty}^{\infty} |f(x)|^2\mathrm{d}x < \infty. \tag{3.130}$$

As an example, functions of the form $p(x)\exp(-x^2/2)$ with a polynomial $p(x)$ have finite norm and therefore provide a function space with a meaningful inner product. We can determine an orthonormal basis for this space by using our solutions for bound states of the harmonic oscillator, which are of the same form up to a constant rescaling of x. Using the Hermite polynomials $H_n(x)$, an orthonormal basis for the inner product (3.124) is given by

$$e_n(x) = \frac{1}{\sqrt{2^n\sqrt{\pi}n!}}H_n(x)\exp(-x^2/2). \tag{3.131}$$

There are infinitely many basis elements $|e_n\rangle$ labeled by a non-negative integer $n = 0,1,2,\ldots$.

Completeness of the bound-state wave functions implies that every function that obeys the boundary conditions at infinity can be written as an infinite series

$$|f\rangle = \sum_{n=0}^{\infty} f_n|e_n\rangle \tag{3.132}$$

with complex coefficients

$$f_n = \langle e_n|f\rangle = \int_{-\infty}^{\infty} e_n(x)^* f(x)\mathrm{d}x\,. \tag{3.133}$$

Such a function is normalizable if

$$||f||^2 = \langle f|f\rangle = \sum_{n=0}^{\infty} |f_n|^2 \tag{3.134}$$

is finite, even if the f_n are non-zero for infinitely many n such that $f(x)$ is no longer a polynomial times $\exp(-x^2/2)$. The space of all normalizable functions in this space is therefore greater than the originally defined space, given by linear combinations of polynomials times $\exp(-x^2/2)$. Completeness in infinite dimensions often requires us to enlarge the original vector space. A vector space with an inner product that fulfills the finiteness and completeness conditions is called a **Hilbert space**.

As with finite-dimensional vector spaces, we can take tensor products of function spaces. The tensor product of function spaces may be interpreted as introducing additional functional dependence. For instance, if we take two copies of the same basis functions, $e_n(x)$ and $f_m(y) = e_m(y)$, we obtain two Hilbert spaces with bases $|e_n\rangle$ and $|f_m\rangle$, respectively. The tensor product of two functions is given by an abstract vector $|g\rangle \otimes |h\rangle$ that can be written as

$$|g\rangle \otimes |h\rangle = \left(\sum_{n=0}^{\infty} g_n|e_n\rangle\right) \otimes \left(\sum_{m=0}^{\infty} h_m|f_m\rangle\right) = \sum_{n=0}^{\infty}\sum_{m=0}^{\infty} g_n h_m |e_n\rangle \otimes |f_m\rangle\,. \tag{3.135}$$

However, a general two-index object need not be of the product form $g_n h_m$, as we saw by comparing dimensions in finite dimensions on page 94. Rather, the components F_{nm} of a vector $|F\rangle = \sum_{n,m=0}^{\infty} F_{nm}|e_n\rangle \otimes |f_m\rangle$ in the tensor-product space may belong to any 2-variable function $F(x,y)$ because we can unambiguously relate them to the inner products

$$F_{nm} = (\langle e_n| \otimes \langle f_m|)|F\rangle = \int_{-\infty}^{\infty}\int_{-\infty}^{\infty} e_n(x) f_m^*(y) F(x,y)\mathrm{d}x\mathrm{d}y\,. \tag{3.136}$$

Tensor products therefore introduce additional degrees of freedom in the states described by a function space. They can be used to formulate combined systems in which different subsystems may interact with one another.

3.2.2 OPERATORS AS LINEAR TRANSFORMATIONS

Operators previously defined on wave functions, such as $\hat{x}$, $\hat{p} = -i\hbar\mathrm{d}/\mathrm{d}x$ or the Hamilton operator $\hat{H}$, are examples of linear transformations $\hat{T}$ acting on vectors $|\psi\rangle$ in a function space. Their eigenvectors or eigenfunctions $|f_\lambda\rangle$ are defined by

$$\hat{T}|f_\lambda\rangle = \lambda|f_\lambda\rangle \tag{3.137}$$

for constant λ, provided $f_\lambda(x)$ is not zero for all x (such that $|f_\lambda\rangle$ is not the null vector). The mathematical form of this equation depends on the linear transformation $\hat{T}$ and on the basis representation used for $|f_\lambda\rangle$. So far, we have seen only the basis representation $f_\lambda(x)$ for $|f_\lambda\rangle$, which determines the equation $\hat{T}f_\lambda(x) = \lambda f_\lambda(x)$ we have to solve for eigenfunctions of a given $\hat{T}$. For instance, if $\hat{T} = \mathrm{d}/\mathrm{d}x$,

$$\hat{T}f_\lambda(x) = \frac{\mathrm{d}}{\mathrm{d}x}f_\lambda(x) = \lambda f_\lambda(x) \tag{3.138}$$

is solved by

$$f_\lambda(x) = A_\lambda \exp(\lambda x) \tag{3.139}$$

with a constant $A_\lambda \neq 0$ (which may depend on the value λ).

3.2.2.1 Hermitian and self-adjoint operators

Another example is the time-independent Schrödinger equation $\hat{H}\psi_n(x) = E_n\psi_n(x)$ which implies that stationary states $\psi_n(x)$ are eigenfunctions of the Hamilton operator $\hat{H}$ with eigenvalues E_n. As in this case, we are often interested in real eigenvalues because they may be used to describe measurement results. As we have seen in Section 3.1.5, the eigenvalues of a Hermitian transformation are always real. On a function space, we use the general definition

$$\langle f|\hat{T}|g\rangle = \langle \hat{T}^\dagger f|g\rangle \tag{3.140}$$

for all $|f\rangle$ and $|g\rangle$ to define the Hermitian conjugate $\hat{T}^\dagger$ of $\hat{T}$. Using the inner product (3.124), this condition looks like

$$\int_{-\infty}^{\infty} f(x)^*\hat{T}g(x)\mathrm{d}x = \int_{-\infty}^{\infty} (\hat{T}^\dagger f(x))^* g(x)\mathrm{d}x \tag{3.141}$$

for all $f(x)$ and $g(x)$. An operator is Hermitian if

$$\int_{-\infty}^{\infty} f(x)^*\hat{T}g(x)\mathrm{d}x = \int_{-\infty}^{\infty} (\hat{T} f(x))^* g(x)\mathrm{d}x \tag{3.142}$$

for all $f(x)$ and $g(x)$.

There is another subtlety of infinite-dimensional vector spaces. In general, we cannot apply an operator to all functions in a Hilbert space because, for instance, some functions may have finite norm but not be differentiable, or the derivative may not have finite norm. The momentum operator is therefore defined only on a subset of the Hilbert space of finite-norm functions. Similarly, there are finite-norm functions, such as the function defined by $f(x) = 1/x$ for $|x| > 1$ and $f(x) = x$ for $|x| \leq 1$, that no longer have finite norm after multiplying them by x. Nevertheless, we can find a Hilbert-space basis, such as polynomials times a Gaussian, on which both operators, $\hat{p}$ and $\hat{x}$, are well-defined. Operators for which this condition is fulfilled are **densely-defined**. They are not defined on all functions in the Hilbert space, but on a sufficiently large subset such that there are meaningful relationships between $\hat{x}$ and $\hat{p}$, including their commutator or their combination in the Hamilton operator of the harmonic oscillator. If it is not possible to find a basis such that a linear transformation can be applied to each basis vector, the transformation is not considered a valid operator on the Hilbert space. An example is the inverse of $\hat{H} - \frac{1}{2}\hbar\omega\hat{\mathbb{I}}$ with the harmonic-oscillator Hamilton operator $\hat{H}$.

For a densely-defined operator, such as $\hat{p}$, we may encounter additional difficulties related to the range of functions on which we can perform operations such as computing the Hermitian conjugate. Given an operator $\hat{T}$, the subset of functions $|f\rangle$ on which $\hat{T}^\dagger$ is well-defined may have to be restricted. As an example, consider the momentum operator $\hat{p} = (\hbar/i)\mathrm{d}/\mathrm{d}x$ defined as before, but acting on functions that depend only on positive x. We obtain a Hilbert space for such functions if we restrict integrations in the previous inner product to the positive range. This restriction implies that there are now new normalizable functions, such as $\exp(-x)$. Since the definition of the Hermitian conjugate makes use of a condition that refers to all normalizable functions in the Hilbert space, its properties may change if we consider an operator acting on functions defined on a restricted range.

The Hermitian conjugate of the momentum operator can be derived by integrating by parts. For functions on the full range of x, we compute

$$\begin{aligned}
\langle f|\hat{p}|g\rangle &= \int_{-\infty}^{\infty} f(x)^*\frac{\hbar}{i}\frac{\mathrm{d}g(x)}{\mathrm{d}x}\mathrm{d}x = \frac{\hbar}{i}\left(f(x)^*g(x)|_{-\infty}^{\infty} - \int_{-\infty}^{\infty}\frac{\mathrm{d}f(x)^*}{\mathrm{d}x}g(x)\mathrm{d}x\right)\\
&= \int_{-\infty}^{\infty}\left(\frac{\hbar}{i}\frac{\mathrm{d}f(x)}{\mathrm{d}x}\right)^* g(x)\mathrm{d}x = \int_{-\infty}^{\infty}(\hat{p}f(x))^*g(x)\mathrm{d}x = \langle \hat{p}f|g\rangle
\end{aligned} \tag{3.143}$$

for all $f(x)$ and $g(x)$ that obey the finite-norm condition, and therefore vanish at $x \to \pm\infty$. The momentum operator therefore equals its Hermitian conjugate on all admissible functions on the full real line.

For functions on the positive real line, by contrast, the boundary term in

$$\int_0^\infty f(x)^* \hat{p} g(x) \mathrm{d}x = \frac{\hbar}{i} \left(f(0)^* g(0) - \int_0^\infty (\mathrm{d}f(x)/\mathrm{d}x)^* g(x) \right) \mathrm{d}x \tag{3.144}$$

does not always vanish. We conclude that $\hat{p}^\dagger$ as the Hermitian conjugate of $\hat{p}$ is well-defined only if we impose the boundary condition $f(0) = 0$ for functions on which it acts. The original $\hat{p}$-operator, however, may be applied to functions that do not obey the boundary condition, as long as they are differentiable. While $\hat{p}^\dagger$ acts like $\hat{p}$ when it can be applied, it is not the same operator because it does not act on the same set of functions. To emphasize this possibility, we still call $\hat{p}$ Hermitian in this case, but we distinguish it from **self-adjoint operators** which equal their Hermitian conjugate on all functions on which the original operator can be applied. Any self-adjoint operator is Hermitian, but the reverse is not always correct. The momentum operator acting on functions on the full real line is self-adjoint, while the momentum operator acting on functions over the positive numbers is only Hermitian.

When we derived the result that Hermitian operators in finite dimensions have real eigenvalues, we used the equation $\langle \alpha | \hat{T} | \alpha \rangle = \langle \hat{T}^\dagger \alpha | \alpha \rangle$. This equation assumes that both $\hat{T}$ and $\hat{T}^\dagger$ can be applied to the same set of functions. In the infinite-dimensional case, we can use this property only if we know that the operator is self-adjoint and not just Hermitian. In fact, it is not difficult to find examples of Hermitian operators in infinite dimensions with complex eigenvalues. Returning to the momentum operator acting on functions over the positive numbers, $f(x) = \exp(-x/x_0)$ with some non-zero x_0 has finite norm because we do not include negative numbers in the normalization integral. Acting on this function, we have $\hat{p} f(x) = (i\hbar/x_0) f(x)$, which implies the imaginary eigenvalue $i\hbar/x_0$. This result may be surprising, but it is consistent with the theory of function spaces because $\hat{p}$ on functions over positive numbers is Hermitian and not self-adjoint.

With this additional ingredient, we know from the results of Section 3.1.5 that a self-adjoint operator has real eigenvalues with a complete set of eigenfunctions that are orthogonal for distinct eigenvalues.

We have seen several examples of Hermitian conjugates of operators as well as self-adjoint operators:

1. We have used the integral equation $\int_{-\infty}^{\infty} f(x)^* \hat{a}_- g(x) \mathrm{d}x = \int_{-\infty}^{\infty} (\hat{a}_+ f(x))^* g(x) \mathrm{d}x$; see (2.112), which implies that $\hat{a}_-^\dagger = \hat{a}_+$. Similarly, $\hat{a}_+^\dagger = \hat{a}_-$.
2. The momentum operator over the full real line $\hat{p}$ is self-adjoint, as shown in (3.143) The derivative operator in (3.138) is not Hermitian because we get a negative sign if we integrate by parts. Multiplying this operator by i or $1/i$ turns it into a Hermitian operator, which is shown to be self-adjoint by considering the finite-norm condition.
3. The position operator $\hat{x}$ is self-adjoint:

$$\langle f | \hat{x} | g \rangle = \int_{-\infty}^{\infty} f(x)^* x g(x) \mathrm{d}x = \int_{-\infty}^{\infty} (x f(x))^* g(x) \mathrm{d}x = \langle \hat{x} f | g \rangle \tag{3.145}$$

   for all $f(x)$ and $g(x)$ such that they retain finite norm after being multiplied by x.
4. Powers and sums of Hermitian operators are Hermitian. Therefore, any Hamilton operator

$$\hat{H} = \frac{1}{2m} \hat{p}^2 + V(\hat{x}) \tag{3.146}$$

   with real m and $V(x)$ is Hermitian. The question of whether a Hamilton operator is self-adjoint requires a dedicated analysis of its specific expression.
5. Using the relationship between the ladder operators $\hat{a}_\pm$ and the position and momentum operators, we can rederive their Hermitian conjugates by

$$\hat{a}_-^\dagger = \frac{1}{\sqrt{2m\hbar\omega}} \left((i\hat{p})^\dagger + m\omega\hat{x} \right) = \frac{1}{\sqrt{2m\hbar\omega}} \left(\hat{p}^\dagger i^\dagger + m\omega\hat{x} \right)$$

$$= \frac{1}{\sqrt{2m\hbar\omega}}(-i\hat{p}+m\omega\hat{x}) = \hat{a}_+ . \tag{3.147}$$

In the first step, we view i as a linear transformation that multiplies wave functions by i. Therefore, $i^\dagger = -i$.

3.2.2.2 Spectral properties

The position and momentum operators over all real numbers are self-adjoint and should have real eigenvalues. We derive the range of these values and corresponding eigenfunctions by solving eigenfunction equations. For the momentum operator, this equation is

$$\hat{p}f_\lambda(x) = \frac{\hbar}{i}\frac{\mathrm{d}}{\mathrm{d}x}f_\lambda(x) = \lambda f_\lambda(x) \tag{3.148}$$

and solved by

$$f_\lambda(x) = A_\lambda \exp(i\lambda x/\hbar) \tag{3.149}$$

for a given λ, with a constant $A_\lambda \neq 0$. However, none of these functions are normalizable: We have

$$|f_\lambda(x)|^2 = |A_\lambda|^2|\exp(ix(\mathrm{Re}(\lambda)+i\mathrm{Im}(\lambda))/\hbar)|^2 = |A_\lambda|^2\exp(-2x\mathrm{Im}(\lambda)/\hbar) \tag{3.150}$$

and therefore

$$||f_\lambda||^2 = \int_{-\infty}^{\infty}|f_\lambda|^2\mathrm{d}x = \infty \tag{3.151}$$

for any complex number λ.

The position operator has the eigenfunction equation

$$\hat{x}g_\lambda(x) = xg_\lambda(x) = \lambda g_\lambda(x) . \tag{3.152}$$

This is an unusual equation which requires a solution $g_\lambda(x)$ on which multiplication by the variable x is the same as multiplication by a constant, λ. We first check whether a solution can have a Taylor expansion around $x = 0$, by assuming that $g_\lambda(x)$ is of the form $g_\lambda(x) = a_0 + a_1x + a_2x^2 + \cdots$ with some complex coefficients a_i. The resulting eigenfunction equation

$$x(a_0 + a_1x + a_2x^2 + \cdots) = \lambda(a_0 + a_1x + a_2x^2 + \cdots) \tag{3.153}$$

can be solved by comparing coefficients in the two power series. We obtain an infinite set of linear equations,

$$\lambda a_0 = 0 \quad , \quad \lambda a_1 = a_0 \quad , \quad \lambda a_2 = a_1 \quad , \quad \cdots \tag{3.154}$$

or in general $\lambda a_i = a_{i-1}$ for $i = 0, 1, 2, \ldots$ if we define $a_{-1} = 0$. We now have two cases, depending on whether λ is zero or non-zero. If $\lambda = 0$, the first equation in (3.154) is always valid, the second equation gives us $0 = a_0$, which implies $0 = a_1$ in the next equation and so on such that $a_i = 0$ for all i. If $\lambda \neq 0$, the first equation in (3.154) implies $a_0 = 0$, which then means $a_1 = 0$ from the second equation, and in general $a_i = 0$ for all i. In both cases, we obtain the function $g_\lambda(x) = 0$ as a solution, but the null vector is not allowed as an eigenvector or eigenfunction. Therefore, eigenfunctions of the position operator do not have a Taylor expansion around $x = 0$.

We can see how the eigenfunction equation (3.152) may be solved by looking at positions x such that $g_\lambda(x) \neq 0$. Such positions must exist because an eigenfunction cannot be completely zero. Dividing (3.152) by $g_\lambda(x)$ at values of x where the function does not vanish, we obtain $x = \lambda$. Since λ is a constant, this result implies that $g_\lambda(x)$ can be non-zero only at one point, $x = \lambda$. Such a function is given by the delta function shifted by λ, and indeed

$$g_\lambda(x) = B_\lambda \delta(x-\lambda) \tag{3.155}$$

with a constant $B_\lambda \neq 0$ solves (3.152) as an example of the general equation

$$f(x)\delta(x-x_0) = f(x_0)\delta(x-x_0) \tag{3.156}$$

derived in (2.186). However, the delta function is not normalizable because

$$\int_{-\infty}^{\infty} |\delta(x-\lambda)|^2 \mathrm{d}x = \int_{-\infty}^{\infty} \delta(x-\lambda)\delta(x-\lambda)\mathrm{d}x = \delta(0) = \infty \tag{3.157}$$

if we use (2.186) with $f(x) = \delta(x-\lambda)$.

We conclude that self-adjoint operators in a function space do not always have normalizable eigenfunctions. Completeness of eigenfunctions therefore looks difficult, even while our other results remain true: the eigenvalues of a self-adjoint transformation are always real, and eigenfunctions with distinct eigenvalues are orthogonal. Another example is given by the time-independent Schrödinger equation $\hat{H}\psi_n(x) = E_n\psi_n(x)$ with solutions given by stationary states $\psi_n(x)$ with real energy eigenvalues E_n. If all stationary states are bound states, they are normalizable and provide a complete set of normalizable eigenfunctions, or an eigenfunction basis. Moreover, they are orthogonal for distinct eigenvalues E_n. However, if there are scattering states, they are non-normalizable and a restriction to normalizable eigenfunctions does not give us a complete set or a basis. As we have seen in the example of the free particle, non-normalizable eigenfunctions of $\hat{H}$ are required for completeness, which in this case can be shown explicitly using Fourier transformation.

An indication of a useful treatment of non-normalizable eigenfunctions can be obtained by looking at orthogonality of eigenfunctions for distinct eigenvalues. Returning to our eigenfunctions of position and momentum, we have

$$\int_{-\infty}^{\infty} g_\lambda(x)^* g_\mu(x)\mathrm{d}x = B_\lambda^* B_\mu \int_{-\infty}^{\infty} \delta(x-\lambda)\delta(x-\mu)\mathrm{d}x = B_\lambda^* B_\mu \delta(\mu-\lambda) = |B_\mu|^2\delta(\mu-\lambda) \tag{3.158}$$

using (2.186) twice, once for $f(x)\delta(x-x_0) = \delta(x-\lambda)\delta(x-\mu)$ and once for $f(\lambda)\delta(\lambda-\lambda_0) = B_\lambda^* B_\mu \delta(\lambda-\mu)$. For $\lambda = \mu$, we obtain our previous non-normalizability property because $\delta(0) = \infty$. For $\lambda \neq \mu$ we obtain the finite value $\delta(\mu-\lambda) = 0$, which shows orthogonality.

Similarly, for momentum eigenfunctions we have

$$\int_{-\infty}^{\infty} f_\lambda(x)^* f_\mu(x)\mathrm{d}x = A_\lambda^* A_\mu \int_{-\infty}^{\infty} \exp(-i(\lambda-\mu)x/\hbar)\mathrm{d}x = \begin{cases} 0 & \text{if } \lambda \neq \mu \\ \infty & \text{if } \lambda = \mu \end{cases} \tag{3.159}$$

where the zero value for $\lambda \neq \mu$ is suggested by the oscillations of sine and cosine in the complex exponential function canceling out when they are integrated over an infinite range. If $\lambda = \mu$, however, we integrate the constant function one over an infinite range, which is infinite. The result therefore looks like the delta function, or is at least proportional to the delta function.

A precise expression is given by

$$\int_{-\infty}^{\infty} f_\lambda(x)^* f_\mu(x)\mathrm{d}x = 2\pi\hbar|A_\lambda|^2\delta(\mu-\lambda) \tag{3.160}$$

if we use Fourier transformation, which explains the factor of 2π: The Fourier transform

$$\tilde{F}(k) = \frac{1}{\sqrt{2\pi}}\int_{-\infty}^{\infty} F(y)\exp(-iky)\mathrm{d}y \tag{3.161}$$

and its inverse

$$F(y) = \frac{1}{\sqrt{2\pi}}\int_{-\infty}^{\infty} \tilde{F}(k)\exp(iky)\mathrm{d}k \tag{3.162}$$

can be applied to the delta function. The Fourier transform of the delta function is constant:

$$\tilde{\delta}(k) = \frac{1}{\sqrt{2\pi}}\int_{-\infty}^{\infty} \delta(y)\exp(-iky)\mathrm{d}y = \frac{1}{\sqrt{2\pi}}\exp(ik\cdot 0) = \frac{1}{\sqrt{2\pi}}\,. \tag{3.163}$$

The delta function is therefore equal to the inverse Fourier transform of $1/\sqrt{2\pi}$, which implies the integral formula

$$\delta(y) = \frac{1}{\sqrt{2\pi}} \int_{-\infty}^{\infty} \frac{1}{\sqrt{2\pi}} \exp(iky)\mathrm{d}k = \frac{1}{2\pi} \int_{-\infty}^{\infty} \exp(iky)\mathrm{d}k. \tag{3.164}$$

The final integral in (3.159) is obtained from (3.164) if we substitute $k = x/\hbar$ and change the free variable y to $\mu - \lambda$:

$$\int_{-\infty}^{\infty} \exp(i(\mu-\lambda)x/\hbar)\mathrm{d}x = \hbar \int_{-\infty}^{\infty} \exp(i(\mu-\lambda)k)\mathrm{d}k = 2\pi\hbar\delta(\mu-\lambda). \tag{3.165}$$

Both the position and momentum eigenfunctions give us a delta function for the inner product of two eigenfunctions. This result is a continuous version of the orthonormality condition

$$\langle \psi_m | \psi_n \rangle = \delta_{mn} \tag{3.166}$$

for normalizable wave functions. In our results for f_λ and g_λ, we did not obtain just the delta function but rather the delta function times $|B_\lambda|^2$ and $2\pi\hbar|A_\lambda|^2$, respectively. We set these coefficients equal to one by choosing the free constants A_λ and B_λ as $A_\lambda = 1/\sqrt{2\pi\hbar}$ and $B_\lambda = 1$, respectively. The eigenfunctions are then

$$g_\lambda(x) = \delta(x-\lambda) \tag{3.167}$$

for $\hat{x}$ and

$$f_\lambda(x) = \frac{1}{\sqrt{2\pi\hbar}} \exp(i\lambda x/\hbar) \tag{3.168}$$

for $\hat{p}$, such that

$$\langle f_\lambda | f_\mu \rangle = \delta(\lambda-\mu) \quad \text{and} \quad \langle g_\lambda | g_\mu \rangle = \delta(\lambda-\mu). \tag{3.169}$$

The eigenfunctions remain non-normalizable, but by requiring (3.169) to hold with coefficients one of the delta function, mimicking (3.166), we have fixed the free constants A_λ and B_λ. This outcome is exactly what (3.166) does in the case of normalizable eigenfunctions of stationary states. We can therefore consider (3.169) as a normalization condition for non-normalizable eigenfunctions, which is called **delta-function normalization**. This condition does not turn non-normalizable wave functions into normalizable ones as determined by the Hilbert-space norm, but it helps us to fix the values of free parameters in the solutions of eigenfunction equations.

For stationary states, we saw that there is a correspondence between normalizability of eigenfunctions and discreteness of the eigenvalues or, in this case, the energies. The same mathematical property is realized for any linear transformation on a function space. The set of all eigenvalues, or the **spectrum**, in general has two different components, a continuous spectrum from non-normalizable eigenfunctions and a discrete spectrum from normalizable ones. Some linear transformations have only a discrete part of the spectrum, and some have only a continuous part, but there are also mixed cases that have discrete and continuous parts for the same operator. An example is the Hamilton operator $\hat{H}$ for a potential that has negative values and is bounded from above such that $\lim_{x\to\pm\infty} V(x) = 0$. There are then bound states with discrete negative energies and scattering states with continuous positive energies for the same Hamilton operator.

With the new concept of delta-function normalization, we have the following properties of self-adjoint operators on a function space:

1. All eigenvalues are real.
2. The eigenfunctions are orthonormal if delta-function normalization is used in the continuous part of the spectrum.

3. The eigenfunctions are complete. Every vector $|\alpha\rangle$ in the function space can be written as

$$|\alpha\rangle = \sum_n \alpha_n |f_n\rangle + \int \alpha_\lambda |f_\lambda\rangle \mathrm{d}\lambda \tag{3.170}$$

if $|f_n\rangle$ are normalizable eigenfunctions in a certain discrete range of values n, and $|f_\lambda\rangle$ are non-normalizable eigenfunctions in the continuous range of eigenvalues λ.

3.2.2.3 Basis change

As usual, completeness is hard to prove in general, but we can check it explicitly in the two cases of the position and momentum eigenfunctions. Each case has a completely continuous spectrum with eigenfunctions $|f_\lambda\rangle$ for $\hat{p}$ and $|g_\lambda\rangle$ for $\hat{x}$, respectively. We are therefore looking for expressions of general functions $|\alpha\rangle$ and $|\beta\rangle$ in the form

$$|\alpha\rangle = \int_{-\infty}^{\infty} \alpha_\lambda |f_\lambda\rangle \mathrm{d}\lambda \quad \text{and} \quad |\beta\rangle = \int_{-\infty}^{\infty} \beta_\lambda |g_\lambda\rangle \mathrm{d}\lambda \tag{3.171}$$

with suitable functions α_λ and β_λ of λ that correspond to components of an infinite-dimensional vector in an orthonormal basis. Writing these vectors as functions, we have

$$\alpha(x) = \int_{-\infty}^{\infty} \alpha_\lambda f_\lambda(x) \mathrm{d}\lambda \quad \text{and} \quad \beta(x) = \int_{-\infty}^{\infty} \beta_\lambda g_\lambda(x) \mathrm{d}\lambda\,. \tag{3.172}$$

Orthonormality of basis vectors always simplifies the derivation of expansion coefficients, such as α_λ and β_λ: We have

$$\langle f_\mu|\alpha\rangle = \int_{-\infty}^{\infty} \alpha_\lambda \langle f_\mu|f_\lambda\rangle \mathrm{d}\lambda = \int_{-\infty}^{\infty} \alpha_\lambda \delta(\mu-\lambda) \mathrm{d}\lambda = \alpha_\mu \tag{3.173}$$

and a similar version for β_μ using $|g_\mu\rangle$.

We now have two sets of numbers that represent the abstract vector $|\alpha\rangle$: the values $\alpha(x)$ and the components α_μ. These two sets correspond to two different basis choices. The new components α_μ just derived in (3.173) are the components of $|\alpha\rangle$ in the orthonormal basis $|f_\mu\rangle$. We can see which basis the values $\alpha(x)$ refer to by applying the general equation (3.173) to the eigenfunction basis of $\hat{x}$, given by $g_\mu(x) = \delta(x-\mu)$. The components of $|\alpha\rangle$ in the $\hat{x}$-eigenbasis are then given by

$$\alpha_\mu^{(x)} = \langle g_\mu|\alpha\rangle = \int_{-\infty}^{\infty} \delta(x-\mu)\alpha(x) \mathrm{d}x = \alpha(\mu)\,, \tag{3.174}$$

which is just the original function α applied to μ. When we wrote $\alpha(x)$ to define a function of x, we were thus referring to the components of an abstract vector $|\alpha\rangle$ in the orthonormal basis $g_\mu(x)$ of eigenfunctions of $\hat{x}$, or in the **position eigenbasis**. For convenience, we write the eigenfunctions of the position operator simply as $|x\rangle$ if $\hat{x}|x\rangle = x|x\rangle$. Therefore,

$$\alpha(x) = \langle x|\alpha\rangle\,. \tag{3.175}$$

Equation (3.173) then transforms this representation to a new orthonormal basis $|f_\mu\rangle$, such as the momentum eigenfunctions. For this example, we compute

$$\alpha_\lambda^{(p)} = \langle f_\lambda|\alpha\rangle = \frac{1}{\sqrt{2\pi\hbar}} \int_{-\infty}^{\infty} \exp(-i\lambda x/\hbar)\alpha(x) \mathrm{d}x = \frac{1}{\sqrt{\hbar}} \tilde{\alpha}(\lambda/\hbar)\,. \tag{3.176}$$

The Fourier transform of $\alpha(x)$ therefore represents the abstract vector $|\alpha\rangle$ in the orthonormal basis of momentum eigenfunctions, or the **momentum eigenbasis**. Fourier transformation is just a basis change in an infinite-dimensional vector space.

In quantum mechanics, we started with position-dependent wave functions $\psi(x)$, which we now recognize as the components of an abstract vector $|\psi\rangle$ in the orthonormal basis of position eigenfunctions. The same abstract vector can be represented in the orthonormal basis of momentum eigenfunctions, in which case the components imply a momentum-dependent wave function

$$\phi(p) = \frac{1}{\sqrt{2\pi\hbar}} \int_{-\infty}^{\infty} \psi(x) \exp(-ipx/\hbar) \mathrm{d}x\,. \tag{3.177}$$

This wave function depends on momentum eigenvalues p, just as the original wave function $\psi(x)$ depends on position eigenvalues x. It defines a probability density $|\phi(p)|^2$ for momentum measurements. If we simplify the notation of momentum eigenfunctions to $|p\rangle$ if $\hat{p}|p\rangle = p|p\rangle$, we have

$$\phi(p) = \langle p|\psi\rangle\,. \tag{3.178}$$

Another example is given by stationary states $\psi_n(x)$ of a bound-state potential such as the harmonic oscillator, which form an orthonormal basis of normalizable eigenfunctions. The wave function in this basis is not a continuous function because it is defined only on the label n of eigenvalues E_n of the Hamilton operator, which form a discrete set. Completeness tells us that any wave function $\psi(x)$ can be written as

$$\psi(x) = \sum_{n=0}^{\infty} c_n \psi_n(x) \tag{3.179}$$

with complex coefficients c_n. These coefficients, interpreted as a discrete function defined on the non-negative integers, are the components of the abstract vector $|\psi\rangle$ in the orthonormal basis of energy eigenfunctions with a bound-state potential. Writing the stationary states as $|n\rangle$, a wave function in the energy eigenbasis is determined by the coefficients $c_n = \langle n|\psi\rangle$.

3.2.3 UNCERTAINTY RELATIONS AND COMMUTATORS

Given two self-adjoint operators $\hat{A}$ and $\hat{B}$, the commutator

$$[\hat{A},\hat{B}] = \hat{A}\hat{B} - \hat{B}\hat{A} \tag{3.180}$$

is in general non-zero. For instance, for position and momentum we have already computed the commutator $[\hat{x},\hat{p}] = i\hbar$. In quantum mechanics, the main implication of non-commutativity is given by uncertainty relations, which place positive lower bounds on the products $\sigma_A^2\sigma_B^2$ of the variances for any pair of non-commuting operators.

3.2.3.1 Uncertainty relations

It is possible to write the variance σ_A^2 as the norm squared of a certain vector obtained by applying the operator $\hat{A}$ to a generic state, $|\Psi\rangle$. Products of vector norms are bounded from below by the Cauchy–Schwarz inequality (3.15), which can be translated into uncertainty relations for a given pair of operators $\hat{A}$ and $\hat{B}$. Using (1.59), we first write

$$\sigma_A^2 = \langle\Psi|(\hat{A} - \langle\hat{A}\rangle)^2|\Psi\rangle = \langle(\hat{A} - \langle\hat{A}\rangle)\Psi|(\hat{A} - \langle\hat{A}\rangle)|\Psi\rangle = \langle f|f\rangle \tag{3.181}$$

where we used the condition that $\hat{A}$, and therefore $\hat{A} - \langle\hat{A}\rangle$ (or, more precisely, $\hat{A} - \langle\hat{A}\rangle\hat{\mathbb{I}}$), are self-adjoint operators, and defined the new state

$$|f\rangle = (\hat{A} - \langle\hat{A}\rangle)|\Psi\rangle\,. \tag{3.182}$$

Similarly, we obtain

$$\sigma_B^2 = \langle g|g\rangle \quad \text{with} \quad |g\rangle = (\hat{B} - \langle\hat{B}\rangle)|\Psi\rangle\,. \tag{3.183}$$

The Cauchy–Schwarz inequality implies

$$\sigma_A^2\sigma_B^2 = \langle f|f\rangle\langle g|g\rangle \geq |\langle f|g\rangle|^2 = \mathrm{Re}(\langle f|g\rangle)^2 + \mathrm{Im}(\langle f|g\rangle)^2 \tag{3.184}$$

with a lower bound given by the sum of two non-negative terms. The lower bound is therefore greater than or equal to either one of them, in particular

$$\sigma_A^2\sigma_B^2 \geq \mathrm{Im}(\langle f|g\rangle)^2 \tag{3.185}$$

which turns out to be the more interesting case. In order to compute this lower bound in terms of properties of $\hat{A}$ and $\hat{B}$, we derive

$$\begin{aligned}\langle f|g\rangle &= \langle(\hat{A}-\langle\hat{A}\rangle)\Psi|(\hat{B}-\langle\hat{B}\rangle)|\Psi\rangle = \langle\Psi|(\hat{A}-\langle\hat{A}\rangle)(\hat{B}-\langle\hat{B}\rangle)|\Psi\rangle\\ &= \langle\Psi|(\hat{A}\hat{B}-\hat{A}\langle\hat{B}\rangle-\hat{B}\langle\hat{A}\rangle+\langle\hat{A}\rangle\langle\hat{B}\rangle)|\Psi\rangle = \langle\hat{A}\hat{B}\rangle-\langle\hat{A}\rangle\langle\hat{B}\rangle\end{aligned} \tag{3.186}$$

and similarly

$$\langle g|f\rangle = \langle\hat{B}\hat{A}\rangle - \langle\hat{A}\rangle\langle\hat{B}\rangle \tag{3.187}$$

by swapping $\hat{A}$ and $\hat{B}$. In this derivation, note that $\langle\Psi|\hat{A}\langle\hat{B}\rangle|\Psi\rangle = \langle\hat{B}\rangle\langle\Psi|\hat{A}|\Psi\rangle = \langle\hat{A}\rangle\langle\hat{B}\rangle$ because we can factor out the number $\langle\hat{B}\rangle$.

The commutator now appears in

$$\mathrm{Im}(\langle f|g\rangle) = \frac{1}{2i}(\langle f|g\rangle - \langle f|g\rangle^*) = \frac{1}{2i}(\langle f|g\rangle - \langle g|f\rangle) = \frac{1}{2i}\langle\hat{A}\hat{B}-\hat{B}\hat{A}\rangle = \frac{1}{2i}\langle[\hat{A},\hat{B}]\rangle\,, \tag{3.188}$$

and therefore the uncertainty relation for the pair $(\hat{A},\hat{B})$ reads

$$\sigma_A^2\sigma_B^2 \geq \left(\frac{1}{2i}\langle[\hat{A},\hat{B}]\rangle\right)^2. \tag{3.189}$$

As an example, we derive the position-momentum uncertainty relation from $[\hat{x},\hat{p}] = i\hbar$:

$$\sigma_x^2\sigma_p^2 \geq \frac{\hbar^2}{4}. \tag{3.190}$$

In this case, the lower bound is a state-independent constant, which is not the case in general. For instance, if we replace p with the kinetic-energy operator $\hat{T} = (2m)^{-1}\hat{p}^2$, we compute $[\hat{x},\hat{T}] = i\hbar m^{-1}\hat{p}$ and therefore $\sigma_x^2\sigma_T^2 \geq \frac{1}{4}(\hbar/m)^2\langle\hat{p}\rangle^2$ with a state-dependent lower bound.

The lower bound in (3.189) is sharp (such that it may be achieved by some states) only if $\mathrm{Re}(\langle f|g\rangle)^2 = 0$. More generally, this real part provides another positive contribution to the lower bound, which we compute as

$$\mathrm{Re}(\langle f|g\rangle) = \frac{1}{2}(\langle f|g\rangle + \langle g|f\rangle) = \frac{1}{2}\langle\hat{A}\hat{B}+\hat{B}\hat{A}\rangle - \langle\hat{A}\rangle\langle\hat{B}\rangle = C_{AB}. \tag{3.191}$$

This expression is always state-dependent and equals the covariance C_{AB} of the operators $\hat{A}$ and $\hat{B}$ in a given state, which we already encountered in (1.128) for the pair $(\hat{x},\hat{p})$. The covariance is another statistical quantity, in addition to the two variances, that tells us how much the measurement results for $\hat{A}$ and $\hat{B}$ are correlated. For instance, if position and momentum measurements are correlated, it may be more likely to find a positive momentum rather than a negative momentum to the right of $\langle x\rangle$. If the covariance is non-zero, we can sharpen the uncertainty relation to

$$\sigma_A^2\sigma_B^2 - C_{AB}^2 \geq \left(\frac{1}{2i}\langle[\hat{A},\hat{B}]\rangle\right)^2. \tag{3.192}$$

An example of a state in which position and momentum are correlated is given by a Gaussian such as

$$\Psi(x) = (\mathrm{Re}(\lambda)/\pi)^{1/4} e^{-\frac{1}{2}\lambda x^2} \tag{3.193}$$

with a complex parameter λ. The imaginary part of λ then determines C_{xp}. A Gaussian with real λ and therefore zero C_{xp} saturates the uncertainty relation (3.189), while a general Gaussian with complex λ and non-zero C_{xp} only saturates the sharpened version, (3.192). For an example, see the properties of the free particle derived after (1.128).

The assumption that $\hat{A}$ and $\hat{B}$ are self-adjoint is not strictly necessary for the derivation of uncertainty relations. However, instead of bounding the product of variances, the general case implies inequalities for statistical quantities such as the covariance $C_{AA^\dagger}$ of $\hat{A}$ and its Hermitian conjugate $\hat{A}^\dagger$.

3.2.3.2 Commutators

We have to compute a commutator in order to derive the uncertainty relation for a given pair of operators. Several identities can be used to simplify the required computations:

Antisymmetry: $[\hat{A},\hat{B}] = -[\hat{B},\hat{A}]$ follows directly from the definition of the commutator. This identity also implies that the commutator of two Hermitian operators is anti-Hermitian because $[\hat{A},\hat{B}]^\dagger = [\hat{B},\hat{A}] = -[\hat{A},\hat{B}]$. Multiplication by i can be used to obtain the Hermitian expression $i[\hat{A},\hat{B}]$ that appears in the lower bound of uncertainty relations.

Product rule: The commutator obeys a version of the product rule if there is a product of two (or more) operators on one side. By adding and subtracting $\hat{A}\hat{C}\hat{B}$ in

$$\begin{aligned} [\hat{A}\hat{B},\hat{C}] &= (\hat{A}\hat{B})\hat{C} - \hat{C}(\hat{A}\hat{B}) = \hat{A}\hat{B}\hat{C} - \hat{A}\hat{C}\hat{B} + \hat{A}\hat{C}\hat{B} - \hat{C}\hat{A}\hat{B} \\ &= \hat{A}[\hat{B},\hat{C}] + [\hat{A},\hat{C}]\hat{B} \end{aligned} \tag{3.194}$$

we are able to factor out $\hat{A}$ to the left in the first two terms, and $\hat{B}$ to the right in the next two terms. As a result, the commutator with $\hat{C}$ is taken individually for each factor in $\hat{A}\hat{B}$, which can be used to break down complicated commutators into simpler expressions. We have to keep the correct ordering with $\hat{A}$ on the left and $\hat{B}$ on the right in each term. A similar version of the product rule holds if there is a product on the right, which can be derived from the previous result by using antisymmetry of the commutator.

Derivatives: The product rule can be applied to each term in a polynomial or power series $f(\hat{A}) = \sum_{n=0}^{\infty} a_n \hat{A}^n$ in the commutator, such as $[f(\hat{x}),\hat{p}]$. In this example, the procedure results in many terms with products of $\hat{x}$ and the simple commutators $[\hat{x},\hat{p}] = i\hbar$, which can be shown to add up to

$$[f(\hat{x}),\hat{p}] = i\hbar f'(\hat{x}) \tag{3.195}$$

where $f'(x) = \sum_{n=0}^{\infty} n a_n \hat{x}^{n-1}$ is obtained from the derivative of the power series. A simpler derivation makes use of the position eigenbasis, in which $f(\hat{x}) = f(x)$ is a multiplication operator, and $\hat{p} = (\hbar/i)\mathrm{d}/\mathrm{d}x$. The commutator is then derived from

$$[f(\hat{x}),\hat{p}]\psi(x) = \frac{\hbar}{i}\left(f(x)\frac{\mathrm{d}\psi}{\mathrm{d}x} - \frac{\mathrm{d}(f\psi)}{\mathrm{d}x}\right) = i\hbar\frac{\mathrm{d}f}{\mathrm{d}x}\psi \tag{3.196}$$

acting on a generic wave function $\psi(x)$. Similarly, we obtain

$$[\hat{x},g(\hat{p})] = i\hbar g'(\hat{p}) \tag{3.197}$$

for any power series $g(\hat{p})$ if we use operators in the momentum eigenbasis.

Commutators with a Hamilton operator $\hat{H}$: A specific commutator appears in the time derivative of the expectation value of some operator $\hat{A}$ in a time-dependent state $\Psi(x,t)$ according to

$$\begin{aligned}\frac{\mathrm{d}\langle\hat{A}\rangle}{\mathrm{d}t} &= \frac{\mathrm{d}}{\mathrm{d}t}\langle\Psi|\hat{A}|\Psi\rangle = \left\langle\frac{\partial\Psi}{\partial t}\middle|\hat{A}\middle|\Psi\right\rangle + \left\langle\Psi\middle|\hat{A}\middle|\frac{\partial\Psi}{\partial t}\right\rangle = -\frac{1}{i\hbar}\langle\hat{H}\Psi|\hat{A}|\Psi\rangle + \frac{1}{i\hbar}\langle\Psi|\hat{A}\hat{H}|\Psi\rangle \\ &= \frac{1}{i\hbar}\left(-\langle\Psi|\hat{H}\hat{A}|\Psi\rangle + \langle\Psi|\hat{A}\hat{H}|\Psi\rangle\right) = \frac{i}{\hbar}\langle[\hat{H},\hat{A}]\rangle \end{aligned} \tag{3.198}$$

using the time-dependent Schrödinger equation $\hat{H}|\Psi\rangle = i\hbar\partial|\Psi\rangle/\partial t$ and Hermiticity of $\hat{H}$. If $\hat{A}$ has an explicit time dependence, we have to add a contribution of $\langle\partial\hat{A}/\partial t\rangle$.

The last result allows us to rederive the Ehrenfest equations (1.98) and (1.117):

$$\frac{\mathrm{d}\langle x\rangle}{\mathrm{d}t} = \frac{i}{\hbar}\left\langle\left[\frac{\hat{p}^2}{2m}+V(\hat{x}),\hat{x}\right]\right\rangle = \frac{\langle p\rangle}{m} \tag{3.199}$$

from $[\hat{p}^2,\hat{x}] = -2i\hbar\hat{p}$, and

$$\frac{\mathrm{d}\langle p\rangle}{\mathrm{d}t} = \frac{i}{\hbar}\left\langle\left[\frac{\hat{p}^2}{2m}+V(\hat{x}),\hat{p}\right]\right\rangle = -\langle V'(x)\rangle \tag{3.200}$$

from $[V(\hat{x}),\hat{p}] = i\hbar V'(\hat{x})$, using (3.196). Also the equations (1.127), (1.133) and (1.134), as well as versions with higher powers of $\hat{x}$ and $\hat{p}$, can now be derived more easily by using commutators.

3.2.3.3 Poisson bracket

The derivative properties $[f(\hat{x}),\hat{p}] = i\hbar f'(\hat{x})$ and $[\hat{x},g(\hat{p})] = i\hbar g'(\hat{p})$ have a classical analog given by the Poisson bracket $\{f(x,p),g(x,p)\}$ of two functions of x and p. The result of the bracket operation is defined as another function of x and p such that

$$\{f(x,p),g(x,p)\} = \frac{\partial f}{\partial x}\frac{\partial g}{\partial p} - \frac{\partial f}{\partial p}\frac{\partial g}{\partial x}. \tag{3.201}$$

Just like the quantum commutator, the classical Poisson bracket is also antisymmetric, obeys the product rule because it involves first-order derivatives, and implies $\{f(x),p\} = f'(x)$ as well as $\{x,g(p)\} = g'(p)$. The commutator therefore resembles the Poisson bracket except for the additional factor of $i\hbar$. It is also possible to confirm the **Jacobi identity**

$$\{f,\{g,h\}\} + \{g,\{h,f\}\} + \{h,\{f,g\}\} = 0 \tag{3.202}$$

(cyclic permutations) for any functions f, g and h, as well as the analogous equation

$$[\hat{A},[\hat{B},\hat{C}]] + [\hat{B},[\hat{C},\hat{A}]] + [\hat{C},[\hat{A},\hat{B}]] = 0 \tag{3.203}$$

for the double commutator. An antisymmetric composition law that obeys the Jacobi identity defines a **Lie algebra**.

The Poisson bracket of different versions of $f(x,p)$ with the Hamilton function of a system is of interest in a general formulation of classical dynamics. For instance, we obtain

$$\{x,H\} = \frac{\partial H}{\partial p} \quad , \quad \{p,H\} = -\frac{\partial H}{\partial x}. \tag{3.204}$$

According to Hamilton's equations in classical mechanics, we can write

$$\frac{\mathrm{d}x}{\mathrm{d}t} = \{x,H\} \quad , \quad \frac{\mathrm{d}p}{\mathrm{d}t} = \{p,H\} \tag{3.205}$$

or, for a general function $f(x,p)$,

$$\frac{\mathrm{d}f}{\mathrm{d}t} = \frac{\mathrm{d}x}{\mathrm{d}t}\frac{\partial f}{\partial x} + \frac{\mathrm{d}p}{\mathrm{d}t}\frac{\partial f}{\partial p} = \frac{\partial H}{\partial p}\frac{\partial f}{\partial x} - \frac{\partial H}{\partial x}\frac{\partial f}{\partial p} = \{f,H\}. \tag{3.206}$$

This equation again resembles a commutator relationship, (3.198), except for a factor of $i\hbar$.

The Ehrenfest equations can be derived from a Poisson bracket for expectation values of $\hat{x}$ and $\hat{p}$ as well as their moments. We define the extended Poisson bracket

$$\{\langle\hat{A}\rangle,\langle\hat{B}\rangle\} = \frac{\langle[\hat{A},\hat{B}]\rangle}{i\hbar} \tag{3.207}$$

for any operators $\hat{A}$ and $\hat{B}$. For $\hat{x}$ and $\hat{p}$, we obtain

$$\{\langle\hat{x}\rangle,\langle\hat{p}\rangle\} = \frac{\langle[\hat{x},\hat{p}]\rangle}{i\hbar} = 1 \tag{3.208}$$

just like the classical Poisson bracket. However, the classical Poisson bracket receives quantum corrections if higher powers of $\hat{x}$ and $\hat{p}$ are used. For instance, if we use a cubic potential $V(x) = \lambda x^3$, we obtain the Poisson bracket

$$\{\langle\hat{p}\rangle,\langle V(\hat{x})\rangle\} = \frac{\langle[\hat{p},\lambda\hat{x}^3]\rangle}{i\hbar} = -3\lambda\langle\hat{x}^2\rangle = -3\lambda\langle\hat{x}\rangle^2 - 3\lambda\sigma_x^2. \tag{3.209}$$

In this way, the Poisson bracket (3.207) can be used for efficient derivations of quantum corrections to Ehrenfest equations, such as fluctuation terms. The Poisson bracket is also defined for quantum fluctuations or variances, such as

$$\begin{aligned}\{\sigma_x^2,\sigma_p^2\} &= \{\langle\hat{x}^2\rangle - \langle\hat{x}\rangle^2, \langle\hat{p}^2\rangle - \langle\hat{p}\rangle^2\} \\ &= \frac{\langle[\hat{x}^2,\hat{p}^2]\rangle}{i\hbar} - 2\langle\hat{p}\rangle\frac{\langle[\hat{x}^2,\hat{p}]\rangle}{i\hbar} - 2\langle\hat{x}\rangle\frac{\langle[\hat{x},\hat{p}^2]\rangle}{i\hbar} + 4\langle\hat{x}\rangle\langle\hat{p}\rangle\frac{\langle[\hat{x},\hat{p}]\rangle}{i\hbar} \\ &= 2\langle\hat{x}\hat{p} - \hat{p}\hat{x}\rangle - 4\langle\hat{x}\rangle\langle\hat{p}\rangle = 4C_{xp},\end{aligned} \tag{3.210}$$

which allows a quick rederivation of (1.127). In the second line, we applied the product rule and linearity in order to bring all Poisson brackets to the basic form (3.207) used in its definition. For the free particle as an example, the expectation value

$$\langle\hat{H}\rangle = \frac{1}{2m}\langle\hat{p}\rangle^2 + \frac{1}{2m}\sigma_p^2 \tag{3.211}$$

depends on the momentum variance. Combining (3.210) with (3.211) directly reproduces equation (1.127).

3.2.4 SYMMETRIES

Commuting operators have special properties because they are not subject to uncertainty relations. If one of them is the Hamilton operator of a given system, another important implication is the existence of conservation laws. As a simple example, any operator commutes with itself. The corresponding uncertainty relation is trivial because $C_{AA} = \sigma_A^2$, and therefore the $(\hat{A},\hat{A})$-version of (3.192) amounts to $0 \geq 0$ for any $\hat{A}$. However, using the commutator $[\hat{H},\hat{H}]$ in (3.198) means that $\mathrm{d}\langle\hat{H}\rangle/\mathrm{d}t = 0$ for any time-independent $\hat{H}$. Therefore, a conservative potential implies energy conservation.

More generally, any operator that commutes with the Hamilton operator, $[\hat{H},\hat{A}] = 0$, has a time-independent expectation value:

$$\frac{\mathrm{d}\langle A\rangle}{\mathrm{d}t} = \frac{i}{\hbar}\langle[\hat{H},\hat{A}]\rangle = 0. \tag{3.212}$$

An operator that commutes with the Hamilton operator therefore represents a conserved quantity. An example is the momentum of a free particle. Using the product rule for the commutator, this result can be extended to any integer power of A, and therefore to the variance σ_A and higher A-moments $\langle(\hat{A}-\langle\hat{A}\rangle\mathbb{I})^n\rangle$. In the example of the free particle, we already saw that not only $\langle\hat{p}\rangle$ is conserved but also σ_p; see (1.134).

3.2.4.1 Simultaneous diagonalizability

Conserved quantities are always important in physics, but in quantum mechanics they have special relevance thanks to a related mathematical feature. If two diagonalizable matrices or operators do not commute, $\hat{A}\hat{B}\neq\hat{B}\hat{A}$, then they cannot have the same eigenbasis. This result follows from the fact that an operator in its eigenbasis is a multiplication operator, and multiplication by real or complex numbers is commutative. In the finite-dimensional case, the two linear transformations are represented by diagonal matrices in the simultaneous eigenbasis, if it exists, and diagonal matrices always commute.

The converse is also valid: If two diagonalizable matrices or operators commute, $\hat{A}\hat{B}=\hat{B}\hat{A}$, then they have the same eigenbasis. To see this, start with an eigenvector $|\alpha\rangle$ of $\hat{A}$, with eigenvalues λ. Then,

$$\hat{A}(\hat{B}|\alpha\rangle)=\hat{B}(\hat{A}|\alpha\rangle)=\lambda\hat{B}|\alpha\rangle \tag{3.213}$$

shows that $\hat{B}|\alpha\rangle$ is also an eigenvector of $\hat{A}$ with eigenvalue λ. If λ has multiplicity one, we must have $\hat{B}|\alpha\rangle\propto|\alpha\rangle$, and $|\alpha\rangle$ is an eigenvector of $\hat{B}$ as well. If λ has multiplicity greater than one, $\hat{B}|\alpha\rangle$ is contained in the span of all eigenvectors of $\hat{A}$ with the same eigenvalue λ. Diagonalizing $\hat{B}$ restricted to this subspace then rearranges the λ-eigenvectors of $\hat{A}$ such that they become eigenvectors of $\hat{B}$ as well. They remain eigenvectors of $\hat{A}$ because $\hat{A}$ restricted to the span of all eigenvectors with the same eigenvalue λ is proportional to the identity transformation.

We conclude that two diagonalizable operators have the same eigenbasis if and only if they commute. If $\hat{A}$ is the Hamilton operator $\hat{H}$ that defines a system, a commuting operator implies a conserved quantity thanks to (3.212), and we now know that $\hat{A}$ and $\hat{H}$ can be diagonalized in the same eigenbasis in this case. This result allows us to attach new properties to eigenstates of $\hat{H}$, in addition to the energy eigenvalues that directly follow from the eigenvalue equation of $\hat{H}$.

We have implicitly used this procedure when we showed that bound states of a Hamilton operator with reflection-symmetric potential, $V(-x)=V(x)$, are either even or odd. The property of being even or odd, also called the parity of the state, is then conserved because a superposition of even bound states remains even, and a superposition of odd bound states remains odd.

The existence of a conserved quantity is closely related to symmetry. As a symmetry operation, the parity operator $\hat{P}$ is given by

$$\hat{P}\psi(x)=\psi(-x) \tag{3.214}$$

in the position basis. It is invertible (being equal to its own inverse) such that we can write $\hat{H}=\hat{P}^{-1}\hat{H}\hat{P}$ as an equivalent version of the commutation property $\hat{P}\hat{H}=\hat{H}\hat{P}$. Acting on wave functions in the position eigenbasis, this equation reads

$$\begin{aligned}\left(\frac{\hat{p}^2}{2m}+V(x)\right)\psi(x) &= \hat{P}^{-1}\left(\frac{\hat{p}^2}{2m}+V(x)\right)\hat{P}\psi(x)=\hat{P}^{-1}\left(\left(\frac{\hat{p}^2}{2m}+V(x)\right)\psi(-x)\right)\\ &= \left(\frac{\hat{p}^2}{2m}+V(-x)\right)\psi(x)\end{aligned} \tag{3.215}$$

for all $\psi(x)$, which is equivalent to $V(x)=V(-x)$. (The first operator $\hat{P}^{-1}$ acts on any x to its right, including the dependence in $V(x)$ and in $\psi(x)$.) Therefore, if $V(-x)=V(x)$, $\hat{H}$ and $\hat{P}$ have simultaneous eigenfunctions. Since the eigenfunctions of the parity operator are all even functions (eigenvalue $+1$) and all odd functions (eigenvalue -1), both with infinite multiplicity, an eigenfunction of $\hat{H}$ is either even or odd.

3.2.4.2 Exponentiation

In general, a conserved quantity is not the same as a symmetry transformation, even though these concepts are closely related. A symmetry transformation maps states to states and therefore preserves the normalization condition, while a conserved quantity may have real eigenvalues other than ± 1 that do not preserve the norm of states in the eigenbasis, where it multiplies by these numbers.

A real number can be mapped to a norm-preserving factor by exponentiating it, including a factor of i: If a is real, $\exp(\pm ia)$ has complex norm one. If we consider a self-adjoint operator, this mapping can be used for each eigenvalue λ, replacing it with $\exp(-i\lambda)$ on the same eigenvector. (It is common to choose the negative sign in this exponent.) The self-adjoint operator $\hat{A}$ with eigenvalues λ is then mapped to a new operator written as $\hat{u}_{\hat{A}} = \exp(-i\hat{A})$. The operator (or matrix) exponential of $\hat{A}$ can be computed easily if we know all eigenvectors and eigenvalues of $\hat{A}$, which we then exponentiate individually. In some cases, it is also possible to compute the exponential by evaluating the Taylor series

$$\hat{u}_{\hat{A}} = \sum_{n=0}^{\infty} \frac{(-i)^n}{n!} \hat{A}^n \tag{3.216}$$

as we will see in several examples.

Since multiplication by the new eigenvalues preserves the norm, $\hat{u}_{\hat{A}}$ is a **unitary** operator. If $\hat{A}$ is conserved and therefore commutes with the Hamilton operator $\hat{H}$, $\hat{u}_{\hat{A}}$ also commutes with the Hamilton operator because its definition implies that it has the same eigenbasis as $\hat{A}$. We interpret the norm-preserving unitary operator $\hat{u}_{\hat{A}}$ as the symmetry transformation corresponding to the conserved quantity $\hat{A}$.

The parity operator is norm-preserving and therefore may be viewed as a symmetry transformation. The corresponding conserved quantity may be defined as the self-adjoint operator $\hat{A}_{\hat{P}}$ such that $\hat{A}_{\hat{P}}|\psi\rangle = 0$ if $\psi(x)$ is even, and $\hat{A}_{\hat{P}}|\psi\rangle = \pi|\psi\rangle$ if $\psi(x)$ is odd, such that $\hat{P} = \hat{u}_{\hat{A}_{\hat{P}}}$. The eigenvalues of $\hat{A}_{\hat{P}}$ are not of complex norm one, but they can be used to distinguish between even and odd states and therefore determine the parity.

The Hamilton operator $\hat{H}$ with a conservative potential is conserved. Since it has units, we cannot directly exponentiate this operator in order to identify the corresponding symmetry. However, if we use a time value t, the product $tE/\hbar$ is unitless for any energy value E. The corresponding symmetry transformation, represented by $\hat{U}_{\hat{H}}(t) = \exp(-it\hat{H}/\hbar)$, corresponds to time translations or time evolution because it multiplies a stationary state $|\psi_n\rangle$ by $\exp(-itE_n/\hbar)$. A state $|\Psi\rangle(t)$ at time t, represented by the wave function $\Psi(x,t)$ in the position eigenbasis, equals

$$|\Psi\rangle(t) = \hat{U}_{\hat{H}}(t-t_0)|\Psi\rangle(t_0) = \exp(-i(t-t_0)\hat{H}/\hbar)|\Psi\rangle(t_0) \tag{3.217}$$

in terms of an initial state $|\Psi\rangle(t_0)$ at time t_0. Since $\hat{U}_{\hat{H}}(t-t_0)$ is unitary, normalization of $|\Psi\rangle(t)$ is preserved in time, as we derived directly from the Schrödinger equation in (1.94).

It is not always possible to derive a unique $\hat{A}$ for a given unitary operator $\hat{U}$ such that $\hat{U}_{\hat{A}} = \hat{U}$, where we follow the conventions used for the Hamilton operator and define

$$\hat{U}_{\hat{A}}(t) = \exp(-it\hat{A}/\hbar) \tag{3.218}$$

for any self-adjoint $\hat{A}$. The units of the parameter, here t, in general depend on the units we wish to assign to the resulting $\hat{A}$ and are not required to represent time in all cases. In contrast to the exponential function or its Taylor series, its inverse, given by the natural logarithm, has poles and is not always uniquely defined for a complex number such as an eigenvalue of $\hat{U}$. For instance, since $\exp(ia)$ is periodic for real a, $\ln(\exp(ia))$ is defined only up to adding an integer multiple of 2π. However, if we have a one-parameter family $\hat{U}(t)$ of unitary operators, as in the example of $\hat{U}_{\hat{H}}(t)$, it is possible to derive a corresponding self-adjoint operator by taking the derivative

$$\hat{A} = i\hbar \left.\frac{d\hat{U}(t)}{dt}\right|_{t=0} . \tag{3.219}$$

If the result turns out to be a densely-defined operator, it follows that $\hat{U}(t) = \exp(-it\hat{A}/\hbar)$.

The one-parameter family $\hat{U}(t)$ must be differentiable at $t=0$, which is not always the case and, in general, presents a tricky question for an operator-valued function that maps $t \in \mathbb{R}$ to an operator $\hat{U}(t)$. However, it is always possible to try computing the derivative and check whether it produces a well-defined operator with finite eigenvalues. A common example in which the derivative can be used is given by a family $\hat{U}(t)$ with known continuous spectrum for any t around $t=0$, in which case we can take the derivative directly on the eigenvalues.

3.2.4.3 Translations

Consider spatial translations represented by the one-parameter family $\hat{U}(\delta)\psi(x) = \psi(x-\delta)$ of unitary operators, where we replaced t by δ because it no longer has units of time. An eigenfunction of $\hat{U}(\delta)$ must turn a shift of x in the argument of ψ into a multiplicative factor of ψ, which is what the exponential function does: $\exp(a(x-\delta)) = e^{-a\delta}\exp(ax)$. We should choose the free constant a such that the exponent is unitless, and such that we have normalizability at least in the delta-function sense. Thus, a should be imaginary and have units of inverse length: We write $a = ik$ with a wave number k, and the eigenfunctions become plane waves with suitable delta-function normalization.

These are the eigenfunctions $|k\rangle$ of the momentum operator, on which $\hat{U}(\delta)$ acts by

$$\hat{U}(\delta)|k\rangle = e^{-ik\delta}|k\rangle\,. \tag{3.220}$$

We derive

$$\hat{A}|k\rangle = i\hbar \left.\frac{\mathrm{d}\hat{U}(\delta)}{\mathrm{d}\delta}\right|_{\delta=0} |k\rangle = \hbar k|k\rangle \tag{3.221}$$

on momentum eigenfunctions $|k\rangle$. The factor $\hbar k$ is the eigenvalue of the momentum operator in $|k\rangle$, and we obtain $\hat{A} = \hat{p}$, or

$$\hat{U}(\delta) = \hat{U}_{\hat{p}}(\delta) = \exp(-i\delta\hat{p}/\hbar)\,. \tag{3.222}$$

It is possible to evaluate the operator exponential $\exp(-i\delta\hat{p}/\hbar)$ in the position eigenbasis, using the Taylor series (3.216):

$$\hat{U}_{\hat{p}}(\delta)\psi(x) = \exp(-i\delta\hat{p}/\hbar)\psi(x) = \sum_{n=0}^{\infty} \frac{(-\delta)^n}{n!}\frac{\mathrm{d}^n\psi(x)}{\mathrm{d}x^n} = \psi(x-\delta)\,, \tag{3.223}$$

applying the Taylor series of $\psi(x-\delta)$ in the last step.

Commutators can also be used here. The position operator is transformed by $\hat{U}_{\hat{p}}(\delta)$ according to a construction similar to (3.215): First, a state $|\psi\rangle$ is mapped to $|\psi\rangle(\delta) = \hat{U}_{\hat{p}}(\delta)|\psi\rangle$ by the symmetry transformation $\hat{U}_{\hat{p}}(\delta)$. Any operator $\hat{B}$ is correspondingly mapped to

$$\hat{B}(\delta) = \hat{U}_{\hat{p}}(\delta)\hat{B}\hat{U}_{\hat{p}}(\delta)^{-1} = \hat{U}_{\hat{p}}(\delta)\hat{B}\hat{U}_{\hat{p}}(-\delta) \tag{3.224}$$

by a similarity transformation because it implies

$$\hat{B}(\delta)|\psi\rangle(\delta) = \left(\hat{U}_{\hat{p}}(\delta)\hat{B}\hat{U}_{\hat{p}}(-\delta)\right)\left(\hat{U}_{\hat{p}}(\delta)|\psi\rangle\right) = \hat{U}_{\hat{p}}(\delta)\left(\hat{B}|\psi\rangle\right)\,. \tag{3.225}$$

Applied to $\hat{B} = \hat{x}$ and $\hat{U}_{\hat{p}}(\delta) = \exp(-i\delta\hat{p}/\hbar)$, we first compute

$$\begin{aligned}
\hat{x}\exp(i\delta\hat{p}/\hbar) &= \hat{x}\sum_{n=0}^{\infty}\frac{(i\delta/\hbar)^n}{n!}\hat{p}^n = \sum_{n=0}^{\infty}\frac{(i\delta/\hbar)^n}{n!}\left(\hat{p}^n\hat{x} + [\hat{x},\hat{p}^n]\right) \\
&= \sum_{n=0}^{\infty}\frac{(i\delta/\hbar)^n}{n!}\left(\hat{p}^n\hat{x} + i\hbar n\hat{p}^{n-1}\right) = \sum_{n=0}^{\infty}\frac{(i\delta/\hbar)^n}{n!}\hat{p}^n\hat{x} + i\hbar\sum_{n=1}^{\infty}\frac{(i\delta/\hbar)^n}{(n-1)!}\hat{p}^{n-1}
\end{aligned} \tag{3.226}$$

using the commutator relationship (3.197). In the first sum, we factor out $\hat{x}$ to the right, such that it equals $\exp(i\delta\hat{p}/\hbar)\hat{x}$. The second sum equals $-\delta\exp(i\delta\hat{p}/\hbar)$ after factoring out $i\delta/\hbar$ and shifting the summation label n by one. Therefore,

$$\hat{x}(\delta) = \exp(-i\delta\hat{p}/\hbar)\hat{x}\exp(i\delta\hat{p}/\hbar) = \hat{x} - \delta\hat{\mathbb{I}}. \tag{3.227}$$

This result shows that any pair of self-adjoint **canonically conjugate** operators, $\hat{A}$ and $\hat{B}$ such that $[\hat{A},\hat{B}] = i\hbar$, must have continuous spectra given by all real numbers. Since we only used the commutator in (3.227), it always applies to canonically conjugate self-adjoint operators. Thus, $\exp(-i\delta\hat{B}/\hbar)$ represents translations on the spectrum of $\hat{A}$. For any eigenfunction $|\psi\rangle$ of $\hat{A}$ with eigenvalue λ there is then an eigenfunction $\exp(-i\delta\hat{B}/\hbar)|\psi\rangle$ with eigenvalue $\lambda-\delta$. Similarly, $\exp(i\varepsilon\hat{A}/\hbar)$ represents translations on the spectrum of $\hat{B}$, such that any eigenfunction $|\phi\rangle$ of $\hat{B}$ with eigenvalue κ has a corresponding eigenfunction $\exp(i\varepsilon\hat{A}/\hbar)|\phi\rangle$ with eigenvalue $\kappa-\varepsilon$. Because δ and ε can take any real value, the spectra of $\hat{A}$ and $\hat{B}$ must include all real numbers.

3.2.4.4 Evolution

In the example of the (inverse) evolution operator $\hat{U}_{\hat{H}}(-t) = \exp(it\hat{H}/\hbar)$, the time-dependent operator

$$\hat{B}(t) = \exp(it\hat{H}/\hbar)\hat{B}\exp(-it\hat{H}/\hbar) \tag{3.228}$$

is called the **Heisenberg operator** corresponding to $\hat{B}$. According to (3.225), acting with $\hat{B}(-t)$ on a time-dependent wave function is equivalent to acting with $\hat{B} = \hat{B}(0)$ on an initial wave function and then evolving the result. Alternatively, if we act with $\hat{B}(t)$ on a time-independent wave function,

$$\hat{B}(t)|\psi\rangle = \hat{U}_{\hat{H}}(-t)(\hat{B}|\psi\rangle(t)), \tag{3.229}$$

the result equals the application of $\hat{B}$ to the wave function at time t, and then evolved back to time zero. In this form, the wave function is always evaluated at time $t=0$, but evolution is expressed by an evolving operator that takes into account dynamical changes of expectation values and other statistical properties.

Instead of solving differential equations for vectors or wave functions, we can then solve differential equations for matrices or operators. The Heisenberg operator $\hat{B}(t)$ is subject to the **Heisenberg equation**

$$\frac{\mathrm{d}\hat{B}(t)}{\mathrm{d}t} = \frac{\mathrm{d}}{\mathrm{d}t}\left(\exp(it\hat{H}/\hbar)\hat{B}\exp(-it\hat{H}/\hbar)\right) = \frac{[\hat{B},\hat{H}(t)]}{i\hbar}. \tag{3.230}$$

The derivative of an operator requires a limiting procedure or a topology. Here, it is sufficient to note that an equation such as $\mathrm{d}\hat{B}/\mathrm{d}t = \hat{A}$ means that $\mathrm{d}\langle\psi|\hat{B}(t)|\psi\rangle/\mathrm{d}t = \langle\psi|\hat{A}|\psi\rangle$ for any time-independent state $|\psi\rangle$. The expression $\langle\psi|\hat{B}(t)|\psi\rangle$ for a given $|\psi\rangle$ is a standard time-dependent function, for which the usual notion of derivatives can be used.

In general, equation (3.230) does not provide a solvable differential equation for $\hat{B}(t)$ because the commutator $[\hat{B}(t),\hat{H}]$ is not guaranteed to be related to B. A solvable system requires another differential equation for $\mathrm{d}[\hat{B},\hat{H}]/\mathrm{d}t$, which introduces another operator $[[\hat{B},\hat{H}],\hat{H}]$ with its own differential equation and so on. However, there are examples in which (3.230) is useful, such as the free particle where $[\hat{x},\hat{H}] = i\hbar\hat{p}/m$ is proportional to the momentum operator, which obeys $[\hat{p},\hat{H}] = 0$. Therefore, $\mathrm{d}\hat{x}(t)/\mathrm{d}t = \hat{p}/(2m)$ and $\mathrm{d}\hat{p}/\mathrm{d}t = 0$, or $\hat{x}(t) = \hat{x}(0) + t\hat{p}/m$ just like the classical solution.

3.2.4.5 Harmonic oscillator

The harmonic oscillator has several special mathematical features that are relevant for its solvability properties. The commutators of $\hat{x}$, $\hat{p}$ and $\hat{H}$ are all linear expressions in these operators and the identity, $\hat{\mathbb{I}}$:

$$[\hat{x},\hat{p}] = i\hbar\hat{\mathbb{I}} \quad , \quad [\hat{x},\hat{H}] = i\hbar\frac{\hat{p}}{m} \quad , \quad [\hat{H},\hat{p}] = i\hbar m\omega^2\hat{x}. \tag{3.231}$$

Since $[\hat{\mathbb{I}},\hat{A}]=0$ for any operator $\hat{A}$, the set of four operators $\{\hat{\mathbb{I}},\hat{x},\hat{p},\hat{H}\}$ is preserved by taking commutators, up to numerical factors. Linearity of the commutator on each side implies that the 4-dimensional vector space of linear combinations $\delta_{\mathbb{I}}\hat{\mathbb{I}}+\delta_x\hat{x}+\delta_p\hat{p}+\delta_H\hat{H}$ is also preserved by the commutator, which therefore defines a **Lie algebra**. Lie algebras and the closely related concept of Lie groups describe symmetries of a physical system and can often be employed to develop algebraic solution methods.

The specific example defined here is the harmonic-oscillator algebra. If we first exclude $\hat{H}$, we obtain a subalgebra from linear combinations of $\hat{x}$, $\hat{p}$ and $\hat{\mathbb{I}}$, called the **Heisenberg algebra**. These operators have commutator relations

$$[\hat{x},\hat{p}]=i\hbar\hat{\mathbb{I}} \quad , \quad [\hat{\mathbb{I}},\hat{x}]=0=[\hat{\mathbb{I}},\hat{p}] \tag{3.232}$$

such that

$$[\delta_x\hat{x}+\delta_p\hat{p}+\delta_{\mathbb{I}}\hat{\mathbb{I}},\varepsilon_x\hat{x}+\varepsilon_p\hat{p}+\varepsilon_{\mathbb{I}}\hat{\mathbb{I}}]=i\hbar(\delta_x\varepsilon_p-\delta_p\varepsilon_x)\hat{\mathbb{I}} \tag{3.233}$$

for linear combinations. This relationship can be obtained in the same form from upper-triangular matrices

$$A_{\delta_x,\delta_p,\delta_{\mathbb{I}}}=\begin{pmatrix} 0 & \delta_p & -i\delta_{\mathbb{I}}/\hbar \\ 0 & 0 & \delta_x \\ 0 & 0 & 0 \end{pmatrix}. \tag{3.234}$$

The set of matrices $A_{\delta_x,\delta_p,\delta_{\mathbb{I}}}$ is called a **matrix representation** of the Heisenberg algebra. Since

$$[A_{\delta_x,\delta_p,\delta_{\mathbb{I}}},A_{\varepsilon_x,\varepsilon_p,\varepsilon_{\mathbb{I}}}]=i\hbar A_{0,0,\delta_p\varepsilon_x-\delta_x\varepsilon_p} \tag{3.235}$$

and $[A_{\delta_x,\delta_p,\delta_{\mathbb{I}}},A_{0,0,\varepsilon_{\mathbb{I}}}]=0$, the algebra is reduced to zero by iterating commutators. Such a Lie algebra is called **nilpotent**.

The matrix product has a similar effect:

$$A_{\delta_x,\delta_p,\delta_{\mathbb{I}}}A_{\varepsilon_x,\varepsilon_p,\varepsilon_{\mathbb{I}}}=\begin{pmatrix} 0 & 0 & \delta_p\varepsilon_x \\ 0 & 0 & 0 \\ 0 & 0 & 0 \end{pmatrix}=i\hbar A_{0,0,\delta_p\varepsilon_x} \tag{3.236}$$

and $A_{\delta_x,\delta_p,\delta_{\mathbb{I}}}A_{0,0,\varepsilon_{\mathbb{I}}}=A_{0,0,0}=0$. The matrix product is not an operation considered for a Lie algebra, in which we only add multiples of matrices and take commutators. The matrix product is an operation in a group. Addition in an algebra can be turned into group multiplication by applying the exponential function: $e^{a+b}=e^ae^b$. The exponential can also be defined for a matrix by using the Taylor series, but we have to be careful if the matrix analogs of a and b in e^{a+b} do not commute, and therefore the matrix analogs of e^a and e^b do not commute. Non-zero commutators may then contribute to the product formula, but the sum in e^{a+b} does not suggest a specific ordering of a and b. We will soon encounter a formula that provides additional terms for matrix exponentials.

In the matrix representation of the Heisenberg algebra, exponentiation does not lead to unitary matrices because the $A_{\delta_x,\delta_p,\delta_{\mathbb{I}}}$ are not Hermitian. The product version of nilpotent behavior, based on (3.236), simplifies direct computations of the matrix exponential because it implies $A^k=0$ if $k\geq 3$, which holds for any upper triangular matrix with zero diagonal. The Taylor series therefore contains only three terms,

$$\begin{aligned}\exp(A_{\delta_x,\delta_p,\delta_{\mathbb{I}}}) &= \sum_{k=0}^{\infty}\frac{1}{k!}A^k_{\delta_x,\delta_p,\delta_{\mathbb{I}}}=\mathbb{I}+A_{\delta_x,\delta_p,\delta_{\mathbb{I}}}+\frac{1}{2}A^2_{\delta_x,\delta_p,\delta_{\mathbb{I}}} \\ &= \begin{pmatrix} 1 & \delta_p & -i\delta_{\mathbb{I}}/\hbar+\frac{1}{2}\delta_x\delta_p \\ 0 & 1 & \delta_x \\ 0 & 0 & 1 \end{pmatrix}.\end{aligned} \tag{3.237}$$

Matrix multiplication shows that

$$\begin{aligned}
&\exp(A_{\delta_x,\delta_p,\delta_{\mathbb{I}}})\exp(A_{\varepsilon_x,\varepsilon_p,\varepsilon_{\mathbb{I}}})\\
&= \begin{pmatrix} 1 & \delta_p+\varepsilon_p & -i(\delta_{\mathbb{I}}+\varepsilon_{\mathbb{I}})/\hbar+\frac{1}{2}(\delta_x\delta_p+2\delta_p\varepsilon_x+\varepsilon_x\varepsilon_p) \\ 0 & 1 & \delta_x+\varepsilon_x \\ 0 & 0 & 1 \end{pmatrix}\\
&= \begin{pmatrix} 1 & \delta_p+\varepsilon_p & -i(\delta_{\mathbb{I}}+\varepsilon_{\mathbb{I}})/\hbar+\frac{1}{2}(\delta_x+\varepsilon_x)(\delta_p+\varepsilon_p) \\ 0 & 1 & \delta_x+\varepsilon_x \\ 0 & 0 & 1 \end{pmatrix}\begin{pmatrix} 1 & 0 & \frac{1}{2}(\delta_p\varepsilon_x-\delta_x\varepsilon_p) \\ 0 & 1 & 0 \\ 0 & 0 & 1 \end{pmatrix}\\
&= \exp(A_{\delta_x,\delta_p,\delta_{\mathbb{I}}}+A_{\varepsilon_x,\varepsilon_p,\varepsilon_{\mathbb{I}}})\exp(\tfrac{1}{2}[A_{\delta_x,\delta_p,\delta_{\mathbb{I}}},A_{\varepsilon_x,\varepsilon_p,\varepsilon_{\mathbb{I}}}])\,. \qquad (3.238)
\end{aligned}$$

The exponential function applied to the sum of two matrices therefore does not always equal the product of the two matrix exponentials. However, it is correct to write

$$\exp(X)\exp(Y)=\exp(X+Y)\exp(\tfrac{1}{2}[X,Y]) \qquad (3.239)$$

for matrices X and Y such that $[[X,Y],X]=0=[[X,Y],Y]$. This equation is a special case of the **Baker–Campbell–Hausdorff** formula. Inserting (3.235) in (3.238), we obtain the **Weyl relation**

$$\exp(A_{\delta_x,\delta_p,\delta_{\mathbb{I}}})\exp(A_{\varepsilon_x,\varepsilon_p,\varepsilon_{\mathbb{I}}})=\exp(A_{\delta_x,\delta_p,\delta_{\mathbb{I}}}+A_{\varepsilon_x,\varepsilon_p,\varepsilon_{\mathbb{I}}})\exp(\tfrac{1}{2}i\hbar A_{0,0,\delta_p\varepsilon_x-\delta_x\varepsilon_p})\,. \qquad (3.240)$$

Returning to the harmonic-oscillator algebra, the commutation relations are represented by $\hat{x}$ and $\hat{p}$ through ladder operators, and diagonal $\hat{H}$ and $\hat{\mathbb{I}}$ in the infinite-dimensional Hilbert space spanned by stationary states. However, as in the case of the Heisenberg algebra, there are finite-dimensional matrix representations as well.[1] The correct commutators can be confirmed if we represent the basic operators by matrices

$$\begin{aligned}
x &= \begin{pmatrix} 0 & \sqrt{\hbar/(m\omega)} & 0 & 0 \\ 0 & 0 & 0 & 0 \\ 0 & 0 & 0 & -\sqrt{\hbar/(m\omega)} \\ 0 & 0 & 0 & 0 \end{pmatrix}\,,\\
p &= \begin{pmatrix} 0 & 0 & \sqrt{m\hbar\omega} & 0 \\ 0 & 0 & 0 & \sqrt{m\hbar\omega} \\ 0 & 0 & 0 & 0 \\ 0 & 0 & 0 & 0 \end{pmatrix}\\
H &= \begin{pmatrix} 0 & 0 & 0 & 0 \\ 0 & 0 & i\hbar\omega & 0 \\ 0 & -i\hbar\omega & 0 & 0 \\ 0 & 0 & 0 & 0 \end{pmatrix}\,,\quad I=\begin{pmatrix} 0 & 0 & 0 & -2i \\ 0 & 0 & 0 & 0 \\ 0 & 0 & 0 & 0 \\ 0 & 0 & 0 & 0 \end{pmatrix} \qquad (3.241)
\end{aligned}$$

acting on a 4-dimensional vector space. Notice that the identity operator is not represented by the identity matrix because we are interested in commutator relationships rather than matrix products. Unlike H, the matrices x, p, and I are not Hermitian if we use an inner product in which the standard basis is orthonormal.

Each of these matrices can be exponentiated by using the Taylor-series method. The three matrices x, p and I square to zero and therefore only require the constant and linear terms in the exponential series:

$$\exp(-i\delta_x x/\hbar) = 1-i\frac{\delta_x}{\hbar}x=\begin{pmatrix} 1 & -i\delta_x/\sqrt{m\hbar\omega} & 0 & 0 \\ 0 & 1 & 0 & 0 \\ 0 & 0 & 1 & i\delta_x/\sqrt{m\hbar\omega} \\ 0 & 0 & 0 & 1 \end{pmatrix} \qquad (3.242)$$

[1] R. Streater: The representations of the oscillator group, *Commun. Math. Phys.* 4 (1967) 217–236.

$$\exp(-i\delta_p p/\hbar) = 1 - i\frac{\delta_p}{\hbar}p = \begin{pmatrix} 1 & 0 & -i\sqrt{m\omega}\delta_p/\sqrt{\hbar} & 0 \\ 0 & 1 & 0 & -i\sqrt{m\omega}\delta_p/\sqrt{\hbar} \\ 0 & 0 & 1 & 0 \\ 0 & 0 & 0 & 1 \end{pmatrix}$$

$$\exp(-i\delta_I I/\hbar) = 1 - i\frac{\delta_I}{\hbar}I = \begin{pmatrix} 1 & 0 & 0 & 2\delta_I/\hbar \\ 0 & 1 & 0 & 0 \\ 0 & 0 & 1 & 0 \\ 0 & 0 & 0 & 1 \end{pmatrix}.$$

These matrices are not unitary and therefore do not correspond to symmetry transformations in the strict sense. Nevertheless, we will see interesting relationships based on the original commutators.

For H, we have $(iH/\hbar)^2 = -\omega^2 I_2$ with the matrix

$$I_2 = \begin{pmatrix} 0 & 0 & 0 & 0 \\ 0 & 1 & 0 & 0 \\ 0 & 0 & 1 & 0 \\ 0 & 0 & 0 & 0 \end{pmatrix}. \tag{3.243}$$

Therefore, $(iH/\hbar)^{2k} = (-1)^k \omega^{2k} I_2$ and $(iH/\hbar)^{2k+1} = (-1)^k \omega^{2k} iH/\hbar$. These equations allow us to split the Taylor series $\exp(-i\delta_H H/\hbar)$ into even terms (proportional to I_2) and odd terms (proportional to H), in which the coefficients form the Taylor series of $\cos(\omega\delta_H)$ and $\sin(\omega\delta_H)$, respectively:

$$\exp(-i\delta_H H/\hbar) = \begin{pmatrix} 0 & 0 & 0 & 0 \\ 0 & \cos(\omega\delta_H) & -\sin(\omega\delta_H) & 0 \\ 0 & \sin(\omega\delta_H) & \cos(\omega\delta_H) & 0 \\ 0 & 0 & 0 & 0 \end{pmatrix}. \tag{3.244}$$

The product of all four exponentials equals

$$\exp(-i\delta_I I/\hbar)\exp(-i\delta_x x/\hbar)\exp(-i\delta_p p/\hbar)\exp(-i\delta_H H/\hbar) \tag{3.245}$$

$$= \begin{pmatrix} 1 & -i(\delta_x C + m\omega\delta_p S)/\sqrt{m\hbar\omega} & i(\delta_x S - m\omega\delta_p C)/\sqrt{m\hbar\omega} & 2(\delta_I - \frac{1}{2}\delta_x\delta_p)/\hbar \\ 0 & C & -S & -i\sqrt{m\omega}\delta_p/\sqrt{\hbar} \\ 0 & S & C & i\delta_x/\sqrt{m\hbar\omega} \\ 0 & 0 & 0 & 1 \end{pmatrix}$$

where we defined $C = \cos(\omega\delta_H)$ and $S = \sin(\omega\delta_H)$. Setting $\delta_I = 0$ and using two versions of the exponentiated H, we obtain the more symmetric expression

$$\exp(i\delta_H H/\hbar)\exp(-i\delta_x x/\hbar)\exp(-i\delta_p p/\hbar)\exp(-i\delta_H H/\hbar) \tag{3.246}$$

$$= \begin{pmatrix} 1 & -i(\delta_x C + m\omega\delta_p S)/\sqrt{m\hbar\omega} & i(\delta_x S - m\omega\delta_p C)/\sqrt{m\hbar\omega} & -\delta_x\delta_p/\hbar \\ 0 & 1 & 0 & i(\delta_x S - m\omega\delta_p C)/\sqrt{m\hbar\omega} \\ 0 & 0 & 1 & i(\delta_x C + m\omega\delta_p S)/\sqrt{m\hbar\omega} \\ 0 & 0 & 0 & 1 \end{pmatrix}.$$

If we rename $\delta_H = -t$, we recognize the evolution operator $\exp(-itH/\hbar)$ on the left of the product, and its inverse on the right, which on a generic operator in the middle, here $\exp(-i\delta_x x/\hbar)\exp(-i\delta_p p/\hbar)$, should act as in (3.224). We can therefore use the matrix components of (3.246) in order to identify time-dependent $\delta_x(t)$ and $\delta_p(t)$. For unitless exponents, $\delta_x(t) = p(t)$ is a momentum, while $\delta_p(t) = x(t)$ is a position. (These properties also follow from our discussion of translation operators, in particular (3.227), but here we have a different representation.)

The evolution operator $\exp(i\delta_H H/\hbar) = \exp(-itH/\hbar)$ then implements the time dependence of x and p in the components of (3.246), where we recognize

$$x(t) = \delta_p(t) = \delta_p C - \frac{1}{m\omega}\delta_x S = x_0\cos(\omega t) + \frac{1}{m\omega}p_0\sin(\omega t) \tag{3.247}$$

and

$$p(t) = \delta_x(t) = \delta_x C + m\omega\delta_p S = p_0\cos(\omega t) - m\omega x_0 \sin(\omega t)\,, \tag{3.248}$$

observing $S = \sin(\omega\delta_H) = -\sin(\omega t)$, as the correct solutions for the equations of motion of the harmonic oscillator. These are classical solutions, but the Ehrenfest theorem tells us that they also hold for expectation values of $\hat{x}$ and $\hat{p}$ in quantum mechanics, or for Heisenberg operators. For the harmonic oscillator, they can be derived within a finite-dimensional group.

EXERCISES

1. A 3-dimensional complex vector space has an orthonormal basis given by the vectors $|e_1\rangle$, $|e_2\rangle$ and $|e_3\rangle$.
   (i) Confirm that the following vectors also form an orthonormal basis of the same vector space:
   $$|f_1\rangle = i|e_3\rangle \quad , \quad |f_2\rangle = \frac{1}{\sqrt{2}}(|e_1\rangle + |e_2\rangle) \quad , \quad |f_3\rangle = \frac{1}{\sqrt{2}}(|e_1\rangle - |e_2\rangle)\,.$$
   (ii) If the vector $|\alpha\rangle$ has components $(1,0,0)$ in the e-basis and $|\beta\rangle$ has components $(0,0,1)$ in the e-basis, what are their components in the f-basis?
   Compute $\langle\alpha|\beta\rangle$, $||\alpha||$ and $||\beta||$ in the e-basis and in the f-basis.
   (iii) A linear transformation $\hat{T}$ is defined by $\hat{T}|e_i\rangle = |f_i\rangle$ for $i = 1,2,3$ with the $|f_i\rangle$ from part (i). Compute the matrix T that represents $\hat{T}$ in the e-basis.
2. Compute the Hermitian conjugate of the matrix
   $$T = \begin{pmatrix} 0 & 1/\sqrt{2} & 1/\sqrt{2} \\ 0 & 1/\sqrt{2} & -1/\sqrt{2} \\ i & 0 & 0 \end{pmatrix}$$
   Show that this matrix is unitary.
3. Show that the matrix
   $$U = \begin{pmatrix} \cos(\psi) - i\sin(\psi)\cos(\vartheta) & -i\sin(\psi)\sin(\vartheta)e^{-i\varphi} \\ -i\sin(\psi)\sin(\vartheta)e^{i\varphi} & \cos(\psi) + i\sin(\psi)\cos(\vartheta) \end{pmatrix}$$
   with real ψ, ϑ, φ is unitary and has $\det U = 1$.
4. Write down all the conditions on T_{ij} if they are matrix elements of a real and antisymmetric matrix.
   Compute the eigenvalues of a generic real and antisymmetric 3×3 matrix, using the conditions on its elements.
5. Show that a 2×2-matrix $S = \begin{pmatrix} S_{11} & S_{12} \\ S_{21} & S_{22} \end{pmatrix}$ is unitary if and only if its two columns form an orthonormal set.
6. Show that a 2×2-matrix S is unitary if it preserves the norm of any 2-vector.
7. A matrix S is defined by
   $$S = \begin{pmatrix} \cos(\beta) & 0 & \sin(\beta)e^{-i\delta} \\ -\sin(\beta)\sin(\gamma)e^{i\delta} & \cos(\gamma) & \cos(\beta)\sin(\gamma) \\ -\sin(\beta)\cos(\gamma)e^{i\delta} & -\sin(\gamma) & \cos(\beta)\cos(\gamma) \end{pmatrix}$$
   with real numbers β, γ, and δ.
   (i) Compute the Hermitian conjugate of S.
   (ii) Show that S is unitary.
   (iii) Confirm that the vector with components $\alpha_1 = \cot(\frac{1}{2}\beta)e^{-i\delta}$, $\alpha_2 = -\cot(\frac{1}{2}\gamma)$, $\alpha_3 = 1$ is an eigenvector of S with eigenvalue one.

8. Derive an exact expression for the 10^{100}-th power of the matrix $T = \begin{pmatrix} 0 & -2 \\ 1 & 3 \end{pmatrix}$, defined as the product of T with itself 10^{100} times. Use the following method:
   a. Diagonalize T by writing it in the form $T = S^{-1}DS$ with a diagonal matrix D.
   b. Show that $(S^{-1}DS)^N = S^{-1}D^N S$ for a matrix D, an invertible matrix S and an integer N.
   c. Apply this equation to $T^{10^{100}}$.
9. The electron has two spin states depending on whether its spin points up or down, which can be described by a 2-vector $\alpha = \begin{pmatrix} \alpha_1 \\ \alpha_2 \end{pmatrix}$. If the electron is in a constant magnetic field pointing in the x-direction, the energy or Hamilton operator of the spin states is given by the matrix
$$H = a \begin{pmatrix} 0 & 1 \\ 1 & 0 \end{pmatrix} \tag{3.249}$$
with a real constant a.
Solve the evolution equation $i\hbar d\alpha/dt = H\alpha$ by diagonalizing the matrix H.
10. Diagonalization of a matrix can be used to decouple potentials for multiple degrees of freedom. For instance, the two coordinates (x,y) of a particle with mass m in the two-dimensional potential $V(x,y) = \frac{1}{2}m\omega^2(3x^2 - 2xy + 3y^2)$ have coupled equations of motion in classical mechanics, and the 2-dimensional Schrödinger equation with this potential is not separable. The time-independent Schrödinger equation is not separable in the original form because of the cross-term xy. These equations simplify if the potential can be transformed to some $V(x',y') = V_1(x') + V_2(y')$. To eliminate the xy-term, transform to new coordinates as follows:
   (i) Write the potential as $V(x,y) = \frac{1}{2}m\omega^2\vec{v}^\dagger T\vec{v}$, where $\vec{v}$ is the 2-vector $\vec{v} = \begin{pmatrix} x \\ y \end{pmatrix}$ and $T = \begin{pmatrix} 3 & -1 \\ -1 & 3 \end{pmatrix}$. Diagonalize $T = S^{-1}DS$, where D is diagonal and show that S is unitary if you normalize the eigenvectors before choosing them as columns of S^{-1}.
   (ii) Use the unitary S to introduce new coordinates $\begin{pmatrix} x' \\ y' \end{pmatrix} = \vec{v}' = S\vec{v}$ and compute the new potential
$$V(x',y') = \frac{1}{2}m\omega^2(S^{-1}\vec{v}')^\dagger TS^{-1}\vec{v}' = \frac{1}{2}m\omega^2(\vec{v}')^\dagger STS^{-1}\vec{v}'.$$
   Confirm that $V(x',y')$ is of the form $V_1(x') + V_2(y')$.
   (iii) We also have to transform the two components of the momentum operator to our new coordinates. Use the chain rule to derive the transformation of derivatives: $\nabla = S^\dagger\nabla'$, where $\nabla' = (\frac{\partial}{\partial x'}, \frac{\partial}{\partial y'})^\dagger$. Conclude that the derivative part of the Hamilton operator is unchanged: $\nabla^\dagger\nabla = \nabla'^\dagger\nabla'$.
   (iv) The time-independent Schrödinger equation in the coordinates (x',y') is separable by using wave functions of the form $\psi(x',y') = X(x')Y(y')$ with arbitrary functions X and Y. Show that X and Y must both be solutions of the Schrödinger equation for the one-dimensional harmonic oscillator (with different frequencies). Use the harmonic-oscillator results to find all allowed energy eigenvalues and the stationary states for this system.

11. Consider the parity operator $\hat{P}$, defined by its action $(\hat{P}f)(x) = f(-x)$ for any function $f(x)$.
    (i) Is $\hat{P}$ a linear transformation? Why/why not?
    (ii) Show that $\hat{P}^2 = 1$ is the identity operator, and that this result implies $\hat{P}^{-1} = \hat{P}$.
    (iii) Prove $\hat{P}^\dagger = \hat{P}$. Therefore, the operator $\hat{P}$ is Hermitian and, combined with the result of part (ii), unitary.
    (iv) Use the result of part (ii) to show that $\hat{P}$ has two different eigenvalues, $\lambda_\pm = \pm 1$. What are the eigenfunctions?
12. Define the dilation operator $\hat{D}$ by $(\hat{D}f)(x) = \sqrt{2}f(2x)$ for all functions $f(x)$.
    (i) Show that $\hat{D}$ is a linear transformation.
    (ii) Compute $\hat{D}^\dagger$ and show that $\hat{D}$ is unitary.
    (iii) Determine all eigenfunctions of $\hat{D}$.
    [Hint: Try $f_\lambda(x) = g(\lambda)^{\log|x|}/\sqrt{|x|}$ and solve for the function $g(\lambda)$.]
13. Derive and solve the Heisenberg equations (3.230) for $\hat{x}(t)$ and $\hat{p}(t)$ of the harmonic oscillator.
14. Demonstrate that the harmonic-oscillator group is not exponential in the following sense: Compute $\exp(-i(\delta_I I + \delta_x x + \delta_p p + \delta_H H)/\hbar)$ in the matrix representation (3.241) and show that the result does not include matrices of the form (3.245).

4 Angular Momentum and Spin

If we include all three spatial dimensions, the time-dependent wave function $\Psi(x,y,z,t)$ depends on four coordinates. A wave function of this form determines the probability density $\rho_t(x,y,z) = |\Psi(x,y,z,t)|^2$ for measurements of the spatial position (x,y,z). The product $|\Psi(x,y,z,t)|^2 \mathrm{d}x\mathrm{d}y\mathrm{d}z$ therefore equals the probability of finding a particle in a small cubic volume with side lengths $(\mathrm{d}x, \mathrm{d}y, \mathrm{d}z)$ around (x,y,z) at time t. The normalization condition for this probability density is given by

$$\int_{-\infty}^{\infty}\int_{-\infty}^{\infty}\int_{-\infty}^{\infty} |\Psi(x,y,z,t)|^2 \mathrm{d}x\mathrm{d}y\mathrm{d}z = 1 \qquad \text{for all } t\,. \tag{4.1}$$

Wave functions $\Psi(x,y,z,t)$ represent states in the position eigenbasis. Therefore, the three position operators for individual Cartesian coordinates are multiplication operators,

$$\hat{x} = x \quad , \quad \hat{y} = y \quad , \quad \hat{z} = z\,, \tag{4.2}$$

and the momentum components are derivative operators:

$$\hat{p}_x = \frac{\hbar}{i}\frac{\partial}{\partial x} \quad , \quad \hat{p}_y = \frac{\hbar}{i}\frac{\partial}{\partial y} \quad , \quad \hat{p}_z = \frac{\hbar}{i}\frac{\partial}{\partial z}\,. \tag{4.3}$$

For pairs of these operators we have the commutators

$$[\hat{x}, \hat{p}_x] = [\hat{y}, \hat{p}_y] = [\hat{z}, \hat{p}_z] = i\hbar \tag{4.4}$$

as in one dimension, while all other commutators, including for instance $[\hat{x}, \hat{p}_y] = 0$, are zero because the momentum operators use partial derivatives. These commutators imply the uncertainty relations

$$\sigma_x \sigma_{p_x} \geq \frac{1}{2}\hbar \quad , \quad \sigma_y \sigma_{p_y} \geq \frac{1}{2}\hbar \quad , \quad \sigma_z \sigma_{p_z} \geq \frac{1}{2}\hbar\,. \tag{4.5}$$

The classical Hamilton function for a particle in three dimensions with a conservative potential $V(x,y,z)$ is given by

$$H = \frac{1}{2m}(p_x^2 + p_y^2 + p_z^2) + V(x,y,z)\,. \tag{4.6}$$

Using the position and momentum operators, we turn this expression into the Hamilton operator

$$\hat{H} = -\frac{\hbar^2}{2m}\left(\frac{\partial^2}{\partial x^2} + \frac{\partial^2}{\partial y^2} + \frac{\partial^2}{\partial z^2}\right) + V(x,y,z) = -\frac{\hbar^2}{2m}\nabla^2 + V(x,y,z) \tag{4.7}$$

with the Laplacian

$$\nabla^2 = \Delta = \frac{\partial^2}{\partial x^2} + \frac{\partial^2}{\partial y^2} + \frac{\partial^2}{\partial z^2} \tag{4.8}$$

in Cartesian coordinates. This operator determines the Schrödinger equation in three dimensions:

$$i\hbar\frac{\partial \Psi}{\partial t} = \hat{H}\Psi = -\frac{\hbar^2}{2m}\nabla^2\Psi + V\Psi\,. \tag{4.9}$$

DOI: 10.1201/9781003588702-5

As in our 1-dimensional examples, a conservative potential allows us to separate the time coordinate by using stationary states

$$\Psi_{\vec{n}}(x,y,z,t) = \psi_{\vec{n}}(x,y,z)\exp(-iE_{\vec{n}}t/\hbar) \qquad (4.10)$$

where $\psi_{\vec{n}}(x,y,z)$ is a solution of the 3-dimensional time-independent Schrödinger equation

$$\hat{H}\psi_{\vec{n}}(x,y,z) = -\frac{\hbar^2}{2m}\nabla^2\psi_{\vec{n}}(x,y,z) + V\psi_{\vec{n}}(x,y,z) = E_{\vec{n}}\psi_{\vec{n}}(x,y,z)\,. \qquad (4.11)$$

In general, we need three quantum numbers $\vec{n} = (n_1,n_2,n_3)$, or a continuous version $\vec{k} = (k_1,k_2,k_3)$ for scattering states, in order to obtain a complete set of solutions in three dimensions, or an eigenfunction basis. The energy eigenvalues then also have three independent quantum numbers, $\vec{n}$ or $\vec{k}$. Using completeness, the general time-dependent solution of the Schrödinger equation can be obtained as a superposition

$$\Psi(x,y,z,t) = \sum_{\vec{n}} c_{\vec{n}}\psi_{\vec{n}}(x,y,z)\exp(-iE_{\vec{n}}t/\hbar) \qquad (4.12)$$

with constants $c_{\vec{n}}$ determined by an initial wave function $\Psi(x,y,z,0)$.

We are still able to separate time as in one dimension, but the time-independent Schrödinger equation in three dimensions is a partial differential equation and much harder to solve than an ordinary differential equation in one dimension (which already required quite a bit of work). The time coordinate can be separated because a conservative potential has the property of time independence, a result of the symmetry of time translation invariance. Similarly, additional spatial symmetry implies that the potential depends on a smaller number of independent coordinates instead of all three Cartesian ones, (x,y,z). For instance, if the potential depends only on x, such we have translation symmetry in the y-z plane, an eigenbasis of the 3-dimensional Hamilton operator is given by products of stationary states for the 1-dimensional potential $V(x)$, labeled by an integer n or a wave number k, times plane waves in y and z with independent wave numbers, k_y and k_z.

If there is spatial symmetry, we are able to use separation of variables even in the time-independent 3-dimensional Schrödinger equation and reduce it to ordinary differential equations if there are sufficiently many independent symmetry transformation. However, depending on the symmetry, we may have to transform to non-Cartesian coordinates in order to obtain a potential that depends on fewer than three coordinates in an explicit form. Alternatively, if we can represent the symmetry transformations by parameterized families of unitary operators $\hat{U}(\delta)$, we can look for simultaneous eigenfunctions of $\hat{H}$ together with an independent set the self-adjoint operators derived from $i\hbar \mathrm{d}\hat{U}(\delta)/\mathrm{d}\delta|_{\delta=0}$.

4.1 ANGULAR MOMENTUM

An important example is given by rotational or spherical symmetry, such that the potential depends only on the radial distance from the origin. Such systems are best analyzed in spherical coordinates. We assume that the potential is not only time independent but also spherically symmetric, such that the general potential $V(x,y,z)$ is replaced by a function $V(r)$ that depends on (x,y,z) only through the radius $r = \sqrt{x^2+y^2+z^2}$. Our main example will be Coulomb's potential, where $V(r)$ is proportional to $-1/r$, as a model of the attractive electric force between a proton and an electron in the hydrogen atom. Mathematically, the fact that a spherically symmetric potential depends on only one spatial coordinate in a spherical coordinate system implies that the partial differential equation (4.11) can be separated further and be reduced to ordinary differential equations. These three equations correspond to simultaneous eigenfunction equations of a suitable set of commuting operators.

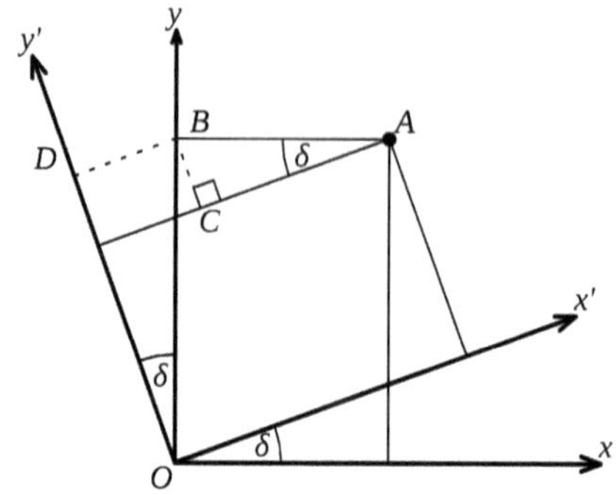

Figure 4.1 Rotation of a coordinate system. The rotation angle δ determines relationships between the coordinates in different frames. The dotted auxiliary lines DB and BC, the latter with a right angle at C, make it possible to derive the coordinate x' of a point A as a function of the coordinates x and y as well as the rotation angle δ, which appears three times in the diagram.

4.1.1 SYMMETRY TRANSFORMATIONS

Spherical symmetry implies three independent transformations from which a general rotation can be constructed by repeated application: rotations around the x, y and z axes, respectively. The traditional choice of spherical coordinates (r, ϑ, φ), expressed by

$$x = r\sin(\vartheta)\cos(\varphi) \quad , \quad y = r\sin(\vartheta)\sin(\varphi) \quad , \quad z = r\cos(\vartheta)\,, \tag{4.13}$$

singles out the z-axis because z depends only on two of the new coordinates. This choice is arbitrary, but it is convenient because it simplifies some of the general expressions for rotations. In particular, a rotation around the z-axis is simply represented by changing the angle φ. These rotations therefore resemble spatial translations, even though φ takes values in a bounded range (from 0 to 2π) and changing it does not move the corresponding point in a linear fashion. Nevertheless, we will see that the (angular) momentum operator associated with the symmetry of rotating around the z-axis shares some features with the momentum operator $\hat{p}$ in one spatial dimension. This is not the case for operators associated with the symmetries of rotating around the x and y axes, respectively, when they are expressed in the common choice of spherical coordinates.

The Cartesian coordinates of a point expressed in two coordinate systems with a relative rotation angle δ around the z-axis are related by

$$x' = x\cos(\delta) + y\sin(\delta) \tag{4.14}$$
$$y' = -x\sin(\delta) + y\cos(\delta) \tag{4.15}$$
$$z' = z\,. \tag{4.16}$$

Using Figure 4.1 we recognize the x' coordinate of point A as the sum of the lengths of the two lines AC and BD. Noting the right angle at point C, the length of AC equals $x\cos(\delta)$ based on the triangle ABC, while the length of BD equals $y\sin(\delta)$ based on the right angle at D in the triangle OBD. The sum equals (4.14), and a similar construction implies (4.15).

In Cartesian coordinates, a rotation around the z-axis by an angle δ is therefore represented by the matrix

$$R_z(\delta) = \begin{pmatrix} \cos(\delta) & \sin(\delta) & 0 \\ -\sin(\delta) & \cos(\delta) & 0 \\ 0 & 0 & 1 \end{pmatrix}\,. \tag{4.17}$$

This one-parameter family of matrices consists of **orthogonal** transformations that preserve the norm of vectors $R_z(\delta)\vec{x}$. It also acts on wave functions $\psi(\vec{x})$ by

$$\hat{U}_z(\delta)\psi(\vec{x}) = \psi(R_z(\delta)\vec{x}) \tag{4.18}$$

These transformations are unitary because

$$\begin{aligned}\int_{-\infty}^{\infty}\int_{-\infty}^{\infty}\int_{-\infty}^{\infty}|\psi(R_z(\delta)\vec{x})|^2\mathrm{d}x\mathrm{d}y\mathrm{d}z &= \int_{-\infty}^{\infty}\int_{-\infty}^{\infty}\int_{-\infty}^{\infty}|\psi(R_z(\delta)\vec{x})|^2\det(J(\delta))\mathrm{d}x\mathrm{d}y\mathrm{d}z\\ &= \int_{-\infty}^{\infty}\int_{-\infty}^{\infty}\int_{-\infty}^{\infty}|\psi(\vec{x}')|^2\mathrm{d}x'\mathrm{d}y'\mathrm{d}z'\end{aligned} \tag{4.19}$$

with the Jacobian

$$J(\delta) = \begin{pmatrix} \partial x'/\partial x & \partial x'/\partial y & \partial x'/\partial z\\ \partial y'/\partial x & \partial y'/\partial y & \partial y'/\partial z\\ \partial z'/\partial x & \partial z'/\partial y & \partial z'/\partial z\end{pmatrix} = R_z(\delta) \tag{4.20}$$

of the coordinate transformation $\vec{x}' = R_z(\delta)\vec{x}$. Since $J(\delta)$ has determinant one for any δ, we can insert it in the second step of (4.19) and transform the integral to the new coordinates $\vec{x}'$, where it takes the standard form.

4.1.2 ANGULAR-MOMENTUM OPERATORS

The unitary transformation now maps $\psi(x,y,z)$ to

$$\hat{U}_z(\delta)\psi(x,y,z) = \psi(x'(\delta),y'(\delta),z'(\delta)) = \psi(\cos(\delta)x+\sin(\delta)y, -\sin(\delta)x+\cos(\delta y),z)\,. \tag{4.21}$$

Its associated self-adjoint operator is given by

$$\begin{aligned}\hat{L}_z\psi(x,y,z) &= i\hbar\left.\frac{\mathrm{d}\hat{U}_z(\delta)}{\mathrm{d}\delta}\right|_{\delta=0}\psi(x,y,z) = i\hbar\left.\frac{\mathrm{d}\psi(x'(\delta),y'(\delta),z'(\delta))}{\mathrm{d}\delta}\right|_{\delta=0}\\ &= i\hbar\left.\left(\frac{\mathrm{d}x'(\delta)}{\mathrm{d}\delta}\frac{\partial\psi}{\partial x'}+\frac{\mathrm{d}y'(\delta)}{\mathrm{d}\delta}\frac{\partial\psi}{\partial y'}+\frac{\mathrm{d}z'(\delta)}{\mathrm{d}\delta}\frac{\partial\psi}{\partial z'}\right)\right|_{\delta=0}\\ &= \frac{\hbar}{i}\left(x\frac{\partial}{\partial y}-y\frac{\partial}{\partial x}\right)\psi\,.\end{aligned} \tag{4.22}$$

Using the linear momentum operators $\hat{p}_x = -i\hbar\partial/\partial x$ and $\hat{p}_y = -i\hbar\partial/\partial y$, we recognize this operator as the quantum version of the classical z-component of angular momentum,

$$L_z = xp_y - yp_x = (\vec{x}\times\vec{p})_z\,. \tag{4.23}$$

We expect that a rotation around the z-axis acts like a translation in φ, such that the momentum operator should be

$$\hat{L}_z = \frac{\hbar}{i}\frac{\partial}{\partial\varphi}\,. \tag{4.24}$$

This expression is indeed equivalent to (4.22), based on the chain rule for multiple variables. We have

$$\frac{\partial}{\partial x} = \frac{\partial r}{\partial x}\frac{\partial}{\partial r}+\frac{\partial\vartheta}{\partial x}\frac{\partial}{\partial\vartheta}+\frac{\partial\varphi}{\partial x}\frac{\partial}{\partial\varphi} = \frac{x}{r}\frac{\partial}{\partial r}+\frac{xz}{\sqrt{x^2+y^2}}\frac{\partial}{\partial\vartheta}-\frac{y}{\sqrt{x^2+y^2}}\frac{\partial}{\partial\varphi} \tag{4.25}$$

and

$$\frac{\partial}{\partial y} = \frac{\partial r}{\partial y}\frac{\partial}{\partial r}+\frac{\partial\vartheta}{\partial y}\frac{\partial}{\partial\vartheta}+\frac{\partial\varphi}{\partial y}\frac{\partial}{\partial\varphi} = \frac{y}{r}\frac{\partial}{\partial r}+\frac{yz}{\sqrt{x^2+y^2}}\frac{\partial}{\partial\vartheta}+\frac{x}{\sqrt{x^2+y^2}}\frac{\partial}{\partial\varphi} \tag{4.26}$$

derived from the inverse

$$r = \sqrt{x^2+y^2+z^2} \quad , \quad \vartheta = \arctan\left(\frac{\sqrt{x^2+y^2}}{z}\right) \quad , \quad \varphi = \arctan\left(\frac{y}{x}\right) \tag{4.27}$$

of (4.13). Therefore, $x\partial/\partial y - y\partial/\partial x = \partial/\partial\varphi$.

The same procedure can be applied to rotations around the x and y axis, respectively. These transformations are not simple translations in φ or ϑ, and therefore the form of corresponding momentum operators is hard to guess. Following the same steps as before shows that

$$\hat{L}_x = \frac{\hbar}{i}\left(y\frac{\partial}{\partial z} - z\frac{\partial}{\partial y}\right) \tag{4.28}$$

$$= \frac{\hbar}{i}\left(-\sin(\varphi)\frac{\partial}{\partial\vartheta} - \cot(\vartheta)\cos(\varphi)\frac{\partial}{\partial\varphi}\right) \tag{4.29}$$

and

$$\hat{L}_y = \frac{\hbar}{i}\left(z\frac{\partial}{\partial x} - x\frac{\partial}{\partial z}\right) \tag{4.30}$$

$$= \frac{\hbar}{i}\left(\cos(\varphi)\frac{\partial}{\partial\vartheta} - \cot(\vartheta)\sin(\varphi)\frac{\partial}{\partial\varphi}\right) \tag{4.31}$$

are the corresponding self-adjoint operators for angular-momentum components.

4.1.3 ADJOINTNESS PROPERTIES

Position and momentum operators do not always commute, and therefore their ordering in composite operators such as the $\hat{L}_j$ is important. If we first look at angular-momentum operators in Cartesian coordinates, the ordering of coordinate factors and derivatives is unambiguous because we always have products of components in different directions. For instance $[\hat{y},\hat{p}_z] = 0$ in the first term of $\hat{L}_x$ according to (4.28). In spherical coordinates, the two terms in $\cos(\varphi)\partial/\partial\varphi$ from (4.29) do not commute with each other. In this case, the derivation from rotation matrices and transformation to spherical coordinates shows that the specific ordering with $\partial/\partial\varphi$ to the right of $\cos(\varphi)$ should be used.

For the same reason, the operators $\hat{L}_j$ are easily seen to be Hermitian in the Cartesian form: They are sums of products of commuting Hermitian operators. For instance, $(\hat{y}\hat{p}_z)^\dagger = \hat{p}_z^\dagger\hat{y}^\dagger = \hat{p}_z\hat{y} = \hat{y}\hat{p}_z$. In spherical coordinates, showing Hermiticity is more tricky. Referring again to (4.29), the two terms in $\cos(\varphi)(i\partial/\partial\varphi)$ do not commute with each other. Therefore, the Hermitian conjugate of their product, acting on a generic wave function ψ, equals

$$\left(\cos(\varphi)i\frac{\partial}{\partial\varphi}\right)^\dagger\psi = \left(i\frac{\partial}{\partial\varphi}\right)^\dagger\cos(\varphi)^\dagger\psi = i\frac{\partial}{\partial\varphi}(\cos(\varphi)\psi) = \cos(\varphi)i\frac{\partial}{\partial\varphi}\psi - i\sin(\varphi)\psi \tag{4.32}$$

using the product rule. The extra term of $-i\sin(\varphi)$ must cancel out with a corresponding term from the Hermitian conjugate of $\sin(\varphi)(i\partial/\partial\vartheta)$ for $\hat{L}_x$ to be Hermitian. These two factors do commute, and each one appears to be Hermitian at first sight. However, it turns out that $i\partial/\partial\vartheta$ is *not* Hermitian.

We should carefully compute the Hermitian conjugate of $i\partial/\partial\vartheta$. In three spatial dimensions, the inner product of wave functions is defined by

$$\langle f|g\rangle = \int_{-\infty}^{\infty}\int_{-\infty}^{\infty}\int_{-\infty}^{\infty} f(x,y,z)^* g(x,y,z)\mathrm{d}x\mathrm{d}y\mathrm{d}z\,. \tag{4.33}$$

The Jacobian of the transformation to spherical coordinates allows us to express the integral in terms of r, ϑ and φ over their respective ranges:

$$\langle f|g\rangle = \int_0^r \int_0^\pi \int_0^{2\pi} f(r,\vartheta,\varphi)^* g(r,\vartheta,\varphi) r^2 \sin(\vartheta) \mathrm{d}\varphi \mathrm{d}\vartheta \mathrm{d}r \tag{4.34}$$

for two generic normalizable functions depending on spherical coordinates. The Jacobian $r^2\sin(\vartheta)$ now presents an additional function that must be taken into account when we compute the Hermitian conjugate of $i\partial/\partial\vartheta$ by integrating by parts (as well as $i\partial/\partial r$, which is not required for our current purpose). Since there is no additional factor depending on φ, $i\partial/\partial\varphi$ (and therefore $\hat{L}_z$) is guaranteed to be Hermitian, as used in (4.32).

For $\hat{T} = i\partial/\partial\vartheta$, we compute

$$\begin{aligned}
\langle f|\hat{T}g\rangle &= i\int_0^r \int_0^\pi \int_0^{2\pi} f(r,\vartheta,\varphi)^* \frac{\partial g(r,\vartheta,\varphi)}{\partial\vartheta} r^2 \sin(\vartheta) \mathrm{d}\varphi \mathrm{d}\vartheta \mathrm{d}r \\
&= -i\int_0^r \int_0^\pi \int_0^{2\pi} g(r,\vartheta,\varphi) \frac{\partial(f(r,\vartheta,\varphi)^* \sin(\vartheta))}{\partial\vartheta} r^2 \mathrm{d}\varphi \mathrm{d}\vartheta \mathrm{d}r \\
&= -i\int_0^r \int_0^\pi \int_0^{2\pi} g(r,\vartheta,\varphi) \frac{\partial f(r,\vartheta,\varphi)^*}{\partial\vartheta} r^2 \sin(\vartheta) \mathrm{d}\varphi \mathrm{d}\vartheta \mathrm{d}r \\
&\quad -i\int_0^r \int_0^\pi \int_0^{2\pi} g(r,\vartheta,\varphi) f(r,\vartheta,\varphi)^* r^2 \cos(\vartheta) \mathrm{d}\varphi \mathrm{d}\vartheta \mathrm{d}r \\
&= \langle (\hat{T} - i\cot(\vartheta)) f|g\rangle .
\end{aligned} \tag{4.35}$$

In the second step, integrating by parts, we used the fact that $\sin(\vartheta) = 0$ at the boundary values $\vartheta = 0$ and $\vartheta = \pi$. Therefore, $\hat{T}^\dagger = \hat{T} - i\cot(\vartheta)$ for $\hat{T} = i\partial/\partial\vartheta$. Combining the Hermitian conjugates of both terms in (4.29), using (4.32) and (4.35), now shows that this expression is Hermitian in spherical coordinates. A similar result follows for (4.31). As a useful lesson, we saw that it is important to consider the specific form of an inner product when we compute a Hermitian conjugate.

4.1.4 ALGEBRAIC PROPERTIES

The commutators of angular-momentum operators have interesting properties which are closely related to their origin, derived from rotations as symmetry transformations. There are three independent ways to rotate 3-dimensional space, given by rotations around the three axes of a Cartesian coordinate system. Any other rotation can be obtained as a combination of these three **generating** rotations: For a given angle δ and a rotation axis, we can always write the corresponding unitary operator as $\hat{U}(\delta) = \hat{U}_x(\delta_x)\hat{U}_y(\delta_y)\hat{U}_z(\delta_z)$ with suitable angles δ_x, δ_y and δ_z. Here, we represent the three generating rotations by $\hat{U}_j(\delta_j)$, $j = 1,2,3$. The three $\hat{U}_j(\delta_j)$ do not commute with one another, but we can always choose a specific ordering as an initial step to generate other transformations.

The unitary operators are exponentiated angular-momentum operators,

$$\hat{U}_j(\delta_j) = \exp(-i\delta_j\hat{L}_j/\hbar) . \tag{4.36}$$

Angular momentum, defined classically as $\vec{x} \times \vec{p}$, has units of position times momentum which, as demonstrated by a quick look at the basic uncertainty relation, happens to be the same as the units of $\hbar$. Therefore, the parameters δ_j are unitless in the present case, and can indeed be interpreted as angles.

Mathematically, the condition that the unitary operator for any rotation can be expressed as a product of generating unitary operators of the form (4.36) has implications for operations that can be performed with the $\hat{L}_j$. These operations can be derived by looking at small angles and using Taylor expansions of the exponential function. First, the product of two operators of the form (4.36), such as $\hat{U}_x(\delta_x)\hat{U}_y(\delta_y)$, has the expression

$$\hat{U}_x(\delta_x)\hat{U}_y(\delta_y) \approx (\mathbb{I} - i\delta_x\hat{L}_x/\hbar)(\mathbb{I} - i\delta_y\hat{L}_y/\hbar) \approx \mathbb{I} - i(\delta_x\hat{L}_x + \delta_y\hat{L}_y)/\hbar \tag{4.37}$$

to leading order in the small δ_x and δ_y. Therefore, multiplication of unitary operators corresponds to addition of the corresponding self-adjoint operators. Addition is commutative while matrix or operator multiplication is not. In the current expression, non-commutativity is seen only at higher orders in the expansion by angles. Since we also multiply the self-adjoint operators by numbers δ_j, these operators form a vector space in which we have addition and scalar multiplication.

Another multiplication of interest is

$$\begin{aligned}\hat{U}_x(\delta_x)\hat{U}_y(\delta_y)\hat{U}_x(-\delta_x)\hat{U}_y(-\delta_y) &\approx (\hat{\mathbb{I}}-i\delta_x\hat{L}_x/\hbar-\tfrac{1}{2}\delta_x^2\hat{L}_x^2/\hbar^2)(\hat{\mathbb{I}}-i\delta_y\hat{L}_y/\hbar-\tfrac{1}{2}\delta_y^2\hat{L}_y^2/\hbar^2)\\ &\quad\times(\hat{\mathbb{I}}+i\delta_x\hat{L}_x/\hbar-\tfrac{1}{2}\delta_x^2\hat{L}_x^2/\hbar^2)(\hat{\mathbb{I}}+i\delta_y\hat{L}_y/\hbar-\tfrac{1}{2}\delta_y^2\hat{L}_y^2/\hbar^2)\\ &\approx \hat{\mathbb{I}}-\delta_x\delta_y[\hat{L}_x,\hat{L}_y]/\hbar^2\,. \end{aligned} \tag{4.38}$$

We have to expand to second order in δ_x and δ_y because first-order terms cancel out completely. The left-hand side can be used as a measure of non-commutativity of the unitary operators because all terms would cancel out if the operators were commuting. Correspondingly, we see the commutator $[\hat{L}_x,\hat{L}_y]=\hat{L}_x\hat{L}_y-\hat{L}_y\hat{L}_x$ of self-adjoint operators on the right. Taking the commutator is therefore another operation that we are able to perform on these self-adjoint operators with a corresponding operation in the one-parameter families of unitary operators. Since the commutator is linear on each side, it can be combined with addition and scalar multiplication. The direct operator product or composition of self-adjoint operators, by contrast, does not have a corresponding operation for the full Taylor series or for the unitary operators.

Commutators as well as composition, applying one after the other, are defined for any pair of operators. In the present case, the commutator of $\hat{L}_x$ and $\hat{L}_y$ plays a special role because we can derive it from (4.38). Since the left-hand side of this equation is a general unitary operator for a rotation around some axis, its Taylor expansion must, to leading order, equal the identity operator plus a linear combination of all the available self-adjoint operators, which are given by $\hat{L}_x$, $\hat{L}_y$ and $\hat{L}_z$. In particular, we may view the product $\hat{U}_x(\delta)\hat{U}_y(\delta)\hat{U}_x(-\delta)\hat{U}_y(-\delta)$, setting $\delta_x=\delta_y$, as another one-parameter family of unitary operators, $\hat{U}(\varepsilon)$ with $\varepsilon=\delta^2$, and compute the corresponding self-adjoint operator

$$\hat{L}=i\hbar\left.\frac{\mathrm{d}\hat{U}(\varepsilon)}{\mathrm{d}\varepsilon}\right|_{\varepsilon=0}=-\frac{i}{\hbar}[\hat{L}_x,\hat{L}_y] \tag{4.39}$$

from the right-hand side of (4.38). Since the product of unitary operators for rotations is a unitary operator for another rotation, the resulting $\hat{L}$ must be an element of the vector space of self-adjoint operators generated by $\hat{L}_x$, $\hat{L}_y$ and $\hat{L}_z$, and therefore be a linear combination of them.

In other words, the operation of performing the commutator preserves the vector space. Such a space defines a **Lie algebra**, and the commutator is an example of a **Lie bracket**, the general name for the operation in a Lie algebra. In general, a Lie bracket is defined as an antisymmetric operation on two elements of a vector space that satisfies the **Jacobi identity**

$$[[\hat{A},\hat{B}],\hat{C}]+[[\hat{B},\hat{C}],\hat{A}]+[[\hat{C},\hat{A}],\hat{B}]=0\,, \tag{4.40}$$

using cyclic permutations, for all $\hat{A}$, $\hat{B}$, and $\hat{C}$ in the vector space. (We use operator notation because the present case implies a vector space of operators.) For commutators, this identity can be confirmed directly based on the definition $[\hat{A},\hat{B}]=\hat{A}\hat{B}-\hat{B}\hat{A}$, without using specific operators. The corresponding set of unitary operators and their multiplication law is called a **Lie group**. As used here, exponentiation and taking derivatives map a Lie algebra to a Lie group, and vice versa.

For angular-momentum operators, it is easy to compute commutators that include $\hat{L}_z$ because it just requires taking a single derivative by φ. For instance, using (4.24) and (4.29), we have

$$[\hat{L}_z,\hat{L}_x]=\hbar^2\frac{\mathrm{d}\sin(\varphi)}{\mathrm{d}\varphi}\frac{\partial}{\partial\vartheta}+\cot(\vartheta)\frac{\mathrm{d}\cos(\varphi)}{\mathrm{d}\varphi}\frac{\partial}{\partial\varphi}=i\hbar\hat{L}_y\,, \tag{4.41}$$

the final step as per (4.31). Using a spherical coordinate system in which rotations around the z-axis are a simple change in one of the angles, φ, was a choice required for a specific coordinate system, but not for the representation of rotations in general. Choosing instead the y or x axis amounts to a rotation of the entire coordinate system by ninety degrees around the $-x$ or y axis, respectively, which can always be expressed as a cyclic permutation of the three axes. The remaining commutators of angular-momentum components must therefore be cyclic permutations of (4.41):

$$[\hat{L}_x,\hat{L}_y] = i\hbar\hat{L}_z \quad , \quad [\hat{L}_y,\hat{L}_z] = i\hbar\hat{L}_x \quad , \quad [\hat{L}_z,\hat{L}_x] = i\hbar\hat{L}_y . \tag{4.42}$$

Unlike the expressions of angular momentum as derivative operators, the commutators (4.42) are independent of choices of coordinates. They define the Lie algebra corresponding to the rotation group.

Commutators imply uncertainty relations for the statistics of measurements, here related to different components of angular momentum. For instance, we have

$$\sigma_{L_x}\sigma_{L_y} \geq \left|\frac{1}{2}i\langle[\hat{L}_x,\hat{L}_y]\rangle\right| = \frac{\hbar}{2}|\langle\hat{L}_z\rangle| \tag{4.43}$$

for measurements of L_x and L_y. Unless $\langle\hat{L}_z\rangle = 0$ in the state realized for the measurement, we cannot have sharp values for both L_x and L_y.

At most one of the components of $\vec{L}$ can be measured sharply at one time: We cannot have a complete set of simultaneous eigenfunctions for two or more angular-momentum components. However, the (squared) magnitude of angular momentum, $||\vec{L}||^2 = L_x^2 + L_y^2 + L_z^2$, can be measured sharply together with any one of the components. In the case of $\hat{L}_z$,

$$\begin{aligned}
[\hat{L}^2,\hat{L}_z] &= [\hat{L}_x^2,\hat{L}_z] + [\hat{L}_y^2,\hat{L}_z] + [\hat{L}_z^2,\hat{L}_z] \\
&= \hat{L}_x[\hat{L}_x,\hat{L}_z] + [\hat{L}_x,\hat{L}_z]\hat{L}_x + \hat{L}_y[\hat{L}_y,\hat{L}_z] + [\hat{L}_y,\hat{L}_z]\hat{L}_y \\
&= -i\hbar\hat{L}_x\hat{L}_y - i\hbar\hat{L}_y\hat{L}_x + i\hbar\hat{L}_y\hat{L}_x + i\hbar\hat{L}_x\hat{L}_y = 0
\end{aligned} \tag{4.44}$$

follows from the commutators (4.42) and the product rule (3.194). This result can also be obtained from the underlying symmetries: The length of a vector, such as L^2, is invariant under rotations. For operators, we therefore have

$$\begin{aligned}
\hat{L}^2 &= \exp(-i\delta\hat{L}_z/\hbar)\hat{L}^2\exp(i\delta\hat{L}_z/\hbar) \approx (\hat{\mathbb{I}} - i\delta\hat{L}_z/\hbar)\hat{L}^2(\hat{\mathbb{I}} + i\delta\hat{L}_z/\hbar) \\
&\approx \hat{L}^2 + i\frac{\delta}{\hbar}[\hat{L}^2,\hat{L}_z]
\end{aligned} \tag{4.45}$$

as a version of a conservation law relative to transformations $\hat{U}_{\hat{L}_z}(\delta)$. This calculation implies $[\hat{L}^2,\hat{L}_z] = 0$, or

$$[\hat{L}^2,\hat{L}_x] = [\hat{L}^2,\hat{L}_y] = [\hat{L}^2,\hat{L}_z] = 0 \tag{4.46}$$

for all the components of angular momentum.

It is therefore possible to find a simultaneous eigenbasis of $\hat{L}^2$ and one of the components, usually chosen as $\hat{L}_z$:

$$\hat{L}^2|f_{\lambda,\mu}\rangle = \lambda|f_{\lambda,\mu}\rangle \quad , \quad \hat{L}_z|f_{\lambda,\mu}\rangle = \mu|f_{\lambda,\mu}\rangle \tag{4.47}$$

for a single set of states $|f_{\lambda,\mu}\rangle$ labeled by two eigenvalues, λ and μ.

4.1.5 EIGENVALUES

The eigenvalue problem of angular-momentum operators can be solved by the algebraic method, defining ladder operators suitable for the commutators of angular-momentum components. We introduce the non-Hermitian ladder operators

$$\hat{L}_\pm = \hat{L}_x \pm i\hat{L}_y \tag{4.48}$$

which have the commutators

$$[\hat{L}_z, \hat{L}_\pm] = \pm\hbar\hat{L}_\pm \quad , \quad [\hat{L}^2, \hat{L}_\pm] = 0 \tag{4.49}$$

with $\hat{L}_z$ and $\hat{L}^2$. If we assume that we know one eigenfunction, $|f\rangle = |f_{\lambda,\mu}\rangle$ for some fixed λ and μ, then $\hat{L}_\pm|f\rangle$ has the same eigenvalue λ of $\hat{L}^2$ because

$$\hat{L}^2(\hat{L}_\pm|f\rangle) = \hat{L}_\pm(\hat{L}^2|f\rangle) = \lambda\hat{L}_\pm|f\rangle\,, \tag{4.50}$$

and it has the eigenvalue $\mu \pm \hbar$ of $\hat{L}_z$ shifted by $\pm\hbar$ because

$$\hat{L}_z(\hat{L}_\pm|f\rangle = \left([\hat{L}_z, \hat{L}_\pm] + \hat{L}_\pm\hat{L}_z\right)|f\rangle = \pm\hbar\hat{L}_\pm|f\rangle + \mu\hat{L}_\pm|f\rangle = (\mu\pm\hbar)\hat{L}_\pm|f\rangle\,. \tag{4.51}$$

This equation is similar to our result (2.97) obtained for ladder operators $\hat{a}_\pm$ and the energy operator instead of $\hat{L}_z$, acting on stationary states of the harmonic oscillator. As in this case, we call $\hat{L}_+$ the raising operator and $\hat{L}_-$ the lowering operator. By acting with these operators multiple times on one eigenfunction of $\hat{L}_z$, we obtain a whole ladder of eigenfunctions with a spacing given by $\hbar$.

All these eigenfunctions of $\hat{L}_z$ have the same eigenvalue of $\hat{L}^2$. Classically, the value of L_z is bounded from above and below by the inequality $L_z^2 \leq L^2$, such that we have a finite range of values of L_z for given L^2. There is a similar bound in quantum mechanics, which can be derived from the equation

$$\hat{L}_-\hat{L}_+ = (\hat{L}_x - i\hat{L}_y)(\hat{L}_x + i\hat{L}_y) = \hat{L}_x^2 + \hat{L}_y^2 + i[\hat{L}_x, \hat{L}_y] = \hat{L}_x^2 + \hat{L}_y^2 - \hbar\hat{L}_z = \hat{L}^2 - \hat{L}_z^2 - \hbar\hat{L}_z \tag{4.52}$$

and a similar version for

$$\hat{L}_+\hat{L}_- = (\hat{L}_x + i\hat{L}_y)(\hat{L}_x - i\hat{L}_y) = \hat{L}_x^2 + \hat{L}_y^2 - i[\hat{L}_x, \hat{L}_y] = \hat{L}_x^2 + \hat{L}_y^2 + \hbar\hat{L}_z = \hat{L}^2 - \hat{L}_z^2 + \hbar\hat{L}_z\,. \tag{4.53}$$

These two equations imply

$$\hat{L}^2 = \hat{L}_-\hat{L}_+ + \hat{L}_z^2 + \hbar\hat{L}_z = \hat{L}_+\hat{L}_- + \hat{L}_z^2 - \hbar\hat{L}_z\,. \tag{4.54}$$

In an eigenstate $|f\rangle$ of $\hat{L}^2$ and $\hat{L}_z$, we therefore have

$$\begin{aligned} \lambda\langle f|f\rangle &= \langle f|\hat{L}^2|f\rangle = \langle f|\hat{L}_-\hat{L}_+ + \hat{L}_z^2 + \hbar\hat{L}_z|f\rangle = \langle\hat{L}_+f|\hat{L}_+f\rangle + \mu^2\langle f|f\rangle + \mu\hbar\langle f|f\rangle \\ &\geq (\mu^2 + \mu\hbar)\langle f|f\rangle \end{aligned} \tag{4.55}$$

using $\hat{L}_-^\dagger = \hat{L}_+$ and $||\hat{L}_+f||^2 = \langle\hat{L}_+f|\hat{L}_+f\rangle \geq 0$ for any $|f\rangle$. We conclude that $\mu^2 + \mu\hbar \leq \lambda$ and, repeating the calculation for the second equality in (4.54), $\mu^2 - \mu\hbar \leq \lambda$. Adding the two inequalities implies

$$\mu^2 \leq \lambda\,. \tag{4.56}$$

Therefore, $\lambda \geq 0$ cannot be negative, and μ is bounded from above and below for a given λ.

For a fixed λ, a ladder of $\hat{L}_z$-eigenfunctions obtained by repeated actions of $\hat{L}_\pm$ must be of finite length in order to respect the inequality (4.56). The ladder is bounded by a top state $|f_{\rm top}\rangle$ with

$$\hat{L}_+|f_{\rm top}\rangle = 0\,, \tag{4.57}$$

and a bottom state $|f_{\rm bottom}\rangle$ with

$$\hat{L}_-|f_{\rm bottom}\rangle = 0\,. \tag{4.58}$$

In $|f_{\rm top}\rangle$, we have the greatest eigenvalue of $\hat{L}_z$ for a given $\hat{L}^2$-eigenvalue λ, which we call $\hbar\ell$, and in $|f_{\rm bottom}\rangle$ we have the smallest eigenvalue of $\hat{L}_z$, which we call $\hbar\bar{\ell}$ with some $\bar{\ell} \leq \ell$. As equations,

$$\hat{L}_z|f_{\rm top}\rangle = \hbar\ell|f_{\rm top}\rangle \quad , \quad \hat{L}_z|f_{\rm bottom}\rangle = \hbar\bar{\ell}|f_{\rm bottom}\rangle \tag{4.59}$$

as well as

$$\hbar\bar{\ell} \leq \mu \leq \hbar\ell\,. \tag{4.60}$$

We defined the states $|f_{\rm top}\rangle$ and $|f_{\rm bottom}\rangle$ as states on the same ladder with a given $\hat{L}^2$-eigenvalue λ. This eigenvalue can be computed independently in two different ways,

$$\hat{L}^2|f_{\rm top}\rangle = (\hat{L}_-\hat{L}_+ + \hat{L}_z^2 + \hbar\hat{L}_z)|f_{\rm top}\rangle = \left((\hbar\ell)^2 + \hbar^2\ell\right)|f_{\rm top}\rangle = \hbar^2\ell(\ell+1)|f_{\rm top}\rangle \tag{4.61}$$

using (4.57) and (4.59), and

$$\begin{aligned}\hat{L}^2|f_{\rm bottom}\rangle &= (\hat{L}_+\hat{L}_- + \hat{L}_z^2 - \hbar\hat{L}_z)|f_{\rm bottom}\rangle = \left((\hbar\bar{\ell})^2 - \hbar^2\bar{\ell}\right)|f_{\rm bottom}\rangle \\ &= \hbar^2\bar{\ell}(\bar{\ell}-1)|f_{\rm bottom}\rangle \end{aligned} \tag{4.62}$$

using (4.58) and (4.59). Since both states have the same eigenvalue of $\hat{L}^2$, we obtain $\lambda = \hbar^2\ell(\ell+1) = \hbar^2\bar{\ell}(\bar{\ell}-1)$ which is possible if $\bar{\ell} = -\ell$ or $\bar{\ell} = \ell+1$. However, only the first solution is consistent with our definitions which implied that $\bar{\ell} \leq \ell$. Therefore, $\bar{\ell} = -\ell$ is strictly related to ℓ. This equation then implies that ℓ cannot be negative in order to obey the condition $\bar{\ell} = -\ell \leq \ell$. For a given value of $\ell \geq 0$, we therefore have the $\hat{L}^2$-eigenvalue $\hbar^2\ell(\ell+1)$, and an $\hat{L}_z$-eigenvalue of the form $\mathrm{m}\hbar$ with $-\ell \leq \mathrm{m} \leq \ell$ obtained from (4.60):

$$\hat{L}^2|f_\ell^{\rm m}\rangle = \hbar^2\ell(\ell+1)|f_\ell^{\rm m}\rangle \quad , \quad \hat{L}_z|f_\ell^{\rm m}\rangle = \hbar\mathrm{m}|f_\ell^{\rm m}\rangle\,. \tag{4.63}$$

The result that ℓ is uniquely determined by λ shows that there is only one ladder of $\hat{L}_z$-eigenstates for fixed λ.

Properties of the ladder imply further restrictions on ℓ and m. By construction, it is possible to move from $|f_\ell^{-\ell}\rangle = |f_{\rm bottom}\rangle$ to $|f_\ell^{\ell}\rangle = |f_{\rm top}\rangle$ by acting with the raising operator $\hat{L}_+$ a certain number of times. Each action raises the value of m by an integer amount because of (4.51) and (4.63). The difference $\ell-(-\ell) = 2\ell$ of the top and bottom values of m equals the integer number of applications of the raising operator. Since 2ℓ is an integer, the allowed values for $\ell \geq 0$ are

$$\ell = 0, \frac{1}{2}, 1, \frac{3}{2}, 2, \ldots \tag{4.64}$$

For each of these values, there are $2\ell+1$ integer values for

$$\mathrm{m} = -\ell, -\ell+1, \ldots, \ell-1, \ell\,. \tag{4.65}$$

These relations determine all properties of the action of $\hat{L}_z$ and $\hat{L}^2$ on eigenfunctions. The action of $\hat{L}_x$ and $\hat{L}_y$ can be obtained from the ladder operators as well. We have $\hat{L}_\pm|f_\ell^{\rm m}\rangle = A^\pm_{\ell,\rm m}|f_\ell^{\rm m\pm1}\rangle$, with constants $A^\pm_{\ell,\rm m}$ determined by normalization coefficients of the two eigenfunctions $|f_\ell^{\rm m}\rangle$ and $|f_\ell^{\rm m\pm1}\rangle$. Using $\hat{L}_\pm^\dagger = \hat{L}_\mp$ as well as (4.52) and (4.53), we compute

$$|A^\pm_{\ell,\rm m}|^2 = \langle \hat{L}_\pm f_\ell^{\rm m}|\hat{L}_\pm f_\ell^{\rm m}\rangle = \langle f_\ell^{\rm m}|\hat{L}_\mp\hat{L}_\pm|f_\ell^{\rm m}\rangle \tag{4.66}$$

$$= \langle f_\ell^{\rm m}|\hat{L}^2 - \hat{L}_z^2 \mp \hbar\hat{L}_z|f_\ell^{\rm m}\rangle = \hbar^2(\ell(\ell+1) - \mathrm{m}^2 \mp \mathrm{m})\,. \tag{4.67}$$

Therefore,

$$\hat{L}_\pm|f_\ell^{\rm m}\rangle = \hbar\sqrt{\ell(\ell+1) - \mathrm{m}(\mathrm{m}\pm1)}\,|f_\ell^{\rm m\pm1}\rangle\,. \tag{4.68}$$

For fixed ℓ, the coefficients in this equation can be interpreted as matrix elements

$$L_{\pm,\rm m\pm1\,m} = \langle f_\ell^{\rm m\pm1}|\hat{L}_\pm|f_\ell^{\rm m}\rangle = \hbar\sqrt{\ell(\ell+1) - \mathrm{m}(\mathrm{m}\pm1)} \tag{4.69}$$

of the operator $\hat{L}_\pm$ represented on the finite-dimensional vector space with an orthonormal basis $\{|f_\ell^{\ell}\rangle, |f_\ell^{\ell-1}\rangle, \ldots, |f_\ell^{-\ell}\rangle\}$. Adding and subtracting these matrices, we obtain representations of $\hat{L}_x = \frac{1}{2}(\hat{L}_+ + \hat{L}_-)$ and $\hat{L}_y = -\frac{1}{2}i(\hat{L}_+ - \hat{L}_-)$ as non-diagonal matrices.

4.1.6 EIGENVECTORS

If we fix ℓ, there are $2\ell+1$ independent eigenvectors of the angular-momentum operators, which can be represented on a finite-dimensional Hilbert space. For $\ell=1$, for instance, we have three abstract basis states, $|f_1^1\rangle$, $|f_1^0\rangle$, and $|f_1^{-1}\rangle$ labeled by the three allowed values for $\mathfrak{m}$. In this case, we are constructing a 3-dimensional representation of angular-momentum operators by 3×3-matrices.

4.1.6.1 Matrix representation

Since the label $\mathfrak{m}$ of basis states determines the eigenvalue $\mathfrak{m}\hbar$ of $\hat{L}_z$, this operator is represented by a diagonal matrix,

$$\hat{L}_z \leftrightarrow L_z = \begin{pmatrix} \hbar & 0 & 0 \\ 0 & 0 & 0 \\ 0 & 0 & -\hbar \end{pmatrix}. \tag{4.70}$$

We next derive the matrices that represent $\hat{L}_\pm$, using (4.68). For $\ell=1$, $\ell(\ell+1)=2$ and therefore

$$\hat{L}_\pm|f_1^{\mathfrak{m}}\rangle = \hbar\sqrt{2-\mathfrak{m}(\mathfrak{m}\pm 1)}|f_1^{\mathfrak{m}\pm 1}\rangle. \tag{4.71}$$

We have the following individual actions of $\hat{L}_\pm$ on the three basis states:

$$\hat{L}_+|f_1^1\rangle = 0 \quad , \quad \hat{L}_-|f_1^1\rangle = \sqrt{2}\hbar|f_1^0\rangle \tag{4.72}$$

$$\hat{L}_+|f_1^0\rangle = \sqrt{2}\hbar|f_1^1\rangle \quad , \quad \hat{L}_-|f_1^0\rangle = \sqrt{2}\hbar|f_1^{-1}\rangle \tag{4.73}$$

$$\hat{L}_+|f_1^{-1}\rangle = \sqrt{2}\hbar|f_1^0\rangle \quad , \quad \hat{L}_-|f_1^{-1}\rangle = 0. \tag{4.74}$$

Matrix elements of a linear transformation $\hat{T}$ are defined by the expansion coefficients T_{ji} in the action

$$\hat{T}|e_i\rangle = \sum_{j=1}^{n} T_{ji}|e_j\rangle \tag{4.75}$$

of $\hat{T}$ on basis states $|e_i\rangle$. Therefore, the i-th column of the matrix T is determined by the expansion coefficients of $\hat{T}|e_i\rangle$ in the basis. In the present case, this procedure implies the matrices

$$\hat{L}_+ \leftrightarrow L_+ = \begin{pmatrix} 0 & \sqrt{2}\hbar & 0 \\ 0 & 0 & \sqrt{2}\hbar \\ 0 & 0 & 0 \end{pmatrix} \tag{4.76}$$

and

$$\hat{L}_- \leftrightarrow L_- = \begin{pmatrix} 0 & 0 & 0 \\ \sqrt{2}\hbar & 0 & 0 \\ 0 & \sqrt{2}\hbar & 0 \end{pmatrix}. \tag{4.77}$$

These matrices then determine

$$\hat{L}_x \leftrightarrow L_x = \frac{1}{2}(L_+ + L_-) = \begin{pmatrix} 0 & \hbar/\sqrt{2} & 0 \\ \hbar/\sqrt{2} & 0 & \hbar/\sqrt{2} \\ 0 & \hbar/\sqrt{2} & 0 \end{pmatrix} \tag{4.78}$$

and

$$\hat{L}_y \leftrightarrow L_y = \frac{1}{2i}(L_+ - L_-) = \begin{pmatrix} 0 & -i\hbar/\sqrt{2} & 0 \\ i\hbar/\sqrt{2} & 0 & -i\hbar/\sqrt{2} \\ 0 & i\hbar/\sqrt{2} & 0 \end{pmatrix}. \tag{4.79}$$

4.1.6.2 Position representation

It is possible to look for eigenfunctions of $\hat{L}_z$ and $\hat{L}^2$ as derivative operators in the position representation, given by wave functions $f_\ell^{\mathfrak{m}}(\vartheta,\varphi)$. A function space is infinite-dimensional, but if we fix ℓ, a finite number of functions suffices to span the space of all functions obtained by acting with the angular-momentum operators. The angle dependence of these wave functions determines the form of atomic orbitals, which are characterized by different values of angular momentum.

In order to derive explicit wave functions for our previous abstract representation, we use the angular-momentum operators in spherical coordinates, given by (4.24), (4.29) and (4.31), or

$$\hat{L}_x = \frac{\hbar}{i}\left(-\sin(\varphi)\frac{\partial}{\partial\vartheta} - \cos(\varphi)\frac{\cos(\vartheta)}{\sin(\vartheta)}\frac{\partial}{\partial\varphi}\right) \quad (4.80)$$

$$\hat{L}_y = \frac{\hbar}{i}\left(\cos(\varphi)\frac{\partial}{\partial\vartheta} - \sin(\varphi)\frac{\cos(\vartheta)}{\sin(\vartheta)}\frac{\partial}{\partial\varphi}\right) \quad (4.81)$$

$$\hat{L}_z = \frac{\hbar}{i}\frac{\partial}{\partial\varphi}. \quad (4.82)$$

The ladder operators in the position representation are

$$\begin{aligned}\hat{L}_\pm &= \hat{L}_x \pm i\hat{L}_y = \frac{\hbar}{i}\left((-\sin(\varphi)\pm i\cos(\varphi))\frac{\partial}{\partial\vartheta} - (\cos(\varphi)\pm i\sin(\varphi))\frac{\cos(\vartheta)}{\sin(\vartheta)}\frac{\partial}{\partial\varphi}\right)\\ &= \pm\hbar e^{\pm i\varphi}\left(\frac{\partial}{\partial\vartheta} \pm i\frac{\cos(\vartheta)}{\sin(\vartheta)}\frac{\partial}{\partial\varphi}\right). \end{aligned} \quad (4.83)$$

The eigenfunction conditions for angular-momentum operators can now be written as differential equations. For simultaneous eigenstates $f_\ell^{\mathfrak{m}}(\vartheta,\varphi)$ of $\hat{L}_z$ and $\hat{L}^2$, we have

$$\hat{L}_z f_\ell^{\mathfrak{m}} = \frac{\hbar}{i}\frac{\partial f_\ell^{\mathfrak{m}}}{\partial\varphi} = \hbar\mathfrak{m} f_\ell^{\mathfrak{m}} \quad (4.84)$$

and therefore

$$f_\ell^{\mathfrak{m}}(\vartheta,\varphi) = g_\ell^{\mathfrak{m}}(\vartheta)e^{i\mathfrak{m}\varphi}. \quad (4.85)$$

For these functions to have unique values everywhere in 3-dimensional space, we require $f_\ell^{\mathfrak{m}}(\vartheta,\varphi+2\pi) = f_\ell^{\mathfrak{m}}(\vartheta,\varphi)$ since $\varphi+2\pi$ and φ describe the same point in space, characterized by the same Cartesian coordinates x, y and z. Thus,

$$e^{\pm i\mathfrak{m}(\varphi+2\pi)} = e^{\pm i\mathfrak{m}\varphi} \quad (4.86)$$

or $e^{\pm 2\pi i\mathfrak{m}} = 1$, which implies that $\mathfrak{m}$ must be an integer. This condition is almost what we obtained by the algebraic method, where (4.65) with (4.64) leads to integer $\mathfrak{m}$ only if ℓ is an integer. The half-integer values $\ell = 1/2, 3/2, \ldots$ do not admit a position representation, while a matrix representation can still be constructed. This distinction gives rise to the phenomena of **spin** to be discussed in due course.

The allowed form of functions $g_\ell^{\mathfrak{m}}(\vartheta)$ is determined by the condition that the state is not only an eigenfunction of $\hat{L}_z$ but also of $\hat{L}^2$. For this operator, we compute

$$\begin{aligned}\hat{L}^2 f_\ell^{\mathfrak{m}} &= (\hat{L}_+\hat{L}_- + \hat{L}_z^2 - \hbar\hat{L}_z)f_\ell^{\mathfrak{m}} \quad (4.87)\\ &= \hbar e^{i\varphi}\left(\frac{\partial}{\partial\vartheta} + i\frac{\cos(\vartheta)}{\sin(\vartheta)}\frac{\partial}{\partial\varphi}\right)\left(-\hbar e^{-i\varphi}\left(\frac{\partial f_\ell^{\mathfrak{m}}}{\partial\vartheta} - i\frac{\cos(\vartheta)}{\sin(\vartheta)}\frac{\partial f_\ell^{\mathfrak{m}}}{\partial\varphi}\right)\right) - \hbar^2\frac{\partial^2 f_\ell^{\mathfrak{m}}}{\partial\varphi^2} - \frac{\hbar^2}{i}\frac{\partial f_\ell^{\mathfrak{m}}}{\partial\varphi}\\ &= -\hbar^2 e^{i\varphi}\left(\frac{\partial}{\partial\vartheta} + i\frac{\cos(\vartheta)}{\sin(\vartheta)}\frac{\partial}{\partial\varphi}\right)\left(e^{i(\mathfrak{m}-1)\varphi}\left(\frac{dg_\ell^{\mathfrak{m}}}{d\vartheta} + \mathfrak{m}\frac{\cos(\vartheta)}{\sin(\vartheta)}\right)\right) + \hbar^2(\mathfrak{m}^2-\mathfrak{m})g_\ell^{\mathfrak{m}}e^{i\mathfrak{m}\varphi}\end{aligned}$$

$$= \hbar^2 e^{im\varphi}\left(-\frac{d}{d\vartheta}\left(\frac{dg_\ell^m}{d\vartheta}+m\frac{\cos(\vartheta)}{\sin(\vartheta)}g_\ell^m\right)\right.$$
$$\left.+(m-1)\frac{\cos(\vartheta)}{\sin(\vartheta)}\left(\frac{dg_\ell^m}{d\vartheta}+m\frac{\cos(\vartheta)}{\sin(\vartheta)}g_\ell^m\right)+m(m-1)g_\ell^m\right)$$

and set the result equal to $\hbar^2\ell(\ell+1)g_\ell^m e^{im\varphi}$:

$$-\frac{d^2g_\ell^m}{d\vartheta^2}-\frac{\cos(\vartheta)}{\sin(\vartheta)}\frac{dg_\ell^m}{d\vartheta}+\frac{m^2}{\sin(\vartheta)^2}g_\ell^m=\ell(\ell+1)g_\ell^m. \tag{4.88}$$

Multiplying this differential equation for eigenfunctions of $\hat{L}^2$ by $\sin(\vartheta)^2$, we obtain

$$\sin(\vartheta)\frac{d}{d\vartheta}\left(\sin(\vartheta)\frac{dg_\ell^m}{d\vartheta}\right)+\left(\ell(\ell+1)\sin(\vartheta)^2-m^2\right)g_\ell^m=0. \tag{4.89}$$

This second-order differential equation should have two independent solutions for given ℓ and m, but it turns out that one of them has singularities at $\sin(\vartheta)=0$ and is not normalizable even over the finite range $0\leq\vartheta\leq\pi$. Solutions of (4.89), multiplied by $e^{im\varphi}$ are called **spherical harmonics** $Y_\ell^m(\vartheta,\varphi)=g_\ell^m(\vartheta)e^{im\varphi}$ because, as eigenfunctions of $\hat{L}^2$ and $\hat{L}_z$, they are spherical analogs of harmonic functions, the eigenfunctions of linear momentum. Unlike sine and cosine, spherical harmonics are not simple functions, but thanks to their importance they are readily available in tables or computer algebra.

In terms of somewhat standard mathematical functions, we write

$$g_\ell^m(\vartheta)=C_{\ell m}P_\ell^m(\cos(\vartheta)) \tag{4.90}$$

with a normalization constant $C_{\ell m}$ and the **associated Legendre functions** defined as

$$P_\ell^m(\xi)=\frac{1}{2^\ell\ell!}(1-\xi^2)^{|m|/2}\left(\frac{d}{d\xi}\right)^{|m|+\ell}(\xi^2-1)^\ell. \tag{4.91}$$

The derivative structure of this expression determines conditions on the allowed values for ℓ, which we can compare with the conditions derived by the algebraic method.

In contrast to φ, from which we obtained a quantization condition on m, there is no periodicity condition in ϑ or $\xi=\cos(\vartheta)$: The two boundary points $\vartheta=0$ and $\vartheta=\pi$ of the range correspond to two different points on the sphere, the North pole and the South pole. Nevertheless, the specific form of the associated Legendre functions implies restrictions on the parameter ℓ in order to have regular functions on the sphere:

1. For P_ℓ^m to be regular at $\xi=1$ and $\xi=-1$, or $\cos(\vartheta)=\pm1$ at the two poles, ℓ cannot be negative in order to avoid a divergence of $(\xi^2-1)^\ell$.
2. Since $\xi^2-1\leq0$ for $\xi=\cos(\vartheta)$, we can only take integer powers of this expression for a real-valued solution of the real equation (4.89). Therefore, ℓ must be an integer, $\ell=0,1,2,\ldots$
3. Given the previous condition on ℓ, $(\xi^2-1)^\ell$ is a polynomial in ℓ of degree 2ℓ. When we compute $P_\ell^m(\xi)$ for given ℓ and m, we take $|m|+\ell$ derivatives of a polynomial $(\xi^2-1)^\ell$ of degree 2ℓ. Therefore, a non-zero solution $P_\ell^m(\xi)$ is obtained only if $|m|\leq\ell$.

These conditions are equivalent to the algebraic relations (4.64) and (4.65), provided ℓ is restricted to integer values.

The constant $C_{\ell m}$ in $g_\ell^m(\vartheta)$ is determined by normalization. In spherical coordinates, we have the volume element

$$dxdydz=r^2\sin(\vartheta)drd\vartheta d\varphi \tag{4.92}$$

such that the angular dependence $Y_\ell^{\mathrm{m}}(\vartheta,\varphi)$ is normalized if

$$\int_0^{2\pi}\int_0^{\pi}|Y_\ell^{\mathrm{m}}(\vartheta,\varphi)|^2\sin(\vartheta)\mathrm{d}\vartheta\mathrm{d}\varphi = 1\,. \tag{4.93}$$

Mathematical results then imply the normalized spherical harmonics

$$Y_\ell^{\mathrm{m}}(\vartheta,\varphi) = \varepsilon\sqrt{\frac{2\ell+1}{4\pi}\frac{(\ell-|\mathrm{m}|)!}{(\ell+|\mathrm{m}|)!}}e^{im\varphi}P_\ell^{\mathrm{m}}(\cos(\vartheta)) \tag{4.94}$$

with a conventional sign factor of $\varepsilon = (-1)^{\mathrm{m}}$ if $\mathrm{m}\geq 0$ and $\varepsilon = 1$ if $\mathrm{m}<0$. These functions are orthonormal in the sense that

$$\int_0^{2\pi}\int_0^{\pi}Y_\ell^{\mathrm{m}}(\vartheta,\varphi)^*Y_{\ell'}^{\mathrm{m}'}(\vartheta,\varphi)\sin(\vartheta)\mathrm{d}\vartheta\mathrm{d}\varphi = \delta_{\ell\ell'}\delta_{\mathrm{m}\mathrm{m}'}\,. \tag{4.95}$$

4.1.7 ADDITIONAL PROPERTIES OF ANGULAR MOMENTUM

The classical inequality $L_z^2\leq L^2$ is saturated if $L_x = 0 = L_y$. In quantum mechanics, the corresponding eigenvalues obey $\mu^2 = \mathrm{m}^2\hbar^2\leq\ell^2\hbar^2 < \ell(\ell+1)\hbar^2 = \lambda$ with a strict inequality unless $\ell = 0$. The fact that μ^2 cannot equal λ is implied by the uncertainty relation (4.43): We cannot have sharp values of L_x and L_y in a state with $\langle\hat{L}_z\rangle\neq 0$, such as an eigenstate of $\hat{L}_z$ that maximizes $|\mu| = \langle\hat{L}_z\rangle$ for given $\lambda\neq 0$.

The expectation value of $\hat{L}^2$ can be expressed as

$$\langle\hat{L}^2\rangle = \langle\hat{L}_x^2+\hat{L}_y^2+\hat{L}_z^2\rangle = \langle\hat{L}_x\rangle^2+\sigma_{L_x}^2+\langle\hat{L}_y\rangle^2+\sigma_{L_y}^2+\langle\hat{L}_z\rangle^2+\sigma_{L_z}^2 \tag{4.96}$$

using the definition of variances. We may assume that the state is an eigenstate of $\hat{L}_z$, in which $\sigma_{L_z} = 0$, and that the expectation values of $\hat{L}_x$ and $\hat{L}_y$ vanish, modeling the classical condition for saturation of $L_z^2\leq L^2$. Since the variances $\sigma_{L_x}^2$ and $\sigma_{L_y}^2$ cannot both be zero unless $\langle\hat{L}_z\rangle = 0$, owing to the uncertainty relation, we have a strict inequality $\lambda = \langle\hat{L}^2\rangle > \langle\hat{L}_z\rangle^2 = \mu^2$, unless $\ell = 0$ which implies $\langle\hat{L}_j\rangle = 0$ and $\sigma_{L_j} = 0$ for $j = x,y,z$.

More precisely, these conditions show that

$$\langle\hat{L}^2\rangle = \sigma_{L_x}^2+\sigma_{L_y}^2+\langle\hat{L}_z\rangle^2\,. \tag{4.97}$$

If we further assume that $\sigma_{L_x} = \sigma_{L_y}$, invoking rotational symmetry around the z-axis, the uncertainty relation (4.43) implies

$$\sigma_{L_x}^2 = \sigma_{L_y}^2 = \sigma_{L_x}\sigma_{L_y}\geq\frac{1}{2}\hbar|\langle\hat{L}_z\rangle| \tag{4.98}$$

such that

$$\langle\hat{L}^2\rangle\geq|\langle\hat{L}_z\rangle|(\hbar+|\langle\hat{L}_z\rangle|) \tag{4.99}$$

from (4.97). In an eigenstate of $\hat{L}^2$ and $\hat{L}_z$, this inequality leads to

$$\hbar^2\ell(\ell+1)\geq\hbar^2|\mathrm{m}|(|\mathrm{m}|+1) \tag{4.100}$$

which is saturated if $|\mathrm{m}| = \ell$. In quantum mechanics, it is therefore possible to maximize the z-component of angular momentum, but only within the confines set by the uncertainty relation.

The components of angular momentum are conserved quantities,

$$\frac{\mathrm{d}\langle\hat{L}_j\rangle}{\mathrm{d}t} = \frac{i}{\hbar}\langle[\hat{H},\hat{L}_j]\rangle = 0\,, \tag{4.101}$$

if the potential is spherically symmetric, given by a function $V(r)$. For $j=z$, for instance, we have

$$[\hat{H},\hat{L}_z]=\frac{1}{2m}[\hat{p}_x^2+\hat{p}_y^2+\hat{p}_z^2,\hat{x}\hat{p}_y-\hat{y}\hat{p}_x]+[V(\hat{r}),\hat{x}\hat{p}_y-\hat{y}\hat{p}_x]\,. \tag{4.102}$$

The individual commutators are

$$\begin{aligned}[\hat{p}_x^2+\hat{p}_y^2+\hat{p}_z^2,\hat{x}\hat{p}_y-\hat{y}\hat{p}_x] &= [\hat{p}_x^2,\hat{x}\hat{p}_y]-[\hat{p}_y^2,\hat{y}\hat{p}_x]=[\hat{p}_x^2,\hat{x}]\hat{p}_y-[\hat{p}_y^2,\hat{y}]\hat{p}_x\\ &= -2i\hbar\hat{p}_x\hat{p}_y+2i\hbar\hat{p}_y\hat{p}_x=0\end{aligned} \tag{4.103}$$

because $\hat{p}_x$ and $\hat{p}_y$ commute, and

$$[V(\hat{r}),\hat{x}\hat{p}_y-\hat{y}\hat{p}_x]=\hat{x}[V(\hat{r}),\hat{p}_y]-\hat{y}[V(\hat{r}),\hat{p}_x]=i\hbar\hat{x}\widehat{\frac{\partial V}{\partial y}}-i\hbar\hat{y}\widehat{\frac{\partial V}{\partial x}}\,. \tag{4.104}$$

A spherically symmetric potential $V(r)$ with $r=\sqrt{x^2+y^2+z^2}$ obeys the condition

$$x\frac{\partial V}{\partial y}-y\frac{\partial V}{\partial x}=x\frac{\partial r}{\partial y}\frac{\mathrm{d}V}{\mathrm{d}r}-y\frac{\partial r}{\partial x}\frac{\mathrm{d}V}{\mathrm{d}r}=0 \tag{4.105}$$

because $\partial r/\partial y=y/r$ and $\partial r/\partial x=x/r$, and thus $[V(\hat{r}),\hat{L}_z]=0$. (This commutator follows immediately if we use the position representation in spherical coordinates, where $\hat{L}_z=i\hbar\partial/\partial\varphi$.) We conclude that $\hat{L}_z$ and similarly $\hat{L}_x$ and $\hat{L}_y$ are conserved quantities, obeying

$$[\hat{H},\hat{L}_x]=[\hat{H},\hat{L}_y]=[\hat{H},\hat{L}_z]=0 \tag{4.106}$$

for any Hamilton operator with a spherically symmetric potential $V(r)$. The operator $\hat{L}^2$ then also commutes with $\hat{H}$ and is conserved.

In the case of a spherical potential $V(r)$, all three operators $\hat{H}$, $\hat{L}^2$ and $\hat{L}_z$ commute with one another and have a simultaneous eigenbasis, labeled by $\mathfrak{m}$, ℓ and a third quantum number n (or k in the case of scattering states) for different eigenvalues of $\hat{H}$. We may write the complete set of simultaneous eigenfunctions of $\hat{L}_z$, $\hat{L}^2$ and $\hat{H}$ as $\psi_{n\ell\mathfrak{m}}(r,\vartheta,\varphi)=R_{n\ell}(r)Y_\ell^{\mathfrak{m}}(\vartheta,\varphi)$ with a function $R_{n\ell}(r)$ that depends only on r since the angular dependence of the 3-dimensional wave function is given by the eigenfunction conditions for $\hat{L}^2$ and $\hat{L}_z$. The radial dependence of simultaneous eigenfunctions is determined by the Hamilton operator, such that

$$\hat{H}\psi_{n\ell\mathfrak{m}}=E_{n\ell}\psi_{n\ell\mathfrak{m}}\quad,\quad \hat{L}^2\psi_{n\ell\mathfrak{m}}=\hbar^2\ell(\ell+1)\psi_{n\ell\mathfrak{m}}\quad,\quad \hat{L}_z\psi_{n\ell\mathfrak{m}}=\hbar\mathfrak{m}\psi_{n\ell\mathfrak{m}}\,. \tag{4.107}$$

One might expect a label for $\mathfrak{m}$ also on $R_{n\ell}(r)$ and $E_{n\ell}$. However, for a spherically symmetric system, the energy, governed by the eigenfunction equation of $\hat{H}$, should not depend on the z-component of angular momentum. We will demonstrate this property in our explicit analysis of the radial equation. The angular part $Y_\ell^{\mathfrak{m}}(\vartheta,\varphi)$ is normalized by (4.93), and solutions for the radial part are normalized by

$$\int_0^\infty |R_{n\ell}(r)|^2r^2\mathrm{d}r=1 \tag{4.108}$$

with a factor of r^2 from (4.92).

4.2 RADIAL WAVE EQUATION

The time-independent Schrödinger equation in 3-dimensional space takes the form

$$-\frac{\hbar^2}{2m}\nabla^2\psi+V(r)\psi=E\psi \tag{4.109}$$

if the potential is spherically symmetric and depends only on $r=\sqrt{x^2+y^2+^2}$. The Laplacian $\nabla^2=\partial^2/\partial x^2+\partial^2/\partial y^2+\partial^2/\partial z^2$ takes a simple form in Cartesian coordinates (x,y,z), but for calculations with a spherically symmetric potential it is often more convenient to use spherical coordinates (r, ϑ, φ), related to Cartesian coordinates by (4.13) or the inverse (4.27). The chain rule for multiple variables makes it possible to transform the Laplacian in Cartesian coordinates to an equivalent expression in spherical coordinates:

$$\nabla^2 = \frac{1}{r^2}\frac{\partial}{\partial r}\left(r^2\frac{\partial}{\partial r}\right) + \frac{1}{r^2\sin(\vartheta)}\frac{\partial}{\partial\vartheta}\left(\sin(\vartheta)\frac{\partial}{\partial\vartheta}\right) + \frac{1}{r^2\sin(\vartheta)^2}\frac{\partial^2}{\partial\varphi^2}. \tag{4.110}$$

In a later chapter in Part II, we will encounter a more efficient method for this transformation, using the notion of a covariant derivative (9.18) from Riemannian geometry; see equation (9.190). The symbol ∇ represents this covariant derivative as a vector operation, and the square indicates taking its norm in spherical geometry.

4.2.1 RADIAL WAVE FUNCTION

The time-independent Schrödinger equation now takes the form

$$-\frac{\hbar^2}{2m}\left(\frac{1}{r^2}\frac{\partial}{\partial r}\left(r^2\frac{\partial\psi}{\partial r}\right) + \frac{1}{r^2\sin(\vartheta)}\frac{\partial}{\partial\vartheta}\left(\sin(\vartheta)\frac{\partial\psi}{\partial\vartheta}\right) + \frac{1}{r^2\sin(\vartheta)^2}\frac{\partial^2\psi}{\partial\varphi^2}\right) + V(r)\psi = E\psi. \tag{4.111}$$

In the new coordinates, wave functions are represented by $\psi(r, \vartheta, \varphi)$ in the position representation, such that r and $\sin(\vartheta)$ are multiplication operators. The momentum operators are replaced by partial derivatives with respect to r, ϑ, and φ. Functions are normalized according to

$$\int_0^\infty \int_0^\pi \int_0^{2\pi} |\psi(r, \vartheta, \varphi)|^2 r^2 \sin(\vartheta) d\varphi d\vartheta dr. \tag{4.112}$$

The factor of $r^2\sin(\vartheta)$ implies that the operators

$$\frac{1}{r^2}\frac{\partial}{\partial r}\left(r^2\frac{\partial}{\partial r}\right) \tag{4.113}$$

and

$$\frac{1}{\sin(\vartheta)}\frac{\partial}{\partial\vartheta}\left(\sin(\vartheta)\frac{\partial}{\partial\vartheta}\right) \tag{4.114}$$

are Hermitian even though r does not commute with $\partial/\partial r$, nor $\sin(\vartheta)$ with $\partial/\partial\vartheta$, and combinations of these operators do not look symmetrically ordered. However, multiplied by r^2 and $\sin(\vartheta)$, respectively, from the integrand in (4.112), these two operators take on symmetric forms and can be shown to be Hermitian by evaluating the required conditions, as in (4.35).

If the potential is spherically symmetric, we can find a complete set of stationary states by looking for simultaneous eigenfunctions of $\hat{H}$, $\hat{L}^2$ and $\hat{L}_z$. Equation (4.111) is the full eigenfunction equation for $\hat{H}$, while the eigenfunction equations for $\hat{L}^2$ and $\hat{L}_z$ are given in (4.89) and

$$i\hbar\frac{\partial\psi}{\partial\varphi} = \mathfrak{m}\hbar\psi \tag{4.115}$$

from (4.24). Solutions of the latter two equations are spherical harmonics $Y_\ell^{\mathfrak{m}}(\vartheta, \varphi)$ with eigenvalues $\ell(\ell+1)\hbar^2$ and $\mathfrak{m}\hbar$, respectively. Including a radial wave function $R_{n\ell}(r)$, the full wave function takes the product form

$$\psi_{n\ell\mathfrak{m}}(r, \vartheta, \varphi) = R_{n\ell}(r)Y_\ell^{\mathfrak{m}}(\vartheta, \varphi). \tag{4.116}$$

The function $R_{n\ell}$ depends only on r because the eigenfunction equation for $\hat{H}$, which determines this function, cannot change the dependence on ϑ and φ without affecting the condition that we have simultaneous eigenfunctions of all three operators.

Given the form

$$\frac{1}{\sin(\vartheta)}\frac{\partial}{\partial\vartheta}\left(\sin(\vartheta)\frac{\partial Y_\ell^{\mathfrak{m}}}{\partial\vartheta}\right)=\left(-\ell(\ell+1)+\frac{\mathfrak{m}^2}{\sin(\vartheta)^2}\right)Y_\ell^{\mathfrak{m}}(\vartheta,\varphi) \tag{4.117}$$

equivalent to (4.89) as well as (4.115), the angle dependence indeed disappears from (4.111), which turns into

$$-\frac{\hbar^2}{2m}\frac{1}{r^2}\frac{\mathrm{d}}{\mathrm{d}r}\left(r^2\frac{\mathrm{d}R_{n\ell}}{\mathrm{d}r}\right)+\left(V(r)+\frac{\hbar^2}{2m}\frac{\ell(\ell+1)}{r^2}\right)R_{n\ell}=E_{n\ell}R_{n\ell}\,. \tag{4.118}$$

The dependence on $\mathfrak{m}$ cancels out, which confirms our expectation that the radial eigenfunctions $R_{n\ell}(r)$ and the energy values $E_{n\ell}$ only depend on the first two quantum numbers, n and ℓ. However, the full wave function, $\psi_{n\ell\mathfrak{m}}(r,\vartheta,\varphi)$, and the resulting probability density depend on all three quantum numbers. The relationship $\lambda=\hbar^2\ell(\ell+1)$ for eigenvalues of $\hat{L}^2$ shows that the contribution $\hbar^2\ell(\ell+1)/(2mr^2)$ to the effective potential

$$V_{\mathrm{eff}}(r)=V(r)+\frac{\hbar^2\ell(\ell+1)}{2mr^2}\,. \tag{4.119}$$

in (4.118) is the centrifugal potential $L^2/(2mr^2)$, where eigenvalues of $\hat{L}^2$ replace the classical square L^2 of angular momentum.

Based on our results for angular momentum in the position representation, we may assume that ℓ is a fixed non-negative integer when we try to solve equation (4.118). Introducing $u(r)=rR(r)$ instead of $R(r)$, we have

$$r^2\frac{\mathrm{d}R}{\mathrm{d}r}=r^2\frac{\mathrm{d}(u/r)}{\mathrm{d}r}=r\frac{\mathrm{d}u}{\mathrm{d}r}-u \tag{4.120}$$

and therefore

$$\frac{1}{r}\frac{\mathrm{d}}{\mathrm{d}r}\left(r^2\frac{\mathrm{d}R}{\mathrm{d}r}\right)=\frac{1}{r}\left(\frac{\mathrm{d}}{\mathrm{d}r}\left(r\frac{\mathrm{d}u}{\mathrm{d}r}\right)-\frac{\mathrm{d}u}{\mathrm{d}r}\right)=\frac{\mathrm{d}^2u}{\mathrm{d}r^2}\,. \tag{4.121}$$

Substituting $u(r)$ for $R(r)$ results in the radial equation

$$-\frac{\hbar^2}{2m}\frac{\mathrm{d}^2u}{\mathrm{d}r^2}+\left(V(r)+\frac{\hbar^2\ell(\ell+1)}{2mr^2}\right)u=Eu \tag{4.122}$$

which has the same form as the time-independent 1-dimensional Schrödinger equation with the effective potential (4.119). This correspondence allows us to use methods for solving the 1-dimensional Schrödinger equation, in particular the power-series method. We can also use the same classification of solutions into bound states and scattering states.

Solving the 1-dimensional Schrödinger equation and finding possible restrictions on the energy values E for bound states requires a normalizability condition for the function $u(r)$. Since the original $R(r)$ is normalized by $\int_0^\infty |R(r)|^2r^2\mathrm{d}r=1$ with a factor of r^2 from the spherical volume element (4.92), the function $u(r)=rR(r)$ is subject to the normalization condition

$$\int_0^\infty |u(r)|^2\mathrm{d}r=1 \tag{4.123}$$

applied for the wave function $\psi(x)$ in one dimension. The integrand is just as in 1-dimensional quantum mechanics, but the wave function $u(r)$ is defined only on the half-line $r\geq 0$, and we integrate only over positive values of r in the normalization condition. Instead of the usual fall-off behavior at $x\to-\infty$, we have the strict boundary condition

$$u(0)=0 \tag{4.124}$$

at $r=0$ because $R(r)=u(r)/r$ should remain finite at $r=0$. Normalizable solutions should also approach zero at $r\to\infty$. For given ℓ, there is a unique and non-degenerate bound-state energy $E_{n\ell}$ with quantum number n. However, if $\ell>0$, there are more than one linearly independent wave functions with this energy, labeled by $\mathfrak{m}$. Unlike in one dimension, energy values in three dimensions can have multiplicities greater than one, in which case they are called **degenerate**.

4.2.2 THE HYDROGEN EXAMPLE

An atom is a bound state of a heavy, positively charged nucleus surrounded by light electrons. At low energies, an atom is in its ground state. Transitions between the ground state and excited states can happen if light, a photon with frequency $\omega_{\rm photon}$, is absorbed by the atom in order to supply the required energy $\Delta E = \hbar\omega_{\rm photon}$. Since the energy spectrum of bound states is discrete, only a discrete set of frequencies of light can be absorbed, or emitted when an excited atom moves back to its ground state. Measured absorption and emission spectra can be used to draw conclusions about the energy spectrum of an atom. In practical applications, if the spectrum of an atom or molecule is known, it can be used to identify ingredients of distant objects such as stars through spectroscopy.

The simplest atom is hydrogen, with one proton in the nucleus and one electron. Since the proton is heavier than the electron, we may assume that the nucleus stays fixed at the center while the electron moves around. This assumption constitutes an approximation because momentum conservation requires the nucleus to move a little bit each time the electron moves. A fraction of the energy of the entire atom therefore contributes to the kinetic energy of the nucleus and is not contained in the Hamilton operator of the electron. This fraction is small and can be ignored in a first approximation. But it can be measured by sensitive spectroscopy for instance by comparing the hydrogen spectrum with the deuterium or tritium spectra. Their nuclei are more heavy because they contain neutrons in addition to the proton.

4.2.2.1 Radial wave equation

In this approximation, the only role of the proton (with charge $e > 0$) is to provide the Coulomb potential

$$V(r) = -\frac{e^2}{4\pi\varepsilon_0}\frac{1}{r} \tag{4.125}$$

and the corresponding force acting on the electron. There are additional, magnetic-type forces because both the proton and the electron have an intrinsic angular momentum, called spin to be discussed later, and the electron's moving charge implies an electric current. In a first approximation, we also ignore these forces, as well as relativistic effects.

The potential (4.125) and the force depend only on the distance r between the nucleus and the electron, which makes them spherically symmetric. We can therefore use a radial wave equation in order to compute the energy spectrum of hydrogen. Heavier elements, starting with helium, are more complicated because they have at least two electrons, such that there is electron-electron repulsion in addition to attraction to the nucleus. The combined potential in this case depends on at least two different distance functions and is not spherically symmetric.

Since the nucleus only provides the potential and the force, we do not include a nuclear wave function for states of the entire atom. Wave functions $\psi_{n\ell{\rm m}}(r,\vartheta,\varphi) = r^{-1}u_{n\ell}(r)Y_\ell^{\rm m}(\vartheta,\varphi)$, expressed as in our general spherically symmetric potentials, describe the motion of an electron in a time-independent Coulomb potential. The kinetic energy depends on the electron mass m. Using our previous results, bound states of hydrogen are then described by the radial wave equation

$$-\frac{\hbar^2}{2m}\frac{{\rm d}^2u_{n\ell}}{{\rm d}r^2} + \left(-\frac{e^2}{4\pi\varepsilon_0}\frac{1}{r} + \frac{\hbar^2\ell(\ell+1)}{2mr^2}\right)u_{n\ell} = E_{n\ell}u_{n\ell}\,. \tag{4.126}$$

The effective potential has the limit $\lim_{r\to\infty}V(r) = 0$, such that all bound states have negative energies $E_{n\ell} < 0$. The radial wave function must obey the general boundary condition $u_{n\ell}(0) = 0$ and, for bound states, is normalized according to

$$\int_0^\infty |u_{n\ell}(r)|^2{\rm d}r = 1\,. \tag{4.127}$$

The radial wave equation can be solved for $u_{n\ell}(r)$ by using the analytic method as in the example of the harmonic oscillator in Section 2.3.1. We first reduce the parameters m, e, ε_0, $\hbar$, and E to a single independent and unitless constant,

$$\rho_0 = \frac{me^2}{2\pi\varepsilon_0\hbar^2\kappa}, \tag{4.128}$$

in addition to the quantum number ℓ. Here,

$$\kappa = \frac{\sqrt{-2mE}}{\hbar} > 0 \tag{4.129}$$

is the usual version of the wave number for bound states, $E < 0$. Any discreteness condition on ρ_0 from the normalizability condition will therefore imply discreteness of E.

At the same time, we transform the coordinate r to a unitless variable

$$\rho = \kappa r. \tag{4.130}$$

The radial equation (4.126) for bound states then turns into

$$\begin{aligned}\frac{\mathrm{d}^2u}{\mathrm{d}\rho^2} &= \frac{1}{\kappa^2}\frac{\mathrm{d}^2u}{\mathrm{d}r^2} = -\frac{\hbar^2}{2mE}\frac{\mathrm{d}^2u}{\mathrm{d}r^2} = u + \frac{e^2}{4\pi\varepsilon_0 E}\frac{u}{r} - \frac{\hbar^2\ell(\ell+1)}{2mE}\frac{u}{r^2} \\ &= \left(1 - \frac{me^2}{2\pi\varepsilon_0\hbar^2\kappa^2 r} + \frac{\ell(\ell+1)}{\kappa^2 r^2}\right)u = \left(1 - \frac{\rho_0}{\rho} + \frac{\ell(\ell+1)}{\rho^2}\right)u.\end{aligned} \tag{4.131}$$

We now have a second-order differential equation for $u(\rho)$ with solutions labeled by the two constants ρ_0 and ℓ that appear in the potential-like term on the right-hand side of this equation.

Recalling the main steps of the analytic method, we will

1. determine asymptotic solutions at the boundaries, given by the limiting behavior at $\rho \to \infty$ and $\rho \to 0$, respectively,
2. turn the differential equation for $u(\rho)$ into a recursion formula on coefficients of a general power series multiplied by the asymptotic behaviors at both $\rho \to 0$ and $\rho \to \infty$, and
3. check normalizability of $u(\rho)$ using asymptotic solutions of the recursion formula for power-series coefficients with large exponents.

The final step not only results in normalizable wave functions but also determines the allowed energy values by a condition on the parameter ρ_0.

4.2.2.2 Asymptotic solutions

For large values of ρ, (4.131) implies

$$\frac{\mathrm{d}^2u}{\mathrm{d}\rho^2} = \left(1 - \frac{\rho_0}{\rho} + \frac{\ell(\ell+1)}{\rho^2}\right)u \approx u \tag{4.132}$$

with the general solution

$$u(\rho) \approx Ae^{-\rho} + Be^{\rho}. \tag{4.133}$$

Only $e^{-\rho}$ is normalizable in the range of $\rho \geq 0$, such that $B = 0$ and $u(\rho) \approx Ae^{-\rho}$.

Close to the other boundary, $\rho = 0$, we have

$$\frac{\mathrm{d}^2u}{\mathrm{d}\rho^2} = \left(1 - \frac{\rho_0}{\rho} + \frac{\ell(\ell+1)}{\rho^2}\right)u \approx \frac{\ell(\ell+1)}{\rho^2}u \tag{4.134}$$

if $\ell \neq 0$. The case of $\ell = 0$, in which the $1/\rho^2$-term disappears and we have the asymptotic equation

$$\frac{\mathrm{d}^2u}{\mathrm{d}\rho^2} \approx -\frac{\rho_0}{\rho}u \tag{4.135}$$

near $\rho = 0$, would have to be treated separately. It has more complicated solutions (given by Bessel functions), but fortunately we will not need them because the asymptotic behavior for $\ell > 0$ can be used as a factor in the power-series form for $u(\rho)$ even in the case of $\ell = 0$. The main purpose of determining approximate solutions close to the boundaries of the ρ-range is to eliminate terms that would not be compatible with boundary or normalizability conditions, and to simplify recursion relations for coefficients in a power series.

The differential equation

$$\rho^2\frac{\mathrm{d}^2u}{\mathrm{d}\rho^2} - \ell(\ell+1)u = 0 \tag{4.136}$$

implied by (4.134) is of the general Euler form

$$\sum_k A_k\rho^k\frac{\mathrm{d}^ku}{\mathrm{d}\rho^k} = 0 \tag{4.137}$$

and can be solved by power laws, $u(\rho) \propto \rho^s$ with suitable exponents s: In this type of differential equation, any reduction of the exponent of ρ implied by taking a derivative of order k, $\mathrm{d}^k\rho^s/\mathrm{d}\rho^k \propto \rho^{s-k}$, is compensated for by the same number of factors of ρ multiplying this derivative terms. Inserting $u(\rho) = \rho^s$ in (4.137) therefore implies a polynomial equation for s.

In the specific case of (4.136), inserting $u(\rho) = \rho^s$ such that $\mathrm{d}^2\rho^s/\mathrm{d}\rho^2 = s(s-1)\rho^{s-2}$ we determine the allowed values for s by solving the quadratic equation implied by

$$s(s-1)\rho^s - \ell(\ell+1)\rho^s = (s(s-1) - \ell(\ell+1))\rho^s = 0\,. \tag{4.138}$$

Since this equation must hold for all $\rho \geq 0$, we obtain $s(s-1) = \ell(\ell+1)$ or, as a quadratic equation in standard form, $s^2 - s - \ell(\ell+1) = 0$. The two solutions of this equation are $s = \ell+1$ and $s = -\ell$.

The general solution to the asymptotic equation (4.136) at $\rho = 0$ is given by $u(\rho) \approx C\rho^{\ell+1} + D\rho^{-\ell}$ with two free constants for a given ℓ, as required for the general solution of a second-order differential equation. However, such a solution is compatible with a normalizable wave function only if $D = 0$ because the boundary condition $u(0) = 0$ cannot be imposed on $\rho^{-\ell}$ given that ℓ is non-negative. Therefore, the asymptotic behavior near $\rho = 0$ is given by

$$u(\rho) \approx C\rho^{\ell+1}\,. \tag{4.139}$$

In the next step, we will try to find an exact solution for $u(\rho)$ in the form of a power series multiplied by the two asymptotic functions $\rho^{\ell+1}$ and $e^{-\rho}$, such that a tractable recursion relation is implied for the coefficients. It turns out that this recursion relation is solvable for $\ell > 0$ as well as $\ell = 0$, and thus we do not have to find the precise asymptotic behavior close to $\rho = 0$ in the case of $\ell = 0$. Instead, we simply use (4.139) for all ℓ.

4.2.2.3 Power series

We now substitute $u(\rho) = \rho^{\ell+1}e^{-\rho}v(\rho)$ in (4.131), splitting off the two asymptotic solutions as explicit factors. A polynomial solution for $v(\rho)$ will then be normalizable and obey the boundary condition at $\rho = 0$, but its coefficients must be chosen carefully in order to solve the differential equation. More generally, we will look for solutions $v(\rho)$ in the form of a power series.

Using

$$\frac{\mathrm{d}u}{\mathrm{d}\rho} = \rho^\ell e^{-\rho}\left((\ell+1-\rho)v(\rho) + \rho\frac{\mathrm{d}v}{\mathrm{d}\rho}\right) \tag{4.140}$$

and

$$\frac{\mathrm{d}^2u}{\mathrm{d}\rho^2}=\rho^\ell e^{-\rho}\left(\left(\frac{\ell(\ell+1)}{\rho}-2\ell-2+\rho\right)v+2(\ell+1-\rho)\frac{\mathrm{d}v}{\mathrm{d}\rho}+\rho\frac{\mathrm{d}^2v}{\mathrm{d}\rho^2}\right), \tag{4.141}$$

the radial equation (4.131) implies

$$\rho\frac{\mathrm{d}^2v}{\mathrm{d}\rho^2}+2(\ell+1-\rho)\frac{\mathrm{d}v}{\mathrm{d}\rho}+(\rho_0-2(\ell+1))v=0. \tag{4.142}$$

We now write $v(\rho)=\sum_{j=0}^{\infty}a_j\rho^j$ as a power series with the first-order derivative

$$\frac{\mathrm{d}v}{\mathrm{d}\rho}=\sum_{j=0}^{\infty}ja_j\rho^{j-1}=\sum_{j=1}^{\infty}ja_j\rho^{j-1}=\sum_{k=0}^{\infty}(k+1)a_{k+1}\rho^k \tag{4.143}$$

and the second-order derivative

$$\frac{\mathrm{d}^2v}{\mathrm{d}\rho^2}=\sum_{j=0}^{\infty}j(j-1)a_j\rho^{j-2}=\sum_{j=1}^{\infty}j(j-1)a_j\rho^{j-2}=\sum_{k=0}^{\infty}k(k+1)a_{k+1}\rho^{k-1}. \tag{4.144}$$

In both cases, the shift to $k=j-1$ will be convenient.

Inserting these series expressions in the differential equation and combining terms, we obtain

$$\begin{aligned}&\rho\frac{\mathrm{d}^2v}{\mathrm{d}\rho^2}+2(\ell+1-\rho)\frac{\mathrm{d}v}{\mathrm{d}\rho}+(\rho_0-2(\ell+1))v \\ =&\sum_{j=0}^{\infty}\left(j(j+1)a_{j+1}+2(\ell+1)(j+1)a_{j+1}-2ja_j+(\rho_0-2(\ell+1))a_j\right)\rho^j=0.\end{aligned} \tag{4.145}$$

This final power series is zero for all ρ if and only if all coefficients vanish, which implies the recursion formula

$$a_{j+1}=\frac{2(j+\ell+1)-\rho_0}{(j+1)(j+2\ell+2)}a_j \tag{4.146}$$

for the a_j, given specific values of ℓ and ρ_0.

Equation (4.146) is a first-order or 1-step recursion formula, relating a_{j+1} directly to a_j. A solution is completely determined if we choose a single value, a_0. In general, we expect two free constants for solutions $v(\rho)$ of a second-order differential equation. We see only one here, determined by a_0, because normalizability has partially been implemented. (In the case of the harmonic oscillator, we had a 2-step recursion relation with two free constants expected for a generic solution. However, we were able to use reflection symmetry and restrict attention to even and odd solutions, each of which was determined by a 1-step recursion formula with a single free constant.)

4.2.2.4 Normalizability and energy values

The large-j behavior of solutions of (4.146) gives us conditions on a power series $v(\rho)=\sum_{j=0}^{\infty}a_j\rho^j$ such that we may obtain a normalizable solution $u(\rho)=\rho^{\ell+1}e^{-\rho}v(\rho)$ of the radial equation. In order to find such asymptotic solutions a_j, we solve the simplified approximate equation

$$a_{j+1}\approx\frac{2}{j+1}a_j \tag{4.147}$$

valid for a_j with sufficiently large j such that $j\gg 2$, $j\gg\ell$ and $j\gg\rho_0$.

We could replace $2/(j+1)$ with $2/j$ under these conditions, but it is convenient to keep $j+1$. We can then use the asymptotic form even for $j=0$, starting with a_0 and working our way up to a_j with large j, even though the small-j part of the solution will not be a good approximation to the full equation (4.146). Since each recursion step requires us to multiply by 2 and divide by the integer number $j+1$ of the step starting from $j=1$, the large-j solution for a_j is given by

$$a_j \approx \frac{2^j}{j!}B \tag{4.148}$$

with a constant B that would equal a_0 if we were able to trust small-j solutions.

We can now check whether a full power series with the given large-j behavior of its coefficients can produce a normalizable wave function $u(\rho)$. The asymptotic form for $a_j\rho^j$ is summed explicitly by an exponential function,

$$v(\rho) \approx B\sum_{j=0}^{\infty}\frac{2^j}{j!}\rho^j = B\sum_{j=0}^{\infty}\frac{(2\rho)^j}{j!} = Be^{2\rho}\,. \tag{4.149}$$

However, the radial wave function

$$u(\rho) = \rho^{\ell+1}e^{-\rho}v(\rho) = B\rho^{\ell+1}e^{\rho} \tag{4.150}$$

with such a $v(\rho)$ is not normalizable over the range of $\rho \geq 0$.

The only way to obtain a normalizable solution is to have a power series that terminates at a finite $j_{\max}$ and therefore determines a polynomial of degree $j_{\max}$. We are now looking for conditions on the free parameters ℓ and ρ_0 such that (4.146) has a solution with $a_j = 0$ for $j > j_{\max}$ while $a_{j_{\max}} \neq 0$. Solving (4.146) for $a_{j_{\max}+1}$ for a given non-zero $a_{j_{\max}}$, we obtain zero if and only if

$$2(j_{\max}+\ell+1)-\rho_0 = 0\,. \tag{4.151}$$

Like ℓ, $j_{\max}$ is a non-negative integer. Thus,

$$\rho_0 = 2(j_{\max}+\ell+1) = 2n \tag{4.152}$$

with a positive integer $n = j_{\max}+\ell+1 = 1,2,3,\ldots$ The definition (4.128) of ρ_0 implies that this equation is a condition on allowed energies, which form a discrete spectrum with a single independent quantum number n, called the principal quantum number. From now on, we will use n and ℓ as the independent quantum numbers, from which the polynomial degree

$$j_{\max} = n-\ell-1 \tag{4.153}$$

can be computed.

Using

$$E = -\frac{\hbar^2\kappa^2}{2m} = -\frac{me^4}{8\pi^2\varepsilon_0^2\hbar^2\rho_0^2} \tag{4.154}$$

from (4.129) and (4.128), we obtain **Bohr's formula** for the energy spectrum

$$E_n = -\frac{m}{2\hbar^2}\left(\frac{e^2}{4\pi\varepsilon_0}\right)^2\frac{1}{n^2} = \frac{E_1}{n^2} \qquad \text{for } n = 1,2,3,\ldots, \tag{4.155}$$

as illustrated in Figure 4.2. The smallest energy value is

$$E_1 = -\frac{m}{2\hbar^2}\left(\frac{e^2}{4\pi\varepsilon_0}\right)^2 = -13.6\text{eV} \tag{4.156}$$

for the ground state. Even though the radial equation for $u(r)$ contains two independent parameters, ℓ and ρ_0, the energy spectrum depends only on $\rho_0 = 2n$. This is a special property of the Coulomb

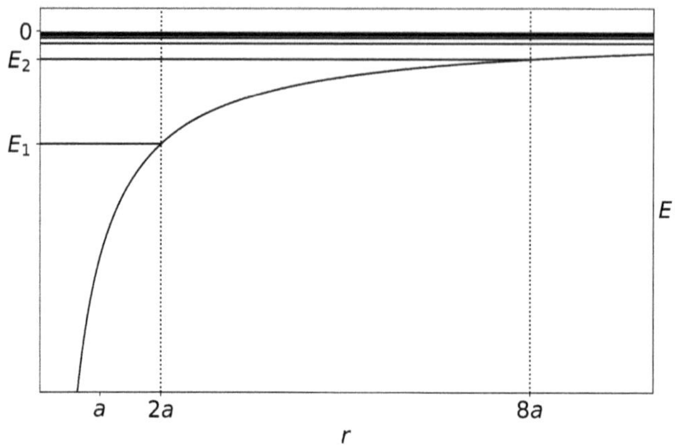

Figure 4.2 Energy spectrum of bound states in the Coulomb potential. As shown by the horizontal lines, infinitely many values E_n lie in a finite range between E_1 and $E = 0$, with an accumulation point at $E = 0$. Units are chosen according to the relationship $E_1 = V(2a)$ in terms of Coulomb's potential and the Bohr radius (4.161).

potential, unlike other spherically symmetric potentials for which the energy values $E_{n\ell}$ do depend on ℓ. However, even for the Coulomb potential, wave functions $u_{n\ell}(r)$ depend on both n and ℓ.

Conservation of angular momentum implies that transitions between different energy levels are restricted because the difference $\Delta\ell$ in an atomic transition must be compensated for by the angular momentum carried away by an electromagnetic wave or a photon. The ℓ-degeneracy of the energy spectrum implies that the energy of an excited state can be realized for different values of ℓ. For the basic hydrogen model, it is then always possible to connect any two energy levels with the correct difference in angular momentum. An energy transition after emitting or absorbing a photon implies a specific final value of ℓ and $\mathfrak{m}$ and the corresponding orbital determined by the wave function. However, additional interactions that we ignored here, such as spin or relativistic effects, lift the ℓ-degeneracy. Energy levels then also split up, which can be seen in fine details of spectroscopy.

Wave functions are determined by specific solutions for the coefficients a_j. For given quantum numbers n and ℓ, which provide $j_{\max} = n - \ell - 1$ and $\rho_0 = 2n$ in (4.146), they can be computed recursively as multiples of the free a_0. The remaining parameter, a_0, will be fixed by normalization. In mathematical terminology, the resulting polynomials have the form

$$v_{n,\ell}(\rho) = L^{2\ell+1}_{n-\ell-1}(2\rho) \tag{4.157}$$

in terms of **associated Laguerre polynomials** $L^m_n(\xi)$. The associated Laguerre polynomials can be derived from the **Laguerre polynomials**

$$L_q(\xi) = e^{\xi}\frac{\mathrm{d}^q(e^{-\xi}\xi^q)}{\mathrm{d}\xi^q} \tag{4.158}$$

by taking additional derivatives,

$$L^p_{q-p}(\xi) = (-1)^p\frac{\mathrm{d}^pL_q(\xi)}{\mathrm{d}\xi^p}\,. \tag{4.159}$$

The radial wave function depends on

$$\rho = \kappa r = \frac{me^2}{2\pi\varepsilon_0\hbar^2\rho_0}r = \frac{me^2}{4\pi\varepsilon_0\hbar^2}\frac{r}{n} = \frac{r}{an} \tag{4.160}$$

if we introduce the **Bohr radius**

$$a = \frac{4\pi\varepsilon_0\hbar^2}{me^2} = 0.529 \cdot 10^{-10}\mathrm{m}\,. \tag{4.161}$$

The radial wave function $R(r) = u(r)/r$ is then given by

$$R_{n\ell}(r) = \frac{1}{r}\rho^{\ell+1}e^{-\rho}v_{n\ell}(\rho) \tag{4.162}$$

with $v_{n\ell}(\rho)$ a polynomial of degree $j_{\max} = n-\ell-1$, depending on $\rho = r/(an)$. This function depends on both n and ℓ. Since $j_{\max} \geq 0$ for a polynomial, we have

$$\ell \leq n-1 \tag{4.163}$$

for a given n, in addition to the general condition $\ell \geq 0$ derived for spherical harmonics.

Complete solutions of the 3-dimensional Coulomb problem for bound states are given by the wave functions

$$\psi_{n\ell\mathfrak{m}}(r,\vartheta,\varphi) = \sqrt{\left(\frac{2}{na}\right)^3 \frac{(n-\ell-1)!}{2n((n+\ell)!)^3}} \exp\left(-\frac{r}{na}\right)\left(\frac{2r}{na}\right)^{\ell} L_{n-\ell-1}^{2\ell+1}\left(\frac{2r}{na}\right) Y_\ell^{\mathfrak{m}}(\vartheta,\varphi) \tag{4.164}$$

which obey orthonormality:

$$\int_0^{2\pi}\int_0^{\pi}\int_0^{\infty} \psi^*_{n\ell\mathfrak{m}}\psi_{n'\ell'\mathfrak{m}'}r^2\sin(\vartheta)\mathrm{d}r\mathrm{d}\varphi\mathrm{d}\vartheta = \delta_{nn'}\delta_{\ell\ell'}\delta_{\mathfrak{m}\mathfrak{m}'}\,. \tag{4.165}$$

We end our discussion of the Coulomb problem with a few explicit examples.

Ground state: The smallest value of the quantum number n, given by $n = 1$, requires $\ell = 0$ because of (4.163). The ranges for quantum numbers of the spherical harmonic then imply $\mathfrak{m} = 0$ because $|\mathfrak{m}| \leq \ell$. The ground state therefore has a unique wave function $\psi_{100}(r,\vartheta,\varphi)$ with energy $E_1 = -13.6\mathrm{eV}$. The absolute value $|E_1|$ is the **binding energy** of hydrogen. The energy required to ionize hydrogen, making it possible to push a ground-state electron into a scattering state that is no longer bound to the nucleus, is $-E_1 = 13.6\mathrm{eV}$.

The quantum numbers ℓ and $\mathfrak{m}$ determine the angular momentum in a state, such as the orbital of an electron in an atom. In the ground state of hydrogen, we have the quantum numbers $\mathfrak{m} = \ell = 0$ and there is no angular momentum or orbital motion. For $\ell = 0$, the effective potential (4.119) equals the Coulomb potential. Unlike our previous bound-state examples with an extended allowed region for negative energies (excluding the delta-function potential), the Coulomb potential is not bounded from below. In classical physics, there is no ground state at zero angular momentum because the electron would fall straight into the nucleus. Even if the electron is orbiting around the nucleus, it keeps loosing energy by electromagnetic radiation as it spirals toward the center. Any classical model of hydrogen would predict that the atom is unstable. The stable ground state in quantum mechanics explains why hydrogen docs not decay even at zero angular momentum, and why it has a size significantly larger than the nuclear size. Heuristically, the electron at zero angular momentum does not fall into the nucleus because the uncertainty principle implies that its wave function must be spread out over a region greater than the nucleus.

The ground-state wave function is given by $\psi_{100}(r,\vartheta,\varphi) = R_{10}(r)Y_0^0(\vartheta,\varphi)$, where

$$R_{10}(r) = r^{-1}u_{10}(r) = r^{-1}\rho e^{-\rho}v_{10}(\rho) \tag{4.166}$$

with $v_{10}(\rho) = a_0^{(10)}$ a constant. The superscript on $a_0^{(10)}$ refers to the quantum numbers $(n\ell)$, while the subscript labels coefficients of the polynomial $v_{n\ell}(\rho)$. Since $\rho = r/a$ for $n = 1$, we have $R_{10}(r) = (a_0^{(10)}/a)e^{-r/a}$. The angular wave function $Y_0^0(\vartheta,\varphi) = (4\pi)^{-1/2}$ equals a constant fixed by the angular normalization integral. Normalization of the radial wave function,

$$1 = \int_0^\infty |R_{10}|^2 r^2 \mathrm{d}r = \frac{|a_0^{(10)}|^2}{a^2}\int_0^\infty r^2 e^{-2r/a}\mathrm{d}r = \frac{1}{4}a|a_0^{(10)}|^2 \tag{4.167}$$

if we integrate by parts twice, implies $a_0^{(10)} = 2/\sqrt{a}$. The normalized ground-state wave function therefore equals

$$\psi_{100}(r,\vartheta,\varphi) = \frac{1}{\sqrt{\pi a^3}}e^{-r/a}\,. \tag{4.168}$$

The expectation value of r in the ground state gives us an estimate of the size of a hydrogen atom. We compute

$$\langle r\rangle = \frac{1}{\pi a^3}\int_0^{2\pi}\int_0^{\pi}\int_0^\infty e^{-2r/a}r^3\sin(\vartheta)\mathrm{d}r\mathrm{d}\vartheta\mathrm{d}\varphi = \frac{4}{a^3}\int_0^\infty r^3 e^{-2r/a}\mathrm{d}r = \frac{3}{2}a\,, \tag{4.169}$$

integrating by parts thrice. The Bohr radius a determines the correct order of magnitude of the expected distance between the electron and the nucleus.

First excited state: For $n=2$, (4.163) allows two possible values of the integer ℓ, $\ell=0$ (in which case $\mathfrak{m}=0$ is unique) or $\ell=1$ (in which case there are three choices, $\mathfrak{m}=-1,0,1$). The first excited state has multiplicity four as an eigenvalue of the Hamilton operator. Its energy value is $E_2 = \frac{1}{4}E_1 = -3.4\text{eV}$.

If $\ell=0$ in the first excited state, $v_{20}(\rho)$ is a polynomial of degree one because $j_{\max} = n-\ell-1 = 1$ in the recursion relation for its coefficients. Given a value of $a_0^{(20)}$, the linear coefficient equals $a_1^{(20)} = -a_0^{(20)}$ according to (4.146), and we obtain $v_{20}(\rho) = a_0^{(20)}(1-\rho)$. Using $\rho = r/(2a)$ for $n=2$, the radial wave function equals

$$R_{20}(r) = \frac{a_0^{(20)}}{2a}\left(1-\frac{r}{2a}\right)e^{-r/(2a)} \tag{4.170}$$

where the first $(2a)^{-1}$ is implied by $\rho^{\ell+1}/r = \rho/r$ in this case. The free constant $a_0^{(20)}$ is again determined by normalization, such that

$$R_{20}(r) = \frac{1}{\sqrt{2a^3}}\left(1-\frac{r}{2a}\right)e^{-r/(2a)}\,. \tag{4.171}$$

If $\ell=1$, $v_{21}(\rho)$ is a polynomial of degree $j_{\max} = n-\ell-1 = 0$ and therefore constant. The radial wave function now equals

$$R_{21}(r) = \frac{a_0^{(21)}}{4a^2}re^{-r/(2a)} \tag{4.172}$$

with a factor of $r/(4a^2)$ from $\rho^{\ell+1}/r = \rho^2/r = r/(2a)^2$. Normalization determines $a_0^{(21)}$ and leads to

$$R_{21}(r) = \frac{1}{\sqrt{6a^3}}re^{-r/(2a)}\,. \tag{4.173}$$

General n: The condition $0\le\ell\le n-1$ for given n, together with $-\ell\le\mathfrak{m}\le\ell$ for given ℓ, implies that there are $2\ell+1$ allowed values of $\mathfrak{m}$ for any ℓ, and

$$d(n) = \sum_{\ell=0}^{n-1}(2\ell+1) = \sum_{k=1}^{n}(2k-1) = 2\cdot\frac{n(n+1)}{2} - n = n^2 \tag{4.174}$$

independent states with the same energy E_n. The hydrogen atom has excited states whose energy eigenvalues have greater multiplicities, or more degeneracy, than general spherically symmetric systems in which $E_{n\ell}$ usually depends on n and ℓ, and therefore has degeneracy $2\ell+1$. The greater degeneracy is related to an additional mathematical symmetry transformation beyond spherical symmetry, which can be constructed for the hydrogen atom.

4.2.2.5 Algebraic properties

Spherically symmetric configurations can be characterized by their dependence on the radius operator $\hat{r}\psi(x,y,z) = \sqrt{x^2+y^2+z^2}\,\psi(x,y,z)$, the momentum-norm operator

$$|\hat{p}|^2\psi(x,y,z) = (\hat{p}_x^2+\hat{p}_y^2+\hat{p}_z^2)\psi(x,y,z)\,, \tag{4.175}$$

and the dot-product operator

$$\widehat{\vec{x}\cdot\vec{p}}\,\psi(x,y,z) = (\hat{x}\hat{p}_x+\hat{y}\hat{p}_y+\hat{z}\hat{p}_z)\psi(x,y,z)\,. \tag{4.176}$$

We use these expressions in order to introduce a set of three basic operators given by

$$\hat{r} \quad , \quad \hat{P} = \hat{r}|\hat{p}|^2 \quad , \quad \hat{Q} = \hat{x}\hat{p}_x+\hat{y}\hat{p}_y+\hat{z}\hat{p}_z - i\hbar\hat{\mathbb{I}}\,. \tag{4.177}$$

They obey a closed set of commutators,

$$[\hat{r},\hat{Q}] = i\hbar\hat{r} \quad , \quad [\hat{r},\hat{P}] = 2i\hbar\hat{Q} \quad , \quad [\hat{Q},\hat{P}] = i\hbar\hat{P}\,, \tag{4.178}$$

with results linear in the original operators. A closed set of this form defines a **Lie algebra**.

Note that neither $\hat{P}$ nor $\hat{Q}$ is Hermitian. The operator $\hat{Q}$ would be Hermitian if the coefficient of $i\hbar$ were equal to $-3/2$ instead of -1. But then the commutators would not close unless we included the identity $\hat{\mathbb{I}}$ in the list of basic operators, as done in the Heisenberg algebra, implying a 4-dimensional Lie algebra. Making $\hat{P}$ Hermitian by subtracting $\frac{1}{2}[\hat{r},|\hat{p}|^2]$ would result in an expression that is no longer compatible with a closed Lie algebra of small dimension. We will see that it is possible to extract much physics information from the 3-dimensional or 4-dimensional Lie algebra (if we include the identity operator), keeping in mind the non-Hermitian behavior of $\hat{Q}$ as well as $\hat{P}$.

The commutators (4.178) share some features with those of the three components of angular momentum, (4.42), but the coefficients are not the same. The corresponding symmetries are therefore different. It is not straightforward to compare Lie algebras determined by different sets of commutators because we may rearrange terms by replacing the original basic operators with linear combinations, changing the basis of the underlying vector space. The commutators then remain closed, but they obtain different coefficients. This situation is similar to comparing geometrical expressions such as the volume element for integrations, written in various coordinate systems. The coefficient of the volume element may look different, for instance in Cartesian and spherical coordinates, even though it amounts to the same geometrical quantity.

The task of comparing expressions obtained in different basis or coordinate choices is greatly simplified if we can find invariant quantities that are independent of such a choice. In the present case, a Lie-algebra invariant is an operator that commutes with all basic operators of the Lie algebra, also called a **Casimir operator**. For angular momentum, an invariant is given by $\hat{L}^2 = \hat{L}_x^2+\hat{L}_y^2+\hat{L}_z^2$. Comparing the coefficients with those of the vector length in Cartesian coordinates tells us the underlying symmetry, given by spatial rotations.

The new commutators (4.178) also admit an invariant quadratic operator, but it has different coefficients,

$$\hat{K} := \frac{1}{2}(\hat{r}\hat{P}+\hat{P}\hat{r}) - \hat{Q}^2\,, \tag{4.179}$$

that are not all positive. As we will see in more detail in our discussion of special relativity, the coefficients of the new invariant correspond to the vector length in Minkowski geometry: While angular-momentum operators correspond to the Lie algebra of the rotation group in three spatial dimensions, the new commutators (4.178) formally correspond to 2-dimensional rotations combined with Lorentz boosts in a 3-dimensional space-time model.

It turns out that the quadratic operator (4.179) does in fact equal $\hat{L}^2$ when expressed in terms of the components of position and linear momentum. However, unlike the $\hat{L}_j$, not all of the basic operators (4.177) that appear in the commutators (4.178) are Hermitian. The difference in reality properties explains the transition from complete rotations to rotations plus Lorentz boosts in the corresponding symmetries.

Hermitian conjugates of the basic operators, in addition to $\hat{r}^\dagger = \hat{r}$, are

$$\hat{Q}^\dagger = \hat{Q} - i\hbar\hat{\mathbb{I}} \tag{4.180}$$

and

$$\hat{P}^\dagger = \hat{P} - 2i\hbar\hat{r}^{-1}\hat{Q} = \hat{P} - 2i\hbar\hat{Q}\hat{r}^{-1} - 2\hbar^2\hat{r}^{-1}\,. \tag{4.181}$$

(Both versions of $\hat{P}^\dagger$ will be useful.) The terms in (4.181) follow from specific commutators that are best derived in the position eigenbasis:

$$[\hat{r}, |\hat{p}|^2] = 2i\hbar\hat{r}^{-1}\hat{Q} \tag{4.182}$$

for the first part of the equation, and

$$[\hat{r}^{-1}, \hat{Q}] = -i\hbar\hat{r}^{-1} \tag{4.183}$$

for the second part.

In the position eigenbasis, the operator $\hat{r}^{-1}$ is defined by multiplication with the function $1/\sqrt{x^2+y^2+z^2}$. It can be applied to the derivation of additional commutators that will be useful, such as

$$
\begin{aligned}
[\hat{Q}, |\hat{p}|^2] &= 2i\hbar\hat{r}^{-1}\hat{P} &(4.184)\\
[\hat{r}^{-1}, |\hat{p}|^2] &= 2i\hbar\hat{r}^{-3}(\hat{Q} + i\hbar\hat{\mathbb{I}}) &(4.185)\\
[\hat{r}^{-1}, \hat{P}] &= 2i\hbar\hat{r}^{-2}(\hat{Q} + i\hbar\hat{\mathbb{I}}) &(4.186)
\end{aligned}
$$

and

$$
\begin{aligned}
[\hat{P}^\dagger, \hat{P}] &= -2i\hbar\hat{r}^{-1}\hat{Q}\hat{P}^\dagger + 2i\hbar\hat{P}(\hat{Q} - i\hbar\hat{\mathbb{I}})\hat{r}^{-1} + 4\hbar^2\hat{r}^{-1}\hat{Q}(\hat{Q} - i\hbar\hat{\mathbb{I}})\hat{r}^{-1} &(4.187)\\
[\hat{P}^\dagger, \hat{r}] &= [\hat{r}, \hat{P}]^\dagger = -i\hbar\hat{r} &(4.188)\\
[\hat{P}^\dagger, \hat{Q}] &= -i\hbar\hat{P}^\dagger = [\hat{P}^\dagger, \hat{Q}^\dagger]\,. &(4.189)
\end{aligned}
$$

The Hermitian conjugates of $\hat{Q}$ and $\hat{P}$ imply the imaginary parts

$$\mathrm{Im}\langle\hat{Q}\rangle = \frac{1}{2i}\langle\hat{Q} - \hat{Q}^\dagger\rangle = \frac{1}{2}\hbar \tag{4.190}$$

and

$$
\begin{aligned}
\mathrm{Im}\langle\hat{r}\hat{P}\rangle &= \frac{1}{2i}\left(\langle\hat{r}\hat{P}\rangle - \langle\hat{P}^\dagger\hat{r}\rangle\right) = \frac{1}{2i}\left(\langle\hat{P}\hat{r} + 2i\hbar\hat{Q}\rangle - \langle\hat{P}\hat{r} - 2i\hbar\hat{Q} - 2\hbar^2\hat{\mathbb{I}}\rangle\right)\\
&= \hbar\langle\hat{Q} + \hat{Q}^\dagger\rangle = 2\hbar\mathrm{Re}\langle\hat{Q}\rangle &(4.191)
\end{aligned}
$$

in any state.

The operator $\hat{Q}$ can be employed for a quick derivation of an important result, the **virial theorem**, valid for any spherically symmetric power-law potential $V(r) = \alpha r^k$ with real numbers α and k (the latter usually of integer form): In any stationary state of the Hamilton operator

$$\hat{H} = \frac{1}{2m}|\hat{p}|^2 + \alpha\hat{r}^k\,, \tag{4.192}$$

the expectation values of kinetic and potential energy are related by

$$\frac{\langle|\hat{p}|^2\rangle}{2m} = \frac{1}{2}k\alpha\langle\hat{r}^k\rangle\,. \tag{4.193}$$

This equation follows directly from the stationarity condition applied to $\langle\hat{Q}\rangle$:

$$0 = \frac{\mathrm{d}\langle\hat{Q}\rangle}{\mathrm{d}t} = -\frac{i}{\hbar}\langle[\hat{x}\hat{p}_x+\hat{y}\hat{p}_y+\hat{z}\hat{p}_z,\hat{H}]\rangle = \langle m^{-1}|\hat{p}|^2 - k\alpha\hat{r}^k\rangle \tag{4.194}$$

using $[\hat{p}_x,\hat{r}] = -i\hbar\widehat{\partial r/\partial x} = -i\hbar\hat{x}\hat{r}^{-1}$ and corresponding versions with $\hat{p}_y$ and $\hat{p}_z$.

A stationary state is also an eigenstate of the Hamilton operator, such that $\langle\hat{H}\rangle = E_n$ and therefore

$$\frac{1}{2m}\langle|\hat{p}|^2\rangle + \alpha\langle\hat{r}^k\rangle = E_n\,. \tag{4.195}$$

We now have two independent equations relating the same expectation values, $\langle|\hat{p}|^2\rangle$ and $\langle\hat{r}^k\rangle$, which thus have to equal

$$\langle|\hat{p}|^2\rangle = \frac{2kmE_n}{k+2} \quad , \quad \alpha\langle\hat{r}^k\rangle = \frac{2E_n}{k+2}\,. \tag{4.196}$$

For the Coulomb potential as a special case,

$$\langle\hat{r}^{-1}\rangle = \frac{2E_n}{\alpha}\,. \tag{4.197}$$

The Hamilton operator is not an element of the Lie algebra because it is not a linear combination of the operators (4.177). However, $\hat{r}\hat{H} = \hat{P}/(2m) - \alpha\hat{\mathbb{I}}$ is an element of the algebra if we also include the identity operator (which has zero commutator with any other operator). The eigenvalue equation $\hat{H}\psi_n = E_n\psi_n$ can therefore be formulated as a condition within the Lie algebra,

$$(\hat{r}\hat{H} - E_n\hat{r})\psi_n = 0\,. \tag{4.198}$$

This equation implies additional conditions on expectation values in stationary states that allow us to derive some statistical properties, such as the radius variance. For expectation values in stationary states, (4.198) implies not only

$$\langle\hat{r}(\hat{H} - E_n\hat{\mathbb{I}})\rangle = 0 \tag{4.199}$$

but also

$$\langle\hat{A}\hat{r}(\hat{H} - E_n\hat{\mathbb{I}})\rangle = 0 \tag{4.200}$$

for any operator $\hat{A}$. Suitable choices of $\hat{A}$ then lead to useful relationships between various expectation values in stationary states.

The operator

$$\hat{C}_n = \hat{r}(\hat{H} - E_n\hat{\mathbb{I}}) = \frac{1}{2m}\hat{P} - E_n\hat{r} - \alpha\hat{\mathbb{I}} \tag{4.201}$$

can be used instead of $\hat{H}$ in an algebraic derivation of energy eigenvalues or, as we do here, of state properties for given energy eigenvalues. Because $\hat{C}_n\psi_n = 0$, it is called a **constraint operator**. Unlike $\hat{H}$, $\hat{C}_n$ is not Hermitian, but values of E_n derived from the constraint equation $\hat{C}_n\psi_n = 0$ are real because this equation is equivalent to the original eigenvalue equation $\hat{H}\psi_n = E_n\psi_n$.

The Hamilton operator $\hat{H}$ is the self-adjoint operator corresponding to the symmetry of time translations. It is exponentiated to the one-parameter family of unitary operators $\hat{U}_{\hat{H}}(t) = \exp(-it\hat{H}/\hbar)$. The expectation value of some operator $\hat{A}$ evolves according to

$$\langle\hat{A}\rangle(t) = \langle\exp(it\hat{H}/\hbar)\hat{A}\exp(-it\hat{H}/\hbar)\rangle(0)\,. \tag{4.202}$$

If the state is stationary, $\langle\hat{A}\rangle(t) = \langle\hat{A}\rangle(0)$, and we obtain $\langle[\hat{A},\hat{H}]\rangle = 0$ from a time derivative.

We derive a related condition from $\hat{C}_n$ by observing that $\hat{C}_n\psi_n = 0$ on stationary states implies $\exp(-i\varepsilon\hat{C}_n/\hbar)\psi_n = \psi_n$, using for instance the Taylor expansion of the exponential function. The operator $\hat{C}_n$ does not determine symmetry transformations by unitary operators because it is not Hermitian, but we can still define the formal one-parameter family of transformations $\exp(-i\varepsilon\hat{C}_n/\hbar)$. For expectation values, the constraint equation $\hat{C}_n\psi_n = 0$ is then equivalent to

$$\langle\exp(i\varepsilon\hat{C}_n^\dagger/\hbar)\hat{A}\exp(-i\varepsilon\hat{C}_n/\hbar\rangle = \langle\hat{A}\rangle \tag{4.203}$$

as a direct analog of (4.202). The one-parameter family on the left, derived from a constraint operator, is called a **gauge transformation**. As derived here, it is mathematically related to time evolution given by a Hamilton operator. Taking a derivative by ε, we obtain the condition

$$\langle\hat{A}\hat{C}_n - \hat{C}_n^\dagger\hat{A}\rangle = 0 \tag{4.204}$$

for any $\hat{A}$, provided the state is stationary. A state that solves this equation is called **gauge-invariant**.

For $\hat{C}_n$ corresponding to the Hamilton operator of hydrogen, we compute

$$\hat{C}_n^\dagger = \hat{C}_n - \frac{i\hbar}{m}\hat{r}^{-1}\hat{Q} = \hat{C}_n - \frac{i\hbar}{m}\hat{Q}\hat{r}^{-1} - \frac{\hbar^2}{m}\hat{r}^{-1}. \tag{4.205}$$

Substituting in (4.204), we have

$$\frac{\langle[\hat{A},\hat{C}_n]\rangle}{i\hbar} = -\frac{\langle(\hat{Q}-i\hbar\hat{\mathbb{I}})\hat{r}^{-1}\hat{A}\rangle}{m} \tag{4.206}$$

for stationary states. This equation is valid for any operator $\hat{A}$, such that we may derive several independent equations by making specific choices for $\hat{A}$.

Choosing $\hat{A} = \hat{r}$ implies $m\langle[\hat{r},\hat{C}_n]\rangle/(i\hbar) = -\langle\hat{Q} - i\hbar\hat{\mathbb{I}}\rangle$. Since $m[\hat{r},\hat{C}_n] = \frac{1}{2}[\hat{r},\hat{P}] = i\hbar\hat{Q}$ based on the definition of $\hat{C}_n$ as well as (4.178), we obtain

$$\langle\hat{Q}\rangle = \frac{1}{2}i\hbar, \tag{4.207}$$

or $\mathrm{Re}\langle\hat{Q}\rangle = 0$ in combination with (4.190).

Choosing $\hat{A} = \hat{Q}$ and (4.178) as well as (4.183) implies

$$\frac{1}{2m}\langle\hat{P}\rangle + E_n\langle\hat{r}\rangle + \frac{1}{m}\langle\hat{Q}^2\hat{r}^{-1}\rangle - \frac{\hbar^2}{m}\langle\hat{r}^{-1}\rangle = 0. \tag{4.208}$$

Here,

$$\frac{1}{2m}\langle\hat{P}\rangle = E_n\langle\hat{r}\rangle + \alpha \tag{4.209}$$

from $\langle\hat{C}_n\rangle = 0$, and $\hat{Q}^2$ in the third term can be eliminated by using the operator $\hat{K}$:

$$\begin{aligned}\hat{Q}^2\hat{r}^{-1} &= \left(-\hat{K} + \frac{1}{2}(\hat{r}\hat{P} + \hat{P}\hat{r})\right)\hat{r}^{-1} = -\hat{K}\hat{r}^{-1} + \frac{1}{2}[\hat{r},\hat{P}]\hat{r}^{-1} + \hat{P}\\ &= -\hat{K}\hat{r}^{-1} + i\hbar\hat{Q}\hat{r}^{-1} + \hat{P}.\end{aligned} \tag{4.210}$$

Expectation values of the first term are related to $\langle\hat{r}^{-1}\rangle$, which we derived from the virial theorem, and eigenvalues of $\hat{K}$, which equals the norm squared of angular momentum and can be diagonalized simultaneously with $\hat{H}$. The middle term can be computed by applying (4.206) to $\hat{A} = \hat{\mathbb{I}}$: $i\hbar\langle\hat{Q}\hat{r}^{-1}\rangle = \hbar^2\langle\hat{r}^{-1}\rangle$. Combining these results, we arrive at the equation

$$3\alpha + 4E_n\langle\hat{r}\rangle - \frac{K}{m}\langle\hat{r}^{-1}\rangle = 0 \tag{4.211}$$

in which $K = \hbar^2\ell(\ell+1)$ is an eigenvalue of $\hat{K}$. Inserting $\langle\hat{r}^{-1}\rangle = 2E_n/\alpha$, we obtain the radius expectation value

$$\langle\hat{r}\rangle = -\frac{1}{2}\ell(\ell+1)\frac{\hbar^2}{m\alpha} - \frac{3}{4}\frac{\alpha}{E_n}\,. \tag{4.212}$$

An equation for $\langle\hat{r}^2\rangle$ is obtained by applying the constraint $\langle\hat{A}\hat{C}_n\rangle = 0$ to the choice of $\hat{A} = \hat{r}$:

$$\langle\hat{r}^2\rangle = \frac{1}{2mE_n}\langle\hat{r}\hat{P}\rangle - \frac{\alpha}{E_n}\langle\hat{r}\rangle\,. \tag{4.213}$$

We do not know $\langle\hat{r}\hat{P}\rangle$ yet, but this expectation value appears in the condition (4.206) applied to $\hat{A} = \hat{r}\hat{Q}$, which reads

$$\frac{1}{2m}\langle\hat{r}\hat{P}\rangle + \frac{2}{m}\langle\hat{Q}^2\rangle + E_n\langle\hat{r}^2\rangle + \frac{\hbar^2}{2m} = 0\,. \tag{4.214}$$

Combining (4.213) and (4.214) and relating $\hat{Q}^2$ to angular momentum as before, we derive

$$\langle\hat{r}^2\rangle = \frac{1}{4}\frac{(3\ell(\ell+1)-1)\hbar^2}{mE_n} + \frac{5}{8}\frac{\alpha^2}{E_n^2}\,. \tag{4.215}$$

The radius variance therefore equals

$$\sigma_r^2 = \langle\hat{r}^2\rangle - \langle\hat{r}\rangle^2 = -\frac{\ell^2(\ell+1)^2\hbar^4}{4m^2\alpha^2} + \frac{\alpha^2}{16E_n^2} - \frac{\hbar^2}{4mE_n}\,. \tag{4.216}$$

The standard deviation σ_r grows as a quadratic function of n for rising energy levels, and as a quadratic function of ℓ for increasing angular momentum.

4.3 SPIN

The algebraic method applied to angular-momentum operators implies a larger spectrum than found by solving differential equations for wave functions in the position representation. If the position representation is assumed, eigenfunctions of the form (4.85) must have unique values at every point in space, and since φ and $\varphi + 2\pi$ imply the same point, a periodicity condition follows that requires the quantum number $\mathfrak{m}$ to be integer. The second quantum number, ℓ, must then also be integer because it equals the largest value of $\mathfrak{m}$ for a given norm squared of angular momentum, L^2. By contrast, the algebraic commutators (4.42) and the properties they imply for ladder operators $\hat{L}_\pm$ are consistent with half-integer ℓ according to (4.64), which corresponds to half-integer $\mathfrak{m}$ in the range $-\ell, -\ell+1, \ldots, \ell-1, \ell$.

The physical implication of these mathematical results is that quantum mechanics leads to a new phenomenon of angular momentum not associated with rotation or orbiting in space, which would require a corresponding φ-dependence. Even a point particle that could not rotate in any classical way may have non-zero quantum-mechanical angular momentum if its quantum state is part of a matrix representation for a given ℓ, such as the example of $\ell = 1$ explicitly constructed by equations (4.70), (4.78) and (4.79). Its quantum state is then a vector that can be rotated in an abstract space. The self-adjoint operators derived from these symmetry transformations have commutators just like the angular-momentum relations (4.42) because the symmetry transformations are closely related. Correspondingly, their quantum numbers are subject to the same conditions.

In order to distinguish this kind of angular momentum from the more intuitive notion of orbital angular momentum, for which a state can be written in the position representation, a particle that has a state in an abstract matrix representation of the angular-momentum operators for some ℓ is said to have **spin** ℓ. In order to emphasize the distinction, we refer to the spin quantum number as s rather than ℓ, with corresponding numbers $\mathfrak{m}_s$.

An example of a particle with non-zero spin is the photon, with spin $s = 1$. This value can be inferred from spectroscopy because angular-momentum conservation implies that every photon emitted by an atomic transition between energy levels requires a change of the orbital quantum number $\Delta\ell$ that cannot be greater than one. There is no heuristic explanation for the spin of a point particle, but spin is a physical property because it determines how the particle interacts with other particles or forces. A spinning electrically charged particle has a magnetic moment that can be measured by testing how the particle interacts with an external magnetic field.

The spin number is therefore a characteristic property of a particle, just like its electric charge, that cannot be reduced to more elementary properties (unless we find new elementary particles, such as the Higgs field in the case of another parameter, a particle's mass): We determine the electric charge by observing how a particle interacts with an electric field, we determine the mass by observing how a particle interacts with a gravitational field, and we determine its spin by observing how a particle interacts with a magnetic field. One mathematical difference between these quantities is that we characterize the spin by a unitless quantum number, while the electric charge and mass have units. For this reason, it might seem more like an abstract mathematical concept rather than a physical parameter. However, if we measure spin or angular momentum, it does have units. The classical units of angular momentum just happen to be the same as those of $\hbar$, which was factored out when we introduced the quantum number $\mathfrak{m}$ through the eigenvalue $\mu = \mathfrak{m}\hbar$.

4.3.1 MATRIX REPRESENTATIONS

Spin, as an abstract finite-dimensional representation of angular-momentum operators, can always be represented by matrices. For spin s, they act on a vector space of dimension $2s+1$ with independent basis states labeled by all allowed $\mathfrak{m}_s$-values, $\mathfrak{m}_s = -s, \ldots, s$ in integer steps. Algebraic properties of these matrices follow from our constructions for angular-momentum operators: We have **spin operators** $\hat{S}_x$, $\hat{S}_y$ and $\hat{S}_z$ acting on abstract vectors $|f_s^{\mathfrak{m}_s}\rangle = |s, \mathfrak{m}_s\rangle$ for a fixed value of $s \in \{0, 1/2, 1, 3/2, \ldots\}$.

The abstract notation $|s, \mathfrak{m}_s\rangle$ of states requires only the quantum numbers but no position representation. A generic state in this representation, called a **spinor**, is given by a superposition

$$\chi = \sum_{\mathfrak{m}_s=-s}^{s} \chi_{\mathfrak{m}_s} |s, \mathfrak{m}_s\rangle \leftrightarrow \begin{pmatrix} \chi_s \\ \vdots \\ \chi_{-s} \end{pmatrix} \tag{4.217}$$

with complex numbers $\chi_{\mathfrak{m}_s}$ that provide vector components of the state in the eigenbasis of $\hat{S}_z$. We introduce an inner product by declaring that the basis is orthonormal, motivated by the previous results that s and $\mathfrak{m}_s$ label different eigenstates of Hermitian operators:

$$\langle s_1, \mathfrak{m}_{s,1} | s_2, \mathfrak{m}_{s,2}\rangle = \delta_{s_1 s_2} \delta_{\mathfrak{m}_{s,1} \mathfrak{m}_{s,2}} \,. \tag{4.218}$$

A spinor is therefore normalized if its components obey the condition

$$\sum_{\mathfrak{m}_s=-s}^{s} |\chi_{\mathfrak{m}_s}|^2 = 1 \,. \tag{4.219}$$

Since $\mathfrak{m}_s$ is identified with the eigenvalues of $\hat{S}_z$ (divided by $\hbar$), the algebraic method implies that

$$\hat{S}_z |s, \mathfrak{m}_s\rangle = \hbar \mathfrak{m}_s |s, \mathfrak{m}_s\rangle \tag{4.220}$$

$$\hat{S}_x |s, \mathfrak{m}_s\rangle = \frac{1}{2}(\hat{S}_+ + \hat{S}_-)|s, \mathfrak{m}_s\rangle \tag{4.221}$$

$$\hat{S}_y |s, \mathfrak{m}_s\rangle = \frac{1}{2i}(\hat{S}_+ - \hat{S}_-)|s, \mathfrak{m}_s\rangle \tag{4.222}$$

with

$$\hat{S}_{\pm}|s,\mathfrak{m}_s\rangle = \hbar\sqrt{s(s+1)-\mathfrak{m}_s(\mathfrak{m}_s\pm 1)}\,|s,\mathfrak{m}_s\pm 1\rangle\,. \tag{4.223}$$

It is possible to check explicitly that the operators $\hat{S}_x$, $\hat{S}_y$ and $\hat{S}_z$, defined in this way, are Hermitian and have commutators (4.42), and that the spin magnitude $\hat{S}^2 = \hat{S}_x^2+\hat{S}_y^2+\hat{S}_z^2 = \hat{S}_-\hat{S}_+ + \hat{S}_z^2 + \hbar\hat{S}_z$ has identical eigenvalues $\hbar^2 s(s+1)$ on all basis states $|s,\mathfrak{m}_s\rangle$ for fixed s. For integer s, the spin operators can be represented as derivative operators (4.80)–(4.82), but this is not possible for half-integer s because no position representation exists in this case.

4.3.2 SPIN 1/2

We obtain a 2-dimensional matrix representation for spin operators with $s = 1/2$, following the earlier example for $\ell = 1$ in Section 4.1.6. An example of a particle with spin 1/2 is the electron, which is relevant for many applications of quantum mechanics to atomic, molecular, condensed-matter, and particle physics, and even in astrophysics. For $s = 1/2$, we have two basis states, $|1/2,1/2\rangle$ and $|1/2,-1/2\rangle$. Intuitively, the state $|1/2,1/2\rangle$ is referred to as "spin up," while $|1/2,-1/2\rangle$ is "spin down." A generic state in this representation is a linear combination of $|1/2,1/2\rangle$ and $|1/2,-1/2\rangle$, given by a spinor

$$\chi = a|1/2,1/2\rangle + b|1/2,-1/2\rangle \leftrightarrow \begin{pmatrix} a \\ b \end{pmatrix} \tag{4.224}$$

with two complex numbers, a and b. Normalization requires that $|a|^2+|b|^2 = 1$. The normalization condition implies that the values $P(\mathrm{up}) = |a|^2$ and $P(\mathrm{down}) = |b|^2$ determine a probability distribution for the two possible outcomes of a single spin measurement, up or down. Therefore, $|a|^2$ is the probability of finding the value $S_z = \frac{1}{2}\hbar$ in a measurement, and $|b|^2$ is the probability of finding the value $S_z = -\frac{1}{2}\hbar$. According to the probability distribution, no other value can be obtained in a single measurement of S_z on a spin-1/2 particle. The expectation value of $\hat{S}_z$ in a given spinor equals $\langle\hat{S}_z\rangle = \frac{1}{2}\hbar(|a|^2-|b|^2)$ and can take any value between $-\frac{1}{2}\hbar$ and $\frac{1}{2}\hbar$.

If we are interested in a probability distribution for measurements of S_x or S_y, we have to transform the spinor to the eigenbasis of one of these operators, just as we Fourier transformed from the position eigenbasis to the momentum eigenbasis in order to derive the probability density $|\phi(p)|^2$ for momentum measurements. Here, the transformation is a standard basis change in a finite-dimensional Hilbert space. We first derive the matrix representations of $\hat{S}_x$ and $\hat{S}_y$ for $s = 1/2$.

Since the states $|1/2,\pm 1/2\rangle$ form an eigenbasis of $\hat{S}_z$, we immediately obtain the action

$$\hat{S}_z|1/2,\pm 1/2\rangle = \pm\frac{1}{2}\hbar|1/2,\pm 1/2\rangle\,. \tag{4.225}$$

The action of $\hat{S}_\pm$ follows from (4.68):

$$\hat{S}_\pm|1/2,1/2\rangle = \hbar\sqrt{3/4-1/2(1/2\pm 1)}|1/2,1/2\pm 1\rangle = \left\{\begin{array}{cl} |0\rangle & \text{for } \hat{S}_+ \\ \hbar|1/2,-1/2\rangle & \text{for } \hat{S}_- \end{array}\right. \tag{4.226}$$

and

$$\hat{S}_\pm|1/2,-1/2\rangle = \hbar\sqrt{3/4+1/2(-1/2\pm 1)}|1/2,-1/2\pm 1\rangle = \left\{\begin{array}{cl} \hbar|1/2,1/2\rangle & \text{for } \hat{S}_+ \\ |0\rangle & \text{for } \hat{S}_-\,. \end{array}\right. \tag{4.227}$$

Using $\hat{S}_x = \frac{1}{2}(\hat{S}_+ + \hat{S}_-)$ and $\hat{S}_y = -\frac{1}{2}i(\hat{S}_+ - \hat{S}_-)$, this result leads to

$$\hat{S}_x|1/2,1/2\rangle = \frac{1}{2}\hbar|1/2,-1/2\rangle \quad , \quad \hat{S}_x|1/2,-1/2\rangle = \frac{1}{2}\hbar|1/2,1/2\rangle \tag{4.228}$$

$$\hat{S}_y|1/2,1/2\rangle = \frac{1}{2}i\hbar|1/2,-1/2\rangle \quad , \quad \hat{S}_y|1/2,-1/2\rangle = -\frac{1}{2}i\hbar|1/2,1/2\rangle\,. \tag{4.229}$$

In matrix notation with the basis $|1/2, \pm 1/2\rangle$, such that

$$|1/2, 1/2\rangle \leftrightarrow \begin{pmatrix} 1 \\ 0 \end{pmatrix} \quad , \quad |1/2, -1/2\rangle \leftrightarrow \begin{pmatrix} 0 \\ 1 \end{pmatrix}, \tag{4.230}$$

we write

$$S_z = \frac{1}{2}\hbar \begin{pmatrix} 1 & 0 \\ 0 & -1 \end{pmatrix} \tag{4.231}$$

$$S_+ = \hbar \begin{pmatrix} 0 & 1 \\ 0 & 0 \end{pmatrix} \quad , \quad S_- = \hbar \begin{pmatrix} 0 & 0 \\ 1 & 0 \end{pmatrix} \tag{4.232}$$

$$S_x = \frac{1}{2}\hbar \begin{pmatrix} 0 & 1 \\ 1 & 0 \end{pmatrix} \quad , \quad S_y = \frac{1}{2}\hbar \begin{pmatrix} 0 & -i \\ i & 0 \end{pmatrix}. \tag{4.233}$$

The spin magnitude is represented by the matrix

$$S^2 = S_x^2 + S_y^2 + S_z^2 = \frac{3}{4}\hbar \begin{pmatrix} 1 & 0 \\ 0 & 1 \end{pmatrix}. \tag{4.234}$$

The matrix representation allows us to see a different reason why half-integer spin quantum numbers cannot be related to a position representation. The spin matrices are self-adjoint and therefore generate unitary symmetry transformations by exponentiation. For the example of the diagonal S_z, we compute the matrix exponential by splitting the Taylor series into even and odd terms:

$$\begin{aligned} U_{S_z}(\delta) &= \exp(-i\delta S_z/\hbar) = \exp\left(-i\frac{\delta}{2}\begin{pmatrix} 1 & 0 \\ 0 & -1 \end{pmatrix}\right) \\ &= \sum_{n=0}^{\infty} \frac{(-i)^n}{n!}\left(\frac{\delta}{2}\right)^n \begin{pmatrix} 1 & 0 \\ 0 & -1 \end{pmatrix}^n \\ &= \begin{pmatrix} 1 & 0 \\ 0 & 1 \end{pmatrix} \sum_{k=0}^{\infty} \frac{(-1)^k}{(2k)!}\left(\frac{\delta}{2}\right)^{2k} - i \begin{pmatrix} 1 & 0 \\ 0 & -1 \end{pmatrix} \sum_{k=0}^{\infty} \frac{(-1)^k}{(2k+1)!}\left(\frac{\delta}{2}\right)^{2k+1}. \end{aligned} \tag{4.235}$$

In the last line, we were able to remove matrices from the summation because $(2S_z/\hbar)^{2k} = \mathbb{I}$ is the identity matrix for any even exponent, and $(2S_z/\hbar)^{2k+1} = 2S_z/\hbar$ for any odd exponent. These relations are also correct for S_x and S_y, for which we therefore obtain analogous exponentials. Recognizing the remaining summations as the Taylor series of cosine and sine, respectively, we write

$$U_{S_j}(\delta_j) = \mathbb{I}\cos(\delta_j/2) - \frac{2i}{\hbar} S_j \sin(\delta_j/2) \tag{4.236}$$

or

$$U_{S_x}(\delta_x) = \begin{pmatrix} \cos(\delta_x/2) & -i\sin(\delta_x/2) \\ -i\sin(\delta_x/2) & \cos(\delta_x/2) \end{pmatrix} \tag{4.237}$$

$$U_{S_y}(\delta_y) = \begin{pmatrix} \cos(\delta_y/2) & -\sin(\delta_y/2) \\ \sin(\delta_y/2) & \cos(\delta_y/2) \end{pmatrix} \tag{4.238}$$

$$U_{S_z}(\delta_z) = \begin{pmatrix} \exp(-i\delta_z/2) & 0 \\ 0 & \exp(i\delta_z/2) \end{pmatrix}. \tag{4.239}$$

We construct a generic transformation by exponentiating a linear combination $-i(\delta_x S_x + \delta_y S_y + \delta_z S_z)/\hbar$. There is a mathematical subtlety: We do not always obtain the full Lie group corresponding to a Lie algebra if we consider individual exponentials of all linear combinations. In general, the Lie group is defined as the group obtained by multiplying all possible exponentials, and in some groups there may be products of exponentials that are not an exponential of a linear combination of Lie algebra elements. For the relations considered here, however, it is known that exponentiation of linear combinations is sufficient to generate the whole group. Whenever this is the case, the resulting Lie group is called **exponential**. The harmonic-oscillator group with Lie algebra determined by the relations (3.231) is an example of a Lie group that is not exponential; see page 133.

The general matrix for a linear combination of spin-1/2 matrices looks rather complicated, but if we write the vector $\vec{\delta}=(\delta_x,\delta_y,\delta_z)=\delta\vec{r}$ with $\delta=\sqrt{\delta_x^2+\delta_y^2+\delta_z^2}$ and the unit vector $\vec{r}=(\sin(\vartheta)\cos(\varphi),\sin(\vartheta)\sin(\varphi),\cos(\vartheta))$ in spherical coordinates, we still have the useful identity

$$\left(\frac{2}{\hbar}(r_xS_x+r_yS_y+r_zS_z)\right)^{2k}=\mathbb{I} \tag{4.240}$$

and

$$\left(\frac{2}{\hbar}(r_xS_x+r_yS_y+r_zS_z)\right)^{2k+1}=\frac{2}{\hbar}(r_xS_x+r_yS_y+r_zS_z), \tag{4.241}$$

which can be shown by induction using $S_jS_k=-S_kS_j$ for $j\neq k$. Therefore, a general unitary transformation generated by spin operators is given by

$$\begin{aligned}\hat{U}(\delta,\vartheta,\varphi) &= \exp(-i\delta(r_xS_x+r_yS_y+r_zS_z)/\hbar)\\ &= \mathbb{I}\cos(\delta/2)-\frac{2i}{\hbar}(r_xS_x+r_yS_y+r_zS_z)\sin(\delta/2)\\ &= \begin{pmatrix}\cos(\delta/2)-i\sin(\delta/2)\cos(\vartheta) & -i\sin(\delta/2)\sin(\vartheta)e^{-i\varphi}\\ -i\sin(\delta/2)\sin(\vartheta)e^{i\varphi} & \cos(\delta/2)+i\sin(\delta/2)\cos(\vartheta)\end{pmatrix}\end{aligned} \tag{4.242}$$

These symmetry transformations are periodic and resemble rotations. However, a full symmetry cycle is obtained only when we reach the value $\delta=4\pi$, while $\delta=2\pi$ would be sufficient for rotations in space if we exponentiate orbital angular momentum given by $\hat{L}_x$, $\hat{L}_y$ and $\hat{L}_z$. Algebraically, the spin matrices behave like rotation generators, but their exponentiations contain additional symmetry transformations. Therefore, they form a group that is not identical with the rotation group in a strict mathematical sense. The latter can be characterized as the group of all norm-preserving (or orthogonal) 3×3 matrices with determinant one, a combined condition that is shortened to the name SO(3) for the Lie group of **special orthogonal matrices** in three dimensions. What makes them "special" is the fact that they have determinant one. Exponentials of spin-1/2 matrices are 2×2 matrices that also have determinant one but are unitary, shortened to the name SU(2) for the corresponding Lie group. These properties can be confirmed directly for the matrices (4.237)–(4.239) as well as the general (4.242), and then also hold for their products.

For $\delta_j=2\pi$, each of these matrices equals $-\mathbb{I}$. If we ignore the overall minus sign (or, in mathematical language, if we consider equivalence classes of pairs $(U,-U)$ of SU(2)-matrices U), there is a one-to-one correspondence with matrices in SO(3): The spherical angles ϑ and φ can be used to define a unique rotation axis, and an angle $0\leq\delta<2\pi$ around this axis produces a general rotation in three dimensions. The group SU(2) is therefore called a double cover of SO(3).

It is common and often useful to write the spin-1/2 matrices as $\hat{S}_j=\frac{1}{2}\hbar\sigma_j$, where

$$\sigma_x=\sigma_1=\begin{pmatrix}0&1\\1&0\end{pmatrix},\quad \sigma_y=\sigma_2=\begin{pmatrix}0&-i\\i&0\end{pmatrix},\quad \sigma_z=\sigma_3=\begin{pmatrix}1&0\\0&-1\end{pmatrix} \tag{4.243}$$

are called **Pauli** matrices. (Using the letter σ is standard notation for these matrices, which are distinct from standard deviations.) These matrices obey the useful identity

$$\sigma_j\sigma_k=\delta_{jk}\mathbb{I}+i\sum_{l=1}^{3}\varepsilon_{jkl}\sigma_l \tag{4.244}$$

with the **Levi–Civita** symbol ε_{jkl} defined by complete antisymmetry in all three indices, together with $\varepsilon_{123} = 1$. (The same symbol can be used in the vector cross product, which has components $(a \times b)_j = \sum_{k,l} \varepsilon_{jkl} a_k b_l$.) For instance, $\sigma_1 \sigma_2 = i\sigma_3$ mimics the commutator of spin operators but in the case of Pauli matrices even applies to the matrix product.

We can now compute expectation values using simple matrix notation, for instance

$$\langle S_x \rangle_+ = \langle 1/2, 1/2 | S_x | 1/2, 1/2 \rangle = \frac{1}{2}\hbar \begin{pmatrix} 1 \\ 0 \end{pmatrix}^\dagger \begin{pmatrix} 0 & 1 \\ 1 & 0 \end{pmatrix} \begin{pmatrix} 1 \\ 0 \end{pmatrix} = 0 \tag{4.245}$$

and similarly $\langle S_x \rangle_- = \langle 1/2, -1/2 | S_x | 1/2, -1/2 \rangle = 0$, while $\langle S_x^2 \rangle_\pm = \frac{1}{4}\hbar^2$. The standard deviations of S_x in the two states are therefore

$$\sigma_{S_x}^\pm = \sqrt{\langle S_x^2 \rangle_\pm - \langle S_x \rangle_\pm^2} = \frac{1}{2}\hbar\,. \tag{4.246}$$

We compute the same value for σ_{S_y}, and see that the uncertainty relation is saturated in both states $|1/2, \pm 1/2\rangle$:

$$\sigma_{S_x}^\pm \sigma_{S_y}^\pm = \frac{1}{4}\hbar^2 = \frac{1}{2}|\hbar \langle S_z \rangle| = \frac{1}{2}|-i\langle [S_x, S_y] \rangle|\,. \tag{4.247}$$

Returning to the probability distribution for measurements of S_x rather than S_z, we need to transform to the eigenbasis of S_x. This transformation is accomplished by computing the eigenvalues from

$$\det(S_x - \lambda \mathbb{I}) = \det \begin{pmatrix} -\lambda & \frac{1}{2}\hbar \\ \frac{1}{2}\hbar & -\lambda \end{pmatrix} = \lambda^2 - \frac{1}{4}\hbar^2 = 0\,, \tag{4.248}$$

resulting in $\lambda_\pm = \pm\frac{1}{2}\hbar$ (as in the case of S_z), and corresponding normalized eigenvectors

$$v_+ = \frac{1}{\sqrt{2}} \begin{pmatrix} 1 \\ 1 \end{pmatrix} \quad , \quad v_- = \frac{1}{\sqrt{2}} \begin{pmatrix} 1 \\ -1 \end{pmatrix} \tag{4.249}$$

from

$$(S_x - \lambda \mathbb{I}) v_\pm = \begin{pmatrix} -\lambda & \frac{1}{2}\hbar \\ \frac{1}{2}\hbar & -\lambda \end{pmatrix} \begin{pmatrix} v_\pm^{(1)} \\ v_\pm^{(2)} \end{pmatrix} = \begin{pmatrix} 0 \\ 0 \end{pmatrix}. \tag{4.250}$$

Expressed in the S_z-eigenbasis, we have $S_x v_+ = \frac{1}{2}\hbar v_+$ with

$$v_+ = \frac{1}{\sqrt{2}} (|1/2, 1/2\rangle + |1/2, -1/2\rangle) \tag{4.251}$$

and $S_x v_- = -\frac{1}{2}\hbar v_-$ with

$$v_- = \frac{1}{\sqrt{2}} (|1/2, 1/2\rangle - |1/2, -1/2\rangle)\,. \tag{4.252}$$

For measurements of S_x, we write a given state, such as $|1/2, 1/2\rangle$, in the eigenbasis (v_+, v_-) of S_x:

$$|1/2, 1/2\rangle = \frac{1}{\sqrt{2}} (v_+ + v_-) = c_+ v_+ + c_- v_- \tag{4.253}$$

with coefficients $c_+ = c_- = 1/\sqrt{2}$. Now, the two values $|c_+|^2$ and $|c_-|^2$ form a probability distribution for measurements of S_x. Therefore, if a particle is in the eigenstate of S_z with eigenvalue $\frac{1}{2}\hbar$, a measurement of S_x returns the value $S_x = \frac{1}{2}\hbar$ with probability $|c_+|^2 = 1/2$, and the value $S_x = -\frac{1}{2}\hbar$ with probability $|c_-|^2 = 1/2$. This result is consistent with the previous observation (4.245) that the expectation value $\langle S_x \rangle_+ = 0$ vanishes in this state.

4.3.3 ELECTRON IN A MAGNETIC FIELD

According to classical physics, an electrically charged spinning object implies circular electric currents on its surface. Like a coil, such currents generate a magnetic dipole moment $\vec{\mu} = \gamma\vec{S}$ proportional to the angular momentum $\vec{S}$. The constant γ, called the gyromagnetic ratio, depends on the shape of the object and its electromagnetic properties. Even though a spinning particle in quantum mechanics may not have a surface on which we could have electric currents, it still implies a magnetic dipole moment if it is electrically charged. An important example is the electron, which is charged and has spin $1/2$.

A magnetic dipole moment $\vec{\mu}$ interacts with a magnetic field. If it is placed in an external magnetic field $\vec{B}$, the dipole has an energy

$$H_{\text{mag}} = -\vec{\mu}\cdot\vec{B} = -\gamma\vec{S}\cdot\vec{B} \tag{4.254}$$

depending on the angle between the spin and the magnetic field. "External" here means that we will not try to turn the magnetic field into an operator for a quantum-mechanical description. If we choose the magnetic field to point in the z-direction, we can write

$$H_{\text{mag}} = -\gamma B S_z \tag{4.255}$$

and use the simple diagonal S_z-matrix in the energy operator,

$$\hat{H}_{\text{mag}} = -\gamma B\hat{S}_z = -\frac{1}{2}\gamma B\hbar\begin{pmatrix} 1 & 0 \\ 0 & -1 \end{pmatrix} \tag{4.256}$$

for a spin-$1/2$ particle such as an electron. This magnetic Hamilton operator is not a differential operator but rather a matrix operator acting on spinors $\chi = \begin{pmatrix} \chi_+ \\ \chi_- \end{pmatrix}$.

If a particle is moving, it can be found at different positions as described statistically by a wave function $\psi(x,y,z,t)$. A spinning particle in addition has the two components of its spinor. The combined behavior of motion and spin is then described by a position-dependent spinor

$$\chi(x,y,z,t) = \begin{pmatrix} \chi_+(x,y,z,t) \\ \chi_-(x,y,z,t) \end{pmatrix} \tag{4.257}$$

with two components $\chi_\pm(x,y,z,t)$. Mathematically, the combination of position and spin degrees of freedom is described by the tensor product, $\chi(x,y,z,t) = \psi(x,y,z,t)\otimes\chi$. We act on this state by 2-dimensional spin matrices as well as differential and multiplication operators for the position. The complete Hamilton operator of the particle, including kinetic and potential energies for the motion and the magnetic energy for its spin, is given by the sum

$$\hat{H} = \hat{H}_{\text{kin}} + \hat{H}_{\text{pot}} + \hat{H}_{\text{mag}} = -\frac{\hbar^2}{2m}\left(\frac{\partial^2}{\partial x^2} + \frac{\partial^2}{\partial y^2} + \frac{\partial^2}{\partial z^2}\right) + V(x,y,z) - \gamma B\hat{S}_z\,. \tag{4.258}$$

If the particle moves slowly, such that we can ignore the Lorentz force from the magnetic field, which is proportional to the speed, the main non-kinetic energy is given by the magnetic term. Acting on a position-dependent spinor χ, we obtain the time-dependent Schrödinger equation

$$\begin{aligned} \hat{H}\chi &= (\hat{H}_{\text{kin}} + \hat{H}_{\text{mag}})\chi & (4.259) \\ &= \left(-\frac{\hbar^2}{2m}\left(\frac{\partial^2}{\partial x^2} + \frac{\partial^2}{\partial y^2} + \frac{\partial^2}{\partial z^2}\right) - \frac{1}{2}\gamma B\hbar\begin{pmatrix} 1 & 0 \\ 0 & -1 \end{pmatrix}\right)\begin{pmatrix} \chi_+ \\ \chi_- \end{pmatrix} = i\hbar\frac{\partial\chi}{\partial t}\,. \end{aligned}$$

This vector equation implies two independent equations for the components χ_+ and χ_-,

$$-\frac{\hbar^2}{2m}\left(\frac{\partial^2}{\partial x^2} + \frac{\partial^2}{\partial y^2} + \frac{\partial^2}{\partial z^2}\right)\chi_+ - \frac{1}{2}\gamma B\hbar\chi_+ = i\hbar\frac{\partial\chi_+}{\partial t} \tag{4.260}$$

$$-\frac{\hbar^2}{2m}\left(\frac{\partial^2}{\partial x^2} + \frac{\partial^2}{\partial y^2} + \frac{\partial^2}{\partial z^2}\right)\chi_- + \frac{1}{2}\gamma B\hbar\chi_- = i\hbar\frac{\partial\chi_-}{\partial t}\,, \tag{4.261}$$

which are decoupled because the matrix S_z is diagonal.

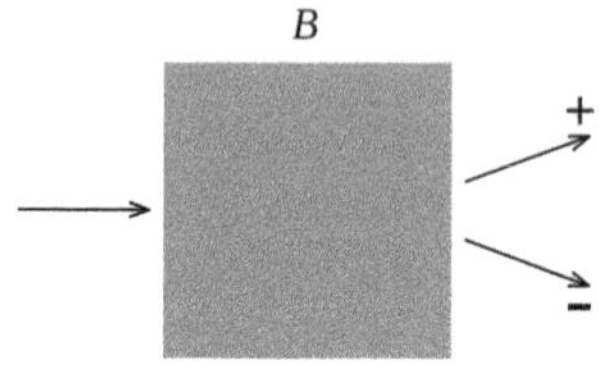

Figure 4.3 A beam of spinning particles sent through a non-constant magnetic field splits up into two beams according to the spin alignment with the magnetic field.

The motion of a charged spinning particle depends on how the spin vector is aligned with the magnetic field. If B is constant, the magnetic energy is just a constant shift for each component of χ and does not affect how the particle moves, but a non-constant field implies different forces acting on χ_+ and χ_-, respectively. A beam of spinning particles that contains both spin-up and spin-down states can therefore be separated into the two spinor components by sending it through a non-constant magnetic field. For instance, for a linear dependence $B(x,y,z) = \alpha z$ with a constant α, the two components experience oppositive forces in the z-direction, determined by linear potentials $V_\pm(z) = \mp\frac{1}{2}\gamma\alpha\hbar z$ for $\chi_\pm$.

The first measurement that indicated the existence of quantized spin was made by the **Stern–Gerlach experiment** in 1922; see Figure 4.3. The experiment used a beam of ions that happened to have a non-zero spin quantum number, and sent it through a magnetic field. Classically, one would expect a widening of the beam because ions with different alignments of a magnetic dipole moment with the magnetic field are deflected up or down in a continuum of directions. However, the experiment resulted in two clearly distinct bands on a screen, showing that the beam contained ions with only two different directions of the magnetic moment. If the ions have spin s, we expect $2s+1$ different projections of the spin on the direction of the magnetic field, given by the different values of $\mathfrak{m}_s$. For two final beams, $2s+1=2$ requires $s=1/2$. The result can therefore be explained only if half-integer spin values are realized physically, as noticed by Uhlenbeck and Goudsmit in 1925.

4.3.4 IDENTICAL PARTICLES

If we have multiple particles at positions $\vec{r}_j = (x_j, y_j, z_j)$, $j = 1,\ldots,n$, the applicability of position and momentum operators for each of them requires a joint wave function $\psi(\vec{r}_1,\ldots,\vec{r}_n)$ that depends on all the positions. It is normalized according to

$$\int\cdots\int \mathrm{d}x^n\mathrm{d}y^n\mathrm{d}z^n|\psi(\vec{r}_1,\ldots,\vec{r}_n)|^2 = 1\,. \tag{4.262}$$

There are independent position and momentum operators for all the particles, $\hat{x}_j$, $\hat{y}_j$, $\hat{z}_j$ and $\hat{p}_{x_j}$, $\hat{p}_{y_j}$, $\hat{p}_{z_j}$. They have the usual commutators such as $[\hat{x}_j, \hat{p}_{x_j}] = i\hbar$ if they refer to the same particle and the same dimension, but $[\hat{x}_j, \hat{p}_{x_k}] = 0$ if they refer to different particles, $j \neq k$. There are therefore no uncertainty relations for a position and a momentum measurement of different particles.

Nevertheless, it turns out that the position dependence of ψ is restricted if the n particles are identical, that is, if they are of the same type, such as n electrons. The wave function for particles of different types, such as an electron and a proton, is in general unrestricted. The reason is the basic property that the state of the particles is described by a wave function, rather than individual time-dependent positions along trajectories as in the classical case.

As an example, a 2-particle state may be represented at two different times, t_1 and t_2, by positions as indicated in Figure 4.4. The particles are identical and therefore do not have any distinguishing information, other than their positions. This is different from the classical case, shown in Figure 4.5, in which we can always keep track of individual particles by observing their trajectories. Even if

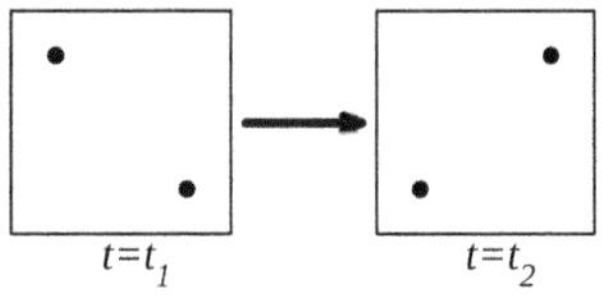

Figure 4.4 Two identical particles at different positions at times t_1 and t_2. Quantum mechanics does not tell us to which position at time t_2 a specific particle moved, starting from a position at time t_1.

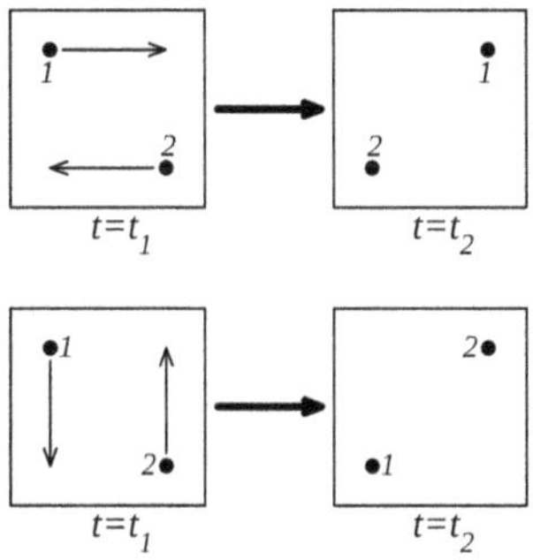

Figure 4.5 A classical picture in which we keep track of individual particles and the trajectories along which they move. We can therefore think of the particles as being labeled by an arbitrary assignment at time t_1, even if they are identical, which can be uniquely maintained at later times by following their trajectories. The two labelings at time t_2 given by $(1,2)$, top, and $(2,1)$, bottom, therefore represent distinguishable states in classical physics.

they are identical, we can label them by taking a picture of their positions at an initial time, and keeping the same labels along their trajectories at later times. Figure 4.5 shows two different states distinguished by labels at different positions at time t_2. Without labels, as in quantum mechanics, we have only one final state shown in Figure 4.4.

The wave function, representing a state, by definition contains all the information we can obtain about the system. It should therefore implement the **indistinguishability** behavior of multi-particle states implied by the absence of individual trajectories or labels. The classical understanding has led us to introducing a list of vectors $\vec{r}_1,\ldots,\vec{r}_n$ for a set of n particles, suggesting that the first particle is at position $\vec{r}_1$, the second particle is at position $\vec{r}_2$ and so on. However, calling a particle "the first particle" or "the second particle" introduces labels that are not implied by physical properties of the particles, assumed to be identical. In quantum mechanics, we cannot distinguish between the two outcomes described as "the first particle at position $\vec{r}_1$ and the second particle at position $\vec{r}_2$" (probability density $|\psi(\vec{r}_1,\vec{r}_2)|^2$) and "the first particle at position $\vec{r}_2$ and the second particle at position $\vec{r}_1$" (probability density $|\psi(\vec{r}_2,\vec{r}_1)|^2$). These probability densities should therefore be the same,

$$|\psi(\vec{r}_1,\vec{r}_2)|^2 = |\psi(\vec{r}_2,\vec{r}_1)|^2\,, \tag{4.263}$$

whenever the two particles are identical. Similar symmetry conditions exist for n particles, and for wave functions in the momentum eigenbasis or some other basis.

We evaluate this condition by introducing the **exchange operator**

$$(\hat{E}\psi)(\vec{r}_1,\vec{r}_2) = \psi(\vec{r}_2,\vec{r}_1)\,. \tag{4.264}$$

An eigenstate of $\hat{E}$ is such that $\psi(\vec{r}_2,\vec{r}_1)$ is proportional to $\psi(\vec{r}_1,\vec{r}_2)$, and is therefore a candidate for a wave function that obeys (4.263).

This operator defines a linear transformation because

$$\begin{aligned}(\hat{E}(a\psi_1+b\psi_2))(\vec{r}_1,\vec{r}_2) &= (a\psi_1+b\psi_2)(\vec{r}_2,\vec{r}_1)\\ &= a\psi_1(\vec{r}_2,\vec{r}_1)+b\psi_2(\vec{r}_2,\vec{r}_1)=(a\hat{E}\psi_1+b\hat{E}\psi_2)(\vec{r}_1,\vec{r}_2). \quad (4.265)\end{aligned}$$

The exchange operator is Hermitian because

$$\begin{aligned}\langle\psi_1|\hat{E}\psi_2\rangle &= \int\cdots\int \psi_1(\vec{r}_1,\vec{r}_2)^*\hat{E}\psi_2(\vec{r}_1,\vec{r}_2)\mathrm{d}^3r_1\mathrm{d}^3r_2 \quad (4.266)\\ &= \int\cdots\int \psi_1(\vec{r}_1,\vec{r}_2)^*\psi_2(\vec{r}_2,\vec{r}_1)\mathrm{d}^3r_1\mathrm{d}^3r_2\\ &= \int\cdots\int \psi_1(\vec{s}_2,\vec{s}_1)^*\psi_2(\vec{s}_1,\vec{s}_2)\mathrm{d}^3s_2\mathrm{d}^3s_1\\ &= \int\cdots\int (\hat{E}\psi_1)(\vec{s}_1,\vec{s}_2)^*\psi_2(\vec{s}_1,\vec{s}_2)\mathrm{d}^3s_1\mathrm{d}^3s_2=\langle\hat{E}\psi_1|\psi_2\rangle\end{aligned}$$

where we substituted $(\vec{s}_1,\vec{s}_2)=(\vec{r}_2,\vec{r}_1)$. This is not an exchange operation because $r\neq s$ are general labels. We further conclude that the operator $\hat{E}$ is self-adjoint because the calculation in (4.266) does not require any specific properties of ψ_1 and ψ_2, other than their normalizability, for the integrals to be finite. Therefore, $\hat{E}^\dagger=\hat{E}$.

The square of the exchange operator is the identity operator, $\hat{E}^2=\hat{\mathbb{I}}$:

$$(\hat{E}^2\psi)(\vec{r}_1,\vec{r}_2)=(\hat{E}(\hat{E}\psi))(\vec{r}_1,\vec{r}_2)=(\hat{E}\psi)(\vec{r}_2,\vec{r}_1)=\psi(\vec{r}_1,\vec{r}_2). \quad (4.267)$$

Replacing one of the $\hat{E}$ in $\hat{E}^2=\hat{\mathbb{I}}$ with $\hat{E}^\dagger=\hat{E}$, we obtain

$$\hat{E}^\dagger\hat{E}=\hat{\mathbb{I}}=\hat{E}\hat{E}^\dagger. \quad (4.268)$$

The Hermitian conjugate therefore equals the inverse, $\hat{E}^\dagger=\hat{E}^{-1}$, and the exchange operator is unitary.

Because the exchange operator is self-adjoint, it has real eigenvalues λ with eigenfunctions ψ_λ, such that $\hat{E}\psi_\lambda=\lambda\psi_\lambda$. The equation $\hat{E}^2=\hat{\mathbb{I}}$ implies

$$\hat{E}^2\psi_\lambda=\hat{E}(\lambda\psi_\lambda)=\lambda^2\psi_\lambda=\hat{\mathbb{I}}\psi_\lambda=\psi_\lambda \quad (4.269)$$

and therefore $\lambda^2=1$ because an eigenfunction cannot vanish everywhere. For real eigenvalues, this equation has only two solutions, $\lambda=\pm1$. This result is consistent with unitarity because $|\lambda|^2=1$.

The corresponding eigenfunctions obey the conditions

$$(\hat{E}\psi_+)(\vec{r}_1,\vec{r}_2)=\psi_+(\vec{r}_2,\vec{r}_1)=\psi_+(\vec{r}_1,\vec{r}_2) \quad (4.270)$$

for $\lambda=+1$, and

$$(\hat{E}\psi_-)(\vec{r}_1,\vec{r}_2)=\psi_-(\vec{r}_2,\vec{r}_1)=-\psi_-(\vec{r}_1,\vec{r}_2) \quad (4.271)$$

for $\lambda=-1$. Wave functions of the type ψ_+ are even under the exchange operation, while wave functions of the type ψ_- are odd. In both cases,

$$|\psi_\pm(\vec{r}_1,\vec{r}_2)|^2=|\psi_\pm(\vec{r}_2,\vec{r}_1)|^2. \quad (4.272)$$

Both types of wave functions are therefore suitable for pairs of identical particles, such as two electrons.

The **spin-statistics theorem** determines which types have to be used for different particles, even or odd wave functions under exchange. The theorem does not have a simple explanation but rather follows from properties of quantum field theory, which is the general framework used to describe particles and their interactions, as well as pair creation and annihilation events. According to this theorem, an even wave function, $\psi(\vec{r}_1,\vec{r}_2) = \psi(\vec{r}_2,\vec{r}_1)$, has to be used for identical particles with integer spin, $s = 0, 1, \ldots$. Particles of this type are called **bosons**. An example is the photon. An odd wave function, $\psi(\vec{r}_1,\vec{r}_2) = -\psi(\vec{r}_2,\vec{r}_1)$, has to be used for identical particles with half-integer spin, $s = 1/2, 3/2, \ldots$. Particles of this type are called **fermions**. Examples are the electron, the proton, and the neutron.

In particular, pairs of electrons, fermions with spin $s = 1/2$, must always be obtained from antisymmetric combinations of individual states. States of two distinguishable particles, such as an electron and a proton in a hydrogen atom, can be combined by simply multiplying their wave functions, $\psi(\vec{r}_1,\vec{r}_2) = \psi_1(\vec{r}_1)\psi_2(\vec{r}_2)$. The product of wave functions implies that their probability densities are products of individual densities, which is suitable for probabilities of two independent measurements given by the two positions of distinguishable particles. However, the product is not suitable for indistinguishable particles because it is not an eigenstate of the exchange operator if $\psi_1 \neq \psi_2$. For indistinguishable particles, the two individual position measurements are not independent because we cannot separate them from each other. If we have one electron in the state $\psi_1(\vec{r}_1)$ and one electron in the state $\psi_2(\vec{r}_2)$, the combined state must be antisymmetric for these fermions, and can be written as

$$\psi(\vec{r}_1,\vec{r}_2) \propto \psi_1(\vec{r}_1)\psi_2(\vec{r}_2) - \psi_1(\vec{r}_2)\psi_2(\vec{r}_1)\,. \tag{4.273}$$

In this way, we combine the product behavior of probabilities with the required symmetry properties for fermions. (The coefficient in this proportionality, which is required for normalization, will be discussed in Section 5.1.)

As a consequence, two identical fermions cannot be in the same state because $\psi_1 = \psi_2$ implies that $\psi_1(\vec{r}_1)\psi_2(\vec{r}_2) - \psi_1(\vec{r}_2)\psi_2(\vec{r}_1) = 0$. This implication of the spin-statistics theorem is called **Pauli's exclusion principle**. It has far-reaching consequences because it determines how electrons can be combined in atoms with nuclear charge greater than one, how different atoms can be combined and bound in molecules, and how macroscopic objects such as condensed-matter systems or even compact stellar objects behave.

If we combine specific electron states, we have to keep in mind the position dependence $\psi(\vec{r})$ and the spinor part χ in a finite-dimensional representation. Since there are always two available spin states for a spin-1/2 particle, given by the two different values of $\mathfrak{m}_s = \pm 1/2$, the position dependence may be the same, $\psi(\vec{r}_1) = \psi(\vec{r}_2)$. A non-zero odd wave function including spin is then given by

$$\chi(\vec{r}_1,\vec{r}_2) = \frac{1}{\sqrt{2}}\psi(\vec{r}_1)\psi(\vec{r}_2)\left(|1/2,1/2\rangle|1/2,-1/2\rangle - |1/2,-1/2\rangle|1/2,1/2\rangle\right) \tag{4.274}$$

where we use the notation $|s,\mathfrak{m}_{s,1}\rangle|s,\mathfrak{m}_{s,2}\rangle$ for combined spin states.

Mathematically, this combination is a tensor product, which is not commutative as shown by the example (3.40) for the tensor product of two matrices. For column vectors χ_j in a basis, the components of their tensor product are obtained from the matrix $\chi_1\chi_2^T$. The two spinors in (4.274) are represented by different expressions,

$$\begin{pmatrix} 1 \\ 0 \end{pmatrix}(0\,1) = \begin{pmatrix} 0 & 1 \\ 0 & 0 \end{pmatrix} \quad \text{and} \quad \begin{pmatrix} 0 \\ 1 \end{pmatrix}(1\,0) = \begin{pmatrix} 0 & 0 \\ 1 & 0 \end{pmatrix}. \tag{4.275}$$

The tensor-product notation $|1/2,1/2\rangle \otimes |1/2,-1/2\rangle$ is useful in order to emphasize this non-commutative nature.

The state (4.274) is therefore non-zero, and it is odd under the combined exchange $\vec{r}_1 \leftrightarrow \vec{r}_2$ and $\mathfrak{m}_{s,1} \leftrightarrow \mathfrak{m}_{s,2}$. It can be used for the ground state of helium (in which case ψ is approximately given by the ground-state wave function of hydrogen with an adjustment for the increased nuclear charge), or for the two electrons forming a covalent bond in the hydrogen molecule.

The exclusion principle has important implications in a wide range of physical applications:

1. The periodic table: The structure of atoms can be determined by starting with a nucleus of central charge Ze with an integer $Z \geq 1$ and adding Z electrons in suitable states in order to form a neutral atom. After hydrogen, which just has one electron, the next example is helium with $Z = 2$. The two electrons are in general at two different positions, $\vec{r}_1$ and $\vec{r}_2$, which implies the complicated Hamilton operator

$$\hat{H} = -\frac{\hbar^2}{2m}(\nabla_1^2 + \nabla_2^2) - \frac{e^2}{4\pi\varepsilon_0}\left(\frac{2}{r_1} + \frac{2}{r_2} - \frac{1}{|\vec{r}_1 - \vec{r}_2|}\right) \tag{4.276}$$

including electron-nucleus as well as electron-electron interactions. The potential is not spherically symmetric, which means that, in general, we cannot use simultaneous eigenfunctions of $\hat{H}$ with angular-momentum operators in order to simplify the time-independent Schrödinger equation and characterize different stationary states.
Nevertheless, as a first approximation, we may use the eigenfunctions $\psi_{n\ell\mathrm{m}}(\vec{r})$ derived for hydrogen, just substituting $2e^2$ for e^2 in order to account for the greater central charge of helium. This procedure ignores electron-electron interactions and therefore does not result in a good quantitative approximation. The two electrons may well come close to each other, implying a large repulsive force greater than the two attractive forces toward the nucleus. Nevertheless, several qualitative properties of helium can be seen in this way.
The ground state of this 2-electron system has to be odd under exchange symmetry, which can be achieved by using two hydrogen ground states with minimal energy and opposite spin directions for the two electrons:

$$\chi(\vec{r}_1,\vec{r}_2) = \frac{1}{\sqrt{2}}\psi_{100}(\vec{r}_1)\psi_{100}(\vec{r}_2)\left(|1/2,1/2\rangle|1/2,-1/2\rangle - |1/2,-1/2\rangle|1/2,1/2\rangle\right) \tag{4.277}$$

just as in (4.274). We can then estimate the ground-state energy of helium by computing $\langle\chi|\hat{H}|\chi\rangle$ in this state. The approximation may be improved by using a variational method, in which we keep the Bohr radius a in ψ_{100} initially as a free parameter, such that $\langle\chi|\hat{H}|\chi\rangle(a)$ is a function of a. This function has a local minimum at a certain value of a, providing a better estimate of the ground-state energy.
Irrespective of the precise value, an important prediction follows from symmetry properties of the state. Minimal energy can be obtained only if both electrons are in a hydrogen-like ground state because assigning one electron to the hydrogen level $n = 2$ would raise the energy. The wave function can then be odd only if the two spins point in opposite directions, which means that a helium atom in its ground state does not have a magnetic dipole moment: It is diamagnetic. In fact, there is a metastable state in which one of the two electrons has a wave function of the type ψ_{200}, while the other electron remains in ψ_{100}. This state has a greater energy than the ground state, but it makes it possible to have aligned spins of the two electrons:

$$\chi_{\mathrm{ortho}}(\vec{r}_1,\vec{r}_2) \propto \left(\psi_{100}(\vec{r}_1)\psi_{200}(\vec{r}_2) - \psi_{200}(\vec{r}_1)\psi_{100}(\vec{r}_2)\right)|1/2,1/2\rangle|1/2,1/2\rangle\,. \tag{4.278}$$

This metastable version of helium, called orthohelium, has a non-vanishing magnetic dipole moment. Helium in its actual ground state is called parahelium.
For an atom with central charge Ze, we have to combine Z hydrogen wave functions in order to construct the same approximation. The available quantum numbers are $n \geq 1$, $0 \leq \ell \leq n-1$, $-\ell \leq \mathrm{m} \leq \ell$ and $\mathrm{m}_s = \pm\frac{1}{2}$. The number of independent states for a given n, called a shell, equals $2n^2$, twice the value derived in (4.174) because we now include the spin degree of freedom. For a hydrogen model with a single Coulomb potential, all states with the same n but different ℓ have the same energy E_n. Electron-electron interactions in heavier atoms

imply a discrete range of energies even for a given n: The electron-electron potential does not commute with the Hamilton operator of hydrogen and therefore cannot have the same eigenstates. Even for hydrogen, there are additional terms such as relativistic corrections or coupling terms between the electron and nuclear spin that imply ℓ-dependent corrections to E_n.

After hydrogen, the next few elements of the periodic table have the following electron configurations:

a. Z=2, helium: The two electrons have quantum numbers $n,\ell,\mathfrak{m},\mathfrak{m}_s = 1,0,0,\pm\frac{1}{2}$. The first shell with $n=1$ is full.
b. Z=3, lithium: Two electrons are in states as in helium, and the third has quantum numbers $n,\ell,\mathfrak{m} = 2,0,0$, starting the second shell.
c. $Z=4$–$Z=9$: The second shell has $2n^2 = 8$ available states. which are being filled up.
d. $Z=10$, neon: Two electrons fill the first shell, and the remaining eight electrons fill the second shell.

Chemical reactions are determined by the outermost electrons in an atom that are most likely to interact with electrons of a different atom. Atoms with the same type of outermost shell, such as hydrogen and lithium or helium and neon, therefore have similar chemical properties, which explains many characteristics of the periodic table.

2. Molecules: Electrons in the outermost shell of an atom determine properties of the covalent bond. For instance, a hydrogen molecule has two electrons in a spinor wave function $\chi(\vec{r}_1,\vec{r}_2)$ similar to helium, but they now interact with two nuclei. The electric force depends on the distance between two charged particles, which we can relate to the position vectors $\vec{r}_1$ and $\vec{r}_2$ of the two electrons relative to one of the nuclei. The distances to this nucleus are then $r_{11} = |\vec{r}_1|$ and $r_{12} = |\vec{r}_2|$, respectively. If the position vector of the second nucleus relative to the first one is $\vec{R}$, the distances between the two electrons and the second nucleus are $r_{21} = |\vec{r}_1 - \vec{R}|$ and $r_{22} = |\vec{r}_2 - \vec{R}|$, respectively. We have the Hamilton operator

$$\hat{H} = -\frac{\hbar^2}{2m}(\nabla_1^2 + \nabla_2^2) - \frac{e^2}{4\pi\varepsilon_0}\left(\frac{1}{r_{11}} + \frac{1}{r_{21}} + \frac{1}{r_{12}} + \frac{1}{r_{22}} - \frac{1}{|\vec{r}_1 - \vec{r}_2|}\right) \tag{4.279}$$

including electron-electron interactions in the last term. There is also nucleus-nucleus repulsion depending on R. However, this expression is independent of $\vec{r}_1$ and $\vec{r}_2$ and therefore does not contribute to electron wave functions.

As in helium, (4.274) can be used as an approximate ground-state wave function,

$$\chi(\vec{r}_1,\vec{r}_2) = \frac{1}{\sqrt{2}}\psi_{100}(\vec{r}_1)\psi_{100}(\vec{r}_2)\left(|1/2,1/2\rangle|1/2,-1/2\rangle - |1/2,-1/2\rangle|1/2,1/2\rangle\right). \tag{4.280}$$

The product $\psi_{100}(\vec{r}_1)\psi_{100}(\vec{r}_2)$ takes the greatest values in a region between the two nuclei, where $|\vec{r}_1|$ and $|\vec{r}_2|$ are close to each other and not sufficiently large for the exponential decrease of $R_{10}(r)$ to be significant. There is therefore a large probability for a negative charge density between the two positively charged nuclei, implying a covalent bond.

If we try to construct a state with aligned spins, a non-vanishing wave function requires that one of the electrons is in a state with $n=2$:

$$\chi(\vec{r}_1,\vec{r}_2) \propto (\psi_{100}(\vec{r}_1)\psi_{200}(\vec{r}_2) - \psi_{200}(\vec{r}_1)\psi_{100}(\vec{r}_2))|1/2,1/2\rangle|1/2,1/2\rangle \tag{4.281}$$

as in orthohelium. Here, $\psi_{100}(\vec{r}_1)\psi_{200}(\vec{r}_2) - \psi_{200}(\vec{r}_1)\psi_{100}(\vec{r}_2)$ is close to zero in the region between the nuclei, where $|\vec{r}_1| \approx |\vec{r}_2|$. There is therefore only a small negative magnetic charge density between the positive nuclei, which in this case do not form a bound state. Based on the spin alignment, the hydrogen molecule should have magnetic properties similar to parahelium, which can be confirmed by experiments.

3. Solid states: Important properties of matter such as conductivity of metals, semiconductors, or superconductivity follow from the fact that electrons are fermions.
4. Compact stellar objects: Stars maintain their temperature and pressure by fusion processes in their centers, starting with primordial hydrogen and helium produced in the big bang and leading up to iron nuclei. Beyond iron, fusion no longer produces energy. When most of the initial nuclei have been fused to iron, the star cools down and loses pressure, leading to stellar collapse. In classical physics, there would be no force that could stop the collapse of cold matter, but eventually the electron distribution in a collapsing star will be compressed so much that all available states are occupied. The star then cannot collapse further, being stabilized to a white dwarf only by the exclusion principle.
   However, at such high densities electrons are close to protons, such that they are likely to react through the weak nuclear force and form a neutron and a neutrino. Neutrons are also fermions, but more massive than electrons. The star can therefore collapse further until the exclusion principle prevents neutrons from getting more compressed based on the exclusion principle. The result is a neutron star. A neutron star can be stable only if its mass is sufficiently small. If it is too heavy, it cannot be supported by space-time, and collapses into a black hole.

EXERCISES

1. Transform the angular-momentum operator
$$\hat{L}_x = \frac{\hbar}{i}\left(y\frac{\partial}{\partial z} - z\frac{\partial}{\partial y}\right)$$
to spherical coordinates, confirming (4.29).
2. For $\ell = 1$, the angular-momentum components $\hat{L}_x$, $\hat{L}_y$ and $\hat{L}_z$ are represented by the matrices
$$L_x = \begin{pmatrix} 0 & \hbar/\sqrt{2} & 0 \\ \hbar/\sqrt{2} & 0 & \hbar/\sqrt{2} \\ 0 & \hbar/\sqrt{2} & 0 \end{pmatrix}$$
$$L_y = \begin{pmatrix} 0 & -i\hbar/\sqrt{2} & 0 \\ i\hbar/\sqrt{2} & 0 & -i\hbar/\sqrt{2} \\ 0 & i\hbar/\sqrt{2} & 0 \end{pmatrix}$$
$$L_z = \begin{pmatrix} \hbar & 0 & 0 \\ 0 & 0 & 0 \\ 0 & 0 & -\hbar \end{pmatrix}$$
   (i) Compute $L^2 = L_x^2 + L_y^2 + L_z^2$ and show that it is a multiple of the identity matrix.
   (ii) Confirm that the commutator $[L_x, L_y]$ has the correct form.
   (iii) Derive the eigenvalues of L_x and L_y and the corresponding eigenvectors.
3. Give a complete demonstration that the operators (4.113) and (4.114) are Hermitian.
4. Derive equations (4.117) and (4.118) by applying separation of variables to (4.111), using $\hat{L}_z\psi = \mathfrak{m}\hbar\psi$.
5. Compute the commutators of $\hat{x}$, $\hat{y}$, $\hat{z}$, $\hat{p}_x$, $\hat{p}_y$, and $\hat{p}_z$ with the Hamilton operator $\hat{H}$ for a particle in an arbitrary three-dimensional potential $V(x,y,z)$. Use the results to generalize Ehrenfest's theorem to three dimensions.

6. A 2-particle system has a wave function $\psi(x_1,y_1,z_1,x_2,y_2,z_2)$, depending on all six coordinates of the two particles with masses m_1 and m_2. Assume that the force between the particles is described by a potential $V(r)$ that depends only on the distance $r = \sqrt{(x_1-x_2)^2+(y_1-y_2)^2+(z_1-z_2)^2}$ between the two particles. The Schrödinger equation for the complete system is obtained from the combined energy, which now has kinetic contributions from both particles and the potential $V(r)$.
In order to reduce the problem to one for a single particle, introduce new coordinates defined by $x = x_1 - x_2$, $y = y_1 - y_2$, $z = z_1 - z_2$ (the particle separation) and

$$X = \frac{m_1 x_1 + m_2 x_2}{M} \quad , \quad Y = \frac{m_1 y_1 + m_2 y_2}{M} \quad , \quad Z = \frac{m_1 z_1 + m_2 z_2}{M}$$

(the center-of-mass coordinates), where $M = m_1 + m_2$ is the combined mass.

(i) Show that the Schrödinger equation in these coordinates is separable by using wave functions $\psi(x_1,y_1,z_1,x_2,y_2,z_2) = \psi_1(x,y,z)\psi_2(X,Y,Z)$, such that ψ_2 satisfies the Schrödinger equation for a free particle with mass M (describing the center-of-mass motion) and ψ_1 solves the Schrödinger equation for a single particle with the reduced mass $\mu = m_1 m_2/M$ in a spherically symmetric potential.
(ii) Identify symmetry transformations of the system according to which the separation procedure used in part (i) can be viewed as looking for simultaneous eigenfunctions of $\hat{H}$ and suitable self-adjoint operators.
(iii) The time-independent Schrödinger equation for ψ_1 has a spherically symmetric potential $V(r)$. Its solutions can therefore be found as simultaneous eigenfunctions with angular-momentum operators, such that $\psi_1(x,y,z) = R(r)Y_\ell^m(\vartheta,\varphi)$. Derive the corresponding radial equation

$$-\frac{\hbar^2}{2\mu}\frac{\mathrm{d}^2 u}{\mathrm{d}r^2} + V(r)u + \frac{\hbar^2 \ell(\ell+1)}{2\mu r^2}u = Eu$$

for $u(r) = rR(r)$.

7. A deuteron is a bound state of a proton (mass m_p) and a neutron (mass $m_\mathrm{n} \approx m_\mathrm{p}$). According to the preceding problem, its bound states can be obtained from a 1-particle Schrödinger equation for ψ_1 with the reduced mass μ.

(i) Write down the radial equation with a potential modeled as a finite square well, $V(r) = -V_0$ if $r < a$ and $V(r) = 0$ otherwise.
(ii) Note that $R(r)$ must be finite at $r = 0$, and thus $u(r)$ must vanish at $r = 0$. Use this boundary condition and continuity of $u(r)$ to find an equation for the energies of bound state in a deuteron, assuming vanishing angular momentum ($\ell = 0$).
(iii) How many bound states with $\ell = 0$ are there? Determine the ground-state energy by solving the energy equation graphically, and compare with the experimental result for the binding energy E_b. Relevant numerical values can be found on page 90.

8. A small modification of Bohr's formula for bound-state energies,

$$E_n = -\frac{m}{2\hbar^2}\left(\frac{Ze^2}{4\pi\varepsilon_0}\right)^2 \frac{1}{n^2} = \frac{E_1}{n^2} \qquad \text{for } n = 1,2,3,\ldots \tag{4.282}$$

can be used for atoms of nuclear charge Ze (Z protons) if they are almost completely ionized such that there is only one remaining electron. Wave functions are obtained by using the same replacement of Ze^2 for e^2 in the modified Bohr radius

$$a_Z = \frac{4\pi\varepsilon_0\hbar^2}{mZe^2}\,. \tag{4.283}$$

Suppose that a tritium atom T ($Z = 1$) in its ground state decays into a He^+ ion ($Z = 2$). The following steps indicate how the probabilities of finding He^+ in its ground state and in the first excited states, respectively, can be derived:

(i) Use the general formulas for $\psi_{n\ell\mathrm{m}}$ and Y_ℓ^{m} in terms of Laguerre and Legendre polynomials in order to find the ground-state wave function and the wave functions of the first excited states of He^+.

(ii) Assume that the electron wave function is unchanged by the decay and is still given by the tritium ground state $\psi_{100}(\mathrm{T})$. Compute the coefficients $c_{1\ell\mathrm{m}} = \langle\psi_{1\ell\mathrm{m}}(\mathrm{He}^+)|\psi_{100}(\mathrm{T})\rangle$ and $c_{2\ell\mathrm{m}} = \langle\psi_{2\ell\mathrm{m}}(\mathrm{He}^+)|\psi_{100}(\mathrm{T})\rangle$ that appear in the linear combination

$$\psi_{100}(\mathrm{T}) = \sum_{n\ell\mathrm{m}} c_{n\ell\mathrm{m}}\,\psi_{n\ell\mathrm{m}}(\mathrm{He}^+)\,.$$

(iii) The coefficients in $\psi_{100}(\mathrm{T})$ provide a probability distribution for energy measurements. Accordingly, the probabilities of finding He^+ in the ground state and the first excited state, respectively, are given by $P_1 = \sum_{\ell\mathrm{m}} |c_{1\ell\mathrm{m}}|^2$ and $P_2 = \sum_{\ell\mathrm{m}} |c_{2\ell\mathrm{m}}|^2$, where the sums are over all allowed values of ℓ and m for a given $n = 1$ or $n = 2$. Compute P_1 and P_2.

5 Quantum Information

There is more information in a quantum state than in the classical initial values (x_0, p_0) that determine the complete 1-dimensional motion of a point particle for a given force. The initial wave function $\Psi(x,0)$ requires an infinite number of parameters, which can be expressed in observational terms as the two basic expectation values $\langle\hat{x}\rangle$ and $\langle\hat{p}\rangle$ together with the two variances (σ_x^2, σ_p^2) and the covariance

$$C_{xp} = \frac{1}{2}\langle\hat{x}\hat{p} + \hat{p}\hat{x}\rangle - \langle\hat{x}\rangle\langle\hat{p}\rangle\,, \tag{5.1}$$

as well as an extension of these quadratic expressions in x and p to arbitrary polynomial orders, defining the statistical moments. All these values are independent degrees of freedom that can vary in an initial state given by a specific wave function, and they are subject to dynamics. An important question is how different kinds of this non-classical information can be organized or generated in a specific state. Two important concepts related to quantum information are correlations and entanglement.

5.1 QUANTUM CORRELATIONS

Correlations of two degrees of freedom, x_1 and x_2, provide a statistical concept that tells us whether measuring x_1 can make it more likely to measure x_2 in a certain range, such as less than x_1 or greater than x_1, in a second measurement. From medical studies it is well-known that identifying correlations in a sample is an important and, in general, difficult problem.

In quantum mechanics, correlations between x_1 and x_2 are described by a 2-variable wave function $\psi(x_1,x_2)$. As a first example, consider the simple product state $\psi(x_1,x_2) = \psi_1(x_1)\psi_2(x_2)$ of two 1-variable states, ψ_1 and ψ_2. We compute the (x_1,x_2)-version of the covariance (5.1), in which we do not need to use the symmetrized product because $\hat{x}_1$ and $\hat{x}_2$ commute:

$$\begin{aligned} C_{x_1x_2} &= \langle\hat{x}_1\hat{x}_2\rangle - \langle\hat{x}_1\rangle\langle\hat{x}_2\rangle \\ &= \int_{-\infty}^{\infty}\int_{-\infty}^{\infty} x_1x_2|\psi_1(x_1)\psi_2(x_2)|^2 \mathrm{d}x_1\mathrm{d}x_2 - \left(\int_{-\infty}^{\infty} x_1|\psi_1(x_1)|^2\mathrm{d}x_1\right)\left(\int_{-\infty}^{\infty} x_2|\psi_2(x_2)|^2\mathrm{d}x_2\right) \\ &= \left(\int_{-\infty}^{\infty} x_1|\psi_1(x_1)|^2\mathrm{d}x_1\right)\left(\int_{-\infty}^{\infty} x_2|\psi_2(x_2)|^2\mathrm{d}x_2\right) - \langle\hat{x}_1\rangle\langle\hat{x}_2\rangle = 0\,. \end{aligned} \tag{5.2}$$

In the first integral, the dependences on x_1 and x_2 factorize and produce the same product of expectation values that is subtracted from this expression. Quantum correlations therefore vanish in a product state.

As another example, we may use symmetric or antisymmetric combinations of the two states as they appeared in atomic and molecular models, such that $\psi(x_1,x_2)$ is proportional to $\psi_1(x_1)\psi_2(x_2) \pm \psi_2(x_1)\psi_1(x_2)$. We first have to normalize this state by setting the integral of

$$\begin{aligned} |\psi_1(x_1)\psi_2(x_2) \pm \psi_2(x_1)\psi_1(x_2)|^2 &= |\psi_1(x_1)|^2|\psi_2(x_2)|^2 + |\psi_2(x_1)|^2|\psi_1(x_2)|^2 \\ &\quad \pm\psi_1(x_1)^*\psi_2(x_1)\psi_2(x_2)^*\psi_1(x_2) \\ &\quad \pm\psi_1(x_1)\psi_2(x_1)^*\psi_2(x_2)\psi_1(x_2)^* \end{aligned} \tag{5.3}$$

equal to one. Integrating the first line results in the number two, using normalization of the two 1-variable wave functions. The remaining terms factorize in x_1 and x_2-versions of a single integral,

$$O = \int_{-\infty}^{\infty} \psi_1(x)^*\psi_2(x)\mathrm{d}x = \langle\psi_1|\psi_2\rangle\,, \tag{5.4}$$

DOI: 10.1201/9781003588702-6

or its complex conjugate. This expression is called the **overlap integral** because it receives most of its contributions from regions, where both $\psi_1(x)$ and $\psi_2(x)$ have large real or imaginary parts. The integral of (5.3) then equals $2(1\pm|O|^2)$, such that we have a normalized wave function

$$\psi(x_1,x_2) = \frac{\psi_1(x_1)\psi_2(x_2)\pm\psi_2(x_1)\psi_1(x_2)}{\sqrt{2}\sqrt{1\pm|O|^2}}\,. \tag{5.5}$$

For simplicity, we assume that $O=0$, for instance by using two stationary states of the harmonic oscillator or of the radial wave functions of hydrogen with two different energies, which implies orthogonality. The latter case could be a model wave function for helium or the hydrogen molecule. We can therefore estimate quantum correlations in these states. With this assumption, we have

$$\psi(x_1,x_2) = \frac{\psi_1(x_1)\psi_2(x_2)\pm\psi_2(x_1)\psi_1(x_2)}{\sqrt{2}}\,. \tag{5.6}$$

We first compute the expectation values $\langle\hat{x}_1\rangle$ and $\langle\hat{x}_2\rangle$ in the combined state:

$$\begin{aligned}\langle\hat{x}_1\rangle &= \frac{1}{2}\int_{-\infty}^{\infty}\int_{-\infty}^{\infty} x_1\left(|\psi_1(x_1)|^2|\psi_2(x_2)|^2+|\psi_2(x_1)|^2|\psi_1(x_2)|^2\right)\mathrm{d}x_1\mathrm{d}x_2\\ &\pm\frac{1}{2}\int_{-\infty}^{\infty}\int_{-\infty}^{\infty} x_1\left(\psi_1(x_1)^*\psi_2(x_1)\psi_2(x_2)^*\psi_1(x_2)+\psi_1(x_1)\psi_2(x_1)^*\psi_2(x_2)\psi_1(x_2)^*\right)\mathrm{d}x_1\mathrm{d}x_2\\ &= \frac{1}{2}(\langle\hat{x}\rangle_1+\langle\hat{x}\rangle_2)\,.\end{aligned} \tag{5.7}$$

In the first line, the x_2-integrations produce factors of one from normalization, while the x_1-integrations result in the individual expectation values $\langle\hat{x}\rangle_1 = \int_{-\infty}^{\infty} x_1|\psi_1(x_1)|^2\mathrm{d}x_1$ and $\langle\hat{x}\rangle_2 = \int_{-\infty}^{\infty} x_1|\psi_2(x_1)|^2\mathrm{d}x_1$ in the two 1-variable states. In the second line, the x_1-integration are in general non-zero, but the x_2-integrations give us factors of O which vanish by assumption. The same value is obtained for $\langle\hat{x}_2\rangle$.

In $\langle\hat{x}_1\hat{x}_2\rangle$, we no longer obtain factors of O, resulting in a longer expression

$$\begin{aligned}\langle\hat{x}_1\hat{x}_2\rangle &= \langle\hat{x}\rangle_1\langle\hat{x}\rangle_2\pm\frac{1}{2}\langle\psi_1|\hat{x}|\psi_2\rangle\langle\psi_2|\hat{x}|\psi_1\rangle\pm\frac{1}{2}\langle\psi_2|\hat{x}|\psi_1\rangle\langle\psi_1|\hat{x}|\psi_2\rangle\\ &= \langle\hat{x}\rangle_1\langle\hat{x}\rangle_2\pm|\langle\psi_1|\hat{x}|\psi_2\rangle|^2\,.\end{aligned} \tag{5.8}$$

Therefore,

$$C_{x_1x_2} = \pm|\langle\psi_1|\hat{x}|\psi_2\rangle|^2-\frac{1}{4}(\langle\hat{x}\rangle_1-\langle\hat{x}\rangle_2)^2\,. \tag{5.9}$$

The second term agrees with the classical understanding of statistical correlations: If $\langle\hat{x}\rangle_1\neq\langle\hat{x}\rangle_2$, measuring x_2 after x_1 is more likely to result in a different value. Measurements of x_1 and x_2 are therefore correlated. The first term is an additional quantum contribution that depends on the relative phase of the two product states in the superposition (5.6). As seen in the explicit expression, the covariance may be negative, unlike the two variances of x_1 and x_2, respectively.

The covariance has units of length squared if $\hat{x}$ is a position operator. A unitless expression is given by **quantum correlations** defined as

$$c_{x_1x_2} = \frac{C_{x_1x_2}}{\sigma_{x_1}\sigma_{x_2}}\,. \tag{5.10}$$

For two commuting operators such as x_1 and x_2, quantum correlations are bounded from above and below by ±1. This general result is a consequence of the uncertainty relation (3.192),

$$\sigma_A^2\sigma_B^2-C_{AB}^2\geq\frac{1}{4}|\langle[\hat{A},\hat{B}]\rangle|^2\,. \tag{5.11}$$

For a vanishing commutator, $|C_{AB}|$ cannot be greater than $\sigma_A\sigma_B$, such that $|c_{AB}| \leq 1$.

It is also of interest to derive quantum correlations in spin states that are combined with position states in atomic or molecular wave functions. For two spin-$\frac{1}{2}$ particles, we use the shorter notation

$$|1/2,1/2\rangle \otimes |1/2,1/2\rangle = |\uparrow\uparrow\rangle \quad (5.12)$$

$$|1/2,1/2\rangle \otimes |1/2,-1/2\rangle = |\uparrow\downarrow\rangle \quad (5.13)$$

$$|1/2,-1/2\rangle \otimes |1/2,1/2\rangle = |\downarrow\uparrow\rangle \quad (5.14)$$

$$|1/2,-1/2\rangle \otimes |1/2,-1/2\rangle = |\downarrow\downarrow\rangle \quad (5.15)$$

for eigenstates of $\hat{S}_z$. We have the individual spin operators $\hat{Z}_1 = \hat{S}_z \otimes \hat{\mathbb{I}}$ and $\hat{Z}_2 = \hat{\mathbb{I}} \otimes \hat{S}_z$ as well as the tensor product $\hat{Z}_1\hat{Z}_2 = \hat{S}_z \otimes \hat{S}_z$. For instance,

$$\hat{Z}_1\hat{Z}_2|\uparrow\downarrow\rangle = (\hat{S}_z|1/2,1/2\rangle) \otimes (\hat{S}_z|1/2,-1/2\rangle) = -\frac{\hbar^2}{4}|\uparrow\downarrow\rangle . \quad (5.16)$$

The tensor product of operators is defined by applying the operators in its factors according to their position, pairing them with corresponding states.

A product state such as $|\uparrow\uparrow\rangle$ again has vanishing correlations because

$$\langle\hat{Z}_1\hat{Z}_2\rangle = \frac{\hbar^2}{4} = \langle\hat{Z}_1\rangle\langle\hat{Z}_2\rangle . \quad (5.17)$$

For $\chi_\pm = 2^{-1/2}(|\uparrow\downarrow\rangle \pm |\downarrow\uparrow\rangle)$, we compute

$$\langle\hat{Z}_1\rangle_\pm = \frac{1}{2}(\langle\uparrow\downarrow|\hat{Z}_1|\uparrow\downarrow\rangle \pm 2\mathrm{Re}\langle\uparrow\downarrow|\hat{Z}_1|\downarrow\uparrow\rangle + \langle\downarrow\uparrow|\hat{Z}_1|\downarrow\uparrow\rangle) \quad (5.18)$$

$$= \frac{1}{2}\left(\frac{\hbar}{2} - \frac{\hbar}{2}\right) = 0 . \quad (5.19)$$

The individual inner products can easily be derived using the fact that the up and down states are eigenstates of $\hat{S}_z$ and orthogonal to each other. We obtain the same value for $\langle\hat{Z}_2\rangle_\pm = 0$.

The covariance is therefore given by

$$C_{Z_1Z_2} = \frac{1}{2}(\langle\uparrow\downarrow|\hat{Z}_1\hat{Z}_2|\uparrow\downarrow\rangle \pm 2\mathrm{Re}\langle\uparrow\downarrow|\hat{Z}_1\hat{Z}_2|\downarrow\uparrow\rangle + \langle\downarrow\uparrow|\hat{Z}_1\hat{Z}_2|\downarrow\uparrow\rangle) \quad (5.20)$$

$$= \frac{1}{2}\left(-\frac{\hbar^2}{4} - \frac{\hbar^2}{4}\right) = -\frac{\hbar^2}{4} . \quad (5.21)$$

Similarly, we compute $\sigma_{Z_1}^2 = \hbar^2/4 = \sigma_{Z_2}^2$. These states are maximally correlated:

$$c_{Z_1Z_2} = \frac{C_{Z_1Z_2}}{\sigma_{Z_1}\sigma_{Z_2}} = -1 . \quad (5.22)$$

5.2 DENSITY OPERATORS

In our discussion of the inner product $\langle\psi|\psi\rangle$, we motivated the "reflected" version $\langle\psi|$ of the abstract state $|\psi\rangle$ as an indication of Hermitian conjugation from a column-vector representation of $|\psi\rangle$ to a row vector $\langle\psi| = (|\psi\rangle)^\dagger$, such that $\langle\psi|\psi\rangle$ has a meaningful matrix product that results in a single complex number. Mathematically, we interpret $\langle\psi|$ as a function, or a dual vector, that can be applied to any state $|\phi\rangle$, including $|\psi\rangle$ itself, and results in a complex number $\langle\psi|\phi\rangle$. The object $\langle\psi|$ is a linear map because the inner product is linear on its right.

Assuming a finite-dimensional Hilbert space, we have

$$\langle\psi|\psi\rangle = (|\psi\rangle)^\dagger|\psi\rangle \leftrightarrow \begin{pmatrix} \psi_1^* & \cdots & \psi_n^* \end{pmatrix}\begin{pmatrix} \psi_1 \\ \vdots \\ \psi_n \end{pmatrix} = |\psi_1|^2+\cdots+|\psi_n|^2 \tag{5.23}$$

in components. If we flip the two abstract expressions, $\langle\psi|$ and $|\psi\rangle$, we obtain another meaningful matrix product

$$|\psi\rangle\langle\psi| \leftrightarrow \begin{pmatrix} \psi_1 \\ \vdots \\ \psi_n \end{pmatrix}\begin{pmatrix} \psi_1^* & \cdots & \psi_n^* \end{pmatrix} = \begin{pmatrix} \psi_1\psi_1^* & \psi_1\psi_2^* & \cdots & \psi_1\psi_n^* \\ \psi_2\psi_1^* & \psi_2\psi_2^* & \cdots & \psi_2\psi_n^* \\ \vdots & \vdots & \ddots & \vdots \\ \psi_n\psi_1^* & \psi_n\psi_2^* & \cdots & \psi_n\psi_n^* \end{pmatrix} \tag{5.24}$$

that results in an $n\times n$-matrix. The combination $|\psi\rangle\langle\psi|$ may also be written as $|\psi\rangle\otimes\langle\psi|$ as another example of a tensor product. The resulting matrix can then act on other column vectors representing abstract states $|\phi\rangle$:

$$|\psi\rangle\langle\psi|\cdot|\phi\rangle = |\psi\rangle\langle\psi|\phi\rangle = c|\psi\rangle \tag{5.25}$$

with the complex number $c=\langle\psi|\phi\rangle$. In this derivation, we merely have to reorder the product of a state $|\psi\rangle$ with the complex number $\langle\psi|\cdot|\phi\rangle = \langle\psi|\phi\rangle$. The dot is used here in order to separate the two original expressions and is not related to the dot product of vectors.

5.2.1 PROPERTIES

Interpreted in this way, the expression

$$\hat{\rho}_\psi = |\psi\rangle\langle\psi| \tag{5.26}$$

is called the **density operator** of the state $|\psi\rangle$. It has the following properties:

1. $\hat{\rho}_\psi$ is a linear transformation for a given $|\psi\rangle$. To see this, we write

$$\begin{aligned} \hat{\rho}_\psi(a|\phi_1\rangle+b|\phi_2\rangle) &= |\psi\rangle(a\langle\psi|\phi_1\rangle+b\langle\psi|\phi_2\rangle) \\ &= a|\psi\rangle\langle\psi|\phi_1\rangle+b|\psi\rangle\langle\psi|\phi_2\rangle = a\hat{\rho}_\psi|\phi_1\rangle+b\hat{\rho}_\psi|\phi_2\rangle\,. \end{aligned} \tag{5.27}$$

   Mapping $|\psi\rangle$ to $\hat{\rho}_\psi$, however, is not linear.
2. We can now apply our methods for linear operators. If we choose a Hermitian operator $\hat{Q}$ with eigenbasis $|e_\lambda\rangle$, $\hat{\rho}_\psi$ has the expectation values

$$\langle e_\lambda|\hat{\rho}_\psi|e_\lambda\rangle = \langle e_\lambda|\psi\rangle\langle\psi|e_\lambda\rangle = |\langle e_\lambda|\psi\rangle|^2 = |c_\lambda|^2 \tag{5.28}$$

   if $|\psi\rangle = \sum_\lambda c_\lambda|e_\lambda\rangle$ in the eigenbasis of $\hat{Q}$. For any operator, the expectation values of $\hat{\rho}_\psi$ in the eigenstates therefore determine the probability distribution or probability density for measurements of Q in the state $|\psi\rangle$. This is the reason why we call $\hat{\rho}_\psi$ the *density* operator of $|\psi\rangle$.
3. The density operator is Hermitian:

$$\hat{\rho}_\psi^\dagger = (|\psi\rangle\langle\psi|)^\dagger = \langle\psi|^\dagger|\psi\rangle^\dagger = |\psi\rangle\langle\psi| = \hat{\rho}_\psi\,. \tag{5.29}$$

4. The state $|\psi\rangle$ is an eigenstate of its own density operator $\hat{\rho}_\psi$ with eigenvalue one:

$$\hat{\rho}_\psi|\psi\rangle = |\psi\rangle\langle\psi|\psi\rangle = |\psi\rangle \tag{5.30}$$

using normalization. If we complete the single state $|\psi\rangle$ to an orthonormal basis $(|\psi\rangle, |\phi_i\rangle)$ of our Hilbert space, the remaining basis states $|\phi_i\rangle$ are eigenstates of $\hat{\rho}_\psi$ with eigenvalue zero:

$$\hat{\rho}_\psi|\phi_i\rangle = |\psi\rangle\langle\psi|\phi_i\rangle = 0\,. \tag{5.31}$$

The eigenvalue zero therefore has multiplicity given by the dimension of the Hilbert space minus one, or infinite multiplicity in infinite dimensions. The spectrum is discrete.

5. Trace operations turn out to be important for density operators. For a matrix, the trace is the sum of its diagonal components, which is independent of the basis choice because we always have $\mathrm{tr}(AB) = \mathrm{tr}(BA)$. Therefore, the trace is not changed by a similarity transformation. For the density operator in its eigenbasis, we obtain the value $\mathrm{tr}(\hat{\rho}_\psi) = 1$, the sum of all its eigenvalues.
   We use this result in order to test basis independence. Choosing a basis $|e_\lambda\rangle$ in our Hilbert space, we compute the trace of the density operator as

$$\mathrm{tr}(\hat{\rho}_\psi) = \sum_\lambda \langle e_\lambda|\hat{\rho}_\psi|e_\lambda\rangle = \sum_\lambda \langle e_\lambda|\psi\rangle\langle\psi|e_\lambda\rangle = \sum_\lambda |\langle e_\lambda|\psi\rangle|^2 = \sum_\lambda |c_\lambda|^2 = 1\,. \tag{5.32}$$

   Here, the result follows from normalization of the state $|\psi\rangle$ represented by its components c_λ in the chosen basis.
6. Using the eigenbasis of $\hat{\rho}_\psi$, it directly follows that $\mathrm{tr}(\hat{\rho}_\psi^2) = 1$, which is given by the sum of the squares of the eigenvalues. More generally, $\mathrm{Tr}(\hat{\rho}_\psi^n) = 1$ for any positive integer n.
7. The density operator of a state $|\psi\rangle$ equals its own square,

$$\hat{\rho}_\psi^2 = |\psi\rangle\langle\psi|\cdot|\psi\rangle\langle\psi| = |\psi\rangle\langle\psi| = \hat{\rho}_\psi\,. \tag{5.33}$$

   Such an operator is called a **projection operator**.

5.2.2 ENTROPY

The diagonal components of a density operator $\hat{\rho}_\psi$ always provide a probability density or distribution because $\mathrm{tr}(\hat{\rho}_\psi) = 1$. A useful way of evaluating a classical probability distribution $P(i)$ is given by the **Shannon entropy**, defined as

$$S = -\sum_{i=1}^{n} P(i)\log P(i) \tag{5.34}$$

where any logarithm may be used. In classical statistics, base two is a common choice. The entropy is non-negative and bounded from above by $\log n$, where n is the number of possible outcomes. This maximum is realized for a uniform probability distribution where the $P(i) = 1/n$ are equal. The entropy vanishes if one outcome is certain, such as $P(1) = 1$ and $P(i) = 0$ for $i > 1$.

The definition of entropy can be applied to density operators, in which case it is called the **von Neumann entropy**

$$S(\hat{\rho}_\psi) = -\mathrm{tr}(\hat{\rho}_\psi \log\hat{\rho}_\psi) = -\sum_\lambda \langle e_\lambda|\psi\rangle\langle\psi|\log(|\psi\rangle\langle\psi|)|e_\lambda\rangle\,. \tag{5.35}$$

The trace is computed in a given basis, as indicated, but we also have to know how to evaluate the logarithm of an operator. This object can be defined by using the Taylor expansion of the logarithm around the value one, provided the result is a well-defined operator. For the natural log, the Taylor expansion is given by $\log(x) = -\sum_{k=1}^{\infty}(-1)^k(x-1)^k/k$. Applied to a density operator $\hat{\rho}_\psi$, we have

$$\log(|\psi\rangle\langle\psi|) = -\sum_{k=1}^{\infty}\frac{(-1)^k}{k}(|\psi\rangle\langle\psi| - \hat{\mathbb{I}})^k\,. \tag{5.36}$$

A crucial factor in each term is given by $|\psi\rangle\langle\psi| - \hat{\mathbb{I}}$. In the entropy expression, this factor is mapped to zero by the preceding $\langle\psi|$ in (5.35):

$$\langle\psi|(|\psi\rangle\langle\psi| - \hat{\mathbb{I}}) = \langle\psi| - \langle\psi| = 0\,. \tag{5.37}$$

The entropy of any $\hat{\rho}_\psi$ therefore vanishes.

This result is consistent with our previous derivation of the eigenvalues of $\hat{\rho}_\psi$, which form a probability distribution $(1,0,\ldots)$ with one certain outcome, implying zero entropy. Since the entropy, like any trace, is basis-independent, the same value must be obtained in any basis $|e_\lambda\rangle$. The surprising property is that we always obtain zero entropy of $\hat{\rho}_\psi$ even though the probability distribution or density of the state $|\psi\rangle$ for measurements of a Hermitian operator $\hat{Q}$ in general does not have a certain outcome. Interpreted classically, such a probability distribution or density would have non-zero entropy. However, in quantum mechanics, phase relationships between the coefficients c_λ in the expansion of $|\psi\rangle$ using the eigenbasis of $\hat{Q}$ have additional information that is not represented in a classical probability distribution. The density operator contains this additional information, which makes the entropy zero.

We will soon discuss generalizations of density operators $\hat{\rho}_\psi$ that have non-zero entropy. But first we familiarize ourselves with density matrices by evaluating them for a particle of spin $\frac{1}{2}$, using a 2-dimensional Hilbert space.

5.2.3 BLOCH SPHERE

A spin-$\frac{1}{2}$ particle is described by a 2-vector spinor state $\chi = (\chi_1, \chi_2)^T$ with two complex numbers χ_1 and χ_2 such that $|\chi_1|^2 + |\chi_2|^2 = 1$. Since an overall phase factor $e^{i\varepsilon}$ does not change any expectation values and probabilities, we may assume that χ_1 is real. With these conditions, the space of 2-spinors is uniquely parameterized by

$$\chi(\vartheta,\varphi) = \begin{pmatrix} \cos(\vartheta/2) \\ e^{i\varphi}\sin(\vartheta/2) \end{pmatrix} \tag{5.38}$$

with two real angles, $0 \leq \varphi < 2\pi$ and $0 \leq \vartheta \leq \pi$.

The factors of $\frac{1}{2}$ multiplying ϑ are chosen so as to obtain the standard ranges for these two angles, familiar from spherical coordinates. The restriction on the range of ϑ can be seen by making sure that every state corresponds to a unique pair (ϑ,φ). Had we allowed values of ϑ greater than π, as in $\pi + \Delta\vartheta$ with $\Delta\vartheta > 0$, we would be able to construct an equivalent state with a new angle $\pi - \Delta\vartheta$ less than π because

$$\begin{pmatrix} \cos((\pi+\Delta\vartheta)/2) \\ e^{i\varphi}\sin((\pi+\Delta\vartheta)/2) \end{pmatrix} = \begin{pmatrix} -\cos(\pi-\Delta\vartheta)/2) \\ e^{i\varphi}\sin((\pi-\Delta\vartheta)/2) \end{pmatrix} = -\begin{pmatrix} \cos((\pi-\Delta\vartheta)/2) \\ e^{i(\varphi\pm\pi)}\sin((\pi-\Delta\vartheta)/2) \end{pmatrix} \tag{5.39}$$

follows from symmetry properties of sine and cosine around $\pi/2$. The sign in $\varphi \pm \pi$ can be chosen such that this angle stays in the allowed range: We choose the plus sign if $\varphi < \pi$ and the minus sign if $\varphi \geq \pi$. The factor of -1 in the final expression does not affect physical outcomes such as expectation values. The two states are therefore equivalent, and we can restrict ϑ to values less than or equal to π.

The $\frac{1}{2}$-factors in the spinor imply that the direction of the spin-expectation vector $\langle\hat{\vec{S}}\rangle$ is determined by standard spherical coordinates: Expressing the spin matrices as Pauli matrices times $\hbar/2$, we compute

$$\langle\hat{S}_x\rangle = \frac{\hbar}{2}\sin(\vartheta/2)\cos(\vartheta/2)(e^{i\varphi}+e^{-i\varphi}) = \frac{\hbar}{2}\sin(\vartheta)\cos(\varphi) \tag{5.40}$$

$$\langle\hat{S}_y\rangle = \frac{\hbar}{2}i\sin(\vartheta/2)\cos(\vartheta/2)(e^{-i\varphi}-e^{i\varphi}) = \frac{\hbar}{2}\sin(\vartheta)\sin(\varphi) \tag{5.41}$$

$$\langle\hat{S}_z\rangle = \frac{\hbar}{2}(\cos(\vartheta/2)^2 - \sin(\vartheta/2)^2) = \frac{\hbar}{2}\cos(\vartheta)\,. \tag{5.42}$$

The spin-expectation vector is therefore $\frac{1}{2}\hbar$ times a unit vector.

5.2.3.1 Density operator of a 2-spinor

The density operator of a general 2-spinor χ is given by

$$\begin{aligned}\hat{\rho}_\chi &= |\chi\rangle\langle\chi| = \begin{pmatrix} \cos(\vartheta/2) \\ e^{i\varphi}\sin(\vartheta/2)\end{pmatrix}(\cos(\vartheta/2), e^{-i\varphi}\sin(\vartheta/2)) \\ &= \begin{pmatrix} \cos(\vartheta/2)^2 & e^{-i\varphi}\sin(\vartheta/2)\cos(\vartheta/2) \\ e^{i\varphi}\sin(\vartheta/2)\cos(\vartheta/2) & \sin(\vartheta/2)^2 \end{pmatrix}. \end{aligned} \quad (5.43)$$

This expression is Hermitian and has trace one.

Conversely, we can derive the general form of a 2-spinor density operator $\hat{\rho}_\chi$ from the conditions that it be Hermitian and have trace one, as well as $\hat{\rho}_\chi^2 = \hat{\rho}_\chi$: A Hermitian trace-one 2×2-matrix has the form

$$\hat{\rho}_\chi = \begin{pmatrix} a & c+di \\ c-di & 1-a \end{pmatrix} \quad (5.44)$$

with three real numbers, a, c and d. Its square is given by

$$\hat{\rho}_\chi^2 = \begin{pmatrix} a^2+c^2+d^2 & c+di \\ c-di & (1-a)^2+c^2+d^2 \end{pmatrix}. \quad (5.45)$$

This matrix equals $\hat{\rho}_\chi$ if and only if $a^2+c^2+d^2 = a$, such that $|c+di| = c^2+d^2 = a(1-a)$. In particular, this condition implies $a(1-a)\geq 0$, from which it follows that a must be in the range $0\leq a\leq 1$. We write $c+di = e^{-i\varphi}\sqrt{a(1-a)}$ in polar coordinates of a complex number. The resulting density operator,

$$\hat{\rho}_\chi = \begin{pmatrix} a & e^{-i\varphi}\sqrt{a(1-a)} \\ e^{i\varphi}\sqrt{a(1-a)} & 1-a \end{pmatrix} \quad (5.46)$$

is of the form (5.43): The required range of a allows us to set $a = \cos(\vartheta/2)^2$ with a unique ϑ in the range $0\leq\vartheta\leq\pi$.

Using trigonometric identities, the density operator (5.43) can be written as

$$\hat{\rho}_\chi = \frac{1}{2}\begin{pmatrix} 1+\cos(\vartheta) & e^{-i\varphi}\sin(\vartheta) \\ e^{i\varphi}\sin(\vartheta) & 1-\cos(\vartheta)\end{pmatrix} = \frac{1}{2}(\mathbb{I}+r_x\sigma_1+r_y\sigma_2+r_z\sigma_3) \quad (5.47)$$

with Pauli matrices σ_1, σ_2 and σ_3, and components r_x, r_y and r_z of the unit vector

$$\vec{r} = \begin{pmatrix} \sin(\vartheta)\cos(\varphi) \\ \sin(\vartheta)\sin(\varphi) \\ \cos(\vartheta) \end{pmatrix}. \quad (5.48)$$

In this way, density operators $\hat{\rho}_\chi$ are mapped to the **Bloch sphere**.

To summarize, we can derive the general form of $\hat{\rho}_\chi$ without using a specific χ if we require three properties: $\hat{\rho}_\chi$ is (i) Hermitian ($\hat{\rho}_\chi^\dagger = \hat{\rho}_\chi$), (ii) trace-one ($\mathrm{tr}(\hat{\rho}_\chi) = 1$), and (iii) a projection operator ($\hat{\rho}_\chi^2 = \hat{\rho}_\chi$). We can now try to relax some of these conditions while making sure that we are still able to define meaningful state properties. The final parameterization (5.47) is a good starting point because we can see which conditions are preserved if we allow $\vec{r}$ to be non-normalized. Since the Pauli matrices are Hermitian and have trace zero, the first two conditions on $\hat{\rho}_\chi$ are preserved. However, the last condition is violated because $\hat{\rho}_\chi^2 = \hat{\rho}_\chi$ is true only if $|\vec{r}| = 1$, as follows from our derivation.

If we drop the condition that a density operator be a projection operator, we obtain a more general parameterization. As we will show, the resulting matrices still have all the required properties for statistical predictions of measurement results, provided we impose a **positivity** condition that the eigenvalues be non-negative. If we compute eigenvalues for the matrices derived so far, it turns out that positivity is equivalent to the property that $0\leq a\leq 1$ which we obtained from $\hat{\rho}_\chi^2 = \hat{\rho}_\chi$. We may then replace this condition with the weaker requirement of positivity, resulting in a step toward generalizing the notion of quantum states.

5.2.3.2 General density operator

The density operator $\hat{\rho}_\chi$ of a spinor χ always obeys the condition $\hat{\rho}_\chi^2 = \hat{\rho}_\chi$. If we do not impose this condition, the resulting expression $\hat{\rho}$ is still called a density operator if it is positive. In general, a **density operator** is defined by the conditions that it be Hermitian, $\hat{\rho}^\dagger = \hat{\rho}$, have trace one, $\mathrm{tr}(\hat{\rho}) = 1$, and have non-negative eigenvalues.

The last condition is motivated by a reference to probability distributions. In the eigenbasis $|e_i\rangle$ of a Hermitian operator $\hat{Q}$ with discrete spectrum, we have

$$\mathrm{tr}(\hat{\rho}) = \sum_i \langle e_i|\hat{\rho}|e_i\rangle = 1 \tag{5.49}$$

for any density operator $\hat{\rho}$. Since $\hat{\rho}$ is Hermitian, the numbers $P(i) = \langle e_i|\hat{\rho}|e_i\rangle$ are real and can be interpreted as a probability distribution provided they are non-negative. In particular, if we choose the eigenbasis of $\hat{\rho}$ itself, the $P(i)$ are its eigenvalues, which must therefore be non-negative.

If we continue to work in the eigenbasis of $\hat{\rho}$, we can compute the trace of its square,

$$\mathrm{tr}(\hat{\rho}^2) = \sum_i \sum_j \langle e_i|\hat{\rho}|e_j\rangle\langle e_j|\hat{\rho}|e_i\rangle \,. \tag{5.50}$$

The j-sum represents the matrix product of the density operator with itself, written in the eigenbasis. The i-sum then determines the sum over diagonal components as required for the trace. Since we use the eigenbasis of $\hat{\rho}$, we can use orthogonality, $\langle e_i|\hat{\rho}|e_j\rangle \propto \langle e_i|e_j\rangle = 0$ if $i \neq j$. The double-sum is therefore reduced to a single sum,

$$\mathrm{tr}(\hat{\rho}^2) = \sum_i \langle e_i|\hat{\rho}|e_i\rangle^2 = \sum_i P(i)^2 \leq 1 \,. \tag{5.51}$$

The last inequality follows from the fact that $P(i)$ is a probability distribution, such that $0 \leq P(i) \leq 1$ and therefore $P(i)^2 \leq P(i)$. Since the $P(i)$ sum to one, the $P(i)^2$ sum to a number less than or equal to one.

We obtain the upper limit, $\mathrm{tr}(\hat{\rho}^2) = 1$, if and only if one of the $P(i)$ equals one and the rest are zero, implying one certain outcome. This is the case for any density operator of the form $\hat{\rho}_\chi = |\chi\rangle\langle\chi|$ defined by a wave function or spinor χ. A state represented by a density operator $\hat{\rho}$ such that $\mathrm{tr}(\hat{\rho}^2) = 1$ is called a **pure state**. If $\mathrm{tr}(\hat{\rho}^2) < 1$, the state is called a **mixed state**. The value of $\mathrm{tr}(\hat{\rho}^2)$ defines the **purity** of the state. If the state is mixed, it is not possible to find a corresponding wave function or spinor χ. Instead, statistical properties relevant for measurements are directly determined by $\hat{\rho}$ and its eigenvalues $P(i)$, which then form a generic probability distribution without a certain outcome.

The general density operator of a spin-$\frac{1}{2}$ particle in a mixed state is given by

$$\hat{\rho} = \begin{pmatrix} a & c+di \\ c-di & 1-a \end{pmatrix} \tag{5.52}$$

with real numbers a, c and d, provided it has non-negative eigenvalues. Compared with the density operator of a pure state, there is one additional free parameter because $|c+di|$ is no longer related to a. A convenient parameterization that simplifies the condition of non-negative eigenvalues is given by

$$\hat{\rho} = \frac{1}{2}\begin{pmatrix} 1+r\cos(\vartheta) & re^{-i\varphi}\sin(\vartheta) \\ re^{i\varphi}\sin(\vartheta) & 1-r\cos(\vartheta) \end{pmatrix} . \tag{5.53}$$

Using trigonometric identities, it follows that this density operator belongs to a pure state if and only if $r = 1$. The parameters (r, ϑ, φ) have the usual ranges for spherical coordinates. In particular, we have $r \geq 0$ because a negative value does not give us a new operator not already contained in the positive range: For negative r, we obtain the same matrix if we replace r with $|r|$ and ϑ with $\pi - \vartheta$.

As before, the same expression can be written in terms of Pauli matrices σ_1, σ_2 and σ_3:

$$\hat{\rho} = \frac{1}{2}(\mathbb{I} + r_x\sigma_1 + r_y\sigma_2 + r_z\sigma_3) \tag{5.54}$$

with the position vector

$$\vec{r} = \begin{pmatrix} r\sin(\vartheta)\cos(\varphi) \\ r\sin(\vartheta)\sin(\varphi) \\ r\cos(\vartheta) \end{pmatrix}. \tag{5.55}$$

For a pure state, we have $r = 1$ and the position is on the unit sphere. It is not possible to have r greater than one because such values would violate the non-negativity condition imposed on the eigenvalues of $\hat{\rho}$:

$$\frac{1}{4}\det\begin{pmatrix} 1+r\cos(\vartheta)-2\lambda & re^{-i\varphi}\sin(\vartheta) \\ re^{i\varphi}\sin(\vartheta) & 1-r\cos(\vartheta)-2\lambda \end{pmatrix} = \frac{1}{4}\left((1-2\lambda)^2 - r^2\cos(\vartheta)^2 - r^2\sin(\vartheta)^2\right) = 0 \tag{5.56}$$

implies the eigenvalues

$$\lambda_\pm = \frac{1\pm r}{2}. \tag{5.57}$$

The eigenvalues are non-negative if and only if $r \leq 1$.

Using (5.45), we see that the purity

$$\mathrm{tr}(\hat{\rho}^2) = a^2 + (1-a)^2 + 2|c+di|^2 = \frac{1}{2}(1+r^2) \leq 1 \tag{5.58}$$

is maximized in a pure state. Its minimum is given by $1/2$ (or, more generally, $1/n$ in an n-dimensional Hilbert space). This lower limit is obtained for the density operator

$$\hat{\rho}_0 = \begin{pmatrix} \frac{1}{2} & 0 \\ 0 & \frac{1}{2} \end{pmatrix} \tag{5.59}$$

where $r = 0$.

The entropy of $\hat{\rho}$ can be computed in its eigenbasis:

$$S = -(\lambda_+\log\lambda_+ + \lambda_-\log\lambda_-) = \log 2 - \frac{1}{2}(1+r)\log(1+r) - \frac{1}{2}(1-r)\log(1-r). \tag{5.60}$$

The entropy vanishes for $r = 1$ (pure states) and equals the maximal value $\log 2$ if $r = 0$; see Figure 5.1.

5.2.4 EXPECTATION VALUES

Given an operator $\hat{A}$ with an eigenvalue λ and corresponding eigenstate $|e_\lambda\rangle$, we obtain the pure density operator $\hat{\rho}_{|e_\lambda\rangle} = |e_\lambda\rangle\langle e_\lambda|$. This density operator satisfies $\hat{A}\hat{\rho}_{|e_\lambda\rangle} = \lambda\hat{\rho}_{|e_\lambda\rangle}$. Using $\mathrm{tr}(\hat{\rho}_{|e_\lambda\rangle}) = 1$, we express the eigenvalue as

$$\lambda = \mathrm{tr}(\hat{A}\hat{\rho}_{|e_\lambda\rangle}) = \langle e_\lambda|\hat{A}|e_\lambda\rangle. \tag{5.61}$$

For a general density operator $\hat{\rho}$ of a mixed state, we still have the trace condition, which gives us a probability distribution $P(i) = \langle e_i|\hat{\rho}|e_i\rangle$ in the eigenbasis of $\hat{\rho}$. The expression in (5.61) now reads

$$\mathrm{tr}(\hat{A}\hat{\rho}) = \sum_i\sum_j \langle e_i|\hat{A}|e_j\rangle\langle e_j|\hat{\rho}|e_i\rangle \tag{5.62}$$

in a generic basis $|e_i\rangle$. If we use the eigenbasis of $\hat{\rho}$, we have $\langle e_j|\hat{\rho}|e_i\rangle = 0$ if $i \neq j$ and obtain

$$\mathrm{tr}(\hat{A}\hat{\rho}) = \sum_i \langle e_i|\hat{A}|e_i\rangle\langle e_i|\hat{\rho}|e_i\rangle = \sum_i \langle e_i|\hat{A}|e_i\rangle P(i) = \langle\hat{A}\rangle_{\hat{\rho}}. \tag{5.63}$$

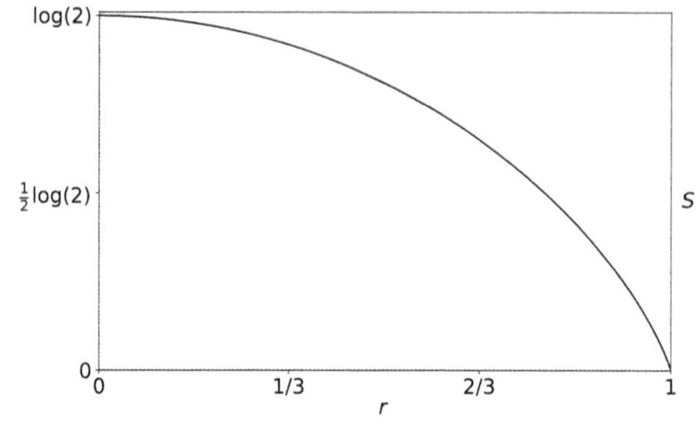

Figure 5.1 Entropy (5.60) of a spinor as a function of the radius parameter r.

This expression defines the expectation value of an observable $\hat{A}$ in a mixed state. The state is mixed because the final result, $\sum_i \langle e_i|\hat{A}|e_i\rangle P(i)$, is a combination of quantum expectation values $\langle e_i|\hat{A}|e_i\rangle = A_i$ and the classical derivation of an expectation value, $\sum_i A_i P(i)$. As an application of this result, we write the entropy as the expectation value

$$S = -\langle \log \hat{\rho} \rangle_{\hat{\rho}}\,. \tag{5.64}$$

In our spin-$\frac{1}{2}$ system, we can now compute the spin-expectation vector $\langle \hat{\vec{S}} \rangle_{\hat{\rho}}$ in a mixed state. We obtain

$$\langle \hat{S}_x \rangle_{\hat{\rho}} = \mathrm{tr}(\hat{S}_x\hat{\rho}) = \frac{1}{2}\hbar\,\mathrm{tr}\begin{pmatrix} \frac{1}{2}re^{i\varphi}\sin(\vartheta) & \frac{1}{2}(1-r\cos(\vartheta)) \\ \frac{1}{2}(1+r\cos(\vartheta)) & \frac{1}{2}re^{-i\varphi}\sin(\vartheta) \end{pmatrix} = \frac{1}{2}\hbar r\sin(\vartheta)\cos(\varphi) \tag{5.65}$$

and similarly $\langle \hat{S}_y \rangle_{\hat{\rho}} = \frac{1}{2}\hbar r \sin(\vartheta)\sin(\varphi)$ and $\langle \hat{S}_z \rangle_{\hat{\rho}} = \frac{1}{2}\hbar r\cos(\vartheta)$. The direction is determined by (ϑ, φ) as before, while the magnitude equals

$$|\langle \hat{\vec{S}} \rangle_{\hat{\rho}}| = \sqrt{\langle \hat{S}_x \rangle_{\hat{\rho}}^2 + \langle \hat{S}_y \rangle_{\hat{\rho}}^2 + \langle \hat{S}_z \rangle_{\hat{\rho}}^2} = \frac{1}{2}\hbar r\,. \tag{5.66}$$

Using the eigenvalues $\lambda_\pm = P_\pm = \frac{1}{2}(1\pm r)$ from (5.57) as a probability distribution, this value can be written as

$$\frac{1}{2}\hbar r = \frac{1}{2}\hbar(\lambda_+ - \lambda_-) = S_+P_+ + S_-P_- \tag{5.67}$$

where $S_\pm = \pm\frac{1}{2}\hbar$ are the two possible outcomes. The magnitude of the spin-expectation vector is therefore the statistical average of the two possible sharp outcomes of single measurements of a spin component.

5.2.5 TIME DEPENDENCE

A pure state evolves in time according to

$$i\hbar\frac{\mathrm{d}|\psi\rangle}{\mathrm{d}t} = \hat{H}|\psi\rangle \tag{5.68}$$

with the Hamilton operator $\hat{H}$ of the system. The Hermitian conjugate $\langle\psi| = (|\psi\rangle)^\dagger$ therefore evolves according to

$$i\hbar\frac{\mathrm{d}\langle\psi|}{\mathrm{d}t} = \left(-i\hbar\frac{\mathrm{d}|\psi\rangle}{\mathrm{d}t}\right)^\dagger = -(\hat{H}|\psi\rangle)^\dagger = -\langle\psi|\hat{H}\,. \tag{5.69}$$

These two equations can be combined for the density operator of a pure state:

$$i\hbar\frac{\mathrm{d}\hat{\rho}_\psi}{\mathrm{d}t} = i\hbar\frac{\mathrm{d}|\psi\rangle\langle\psi|}{\mathrm{d}t} = i\hbar\frac{\mathrm{d}|\psi\rangle}{\mathrm{d}t}\langle\psi| + i\hbar|\psi\rangle\frac{\mathrm{d}\langle\psi|}{\mathrm{d}t} = (\hat{H}|\psi\rangle)\langle\psi| - |\psi\rangle(\langle\psi|\hat{H}) = [\hat{H},\hat{\rho}_\psi]. \quad (5.70)$$

In this form, quantum evolution is referred to as the **von Neumann equation**.

5.2.5.1 Unitary evolution

The von Neumann equation can directly be generalized to the density operator of a mixed state,

$$i\hbar\frac{\mathrm{d}\hat{\rho}}{\mathrm{d}t} = [\hat{H},\hat{\rho}]. \quad (5.71)$$

However, this definition is meaningful only if the conditions imposed on a general density operator are preserved by the differential equation: If we choose an initial density operator $\hat{\rho}(0)$ that is Hermitian, trace-one, and non-negative, a time-dependent solution $\hat{\rho}(t)$ must obey these conditions for any t. We now demonstrate that these properties are indeed preserved in time:

1. Evolution preserves Hermiticity: We have

$$i\hbar\frac{\mathrm{d}(\hat{\rho}-\hat{\rho}^\dagger)}{\mathrm{d}t} = i\hbar\frac{\mathrm{d}\hat{\rho}}{\mathrm{d}t} + \left(i\hbar\frac{\mathrm{d}\hat{\rho}}{\mathrm{d}t}\right)^\dagger = [\hat{H},\hat{\rho}] + [\hat{H},\hat{\rho}]^\dagger = \hat{H}(\hat{\rho}-\hat{\rho}^\dagger) - (\hat{\rho}-\hat{\rho}^\dagger)\hat{H} \quad (5.72)$$

   where we used $[\hat{A},\hat{B}]^\dagger = (\hat{A}\hat{B}-\hat{B}\hat{A})^\dagger = -\hat{A}^\dagger\hat{B}^\dagger + \hat{B}^\dagger\hat{A}^\dagger$ and $\hat{H}^\dagger = \hat{H}$. Therefore, if $\hat{\rho}-\hat{\rho}^\dagger = 0$, this expression has a vanishing time derivative and stays zero.
2. Evolution preserves trace-one: We have

$$i\hbar\frac{\mathrm{d}\,\mathrm{tr}(\hat{\rho})}{\mathrm{d}t} = \mathrm{tr}([\hat{H},\hat{\rho}]) = 0 \quad (5.73)$$

   because $\mathrm{tr}(\hat{A}\hat{B}) = \mathrm{tr}(\hat{B}\hat{A})$ and therefore $\mathrm{tr}([\hat{A},\hat{B}]) = 0$ for any commutator.
   The same property can be used to show that purity is preserved, even if $\mathrm{tr}(\hat{\rho}^2) < 1$: We have

$$i\hbar\frac{\mathrm{d}\hat{\rho}^2}{\mathrm{d}t} = \hat{\rho}[\hat{H},\hat{\rho}] + [\hat{H},\hat{\rho}]\hat{\rho} = \hat{\rho}\hat{H}\hat{\rho} - \hat{\rho}^2\hat{H} + \hat{H}\hat{\rho}^2 - \hat{\rho}\hat{H}\hat{\rho} = [\hat{H},\hat{\rho}^2] \quad (5.74)$$

   and therefore

$$i\hbar\frac{\mathrm{d}\,\mathrm{tr}(\hat{\rho}^2)}{\mathrm{d}t} = \mathrm{tr}([\hat{H},\hat{\rho}^2]) = 0. \quad (5.75)$$

3. Evolution preserves non-negativity: This property follows from the fact that evolution determined by a self-adjoint Hamilton operator is unitary, which means that the norm of any pure-state wave function is preserved by evolution. The linear transformation mapping $|\psi\rangle(0)$ to $|\psi\rangle(t)$ is therefore given by a unitary symmetry transformation, for any t. For a time-independent Hamilton operator $\hat{H}$, the one-parameter family of unitary operators $\hat{U}_{\hat{H}}(t) = \exp(-it\hat{H}/\hbar)$ determines solutions $|\psi\rangle(t) = \hat{U}_{\hat{H}}(t)|\psi\rangle(0)$.
   For a pure state, we have the time-dependent density operator

$$|\psi\rangle\langle\psi|(t) = \hat{U}_{\hat{H}}(t)|\psi\rangle\langle\psi|(0)\hat{U}_{\hat{H}}(t)^{-1} \quad (5.76)$$

   which suggests the time dependence

$$\hat{\rho}(t) = \hat{U}_{\hat{H}}(t)\hat{\rho}(0)\hat{U}_{\hat{H}}(t)^{-1}. \quad (5.77)$$

This expression indeed solves (5.71), even if the state is not pure. Therefore,

$$\begin{aligned}\det(\hat{\rho}(t)-\lambda\hat{\mathbb{I}}) &= \det(\hat{U}_{\hat{H}}(t)\hat{\rho}(0)\hat{U}_{\hat{H}}(t)^{-1}-\lambda\hat{\mathbb{I}}) = \det(\hat{U}_{\hat{H}}(t)(\hat{\rho}(0)-\lambda\mathbb{I})\hat{U}_{\hat{H}}(t)^{-1}) \\ &= \det(\hat{\rho}(0)-\lambda\hat{\mathbb{I}})\frac{\det\hat{U}_{\hat{H}}(t)}{\det\hat{U}_{\hat{H}}(t)}\,. \end{aligned} \tag{5.78}$$

We conclude that the characteristic polynomial and its zeros, the eigenvalues of a density operator, are preserved by unitary evolution, including their signs.

As an example, consider a Hamilton operator for a spin-$\frac{1}{2}$ particle in a magnetic field with strength B in the z-direction, such that

$$\hat{H} = -\gamma B\hat{S}_z \tag{5.79}$$

with the gyromagnetic ratio γ. For this Hamilton operator, we have

$$\begin{aligned}[\hat{H},\hat{\rho}] &= -\frac{1}{4}\gamma B\hbar\left[\begin{pmatrix}1 & 0\\ 0 & -1\end{pmatrix},\begin{pmatrix}1+r\cos(\vartheta) & re^{-i\varphi}\sin(\vartheta)\\ re^{i\varphi}\sin(\vartheta) & 1-r\cos(\vartheta)\end{pmatrix}\right] \\ &= -\frac{1}{2}\gamma B\hbar\begin{pmatrix}0 & re^{-i\varphi}\sin(\vartheta)\\ -re^{i\varphi}\sin(\vartheta) & 0\end{pmatrix}.\end{aligned} \tag{5.80}$$

This expression should equal

$$i\hbar\frac{\mathrm{d}\hat{\rho}}{\mathrm{d}t} = \frac{1}{2}i\hbar\begin{pmatrix}\frac{\mathrm{d}(r\cos(\vartheta))}{\mathrm{d}t} & \frac{\mathrm{d}(r\sin(\vartheta))}{\mathrm{d}t}e^{-i\varphi}-i\frac{\mathrm{d}\varphi}{\mathrm{d}t}e^{-i\varphi}r\sin(\vartheta)\\ \frac{\mathrm{d}(r\sin(\vartheta))}{\mathrm{d}t}e^{i\varphi}+i\frac{\mathrm{d}\varphi}{\mathrm{d}t}e^{i\varphi}r\sin(\vartheta) & -\frac{\mathrm{d}(r\cos(\vartheta))}{\mathrm{d}t}\end{pmatrix}. \tag{5.81}$$

From the diagonal components, it follows that $r\cos(\vartheta)$ is constant. The time derivative of $r\sin(\vartheta)$ is proportional to the sum of the real parts of the off-diagonal components of $[\hat{H},\hat{\rho}]$, and also vanishes. Both r and ϑ are therefore constant, and so is the purity $\frac{1}{2}(1+r^2)$. The non-zero components of $[\hat{H},\hat{\rho}]$ then determine the time derivative $\mathrm{d}\varphi/\mathrm{d}t = -\gamma B$. The resulting time-dependent density matrix describes a spin-expectation vector rotating around the axis of the magnetic field with frequency γB.

5.2.5.2 Lindblad evolution

Evolution implied by a Hamilton operator never changes the purity because it preserves $\mathrm{tr}(\hat{\rho}^2)$. More general time dependence is obtained if additional terms are added to $[\hat{H},\hat{\rho}]$, provided they maintain the conditions that define a density operator: Hermiticity, trace-one, and non-negativity. A suitable version is given by the **Lindblad equation**

$$\begin{aligned}i\hbar\frac{\mathrm{d}\hat{\rho}}{\mathrm{d}t} &= [\hat{H},\hat{\rho}]+i\hbar\sum_j\varepsilon_j\left(\hat{L}_j\hat{\rho}\hat{L}_j^\dagger-\frac{1}{2}(\hat{L}_j^\dagger\hat{L}_j\hat{\rho}+\hat{\rho}\hat{L}_j^\dagger\hat{L}_j)\right) \\ &= [\hat{H},\hat{\rho}]+\frac{1}{2}i\hbar\sum_j\varepsilon_j([\hat{L}_j\hat{\rho},\hat{L}_j^\dagger]-[\hat{\rho}\hat{L}_j^\dagger,\hat{L}_j])\,.\end{aligned} \tag{5.82}$$

The two lines are equivalent, but depending on the application one may be more convenient than the other. The sum may include any number of terms with different constants $\varepsilon_j \geq 0$, called **damping factors** with units of one over time, and unitless **Lindblad operators** $\hat{L}_j$. The Hamiltonian term $[\hat{H},\hat{\rho}]$ may also include several contributions, as we know from combinations of the kinetic energy with different potentials. But since this term is linear in $\hat{H}$, different contributions can simply be combined in a single Hamilton operator $\hat{H}$. The Lindblad terms, by contrast, are quadratic in $\hat{L}_j$ and its Hermitian conjugate, such that adding different contributions to a given $\hat{L}_j$ is not the same as adding different terms in the j-sum.

The operators $\hat{L}_j$ are not required to be Hermitian, but the j-sum is always Hermitian for a Hermitian $\hat{\rho}$. Multiplied by i, we therefore have a skew-Hermitian term, just like $[\hat{H},\hat{\rho}]$. This property implies that evolution preserves Hermiticity of $\hat{\rho}$. It also preserves $\mathrm{tr}(\hat{\rho})=1$ because $\mathrm{d}\hat{\rho}/\mathrm{d}t$ is a sum of commutators, as seen in the second version of the Lindblad equation (5.82). However, Lindblad evolution is not unitary: The mapping from $\hat{\rho}(0)$ to $\hat{\rho}(t)$ is not a unitary transformation. As a consequence, evolution in general does not preserve the eigenvalues of $\hat{\rho}$, and the purity may change: The time derivative of $\hat{\rho}^2$ can no longer be written as a sum of commutators, and its trace does not vanish. Nevertheless, it is still possible to show that non-negativity of $\hat{\rho}$ is preserved. The equation therefore provides a valid generalization of evolution for density operators.

As an example, consider the Hermitian choice $\hat{L}=2\hat{S}_z/\hbar=\hat{L}^\dagger$ as a single Lindblad operator, with zero Hamilton operator $\hat{H}=0$. In this case, we have

$$[\hat{L}\hat{\rho},\hat{L}^\dagger]-[\hat{\rho}\hat{L}^\dagger,\hat{L}]=[[\hat{L},\hat{\rho}],\hat{L}] \tag{5.83}$$

and compute

$$\begin{aligned}[[2\hat{S}_z/\hbar,\hat{\rho}],2\hat{S}_z/\hbar] &= \left[\left(\begin{array}{cc} 0 & re^{-i\varphi}\sin(\vartheta) \\ -re^{i\varphi}\sin(\vartheta) & 0 \end{array}\right),\left(\begin{array}{cc} 1 & 0 \\ 0 & -1 \end{array}\right)\right] \\ &= -2\left(\begin{array}{cc} 0 & re^{-i\varphi}\sin(\vartheta) \\ re^{i\varphi}\sin(\vartheta) & 0 \end{array}\right).\end{aligned} \tag{5.84}$$

This expression equals $2\varepsilon^{-1}\mathrm{d}\hat{\rho}/\mathrm{d}t$ if (5.82) holds, which implies that $r\cos(\vartheta)=c_1$ is constant because the diagonal terms vanish. Unlike in our Hamiltonian example, the off-diagonal terms now have the same sign, such that φ must be constant. The non-zero terms in (5.84) are then assigned to the time derivative of $r\sin(\vartheta)$, which implies the differential equation

$$\frac{\mathrm{d}(r\sin(\vartheta))}{\mathrm{d}t}=-\varepsilon r\sin(\vartheta) \tag{5.85}$$

solved by

$$r(t)\sin(\vartheta(t))=c_2e^{-\varepsilon t}. \tag{5.86}$$

Combining this solution with the constant $r\cos(\vartheta)$, we obtain

$$r(t)=\sqrt{r(t)^2\cos(\vartheta(t))^2+r(t)^2\sin(\vartheta(t))^2}=\sqrt{c_1^2+c_2^2e^{-2\varepsilon t}}. \tag{5.87}$$

We relate the two constants to the initial value $r_0=r(0)=\sqrt{c_1^2+c_2^2}$ and the asymptotic limit $r_\infty=\lim_{t\to\infty}r(t)=|c_1|$, such that

$$r(t)=\sqrt{r_0^2e^{-2\varepsilon t}+r_\infty^2(1-e^{-2\varepsilon t})}. \tag{5.88}$$

The purity changes from $\frac{1}{2}(1+r_0^2)$ to $\frac{1}{2}(1+r_\infty^2)$. From (5.86), we obtain

$$\sin(\vartheta(t))=\frac{c_2}{\sqrt{r_0^2+r_\infty^2(e^{2\varepsilon t}-1)}}=\frac{1}{\sqrt{1+e^{2\varepsilon t}\,r_\infty^2/(r_0^2-r_\infty^2)}}. \tag{5.89}$$

For large t, the spin-expectation vector therefore points toward the north pole (positive c_1) or south pole (negative c_1) if c_1 is non-zero. If $c_1=0$, the final spin-expectation vector is the null vector.

An example with non-Hermitian $\hat{L}$ is given by $\hat{L} = \hat{S}_+/\hbar$ with the raising operator $\hat{S}_+ = \hat{S}_x + i\hat{S}_y$, and thus $\hat{L}^\dagger = \hat{S}_-/\hbar$. In this case, we have

$$\frac{1}{2}\left([\hbar^{-1}\hat{S}_+\hat{\rho},\hbar^{-1}\hat{S}_-] - [\hat{\rho}\hbar^{-1}\hat{S}_-,\hbar^{-1}\hat{S}_+]\right) = \begin{pmatrix} \frac{1}{2}(1-r\cos(\vartheta)) & -\frac{1}{4}re^{-i\varphi}\sin(\vartheta) \\ -\frac{1}{4}re^{i\varphi}\sin(\vartheta) & -\frac{1}{2}(1-r\cos(\vartheta)) \end{pmatrix} \tag{5.90}$$

which implies

$$\frac{\mathrm{d}(r\cos(\vartheta))}{\mathrm{d}t} = \varepsilon(1-r\cos(\vartheta)) \tag{5.91}$$

and

$$\frac{\mathrm{d}(r\sin(\vartheta))}{\mathrm{d}t} = \frac{1}{2}\varepsilon r\sin(\vartheta) \tag{5.92}$$

while φ is constant. These equations are solved by

$$r(t)\cos(\vartheta(t)) = 1 - c_1e^{-\varepsilon t} \quad , \quad r(t)\sin(\vartheta(t)) = c_2e^{-\frac{1}{2}\varepsilon t} \tag{5.93}$$

such that

$$r(t) = \sqrt{1+(c_2^2-2c_1)e^{-\varepsilon t}+c_1^2e^{-2\varepsilon t}}\,. \tag{5.94}$$

For instance, if we choose $r(0) = 1$ for a pure initial state, we have $c_2 = \sqrt{2c_1 - c_1^2}$ and

$$r(t) = \sqrt{1-c_1^2(e^{-\varepsilon t}-e^{-2\varepsilon t})}\,. \tag{5.95}$$

If we then choose $c_1 = 2$, we obtain $c_2 = 0$ and therefore $\sin(\vartheta(t)) = 0$ for all t, while $r(t) = |1-2e^{-\varepsilon t}|$. With this choice, $\cos(\vartheta(0)) = -1$ and $\lim_{t\to\infty}\cos(\vartheta(t)) = 1$. The spin-expectation vector starts at the south pole and moves to the north pole along a straight line through the center.

5.2.6 PARTIAL TRACE AND ENTANGLEMENT

Non-unitary evolution can change the purity of a state. Another procedure with this outcome is the restriction of a state to a subsystem, for instance because there may be, and usually are, uncontrolled degrees of freedom that weakly interact with the system of interest but are not measurable. Examples include interactions with air molecules or, in vacuum, thermal radiation or various kinds of cosmic background radiation. The system of interest is then a subsystem of a larger interacting system, and its properties should depend on the unknown interactions. Such interactions in general imply that the subsystem is not isolated but rather entangled with its environment.

As an example, consider two spin-$\frac{1}{2}$ particles described by a basis of states $|\uparrow\uparrow\rangle$, $|\uparrow\downarrow\rangle$, $|\downarrow\uparrow\rangle$, $|\downarrow\downarrow\rangle$. We interpret the first spin as the subsystem and the second spin as its environment. If we can only measure the first spin, the second spin states should not provide independent parameters of the probability distribution for our measurements. We can eliminate these values by summing over their contributions to the original joint probability distribution, resulting in what is called the marginal distribution in classical probability theory. In quantum mechanics, we need the density operator for this procedure because its expectation values provide the probability distribution. The procedure is also called a **partial trace** because we sum over diagonal components in the basis of only one ingredient of a tensor-product space. If the full Hilbert space $\mathscr{H} = \mathscr{H}_1 \otimes \mathscr{H}_2$ has a tensor-product factor $\mathscr{H}_2$ with basis $|e_j\rangle$, the partial trace of a density operator $\hat{\rho}$ acting on $\mathscr{H}$ is defined as

$$\mathrm{tr}_{\mathscr{H}_2}(\hat{\rho}) = \sum_j (\hat{\mathbb{I}}_1 \otimes \langle e_j|)\,\hat{\rho}\,(\hat{\mathbb{I}}_1 \otimes |e_j\rangle) \tag{5.96}$$

with the identity operator $\hat{\mathbb{I}}_1$ on $\mathscr{H}_1$. As in the case of the complete trace, this definition is independent of the basis choice.

For instance, if the original state is $\chi_1 = |\uparrow\uparrow\rangle$, we have the density operator

$$\hat{\rho}_{\chi_1} = |\uparrow\uparrow\rangle\langle\uparrow\uparrow| .$$

The partial trace over the second spin equals

$$\mathrm{tr}_2(\hat{\rho}_{\chi_1}) = (\hat{\mathbb{I}} \otimes \langle\uparrow|)\hat{\rho}_{\chi_1}(\hat{\mathbb{I}} \otimes |\uparrow\rangle + (\hat{\mathbb{I}} \otimes \langle\downarrow|)\hat{\rho}_{\chi_1}(\hat{\mathbb{I}} \otimes |\downarrow\rangle . \tag{5.97}$$

We evaluate the operators by using

$$(\hat{\mathbb{I}} \otimes \langle\uparrow|)|\uparrow\uparrow\rangle = |\uparrow\rangle \tag{5.98}$$

from normalization and

$$(\hat{\mathbb{I}} \otimes \langle\downarrow|)|\uparrow\uparrow\rangle = |0\rangle \tag{5.99}$$

from orthogonality. Therefore,

$$\mathrm{tr}_2(\hat{\rho}_{\chi_1}) = |\uparrow\rangle\langle\uparrow| = \hat{\rho}_{|\uparrow\rangle} \tag{5.100}$$

is a pure state with zero entropy. In this case, the original 2-spin state, which is a product state, is called unentangled.

If we choose the original state as

$$\chi_2 = \frac{1}{\sqrt{2}}(|\uparrow\downarrow\rangle \pm |\downarrow\uparrow\rangle) \tag{5.101}$$

we have the density operator

$$\hat{\rho}_{\chi_2} = \frac{1}{2}(|\uparrow\downarrow\rangle\langle\uparrow\downarrow| \pm |\uparrow\downarrow\rangle\langle\downarrow\uparrow| \pm |\downarrow\uparrow\rangle\langle\uparrow\downarrow| + |\downarrow\uparrow\rangle\langle\downarrow\uparrow|) \tag{5.102}$$

with partial trace

$$\mathrm{tr}_2(\hat{\rho}_{\chi_2}) = \frac{1}{2}(|\uparrow\rangle\langle\uparrow| + |\downarrow\rangle\langle\downarrow|) \tag{5.103}$$

In the $\hat{S}_z$-eigenbasis, this density operator has the matrix

$$\mathrm{tr}_2(\hat{\rho}_{\chi_2}) \leftrightarrow \begin{pmatrix} 1/2 & 0 \\ 0 & 1/2 \end{pmatrix} \tag{5.104}$$

with a uniform probability distribution and entropy $\log(2)$. This state is mixed, and the original state χ_2 (which is pure) is an **entangled** state.

5.3 OPERATOR-VALUED MEASURES

Just like states, operators must be generalized for realistic descriptions of measurements. The application of a self-adjoint operator $\hat{A}$ to measurements assumes that they can be arbitrarily precise, or sharp, because $\hat{A}$ provides a basis of eigenstates in which the standard deviation σ_A is zero. Realistic measurements are never sharp. This general limitation does not necessarily preclude the application of a self-adjoint operator, in particular if it has a discrete spectrum with a spacing greater than the measurement precision. For instance, it is possible for a measurement to give a sharp answer to the question of whether an electron has spin up or spin down. However, there are cases in which it is difficult to resolve the discrete spacing, and an observable represented by an operator with a continuous spectrum can, in a realistic description, never result in a sharp outcome.

5.3.1 SPECTRAL DECOMPOSITION

The sharp nature of self-adjoint operators can be expressed by using the notion of projection operators. A **projection operator** is defined as a self-adjoint operator $\hat{P}$ such that $\hat{P}^2 = \hat{P}$, and therefore has eigenvalues zero and one. Applying $\hat{P}$ to all vectors in the Hilbert space in general results in vectors contained in a subspace of the Hilbert space, the image of $\hat{P}$. These vectors are the eigenstates with eigenvector one, while all vectors orthogonal to the image subspace have eigenvalue zero. The density operator $|\psi\rangle\langle\psi|$ of a pure state $|\psi\rangle$ is a projection operator with image given by the subspace spanned by $|\psi\rangle$.

Given a self-adjoint operator $\hat{A}$ with discrete eigenvalues λ_j and normalized eigenstates $|e_j\rangle$, we introduce the countable set of projection operators $|e_j\rangle\langle e_j|$. The operator sum

$$\sum_j \lambda_j |e_j\rangle\langle e_j| = \hat{A} \tag{5.105}$$

can easily be seen to have the same eigenvectors and eigenvalues of $\hat{A}$, and is therefore the same as $\hat{A}$. Equation (5.105) is called the **spectral decomposition** of the self-adjoint operator $\hat{A}$. A similar version exists for operators with continuous spectrum, but they require a notion of integration on a Hilbert space, or a measure.

If we only consider the family of projection operators $\{|e_j\rangle\langle e_j|\}$ and ignore the eigenvalues, we obtain the projection operator

$$\hat{P}_{\hat{A}} = \sum_j |e_j\rangle\langle e_j| \,. \tag{5.106}$$

Applying this operator to a vector $|\psi\rangle$ yields its eigenfunction decomposition in the $\hat{A}$-eigenbasis,

$$\hat{P}_{\hat{A}}|\psi\rangle = \sum_j \langle e_j|\psi\rangle |e_j\rangle \,. \tag{5.107}$$

Since the vector $|\psi\rangle$ is not changed by $\hat{P}_{\hat{A}}$ but merely expressed in the eigenvector basis, $\hat{P}_{\hat{A}} = \hat{\mathbb{I}}$ is the identity operator. If we restrict the summation range in $\hat{P}_{\hat{A}}$ to a subset X of the spectrum, we obtain a projection operator

$$\hat{P}_{\hat{A}}(X) = \sum_{j\in X} |e_j\rangle\langle e_j| \tag{5.108}$$

that is not the identity operator.

5.3.2 POSITIVE-OPERATOR VALUED MEASURES

The family of all operators of the form $\hat{P}_{\hat{A}}(X)$ for a given $\hat{A}$ has the following general properties that can be specified for a map $\hat{E}_Z$ from subsets X of a given set Z to operators $\hat{E}_Z(X)$:

1. Positivity: The spectrum of $\hat{E}_Z(X)$ is non-negative for any X. This property is implied if $\hat{E}_Z(X) \geq \hat{0}$: $\hat{E}_Z(X)$ is a **positive operator** in the sense that $\langle\psi|\hat{E}_Z(X)|\psi\rangle \geq 0$ for any $|\psi\rangle$. From this latter condition it also follows that $\hat{E}_Z(X)$ is self-adjoint: First, because a positive number is real, we have

$$\langle\psi|\hat{E}_Z(X)|\psi\rangle = \langle\psi|\hat{E}_Z(X)|\psi\rangle^* = \langle\hat{E}_Z(X)\psi|\psi\rangle \,. \tag{5.109}$$

   The Hermiticity condition for $\langle\psi_1|\hat{E}_Z(X)|\psi_2\rangle$ with $|\psi_1\rangle \neq |\psi_2\rangle$ can be obtained from a combination of (5.109) applied to $|\psi_1\rangle \pm |\psi_2\rangle$ and $|\psi_1\rangle \pm i|\psi_2\rangle$. In this case, Hermiticity implies self-adjointness because the operator $\hat{E}_Z(X)$ is required to be well-defined on any state in the Hilbert space.
2. Boundedness: The spectrum of $\hat{E}_Z(X)$ is bounded from above by the value one. This condition is also written as $\hat{E}_Z(X) \leq \hat{\mathbb{I}}$, which means that $\hat{\mathbb{I}} - \hat{E}_Z(X) \geq \hat{0}$ is a positive operator.

3. Additivity: For two disjoint subsets, $X_1 \cap X_2 = \emptyset$,

$$\hat{E}_Z(X_1 \cup X_2) = \hat{E}_Z(X_1) + \hat{E}_Z(X_2)\,. \tag{5.110}$$

4. Normalization: For the full set Z, we obtain the identity operator, $\hat{E}_Z(Z) = \hat{\mathbb{I}}$.

A mapping from subsets to operators with these properties is called a **positive-operator valued measure**, also referred to as a POVM.

Consistency of the definition requires that all subsets to which the conditions are applied are available: The subsets $X \subset Z$ to which operators are assigned must form a set of sets, Σ, such that: (i) Z is an allowed set, $Z \in \Sigma$. (ii) If $X \in \Sigma$, $Z \backslash X \in \Sigma$ for the complement of X in Z. (iii) For a countable number X_j of sets $X_j \in \Sigma$, their union $X_1 \cup X_2 \cup \cdots \in \Sigma$ is included in Σ. Properties (ii) and (iii) imply that the intersection of a countable number of subsets $X_i \in \Sigma$ is contained in Σ because $X_1 \cap X_2 = Z \backslash ((Z \backslash X_1) \cup (Z \backslash X_2))$. A set of sets Σ obeying these conditions is called a **σ-algebra**.

A positive-operator valued measure assigns an operator to each element of a σ-algebra. This definition can also be applied to continuous sets Z, generalizing self-adjoint operators with continuous spectra. A σ-algebra is a familiar concept from measure theory, in which case numbers are assigned to its elements which may be interpreted as their lengths, areas or volumes. This origin explains the name "measure."

The first two properties required for a positive-operator valued measure are satisfied by any $\hat{P}_{\hat{A}}(X)$ as defined in (5.108) because such a projection operator has only the two eigenvalues zero and one. In general, however, these two properties do not imply that $\hat{E}_Z(X)$ is a projection operator. An operator with these two properties is called an **effect**.

As an example, assume that a measurement device with an analog dial has been constructed in order to determine a discrete value, such as the z-component of some spinning system. To be specific, we assume that the set of all outcomes is $\mathbb{Z}$ which may represent half-integer spin values $\mathtt{m}$ by integer outcomes $2\mathtt{m}$. For a fixed spin quantum number s, all but a finite number of effect operators are zero.

We assume that the measurement is not sharp because two of the dial markers, say numbers one and two, have been placed too close together, such that they do not allow a unique reading: A result of one is sometimes, with probability $\varepsilon > 0$, mistaken for a reading of two, and vice versa. We can describe such a measurement by using the effect operators

$$\hat{E}_{\mathbb{Z}}(1) = (1-\varepsilon)|1\rangle\langle 1| + \varepsilon|2\rangle\langle 2| \quad , \quad \hat{E}_{\mathbb{Z}}(2) = \varepsilon|1\rangle\langle 1| + (1-\varepsilon)|2\rangle\langle 2| \tag{5.111}$$

while $\hat{E}_{\mathbb{Z}}(j) = |j\rangle\langle j|$ for all other outcomes. If we are not interested in distinguishing between the outcomes one and two, but would still like to separate them from the other outcomes, we only consider the combined effect

$$\hat{E}_{\mathbb{Z}}(1,2) = \hat{E}_{\mathbb{Z}}(1) + \hat{E}_{\mathbb{Z}}(2) = |1\rangle\langle 1| + |2\rangle\langle 2|\,. \tag{5.112}$$

This effect is independent of the precision parameter ε.

If we already know from a sharp measurement of the same quantum number that the system is in the eigenstate $|1\rangle$, a measurement with the unsharp device returns the value one with probability $1-\varepsilon$, and the value two with probability ε. We can write these probabilities as

$$P_{|1\rangle}(1) = \langle 1|\hat{E}_{\mathbb{Z}}(1)|1\rangle \quad , \quad P_{|1\rangle}(2) = \langle 1|\hat{E}_{\mathbb{Z}}(2)|1\rangle\,. \tag{5.113}$$

In general, the conditions for a positive-operator valued measure imply that the numbers defined by

$$P_{|\psi\rangle}(j) = \langle \psi|\hat{E}_{\mathbb{Z}}(j)|\psi\rangle \tag{5.114}$$

for a given state $|\psi\rangle$ form a probability distribution. We interpret $P_{|\psi\rangle}(j)$ as the probability of obtaining the outcome j in a measurement described by the positive-operator valued measure $\hat{E}_{\mathbb{Z}}(j)$. If the possible values for each outcome are given by v_j, we obtain the expectation value

$$\langle v\rangle_{|\psi\rangle} = \sum_{j\in Z} v_j P(j) = \sum_{j\in Z} v_j \langle\psi|\hat{E}_{\mathbb{Z}}(j)|\psi\rangle \tag{5.115}$$

for measurements on a system in the state $|\psi\rangle$. This expression generalizes our previous results for a self-adjoint operator $\hat{A}$, in which case

$$\langle\hat{A}\rangle_{|\psi\rangle} = \sum_j \lambda_j |\langle e_j|\psi\rangle|^2 = \sum_j \lambda_j \langle\psi|\hat{P}_{\hat{A}}(j)|\psi\rangle \tag{5.116}$$

if the operator has eigenvalues λ_j and eigenfunctions $|e_j\rangle$.

If $v_j = j^n$ with some positive integer n, we define the n-th **moment**

$$\hat{M}_n(\hat{E}_{\mathbb{Z}}) = \sum_j j^n \hat{E}_{\mathbb{Z}}(j) \tag{5.117}$$

of a positive-operator valued measure over an integer subset $Z \subset \mathbb{Z}$. The expectation value of j then equals

$$\langle j\rangle_{|\psi\rangle} = \langle\psi|\hat{M}_1(\hat{E}_{\mathbb{Z}})|\psi\rangle \tag{5.118}$$

and we define the variance employing the second moment. In our present example, if the values are $v_j = j$, we have the expectation values $\langle j\rangle_{|1\rangle} = 1+\varepsilon$ and $\langle j\rangle_{|2\rangle} = 2-\varepsilon$. We have $\langle j^2\rangle_{|1\rangle} = 1+3\varepsilon$ and $\langle j^2\rangle_{|2\rangle} = 4-3\varepsilon$. Thus, $\sigma_{|1\rangle} = \sqrt{\varepsilon(1-\varepsilon)} = \sigma_{|2\rangle}$, and measurements are not sharp.

5.3.3 CONTINUOUS SPECTRUM

If the spectrum of possible outcomes is continuous, no realistic measurement can be sharp. We should therefore replace self-adjoint operators that have continuous spectra with suitable positive-operator valued measures over a continuous set Z. The general conditions did not make use of discrete values and can therefore be applied in the continuous case as well. Also the notion of σ-algebras can be applied to continuous sets.

On the level of positive-operator valued measures, the difference between discrete and continuous cases is therefore mainly notational. An effect $\hat{E}_{\mathbb{R}}(X)$ corresponding to the subset $X \subset \mathbb{R}$ is written as

$$\hat{E}_{\mathbb{R}}(X) = \int_X |x\rangle\langle x| \mathrm{d}x \tag{5.119}$$

such that

$$\hat{E}_{\mathbb{R}}(\mathbb{R})|\psi\rangle = \int_{-\infty}^{\infty} \psi(x)|x\rangle \mathrm{d}x = |\psi\rangle \tag{5.120}$$

and the normalization condition is fulfilled. We have positivity and boundedness because $\langle\psi|\hat{E}_{\mathbb{R}}(X)|\psi\rangle = \int_X |\psi(x)|^2 \mathrm{d}x$ is non-negative for any $|\psi\rangle$, and bounded from above by one because $\int_X |\psi(x)|^2 \mathrm{d}x \leq \int_{-\infty}^{\infty} |\psi(x)|^2 \mathrm{d}x = 1$.

Note that normalization of a positive-operator valued measure does not require a factor related to the inverse length of the integration range, which would be problematic when the integration is extended to all of $\mathbb{R}$. Such a factor would also be in conflict with additivity. The integration in (5.119) can be interpreted as a regularization of delta functions that we previously employed for the construction of eigenstates $|x\rangle$ of the position operator. The same expressions can be used for any outcome taking values in the full real line, such as the momentum, if we just replace x with a different variable. For now, we continue to interpret x as the outcome of position measurements.

Statistical properties of position measurements are obtained from moments of (5.119),

$$\hat{M}_n(\hat{E}_{\mathbb{R}}) = \int_{-\infty}^{\infty} x^n |x\rangle\langle x| \mathrm{d}x \,. \tag{5.121}$$

We obtain the usual statistical moments

$$\langle x^n \rangle_{|\psi\rangle} = \langle \psi | \hat{M}_n(\hat{E}_{\mathbb{R}}) | \psi \rangle = \int_{-\infty}^{\infty} x^n |\langle x|\psi\rangle|^2 \mathrm{d}x \tag{5.122}$$

determined by the probability density $\rho_{|\psi\rangle}(x) = |\langle x|\psi\rangle|^2 = |\psi(x)|^2$. These equations do not make use of a position operator $\hat{x}$ but only insert the values x of possible outcomes, while effect operators determine the probabilities

$$P_{|\psi\rangle}(X) = \langle \psi | \hat{E}_{\mathbb{R}}(X) | \psi \rangle = \int_X |\psi(x)|^2 \mathrm{d}x \tag{5.123}$$

for subsets of the real line. The position operator acts at any single point and changes the wave function by multiplication, while the limit of $\hat{E}_{\mathbb{R}}(X)$ for $X \to \{x\}$ approaching a single point is zero. Therefore, there is no effect for a single point in a continuous spectrum, and measurements of continuous outcomes cannot be sharp.

5.4 MEASUREMENTS

The first statistics defined in the course of our developments was derived from a wave function $\psi(x)$, which determines the probability density $|\psi(x)|^2$ for measurements of x. In the meantime, we recognized $\psi(x)$ as a basis-dependent representation of a pure state $|\psi\rangle$ of the system. If time changes, the state evolves and provides the time-dependent wave function $\Psi(x,t)$ in the position representation, or a time-dependent vector $|\Psi\rangle(t)$ in an infinite-dimensional Hilbert space.

The initial formulation of quantum mechanics based on position-dependent wave functions led to a preliminary list of ingredients for a statistical description of measurements:

1. The wave function $\Psi(x,t)$ implies the probability density $|\Psi(x,t)|^2$ for measurements of x at time t. The wave function contains all the information we could possibly obtain about the system and therefore represents its state.
2. For meaningful statistics, the wave function must be normalized, such that

$$\int_{-\infty}^{\infty} |\Psi(x,t)|^2 \mathrm{d}x = 1 \tag{5.124}$$

   at all times t.
3. Predictions for measurements other than x can be made by finding the corresponding classical expression $A(x,p)$, called an observable, and turning it into a self-adjoint operator of the form $\hat{A} = A(\hat{x},\hat{p}) = A(x,(\hbar/i)\partial/\partial x)$. The time-dependent expectation value for multiple measurements of the observable A in the state $\Psi(x,t)$ at time t is given by

$$\langle \hat{A} \rangle(t) = \int_{-\infty}^{\infty} \Psi(x,t)^* \hat{A} \Psi(x,t) \mathrm{d}x = \langle \Psi | \hat{A} | \Psi \rangle(t) \,. \tag{5.125}$$

   In order to measure this value for a given state at a fixed time t, its statistical nature requires us to measure A multiple times at the same t (or at the same time difference from the initial time) in copies of the system prepared in identical ways, such that they are in the same state $\Psi(x,t)$ at the time t of the measurements. Measuring the function $\langle \hat{A} \rangle(t)$ requires us to repeat this procedure several times in order to sample it at different values of t.

4. If we perform a single repeatable measurement of x in the state $\Psi(x,t)$ at a fixed time t, an outcome of x in the range $a \leq x \leq b$ is obtained with probability $\int_a^b |\Psi(x,t)|^2 dx$. In terms of a single function, these values are represented by the probability density $\rho_t(x) = |\Psi(x,t)|^2$. Right after the measurement at time t_0, we know the value of $x = x_0 \pm \sigma$ within some measurement uncertainty σ, and therefore the new state after the measurement is described by a wave function spread around the measured value x_0 with standard deviation given by the measurement precision. The precise functional form depends on details of the measurement device and error sources that lead to the standard deviation σ.

These original statements have been generalized in several ways by our additional developments. Quantum mechanics is not restricted to statements about single measurements only for the position variable x. By changing the representation in a Hilbert space, we obtain probability densities or distributions for any observable other than x, provided it is represented by a self-adjoint operator $\hat{A}$ such that we obtain a new basis from its eigenfunctions. Individual measurement outcomes are then given by the real eigenvalues of $\hat{A}$. The probability density or distribution derived from the state in the new basis is consistent with the definition (5.125) of expectation values in the position representation.

In addition, the concept of wave functions should be generalized to density operators if we are interested in describing not only pure states of isolated systems but also mixed states of an **open quantum system** that interacts with an environment under incomplete control. Individual measurement outcomes can be sharp only if they correspond to an observable with discrete spectrum, but even in this case there may be reasons that limit the sharpness of outcomes. General measurements are not described by self-adjoint operators on a Hilbert space, but rather by positive-operator valued measures.

5.4.1 DEFINITIONS

The statistical content of quantum mechanics has three ingredients given by the system in some state at time t, an observable, and their combination in a measurement result.

5.4.1.1 States

The state of an isolated system at time t is represented by a pure state, given by a normalized vector $|\Psi\rangle(t)$ in a Hilbert space. The state of an open system at time t is represented by a mixed state, given by a density operator $\hat{\rho}(t)$ with purity $\mathrm{tr}(\hat{\rho}^2)$ less than one, acting on a Hilbert space.

A Hilbert space is a vector space, which may be finite-dimensional or infinite-dimensional, with an inner product such that all vectors of finite norm are included in the space. If the Hilbert space is finite-dimensional, a pure state is an ordinary normalized vector, which does not look much like a wave. For spin systems, a pure state is called a spinor.

An important implication is that we can apply basis changes in order to simplify some calculations. Our previous statements about measurements of x can be recognized as special cases of statements made in a specific basis. Changing the basis shows how statements about single measurements can be obtained for observables other than x.

5.4.1.2 Observables

An observable A that can take values in the range Z is represented by a positive-operator valued measure $\hat{E}_Z$. Averages and other statistical quantities of multiple measurements of A on a system in the state $\hat{\rho}$ are given by the expectation values

$$\langle A^n \rangle_{\hat{\rho}} = \mathrm{tr}(\hat{M}_n(\hat{E}_Z)\hat{\rho}) \tag{5.126}$$

of the moments

$$\hat{M}_n(\hat{E}_Z) = \sum_j j^n \hat{E}_Z(j) \tag{5.127}$$

of $\hat{E}_Z$, where an integral is used for continuous parts of the range of outcomes. If the state is pure,

$$\langle A^n \rangle_{|\Psi\rangle} = \langle \Psi | \hat{M}_n(\hat{E}_Z) | \Psi \rangle \,. \tag{5.128}$$

Expectation values are real because any $\hat{E}_Z$ takes self-adjoint effect operators as its values $\hat{E}_Z(j)$.

An observable A subject to **sharp measurements** is represented by a self-adjoint operator $\hat{A}$ with discrete spectrum, a special case of a positive-operator valued measure determined by the spectral decomposition of $\hat{A}$. The expectation value is the average of multiple measurements that in general have different values. A specific measurement of an observable A on a system in a given state is sharp if A has a vanishing standard deviation in the state, $\sigma_A = \sqrt{\langle A^2 \rangle - \langle A \rangle^2} = 0$. An observable A is sharp if there is a Hilbert-space basis of pure states such that A has a sharp measurement result in each state of the sharp basis.

If the observable is represented by a positive-operator valued measure $\hat{E}_Z(j)$ and the state $|\Psi\rangle$ is pure, we compute

$$\begin{aligned} \sigma_A^2 &= \sum_j j^2 \langle \Psi | \hat{E}_Z(j) | \Psi \rangle - \left(\sum_j j \langle \Psi | \hat{E}_Z(j) | \Psi \rangle \right)^2 \qquad (5.129) \\ &= \sum_j j^2 (\langle \Psi | \hat{E}_Z(j) | \Psi \rangle - \langle \Psi | \hat{E}_Z(j) | \Psi \rangle^2) - \sum_{j \neq k} jk \langle \Psi | \hat{E}_Z(j) | \Psi \rangle \langle \Psi | \hat{E}_Z(k) | \Psi \rangle \,. \end{aligned}$$

In general, there may be cancellations between the two final sums for specific states, but this possibility cannot apply to an entire basis of pure states. Generically, therefore, the first sum must vanish in each state of the sharp basis, which requires that $\langle \Psi | \hat{E}_Z(j) | \Psi \rangle$ equals zero or one for each basis state unless $j = 0$. Therefore, all $\hat{E}_Z(j)$ for non-zero j must be projection operators which, using normalization, implies that $\hat{E}_Z(0) = \hat{\mathbb{I}} - \sum_{j \neq 0} \hat{E}_Z(j)$ is also a projection operator. (Note that $(\hat{\mathbb{I}} - \hat{P})^2 = \hat{\mathbb{I}} - 2\hat{P} + \hat{P}^2 = \hat{\mathbb{I}} - \hat{P}$ for any projection operator $\hat{P}$.) In this case, the positive-operator valued measure equals the spectral decomposition of a self-adjoint operator. Normalization implies that it is not possible for both $\langle \Psi | \hat{E}_Z(j) | \Psi \rangle$ and $\langle \Psi | \hat{E}_Z(k) | \Psi \rangle$ to equal one in the same state if $j \neq k$. Therefore, at least one factor is zero in each term of the last sum of (5.129), and $\sigma_A^2 = 0$.

We conclude that measurements cannot be sharp for an observable represented by a positive-operator valued measure unless the measure is the spectral decomposition of a self-adjoint operator $\hat{A}$. A specific measurement of A is then sharp if and only if the state $|\Psi\rangle$ is an eigenstate of $\hat{A}$. To see this, we first compute

$$\langle (\hat{A} - \langle \hat{A} \rangle)^2 \rangle = \langle \hat{A}^2 - 2\langle \hat{A} \rangle \hat{A} + \langle \hat{A} \rangle^2 \rangle = \langle \hat{A}^2 \rangle - \langle \hat{A} \rangle^2 = \sigma_A^2 \tag{5.130}$$

where expectation values within expectation values are numbers and can be moved out. For $\sigma_A^2 = 0$, we therefore need

$$\begin{aligned} 0 &= \langle (\hat{A} - \langle \hat{A} \rangle)^2 \rangle = \langle \Psi | (\hat{A} - \langle \hat{A} \rangle)^2 | \Psi \rangle = \langle (\hat{A} - \langle \hat{A} \rangle) \Psi | (\hat{A} - \langle \hat{A} \rangle) | \Psi \rangle \\ &= ||(\hat{A} - \langle \hat{A} \rangle) | \Psi \rangle ||^2 \qquad (5.131) \end{aligned}$$

where we used self-adjointness of $\hat{A}$ and the reality of its expectation values. The norm squared derived from an inner product is zero if and only if it is applied to the null vector, and therefore $(\hat{A} - \langle \hat{A} \rangle)|\Psi\rangle = |0\rangle$ or

$$\hat{A}|\Psi\rangle = \langle \hat{A} \rangle |\Psi\rangle \tag{5.132}$$

such that $|\Psi\rangle$ is an eigenstate of $\hat{A}$ with eigenvalue $\langle \Psi | \hat{A} | \Psi \rangle$. A system can be in an eigenstate $|\Psi\rangle$ of a self-adjoint $\hat{A}$ only if the state is normalizable, which requires an operator $\hat{A}$ with discrete spectrum.

Measurement results of an observable A are certain if the observable is sharp and the system is in an eigenstate of the corresponding $\hat{A}$. If these conditions on observables and states are not fulfilled, we do not obtain certain outcomes but rather a probability distribution or density for possible measurement results, as in the case of position measurements where the probability of finding a value of x in the range $x_0 - \varepsilon < x < x_0 + \varepsilon$ at time t is given by $\int_{x_0-\varepsilon}^{x_0+\varepsilon} |\Psi(x,t)|^2 \mathrm{d}x$ for any $\varepsilon > 0$, provided the state is pure.

A probabilistic statement is necessary if we perform the measurement on a generic state. Having a probability distribution or density allows us to make statements about the likelihood of measuring a specific value in a single measurement. The original position-dependent wave function determines the probability density for position measurements. Writing $\Psi(x,t_0)$ at a given time t_0 as a superposition of position eigenstates $\delta(x-\lambda)$ (with delta-function normalization), we have

$$\Psi(x,t_0) = \int_{-\infty}^{\infty} \Psi(\lambda,t_0)\delta(x-\lambda)\mathrm{d}x \tag{5.133}$$

with the same function Ψ on the left and on the right because $\Psi(x,t_0)$ is already in the basis of position eigenstates.

For a general observable A represented by a self-adjoint operator which, like $\hat{x}$, may have continuous spectrum, we have discrete or continuous eigenvalues λ with eigenfunctions $|e_\lambda\rangle$. Any state can then be written as

$$|\Psi\rangle = \sum\!\!\!\!\!\!\int_{\lambda} c_\lambda |e_\lambda\rangle \tag{5.134}$$

where $\sum\!\!\!\!\!\!\int_\lambda$ indicates that we sum over the discrete part of the spectrum and integrate over the continuous part. The coefficients c_λ in (5.134) are analogous to the coefficients $\Psi(\lambda)$ in (5.133), which give us the probability density for a single position measurement. Similarly, the values $|c_\lambda|^2 = |\langle e_\lambda|\Psi\rangle|^2$ provide the probability distribution or probability density for measurements of A. If the measurement of A is not sharp and represented by a positive-operator valued measure $\hat{E}_Z$, the probability distribution or probability density for measurement results in the range X is given by the expectation values

$$P_{\hat{\rho}}(X) = \mathrm{tr}(\hat{E}_Z(X)\hat{\rho}) \tag{5.135}$$

in a state $\hat{\rho}$ that may in general be mixed.

5.4.1.3 Single measurements

Right after a single *sharp* measurement of an observable A, we know the value of A. Therefore, the post-measurement state, which contains all the information we have about the system, must imply a certain outcome of measuring A, given by the value we just obtained. Since a sharp measurement is represented by a self-adjoint operator $\hat{A}$ with discrete spectrum, and since A-measurements are certain in $\hat{A}$-eigenstates $|e_\lambda\rangle$, the post-measurement state must be one of the eigenstates of $\hat{A}$, and the sharp value obtained in the measurement is the corresponding eigenvalue λ. Before the measurement, we do not know which value we will find unless the system is already in an eigenstate. We will find a specific value λ with probability $|c_\lambda|^2$, where c_λ is the coefficient of $|e_\lambda\rangle$ in the eigenbasis expansion $|\Psi\rangle = \sum_\lambda c_\lambda |e_\lambda\rangle$ of the original state.

A self-adjoint operator $\hat{A}$ may have eigenvalues with multiplicities greater than one, which can easily happen in symmetric systems where the same energy may be obtained for different quantum numbers such as the angular momentum of an electron. If we have an eigenvalue λ with multiplicity $n > 1$, we can always arrange the eigenvalues such that $\lambda_1 = \cdots = \lambda_n = \lambda$ while $\lambda_{n+1} \neq \lambda, \ldots$ in the discrete spectrum. A generic state then has the expansion

$$|\Psi\rangle = c_1|e_1\rangle + \cdots + c_n|e_n\rangle + \sum_{m=n+1}^{\infty} c_m|e_m\rangle \tag{5.136}$$

and it is not clear which of the first n numbers $|c_i|^2$ we should use for the probability of measuring λ. Moreover, the question is mathematically unclear because a linear combination of eigenvectors with the same eigenvalue is also an eigenvector with this eigenvalue.

The question is simplified if we put the basic condition not as having measured λ, but as *not* having measured any of the other values. The state right after the measurement, which takes this information into account, therefore cannot contain the contributions $\sum_{m=n+1}^{\infty} c_m|e_m\rangle$, but it maintains the first part, $\sum_{m=1}^{n} c_m|e_m\rangle$, because the measurement does not provide any information that could restrict the first n eigenstates $|e_m\rangle$. The state right after the measurement cannot be equal to $\sum_{m=1}^{n} c_m|e_m\rangle$ because this linear combination, in general, is not normalized. But using orthonormality of the $|e_m\rangle$, we can easily normalize the new state to obtain

$$|e_\lambda\rangle = \frac{1}{\sqrt{|c_1|^2+\cdots+|c_n|^2}} \sum_{m=1}^{n} c_m|e_m\rangle \tag{5.137}$$

right after the measurement. The initial state then has the form

$$|\Psi\rangle = c_\lambda|e_\lambda\rangle + \sum_{m=n+1}^{\infty} c_m|e_m\rangle \tag{5.138}$$

with $c_\lambda = \sqrt{|c_1|^2+\cdots+|c_n|^2}$. There is now a single eigenstate $|e_\lambda\rangle$ in this eigenvector expansion of $|\Psi\rangle$, with a unique coefficient c_λ that determines the probability

$$|c_\lambda|^2 = |c_1|^2+\cdots+|c_n|^2\,, \tag{5.139}$$

just summing the individual probabilities for an eigenvalue with multiplicity.

Gaining information about a system is an abrupt change that implies a corresponding transition of its state. The abruptness is expressed by saying that the pre-measurement state $|\Psi_1\rangle$ **collapses** into an eigenstate $|\Psi_2\rangle$ of the observable during a sharp measurement. Collapse in this form makes use of the projection operator $\hat{P}_\lambda = |e_\lambda\rangle\langle e_\lambda|$ such that the new state $|\Psi_2\rangle$ equals

$$|\Psi_2\rangle = \frac{\hat{P}_\lambda|\Psi_1\rangle}{||\hat{P}_\lambda|\Psi_1\rangle||} = \frac{\hat{P}_\lambda|\Psi_1\rangle}{\langle\Psi_1|\hat{P}_\lambda|\Psi_1\rangle}\,. \tag{5.140}$$

Any projection operator obeys the equations $\hat{P}_\lambda^2 = \hat{P}_\lambda = \hat{P}_\lambda^\dagger$. We can therefore generalize the collapse to a mixed state $\hat{\rho}_1$, which collapses to

$$\hat{\rho}_2 = \frac{\hat{P}_\lambda\hat{\rho}_1\hat{P}_\lambda}{\mathrm{Tr}(\hat{P}_\lambda\hat{\rho}_1)}\,. \tag{5.141}$$

This operator is self-adjoint and trace-one thanks to the defining relations of a projection operator, and $\hat{P}_\lambda\hat{\rho}_1\hat{P}_\lambda$ is non-negative because $\langle\psi|\hat{P}_\lambda\hat{\rho}_1\hat{P}_\lambda|\psi\rangle = \langle\phi|\hat{\rho}_1|\phi\rangle \geq 0$ for any $|\psi\rangle$, using non-negativity of ρ_1 and defining $|\phi\rangle = \hat{P}_\lambda|\psi\rangle$.

If the measurement is not sharp and A is represented by a positive-operator valued measure $\hat{E}_Z$, there are in general no projection operators that could be used to define the new state. Several properties of projection operators were relevant when we defined state collapse for sharp measurements. The property $\hat{P} = \hat{P}^2$ allowed us to multiply a single projection operator into two copies, such that each of them can act on one side of $\hat{\rho}$ in (5.141), preserving Hermiticity. An operator-valued measure, in general, does not imply a preferred factorization or a square root of an effect such that

$$\hat{E}_Z(j) = \hat{R}_Z(j)^\dagger\hat{R}_Z(j)\,. \tag{5.142}$$

Nevertheless, such factorizations may exist even if they are not unique, and depending on the circumstances there may be physical reasons that suggest a specific $\hat{R}_Z(j)$ for a given effect. If $\hat{R}_Z(j)$

is not unique for a given $\hat{E}_Z(j)$, there are measurements of an observable A that result in the same expectation values, measurement results, and probabilities, but imply different collapse states after a measurement.

In fact, a factorization is never unique: If (5.142) is realized for some $\hat{R}_Z(j)$, it is true for any $\hat{U}(j)\hat{R}_Z(j)$ with a unitary operator $\hat{U}(j)$ that may also depend on j. This ambiguity exists even in the case of projection operators associated to a self-adjoint operator, where $\hat{P}_\lambda$ may be replaced with $\hat{P}'_\lambda = \hat{U}(\lambda)\hat{P}_\lambda$, such that $(\hat{P}'_\lambda)^\dagger \hat{P}'_\lambda = \hat{P}_\lambda$. If a measurement procedure is described by $\hat{P}'_\lambda$ instead of $\hat{P}_\lambda$, a state $\hat{\rho}_1$ would collapse into

$$\hat{\rho}'_2 = \frac{\hat{P}'_\lambda \hat{\rho}_1 \hat{P}'^\dagger_\lambda}{\mathrm{tr}(\hat{P}_\lambda \hat{\rho}_1)} \tag{5.143}$$

instead of (5.141).

If an effect is given in the factorized form (5.142), we are able to define the post-measurement density matrix as

$$\hat{\rho}_2(j) = \frac{\hat{R}_Z(j)\hat{\rho}_1\hat{R}_Z(j)^\dagger}{\mathrm{tr}(\hat{E}_Z(j)\hat{\rho}_1)} \tag{5.144}$$

if we measured the value j for A. This definition fulfills two important conditions: First, it is equivalent to our previous statements about collapse if the positive-operator valued measure is given by the spectral decomposition of a self-adjoint operator $\hat{A}$, in which case we may choose $\hat{R}_Z(j) = |j\rangle\langle j| = \hat{R}_Z(j)^\dagger$. Secondly, if $\hat{\rho}_1$ is a density operator, $\hat{\rho}_2$ is also a density operator. This statement can be demonstrated by the same arguments as used for (5.141)

Sharp measurements are always repeatable in the sense that a second measurement of the same observable right after the first one produces the same value. During the first measurement the state collapses into an eigenstate of the self-adjoint operator corresponding to the measured eigenvalue. The second measurement then has a certain outcome equal to the same eigenvalue. If the measurement is not sharp, however, a second measurement after collapse into (5.144) results in a value k with probability

$$P_{\rho_1}(k|j) = \mathrm{tr}(\hat{E}_Z(k)\hat{\rho}_2(j)) = \frac{\mathrm{tr}(\hat{R}_Z(k)\hat{R}_Z(j)\hat{\rho}_1\hat{R}_Z(j)^\dagger\hat{R}_Z(k)^\dagger)}{\mathrm{tr}(\hat{E}_Z(j)\hat{\rho}_1)}. \tag{5.145}$$

In general, this expression is non-zero for $k \neq j$. Such a measurement is **non-repeatable**.

5.4.1.4 Degeneracy vs. unsharpness

Self-adjoint operators with degenerate spectra share some features with unsharp positive-operator valued measures because they cannot distinguish between some orthogonal states. For instance, a matrix operator A on a 3-dimensional Hilbert space with eigenvalues $\lambda_1 = \lambda_2$ in basis states $|1\rangle$ and $|2\rangle$, and eigenvalue λ_3 in basis state $|3\rangle$ has the projection operator

$$P_1 = \begin{pmatrix} 1 & 0 & 0 \\ 0 & 1 & 0 \\ 0 & 0 & 0 \end{pmatrix} \tag{5.146}$$

onto the eigenspace with eigenvalue λ_1. This operator is identical with the combined effect (5.112) we obtained in our previous example of an unsharp positive-operator valued measure. However, in the case of A, the inability to distinguish between vectors in the 2-dimensional eigenspace does not imply unsharpness because each vector is associated with the same eigenvalue λ_1 that determines measurement outcomes. The previous example (5.112), by contrast, was obtained from an unsharp positive-operator valued measure (5.111) in which two different measurement results could be obtained, λ_1 or λ_2.

Continuing with the self-adjoint A and assuming that we have a pure state $|\psi\rangle = 3^{-1/2}(1,1,1)^T$ before a measurement, the state collapses into $|\psi'\rangle = 2^{-1/2}(1,1,0)^T$ if we measure λ_1, and into $(0,0,1)^T$ if we measure λ_3: With the standard root $P_1 = P_1^\dagger P_1$ of the projection operator P_1, the density operator $\rho = |\psi\rangle\langle\psi|$ collapses into

$$\rho' = \frac{P_1 \rho P_1}{\mathrm{tr}(P_1 \rho)} = \frac{1}{2}\begin{pmatrix} 1 & 1 & 0 \\ 1 & 1 & 0 \\ 0 & 0 & 0 \end{pmatrix} = \frac{1}{2}\begin{pmatrix} 1 \\ 1 \\ 0 \end{pmatrix}\begin{pmatrix} 1 & 1 & 0 \end{pmatrix}. \tag{5.147}$$

We measure $\lambda_1 = \lambda_2$ with probability $2/3$, and λ_3 with probability $1/3$. The expectation value of multiple measurement results equals $\langle\lambda\rangle = \frac{2}{3}\lambda_1 + \frac{1}{3}\lambda_3$.

We now consider a positive-operator valued measure on the same Hilbert space, for an observable that can take three values λ_1, λ_2 and λ_3, defined by

$$E_1 = \frac{1}{2}\begin{pmatrix} 1 & 0 & 0 \\ 0 & 1 & 0 \\ 0 & 0 & 0 \end{pmatrix} = E_2 \quad , \quad E_3 = \begin{pmatrix} 0 & 0 & 0 \\ 0 & 0 & 0 \\ 0 & 0 & 1 \end{pmatrix}. \tag{5.148}$$

The expectation value of multiple measurements of the corresponding observable equals

$$\langle\lambda\rangle = \langle\psi|M_1|\psi\rangle = \frac{1}{3}(\lambda_1 + \lambda_2 + \lambda_3) \tag{5.149}$$

using the same state as before and the first moment

$$M_1 = \lambda_1 E_1 + \lambda_2 E_2 + \lambda_3 E_3 = \begin{pmatrix} \frac{1}{2}(\lambda_1+\lambda_2) & 0 & 0 \\ 0 & \frac{1}{2}(\lambda_1+\lambda_2) & 0 \\ 0 & 0 & \lambda_3 \end{pmatrix}. \tag{5.150}$$

If $\lambda_1 = \lambda_2$, the expectation value equals our result for the self-adjoint A. Moreover, we have the same collapse state (5.147) if we obtain the value λ_1 in a single measurement and use the root $R_1 = \sqrt{2}E_1$, even if $\lambda_1 \neq \lambda_2$. However, in this non-degenerate case, the measurement is not repeatable: If we measure λ_1 in a first measurement, a subsequent measurement on the collapse state returns a different value λ_2 with probability $1/2$. In the degenerate case of a self-adjoint operator, we always re-obtain the value we measured in a first measurement.

In addition to being non-repeatable, the measurement described by the positive-operator valued measure is not sharp: If we continue measuring on the collapse state $|\psi'\rangle$, the variance of measurement outcomes is given by

$$\sigma_\lambda^2 = \langle\psi'|M_2|\psi'\rangle - \langle\psi'|M_1|\psi'\rangle^2 = \frac{1}{2}(\lambda_1^2 + \lambda_2^2) - \frac{1}{4}(\lambda_1 + \lambda_2)^2 = \frac{1}{4}(\lambda_1 - \lambda_2)^2 \tag{5.151}$$

using the second moment $M_2 = \lambda_1^2 E_1 + \lambda_2^2 E_2 + \lambda_3^2 E_3$. This value is zero, implying a sharp measurement, if and only if $\lambda_1 = \lambda_2$, which brings us back to the self-adjoint operator with a degenerate spectrum.

5.4.2 STERN–GERLACH EXPERIMENTS

A Stern–Gerlach experiment can be used as a specific example of measurements with finitely many discrete and sharp outcomes. It can be performed several times in a row with magnetic fields in different directions, as illustrated in Figure 5.2, such that we can test implications of state collapse. The intensity of an individual beam after it has passed a magnetic field is proportional to the probability of measuring one of the two values $\mathfrak{m}_s\hbar = \pm\frac{1}{2}\hbar$ of $(\vec{B}/|\vec{B}|)\cdot\vec{S}$. These probabilities in a single measurement, or a single pass through a magnetic field, are determined by the complex norms squared

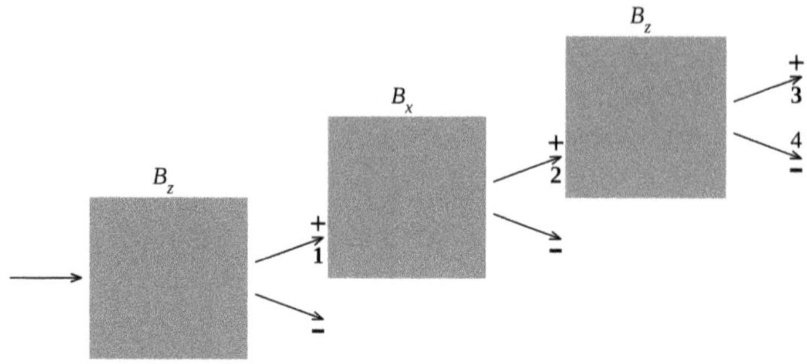

Figure 5.2 A beam of spinning particles sent through a succession of magnetic fields pointing in different directions.

of coefficients of a pure state expanded in the eigenbasis of $(\vec{B}/|\vec{B}|)\cdot\hat{\vec{S}}$. If different directions of the magnetic field are used in successive measurements, eigenbasis transformations must be applied in order to determine the relevant coefficients.

As indicated in Figure 5.2, we will consider magnetic fields in the z and x-direction, respectively, and therefore use the matrices

$$S_z = \frac{1}{2}\hbar \begin{pmatrix} 1 & 0 \\ 0 & -1 \end{pmatrix} \tag{5.152}$$

with eigenstates

$$\{|1/2,1/2\rangle, |1/2,-1/2\rangle\} \leftrightarrow \left\{ \begin{pmatrix} 1 \\ 0 \end{pmatrix}, \begin{pmatrix} 0 \\ 1 \end{pmatrix} \right\} \tag{5.153}$$

and

$$S_x = \frac{1}{2} \begin{pmatrix} 0 & 1 \\ 1 & 0 \end{pmatrix} \tag{5.154}$$

with eigenstates

$$\begin{aligned} \{v_+, v_-\} &= \left\{ \frac{1}{\sqrt{2}}(|1/2,1/2\rangle + |1/2,-1/2\rangle), \frac{1}{\sqrt{2}}(|1/2,1/2\rangle - |1/2,-1/2\rangle) \right\} \\ &\leftrightarrow \left\{ \frac{1}{\sqrt{2}} \begin{pmatrix} 1 \\ 1 \end{pmatrix}, \frac{1}{\sqrt{2}} \begin{pmatrix} 1 \\ -1 \end{pmatrix} \right\}. \end{aligned} \tag{5.155}$$

We have the transformation

$$v_\pm = \frac{1}{\sqrt{2}}(|1/2,1/2\rangle \pm |1/2,-1/2\rangle) \tag{5.156}$$

from the S_z-eigenbasis to the S_x-eigenbasis, and the inverse transformation

$$|1/2,\pm 1/2\rangle = \frac{1}{\sqrt{2}}(v_+ \pm v_-). \tag{5.157}$$

In the experiment illustrated in Figure 5.2, the beam marked as "1" is in the eigenstate $|1/2,1/2\rangle$ of S_z because it has just passed through a magnetic field in the z-direction. It then moves through a magnetic field in the x-direction, which measures the component S_x by splitting up the beam a second time. Since the initial state has the expansion

$$|1/2,1/2\rangle = \frac{1}{\sqrt{2}}(v_+ + v_-) = c_+ v_+ + c_- v_- \tag{5.158}$$

with $c_\pm = 1/\sqrt{2}$, we measure the values of $S_x = \pm\frac{1}{2}\hbar$ with probability $|c_\pm|^2 = 1/2$. Correspondingly, we expect two beams with the same intensity after the second magnetic field.

The beam marked as "2" is then in the eigenstate v_+ of S_x with eigenvalue $\frac{1}{2}\hbar$. The final measurement is again of S_z because the final magnetic field points in the z-direction. The initial state for this magnetic field equals

$$v_+ = \frac{1}{\sqrt{2}}(|1/2,1/2\rangle + |1/2,-1/2\rangle) = d_+|1/2,1/2\rangle + d_-|1/2,-1/2\rangle \tag{5.159}$$

with $d_\pm = 1/\sqrt{2}$. Again, we obtain two beams, "3" and "4" with equal intensities because the probabilities for measuring $S_z = \pm\frac{1}{2}\hbar$ in the last step equal $|d_\pm|^2 = 1/2$.

Even though the first magnetic field was used to eliminate all ions with $S_z = -\frac{1}{2}\hbar$, the second measurement of a non-commuting observable, $[S_x, S_z] \neq 0$, restores some ions with this spin value as measured in the last step. This outcome would also be possible classically because the magnetic field exerts a torque on the magnetic dipole moment, implying precession such that the spin direction in the z-direction is not conserved if the magnetic field points in the x-direction. However, in classical physics the relative intensity of the beams implied by precession depends on how much time the ions spend in the magnetic field as well as the field strength. In quantum mechanics, the intensities in the final step of the setup shown in Figure 5.2 are always equal, and they do not depend on the time spent in the magnetic field.

5.4.3 THE MEASUREMENT PROBLEM

If we measure a sharp observable A and find one of the eigenvalues λ of $\hat{A}$ as a result, the initial state $|\Psi\rangle$ turns into the eigenstate $|e_\lambda\rangle$ because we now know the value. A second measurement right after the first one must reproduce the value λ because knowledge of the value after a sharp measurement is part of the complete state describing the system. A certain outcome in a follow-up experiment is possible only if the system is in the corresponding eigenstate. The state is therefore changed by a measurement, not only because measurements always require interactions with the system but also because they update our information about the system.

The transition from an initial $|\Psi\rangle$ to $|e_\lambda\rangle$ is an example of the collapse of the wave function. The vector components c_λ of the state in the eigenbasis of $\hat{A}$ may initially be non-zero for several λ or a whole range of λ, but after the measurement this support of the function c_λ collapses to a single non-zero value for the measured λ. The state in the eigenbasis of $\hat{A}$ therefore turns into a Kronecker delta, $c_\lambda = \delta_{\lambda\lambda_0}$ if the value λ_0 has been measured. (A state does not collapse into a delta function because only observables with discrete spectra can have sharp measurements.)

The **measurement problem** was identified by John von Neumann in 1932 by attempting to formulate a combined quantum description of system and measurement device. We then have system states $|e_\lambda\rangle$ or their superpositions, and an initial device state $|\Phi\rangle$. Their combination can be written as a product $|e_\lambda\rangle|\Phi\rangle$, given by a tensor product in the mathematical formulation. The product means that the state $|\Psi\rangle|\Phi\rangle$ gives us a combined probability distribution or density $||\Psi||^2 \cdot ||\Phi||^2$ for system and device.

A measurement implies that the device state changes in such a way that we can recognize the measured value. An initial device state $|\Phi\rangle$ is therefore turned into a state $|\Phi_\lambda\rangle$, indicating that the device shows the result λ. If we first assume that the system is initially in an eigenstate $|e_\lambda\rangle$, the measurement process can be described as the transition

$$|e_\lambda\rangle|\Phi\rangle \longrightarrow |e_\lambda\rangle|\Phi_\lambda\rangle \tag{5.160}$$

for an ideal measurement that does not change the system state $|e_\lambda\rangle$.

Details of the transition would mathematically be obtained from a complicated solution of the time-dependent Schrödinger equation that includes all interactions between system and device relevant for the measurement. Here, we only need to know that this transition is linear because the

Schrödinger equation is a linear partial differential equation. Therefore, if the system is in a superposition of eigenstates, the measurement transition looks like

$$\sum_{\lambda} c_\lambda |e_\lambda\rangle|\Phi\rangle \longrightarrow \sum_{\lambda} c_\lambda |e_\lambda\rangle|\Phi_\lambda\rangle \,. \tag{5.161}$$

However, the result is *not* a device showing a unique value λ and a system in a single collapsed state $|e_\lambda\rangle$. Rather, both system and device remain in a superposition, the device showing several possible outcomes at the same time. Schrödinger's cat is an example of such a measurement device in limbo. The collapse of the wave function therefore cannot be a result of linear Schrödinger evolution, and an additional ingredient is necessary.

Hugh Everett in 1957 extended the quantum description from system and device to the whole universe, arguing that, at least in principle, everything in the universe interacts in some way and indirectly with any system we measure. The result of a measurement should therefore be multiple universes in superposition. The collapse of the wave function happens only because we are forced to stay within one universe. Unfortunately, the same limitation makes it impossible to test the proposal by observations.

A middle way is given by **decoherence**, which goes beyond system and device by including an "environment" of a large number of weakly interacting objects, such as air molecules, heat radiation, or cosmic microwaves, neutrinos, or gravitational waves. We cannot remove their influence on the system, which may be weak for a single ingredient in, say, cosmic neutrinos, but can nevertheless be relevant because of their sheer number. The system-device state is not isolated but constitutes an open quantum system.

Since it is impossible to keep track of the environment ingredients in a measurement, there is an additional step that amounts to ignoring these degrees of freedom, striking out their state in the extended measurement process

$$\sum_{\lambda} c_\lambda |e_\lambda\rangle|\Phi\rangle|E\rangle \longrightarrow \sum_{\lambda} c_\lambda |e_\lambda\rangle|\Phi_\lambda\rangle|E_\lambda\rangle \longrightarrow \sum_{\lambda} c_\lambda |e_\lambda\rangle|\Phi_\lambda\rangle|\not{E}_\lambda\rangle \tag{5.162}$$

where the first step is linear evolution with the environment state $|E\rangle$ included, and $|\not{E}_\lambda\rangle$ means ignoring this contribution to the state when we read the measurement result.

The mathematical procedure for removing a state from a tensor product, summing over all its vector components in a basis, is the **partial trace**. This step is not part of linear evolution, and it is irreversible because it removes information. Therefore, including this step eliminates the obvious contradiction between linear evolution and the collapse of the wave function. Detailed calculations show that the removal can indeed imply that states collapse to at least a small range of eigenvalues around the measured value, and they also show how long it takes until the state can be considered collapsed. This time can be measured in quantum experiments that have created superpositions of large molecules. Decoherence may therefore be a solution to the measurement problem.

5.4.4 REPRESENTATIONS

Properties of self-adjoint operators are not only important for a description of sharp measurements but also for choosing alternative representations of states. A state can be expressed in any eigenbasis, even if it belongs to a self-adjoint operator with continuous spectrum. Choosing a suitable basis may simplify derivations of probabilities and other quantities. Formal aspects of eigenfunctions related to a continuous spectrum can also be useful in constructions of positive-operator valued measures.

5.4.4.1 Discrete spectrum

If $\hat{A}$ has a discrete spectrum of eigenvalues λ_n with eigenvectors $|e_n\rangle$, such that $\hat{A}|e_n\rangle = \lambda_n|e_n\rangle$ for a discrete range, say $n = 1,2,\ldots$, the completeness property takes the form

$$|\Psi\rangle = \sum_{n=1}^{\infty} c_n|e_n\rangle \tag{5.163}$$

for any state $|\Psi\rangle$. We compute the coefficients

$$c_n = \langle e_n|\Psi\rangle \tag{5.164}$$

if we use orthonormality of the eigenvector basis,

$$\langle e_n|e_m\rangle = \delta_{nm}\,. \tag{5.165}$$

As a positive-operator valued measure, we can represent the corresponding observable A by the family of projection operators

$$\hat{E}_{\mathbb{Z}_+}(n) = |e_n\rangle\langle e_n|\,. \tag{5.166}$$

The probability of finding an eigenvalue λ_n in a measurement of A on a system in the state $|\Psi\rangle$ is equal to

$$P(n) = |c_n|^2 = |\langle e_n|\Psi\rangle|^2 = \langle\Psi|\hat{E}_{\mathbb{Z}_+}(n)|\Psi\rangle\,. \tag{5.167}$$

The set of all probabilities for different outcomes should be a probability distribution, normalized such that $\sum_{n=1}^{\infty} P(n) = 1$. We can show this by using normalization of the system state $|\Psi\rangle$:

$$1 = \langle\Psi|\Psi\rangle = \sum_{n=1}^{\infty}\sum_{m=1}^{\infty} c_n^* c_m \langle e_n|e_m\rangle = \sum_{m=1}^{\infty} |c_n|^2 = \sum_{n=1}^{\infty} P(n) \tag{5.168}$$

if we again apply orthonormality. A direct derivation uses the positive-operator valued measure:

$$\sum_{n=1}^{\infty} P(n) = \left\langle \Psi \left| \sum_{n=1}^{\infty} \hat{E}_{\mathbb{Z}_+}(n) \right| \Psi \right\rangle = \langle\Psi|\hat{\mathbb{I}}|\Psi\rangle = 1\,. \tag{5.169}$$

This statistical property is realized only if the system state $|\Psi\rangle$ is strictly normalized.

We can also show that $P(n)$ is the correct probability distribution for measurements of A or its eigenvalues λ. The A-expectation value in a state $|\Psi\rangle$ is given by $\langle\hat{A}\rangle = \langle\Psi|\hat{A}|\Psi\rangle$, which we can compare with the expectation value $\langle\lambda\rangle = \sum_{n=1}^{\infty}\lambda_n P(n)$. These two expressions are always the same for a probability distribution $P(n)$ derived from the state $|\Psi\rangle$:

$$\begin{aligned}\langle\hat{A}\rangle &= \langle\Psi|\hat{A}|\Psi\rangle = \langle\Psi|\hat{A}\left(\sum_{n=1}^{\infty} c_n|e_n\rangle\right) = \sum_{n=1}^{\infty} c_n\langle\Psi|\hat{A}|e_n\rangle \\ &= \sum_{n=1}^{\infty} c_n\lambda_n\langle\Psi|e_n\rangle = \sum_{n=1}^{\infty} c_n\lambda_n c_n^* - \sum_{n=1}^{\infty}\lambda_n P(n)\,.\end{aligned} \tag{5.170}$$

This result equals the expectation value of the first moment $\hat{M}_1(\hat{E}_{\mathbb{Z}_+}) = \sum_{n=1}^{\infty}\lambda_n\hat{E}_{\mathbb{Z}_+}(n)$ of $\hat{E}_{\mathbb{Z}_+}(n)$ in the state $|\Psi\rangle$.

We now have additional methods to revisit the collapse state in the case of a degenerate discrete spectrum, (5.137). If the first n eigenvalues are degenerate with an eigenstate that cannot be distinguished by a measurement of some observable A, we have a positive-operator valued measure $\hat{E}_{\mathbb{N}}$ with the effect

$$\hat{E}_{\mathbb{N}}((1,\ldots,n)) = \sum_{m=1}^{n} |e_m\rangle\langle e_m|\,, \tag{5.171}$$

a projection operator. For a projection operator, a simple choice of a root is the operator itself, $\hat{R} = \hat{E}_{\mathbb{N}}((1,\ldots,n))$. If the degenerate eigenvalue is measured, an initial pure state $|\Psi\rangle$ collapses into a state with the density operator

$$\hat{\rho}_2 = \frac{\hat{R}\hat{\rho}_{|\Psi\rangle}\hat{R}^\dagger}{\mathrm{Tr}(\hat{E}_{\mathbb{N}}((1,\ldots,n))\hat{\rho}_{|\Psi\rangle})} \tag{5.172}$$

which can be seen to correspond to the normalized pure state (5.137). More generally, a root is given by $\hat{U}\hat{R}$ with a unitary operator $\hat{U}$. After such a generic measurement, the state may collapse into any normalized linear combination of the degenerate eigenvectors.

5.4.4.2 Continuous and mixed spectra

If $\hat{A}$ has continuous spectrum, we write the eigenvector equation as

$$\hat{A}|e_k\rangle = \lambda_k|e_k\rangle\,, \tag{5.173}$$

parameterizing the eigenvalues by a real number $-\infty < k < \infty$. Completeness tells us that any state $|\Psi\rangle$ can be written as

$$|\Psi\rangle = \int_{-\infty}^{\infty} c_k|e_k\rangle \mathrm{d}k \tag{5.174}$$

with coefficients

$$c_k = \langle e_k|\Psi\rangle \tag{5.175}$$

using orthonormality

$$\langle e_k|e_l\rangle = \delta(k-l) \tag{5.176}$$

in the sense of delta-function normalization.

The possible measurement outcomes λ_k in a continuous range are determined by a probability density

$$p_k = |c_k|^2 = |\langle e_k|\Psi\rangle|^2\,. \tag{5.177}$$

Also here, there is a normalization condition for p_k that follows from normalization of the state $|\Psi\rangle$:

$$1 = \langle\Psi|\Psi\rangle = \int_{-\infty}^{\infty}\int_{-\infty}^{\infty} c_k^* c_l \langle e_k|e_l\rangle \mathrm{d}k\mathrm{d}l = \int_{-\infty}^{\infty} |c_k|^2 \mathrm{d}k\,. \tag{5.178}$$

The last integral being finite implies that the superposition (5.174) of non-normalizable $|e_k\rangle$ is normalizable. The normalizable system state $|\Psi\rangle$ is a generalization of a wave packet as a superposition of non-normalizable plane waves as seen in the example of the free particle. Also the two versions of expectation values agree,

$$\langle A\rangle = \langle\Psi|\hat{A}|\Psi\rangle = \int_{-\infty}^{\infty} \lambda_k|c_k|^2 \mathrm{d}k = \int_{-\infty}^{\infty} \lambda_k p_k \mathrm{d}k\,. \tag{5.179}$$

Measurements of an observable that takes values in a continuous range are unsharp because the eigenfunctions are not strictly normalizable and cannot be obtained as a system state. This property is made explicit by using a positive-operator valued measure

$$\hat{E}_{\mathbb{R}}(K) = \int_K |e_k\rangle\langle e_k| \mathrm{d}k\,. \tag{5.180}$$

Unlike for eigenstates of an operator with discrete spectrum, no equality such as the formal relationship "$|e_k\rangle\langle e_k|\cdot|e_k\rangle\langle e_k| = |e_k\rangle\langle e_k|$" holds here because of non-normalizability: The middle term $\langle e_k|e_k\rangle$ is infinite. Thus, $\hat{E}_{\mathbb{R}}(K)$ is not based on projection operators. However, each operator $\hat{E}_{\mathbb{R}}(K)$ for a given open subset K is a well-defined operator.

Since delta-function normalized eigenstates $|e_k\rangle$ appear in (5.180), they can be used in an explicit derivation of suitable positive-operator valued measures. For instance, for the position operator $\hat{x}$, we have delta-function normalized eigenfunctions $|x\rangle$ such that $\langle x|\Psi\rangle = \Psi(x)$. (In the position representation, we have seen that $|x\rangle$ is a delta function: $\langle x|x'\rangle = \delta(x-x')$.) The action of $\hat{E}_{\mathbb{R}}(X)$ on a general normalized state $|\Psi\rangle$ then evaluates to

$$\hat{E}_{\mathbb{R}}(X)|\Psi\rangle = \int_X |x\rangle\langle x|\Psi\rangle \mathrm{d}x = \int_X \Psi(x)|x\rangle\,. \tag{5.181}$$

We express this state in the position representation by computing

$$\langle x_0|\hat{E}_{\mathbb{R}}(X)|\Psi\rangle = \int_X \Psi(x)\langle x_0|x\rangle \mathrm{d}x = \int_X \Psi(x)\delta(x-x_0)\mathrm{d}x = \begin{cases} \Psi(x_0) & \text{if } x_0 \in X \\ 0 & \text{if } x_0 \notin X \end{cases}\,. \tag{5.182}$$

As a function, (5.181) equals the restriction of $\Psi(x)$ to the subset X. It is normalizable for any open subset $X \subset \mathbb{R}$.

Position measurements in a state $|\Psi\rangle$ have the usual expectation value

$$\langle x\rangle = \langle\Psi|\hat{M}_1(\hat{E}_{\mathbb{R}})|\Psi\rangle = \int_{-\infty}^{\infty} x\langle\Psi|x\rangle\langle x|\Psi\rangle \mathrm{d}x = \int_{-\infty}^{\infty} x|\Psi(x)|^2 \mathrm{d}x\,. \tag{5.183}$$

Measuring the position with a measurement uncertainty given by the width of a region X, the state after the measurement may be written as

$$\hat{\rho}_2 = \frac{\hat{R}(X)\hat{\rho}_1\hat{R}(X)^\dagger}{\mathrm{tr}(\hat{E}_{\mathbb{R}}(X)\hat{\rho}_1)} = \frac{\int_X\int_X |x\rangle\langle x|\hat{\rho}_1|x'\rangle\langle x'|\mathrm{d}x\mathrm{d}x'}{\mathrm{tr}(\int_X |x\rangle\langle x|\hat{\rho}_1 \mathrm{d}x)} \tag{5.184}$$

if we factorize $\hat{E}_{\mathbb{R}}(X) = \hat{R}(X)^\dagger\hat{R}(X)$ with $\hat{R}(X) = \int_X |x\rangle\langle x|\mathrm{d}x$. For a pure state $\hat{\rho}_1 = |\Psi\rangle\langle\Psi|$, the state after the measurement remains pure and has the restricted wave function

$$\Psi_2(x) = \begin{cases} \dfrac{\Psi(x)}{\sqrt{\int_X |\Psi(x)|^2 \mathrm{d}x}} & \text{if } x \in X \\ 0 & \text{if } x \notin X \end{cases} \tag{5.185}$$

An alternative choice of the factorization would be any $\hat{E}_{\mathbb{R}}(X) = (\hat{U}\hat{R}(X))^\dagger\hat{U}\hat{R}(X)$ with a unitary operator $\hat{U}$. For instance, if $\hat{U}(\delta) = \exp(-i\delta\hat{p}/\hbar)$ with a real parameter δ is a translation operator, the state collapses to

$$\begin{aligned} \hat{\rho}_2(\delta) &= \frac{\hat{U}(\delta)\int_X\int_X |x\rangle\langle x|\hat{\rho}_1|x'\rangle\langle x'|\mathrm{d}x\mathrm{d}x'\hat{U}(-\delta)}{\mathrm{tr}(\int_X |x\rangle\langle x|\hat{\rho}_1 \mathrm{d}x)} = \frac{\int_X\int_X |x-\delta\rangle\langle x|\hat{\rho}_1|x'\rangle\langle x'-\delta|\mathrm{d}x\mathrm{d}x'}{\mathrm{tr}(\int_X |x\rangle\langle x|\hat{\rho}_1 \mathrm{d}x)} \\ &= \frac{\int_{X-\delta}\int_{X-\delta} |x\rangle\langle x+\delta|\hat{\rho}_1|x'+\delta\rangle\langle x'|\mathrm{d}x\mathrm{d}x'}{\mathrm{tr}(\int_X |x\rangle\langle x|\hat{\rho}_1 \mathrm{d}x)} \end{aligned} \tag{5.186}$$

where $X-\delta$ is the original range X shifted by δ to the left. The new collapse state therefore assigns the components $\langle x|\hat{\rho}_1|x\rangle$ to a translated range: For a pure state, the new wave function is

$$\Psi_{2,\delta}(x) = \frac{\int_{X-\delta}\Psi(x+\delta)|x\rangle\mathrm{d}x}{\sqrt{\int_X |\Psi(x)|^2\mathrm{d}x}} = \begin{cases} \dfrac{\Psi(x+\delta)}{\sqrt{\int_X |\Psi(x)|^2 \mathrm{d}x}} & \text{if } x+\delta \in X \\ 0 & \text{if } x+\delta \notin X \end{cases}\,. \tag{5.187}$$

The new wave function takes values of $\Psi(x)$ from the old range X and assigns them to the shifted range $X-\delta$. (Therefore, the normalization factor does not change compared with (5.185).) Such a root of the operator-valued measure corresponds to a position measurement with a systematic error that changes the results of a second measurement by an amount of $-\delta$.

There are operators with a mixed spectrum that has both discrete and continuous parts. An example is the Hamilton operator $\hat{H}$ for a potential that has both bound states and scattering states, as in the finite square well. A general state can then be written as

$$|\Psi\rangle = \sum_{n=1}^{N} c_n |\psi_n\rangle + \int_0^\infty c_E |\psi_E\rangle \mathrm{d}E \tag{5.188}$$

with a finite number N of bound states and scattering states parameterized by the energy $E \geq 0$. There are also examples with an infinite number of bound states and a continuum of scattering states, such as the hydrogen atom. In these mixed cases, probabilities for measurement outcomes are determined by a combination of a probability distribution and a probability density, depending on the range of the measured values.

5.4.4.3 Position and momentum representation

If we use a basis, such as an eigenvector basis $|e_\lambda\rangle$ of a Hermitian operator $\hat{A}$, an abstract state $|\Psi\rangle$ is represented by its vector components c_λ. Comparing different representations of the same state allows us to derive transformations for a basis change, which can often simplify calculations. We have seen two operators with explicit sets of eigenvalues and eigenfunctions, given by position $\hat{x}$ and momentum $\hat{p}$. We now derive a basis change between these two representations.

The position operator $\hat{x}$ has eigenfunctions

$$e_{x'}(x) = \delta(x - x') \tag{5.189}$$

with real eigenvalues x'. The x'-component of a state $|\Psi\rangle$ in this representation is therefore given by

$$c_{x'} = \langle e_{x'}|\Psi\rangle = \int_{-\infty}^{\infty} \delta(x - x')\Psi(x,t)\mathrm{d}x = \Psi(x',t) \tag{5.190}$$

if we evaluate a general solution of the Schrödinger equation at a fixed time t. The original x-dependent wave functions we obtained were the components of a general state in the position representation. The components determine the usual probability density

$$|c_{x'}|^2 = |\Psi(x',t)|^2 \tag{5.191}$$

at time t.

The momentum operator $\hat{p}$ has eigenfunctions

$$e_p(x) = \frac{1}{\sqrt{2\pi\hbar}} \exp(ipx/\hbar) \tag{5.192}$$

using delta-function normalization, where the eigenvalue p can be any real number. The p-component of a state $|\Psi\rangle$ at time t is

$$c_p = \langle e_p|\Psi\rangle = \frac{1}{\sqrt{2\pi\hbar}} \int_{-\infty}^{\infty} \exp(-ipx/\hbar)\Psi(x,t)\mathrm{d}x = \Phi(p,t), \tag{5.193}$$

a function proportional to the Fourier transformation of $\Psi(x,t)$ in x. The p-dependent function $\Phi(p.t)$ is referred to as the **momentum-space wave function**, or the wave function in the **momentum representation**. Since Fourier transformation is invertible, we can transform back to the original wave function in the **position representation** by

$$\Psi(x,t) = \frac{1}{\sqrt{2\pi\hbar}} \int_{-\infty}^{\infty} \exp(ipx/\hbar)\Phi(p,t)\mathrm{d}p. \tag{5.194}$$

As another example, the operator $\hat{A} = \hat{H}$ as the Hamilton operator of a system that has only bound states $|\psi_n\rangle$ implies the representation

$$|\Psi\rangle = \sum_{n=1}^{\infty} c_n |\psi_n\rangle \tag{5.195}$$

of a general state by a discrete set of components, c_n. The probability distribution for measuring energy values E_n in a state $|\Psi\rangle$ is given by $P(E_n) = |c_n|^2$. The energy operator in this **energy representation** is diagonal, $\hat{H}_E = E_n$ interpreted as a multiplication operator on the states $|\psi_n\rangle$: $\hat{H}|\Psi\rangle = \sum_{n=1}^{\infty} c_n E_n |\psi_n\rangle$.

The probability density for momentum measurements is given by

$$\rho_p = |c_p|^2 = |\Phi(p,t)|^2 \tag{5.196}$$

at time t. We obtain the momentum expectation value

$$\langle p \rangle = \int_{-\infty}^{\infty} p|\Phi(p,t)|^2 \mathrm{d}p = \int_{-\infty}^{\infty} \Phi(p,t)^* p \Phi(p,t) \mathrm{d}p \tag{5.197}$$

directly from this function, without taking a derivative as required in the position representation. In the momentum representation, the momentum operator $\hat{p}$ is therefore represented as multiplication by p, $\hat{p} = p$. We can compute the expectation value

$$\langle f(p) \rangle = \int_{-\infty}^{\infty} \Phi(p,t)^* f(p) \Phi(p,t) \mathrm{d}p \tag{5.198}$$

for any function of p, including p^2 for the variance of p-measurements or for the kinetic energy.

We should be able to compute expectation values of x in the p-representation, just as we are able to compute p-expectation values in the x-representation of wave functions. Since $\Phi(p,t)$ is not represented in the eigenbasis of $\hat{x}$, this operator is not diagonal and not given by multiplication. We compute the operator by first acting with the original expression $\hat{x}\Psi(x,t) = x\Psi(x,t)$ in the position representation and then using Fourier transformation to find the corresponding operation in the momentum representation:

$$\begin{aligned} \frac{1}{\sqrt{2\pi\hbar}} \int_{-\infty}^{\infty} \exp(-ipx/\hbar) x \Psi(x,t) \mathrm{d}x &= \frac{1}{\sqrt{2\pi\hbar}} \int_{-\infty}^{\infty} \left(-\frac{\hbar}{i} \frac{\partial}{\partial p} \exp(-ipx/\hbar) \right) \Psi(x,t) \mathrm{d}x \\ &= -\frac{\hbar}{i} \frac{\partial}{\partial p} \Phi(p,t) \end{aligned} \tag{5.199}$$

where we obtained the factor of x by taking a p-derivative of $\exp(-ipx/\hbar)$. The operation $-(\hbar/i)\partial/\partial p$ is then x-independent and can be moved in front of the integral.

If we represent Fourier transformation by a basis-change operator $\hat{\mathscr{F}}$ that maps $\Psi(x,t)$ to $\Phi(p,t)$, equation (5.199) can be written as $\hat{\mathscr{F}}\hat{x}_x = \hat{x}_p \hat{\mathscr{F}}$, where $\hat{x}_x$ is the x-operator in the position representation, and $\hat{x}_p$ is the x-operator in the momentum representation. Written in this way,

$$\hat{x}_p = \hat{\mathscr{F}} \hat{x}_x \hat{\mathscr{F}}^{-1} \tag{5.200}$$

takes the form of a similarity transformation, where $\hat{\mathscr{F}}$ is a unitary transformation that preserves the inner product, even though it is no longer a finite-dimensional matrix.

To summarize, in the position representation we have the operators

$$\hat{x}_x = x \quad \text{and} \quad \hat{p}_x = \frac{\hbar}{i} \frac{\partial}{\partial x} \tag{5.201}$$

and in the momentum representation we have the operators

$$\hat{x}_p = -\frac{\hbar}{i}\frac{\partial}{\partial p} \quad \text{and} \quad \hat{p}_p = p\,. \tag{5.202}$$

The momentum representation is useful, for instance, if we try to solve the Schrödinger equation with a linear potential $V(x)$. We obtain a second-order differential equation in the position representation, but a first-order differential equation in the momentum representation, which is easier to solve.

In the momentum representation, the positive-operator valued measure for $\hat{p}$ is given by

$$\hat{E}_{\mathbb{R}}(P) = \int_P |p\rangle\langle p|\mathrm{d}p\,. \tag{5.203}$$

Momentum measurements can then be described just like position measurements. We will discuss some of the formal relationships in the context of time as an observable.

5.5 TIME IN QUANTUM MECHANICS

We have made ample use of the position operator but have not encountered a time operator even though we routinely measure time. Relativity might also be used to suggest that space and time should be treated on a more equal footing, but so far we have not included any relativistic effects for instance in the energy expressions used to define Hamilton operators. We will see later that relativity does not substantially change the basic difference between position and time measurements realized in non-relativistic quantum mechanics. The reason is that we do not measure time directly when we say that we measure time. We rather use position measurements, such as the hands on a clock or the position of Earth on its orbit, and apply established conventions to translate them into what we call time measurements. It is also important to note that there are different kinds of time measurements, including the basic distinction between measuring a moment in time (such as the arrival of a particle at a detector) and measuring a time duration.[1]

On a mathematical level, any expression used to describe time measurements must be different from the position operator because the energy of many systems is bounded from below. Since we have seen expressions such as $\hat{E} = i\hbar\partial/\partial t$, time is related to energy as momentum is related to position: These pairs of observables are canonically conjugate. More precisely, the energy or the Hamilton operator $\hat{H}$ is the observable that corresponds to the unitary family $\hat{U}_{\hat{H}}(t) = \exp(-it\hat{H}/\hbar)$ of time translations. Conversely, a time operator $\hat{t}$, if it were self-adjoint, would be an observable that corresponds to translations $\exp(i\eta\hat{t}/\hbar)$ of the energy. If the energy is bounded, there must be a limit to energy translations, with corresponding restrictions on a time operator that generates this symmetry.

For the self-adjoint position and momentum operators, we have a complete correspondence because $\hat{U}_{\hat{p}}(\delta) = \exp(i\delta\hat{p}/\hbar)$, defined in (3.222), represents the unitary symmetry of spatial translations, while $\hat{U}_{\hat{x}}(\varepsilon) = \exp(-i\varepsilon\hat{x}/\hbar)$ implies momentum translations, as can be seen by analogy with spatial translations if we use the momentum representation of wave functions. Only the direction of translations is different in these two cases, owing to the antisymmetry $[\hat{x},\hat{p}] = -[\hat{p},\hat{x}]$. The correspondence between energy and time cannot be as complete as the correspondence between position and momentum. Position and momentum measurements can take any real number. It may be unlikely to measure a position far from an allowed region of a given potential, but the probability is non-zero in general, unless we consider a simplified and unrealistic potential such as the infinite square well. By contrast, there are many systems in which the energy spectrum is bounded from below, a basic requirement for stability, or is discrete at least in part. On such systems, it is impossible to have a continuous and unitary representation $\exp(i\eta\hat{t}/\hbar)$ of unrestricted energy translations. Therefore, in contrast to $\hat{x}$, there can be no universal self-adjoint time operator $\hat{t}$ valid for any system.

[1] P. Busch: The Time-Energy Uncertainty Relation, `https://arxiv.org/abs/quant-ph/0105049`

5.5.1 TIME AS A POSITIVE-OPERATOR VALUED MEASURE

The preceding conclusion specifically refers to a self-adjoint time operator. If a time operator is not self-adjoint, the exponential $\exp(i\eta\hat{t}/\hbar)$ is not unitary. Non-unitary shift operators are compatible with upper or lower bounds, as we have seen with the lowering operator of the harmonic oscillator. However, we have to make sure that we do not predict complex numbers for measurements of time, a non-trivial task if we cannot use a self-adjoint operator. One way to solve this problem is to assume that the time observable is not represented by an operator but rather by a positive-operator valued measure $\hat{E}_{\mathbb{R}}(T)$.

Since the statistics of measurements do not apply to the time dependence of a wave function $\Psi(x,t)$, we cannot represent time by a direct analogy with the positive-operator valued measure (5.119) for position measurements, integrating $|x\rangle\langle x|$: So far, we have not encountered a meaningful state $|t\rangle$ that could be used in such a measure. Deriving suitable $|t\rangle$ would seem to require a time operator, such that $|t\rangle$ are its eigenstates, bringing us back to the question of what a time operator might be.

At this point, we may drop the first assumption made in our conclusion that "there can be no universal self-adjoint time operator $\hat{t}$ valid for any system." Since time measurements can take different forms, we should look for system-specific rather than universal operators that could represent time. They still cannot be self-adjoint, but they might allow us to construct suitable states $|t\rangle$ to be used in a positive-operator valued measure.

As an example, time and position are strictly related for a free particle with mass m, in a way that also depends on the momentum: The classical expression $t(x,p) = mx/p$ equals the time required for a particle with momentum p to move a distance x. We have operators for $\hat{x}$ and $\hat{p}$, but they do not commute, and we need an inverse of the latter. The inverse momentum operator is defined in the momentum representation, where it multiplies by p^{-1}. In general, one can define the inverse of an operator in its own eigenbasis, in which the inverse operator has the inverse eigenvalues. If the original operator has the eigenvalue zero in its discrete spectrum, the inverse is not defined on the corresponding eigenstates, and therefore it does not have a complete basis and is not densely-defined. However, the inverse is meaningful for an operator with zero in its continuous spectrum. The inverse defined in this way is also Hermitian, just like $\hat{p}$.

The non-commutation property of $\hat{x}$ and $\hat{p}$, or $\hat{p}^{-1}$, implies that we have to choose a symmetric ordering when we define a time operator, such as

$$\hat{t} = \frac{1}{2}m(\hat{x}\hat{p}^{-1} + \hat{p}^{-1}\hat{x})\,. \tag{5.204}$$

In the momentum representation,

$$\hat{t}\Phi(p) = \frac{1}{2}i\hbar m\left(\frac{\mathrm{d}}{\mathrm{d}p}\frac{1}{p} + \frac{1}{p}\frac{\mathrm{d}}{\mathrm{d}p}\right)\Phi(p) = i\hbar m\left(\frac{1}{p}\frac{\mathrm{d}}{\mathrm{d}p} - \frac{1}{2p^2}\right)\Phi(p)\,. \tag{5.205}$$

This operator is Hermitian by construction, but it is not self-adjoint. The eigenfunctions of this operator, given by

$$e_\lambda(p) = \sqrt{\frac{|p|}{2\pi m\hbar}}\exp(-\tfrac{1}{2}i\lambda p^2/(m\hbar)) \tag{5.206}$$

are normalizable (and, in the given form, delta-function normalized) for any real $\lambda = t$ but also for eigenvalues $\lambda = -i\tau$ with negative imaginary part. Such an operator cannot be self-adjoint. Note that the eigenfunctions with eigenvalues $\lambda = -i\tau$ are strictly normalizable, although with a different normalization factor than included in the given expression for $e_\lambda(p)$. The operator therefore has a continuous range of eigenvalues in the discrete part of its spectrum, defined as such by strict normalizability of corresponding eigenfunctions.

In spite of the unusual nature of the spectrum of $\hat{t}$, we can use the eigenfunctions with real eigenvalues in order to construct a positive-operator valued measure by following the example of (5.119). We define

$$\hat{E}_{\mathbb{R}}(T) = \int_T |e_t\rangle\langle e_t|\mathrm{d}t \tag{5.207}$$

with components

$$\langle p|\hat{E}_{\mathbb{R}}(T)|p'\rangle = \frac{1}{2\pi m\hbar}\int_T \sqrt{|pp'|}\exp(-\tfrac{1}{2}it(p^2-p'^2)/(m\hbar))\mathrm{d}t \tag{5.208}$$

in the momentum representation. Employing the first moment, the time expectation value in a state $|\phi\rangle$ equals

$$\begin{aligned}\langle t\rangle_{|\phi\rangle,T} &= \langle\phi|\hat{M}_1(\hat{E}_{\mathbb{R}}(T))|\phi\rangle = \frac{1}{2\pi m\hbar}\int_T t\left|\int_{-\infty}^{\infty}\sqrt{|p|}\exp(-\tfrac{1}{2}itp^2/(m\hbar))\phi(p)\mathrm{d}p\right|^2\mathrm{d}t \\ &= \frac{1}{2\pi m\hbar}\int_T t\left|\int_{-\infty}^{\infty}\sqrt{|p|}\Phi(p,t)\mathrm{d}p\right|^2\mathrm{d}t \end{aligned}\tag{5.209}$$

using evolution with the free-particle Hamilton operator in the last step:

$$\Phi(p,t) = \exp(-\tfrac{1}{2}itp^2/(m\hbar))\phi(p)\,. \tag{5.210}$$

The classical meaning of $t(x,p)$ as the time required for a particle with momentum p to move a distance x cannot be maintained directly in quantum mechanics because we never have sharp values for both x and p, owing to the uncertainty relation. Instead, we may interpret $\langle t\rangle_{|\phi\rangle,T}$ as the average time for particles with momentum p to move some distance x, both weighted by the wave function in the respective representations. The time range T represents a cut-off that takes into account the finite duration of measurements.

The expression (5.209) has the formal nature of a time expectation value of a t-dependent wave function defined by

$$\Xi(t) = \frac{1}{\sqrt{2\pi m\hbar}}\int_{-\infty}^{\infty}\sqrt{|p|}\Phi(p,t)\mathrm{d}p\,. \tag{5.211}$$

However, $\Xi(t)$ is, in general, not guaranteed to be normalizable. Using the time-dependent exponential in (5.209), it may be viewed as a Fourier transform, although not of $\Phi(p)$ which is normalized, a property that would be preserved by Fourier transformation, but rather of $|z|^{-1/4}\Phi(\sqrt{|z|})$ after substituting $z=\mathrm{sgn}(p)p^2$. If $\Phi(p)$ has a Gaussian factor in p, it would be replaced by an exponential of a linear function in $|z|$ which falls off more slowly at infinity. The transformation from $\Phi(p)$ to $\Xi(t)$ is therefore formal and does not represent an example of a basis change in the infinite-dimensional function space. For this reason, which is another example of the peculiarity of time operators, we should use only moments with finite time ranges T, as in (5.209).

5.5.2 TIME UNCERTAINTY

In the momentum representation (5.205), we compute the time-energy commutator

$$[\hat{t},\hat{p}^2/(2m)] = i\hbar\hat{\mathbb{I}} \tag{5.212}$$

for the free particle. Our derivation of the uncertainty relation remains valid if the time operator is Hermitian but not self-adjoint, such that we have

$$\sigma_t\sigma_{p^2/(2m)} \geq \frac{1}{2}\hbar \tag{5.213}$$

in states in which the time operator has a meaningful variance. The uncertainty relation requires a time operator and does not have a general analog if the time observable is represented by a positive-operator valued measure. Moreover, the specific form of the time operator depends on the system. It is therefore not guaranteed that the standard form of the uncertainty relation (5.213) holds for an arbitrary Hamilton operator representing the energy observable, together with a suitable time observable.

An alternative definition of an uncertainty relation makes use of the time derivative (3.198), which as a derivative depends on time and also includes a commutator with the Hamilton operator, now valid for any system. The commutator $[\hat{H},\hat{A}]$ for some choice of $\hat{A}$ then implies

$$\sigma_E^2\sigma_A^2 \geq \left(\frac{1}{2i}\langle[\hat{H},\hat{A}]\rangle\right)^2 = \frac{\hbar^2}{4}\left(\frac{\mathrm{d}\langle\hat{A}\rangle}{\mathrm{d}t}\right)^2 . \tag{5.214}$$

We rewrite this inequality as

$$\sigma_E\Delta_A t \geq \frac{\hbar}{2} \tag{5.215}$$

by defining

$$\Delta_A t = \frac{\sigma_A}{|\mathrm{d}\langle\hat{A}\rangle/\mathrm{d}t|} \tag{5.216}$$

for a given $\hat{A}$. This uncertainty is not statistical because there is, in general, no operator and no probability distribution or density for t. Unlike a standard deviation, the expression depends not only on the state but also on the operator $\hat{A}$. We can interpret $\Delta_A t$ as the time it takes for $\langle\hat{A}\rangle$ to change by an amount equal to σ_A. It is therefore the time required for dynamical changes of $\langle\hat{A}\rangle$ to become noticeably different from statistical fluctuations in A-measurements.

Even though the energy-time uncertainty relation is not as universal as other uncertainty relations derived from pairs of operators, it has important applications in particle physics. For an unstable particle, the lifetime τ provides a maximum $\Delta_A t$ available for any measurement. If $\Delta_A t$ cannot be greater than τ, the energy uncertainty σ_E cannot be smaller than $\Delta E = \frac{1}{2}\hbar/\tau$. The energy E or the mass $m = E/c^2$ of an unstable particle therefore is not sharply defined and has a fundamental uncertainty. High-energy accelerators often produce unstable particles, which can be noticed in an increase of the scattering probability, or a component of the scattering matrix, for this type of particle; see the end of Section 2.5.3 on page 88. The scattering probability has a local maximum at an energy E just sufficient to supply the mass E/c^2 of the unstable particle, but the uncertainty ΔE implies that the probability is elevated in an extended range around mc^2. Measuring the width of this peak, we obtain the value of ΔE, from which the lifetime $\tau = \frac{1}{2}\hbar/\Delta E$ can be computed. In many cases, this is the only possible way to measure the lifetime because some particles, such as the top quark or the Higgs particle, are so unstable that they decay almost immediately and do not leave a visible track in the detector from which the lifetime could be derived directly.

5.5.3 TIME AND CONSTRAINTS

If time is treated on an equal footing with the position in space, we should work with wave functions $\bar{\Psi}(x,t)$ in which both x and t are statistical. Such wave functions should be normalized by

$$\int_{-\infty}^{\infty}\int_{-\infty}^{\infty}|\bar{\Psi}(x,t)|^2\mathrm{d}x\mathrm{d}t = 1\,, \tag{5.217}$$

just as we integrate over both x and y if we include two spatial dimensions. In this case, there can be independent self-adjoint operators $\hat{x}$ and $\hat{t}$ which act in the space-time position representation by

$$\hat{x}\bar{\Psi}(x,t) = x\bar{\Psi}(x,t) \quad , \quad \hat{t}\bar{\Psi}(x,t) = t\bar{\Psi}(x,t)\,. \tag{5.218}$$

There are two independent operators, $\hat{p}$ and $\hat{E}$, with the position representation

$$\hat{p}\bar{\Psi}(x,t) = \frac{\hbar}{i}\frac{\partial\bar{\Psi}(x,t)}{\partial x} \quad , \quad \hat{E}\bar{\Psi}(x,t) = i\hbar\frac{\partial\bar{\Psi}(x,t)}{\partial t} \tag{5.219}$$

such that we have the commutators

$$[\hat{x},\hat{p}] = i\hbar \quad , \quad [\hat{t},\hat{E}] = -i\hbar\,. \tag{5.220}$$

The sign difference may be removed by absorbing a factor of -1 in the definition of $\hat{E}$, but the choice used here is more suitable from the perspective of special relativity; see Section 12.1.1. These commutators imply standard uncertainty relations for both (σ_x,σ_p) and (σ_t,σ_E).

While states $\bar{\Psi}(x,t)$ depend on both space and time, there is no specific relationship between these dependencies. Therefore, such states do not describe evolving or dynamical systems in which the relationship between x and t is determined by equations of motion or a Hamilton operator. The Hilbert space of these wave functions is called the **kinematical Hilbert space**, in contrast to a **physical Hilbert space** in which specific evolution equations hold for a given Hamilton operator. Even though the time operator $\hat{t}$ is self-adjoint on the kinematical Hilbert space, it does not represent an observable because all measurements are performed on dynamical systems and must therefore correspond to operators on a physical Hilbert space.

On the kinematical Hilbert space, the energy operator $\hat{E}$ is completely independent of $\hat{x}$ and $\hat{p}$ because it is represented by a partial derivative with respect to t. A Hamilton operator $\hat{H}$, by contrast, is always defined as a combination of $\hat{x}$ and $\hat{p}$, such as $\hat{p}^2/(2m)+V(\hat{x})$. Dynamical restrictions are imposed on $\bar{\Psi}(x,t)$ by introducing the constraint that the operator $\hat{E}$, acting on suitable states, should always produce the same result as the system's Hamilton operator $\hat{H}$:

$$\hat{C}\bar{\Psi}(x,t) = (\hat{E}-\hat{H})\bar{\Psi}(x,t) = 0 \tag{5.221}$$

with the **constraint operator** $\hat{C}$. This equation is equivalent to

$$i\hbar\frac{\partial\bar{\Psi}(x,t)}{\partial t} = \hat{H}\bar{\Psi}(x,t) \tag{5.222}$$

which is nothing but the time-dependent Schrödinger equation. In the language introduced in the context of (4.203), where we already encountered a constraint operator, solutions to the constraint equation $\hat{C}\bar{\Psi}(x,t) = 0$ are gauge-invariant.

States that obey the constraint equation form the physical Hilbert space. However, since Hamiltonian evolution is unitary, these states are not normalizable elements of the kinematical Hilbert space: They evolve according to $\bar{\Psi}(x,t) = \exp(-it\hat{H}/\hbar)\bar{\Psi}(x,0)$ such that $\int_{-\infty}^{\infty}|\bar{\Psi}(x,t)|^2\mathrm{d}x = \int_{-\infty}^{\infty}|\bar{\Psi}(x,0)|^2\mathrm{d}x$ is independent of t. The t-integration required for the normalization condition (5.217) of the kinematical Hilbert space then results in an infinite value.

It is easy to solve this problem because we already know how to represent normalizable evolving states: We do *not* integrate over t. The physical Hilbert space can therefore be defined as the set of functions $\Psi(x,t_0) = \bar{\Psi}(x,t_0)$ obtained by evaluating all functions of the kinematical Hilbert space at a fixed time, $t = t_0$. Their inner product corresponds to the normalization integral $\int_{-\infty}^{\infty}|\Psi(x,t_0)|^2\mathrm{d}x$. Unitary evolution implies that we obtain the same value for any choice of t_0, which was initially arbitrary but always results in the same physical Hilbert space. The physical Hilbert space is therefore the Hilbert space we worked with for solutions of the time-independent Schrödinger equation. Unfortunately, this straightforward transition is deceptive: In general, constructing a physical Hilbert space for a given constraint operator is a challenging task. The procedure looks easy here because the constraint is linear in $\hat{E}$ and we assume that the Hamilton operator is independent of $\hat{t}$.

On the physical Hilbert space defined with a fixed choice of $t_0 \neq 0$, the kinematical time operator simply multiplies a wave function by the constant t_0. This factor is removed if we normalize the state after acting with this operator. Therefore, the kinematical time operator is reduced to the identity operator on the physical Hilbert space. The construction does not imply a self-adjoint time operator that could represent a measurable observable, in accordance with our previous considerations about the general unavailability of self-adjoint time operators. There is no such issue for the position operator because it retains its non-trivial action when we restrict multiplication by x to a wave function at constant $t = t_0$.

If we compare the actions with $\hat{x}$ on wave functions $\bar{\Psi}(x,t)$ at two different values of t, the results differ because the wave function evolves. This result is expressed by the fact that $[\hat{x},\hat{H}] \neq 0$, and therefore $[\hat{x},\hat{C}] \neq 0$. A non-zero commutator makes it impossible to derive a physical position operator directly from the kinematical $\hat{x}$ by restricting it to the solution space of the constraint equation: A solution satisfies the equation $\hat{C}\bar{\Psi}(x,t) = 0$, but if we first act with the kinematical position operator, we no longer have a solution because

$$\hat{C}\hat{x}\bar{\Psi}(x,t) = \hat{x}\hat{C}\bar{\Psi}(x,t) - [\hat{x},\hat{C}]\bar{\Psi}(x,t) = -[\hat{x},\hat{C}]\bar{\Psi}(x,t) \neq 0\,. \tag{5.223}$$

The same equation shows which kinematical operators can be restricted to the physical Hilbert space: If a kinematical operator $\hat{D}$ obeys $[\hat{D},\hat{C}] = 0$, $\hat{D}\bar{\Psi}(x,t)$ is a solution whenever $\bar{\Psi}(x,t)$ is a solution. Kinematical operators that commute with the constraint operator therefore have a well-defined action on the physical Hilbert space and can represent meaningful observables. Operators that obey this condition are called **Dirac observables**.

For the free particle, examples of Dirac observables are given by $\hat{p}$ and

$$\hat{Q} = \hat{x} - \hat{t}\hat{p}/m\,, \tag{5.224}$$

as shown by a direct calculation. A Dirac observable represents a gauge-invariant quantity with respect to the constraint operator because $[\hat{D},\hat{C}] = 0$ implies that

$$\exp(i\varepsilon\hat{C}/\hbar)\hat{D}\exp(-i\varepsilon\hat{C}/\hbar) = \hat{D}\,. \tag{5.225}$$

It follows that $\hat{D}$ has the same expectation value, variance, and higher moments on all states in the 1-parameter family $\exp(-i\varepsilon\hat{C}/\hbar)\bar{\Psi}(x,t)$, and in this sense it is a conserved quantity.

For the Dirac observable $\hat{Q}$ of the free particle, we may consider its value as the initial-position operator because $\hat{Q} = \hat{x}$ for any state that is annihilated by the kinematical time operator $\hat{t}$. Rewriting equation (5.224) for the Dirac observable, the operator relation

$$\hat{x} = \hat{Q} + \hat{t}\hat{p}/m \tag{5.226}$$

then resembles the classical solution. We can use the same relation in order to derive a time operator from the Dirac observable and the kinematical $\hat{x}$ and $\hat{p}$. We divide (5.226) by $\hat{p}$, but this should be done in a symmetric fashion: We multiply the equation by $|\hat{p}|^{-1/2}$ from the left and from the right, and then multiply by $\mathrm{sgn}(\hat{p})$. The inverse square root and the sign function of an operator are defined by assigning the inverse square root and the sign, respectively, of each eigenvalue to the corresponding eigenfunction, just as we earlier defined the inverse of $\hat{p}$. We obtain

$$\hat{t} = m\,\mathrm{sgn}(\hat{p})|\hat{p}|^{-1/2}(\hat{x} - \hat{Q})|\hat{p}|^{-1/2} \tag{5.227}$$

which, for vanishing initial position $\hat{Q}$, can be shown to be equivalent to (5.204) by using the momentum representation, as in (5.205). The time operator (5.204) can therefore be derived from the restriction of the Dirac observable $\hat{Q}$ to the physical Hilbert space of the free particle.

The constraint operator, which is self-adjoint on the kinematical Hilbert space, can formally be used to define the symmetry transformation $\hat{U}_{\hat{C}}(\delta) = \exp(-i\delta\hat{C}/\hbar)$. Acting on states in the physical Hilbert space, the constraint equation implies that $\hat{U}_{\hat{C}}(\delta) = \hat{\mathbb{I}}$ for all δ. A constraint operator therefore does not imply physical symmetry that could be measured by changes of observables.

Nevertheless, $\hat{U}_{\hat{C}}(\delta)$ implies non-trivial unitary transformations on the kinematical Hilbert space. For this reason, the 1-parameter family of unitary operators obtained by exponentiating a constraint operator are called **gauge transformations** rather than symmetry transformations in the strict sense. In this context, "gauging" means that we change reference variables rather than observables.

Kinematical operators can now be gauge transformed by the general equation (3.224):

$$\hat{A}(\delta) = \exp(-i\delta\hat{C}/\hbar)\hat{A}\exp(i\delta\hat{C}/\hbar)\,. \tag{5.228}$$

The momentum $\hat{p}$ and the energy $\hat{E}$ commute with $\hat{C}$ and are therefore independent of δ, or gauge invariant. Following derivations such as (3.226), we compute

$$\hat{t}(\delta) = \exp(-i\delta\hat{E}/\hbar)\hat{t}\exp(i\delta\hat{E}/\hbar) = \hat{t} + \delta\hat{\mathbb{I}}\,. \tag{5.229}$$

In order to derive the corresponding $\hat{x}(\delta)$, we use the free-particle Hamilton operator and compute

$$\begin{aligned} \hat{x}\exp(-i\delta\hat{H}/\hbar) &= \hat{x}\sum_{n=0}^{\infty}\frac{(-i\delta/(2m\hbar))^n}{n!}\hat{p}^{2n} = \sum_{n=0}^{\infty}\frac{(-i\delta/(2m\hbar))^n}{n!}\left(\hat{p}^{2n}\hat{x} + [\hat{x},\hat{p}^{2n}]\right) \\ &= \sum_{n=0}^{\infty}\frac{(-i\delta/(2m\hbar))^n}{n!}\left(\hat{p}^{2n}\hat{x} + 2i\hbar n\hat{p}^{2n-1}\right) \\ &= \sum_{n=0}^{\infty}\frac{(-i\delta/(2m\hbar))^n}{n!}\hat{p}^{2n}\hat{x} + 2i\hbar\hat{p}\sum_{n=1}^{\infty}\frac{(-i\delta/\hbar)^n}{(n-1)!}\hat{p}^{2(n-1)} \\ &= \exp(-i\delta\hat{p}^2/(2m\hbar))\left(\hat{x} + \frac{\hat{p}}{m}\delta\right)\,. \end{aligned} \tag{5.230}$$

Thus,

$$\hat{x}(\delta) = \exp(i\delta\hat{H}/\hbar)\hat{x}\exp(-i\delta\hat{H}/\hbar) = \hat{x} + \frac{\delta}{m}\hat{p} \tag{5.231}$$

equals the evolved position after a time increment of δ, in agreement with the Heisenberg operator.

Equations (5.229) and (5.231) look just like the classical evolution equations: For a given x at time t, x at time $t+\delta$ equals $x + \delta\hat{p}/m$. However, these equations do not describe evolution of observables because they are formulated on the kinematical rather than physical Hilbert space. It is useful to interpret the kinematical x and t as reference variables chosen to determine the position and time relative to their zero values, such as the origin of a coordinate system. This choice is arbitrary and does not provide information about the physical system that is evolving. Therefore, these quantities as independent parameters are available only on the kinematical Hilbert space. Instead of evolution, the equations (5.229) and (5.231) represent changes of the arbitrary choice of the zero value of t: If we decide to shift the time axis by an amount of δ, we should correspondingly change the position values assigned to a moving system. Equations (5.229) and (5.231) determine transformations of a **quantum reference frame**.[2] The relation to reference parameters distinguishes a gauge transformation from a symmetry transformation of physical observables. Any Dirac observable $\hat{D}$, such as (5.224) for the free particle, is independent of reference choices because it commutes with $\hat{C}$ and also with $\hat{U}_{\hat{C}}(\delta)$.

Time and space coordinates cannot only be changed by translations but also by rotations and Lorentz boosts. Such transformations imply important changes of reference frames, as analyzed in relativity theory.

[2]S. D. Bartlett, T. Rudolph and R. W. Spekkens, Reference frames, superselection rules, and quantum information, *Rev. Mod. Phys.* **79** (2007) 555, quant-ph/0610030

EXERCISES

1. A quantum system with a 3-dimensional Hilbert space has the Hamilton operator

$$H = -E \begin{pmatrix} 1 & 1 & 0 \\ 1 & 1 & 1 \\ 0 & 1 & 1 \end{pmatrix}$$

with a positive energy constant E.

   (i) Compute the eigenvalues and normalized eigenvectors of H.
   (ii) If the state of the system is given by the vector $\psi = (1,0,0)^T$, what is the probability of obtaining the ground-state value when measuring the energy?

2. The potential of a particle moving in one dimension and subject to a constant force F is $V(x) = -Fx$.
What is the Hamilton operator $\hat{H}$ acting on a momentum-space wave function $\phi(p)$ for a particle in this potential? Use this operator to solve the time-independent Schrödinger equation $\hat{H}\phi(p) = E\phi(p)$ in the momentum representation.
Show that the wave function in the position representation is of the form

$$\psi(x) \propto \int_{-\infty}^{\infty} e^{i(f(x)p+gp^3)} \mathrm{d}p$$

and determine the function $f(x)$ and the constant g.

3. Define a state by the wave function

$$\psi(x) = \begin{cases} 0 & \text{if } |x| > a \\ c(1+x/a) & \text{if } -a \le x \le 0 \\ c(1-x/a) & \text{if } 0 < x \le a \end{cases}$$

with two real constants a and c.

   (i) Determine c.
   (ii) Compute the momentum-space wave function $\phi(p)$ corresponding to $\psi(x)$.
   (iii) What are the probability densities for measurements of position and momentum, respectively?
   (iv) Discuss qualitatively how the uncertainty relation is realized by comparing $\psi(x)$ and $\phi(p)$ for different values of a.

4. For a stationary state $\psi_n(x)$ of the infinite square well with width a, compute the function

$$\phi_n(p) = \frac{\tilde{\psi}_n(p/\hbar)}{\sqrt{\hbar}}$$

where $\tilde{\psi}_n(k)$ is the Fourier transform of $\psi_n(x)$.

5. Compute $\langle\lambda\rangle$ and σ_λ as in (5.149) and (5.151), but with a positive-operator valued measure given by replacing E_1 and E_2 with (5.111). What is the collapse state of an initial state $(1,1,1)^T/\sqrt{3}$ for the root

$$R_1 = \begin{pmatrix} \sqrt{1-\varepsilon} & 0 & 0 \\ 0 & \sqrt{\varepsilon} & 0 \\ 0 & 0 & 0 \end{pmatrix}?$$

6. Derive the collapse state after measuring λ_1 in the context of the preceding problem, choosing a root with first 2×2-block given by $R_1' = U(\delta,\vartheta,\varphi)(R_1)_{2\times 2}$, using the general unitary matrix (4.242).

7. Write down a positive-operator valued measure for momentum measurements. What is the first moment and its expectation value in a general state $|\psi\rangle$?
8. Write down a positive-operator valued measure for energy measurements of a free particle. What is the first moment and the energy expectation value in a general state $|\psi\rangle$? If an energy is measured in some range $\mathscr{E} \subset \mathbb{R}_+$, what is the collapse state of a general state? Choose a suitable root of the effect $\hat{E}_{\mathbb{R}_+}(\mathscr{E})$.

Section II

Relativity

6 Basics of Gravity and Relativity

Relativity describes how physical laws are applied by observers using different reference frames, for instance by being placed at different positions and times, by being rotated with respect to one another, or by being in different states of relative motion. All these reference values and transformations between them depend on space and time or on relationships between space and time intervals. Well-defined symmetry transformations therefore require a fundamental understanding of how we measure distances in space and time. Properties of distance measurements are described by geometry. Compared with basic Euclidean geometry, a fundamental description must be able to include the time dimension and, as it turns out, the possibility of non-vanishing curvature.

Special relativity determines how the time dimension can be included in a suitable geometry, which is then used to formulate a complete set of symmetry transformations between reference frames of different observers. Geometry is a fundamental ingredient of physical theories. For instance, describing how a particle reacts to a force requires a well-defined meaning of a direction in space or in space-time, which is one example of a geometrical concept: We use angles to specify the direction, and we need symmetry transformations to make sure that we correctly adjust the angles if we rotate the reference system of an observer, or if the reference system gets rotated by the motion of Earth. According to special relativity, such transformations also happen when we change the speed of the observer, but they are much less intuitive than rotations because, like the speed, they involve space and time intervals and therefore transform space and time.

Geometry, in general, allows for non-vanishing curvature. The intuitive understanding of curvature shows that lines must be curved in order to stay on a curved surface. This basic notion implies a physical description that includes deflection, usually described by the presence of a force. If space-time is curved, this force turns out be gravity. General relativity implements curvature in the space-time description derived by special relativity and thereby explains the relationship with gravity. It also explains phenomena such as the expanding universe and black holes.

6.1 NON-RELATIVISTIC GRAVITY

Dating back to ancient observations, it is well-known that distant objects are able to orbit around one another. This behavior has been seen for the Moon orbiting around the Earth, planets around the sun, or more recently stars around Sagittarius A*, a black hole at the center of the Milky Way.

Isaac Newton explained this motion by the action of the gravitational force

$$F = -G\frac{m_1 m_2}{r^2} \tag{6.1}$$

between two masses m_1 and m_2 separated by a distance r, with a coefficient G called Newton's constant. By unifying a large variety of different phenomena, Newton's description of the gravitational force was very successful and remains in use today. However, it gives rise to several conceptual problems.

6.1.1 GRAVITATIONAL AND INERTIAL MASS

A mass parameter, the **inertial mass** m', also appears in Newton's second law $F = m'a$ with the acceleration a. We usually do not distinguish between the inertial mass and the **gravitational mass**

DOI: 10.1201/9781003588702-8

in (6.1) and set $m = m'$. Nevertheless, the two values can be measured independently: We determine the gravitational mass by weighing an object, and the inertial mass, for instance, by measuring how an electrically charged objects is accelerated by an electric field in a horizontal direction such that gravity does not play a role. These two procedures make use of independent phenomena. They could, in general, lead to different values for m and m'.

Also in the theoretical description, Newtonian mechanics remains consistent if we use two different values for the mass parameters m and m' of a given object. For a single object, we can always adjust units or Newton's constant such that $m = m'$, but this can only be done once and would allow for non-unit ratios of m'/m for other objects. The ratio could even depend on the material decomposition of a single type of object. Nevertheless, measurements have shown that $m = m'$ is always satisfied within the experimental precision.

Galileo Galilei was first to perform such measurements in a systematic fashion, in his case with a relative precision of about 10^{-2}. Lorand Eötvös invented the torsion balance in order to address this question and by 1922 was able to improve the precision to 10^{-9}. Current experiments of this type have reached a precision of about 10^{-13}. Newtonian mechanics does not explain this universal equality of the gravitational and inertial mass.

6.1.2 LOCAL DESCRIPTION

Newton's law is similar to Charles-Augustin de Coulomb's law

$$F = \frac{1}{4\pi\varepsilon_0}\frac{q_1 q_2}{r^2} \tag{6.2}$$

for the electric force between two charges q_1 and q_2. There is a strength difference, for instance for two protons we have

$$\frac{F_G}{F_{EM}} \sim -10^{-36}, \tag{6.3}$$

and a sign difference because gravity is always attractive while the electric force is repulsive for two charges of equal signs.

Coulomb's law has a **local description** given by James Clerk Maxwell's equations: A class of solutions to these equations describes electromagnetic waves. If one charge starts moving, the electric force at the other charge does not change instantaneously. Rather, the accelerated charge emits electromagnetic waves which travel to the other charge at the speed of light and exert a force sometime later. Only then does the second charge react.

Newtonian mechanics does not provide such a local description for the gravitational force, a problem that had been recognized even at Newton's times, before the analog of Coulomb's law (1780s) and the local description of Maxwell's equations (1865) became available. As Newton's confrontational contemporary, Gottfried Wilhelm Leibniz, said in a letter to Samuel Clarke,[1] "I objected that an attraction—properly so called or in the scholastic sense—would be an operation at a distance without any means of intervening. The author [Newton] answers here that an attraction without any means of intervening would be indeed a contradiction. Very well! But then what does he mean, when he will have the Sun to attract the globe of the Earth through an empty space?"

6.1.3 GRAVITATIONAL COLLAPSE

Masses cannot be negative and always attract one another. Even though the force between two masses is mathematically similar to the force between two charges, mass accumulations behave very differently from charge accumulations. A large charge attracts charges of the opposite sign, which partially neutralize, or screen, the initial charge. It is therefore hard to generate significant electric charges comparable with heavy masses. For instance, astrophysical objects have negligible net electric charges.

[1] Samuel Clarke: *A Collection of Papers*, London: Knapton, 1717.

A large mass attracts more mass, which increases its value. This basic process explains how large astrophysical objects can form on cosmological and galactic scales. Newton's law of gravity does not give a unique result for the end stage of this process applied to the complicated many-body problem of all matter in the universe or at least in a sizeable region: a single, point-like collapsed object or some other equilibrium configuration in perpetual orbiting motion.

In addition, we should expect that Newton's law has limited validity when matter is dense, the gravitational force is large, and individual constituents move at high speeds close to the center of a spinning distribution. Basic results from special relativity are expected to lead to two kinds of modifications: (i) If we use $E = mc^2$, energy should contribute to the gravitational force, just like the mass. Newton's gravitational force does not depend on the energy and should therefore be modified in relativistic physics. We expect a more complicated set of evolution equations because the binding energy that is released when a mass enters into an orbit around another object modifies the gravitational force between them. (ii) In special relativity, velocities are limited by the speed of light. In Newtonian gravity, a collapsing mass distribution spins up due to its conserved angular momentum, without an upper limit on the speeds of its parts. The upper limit of the speed of light required by special relativity implies a slow-down of gravitational collapse at least in advanced stages. Moreover, related to the preceding concern, a consistent extension of Newton's law that includes special relativity should result in a local description with a finite transmission time of the gravitational force.

A relativistic extension of Newton's law is important for various astrophysical and fundamental questions, but it has to obey various consistency conditions such as conservation laws. Geometry will help us to handle these challenging questions. The typical scale of relativistic gravity can be estimated by the unitless ratio

$$\rho = \frac{GM}{Rc^2} \tag{6.4}$$

with the mass M and radius R of a stellar object. If this ratio is much less than one, Newton's law may be used to an excellent approximation. If ρ is close to one or greater, we must consider relativity. Characteristic examples are $\rho \sim 10^{-9}$ for Earth, $\rho \sim 10^{-6}$ for the sun, $\rho \sim 10^{-2}$ for white dwarfs, and $\rho \sim 1$ for neutron stars.

6.2 NEWTONIAN PHYSICS AND THE RELATIVITY PRINCIPLE

Relativity theory is required for a fundamental discussion of the gravitational force. As a general definition, relativity determines the laws that govern changes of reference frames in space and time. Formulating the basic condition that we should be able to compare measurement results obtained by different observers, the **relativity principle** states that all physical laws must equally apply in systems that are moving uniformly with respect to one another, in addition to being rotated relative to one another or translated in space and time. The latter conditions mean that we can repeat experiments at various locations and moments in time, with axes of coordinate systems pointing in different directions. The first condition makes sure that physical laws can be applied while standing on the ground or being on a moving airplane or the space station. There should be only one set of laws for all these situations. Without the relativity principle, we would not be able to develop physics theories because it would be impossible to include a special description for any reference frame that may become relevant.

For now, we restrict the relativity principle to *uniform* relative motion, that is unaccelerated and non-rotating motion. We know that some physical laws do change by necessitating new force terms, such as the centrifugal force, if the system is accelerating. Considering uniform relative motion as a first step includes a subset of frame transformations that reveal important properties of space and time. For non-uniform motion, we have to accompany the laws by additional recipes to derive the new forces or other ingredients that may be required. In fact, it is important to generalize the relativity principle to non-uniform motion because our experiments are always done in an accelerating

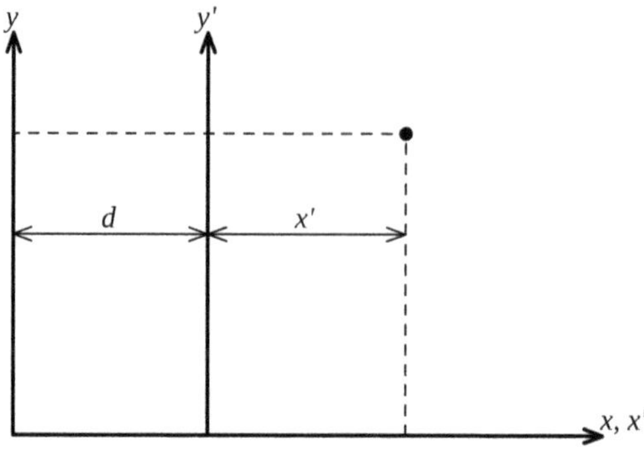

Figure 6.1 Translation of a coordinate system. Translating the system to the right, we subtract the distance d from the x-coordinate of a point.

system, as they are based on Earth orbiting around the sun or, for satellites at a Lagrange point, in the solar system orbiting around the center of the Milky Way.

6.2.1 INERTIAL FRAMES

A specific **reference frame** is introduced by choosing a set of coordinates in space, for instance Cartesian coordinates (x,y,z) measured from the origin at some point, and a time coordinate t. The frame specifications include the directions of the spatial axes as well as the system's motion if the origin is not a fixed point in space.

A reference frame is called **inertial** if it is moving uniformly. This statement introduces the special case of unaccelerated motion, but now in the context of a single reference frame that does not allow us to refer to relative motion. A single inertial frame does not provide a reference point with respect to which it is moving uniformly. This problem, which has played an important role in the historical developments that led to special relativity, was initially addressed by postulating a distinguished reference frame, or absolute space and time. Distinguishing one reference frame makes it possible to determine the inertial nature of other reference frames by uniform relative motion with respect to the distinguished frame.

In practice, it is possible to avoid conceptual questions related to the notion of a distinguished frame by defining inertial frames as those reference frames that have a specific form of their physical laws. In particular, there should be no centrifugal forces in the dynamical laws of inertial frames, and all external effects such as tidal forces as a consequence of gravitational attraction in the Earth-moon-sun system should be negligible. In inertial frames, there are only forces between objects but no forces that act on all objects without an identifiable source. As the conditions suggest, the notion of an inertial frame is approximate or idealized.

Once we have introduced coordinates that parameterize reference frames, different inertial frames are related by **coordinate transformations** that map (x,y,z) and t to (x',y',z') and t'. In Newtonian mechanics, the most common ones are combinations of:

1. **Translations.** In the example of a translation along the x-axis by a distance d, as visualized in Figure 6.1, we have the transformation equations $x' = x-d$, $y' = y$, $z' = z$, $t' = t$.
2. **Rotations.** In the example of a rotation around the z-axis by a fixed, time-independent angle δ, visualized in Figure 4.1 on page 136, we have the transformation equations $x' = x\cos(\delta) + y\sin(\delta)$, $y' = -x\sin(\delta) + y\cos(\delta)$, $z' = z$, $t' = t$.
3. **Speed boosts.** In the example of an instantaneous speed change along the x-axis with velocity v, we have the transformation equations $x' = x - vt$, $y' = y$, $z' = z$, $t' = t$. This **Galilean transformation** is the same as a translation by a linear time-dependent distance vt.

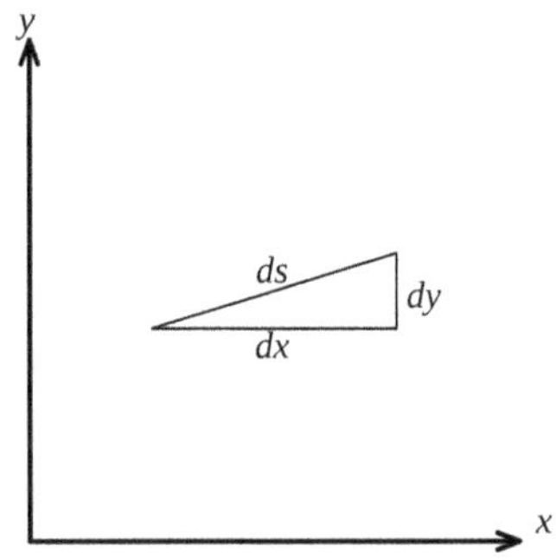

Figure 6.2 The Pythagorean theorem, used to compute the line element of 3-dimensional Euclidean space.

In these examples, the Newtonian relativity principle can be tested explicitly: We have $\mathrm{d}^2\vec{x}'/\mathrm{d}t'^2 = \mathrm{d}^2\vec{x}/\mathrm{d}t^2$ in all cases, and therefore the transformations do not introduce new forces in Newton's second law $\vec{F} = m'\vec{a} = m'\mathrm{d}^2\vec{x}/\mathrm{d}t^2$. Newton's second law is invariant under translations, rotations, and Galilean transformations.

A characteristic feature of Newtonian mechanics is that it postulates the absolute nature of time: In all transformations, we have $t' = t$. In particular, two events that are simultaneous in one frame are simultaneous in any other frame. The time separation between two events is given by the difference of the two real time values assigned to the events and takes the same value in all reference frames.

6.2.2 LINE ELEMENT OF THREE-DIMENSIONAL EUCLIDEAN SPACE

In contrast to the 1-dimensional absolute time direction of Newtonian mechanics, space is 3-dimensional such that we can change our viewpoint by rotating a spatial reference frame. Moreover, two separate points are not connected by a unique curve, and therefore specifying their distance requires additional concepts. This is the subject of geometry, which is assumed to be Euclidean in Newtonian mechanics: We use Euclid's axioms and measure the distance along the straight line connecting two points. If we know the Cartesian coordinates of the points, the Pythagorean theorem allows us to compute the distance.

When we look for modifications or extensions of geometrical laws, for instance by including time measurements along with space measurements, it is inconvenient to work with a list of axioms, which would require us to demonstrate their mutual consistency of any attempted modification. Instead, it is possible to invert the question and define geometry by an expression for the distance function assigned to the separation between points. In order to avoid having to choose a curve connecting the points, we assume that they are infinitesimally close to each other. A simple straight line segment, or a linear function in local Cartesian coordinates, can then be used to connect the points without worrying how curvature might deflect the line if it is extended to a longer distance. If we specify this infinitesimal distance, any extended distance along a specific curve can be obtained by integration. The length of an infinitesimal displacement between two points is called the **line element**.

In 3-dimensional Euclidean space, we compute the line element using the Pythagorean theorem as in Figure 6.2. An infinitesimal displacement is characterized by three Cartesian coordinate increments $(\mathrm{d}x, \mathrm{d}y, \mathrm{d}z)$. Since the axes are orthogonal to one another, the displacement (squared) is given by

$$\mathrm{d}s^2 = \mathrm{d}x^2 + \mathrm{d}y^2 + \mathrm{d}z^2 \,. \tag{6.5}$$

We usually leave the expression in this form and do not take the square root of $\mathrm{d}s^2$. This practice is useful in generalizations in which $\mathrm{d}s^2$ may be negative. We can therefore avoid using complex numbers in geometrical discussions.

We have used results from Euclidean geometry in order to compute its line element. It is possible to reverse the procedure and derive all Euclidean laws from the expression (6.5). The validity of this statement is not obvious at all, but we will see several examples when we discuss geometry in the presence of curvature.

It may seem surprising that a large number of axioms and theorems should be contained in a single equation. However, the line element in fact represents an infinite number of equations because it provides the infinitesimal distance between two points in any direction and anywhere in space. Moreover, in full-fledged geometry, the line element is accompanied by additional concepts that are used in the derivation of properties such as curvature. These additional concepts are provided by the broad framework of Riemannian geometry, in which the line element is used as a compact way of specifying a specific example of a geometry. It is then possible to study different versions of Riemannian geometry by modifying the coefficients in (6.5), as we do in special and general relativity.

The line element is an example of a powerful notion often used in compact derivations: It is **frame independent** or **invariant**. While the components of $(\mathrm{d}x, \mathrm{d}y, \mathrm{d}z)$ depend on the choice of the Cartesian frame, the value of $\mathrm{d}s^2$ remains the same if we translate, rotate, or boost the frame. This property follows from the definition of $\mathrm{d}s^2$ as the square of a distance, which can be measured directly by a ruler and does not depend on where we place the origin or how we orient the axes of a Cartesian system. Geometric derivations are usually most elegant when they can be expressed in terms of invariants. Riemannian geometry, as we will see, provides the means to derive several different invariants that can be used to draw physics conclusions.

We practice differential calculus by explicitly confirming the invariance of $\mathrm{d}s^2$ under translations, rotations and Galilean transformations. The calculation is easy for translations and Galilean transformations because x changes only by a constant. The constant may depend on t, and t is indeed constant in a given 3-dimensional Euclidean space. Therefore, $\mathrm{d}x' = \mathrm{d}x$ and the line element does not change. For a rotation, we compute

$$\begin{aligned} \mathrm{d}x' &= \frac{\partial x'}{\partial x}\mathrm{d}x + \frac{\partial x'}{\partial y}\mathrm{d}y + \frac{\partial x'}{\partial z}\mathrm{d}z = \cos(\delta)\mathrm{d}x + \sin(\delta)\mathrm{d}y \\ \mathrm{d}y' &= \frac{\partial y'}{\partial x}\mathrm{d}x + \frac{\partial y'}{\partial y}\mathrm{d}y + \frac{\partial y'}{\partial z}\mathrm{d}z = -\sin(\delta)\mathrm{d}x + \cos(\delta)\mathrm{d}y \\ \mathrm{d}z' &= \mathrm{d}z \end{aligned} \tag{6.6}$$

using the chain rule of multi-variable calculus, illustrated in Figure 6.3. Here, both $\mathrm{d}x$ and $\mathrm{d}y$ change, but in the line element

$$(\mathrm{d}x')^2 + (\mathrm{d}y')^2 + (\mathrm{d}z')^2 = (\cos(\delta)\mathrm{d}x + \sin(\delta)\mathrm{d}y)^2 + (-\sin(\delta)\mathrm{d}x + \cos(\delta)\mathrm{d}y)^2 + \mathrm{d}z^2 = \mathrm{d}s^2 \tag{6.7}$$

all transformations cancel out.

The chain rule is an important tool in the context of coordinate transformations. In general relativity, it is customary to write it compactly as

$$(\mathrm{d}x')^i = \sum_{j=1}^{3} \frac{\partial x'^i}{\partial x^j}(\mathrm{d}x)^j = \frac{\partial x'^i}{\partial x^j}(\mathrm{d}x)^j\,. \tag{6.8}$$

The superscript is not an exponent but labels different components of the coordinate vector. The same general transformation therefore describes the mapping of $\mathrm{d}x^1 = \mathrm{d}x$, $\mathrm{d}x^2 = \mathrm{d}y$ and $\mathrm{d}x^3 = \mathrm{d}z$. Whenever an index appears twice in a product on the same side of an equation, as j here, it is automatically summed over according to **Einstein's summation convention**. Only the unrepeated index i in (6.8) is free and gives rise to three equations for the components. This notation will be convenient when we talk about curvature in its most general form, which is related to tensorial objects with up to four indices. There are also internal consistency conditions which are automatically

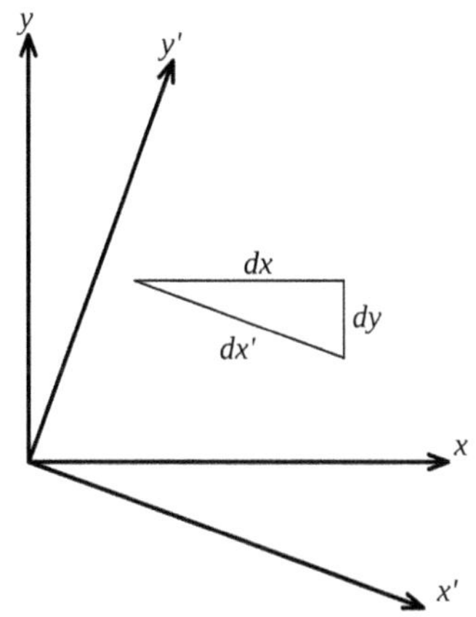

Figure 6.3 Visualization of the chain rule: A coordinate difference such as dx' is the distance between two points separated by a difference dx' in the x'-coordinate while all other coordinates are held fixed. In a rotated coordinate system, the coordinate difference dx' depends on both coordinate differences, dx and dy, in the original system. The old coordinates contribute with different weights, depending on the slopes of the new axes with respect to the old ones. They can be computed from partial derivatives as in the chain rule of multi-variable calculus.

fulfilled by the notation, as we will see later. For instance, if an index appears more than twice in a product on one side of an equation, there has been a mistake in its derivation. For now, we will use this convention sparingly as we continue to introduce more of its rules.

7 Special Relativity

Changes of an inertial frame correspond to mathematical transformations, but they also have physical implications. In particular, we can derive the composition law for velocities, which tells us how the velocity components of a moving object, measured by two observers, differ from each other in a specific way. For a Galilean transformation in the x-direction, $x' = x - vt$ with the frame velocity v, we derive the composition law for velocity components by taking a derivative with respect to $t' = t$:

$$V^{x\prime} = \frac{\mathrm{d}x'}{\mathrm{d}t'} = \frac{\mathrm{d}x}{\mathrm{d}t} - v = V^x - v\,. \tag{7.1}$$

The velocity of an object, seen from a moving frame, is obtained by subtracting the frame velocity from the velocity measured in the original frame.

7.1 GALILEAN TRANSFORMATIONS

Light does not obey the composition law (7.1). Light is an electromagnetic wave described by Maxwell's equations. These are first-order linear partial differential equations for the electric and magnetic fields. If there are no charges and currents, they take the form

$$\nabla\cdot\vec{E} \quad \text{and} \quad \nabla\cdot\vec{B} = 0 \tag{7.2}$$

as well as

$$\nabla\times\vec{E} = -\frac{\partial\vec{B}}{\partial t} \tag{7.3}$$

$$\nabla\times\vec{B} = \varepsilon_0\mu_0\frac{\partial\vec{E}}{\partial t} \tag{7.4}$$

with the vacuum permittivity ε_0 and permeability μ_0, using the derivative operator $\nabla = (\partial/\partial x, \partial/\partial y, \partial/\partial z)$.

If we take a partial time derivative of (7.4), we obtain

$$\nabla\times\frac{\partial\vec{B}}{\partial t} = \varepsilon_0\mu_0\frac{\partial^2\vec{E}}{\partial t^2}\,, \tag{7.5}$$

noting that partial derivatives commute. We can now use (7.3) on the left and obtain a partial differential equation only for $\vec{E}$,

$$-\nabla\times(\nabla\times\vec{E}) = \varepsilon_0\mu_0\frac{\partial^2\vec{E}}{\partial t^2}\,. \tag{7.6}$$

The double vector product can be simplified using the "curl of curl" identity

$$\nabla\times(\nabla\times\vec{E}) = \nabla(\nabla\cdot\vec{E}) - (\nabla\cdot\nabla)\vec{E} \tag{7.7}$$

which holds for any vector $\vec{E}$. For the electric field in vacuum, we have $\nabla\cdot\vec{E} = 0$ in the first term, according to (7.2). Therefore,

$$\nabla\times(\nabla\times\vec{E}) = -(\nabla\cdot\nabla)\vec{E} = -\Delta\vec{E} \tag{7.8}$$

with the Laplacian

$$\Delta = \nabla\cdot\nabla = \frac{\partial^2}{\partial x^2} + \frac{\partial^2}{\partial y^2} + \frac{\partial^2}{\partial z^2}\,. \tag{7.9}$$

DOI: 10.1201/9781003588702-9

Equation (7.6) is then equivalent to the wave equation

$$-\frac{1}{c^2}\frac{\partial^2\vec{E}}{\partial t^2}+\Delta\vec{E}=0 \tag{7.10}$$

with

$$c=\frac{1}{\sqrt{\varepsilon_0\mu_0}}. \tag{7.11}$$

The constant c is a fundamental constant because it is obtained as a combination of the vacuum permittivity ε_0 and permeability μ_0. Solutions of (7.10) are waves moving with phase velocity c, which can be seen by computing the **dispersion relation** $\omega(\vec{k})=c\vec{k}$ after inserting a plane wave $\exp(-i\omega t+i\vec{k}\cdot\vec{x})$ in the wave equation. The fundamental constant c therefore represents the **speed of light**.

7.1.1 PARTIAL DERIVATIVES

Since the speed of light c is a fundamental constant that follows from physical laws, it must take the same value in all frames provided the relativity principle holds. Light, then, cannot obey the composition law (7.1) of Newtonian mechanics. In mathematical terms, Maxwell's equations are not invariant under Galilean transformations $t'=t$ and $x'=x-vt$: The latter imply the transformed derivatives

$$\frac{1}{c^2}\frac{\partial^2}{\partial t^2}=\frac{1}{c^2}\left(\frac{\partial t'}{\partial t}\frac{\partial}{\partial t'}+\frac{\partial x'}{\partial t}\frac{\partial}{\partial x'}\right)^2=\frac{1}{c^2}\frac{\partial^2}{\partial t'^2}-2\frac{v}{c}\frac{\partial^2}{\partial t'\partial x'}+\frac{v^2}{c^2}\frac{\partial^2}{\partial x'^2} \tag{7.12}$$

and

$$\frac{\partial^2}{\partial x^2}=\left(\frac{\partial t'}{\partial x}\frac{\partial}{\partial t'}+\frac{\partial x'}{\partial x}\frac{\partial}{\partial x'}\right)^2=\frac{\partial^2}{\partial x'^2}. \tag{7.13}$$

Here, we are allowed to take simple squares of the transformed first-order derivatives because they have constant coefficients. If we combine these terms in the wave equation, the cross-term does not cancel out and the remaining coefficients change. In this combination of Newtonian mechanics with Maxwell's equations, the relativity principle is violated because these laws do not obey the same transformations.

Incidentally, it may be surprising to see in (7.12) and (7.13) that $\partial/\partial x'=\partial/\partial x$ even though $x'=x-vt\neq x$, while $\partial t\neq\partial t'$ even though $t'=t$. These results are clearly implied by the chain rule, but an explanation is nevertheless warranted. A geometrical understanding can be obtained by keeping in mind the complete definition of a partial derivative: The partial derivative of a function $f(t,x)$ by x is the derivative $\mathrm{d}f(t_0,x)/\mathrm{d}x$ of the single-variable function $f(t_0,x)$ defined by keeping $t=t_0$ fixed in $f(t,x)$. In the more-complete notation used for instance in statistical physics, the constant variable would be indicated as well, as in $(\partial/\partial x)_t$. Therefore, the partial derivative by x depends on the definition of t. Since a Galilean transformation does not change t, keeping t fixed is the same as keeping t' fixed. And since x' differs from x only by adding a constant if t is fixed, we have $\mathrm{d}/\mathrm{d}x=\mathrm{d}/\mathrm{d}x'$ at $t=t_0$. Thus, $\partial/\partial x'=\partial/\partial x$.

Applying the same reasoning to $\partial/\partial t$, the t-dependent function $f(t,x_0)$ is not the same as the t'-dependent function $f(t',x'_0)=f(t',x_0-vt')$, using $t=t'$: Keeping x_0 fixed is not the same as keeping x'_0 fixed. In this new function, t' appears in two places, and correspondingly the chain rule implies two terms in the partial derivative by t'. A geometrical interpretation can be seen in Figure 7.1. Since partial derivatives are defined by keeping all but one coordinate fixed, they correspond to vectors whose components change by coordinate transformations. This observation plays an important role in the understanding of differential geometry, which underlies the formulation of special and general relativity.

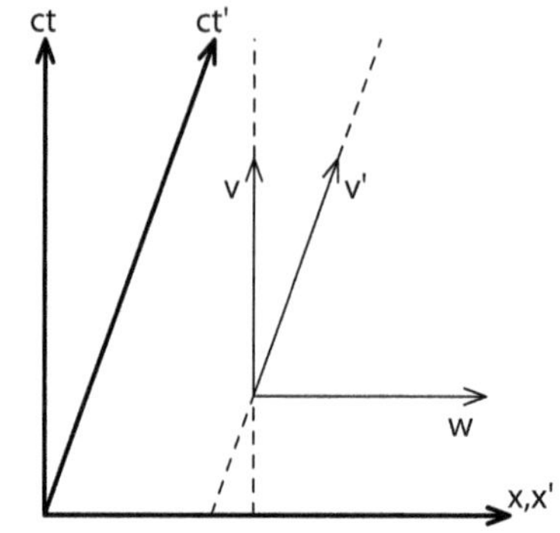

Figure 7.1 Partial derivatives represented by vectors. In anticipation of conventions in special relativity, we have drawn the time axis in the vertical direction and used the coordinate ct with length units. The vector $v = \partial/\partial(ct)$ determines the change of a function in the ct-direction, keeping x fixed. The vector $w = \partial/\partial x$ determines the change of a function in the x-direction, keeping ct fixed. After a Galilean transformation, the partial derivative $\partial/\partial(ct')$ is represented by the vector v', which points in a different direction than v because now x' is to be held fixed. The equation $x' = x'_0$ constant implies the linear relationship $ct(x) = (c/v)(x - x'_0)$, or a straight line with slope c/v. If we draw an axis for ct', defined by $ct(x)$ with $x'_0 = 0$, it is parallel to v', while the x'-axis coincides with the x-axis.

7.1.2 LORENTZ TRANSFORMATIONS

A combination of Newtonian mechanics with Maxwell's equations violates the relativity principle. Different solutions had been proposed when this problem first became apparent. One could modify Maxwell's equations and try to make them invariant under Galilean transformations, but no version could be found that was as successful as Maxwell's equations in correctly describing observations of electromagnetic phenomena. Alternatively, there could be a distinguished frame, called the ether, in which Maxwell's equations are correct. In any other frame moving with respect to the ether, one would have different equations for electromagnetic phenomena obtained by applying a Galilean transformation to Maxwell's equations. However, the existence of an ether is inconsistent with observations as shown by the Michelson–Morley experiment in 1887. If there were an ether, Earth orbiting around the sun should have different relative velocities depending on the seasons. In an interferometer, light would then move with different speeds along the two arms, and interference patterns would change in time. No such changes were seen in the experiment.

At about the same time, Hendrik Lorentz noticed that Maxwell's equations are invariant under a new set of transformations,

$$x' = \frac{x - vt}{\sqrt{1 - v^2/c^2}} \quad , \quad y' = y \quad , \quad z' = z \quad , \quad t' = \frac{t - vx/c^2}{\sqrt{1 - v^2/c^2}} \tag{7.14}$$

for a **Lorentz boost** in the x-direction. They can be combined with rotations in order to boost the observer's speed in any direction. For frame velocities v much smaller than the speed of light, $v \ll c$, Galilean transformations are good approximations of Lorentz boosts. Newtonian mechanics, which has traditionally been applied to objects much slower than light, is therefore approximately invariant under Lorentz transformations, while Maxwell's electromagnetism is exactly invariant under these transformations.

Given these results, Lorentz transformations should be preferred in fundamental physics. However, they have the counter-intuitive consequence that not just space coordinates but also time must be transformed if we use a moving frame. Time is no longer absolute, and there is no universal simultaneity: If we have two events at different positions $x_1 \neq x_2$, happening at the same time $t_1 = t_2$ in one frame, they are not simultaneous in a moving frame, where $t'_1 \neq t'_2$.

In special relativity, the value of the speed of light, as part of a set of physical laws, is required to be the same in all frames. The correct transformations are therefore those found by Lorentz. As a consequence, Newtonian mechanics must be modified if it should apply to objects with high speeds. The linear addition (7.1) of velocities, derived from Galilean transformations, is no longer correct. Moreover, we have to make sense of space-time without an absolute time and understand observations in the absence of universal simultaneity at a distance.

7.2 SPACE-TIME GEOMETRY

If there is no absolute time and time is combined with space in Lorentz transformations, we should keep all four space-time coordinates on the same footing. We call them $(t,x,y,z) = x^{\alpha}$, where the Greek index $\alpha = 0,1,2,3$ labels the four dimensions. Since we do not have an intuitive understanding of geometry in the time dimension, it is useful to develop visualization tools by drawing the main effects in space-time diagrams.

7.2.1 SPACE-TIME DIAGRAMS

A **space-time diagram** has the following ingredients, see Figure 7.2:

1. **Coordinate axes** of an inertial frame. We usually draw 2-dimensional cross-sections, such as ct and x. The speed of light is included in order to have the same units in both directions. In special relativity, the speed of light is the same in all frames, and we do not have to worry about transforming c. In a relativistic context, it is customary to reserve the vertical direction for ct.
2. A point P in space-time is called an **event**.
   It is easy to draw coordinate axes as straight lines, but they represent a large amount of physics requirements if they are used to describe an actual experiment. In particular, drawing the axes implies the understanding that we are able to assign unique position and time coordinates to an event, anywhere in space-time because the axes are supposed to extend to infinity. Even if we restrict the coordinate system to a finite region, a correspondence with an experimental context requires us to make sure that we place a clock at each event relevant for a measurement, and that all clocks used are synchronized and can be compared in a meaningful way. This task may be feasible in a small lab, but it is impossible if we use a space-time diagram to represent astronomical observations such as the redshift of light emitted by a distant star and measured on Earth. Coordinate axes of a **global inertial frame** therefore constitute a highly idealized picture that will have to be weakened considerably for a valid description of most observations, keeping only the essential ingredients required for reference directions. We will

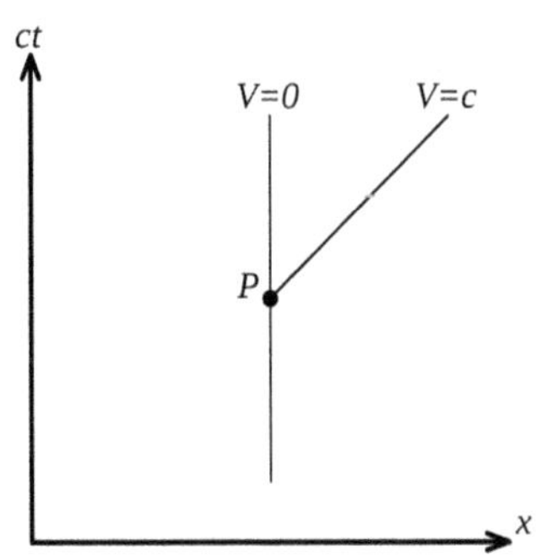

Figure 7.2 Space-time diagram with two worldlines: An object at rest and a light ray emitted at an event P.

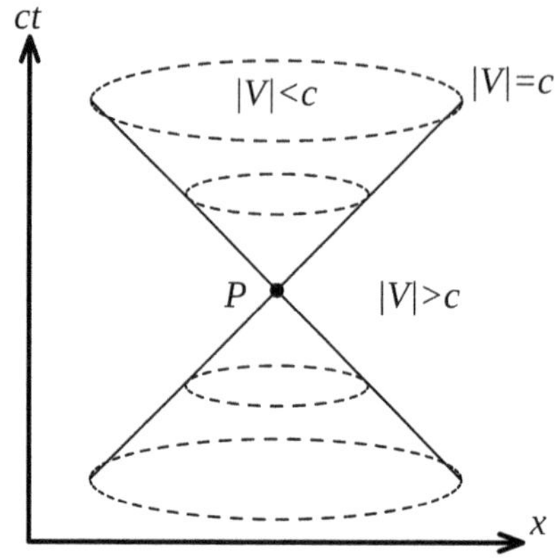

Figure 7.3 The light cone through an event P. A 2-dimensional space-time diagram only shows a "light cross" as a cross-section of the full light cone. In full 4-dimensional space-time, the light cone at a given event is a 3-dimensional hypersurface. The dashed ellipses indicate only one additional dimension.

postpone this task and for now use coordinate axes mainly as a mathematical tool to analyze basic effects of special relativity.

3. A curve followed by an object in space-time is called a **worldline**, parameterized for instance as $x(t)$. Even though t changes in the vertical direction, it is convenient to use this variable as an independent parameter of $x(t)$ because t, unlike x in general, changes in a monotonic fashion.
4. The slope of a worldline is related to the velocity of the object. We have $\mathrm{d}(ct)/\mathrm{d}x = c(\mathrm{d}x/\mathrm{d}t)^{-1} = c/V^x$. A particle at rest in the inertial frame, $V^x = 0$, has a vertical worldline with infinite slope. A light ray moving in the x-direction, $V^x = \pm c$, always has a diagonal direction in our drawing.
5. The fact that the speed of light is a constant independent of the frame used allows us to define three distinct regions in space-time for any given event P: The property that a second event is connected to P by motion at speed $|V| < c$, $|V| > c$ or $|V| = c$, respectively, holds in any frame if it holds in one frame: Lorentz transformations are continuous, and since they preserve the speed of light, they cannot map motion at $|V| < c$ to motion at $|V| = c$ or $|V| > c$. The region with $|V| = c$ around P, shown in Figure 7.3, is called the **light cone** of P. It is spanned by light rays starting at P or ending at P.

Since the light cone as a subset of space-time is frame independent or invariant, it is part of the geometry of space-time. It should be possible to derive its form from a suitable line element. Space-time geometry is therefore different from 4-dimensional Euclidean geometry because there is no such structure in the Euclidean laws.

7.2.2 DISTANCES IN SPACE-TIME

Newtonian physics uses Euclidean geometry of space, described by the line element $\mathrm{d}s^2 = \mathrm{d}x^2 + \mathrm{d}y^2 + \mathrm{d}z^2$. It is invariant under Galilean transformations if t is regarded as an absolute parameter rather than one of the coordinates of geometry. However, the Euclidean line element is not invariant under Lorents transformations, in which we change t and therefore cannot keep it as a parameter separate from the spatial coordinates. We need a new line element to describe the geometry of space-time, one that is invariant under Lorentz transformations.

We derive the new line element by implementing the basic physical feature of Lorentz transformations rather than their mathematical form: the fact that they leave the speed of light invariant.

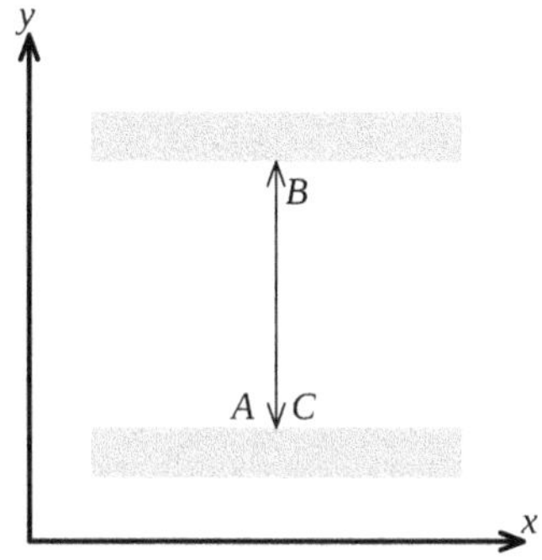

Figure 7.4 Spatial diagram of the light clock. The z-axis (not shown) is parallel to the mirror surfaces (gray) just like the x-axis.

Constructing a clock based on light rays or properties of electromagnetic waves allows us to measure durations in different frames, without assuming Newtonian mechanics beyond basic kinematical features such as the definition of velocity. Durations are distances in space-time, and they give us information about the geometry.

The light clock is based on electromagnetism rather than mechanics in order to make sure that it is described in different frames according to Lorentz transformations. As shown in Figure 7.4, it has the following mechanism:

1. In the rest frame of the clock, there are two parallel mirror surfaces in the x-z plane, separated by a distance L in the y-direction.
2. A light ray is bouncing back and forth between the mirrors. A detection and counting mechanism allows us to measure time with a resolution given by the time required for light to complete one cycle of the clock. For one cycle, the light ray has to travel a distance $2L$. The resolution is therefore given by $c\Delta t = 2L$.
3. Figure 7.5 shows a space-time diagram of the light clock, with the two vertical worldlines of the mirror surfaces in their rest frame. The space-time separation between events A and C, at the beginning and the end of one cycle, is given by the **displacement vector** $(c\Delta t, \Delta x, \Delta y, \Delta z) = (2L, 0, 0, 0)$.
4. We now consider the clock in an inertial frame moving with constant speed v in the x-direction. Our first task is to find the space-time separation $(c\Delta t', \Delta x')$ between events A and C as seen in this frame. Figure 7.6 shows a spatial diagram of the clock in the moving frame. In space, we can use Euclidean geometry to compute the new intervals $c\Delta t'$ and $\Delta x'$, or rather a relationship between them. The triangle ABC can be split in two right-angled ones, as shown in the figure. Since the spatial distance between A and C in the moving frame is $\Delta x' = -v\Delta t'$, equal to the distance traveled by the mirrors as seen from the moving frame, the horizontal side of each right-angled triangle has length $\frac{1}{2}\Delta x'$. The vertical side has length L if we assume that a transformation in the x-direction does not change spatial distances in perpendicular directions. The remaining side length follows from the Pythagorean theorem, which is valid in a spatial diagram, and is given by the square root of $L^2 + \frac{1}{4}(\Delta x')^2$. A light ray travels twice this distance in order to complete one cycle of the clock from A to C. The time between these events, as measured by the clock, is therefore given by

$$\Delta t' = \frac{2}{c}\sqrt{L^2 + \frac{1}{4}(\Delta x')^2}\,. \tag{7.15}$$

In this step, we have used the fact that the speed of light has the same value c in all frames. The equation is therefore a result of special relativity, even though we have mainly used Euclidean

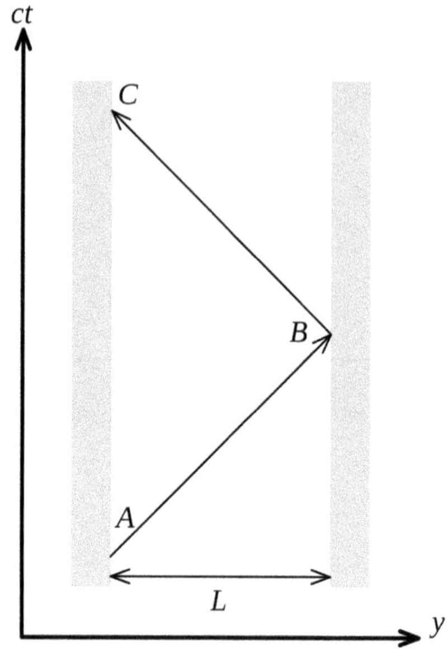

Figure 7.5 Space-time diagram of the light clock.

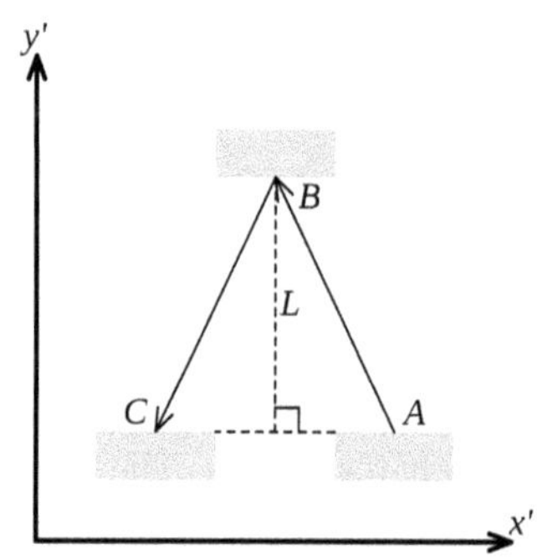

Figure 7.6 Spatial diagram of the light clock in a moving frame. The mirrors, shown at three different times, move to the left if the observer is moving to the right. The dashed auxiliary lines are used to compute (7.15).

geometry in space.

We rewrite our relationship (7.15) between $\Delta t'$ and $\Delta x'$ as

$$-(c\Delta t')^2 + (\Delta x')^2 = -4L^2 = -(c\Delta t)^2\,, \tag{7.16}$$

combining it with the rest-frame result $c\Delta t = 2L$. The expression on the left has the form of a line element, with linear combinations of squared coordinate differences. Using the fact that Δx, Δy, Δz, $\Delta y'$ and $\Delta z'$ are all zero for the two frames used here, we can write our result more symmetrically with respect to the spatial coordinates:

$$-(c\Delta t')^2 + (\Delta x')^2 + (\Delta y')^2 + (\Delta z')^2 = -(c\Delta t)^2 + (\Delta x)^2 + (\Delta y)^2 + (\Delta z)^2\,. \tag{7.17}$$

Although some of the coordinate differences change by a Lorentz transformation, the combination shown in (7.17) is invariant. It contains the Euclidean part if we restrict it to spatial coordinates. But also the time differences may be non-zero in general, such that we have an extension to space-time geometry.

Written with differentials, or infinitesimal differences, our result

$$ds^2 = -c^2 dt^2 + dx^2 + dy^2 + dz^2 \tag{7.18}$$

is called the line element of **Minkowski space-time** or, equivalently, of **flat space-time**. The most important difference with Euclidean space is that the time difference squared enters with a negative sign. Formally, all the implications of special relativity follow from this property of the Minkowski line element.

7.2.3 CONSEQUENCES

The negative sign of $c^2 dt^2$ in the Minkowski line element has several direct consequences.

7.2.3.1 Characterization of separations between two events

Since (7.18) is not positive definite, two distinct events in space-time can be separated by positive, negative, or vanishing line elements in space-time. We characterize these three cases as follows:

$ds^2 > 0$: The two events are **spacelike separated.**

$ds^2 = 0$: The two events are **lightlike separated** (or **null separated**).

$ds^2 < 0$: The two events are **timelike separated.**

In Euclidean space, for comparison, ds^2 cannot be negative, and it is zero only if $dx = dy = dz = 0$, in which case we do not have separate points. For two distinct points in Euclidean space, we always have $ds^2 > 0$.

7.2.3.2 The light cone and causality

The light cone of an event P can now be defined as the set of all events that are lightlike separated from P. As shown in Figure 7.7, an event A outside the light cone of P is spacelike separated from P, an event B on the light cone is lightlike separated, and an event inside the light cone is timelike separated. Physical motion allows massless particles, such as light or gravitational waves, to move from P to B, and massive particles to move from P to C. A straight line connecting P with B is a light

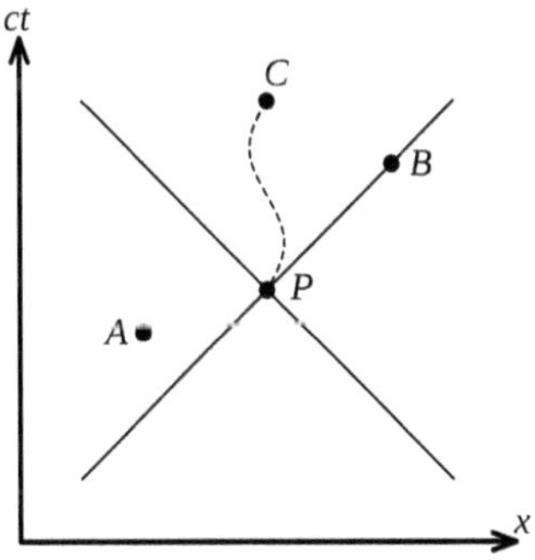

Figure 7.7 The light cone of P with a spacelike separated event (A), a lightlike separated event (B) and a timelike separated event (C). Only B and C can be connected to P by motion, with speed equal to c for B and (possibly varying) speed less than c for C. The light cone therefore introduces causal relationships between events.

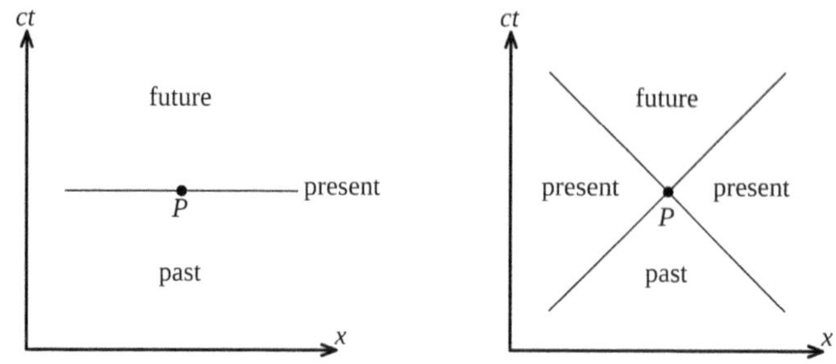

Figure 7.8 The non-relativistic and relativistic descriptions of causality. The absolute time in Newtonian mechanics implies a much smaller present as a subset of space-time.

ray, and a line (not necessarily straight) connecting P with C is a timelike worldline if its tangent vector always points up within the light cone. Moving from A to P would require speeds faster than light.

The light cone of P introduces a **causal structure** in space-time. If we remove P from the set, the light cone and its interior both have two parts, while the exterior remains connected if at least two spatial dimensions are included. We call the top half of the light cone the **future light cone** and the bottom half the **past light cone**. Causal relationships are described as follows: P can influence events in or on the future light cone, and it can be influenced by events in or on the past light cone. A comparison with the Newtonian view is shown in Figure 7.8.

7.2.3.3 Proper time

Consider two timelike related events A and B. Using the line element $\mathrm{d}s^2$ between A and B, we define the positive quantity $\mathrm{d}\tau^2 = -\mathrm{d}s^2/c^2$ with units of time squared. Since it is positive, we can take the square root and compute a real number

$$\mathrm{d}\tau = \frac{\sqrt{-\mathrm{d}s^2}}{c}. \tag{7.19}$$

For two timelike related events, $\mathrm{d}\tau$ is called the **proper-time interval**. It is the time measured by a clock moving along with an object that moves from A to B: Relative to such a clock, the position of the object does not change, and therefore $\mathrm{d}x = \mathrm{d}y = \mathrm{d}z = 0$. The line element is then $\mathrm{d}s^2 = -c^2\mathrm{d}t^2$, and $\mathrm{d}\tau = \mathrm{d}t$. The proper-time difference agrees with the coordinate-time difference of a corresponding observer.

For an observer moving with a different speed, the time difference $\mathrm{d}t'$ does not agree with $\mathrm{d}t$ or $\mathrm{d}\tau$. However, since the line element and the speed of light, and therefore $\mathrm{d}\tau$, are invariant, the result

$$\mathrm{d}\tau = \frac{\sqrt{-\mathrm{d}s^2}}{c} = \sqrt{\mathrm{d}t'^2 - \frac{1}{c^2}(\mathrm{d}x'^2 + \mathrm{d}y'^2 + \mathrm{d}z'^2)} \tag{7.20}$$

obtained in this frame should take the same value as in the previous frame. The invariant proper time is a powerful concept because it allows us to compute time differences experienced by a moving object without transforming to its rest frame. Since we do not need the rest frame of the object, we can use the definition of proper time in order to describe accelerating objects. Only observers must, for now, remain in global inertial frames because they have to ensure synchronization of their clocks.

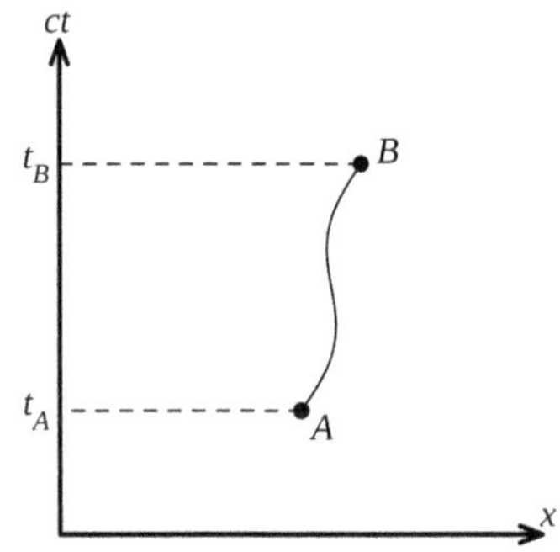

Figure 7.9 Two timelike events, connected by a timelike worldline.

If the separation between two timelike related events is not small, as illustrated in Figure 7.9, we obtain the proper-time duration of motion from one to the other by integrating $d\tau$ along the corresponding timelike worldline:

$$\begin{aligned} \tau_{AB} &= \int_A^B d\tau = \int_A^B \sqrt{dt^2 - \frac{dx^2+dy^2+dz^2}{c^2}} \\ &= \int_{t_A}^{t_B} \sqrt{1 - \frac{1}{c^2}\left(\left(\frac{dx}{dt}\right)^2 + \left(\frac{dy}{dt}\right)^2 + \left(\frac{dz}{dt}\right)^2\right)} dt = \int_{t_A}^{t_B} \sqrt{1 - \frac{V(t)^2}{c^2}} dt \end{aligned} \tag{7.21}$$

with the speed

$$V(t) = \sqrt{\left(\frac{dx}{dt}\right)^2 + \left(\frac{dy}{dt}\right)^2 + \left(\frac{dz}{dt}\right)^2} = \left|\begin{pmatrix} V^x \\ V^y \\ V^z \end{pmatrix}\right| \tag{7.22}$$

along the worldline. Here, $dx^2/dt^2 = (dx/dt)^2$ is the square of a first derivative as inherited from the line element or the Pythagorean theorem, and not a second derivative d^2x/dt^2.

On the worldline,

$$d\tau = \sqrt{1 - \frac{V(t)^2}{c^2}}\, dt \tag{7.23}$$

is always less than dt if there is motion, that is if $V(t) \neq 0$. The proper-time interval is always shorter than the coordinate time interval measured by any observer not simultaneously at rest with the moving object. This effect is called **time dilation**: A clock on a moving object shows a time difference $d\tau$ less than the time difference measured by an observer's clock. To the observer, processes on the moving object seem to be slowed down.

It is possible to place an atomic clock on an airplane and measure the effect of time dilation when it is moving, but as we will see later there is a competing altitude effect. A direct demonstration of relativistic time dilation is given by the observation that muons, unstable elementary particles produced when high-energy cosmic rays hit oxygen or nitrogen nuclei in the upper atmosphere, can be detected on the ground. With a life time of $2.2\mu s = 2.2 \cdot 10^{-6}$s they would not seem to survive a trip through the entire height of the atmosphere. Even though these light particles can move at speeds close to the speed of light, say $V = 0.999c$, they could travel only a distance of 660m before they decay.

We can interpret the life time as the resolution of a simple clock, to which time dilation should be applied. The life time $\Delta\tau = 2.2 \cdot 10^{-6}$s shown in the space-time diagram of Figure 7.10, is the proper-time duration between the creation and decay of a muon, measured in its rest frame. A detector on the ground is not in the muon's rest frame. The life time of a muon corresponds to a duration of

$$\Delta t = \Delta\tau/\sqrt{1-0.999^2} \approx 22\Delta\tau \approx \frac{14,500\text{m}}{c} \tag{7.24}$$

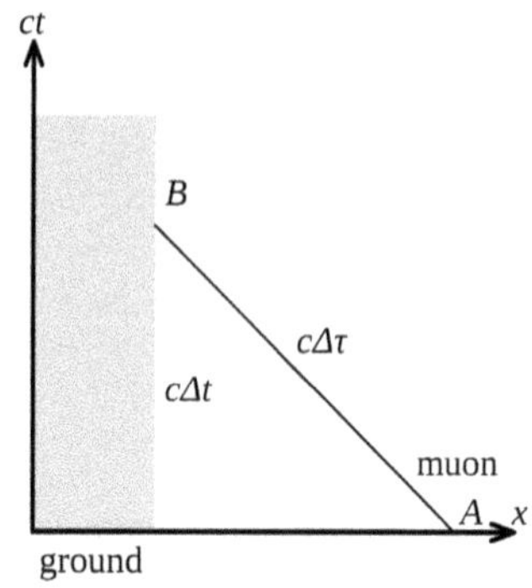

Figure 7.10 Space-time diagram of a muon travelling from the upper atmosphere (event A) to the ground (event B). In Minkowski geometry, the distance $c\Delta\tau$ is shorter than $c\Delta t$, even though it looks longer in this drawing on the Euclidean plane.

in the detector frame. In this frame, the muon can travel a distance of 14,500m before it decays. The entire atmosphere is several hundred kilometers high. (For comparison, the altitude record set by balloons is about 40km, while commercial airplanes fly at an altitude of up to 12km.) But its densest part, the troposphere which contains about 80% of its mass, extends up to 7km at the poles and 17km at the equator. The distance computed in (7.24) is in this range. We should, statistically speaking, be able to detect some of the muons before they decay. This expectation is confirmed by direct observations. Cosmic muons can even be a nuisance in high-energy particle detectors, and they could disturb sensitive qubits in quantum computers.

7.2.3.4 Distances in space-time and the twin paradox

In Euclidean space, the straight line has the shortest distance between two given points. In Minkowski space-time, the shortest distance between two events is zero: As shown in Figure 7.11, we can always connect two events by a series of lightlike worldlines, all of which have zero distance ds^2. (For this mathematical statement, we may use lightlike worldlines going back in time, if necessary.) If the events are not lightlike related, we have to use a combination of left-moving and right-moving light rays. The event where a right-moving light ray meets a left-moving one, or vice versa, is nothing but reflection at a mirror placed at a suitable position.

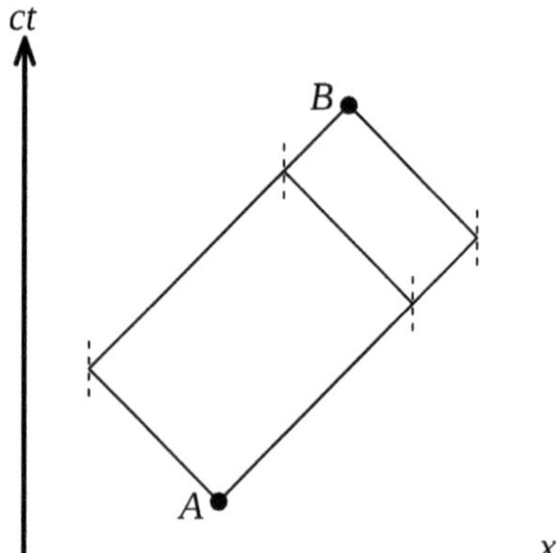

Figure 7.11 Several zero-distance worldlines connecting events A and B. Dotted lines indicate worldlines of mirrors at rest where light rays are reflected in order to reach B.

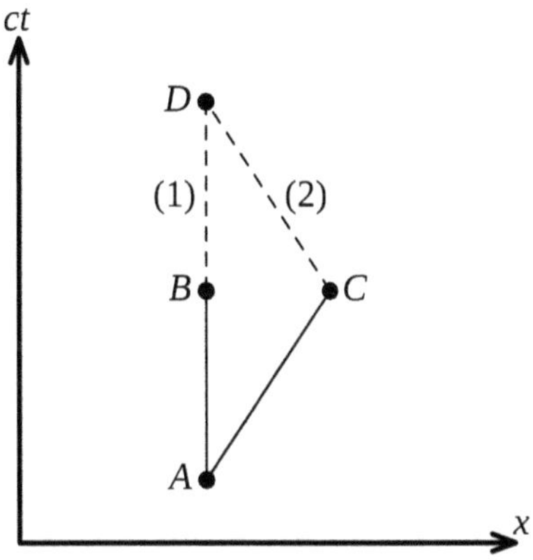

Figure 7.12 Longest distance in space-time and the twin paradox.

Since we can place mirrors at various positions, there is no unique worldline of this form for two given events. Therefore, the question of shortest-distance curves in space-time is not very interesting. However, the question of the *longest* distance between two timelike separated events is of interest. It has a unique answer: the straight worldline connecting the two events. This property is illustrated by the following example.

Define three events, A with coordinates $(ct,x) = (1,4)$, B with coordinates $(ct,x) = (6,4)$, and C with coordinates $(ct,x) = (6,7)$. Events B and C are timelike separated from A. We compute the proper-time differences

$$c\Delta\tau_{AB} = \sqrt{c^2\Delta t_{AB}^2} = 5 \qquad \text{and} \qquad c\Delta\tau_{AC} = \sqrt{c^2\Delta t_{AC}^2 - \Delta x_{AC}^2} = 4\,. \tag{7.25}$$

The straight line from A to C is shorter than the straight line from A to B, even though it looks longer in the Euclidean geometry we intuitively assume when we contemplate a space-time diagram such as Figure 7.12.

In order to test the statement about longest distances in space-time, we extend the worldlines from A to B and C, respectively, by introducing a fourth event, D, with coordinates $(ct,x) = (11,4)$. The two worldlines from A to D, one via B and one through C, are mirror symmetric along the line connecting B with C. Therefore, the two distances from A to D are given by

$$c\Delta\tau_{AD}^{(1)} = c\Delta\tau_{AB} + c\Delta\tau_{BD} = 10 \qquad \text{and} \qquad c\Delta\tau_{AD}^{(2)} = c\Delta\tau_{AC} + c\Delta\tau_{CD} = 8\,. \tag{7.26}$$

The distance through C is shorter than the distance on the straight line through B.

We interpret this geometrical result as an example of the **twin paradox**. One twin stays at rest on the straight worldline from A to D via B, while the other twin first moves to the right with some constant speed v, then turns around at C and meets the first twin again at D. The proper-time differences along the two worldlines are different, the twin at rest ageing faster than the moving twin. The apparent paradox about this situation consists in our expectation that we should be able to consider the second twin as being at rest while the first twin is moving. The age difference should then be reversed, but the age difference we see when the twins re-unite at D cannot depend on the frame used.

Like most paradoxes in special relativity, the twin paradox is resolved by taking into account the fact that simultaneity is relative. If we put ourselves in the frame of the second twin moving away from twin 1, the event C when we start returning is not simultaneous with event B, but rather with an event B' with the same ct'-coordinate as C. This event happens earlier than B or C in the original rest frame, see Figure 7.13.

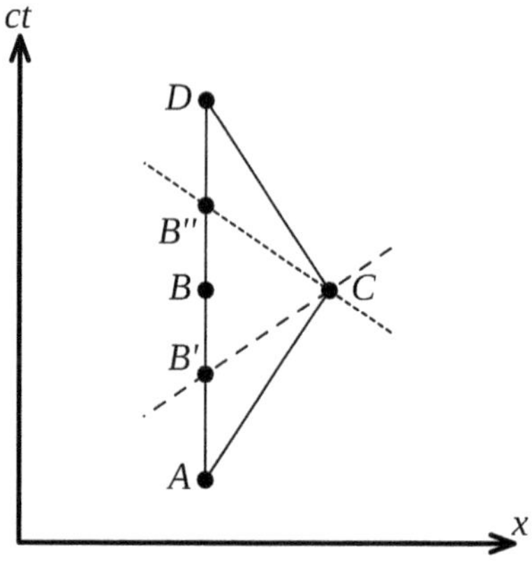

Figure 7.13 In the rest frame of twin 1, event B is simultaneous with C. In the frame of the twin moving away from twin 1, event B' is simultaneous with C, while B'' is simultaneous with C in the frame of the moving twin travelling back to twin 1. All events simultaneous with C while twin 2 is departing are indicated by the dashed line, while all events simultaneous with C while twin 2 is returning are indicated by the dotted line. The slope of these lines is computed by setting ct' equal to a constant in the boost equation with speed v and $-v$, respectively.

The line of all events simultaneous with C in the moving frame is described by the equation $ct' = \gamma(ct - vx/c)$ with a constant ct', or $ct(x) = \text{const}' + vx/c$. This linear relationship describes a straight line with positive slope. If the second twin moves back, the speed is $-v$ and the set of events simultaneous with C in this frame is a straight line $ct(x) = \text{const}'' - vx/c$ with negative slope. In this frame, the event simultaneous with C is B'', which happens later than B in the original rest frame. If we put ourselves in the frame of the moving twin, we cover only the portions from A to B' and from B'' to D while the speed is constant. The missing piece from B' to B'', which corresponds to the brief non-inertial change of velocity at C, accounts for the more-advanced age of twin 1 at D.

Boost equations and relevant parameters such as the speed v allow us to compute $\Delta\tau_{AB'}$ in relation to $\Delta\tau_{AB}$ as well as $\Delta\tau_{AC}$. We first express the condition $t'_{B'} = t'_C$ that defines B' in the original coordinate system:

$$\frac{t_{B'} - (v/c^2)x_{B'}}{\sqrt{1 - v^2/c^2}} = \frac{t_C - (v/c^2)x_C}{\sqrt{1 - v^2/c^2}} \tag{7.27}$$

or

$$t_C - t_{B'} = \frac{v}{c^2}(x_C - x_{B'}). \tag{7.28}$$

Since $x_{B'} = x_B$ and twin 2 moves with speed v, the position difference equals

$$x_C - x_{B'} = v(t_C - t_A) = v(t_B - t_A). \tag{7.29}$$

With these results, we compute

$$\Delta t_{B'B} = t_B - t_{B'} = t_C - t_{B'} = \frac{v}{c^2}(x_C - x_{B'}) = \frac{v^2}{c^2}\Delta t_{AB} \tag{7.30}$$

or

$$\Delta t_{AB'} = \Delta t_{AB} - \Delta t_{BB'} = \left(1 - \frac{v^2}{c^2}\right)\Delta t_{AB}. \tag{7.31}$$

We write (7.31) as

$$\Delta\tau_{AB'} = \left(1 - \frac{v^2}{c^2}\right)\Delta\tau_{AB} \tag{7.32}$$

TABLE 7.1
Comparison of the Two Perspectives in the Twin Paradox, Implying a Frame-Independent Statement in the Last Line

twin at rest	(1)	(2)
event simultaneous with C	B	B'
age of twin at rest when simultaneous with C	$\Delta\tau_{AB}$	$\Delta\tau_{AC}$
age of moving twin when simultaneous with C	$\Delta\tau_{AC}=\sqrt{1-v^2/c^2}\,\Delta\tau_{AB}$	$\Delta\tau_{AB'}=\sqrt{1-v^2/c^2}\,\Delta\tau_{AC}$
frame-independent	$\Delta\tau_{\text{moving}}=\sqrt{1-v^2/c^2}\,\Delta\tau_{\text{at rest}}$	

because all three events referred to in this equation are on a vertical worldline in the rest frame with time coordinate t. We also have

$$\Delta\tau_{AC}=\sqrt{1-\frac{v^2}{c^2}}\,\Delta t_{AB}=\sqrt{1-\frac{v^2}{c^2}}\,\Delta\tau_{AB} \tag{7.33}$$

using our derivation of time dilation. Therefore,

$$\Delta\tau_{AB'}=\left(1-\frac{v^2}{c^2}\right)\Delta\tau_{AB}=\sqrt{1-\frac{v^2}{c^2}}\,\Delta\tau_{AC}\,. \tag{7.34}$$

These equations allow us to compare the two perspectives from the twin at rest and the departing twin as summarized in Table 7.1.

The final equation in Table 7.1 shows that the age of the moving twin and the age of the twin at rest are related in the same way in both frames. However, owing to relative simultaneity, the two frames do not refer to the same worldlines in order to determine the age of twin 1. In the rest frame of twin 2 while departing, the events between B' and B on the worldline of twin 1 are not covered, and similarly in the rest frame of twin 2 returning the events between B and B'' are not covered. The multitude of events between B' and B'' in the life of twin 1 correspond to the brief acceleration event C of twin 2. They account for the visible age difference when the twins re-unite at D. From the perspective of twin 2 during the acceleration event, we have a fast-forward view of the life of twin 1 between events B' and B'': The lines of simultaneity through C sweep through the entire area in the triangle $B'CB''$. Because the spatial distance between C and B may be large, a brief acceleration event can be related to a long duration between B' and B''.

7.2.3.5 Minkowski geometry

The circle is an important object in Euclidean geometry. It is used for instance in the definition of an angle and gives rise to the name of *arc*seconds. Minkowski geometry, as defined by the Minkowski line element, can be developed by analogy with familiar properties of circles and rotations in Euclidean geometry.

In Euclidean space, the circle is characterized by its invariance under rotations around a given point P, its center. We describe rotations by an angle φ in the x-y plane by

$$x'=x\cos(\varphi)+y\sin(\varphi)\qquad,\qquad y'=-x\sin(\varphi)+y\cos(\varphi)\,. \tag{7.35}$$

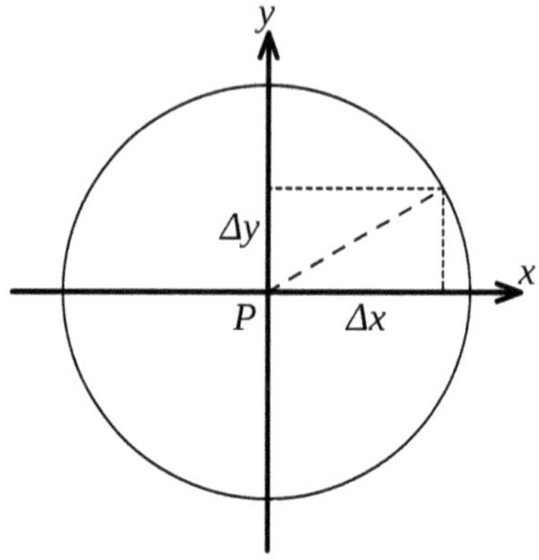

Figure 7.14 A unit circle on the Euclidean plane: $(\Delta x)^2 + (\Delta y)^2 = 1$.

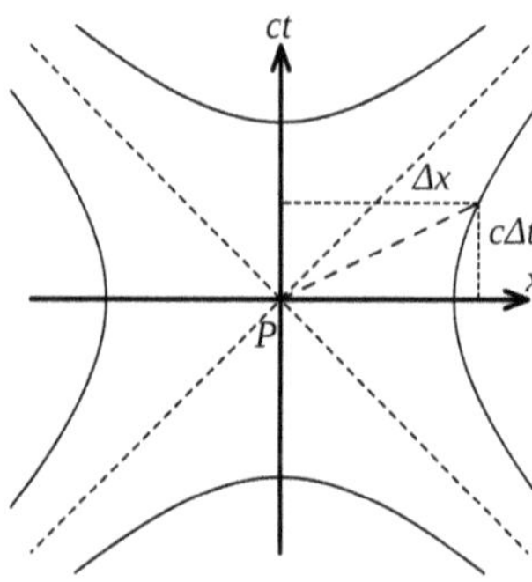

Figure 7.15 The four branches of a unit hyperbola in Minkowski space-time, which asymptotically approach the light cone: $-(c\Delta t)^2 + (\Delta x)^2 = \pm 1$.

The coefficients in this transformation satisfy the equation $\cos(\varphi)^2 + \sin(\varphi)^2 = 1$. At the same time, these coefficients are the coordinates of a point on the unit circle, which is defined by the equation $(\Delta x)^2 + (\Delta y)^2 = 1$, where coordinate differences are taken from the origin; see Figure 7.14.

We do not possess the same intuition about Lorentz boosts in Minkowski space-time, but we can use algebraic equations to derive the corresponding results. In space-time, we describe a Lorentz boost with velocity v in the x-direction by the equations

$$ct' = \gamma ct - \alpha x \quad , \quad x' = -\alpha ct + \gamma x . \tag{7.36}$$

The coefficients $\gamma = (1 - v^2/c^2)^{-1/2}$ and $\alpha = \gamma v/c$ obey the equation $\gamma^2 - \alpha^2 = 1$. For two such coefficients, there always exists a real number ρ such that $\gamma = \cosh(\rho)$ and $\alpha = \sinh(\rho)$. Compared with the angle in Euclidean geometry, we replace trigonometric functions with their hyperbolic versions, $\cosh(x) = \frac{1}{2}(e^x + e^{-x})$ and $\sinh(x) = \frac{1}{2}(e^x - e^{-x})$. We call the new parameter ρ the **rapidity** of the transformation, or the **hyperbolic angle**.

By analogy with the unit circle, we introduce the set of all events with unit space-time distance Δs^2 from a given event, called the **unit hyperbola**. In contrast to Euclidean geometry, unit distance can refer to spacelike or timelike separation, corresponding to $\Delta s^2 = +1$ or $\Delta s^2 = -1$. The unit hyperbola is not a connected set but has several branches described by the equation $-(c\Delta t)^2 + (\Delta x)^2 = \pm 1$; see Figure 7.15.

Invariant unit objects can be used to determine important properties of transformations of a coordinate system. When we rotate a coordinate system in Euclidean space, we move the units on

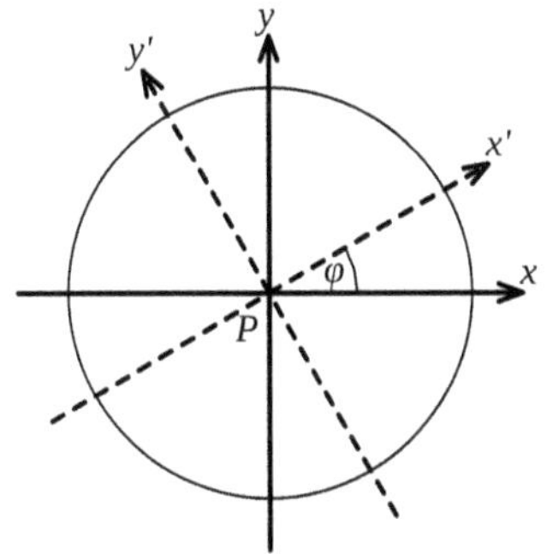

Figure 7.16 Rotation of the coordinate system by an angle φ.

the axes along the unit circle in order to ensure that the distances between the origin and the units remain of unit length. How far we move the units depends on the angle of the rotation. Only one question remains to be answered: Do we move the axes in the same way, or in opposite ways toward each other? In the case of a rotation, intuition tells us that we should move the axes in the same way as shown in Figure 7.16, ensuring that the right angle between them is preserved.

We do not have intuition about Lorentz boosts. Moreover, we have not yet defined geometrical angles in Minkowski geometry, and therefore cannot use the argument referring to right angles being preserved. We know that we should move the units along the unit hyperbolae in order to ensure unit distance between the origin and the units according to Minkowski geometry. But this fact does not tell us whether we should move the axes in the same way or in opposite ways.

In the absence of intuition, it is necessary to derive the correct position of the transformed axes. We start with one of the axes, placed in a new position by a Lorentz boost. Since we cannot refer to angles as in Euclidean geometry, we must find a different way of drawing a new axis determined by the velocity or rapidity of a Lorentz boost. In the original frame (ct, x), the time axis is characterized as the worldline of an observer who stays at $x = 0$. Similarly, the time axis of a new frame (ct', x') is the worldline of an observer who stays at $x' = 0$ and moves at the frame speed v relative to the original frame. The latter condition can be implemented in the original coordinate system by drawing a timelike worldline through the origin with slope c/v (one unit to the right, c/v units up).

The worldline through the origin of an observer at speed v is our new ct'-axis. In this construction, the speed implicitly determines the angle between the axes according to Minkowski geometry. The direction of the new x'-axis should be determined by moving the direction of the x-axis by the same amount as we changed the direction of the t-axis, but should we move it up or down? We cannot appeal to preserving the right angle we used when drawing the original coordinate system because this is a property of the Euclidean plane but not of Minkowski geometry.

Again, since intuition tells us nothing about Lorentz boosts and how they might change angles, we have to derive the correct result. Looking back at the ct'-axis, we characterized it algebraically as the set of all events with $x' = 0$ (or all coordinates other than ct' equal to zero). In the old coordinates, $x' = -\alpha ct + \gamma x$. If $x' = 0$, we have a linear relation $ct = (\gamma/\alpha)x$ between ct and x on the axis. A linear relation between Cartesian coordinates determines a straight line, in this case with slope $\gamma/\alpha = c/v$. This is exactly the straight line we have already obtained as the worldline through the origin of an observer moving at speed v.

In the same way, the x'-axis is the set of all events with $ct' = 0$. According to Lorentz transformations, $ct' = \gamma ct - \alpha x$. The linear relation $\gamma ct - \alpha x = 0$ determines a straight line $ct = (\alpha/\gamma)x$ with slope $\alpha/\gamma = v/c$. For $v > 0$, this parameter is positive, and we have to move the x'-axis upward, into the same quadrant in which we drew the ct'-axis. Figure 7.17 shows the result. The new coordinate system does not look right-angled, but this is a Euclidean notion and does not apply in Minkowski

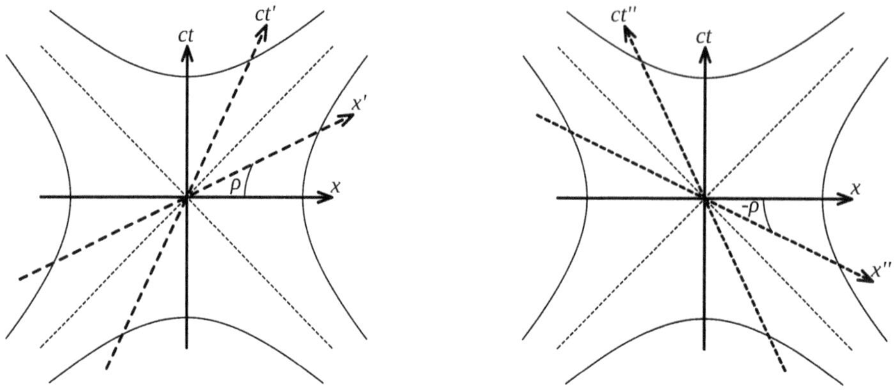

Figure 7.17 Lorentz boosts of a coordinate system with rapidity $\rho = \tanh(v/c)$ with $v > 0$ (left) and $v < 0$ (right). Minkowski angles are indicated by hyperbola segments.

geometry. (In fact, we will see soon that the Minkowski angle between the x and ct axes is infinite.) If $v < 0$, both new axes have negative slopes and therefore move away from each other.

The equations we used to derive the new axes also give us information about the new units. In our diagram of the new system, the units have moved outwards along the unit hyperbola, and their distance from the origin now looks greater than one. But they still have unit distance in Minkowski geometry. In terms of equations, we can characterize the x'-unit P' as follows: It has coordinates $ct'_{P'} = 0$ and $x'_{P'} = 1$ in the new system. Using these values in a Lorentz transformation shows that the new unit has coordinates $ct_{P'} = \alpha$ and $x_{P'} = \gamma$. In Minkowski geometry, this event has unit distance $-(ct_{P'})^2 + x_{P'}^2 = -\alpha^2 + \gamma^2 = 1$ from the origin. In Euclidean geometry as used when we draw and measure distances in the diagram, the length equals

$$\overline{OP'} = \sqrt{(ct_{P'})^2 + x_{P'}^2} = \sqrt{\frac{1+v^2/c^2}{1-v^2/c^2}}\,. \tag{7.37}$$

This number is greater than one for non-zero v. It can be used to rescale distances along the ct' or x'-axis to the correct Minkowski distance if they have been measured in the diagram with a Euclidean ruler. (It is impossible to construct a Minkowski-geometry ruler because it would have to stretch out if we place it in a position that is not horizontal or vertical.)

The coordinates of the new unit allow us to relate the angle between the old and new axes to the rapidity. We introduced the rapidity ρ by the relations $\gamma = \cosh\rho$ and $\alpha = \sinh\rho$. Therefore,

$$\frac{ct_{P'}}{x_{P'}} = \frac{\alpha}{\gamma} = \frac{\sinh\rho}{\cosh\rho} = \tanh\rho\,. \tag{7.38}$$

The ratio of the vertical coordinate of the x'-unit to its horizontal coordinate is equal to the hyperbolic tangents of the rapidity, just as the ratio of the vertical (y) coordinate to the horizontal (x) coordinate of the new unit in a rotated Euclidean system is the tangents of the rotation angle. This analogy identifies the rapidity as the **hyperbolic angle** in Minkowski geometry.

The definition of an angle relies on unit objects in geometry. In Euclidean geometry, the angle between two straight lines is the arc length of a segment of the unit circle between the lines, drawn around the point where the lines intersect. In Minkowski geometry, the hyperbolic angle between two straight lines is the arc length of a piece of the unit hyperbola between the lines, drawn around the point where the lines intersect. We can compute this angle if we integrate the line element along the piece of the hyperbola, or more precisely, we integrate $\mathrm{d}s = \sqrt{\mathrm{d}s^2}$ if the hyperbola is a spacelike worldline (in the case of an angle between two timelike straight lines) or $c\,\mathrm{d}\tau = \sqrt{-\mathrm{d}s^2}$ if the hyperbola is a timelike worldline (in the case of an angle between two spacelike straight lines). If one of the straight lines is lightlike, the hyperbolic angle is infinite because the unit hyperbolae do not reach the lightlike line: The angle equals the full length of a hyperbola from a finite point to infinity, which is infinite. Therefore, the angle between a spacelike line and a timelike line is also infinite because the region between them includes a lightlike line. For this reason, the angle between the axes of a space-time diagram is infinite.

7.2.3.6 Coordinate grids

In a right-angled coordinate system, we can use parallel or orthogonal projection to read off the coordinates of a point P, producing the same results. With parallel projection, we draw a straight line parallel to the x-axis through P and determine the ct-coordinate as the value on the ct-axis where it intersects the straight line through P. For orthogonal projection, we draw a line orthogonal to the ct-axis through P and read off the ct-coordinate of P, where the orthogonal line intersects the ct-axis.

Our transformed space-time diagram with coordinates (ct',x') preserves the angle between the axes, but in Minkowski geometry. Its representation drawn on the Euclidean plane, such as a sheet of paper or a computer screen, does not look right-angled if we assume our intuitive understanding based on Euclidean geometry. In this situation, we cannot use orthogonal projection according to Euclidean geometry. Parallel projection, as shown in Figure 7.18, gives us the correct coordinate values because it does not make use of specific angles. It only requires using the same angle for two parallel lines, for with the symmetry transformation of translations is sufficient. Notice how parallel projection gives us a full coordinate grid, while orthogonal projection would not result in a periodic pattern.

The new method for drawing two frames in one diagram allows us to visualize relative simultaneity. Figure 7.19 shows two events, A and B, that are simultaneous in the moving frame (ct',x'): The events lie on a straight line parallel to the x'-axis, and therefore parallel projection gives the same ct'-coordinate for both of them. However, they do not have the same ct-coordinate. Figure 7.20 shows the same events A and B together with light rays sent from A toward B and from B toward A, respectively. The light rays meet at an event that lies on the worldline of an observer O' who is at rest in the moving frame and always stays in the middle between worldlines at rest through A and B in the frame (ct',x'). In the moving frame, the light rays should indeed meet on this worldline because

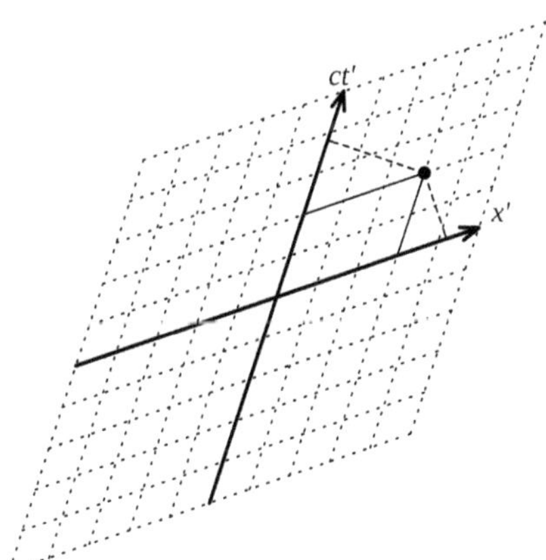

Figure 7.18 Parallel projection in a coordinate grid. Orthogonal projection on the axes, as indicated by the dashed lines, would not result in a periodic pattern.

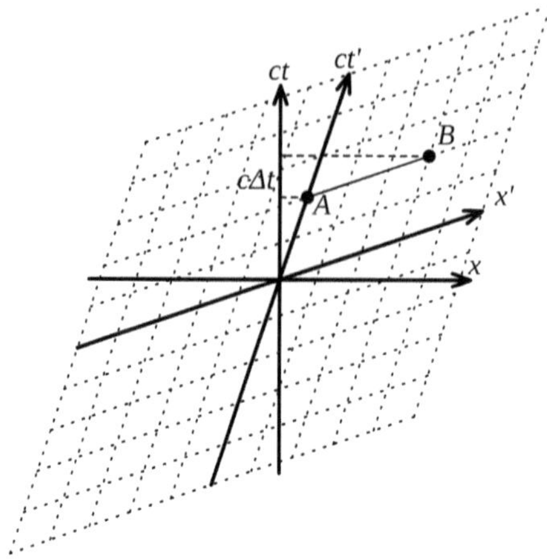

Figure 7.19 Relative simultaneity in a space-time diagram: The two events A and B have the same ct'-coordinate but different ct-coordinates.

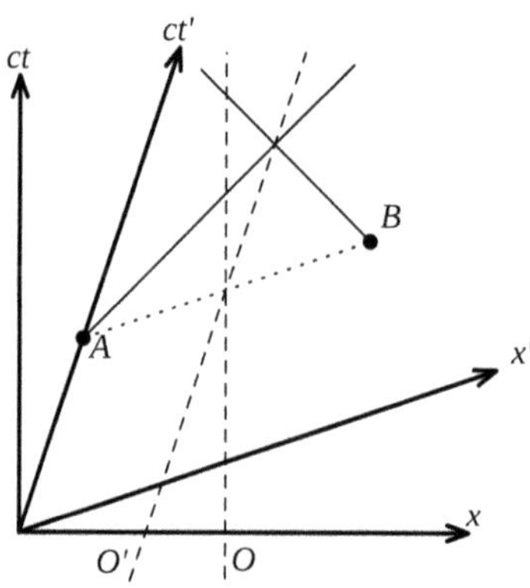

Figure 7.20 Relative simultaneity and the exchange of light signals. In the frame of observer O', A and B are simultaneous events, and the observer in their middle receives their signals at the same time. For an observer at rest, A happens earlier than B and is therefore seen first from the middle position.

they travel the same distance to O' during equal amounts of time from A and B, respectively. An observer O who stays halfway between the positions of A and B in the frame (ct,x) does not receive the light rays at the same time because, in this frame, event B happens later.

In equations, we have $c\Delta t' = 0$ but $\Delta x' \neq 0$. A Lorentz transformation gives

$$c\Delta t = \gamma\left(c\Delta t' + \frac{v}{c}\Delta x'\right) = \gamma\frac{v}{c}\Delta x' \neq 0 \tag{7.39}$$

and

$$\Delta x = \gamma\left(\Delta x' + \frac{v}{c}c\Delta t'\right) = \gamma\Delta x' > \Delta x' \tag{7.40}$$

for these events. The spatial coordinate separation is modified by a Lorentz transformation, which indicates that length measurements change when an observer is boosted, just like time measurements. However, while a single worldline is sufficient to compute the difference between coordinate time and proper time, for a length measurement we need at least two worldlines of the end points. Relative simultaneity must then be taken into account because different observers refer to different

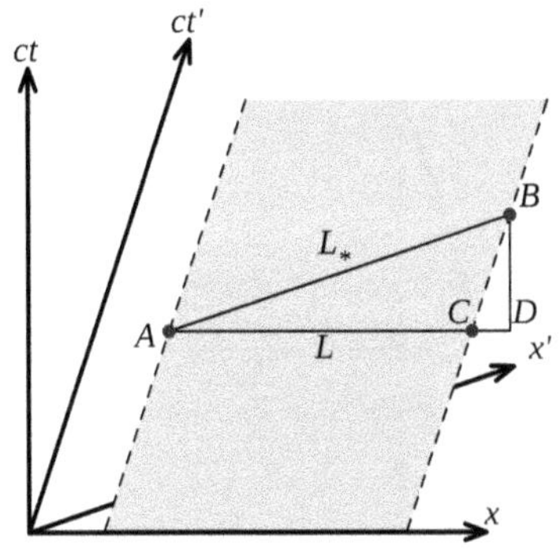

Figure 7.21 Lorentz contraction: moving object. The proper length $L_\star$ in the object's rest frame (ct',x') equals the distance $\overline{AB}$, while the length L measured by an observer at rest in (ct,x) equals the distance $\overline{AC}$. Between the events A and B, the object travels a distance $\overline{CD}$ in the frame (ct,x).

events on the two worldlines for a length measurement at one time. We will also see that the relationship between measured values in different frames depends on the measurement procedure. We should therefore avoid interpreting (7.40) as a direct statement about expanded length measurements until we have developed a proper understanding of spatial length in special relativity.

7.2.3.7 Spatial distances

Equation (7.40) suggests that length measurements might be relative and frame-dependent. However, Δx is a coordinate difference, and we have to see how it may be related to actual measurements. Figure 7.21 is an extended version of Figure 7.19: In addition to two events A and B which are simultaneous in the moving frame (ct',x'), we have worldlines through these events corresponding to endpoints of an object at rest in this frame. The shaded area is swept out by the object as it moves. The space-time distance between A and B is the distance between the two endpoints *when they are simultaneous* in the moving frame. This is the **proper length** L_*, measured in the frame in which the object is at rest. By parallel projection, it equals the coordinate distance $L_* = \Delta x'$ in this frame.

For an observer using the frame (ct,x), the events A and B are not simultaneous. The coordinate difference Δx, computed in (7.40), is *not* the length measured in this frame because the endpoints are not taken at the same time. A length measurement in the observer's rest frame should be based on events A and C in Figure 7.21, which are simultaneous in this frame and lie on the two worldlines of the endpoints. This is the length L measured in the rest frame of the observer. It is smaller than the coordinate difference Δx between A and B because adding Δx to the x-coordinate of A brings us to event D in the figure. The x-difference between C and D is the distance traveled by the moving object during the time difference Δt between events A and B. This travel distance should not be included in a direct length measurement in which we align the endpoints of an object with a ruler at a given time in our frame.

If v is the speed, the travel distance is $v\Delta t$. Adding it to L gives the full distance Δx between A and D:

$$L + v\Delta t = \Delta x\,. \tag{7.41}$$

From our Lorentz transformation with $\Delta t' = 0$ we have $\Delta x = \gamma \Delta x' = \gamma L_*$ in (7.40), as well as $c\Delta t = \gamma(v/c)\Delta x' = \gamma(v/c)L_*$. Inserting these relations, we solve for

$$L = \gamma\left(1 - \frac{v^2}{c^2}\right)L_* = \frac{L_*}{\gamma} \leq L_*\,. \tag{7.42}$$

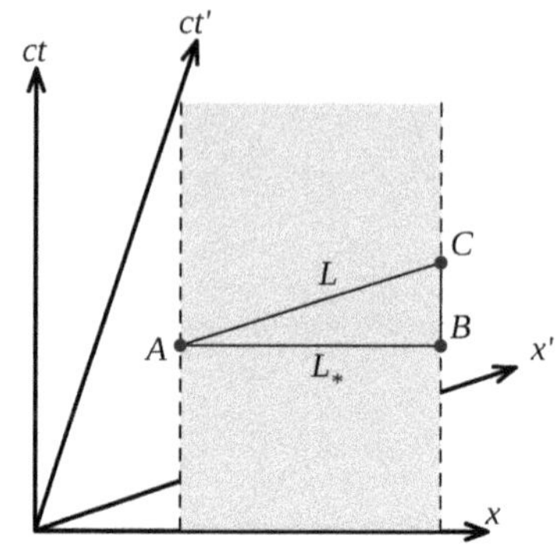

Figure 7.22 Lorentz contraction: object at rest. The events A and C have the same ct' coordinates and are simultaneous in the observer's frame, while A and B have the same ct coordinates and are simultaneous in the object's rest frame. Minkowski geometry implies $L < L_*$ even though a Euclidean interpretation of the drawing suggests the opposite.

The length of a moving object is shorter than the proper length of the object measured in its rest frame. This effect is called **Lorentz contraction.**

Figure 7.22 shows the same situation in the rest frame (ct, x) of the object. The worldlines of the two ends are vertical in this frame, with two simultaneous events A and B on them. The space-time distance between A and B is the proper length L_*, which is equal to the spatial separation, $L_* = \Delta x$, because the events are simultaneous. In the moving frame (ct', x') of an observer, events A and C on the worldlines are simultaneous. The space-time distance between them is the measured length L. We can write this distance in terms of the line element between events A and C, $\Delta s^2 = L^2$. The line element is invariant and has the same value in the rest frame of the object, where it is expressed in coordinates as $\Delta s^2 = -(c\Delta t)^2 + (\Delta x)^2 = -(c\Delta t)^2 + L_*^2$.

We compute the time separation Δt between A and C from the condition $\Delta t' = 0$, together with a Lorentz boost: $\Delta t' = \gamma(\Delta t - (v/c^2)\Delta x) = 0$ implies $\Delta t = (v/c^2)\Delta x = (v/c^2)L_*$. The invariant line element shows that

$$L^2 = -(c\Delta t)^2 + L_*^2 = \left(1 - \frac{v^2}{c^2}\right) L_*^2, \tag{7.43}$$

which is the same result of Lorentz contraction as derived for a moving object.

7.2.3.8 Spaceship paradox

Consider two spaceships which initially are in a common rest frame (ct, x), one at position $x = 0$ and one at position $x = d > 0$. They are programmed to start accelerating in exactly the same way for a brief moment at time $t = 0$. For $t > 0$, after the acceleration event, they are moving at the same constant speed v. Their positions are then described by the equations $x_1(t) = vt$ for spaceship 1, and $x_2(t) = d + vt$ for spaceship 2. At all times, the distance between the spaceships is equal to $\Delta x = x_2(t) - x_1(t) = d$, a constant.

Before the spaceships accelerate, someone has tied a thin string with proper length $L_* = d$ between them. Seen from the original rest frame of the spaceships, the string accelerates with the spaceships and quickly reaches the same speed v. It is then moving and should be Lorentz contracted, such that its length $L = L_*/\gamma = d/\gamma$ is less than the distance d between the spaceships. The string should break. However, since the distance between the spaceships always remains constant, their combination with the string attached can be viewed as a single object. There is then no difference between the accelerating spaceships and the accelerating string. The string should not break. Two perspectives seem to give opposite results, which is a paradox.

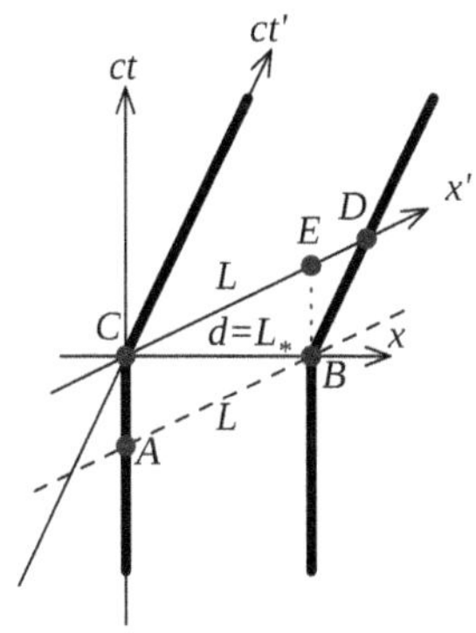

Figure 7.23 Spaceship paradox. After acceleration, the Lorentz-contracted length L of the string only spans the distance from C to E, which is too short to reach spaceship 2 at D. In the moving frame, both spaceships are initially moving and stop at the acceleration events. As the dashed line shows, spaceship 2 stops at a negative ct' in the moving frame (event B), earlier than spaceship 1 at $ct' = 0$ (event C). The distance increases while spaceship 1 is still moving.

The paradox is resolved by drawing a space-time diagram as in Figure 7.23. We have the original rest frame (ct, x) and a moving inertial frame (ct', x') that has the final speed of the spaceships. The two worldlines are drawn thick, and a dashed line has been included to indicate relevant events that are simultaneous in the moving frame. If we follow the worldlines upwards, seen from the moving frame, the two spaceships are initially moving to the left, or to smaller x'. This is as expected because the observer with this frame is moving to the right.

The dashed line, connecting events A and B, contains the event B with coordinates $(0, d)$ in the original frame. Spaceship 2 accelerates at this event, or stops moving as seen from the moving frame: The spaceship is boosted to the observer's speed. However, the worldline of spaceship 1 has not yet encountered its acceleration event C at this value of t'. Spaceship 1 keeps moving to the left until some later time t', when it stops too. Relative simultaneity gives a different picture of the acceleration events in the moving frame. In particular, the two spaceships no longer accelerate at the same time.

We now address the question of whether the string breaks. In the original rest frame, the string is expected to break because of Lorentz contraction. In the moving frame, relative simultaneity implies that spaceship 2 stops first while spaceship 1 keeps moving to the left for sometime. Spaceship 1 moves away from spaceship 2, increasing the distance between them beyond the string length L. The string breaks, as seen also in this frame.

While the distance between the spaceships remains equal to d at all times in the original rest frame, it is changing in the moving frame. Before spaceship 2 accelerates, the distance between the spaceships in the moving frame is equal to the space-time distance between events A and B, which is the Lorentz contracted length

$$L = \frac{d}{\gamma} = \frac{L_*}{\gamma} \tag{7.44}$$

of the string. Also the string is moving at this time, as seen from the moving frame, and it has the same Lorentz contracted length.

After both spaceships have accelerated, the distance between them is equal to the space-time distance between events C and D, which is equal to the coordinate separation $\Delta x'_{CD}$ between these events in the moving frame. As shown in the figure, the previous distance L between the spaceships, or the original length of the string in this frame, extends only from C to E and is much shorter than the distance from C to D between the spaceships. The string is no longer able to connect the spaceships.

The distance between B and C is the proper length of the string, but using the triangle BCD, we can also interpret it as the Lorentz-contracted proper distance between C and D after acceleration. In the original rest frame, the distance between B and C is the proper length L_* of the string. In the moving frame, this space-time distance between the same two events still has the value L_* even though it is now interpreted as the Lorentz contracted version of the distance $\Delta x'_{CD}$ between the spaceships. Therefore, $L_* = \Delta x'_{CD}/\gamma$. Combining this equation with Eq. (7.44), we obtain

$$L = \frac{\Delta x'_{CD}}{\gamma^2}\,. \tag{7.45}$$

Since $\gamma > 1$, the length L of the string in the moving frame is much less than the distance between the spaceships after they accelerate.

7.2.3.9 Optical measurements

Placing a ruler next to an object to measure its length is not practical if the object is moving at high speed or is placed far away, as in astronomical examples. Optical measurements are then more useful, and they can lead to different results than what is expected from Lorentz contraction. In optical measurements, travel time delays of light rays from different parts of an object sometimes have to be taken into account, which can be comparable to the travel time of a moving object if the speed is close to the speed of light.

Consider a cube moving at speed v in a direction normal to one of its sides. If a picture is taken by a camera from an orthogonal direction, one could expect to see a Lorentz-contracted side length of a squashed cube. However, as shown in Figure 7.24, the Lorentz-contracted side is only one part

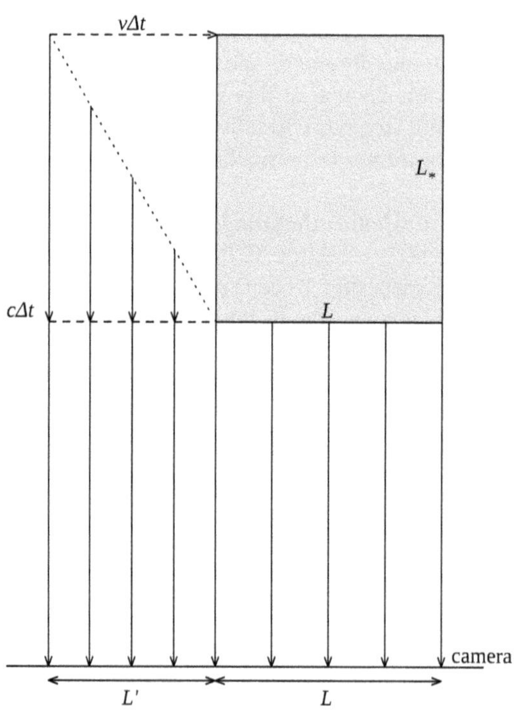

Figure 7.24 The back side of a moving cube (to the right) is visible if the cube moves at high speed.

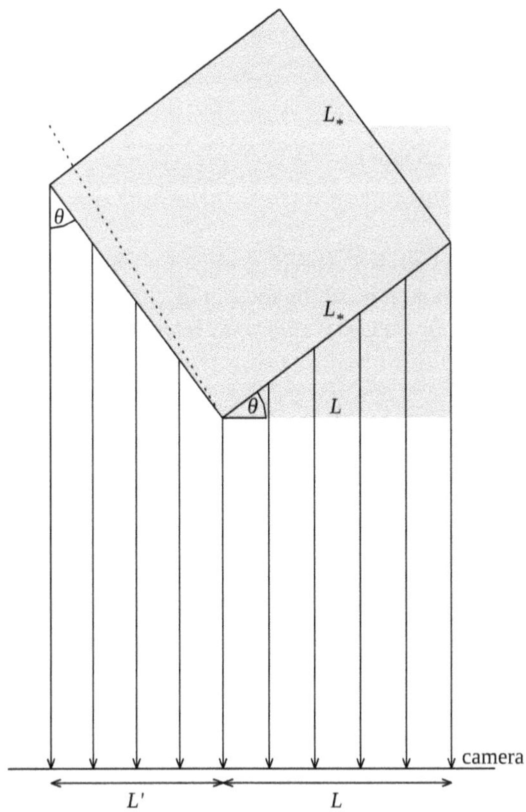

Figure 7.25 Image of a rotated cube compared with the moving cube. The dotted line indicates the emission points from the back side of a moving cube as in Figure 7.24.

of the image seen by the camera. Since the object is moving quickly to the right, it gives way to light rays that had been scattered off its back side such that they travel toward the camera. The picture therefore shows a projected version of the back side in addition to the Lorentz-contracted side.

When we derive the proportions of the image, we should start at the camera and consider the origins of all light rays arriving there at the same time, contributing to a single image. The Lorentz-contracted side is parallel to the camera lens and is therefore directly projected. Light rays from the back side, however, have to travel a longer distance and therefore must start earlier than light rays from the Lorentz-contracted side. If light rays from the upper edge of the back side had been scattered downwards some time Δt earlier, they travelled an additional distance $L_* = c\Delta t$ equal to the proper side length of the cube. (Lengths in directions orthogonal to the velocity are not Lorentz-contracted.) During this time, the cube moved a distance $L' = v\Delta t$ which is equal to the length of the back side projected on the image. Eliminating Δt from these two equations, we obtain $L' = (v/c)L_*$, while the Lorentz-contracted side seen in the image has length $L = \sqrt{1 - v^2/c^2}\, L_*$.

A number v/c less than one can always be written as the sine of some angle θ, such that $L' = L_* \sin(\theta)$. We then have $L = \sqrt{1 - \sin(\theta)^2}\, L_* = L_* \cos(\theta)$. In these relationships, the signs are such that we do not use hyperbolic versions of sine and cosine as in Minkowski geometry. The projected image of the moving cube can therefore be interpreted in terms of Euclidean geometry. In particular, we expect the same projected lengths if we view a cube which is not moving but rotated by the same angle θ; see Figure 7.25. We are used to seeing rotated objects, and would therefore interpret

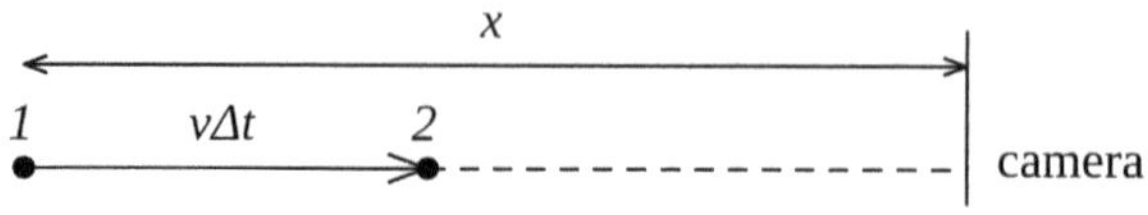

Figure 7.26 Emission of light at high speed.

the image of a cube moving at relativistic speed as one of a rotated cube.[1] In this sense, Lorentz contraction is not visible, even though it can be measured. Since light rays in these two situations have to cross different distances, the images would be different if seen through an absorbing medium such as fog.

7.2.3.10 Apparent velocity

In relativistic physics, velocities of moving objects do not surpass the speed of light. Nevertheless, optical measurements can result in measured speeds greater than c.

Consider a point-like light source moving with speed v toward a camera. During some time Δt, the object moves a distance $v\Delta t$. The speed of the object as seen by the camera is equal to this distance divided by the difference of arrival times of two light rays emitted at the beginning and the end of this period. Since the object is moving, light rays emitted at different times do not travel the same distance. Therefore, the time lag $\Delta\tilde{t}$ seen by the camera is not equal to Δt, and the measured speed V is different from v.

In order to compute V, we consider two emissions, one at time t_1 from the far end and one at time $t_2 = t_1 + \Delta t$ from the near end; see Figure 7.26. If the distance of the object from the camera at time t_1 is x, the light ray from the far end arrives at the camera at time $\tilde{t}_1 = t_1 + x/c$. The light ray emitted from the near end travels a smaller distance $x - v\Delta t$ to the camera. It arrives at the camera at time $\tilde{t}_2 = t_1 + \Delta t + (x - v\Delta t)/c$. We have $\Delta\tilde{t} = \tilde{t}_2 - \tilde{t}_1 = (1 - v/c)\Delta t$, and the measured speed

$$V = \frac{v\Delta t}{\Delta\tilde{t}} = \frac{v}{1 - v/c} \tag{7.46}$$

is always greater than v for $v > 0$. There is no upper limit: If v is close to but still less than the speed of light, the denominator is small, and V can be greater than c. For instance, if $v = \frac{1}{2}c$, we have $V = 2c$. This effect can be observed in relativistic jets, hot matter emitted from active stars at high speeds in a narrow beam. If a jet is pointed almost directly at our telescopes, it may appear to move with a speed faster than light.

7.2.3.11 Composition of velocities

We can compute the velocity of an object in a new frame by applying a Lorentz transformation to the coordinate differentials in $\vec{V} = \mathrm{d}\vec{x}/\mathrm{d}t$. Since the spatial coordinates are not independent along the worldline of a moving object, we have to reparameterize the worldline from $(x(t), y(t), z(t))$ to $(x'(t'), y'(t'), z'(t'))$ where both the dependent and independent variables change.

For the x-component, we write the new velocity component as

$$V^{x'} = \frac{\mathrm{d}x'}{\mathrm{d}t'} = \frac{\gamma(\mathrm{d}x - v\mathrm{d}t)}{\gamma(\mathrm{d}t - (v/c^2)\mathrm{d}x)}. \tag{7.47}$$

Here, the chain rule is formally applied by dividing numerator and denominator by $\mathrm{d}t$:

$$V^{x'} = \frac{\mathrm{d}x/\mathrm{d}t - v}{1 - (v/c^2)\mathrm{d}x/\mathrm{d}t} = \frac{V^x - v}{1 - (v/c^2)V^x}. \tag{7.48}$$

[1] http://www.spacetimetravel.org/fussball/fussball.html

Unlike in Newtonian mechanics, we do not simply subtract the frame velocity from the object's velocity. The relativistic composition of velocities is non-linear. For the remaining components, we proceed in a similar way:

$$V^{y'} = \frac{dy'}{dt'} = \frac{dy}{\gamma(dt - (v/c^2)dx)} = \sqrt{1 - \frac{v^2}{c^2}} \frac{V^y}{1 - (v/c^2)V^x} \tag{7.49}$$

and

$$V^{z'} = \sqrt{1 - \frac{v^2}{c^2}} \frac{V^z}{1 - (v/c^2)V^x} . \tag{7.50}$$

These equations imply that the norms of the velocity vectors are related by

$$|\vec{V}'|^2 = c^2 \left(1 - \frac{(1 - v^2/c^2)(1 - |\vec{V}|^2/c^2)}{(1 - vV^x/c^2)^2} \right) . \tag{7.51}$$

On the right, we always have $1 - v^2/c^2 > 0$ because the frame velocity of an observer cannot equal the speed of light. The denominator on the right is also positive, and therefore the sign of $|\vec{V}'|^2/c^2 - 1$ always equals the sign of $|\vec{V}|^2/c^2 - 1$: We have $|\vec{V}'| < c$ whenever $|\vec{V}| < c$, and $|\vec{V}'| = c$ whenever $|\vec{V}| = c$. Boosting never increases the speed beyond c. Moreover, if an object moves at the speed of light in one frame, it has the same speed in all frames. As a simple example, take $V^x = c$ and $V^y = 0 = V^z$. Then,

$$V^{x'} = \frac{V^x - v}{1 - (v/c^2)V^x} = \frac{c - v}{1 - v/c} = c . \tag{7.52}$$

Since the speed of light is invariant and takes the same value in all frames, we simplify equations by choosing our units such that $c = 1$. Time is then measured in units of length through the distance traveled by light in a certain time. From now on, we therefore write the Minkowski line element as

$$ds^2 = -dt^2 + dx^2 + dy^2 + dz^2 \tag{7.53}$$

and define proper time as $d\tau = \sqrt{-ds^2}$. Velocities then have no units. For evaluations in standard units, it is always possible to re-instate c by replacing any velocity component V with V/c.

7.3 RELATIVISTIC DYNAMICS

The non-linear composition law for velocities seems to make it difficult to apply vector calculus, which is based on linear algebra. However, there is a new notion of vectors in space-time which is compatible with the relativistic composition of velocities. Understanding space-time vectors will help us to derive the correct form of relativistic dynamics.

7.3.1 LINEAR ALGEBRA

From a geometrical perspective, a vector is the mathematical concept used to describe a direction at some point in space, or at an event in space-time. In 3-dimensional Euclidean space, we have the familiar 3-vectors $\vec{a}$, with three components $(a^x, a^y, a^z)^T = (a^i)$ labeled by an index $i = x, y, z$, or $i = 1, 2, 3$ if we simply enumerate the coordinates. Vectors are related to geometry because we can compute their lengths, or angles between them. As everything else in geometry, these definitions are derived from the line element, in the case of Euclidean space given by $ds^2 = dx^2 + dy^2 + dz^2$. We can re-interpret the line element, defined by the length of an infinitesimal displacement, as the squared norm of the displacement vector $d\vec{x} = (dx, dy, dz)^T$ between two points: $ds^2 = |d\vec{x}|^2$. Using the displacement vector as a prototype for any vector in 3-dimensional Euclidean space, we obtain the squared norm

$$|\vec{a}|^2 = (a^x)^2 + (a^y)^2 + (a^z)^2 \tag{7.54}$$

of a generic 3-vector $(a^x, a^y, a^z)^T$.

7.3.1.1 Four-vectors

The same constructions work in four dimensions, even though we then lose all our geometrical intuition. In Minkowski space-time, a **4-vector a** has four components $(a^t, a^x, a^y, a^z)^T = (a^\alpha)$, using Greek labels $\alpha = t, x, y, z$ or $\alpha = 0, 1, 2, 3$ when we include a time component. Sometimes we also write a 4-vector as $\mathbf{a} = (a^t, \vec{a}^T)^T$ when it is useful to keep its spatial part $\vec{a} = (a^x, a^y, a^z)^T$ intact.

In order to derive the norm of a 4-vector, we interpret the Minkowski line element $\mathrm{d}s^2 = -\mathrm{d}t^2 + \mathrm{d}x^2 + \mathrm{d}y^2 + \mathrm{d}z^2$ as the norm (squared) of the displacement vector $\mathbf{dx} = (\mathrm{d}t, \mathrm{d}x, \mathrm{d}y, \mathrm{d}z)^T$: $\mathrm{d}s^2 = ||\mathbf{dx}||^2$. Using the displacement vector as a prototype of any 4-vector, we obtain the **norm** (squared)

$$||\mathbf{a}||^2 = -(a^t)^2 + (a^x)^2 + (a^y)^2 + (a^z)^2 \tag{7.55}$$

of a 4-vector $\mathbf{a} = (a^t, a^x, a^y, a^z)^T$. Unlike in Euclidean space, this norm (squared) may be negative. It is then more convenient to simply work with $||\mathbf{a}||^2$ as the norm, or some kind of length of the vector, without taking the square root.

Given the sign options, we use the norm in order to characterize different types of 4-vectors, just as we used the line element to characterize different types of separations between events. We call a 4-vector **a spacelike** if $||\mathbf{a}||^2 > 0$, **lightlike** if $||\mathbf{a}||^2 = 0$, and **timelike** if $||\mathbf{a}||^2 < 0$.

As always in linear algebra, we use a basis to determine components of a vector. In space-time, we need four independent 4-vectors $\mathbf{e}_0$, $\mathbf{e}_1$, $\mathbf{e}_2$ and $\mathbf{e}_3$ to form a basis. For a given basis, a 4-vector **a** has a unique set of components a^α such that

$$\mathbf{a} = a^0\mathbf{e}_0 + a^1\mathbf{e}_1 + a^2\mathbf{e}_2 + a^3\mathbf{e}_3 = \sum_{\alpha=0}^{3} a^\alpha \mathbf{e}_\alpha = a^\alpha \mathbf{e}_\alpha\,. \tag{7.56}$$

In the last step, we have employed the summation convention which declares that indices that appear twice in a product are automatically summed over their full range. If we draw the basis vectors in a diagram, reading off the components of a vector is just like reading off the coordinates of an event in a coordinate system.

If we change the basis, the components of a vector change. For instance, if we rotate the basis vectors in Euclidean space, the components of a 3-vector should also be rotated, but in the opposite direction: Looking back at Figure 4.1 on page 136, we can interpret the axes as our two sets of basis vectors, such that each vector in the original basis $(\vec{e}_x, \vec{e}_y)$ is rotated counter-clockwise to the corresponding basis vector in $(\vec{e}'_x, \vec{e}'_y)$. From the perspective of $(\vec{e}'_x, \vec{e}'_y)$, the coordinates of the vector $\overrightarrow{OA}$ are rotated by the same angle, but clockwise. This transformation ensures that the summed product of components and basis vectors still corresponds to the same direction after changing the basis.

Similarly, in space-time we must ensure that the components of a 4-vector transform by a Lorentz boost if we boost the basis. We already know how the components of the displacement vector, our prototype 4-vector, have to be transformed. We therefore require the same behavior for the components of any 4-vector. A 4-vector is defined by the following two properties:

1. It has four components $(a^0, a^1, a^2, a^3)^T$.
2. Its components transform by a Lorentz boost if the basis is changed:

$$a^{0'} = \gamma(a^0 - va^1) \tag{7.57}$$

$$a^{1'} = \gamma(a^1 - va^0) \tag{7.58}$$

$$a^{2'} = a^2 \tag{7.59}$$

$$a^{3'} = a^3 \tag{7.60}$$

if the boost is in the x-direction with speed v.

The second requirement ensures that the old and new components express the same direction in two different bases: $\mathbf{a} = a^{\alpha}\mathbf{e}_{\alpha} = a^{\alpha'}\mathbf{e}_{\alpha'}$. Even though we do not have intuition about space-time directions, linear algebra helps us to make sure that the geometrical requirements for a direction in a vector space are satisfied.

Our main example of a 4-vector so far is the displacement vector $\mathbf{dx} = (\mathrm{d}t, \mathrm{d}x, \mathrm{d}y, \mathrm{d}z)^T$. Since we have to observe the correct transformation behavior, we cannot just pick four components to define new 4-vectors. It is more efficient if we start with the known 4-vector $\mathbf{dx}$ and modify it by mathematical operations that do not change the transformation behavior. In particular, we can multiply a 4-vector with an invariant, or divide it by a non-zero invariant, and still have a 4-vector. Since proper time $\mathrm{d}\tau$ is invariant, an important example of a 4-vector derived from the displacement vector is the derivative $\mathbf{dx}/\mathrm{d}\tau$. This 4-vector is the derivative of position by time, and will therefore be discussed in the context of velocity.

The required transformation behavior ensures that we can write meaningful physics equations using 4-vectors. We will derive an equation of the form $\mathbf{f} = m\mathbf{a}$ as a relativistic generalization of Newton's second law, where $\mathbf{f}$ is a 4-vector that describes the force, the mass m is invariant, and $\mathbf{a}$ is a 4-vector for the acceleration of an object. If we apply a Lorentz boost, the components on both sides of the equation change, but they change in the same way. Therefore, the equation holds in all frames if it holds in one frame: It describes an observer-independent physical law. The equation is not invariant in a strict sense because each side changes by a Lorentz boost, but it is **covariant** because its validity is observer independent.

7.3.1.2 Scalar product

In Euclidean geometry, we use the dot product of two vectors to compute the angle between them. The dot product is therefore another geometrical concept and should be derived from the line element, or generalized to non-Euclidean geometries. In its general version, we refer to the underlying concept as a scalar product.

We have already seen that the line element determines the norm $||\mathbf{a}||^2$ of a 4-vector, or its generalized length. Given what we know from the dot product, the scalar product should be related to the norm by the equation $||\mathbf{a}||^2 = \mathbf{a}\cdot\mathbf{a}$. A scalar product is more general because it should be applicable to any pair of vectors, $\mathbf{a}\cdot\mathbf{b}$, where $\mathbf{a}$ and $\mathbf{b}$ are not required to be equal. Nevertheless, since both the norm and the scalar product follow from the line element, it should not be necessary to postulate any additional information. A quick way to see that the scalar product is closely related to the norm is based on the binomial equation

$$||\mathbf{a}+\mathbf{b}||^2 = (\mathbf{a}+\mathbf{b})\cdot(\mathbf{a}+\mathbf{b}) = ||\mathbf{a}||^2 + ||\mathbf{b}||^2 + 2\mathbf{a}\cdot\mathbf{b}. \tag{7.61}$$

If we can compute the norm of any 4-vector, including $||\mathbf{a}||^2$, $||\mathbf{b}||^2$ and $||\mathbf{a}+\mathbf{b}||^2$ for two given vectors $\mathbf{a}$ and $\mathbf{b}$, we can derive the scalar product of two given 4-vectors.

The scalar product should obey certain conditions in order for (7.61) to hold, summarized in the following definition: Given two 4-vectors $\mathbf{a}$ and $\mathbf{b}$, the **scalar product $\mathbf{a}\cdot\mathbf{b}$** is a real number such that

(i) $\mathbf{a}\cdot\mathbf{b} = \mathbf{b}\cdot\mathbf{a}$ (commutativity),
(ii) $\mathbf{a}\cdot(\mathbf{b}+\mathbf{c}) = \mathbf{a}\cdot\mathbf{b} + \mathbf{a}\cdot\mathbf{c}$ (distributivity), and
(iii) $(r\mathbf{a})\cdot\mathbf{b} = r(\mathbf{a}\cdot\mathbf{b})$ (associativity)

if $\mathbf{c}$ is a third 4-vector, and r a real number. Given a scalar product, we relate it to the norm via $\mathbf{a}\cdot\mathbf{a} = ||\mathbf{a}||^2$. We do not require that $\mathbf{a}\cdot\mathbf{a}$ always takes values greater than or equal to zero, as we did for the inner product used in quantum mechanics. Another difference is that we do not use complex numbers in the present case.

The three conditions do not uniquely determine the scalar product. We can parameterize the freedom in choosing different scalar products if we use a basis $\mathbf{e}_{\alpha}$, $\alpha = 0, 1, 2, 3$, and expand two

4-vectors in it: $\mathbf{a} = a^\alpha \mathbf{e}_\alpha$ and $\mathbf{b} = b^\beta \mathbf{e}_\beta$ assuming the summation convention. We use different labels, α and β, because we will combine these two expansions in a scalar product, and we should avoid using the same index more than twice in a product: The expressions $(\sum_\alpha a^\alpha \mathbf{e}_\alpha)(\sum_\beta b^\beta \mathbf{e}_\beta)$ and $\sum_\alpha a^\alpha \mathbf{e}_\alpha b^\alpha \mathbf{e}_\alpha$ are not the same. Properties (ii) and (iii) of a scalar product imply

$$\mathbf{a}\cdot\mathbf{b} = (a^\alpha \mathbf{e}_\alpha)\cdot(b^\beta \mathbf{e}_\beta) = (\mathbf{e}_\alpha\cdot\mathbf{e}_\beta)a^\alpha b^\beta\,. \tag{7.62}$$

We have separated the scalar product into a factor $a^\alpha b^\beta$ which depends on the vector components, and a factor $\mathbf{e}_\alpha\cdot\mathbf{e}_\beta = \mathbf{e}_\beta\cdot\mathbf{e}_\alpha$ which is independent of the vectors chosen for the particular scalar product we just computed. The symmetric 4×4-matrix $\mathbf{e}_\alpha\cdot\mathbf{e}_\beta$ therefore characterizes the scalar product.

So far, we have discussed scalar products in general terms. The version we use for space-time geometry should be derived from the line element. We call the corresponding matrix $\mathbf{e}_\alpha\cdot\mathbf{e}_\beta = \eta_{\alpha\beta}$. The line element is equal to the scalar product of the displacement vector with itself:

$$\begin{aligned} \mathrm{d}s^2 &= \mathrm{d}\mathbf{x}\cdot\mathrm{d}\mathbf{x} = (\mathbf{e}_\alpha\cdot\mathbf{e}_\beta)\mathrm{d}x^\alpha\mathrm{d}x^\beta = \eta_{\alpha\beta}\mathrm{d}x^\alpha\mathrm{d}x^\beta \\ &= \eta_{tt}\mathrm{d}t^2 + \eta_{tx}\mathrm{d}t\mathrm{d}x + \eta_{ty}\mathrm{d}t\mathrm{d}y + \eta_{tz}\mathrm{d}t\mathrm{d}z + \eta_{xt}\mathrm{d}x\mathrm{d}t + \eta_{xx}\mathrm{d}x^2 + \cdots \end{aligned} \tag{7.63}$$

There is not just one displacement vector but infinitely many ones because we can pick arbitrary pairs of events. When we write $\mathrm{d}s^2 = \eta_{\alpha\beta}\mathrm{d}x^\alpha\mathrm{d}x^\beta$, we therefore have many specific equations at our disposal which we can use to compute the components of $\eta_{\alpha\beta}$. It is convenient to make a few simple choices of displacement vectors with many vanishing components and evaluate the equation. If we pick 16 independent such choices, we have 16 independent linear equations from which we can compute all 16 components of $\eta_{\alpha\beta}$. In fact, we only need ten equations because a symmetric matrix $\eta_{\alpha\beta} = \eta_{\beta\alpha}$ has ten independent components in four dimensions.

For instance, if $\mathrm{d}\mathbf{x} = (1,0,0,0)^T$, such that $\mathrm{d}s^2 = ||\mathrm{d}\mathbf{x}||^2 = -1$, we have

$$-1 = \eta_{\alpha\beta}\mathrm{d}x^\alpha\mathrm{d}x^\beta = \eta_{tt} \tag{7.64}$$

because there is only one non-zero term in the sum over α and β, obtained for $\alpha = 0 = \beta$. The three simple spacelike choices for $\mathrm{d}\mathbf{x}$ give $\eta_{xx} = \eta_{yy} = \eta_{zz} = 1$. Using these values, lightlike choices such as $\mathrm{d}\mathbf{x} = (1,1,0,0)^T$ then show that all off-diagonal components of $\eta_{\alpha\beta}$ are zero. We can therefore write the matrix that characterizes our scalar product as

$$(\eta_{\alpha\beta}) = \begin{pmatrix} -1 & 0 & 0 & 0 \\ 0 & 1 & 0 & 0 \\ 0 & 0 & 1 & 0 \\ 0 & 0 & 0 & 1 \end{pmatrix}, \tag{7.65}$$

and the line element is $\mathrm{d}s^2 = \eta_{\alpha\beta}\mathrm{d}x^\alpha\mathrm{d}x^\beta$. When it appears in the line element, the matrix $\eta_{\alpha\beta}$ is called the **metric** of flat space-time, or the **Minkowski metric**.

For two 4-vectors **a** and **b**, we therefore have the Minkowski scalar product

$$\mathbf{a}\cdot\mathbf{b} = \eta_{\alpha\beta}a^\alpha b^\beta = -a^t b^t + a^x b^x + a^y b^y + a^z b^z\,. \tag{7.66}$$

This scalar product, just like the line element from which it is derived, is invariant. Although vector components change by a Lorentz boost (7.57)–(7.60), which we can write in matrix form as

$$\begin{pmatrix} a^{0'} \\ a^{1'} \\ a^{2'} \\ a^{3'} \end{pmatrix} = \begin{pmatrix} \gamma & -\gamma v & 0 & 0 \\ -\gamma v & \gamma & 0 & 0 \\ 0 & 0 & 1 & 0 \\ 0 & 0 & 0 & 1 \end{pmatrix} \begin{pmatrix} a^0 \\ a^1 \\ a^2 \\ a^3 \end{pmatrix}, \tag{7.67}$$

scalar products with **a** remain the same.

As an example, consider $\mathbf{a} = (1,0,0,0)^T$ and $\mathbf{b} = (0,1,0,0)^T$. The Lorentz-boosted vectors are $\mathbf{a}' = (\gamma, -\gamma v, 0, 0)^T$ and $\mathbf{b}' = (-\gamma v, \gamma, 0, 0)^T$. We compute the scalar products

$$\mathbf{a}\cdot\mathbf{a} = -1 = -\gamma^2 + (\gamma v)^2 = \mathbf{a}'\cdot\mathbf{a}' \tag{7.68}$$

$$\mathbf{b}\cdot\mathbf{b} = 1 = -(\gamma v)^2 + \gamma^2 = \mathbf{b}'\cdot\mathbf{b}' \tag{7.69}$$

$$\mathbf{a}\cdot\mathbf{b} = 0 = -\gamma(-\gamma v) + (-\gamma v)\gamma = \mathbf{a}\cdot\mathbf{b}'. \tag{7.70}$$

In particular, the vectors $\mathbf{a}'$ and $\mathbf{b}'$ are orthogonal to each other in the sense that their scalar product is zero, just like **a** and **b**. We can interpret **a** and **b** as unit vectors in the timelike and spacelike direction, respectively, of a coordinate system. The vectors $\mathbf{a}'$ and $\mathbf{b}'$ are then unit vectors in the directions of a transformed coordinate system, which as we have seen do not "look" orthogonal in our Euclidean understanding. But they remain orthogonal in Minkowski geometry in the sense that their scalar product vanishes.

In general, orthogonality, defined by a vanishing scalar product, does not imply a "right" angle. As we have seen, the hyperbolic angle between a timelike and a spacelike direction in Minkowski geometry is infinite. A right angle of $\pi/2$ is implied only if the Cauchy–Schwarz inequality (3.15) is available, which requires a positive inner product.

7.3.1.3 4-Velocity

The motion of a massive particle is described by a timelike worldline. We can parameterize it by specifying four components, the coordinates of the 4-vector $\mathbf{x}(\tau) = (t(\tau), x(\tau), y(\tau), z(\tau))^T$ as functions of proper time along the worldline. We have the relationship

$$\mathrm{d}\tau = \sqrt{1 - |\vec{V}|^2}\,\mathrm{d}t \tag{7.71}$$

between proper time τ and coordinate time t as derived for time dilation, with the 3-velocity

$$\vec{V} = \left(\frac{\mathrm{d}x}{\mathrm{d}t}, \frac{\mathrm{d}y}{\mathrm{d}t}, \frac{\mathrm{d}z}{\mathrm{d}t}\right)^T \tag{7.72}$$

of the moving object.

We need a 4-vector to describe the velocity in covariant equations. Following the definition of the 3-velocity, we define the **4-velocity** of a moving particle as

$$\mathbf{u} = \frac{\mathrm{d}\mathbf{x}}{\mathrm{d}\tau}. \tag{7.73}$$

This object is a 4-vector because $\mathrm{d}\mathbf{x}$ transforms like a 4-vector under Lorentz boosts, while $\mathrm{d}\tau$ is invariant. It is the tangent vector to the worldline in space-time, see Figure 7.27.

Based on the definition, we derive a relationship between the 4-velocity and the 3-velocity. For the time component of **u**, we have

$$u^0 = \frac{\mathrm{d}x^0}{\mathrm{d}\tau} = \frac{\mathrm{d}t}{\mathrm{d}\tau} = \frac{1}{\sqrt{1 - |\vec{V}|^2}} = \gamma_{\vec{V}} \tag{7.74}$$

using (7.71). The final equality defines $\gamma_{\vec{V}}$ in the same way as we defined γ for the frame velocity v in a Lorentz boost. When we use an object's velocity, we indicate it by a subscript on the gamma factor, reserving a plain γ for the frame velocity.

For the spatial components, we use the chain rule and write

$$u^1 = \frac{\mathrm{d}x}{\mathrm{d}\tau} = \frac{\mathrm{d}x}{\mathrm{d}t}\frac{\mathrm{d}t}{\mathrm{d}\tau} = \frac{V^x}{\sqrt{1 - |\vec{V}|^2}} = \gamma_{\vec{V}} V^x \tag{7.75}$$

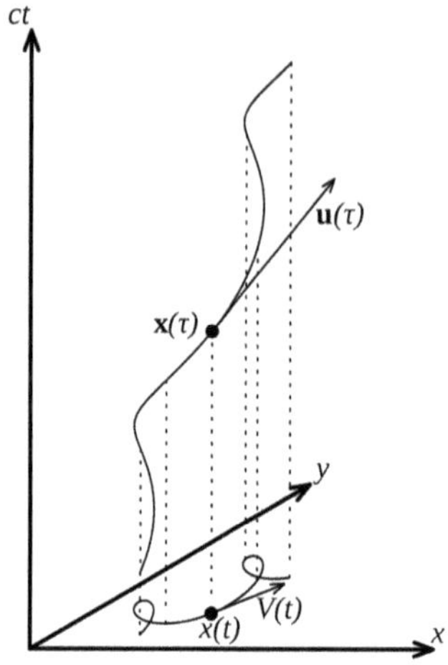

Figure 7.27 The velocity 4-vector $\mathbf{u}$ is tangent to the worldline $\mathbf{x}(\tau)$ of a moving object in space-time, while the 3-velocity $\vec{V}$ is tangent to the trajectory $\vec{x}(t)$ in space obtained by projecting the worldine onto a spatial cross-section.

and similarly for $u^2 = \gamma_{\vec{V}} V^y$ and $u^3 = \gamma_{\vec{V}} V^z$. More compactly, we write the velocity 4-vector as

$$\mathbf{u} = \begin{pmatrix} u^0 \\ \vec{u} \end{pmatrix} = \begin{pmatrix} \gamma_{\vec{V}} \\ \gamma_{\vec{V}} \vec{V} \end{pmatrix}. \tag{7.76}$$

Conversely, we can express the 3-velocity in terms of the 4-velocity components by

$$\vec{V} = \frac{\vec{u}}{u^0}. \tag{7.77}$$

The derivation shows that the 4-velocity contains the same amount of information as the 3-velocity, even though it has an extra component. The fourth component is not independent because there is a relationship between all components: The 4-velocity always has the fixed norm

$$||\mathbf{u}||^2 = -\gamma_{\vec{V}}^2 + \gamma_{\vec{V}}^2 |\vec{V}|^2 = -1 \tag{7.78}$$

independent of the 3-velocity. The 4-velocity is always timelike and normalized. (More precisely, the 4-velocity is timelike for massive particles, as implicitly assumed here by using proper time. Massless particles require a different treatment as we will see.) Its magnitude does not change as the particle speeds up. What changes is the hyperbolic angle of the 4-velocity vector with the time direction. Using our previous equations for the hyperbolic angle in terms of coordinates or vector components, we have

$$\tanh \rho = \frac{u^1}{u^0} = V^x \tag{7.79}$$

using (7.77) and Figure 7.28.

The hyperbolic angle of the space-time direction $\mathbf{u}$ is therefore equal to the rapidity associated with the 3-velocity component V^x. If the object speeds up, the hyperbolic angle of the 4-velocity increases, while its magnitude stays constant. The 4-velocity can be interpreted as a hyperbolic speedometer which measures speeds by an angle. In particular, the 4-velocity is never equal to zero, even for an object at rest. In space-time, there is always motion because the time coordinate must increase even if the spatial coordinates do not change.

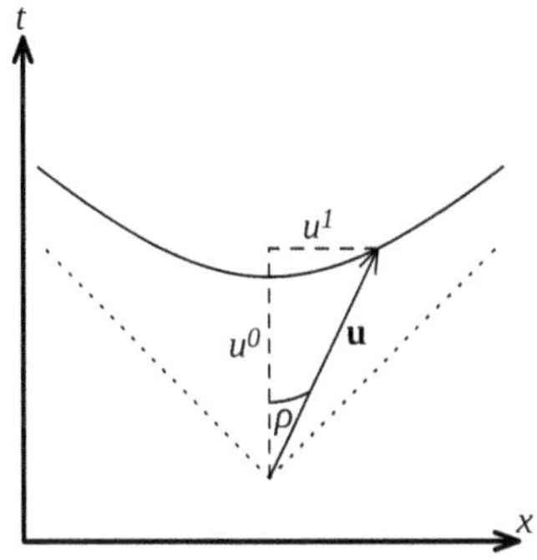

Figure 7.28 Relativistic speedometer: As an object speeds up, the hyperbolic angle ρ of its 4-velocity $\mathbf{u}$ with the time axis increases, while the norm $||\mathbf{u}||^2 = -1$ remains constant. The 4-velocity vector therefore moves along a unit timelike hyperbola centered at the event, where the 4-velocity is drawn. Evaluating triangle relationships in Minkowski geometry confirms equation (7.79).

The velocity 4-vector implies an efficient rederivation of the composition law for velocities. Since $\mathbf{u}$ is a 4-vector, we transform it to a new frame by applying a Lorentz boost. If $\mathbf{u} = (\gamma_{\vec{V}}, \gamma_{\vec{V}}\vec{V}^T)^T$ in the original frame, we have

$$\mathbf{u}' = \begin{pmatrix} \gamma_{\vec{V}'} \\ \gamma_{\vec{V}'}\vec{V}' \end{pmatrix} = \begin{pmatrix} \gamma & -v\gamma & 0 & 0 \\ -v\gamma & \gamma & 0 & 0 \\ 0 & 0 & 1 & 0 \\ 0 & 0 & 0 & 1 \end{pmatrix} \begin{pmatrix} \gamma_{\vec{V}} \\ \gamma_{\vec{V}}V^x \\ \gamma_{\vec{V}}V^y \\ \gamma_{\vec{V}}V^z \end{pmatrix} = \begin{pmatrix} \gamma\gamma_{\vec{V}}(1 - vV^x) \\ \gamma\gamma_{\vec{V}}(V^x - v) \\ \gamma_{\vec{V}}V^y \\ \gamma_{\vec{V}}V^z \end{pmatrix}. \tag{7.80}$$

Notice that there are two different gammas, $\gamma = (1 - v^2)^{-1/2}$ related to the frame velocity, and $\gamma_{\vec{V}} = (1 - |\vec{V}|^2)^{-1/2}$ related to the object's velocity.

From the time components of (7.80), we read off a third gamma,

$$\gamma_{\vec{V}'} = \gamma\gamma_{\vec{V}}(1 - vV^x). \tag{7.81}$$

The new 3-velocity $\vec{V}'$ is then derived by dividing the space components of $\mathbf{u}'$ by this $\gamma_{\vec{V}'}$:

$$\vec{V}' = \frac{1}{\gamma_{\vec{V}'}} \begin{pmatrix} \gamma\gamma_{\vec{V}}(V^x - v) \\ \gamma_{\vec{V}}V^y \\ \gamma_{\vec{V}}V^z \end{pmatrix} = \frac{1}{1 - vV^x} \begin{pmatrix} V^x - v \\ V^y/\gamma \\ V^z/\gamma \end{pmatrix} \tag{7.82}$$

in agreement with the previous result (7.48)–(7.50).

7.3.2 DYNAMICS

The 4-velocity $\mathbf{u}$ combined with an invariant mass m allows us to define the **4-momentum**

$$\mathbf{p} = m\mathbf{u}. \tag{7.83}$$

From the normalization of $\mathbf{u}$ it follows that the norm squared of $\mathbf{p}$ is given by the mass,

$$||\mathbf{p}||^2 = -m^2. \tag{7.84}$$

The components of $\mathbf{u}$ in terms of $\vec{V}$ determine the momentum components

$$p^0 = \frac{m}{\sqrt{1 - |\vec{V}|^2}} \quad , \quad \vec{p} = \frac{m\vec{V}}{\sqrt{1 - |\vec{V}|^2}}. \tag{7.85}$$

These components reveal a close relationship between the mass and possible speeds. If $m \neq 0$, p^0 diverges when $|\vec{V}|$ approaches the speed of light. Therefore, a massive particle cannot reach the speed of light with finite momentum. By contrast, a massless particle, $m = 0$, must move at the speed of light if it is to have non-zero momentum. For massless particles, however, we will have to develop a new formulation because motion at the speed of light makes it impossible to have an increasing proper-time parameter: $\mathrm{d}\tau = 0$ if $|\vec{V}| = 1$.

7.3.2.1 Energy-momentum

For small velocities, $|\vec{V}| \ll 1$, we find a new interpretation of the momentum components by using the Taylor expansion $(1 - |\vec{V}|^2)^{-1/2} = 1 + \frac{1}{2}|\vec{V}|^2 + \cdots$. The spatial components in (7.85) are then close to the Newtonian momentum $\vec{p} = m\vec{V} + \cdots$ up to terms at least cubic in $|\vec{V}|$. The time component equals $p^0 = m + \frac{1}{2}m|\vec{V}|^2 + \cdots$ up to terms at least quartic in $|\vec{V}|$. We recognize the quadratic term as the Newtonian kinetic energy. In relativistic theories, therefore, momentum and energy are combined in a 4-vector, just as space and time are combined in the displacement 4-vector. A Lorentz transformation combines energy and momentum, just as it combines space and time. Sometimes, the 4-momentum is therefore called the **energy-momentum 4-vector**. We write its components as

$$\mathbf{p} = \begin{pmatrix} E \\ \vec{p} \end{pmatrix} = \begin{pmatrix} m\gamma_{\vec{V}} \\ m\gamma_{\vec{V}}\vec{V} \end{pmatrix} . \qquad (7.86)$$

Using (7.84), we compute the energy

$$E = \sqrt{m^2 + |\vec{p}|^2} \qquad (7.87)$$

for a given 3-momentum $\vec{p} = m\gamma_{\vec{V}}\vec{V}$.

In addition to the kinetic energy, $E = p^0$ contains a velocity-independent term equal to m. It is called the **rest energy** because it is present even for a particle at rest. There is no potential energy because it is difficult to make the usual notion of potentials consistent with Lorentz transformations. If a potential, such as Newton's, depends only on spatial coordinates in one frame, it also depends on time in a different frame. Conservative forces are therefore impossible in relativity, at least in an observer-independent way. Instead, forces are usually described in this theory in terms of fields, as in the electromagnetic Lorentz force which we will see soon.

Rewriting (7.87) or (7.84), we obtain the mass as a function of energy and 3-momentum:

$$m = \sqrt{E^2 - |\vec{p}|^2} . \qquad (7.88)$$

The mass is invariant and, unlike E and $\vec{p}$, does not depend on the frame used. Particle-physics experiments measure E and $\vec{p}$ of a charged particle, the former in calorimeters and the latter by the deflection in a strong magnetic field. One can then compute m and identify the particle without having to find a transformation to its rest frame.

7.3.2.2 Force

The momentum is related to the **4-force f** by

$$\mathbf{f} = \frac{\mathrm{d}\mathbf{p}}{\mathrm{d}\tau} . \qquad (7.89)$$

In order to use this definition as an equation of motion, we must determine the components of **f** in terms of the common 3-force $\vec{F}$, which we know for instance from electromagnetic interactions between a charged particle and the electric and magnetic fields.

We begin with the time component f^0 and write it as

$$f^0 = \frac{\mathrm{d}p^0}{\mathrm{d}\tau} = \frac{\mathrm{d}E}{\mathrm{d}\tau} = \gamma_{\vec{V}} \frac{\mathrm{d}E}{\mathrm{d}t}, \tag{7.90}$$

using the time-dilation formula $\mathrm{d}\tau = \sqrt{1-|\vec{V}|^2}\,\mathrm{d}t = \mathrm{d}t/\gamma_{\vec{V}}$ in the last step. The factor $\mathrm{d}E/\mathrm{d}t$ describes the power of an acting mechanical force. If a force $\vec{F}$ moves an object by a distance $\mathrm{d}\vec{x}$, an amount $\mathrm{d}E = \mathrm{d}\vec{x}\cdot\vec{F}$ of work is exercised. The power is therefore equal to

$$\frac{\mathrm{d}E}{\mathrm{d}t} = \vec{V}\cdot\vec{F}. \tag{7.91}$$

Independently, we can relate $\mathrm{d}E/\mathrm{d}t$ to a time derivative of the 3-momentum:

$$\frac{\mathrm{d}E}{\mathrm{d}t} = m\frac{\mathrm{d}\gamma_{\vec{V}}}{\mathrm{d}t} = m\vec{V}\cdot\frac{\mathrm{d}(\gamma_{\vec{V}}\vec{V})}{\mathrm{d}t} = \vec{V}\cdot\frac{\mathrm{d}\vec{p}}{\mathrm{d}t}. \tag{7.92}$$

(The middle equation is best confirmed by fully simplifying the t-derivatives on both sides.) Comparing equations (7.91) and (7.92), both of which have to hold for any 3-velocity $\vec{V}$, we obtain

$$\vec{F} = \frac{\mathrm{d}\vec{p}}{\mathrm{d}t}. \tag{7.93}$$

This equation looks like the corresponding Newtonian one, but here, according to (7.85), $\vec{p} = m\gamma_{\vec{V}}\vec{V}$ is *not* equal to the Newtonian momentum $\vec{P} = m\vec{V}$. Notice that $\gamma_{\vec{V}}$ is usually time dependent through $\vec{V}(t)$. As a differential equation, (7.93) for a given $\vec{F}$ therefore has solutions that differ from the Newtonian ones by more than just a constant factor.

We now combine the previous equations and write the components of the 4-force as

$$f^0 = \gamma_{\vec{V}}\vec{V}\cdot\vec{F} \tag{7.94}$$

and

$$\vec{f} = \frac{\mathrm{d}\vec{p}}{\mathrm{d}\tau} = \gamma_{\vec{V}}\frac{\mathrm{d}\vec{p}}{\mathrm{d}t} = \gamma_{\vec{V}}\vec{F}, \tag{7.95}$$

or

$$\mathbf{f} = \begin{pmatrix} \gamma_{\vec{V}}\vec{V}\cdot\vec{F} \\ \gamma_{\vec{V}}\vec{F} \end{pmatrix}. \tag{7.96}$$

In space-time, the 4-force is always orthogonal to the 4-momentum or the 4-velocity: We have

$$0 = \frac{\mathrm{d}(-m^2)}{\mathrm{d}\tau} = \frac{\mathrm{d}||\mathbf{p}||^2}{\mathrm{d}\tau} = 2\mathbf{p}\cdot\frac{\mathrm{d}\mathbf{p}}{\mathrm{d}\tau} = 2\mathbf{p}\cdot\mathbf{f}. \tag{7.97}$$

This equality can also be confirmed directly for the components of $\mathbf{p}$ and $\mathbf{f}$. In Newtonian mechanics, the situation of a force that is always orthogonal to the momentum corresponds to the special case of circular motion. In this case, the norm $|\vec{P}|^2$ is constant. In relativity, the norm $||\mathbf{p}||^2$ is always constant if the mass does not change. Mathematically, both cases are described by a force orthogonal to the momentum.

The equations of relativistic dynamics are summarized in Table 7.2. In the relativistic expressions, it is important to know both the 4-vector versions as well as how their time and space components are broken up by relating them to corresponding 3-vectors. The 4-vectors determine geometrical directions in 4-dimensional space-time and are crucial ingredients in formulations of covariant equations. The 3-vectors do not have much geometrical meaning in space-time, but they represent how we would measure observables such as the velocity or the force using standard methods of classical mechanics.

TABLE 7.2
The Main Equations of Nonrelativistic and Relativistic Mechanics

	non-relativistic	relativistic
velocity	$\vec{V} = \frac{d\vec{x}}{dt}$	$\mathbf{u} = \frac{d\mathbf{x}}{d\tau} = \begin{pmatrix} \gamma_{\vec{V}} \\ \gamma_{\vec{V}}\vec{V} \end{pmatrix}$ $\|\|\mathbf{u}\|\|^2 = -1$
momentum	$\vec{P} = m\vec{V}$	$\mathbf{p} = m\mathbf{u} = \begin{pmatrix} m\gamma_{\vec{V}} \\ m\gamma_{\vec{V}}\vec{V} \end{pmatrix} = \begin{pmatrix} E \\ \vec{p} \end{pmatrix}$ $\|\|\mathbf{p}\|\|^2 = -m^2$
acceleration	$\vec{A} = \frac{d\vec{V}}{dt}$	$\mathbf{a} = \frac{d\mathbf{u}}{d\tau}$
force	$\vec{F} = \frac{d\vec{P}}{dt}$	$\mathbf{f} = \frac{d\mathbf{p}}{d\tau} = \begin{pmatrix} \gamma_{\vec{V}}\vec{V}\cdot\vec{F} \\ \gamma_{\vec{V}}\vec{F} \end{pmatrix}$ $\mathbf{f}\cdot\mathbf{p} = 0$

A tractable example is given by a constant force vector $\vec{F} = \vec{c}$. We solve $\vec{F} = d\vec{p}/dt$ by $\vec{p}(t) = \vec{p}_0 + \vec{c}t$. This solution is the same as in Newtonian mechanics. In order to find the position as a function of time, we determine $\vec{V}(t)$ from $\vec{p}(t) = m\vec{V}(t)(1-|\vec{V}(t)|^2)^{-1/2}$ and integrate once more. The resulting solution for $\vec{x}(t)$ is different from the Newtonian one because of the factor of $\gamma_{\vec{V}}$ in the momentum. All the required integrals can be computed analytically, but a more elegant solution starts by recognizing that the 4-force components $\mathbf{f} = (\gamma_{\vec{V}}\vec{c}\cdot\vec{V}, \gamma_{\vec{V}}\vec{c}^T)^T$ are proportional to the 4-velocity components: Choosing a spatial coordinate system such that $\vec{c} = ma\vec{e}_x$ points in the x-direction, we see that $f^0 = ma\gamma_{\vec{V}}V^x = mau^1$ and $f^1 = ma\gamma_{\vec{V}} = mau^0$. We obtain linear equations of motion

$$\frac{du^0}{d\tau} = au^1 \quad , \quad \frac{du^1}{d\tau} = au^0 . \tag{7.98}$$

These equations are coupled, but deriving the second-order equations

$$\frac{d^2u^0}{d\tau^2} = a\frac{du^1}{d\tau} = a^2u^0 \quad \text{and} \quad \frac{d^2u^1}{d\tau^2} = a\frac{du^0}{d\tau} = a^2u^1 \tag{7.99}$$

results in separate equations for u^0 and u^1, respectively. Alternatively, we can write the original first-order equations in matrix form and diagonalize the matrix using the methods of Section 3.1.4. With positive coefficients a^2, these equations are hyperbolic versions of the harmonic-oscillator equation of motion. They are solved by

$$u^0(\tau) = A\sinh(a\tau) + B\cosh(a\tau) \quad , \quad u^1(\tau) = B\sinh(a\tau) + A\cosh(a\tau) \tag{7.100}$$

with integration constants A and B. The pair of two second-order equations results in only two integration constants because we still have to make sure that the original first-order equations are satisfied.

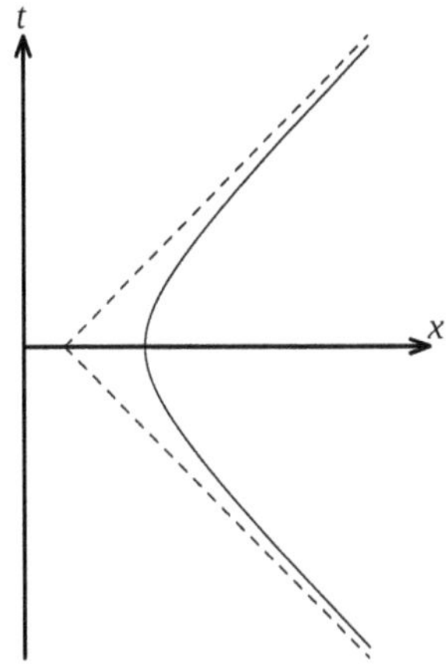

Figure 7.29 The worldline of an object subject to constant acceleration a in an observer's frame is given by a hyperbola that asymptotes to the light cone at $(t,x) = (t(\tau_0), x(\tau_0) - 1/a)$, where τ_0 is the value of proper time when the object is at rest, here on the x-axis.

We obtain the time-dependent space-time position by a direct integration of sinh and cosh:

$$t(\tau) = \frac{B}{a}\sinh(a\tau) + \frac{A}{a}\cosh(a\tau) + t_1 \quad , \quad x(\tau) = \frac{A}{a}\sinh(a\tau) + \frac{B}{a}\cosh(a\tau) + x_1 \tag{7.101}$$

with two additional integration constants, t_1 and x_1, that can be related to initial values. If the object is at rest at time $t = 0$ and we choose this event to correspond to $\tau = 0$, $u^0(0) = 1$ and $u^1(0) = 0$ imply $A = 0$ and $B = 1$, while $t(0) = 0$ implies $t_1 = 0$. Therefore,

$$t(\tau) = \frac{1}{a}\sinh(a\tau) \quad , \quad x(\tau) = \frac{1}{a}(\cosh(a\tau) - 1) + x(0) \tag{7.102}$$

with the initial position $x(0)$ while the object is at rest. These equations parameterize a hyperbola, shown in Figure 7.29.

7.3.2.3 Electromagnetic field

An example of a force given by fields is the Lorentz force

$$\vec{F} = q(\vec{E} + \vec{V} \times \vec{B}) \tag{7.103}$$

exerted on a particle with charge q by an electric field $\vec{E}$ and a magnetic field $\vec{B}$. Since Maxwell's electrodynamics is Lorentz covariant, it is consistent to use this expression in relativistic dynamics. We first assume that magnetic field is zero, implying the 4-force

$$\mathbf{f} = \begin{pmatrix} \gamma_{\vec{V}} q\vec{E}\cdot\vec{V} \\ \gamma_{\vec{V}} qE^x \\ \gamma_{\vec{V}} qE^y \\ \gamma_{\vec{V}} qE^z \end{pmatrix} . \tag{7.104}$$

We know that the 4-force, as a 4-vector, transforms by a Lorentz boost if we change the frame. In a frame moving with speed v in the x-direction, the 4-force is

$$\mathbf{f}' = \begin{pmatrix} \gamma & -v\gamma & 0 & 0 \\ -v\gamma & \gamma & 0 & 0 \\ 0 & 0 & 1 & 0 \\ 0 & 0 & 0 & 1 \end{pmatrix} \begin{pmatrix} \gamma_{\vec{V}} q\vec{E}\cdot\vec{V} \\ \gamma_{\vec{V}} qE^x \\ \gamma_{\vec{V}} qE^y \\ \gamma_{\vec{V}} qE^z \end{pmatrix} = \begin{pmatrix} \gamma\gamma_{\vec{V}} q(E^x(V^x - v) + E^yV^y + E^zV^z) \\ \gamma\gamma_{\vec{V}} q(E^x(1 - vV^x) - vE^yV^y - vE^zV^z) \\ \gamma_{\vec{V}} qE^y \\ \gamma_{\vec{V}} qE^z \end{pmatrix} . \tag{7.105}$$

As seen before, there are two distinct gammas, $\gamma = (1-v^2)^{-1/2}$ depending on the frame velocity and $\gamma_{\vec{V}} = (1-|\vec{V}|^2)^{-1/2}$ depending on the particle's velocity.

The new 4-force $\mathbf{f}'$ should be of the form $\mathbf{f}' = (\gamma_{\vec{V}'}\vec{V}'\cdot\vec{F}', \gamma_{\vec{V}'}\vec{F}'^T)^T$ with the transformed velocity $\vec{V}'$ and the transformed force $\vec{F}'$, as well as $\gamma_{\vec{V}'} \neq \gamma_{\vec{V}}$ belonging to the transformed velocity. We know that the velocity components transform as

$$V^{x'} = \frac{V^x - v}{1 - vV^x} \quad , \quad V^{y'} = \frac{V^y}{\gamma(1-vV^x)} \quad , \quad V^{z'} = \frac{V^z}{\gamma(1-vV^x)} \tag{7.106}$$

and it will be useful to apply the transformation of $\gamma_{\vec{V}}$,

$$\gamma_{\vec{V}'} = u^{0'} = \gamma(u^0 - vu^1) = \gamma\gamma_{\vec{V}}(1-vV^x) \tag{7.107}$$

from (7.76), and the inverse

$$\gamma_{\vec{V}} = \gamma\gamma_{\vec{V}'}(1+vV^{x'}) . \tag{7.108}$$

The inverse can be derived by solving (7.107) for $\gamma_{\vec{V}}$ and replacing V^x with $V^{x'}$, but it is easier to refer to the physical argument that the inverse relationship is obtained by using an inverse Lorentz boost, just replacing v with $-v$ and swapping primed with unprimed $\vec{V}$-components. With these relations and (7.105), we will be able to read off $\vec{F}'$ and determine how the electric and magnetic fields transform.

Substituting all components of $\vec{V}$ and $\gamma_{\vec{V}}$ by components of $\vec{V}'$ and $\gamma_{\vec{V}'}$, we obtain

$$\mathbf{f}' = \begin{pmatrix} \gamma_{\vec{V}'}q(E^xV^{x'} + \gamma E^yV^{y'} + \gamma E^zV^{z'}) \\ \gamma_{\vec{V}'}q(E^x - v\gamma E^yV^{y'} - v\gamma E^zV^{z'}) \\ \gamma_{\vec{V}'}qE^y\gamma(1+vV^{x'}) \\ \gamma_{\vec{V}'}qE^z\gamma(1+vV^{x'}) \end{pmatrix} . \tag{7.109}$$

This 4-vector should be equal to $\mathbf{f}' = (\gamma_{\vec{V}'}\vec{F}'\cdot\vec{V}', \gamma_{\vec{V}'}\vec{F}'^T)^T$ with $\vec{F}' = q(\vec{E}' + \vec{V}'\times\vec{B}')$. For any such $\vec{F}'$, the time component $f^{0'} = q\gamma_{\vec{V}'}\vec{E}'\cdot\vec{V}'$ is independent of $\vec{B}'$ because $\vec{V}'\cdot(\vec{V}'\times\vec{B}') = 0$ for any $\vec{B}'$. This form of $f^{0'}$ is consistent with (7.109) if

$$\vec{E}' = \begin{pmatrix} E^x \\ \gamma E^y \\ \gamma E^z \end{pmatrix} . \tag{7.110}$$

Only the components of $\vec{E}$ orthogonal to the direction of the boost change, which is different from what we have seen for geometrical effects such as Lorentz contraction. Physically, the transformation (7.110) guarantees that the charge is invariant: By Gauss' law, the charge contained in a region is equal to the electric flux through the surface of this region, where the electric flux is the surface integral of the normal component of the electric field. For a small cubic region whose sides of length (L_x, L_y, L_z) are aligned with the spatial coordinate axes, we may assume that the electric field is almost constant on each of the six sides and write the charge/flux as

$$Q = 2(L_xL_yE^z + L_xL_zE^y + L_yL_zE^x) . \tag{7.111}$$

By Lorentz contraction, we have side lengths $L_{x'} = L_x/\gamma$, $L_{y'} = L_y$ and $L_{z'} = L_z$ in a moving frame. Integration as in the electric flux corresponds to a direct measurement of infinitesimal length intervals. Lorentz contraction, rather than some optical length measurement with travel-time delays, is therefore the correct version in this application. The flux in such a frame is then equal to

$$Q' = 2\left(\frac{1}{\gamma}L_xL_yE^{z'} + \frac{1}{\gamma}L_xL_zE^{y'} + L_yL_zE^{x'}\right) . \tag{7.112}$$

We have an invariant charge $Q' = Q$ if the electric field transforms by (7.110).

Given the transformed electric field, we compute

$$\vec{F}' - q\vec{E}' = q \begin{pmatrix} -v\gamma E^y V^{y'} - v\gamma E^z V^{z'} \\ v\gamma E^y V^{x'} \\ v\gamma E^z V^{x'} \end{pmatrix} \tag{7.113}$$

from (7.109), and find $\vec{B}'$ such that $\vec{F}' - q\vec{E}' = q\vec{V}' \times \vec{B}'$. A solution is given by

$$\vec{B}' = v\gamma\vec{E} \times \begin{pmatrix} 1 \\ 0 \\ 0 \end{pmatrix} = \gamma\vec{E} \times \vec{v} \tag{7.114}$$

where $\vec{v} = (v,0,0)^T$ is the velocity vector of the moving frame. Mathematically, the solution for $\vec{B}'$ is not unique because we may add a scalar multiple of $\vec{V}'$ without changing the cross product with $\vec{V}'$. Physically, however, the transformed magnetic field should not depend on the velocity of a specific particle.

In the moving frame, there is a non-zero magnetic field even though there was no magnetic field in the original frame. Electric and magnetic fields are combined by a Lorentz boost, just as space and time or energy and momentum are combined. For this reason, electric and magnetic fields do not have separate observer-independent meanings. They form a single physical concept, the **electromagnetic field**.

The electromagnetic field has six components and cannot be arranged in a 4-vector. Instead, the transformation behavior derived from the 4-force agrees with the components of an antisymmetric 4×4 matrix, the **field-strength tensor**

$$\mathscr{F} = \begin{pmatrix} 0 & E^x & E^y & E^z \\ -E^x & 0 & B^z & -B^y \\ -E^y & -B^z & 0 & B^x \\ -E^z & B^y & -B^x & 0 \end{pmatrix} \tag{7.115}$$

with components $\mathscr{F}_{\alpha\beta}$. As a tensor (a multi-index object) it transforms by a Lorentz boost with matrix Λ according to

$$\mathscr{F}' = (\Lambda^{-1})^T \mathscr{F} \Lambda^{-1} . \tag{7.116}$$

Tensors as well as this transformation law will be discussed in more detail in the context of general relativity.

7.3.3 LIGHT

A massive particle moves with speed $|\vec{V}| < 1$, which ensures that its energy and momentum are finite. Its 4-velocity vector $\mathbf{u} = (\gamma_{\vec{V}}, \gamma_{\vec{V}}\vec{V}^T)^T$ is always timelike: $||\mathbf{u}||^2 = -1$.

If we attempt to derive the 4-velocity of a massless particle moving at speed $|\vec{V}| = 1$ as a limit of the massive case for $|\vec{V}| \to 1$ in $\mathbf{u}$, we do not obtain well-defined components. Also the original definition of the 4-velocity $\mathbf{u} = \mathrm{d}\mathbf{x}/\mathrm{d}\tau$ is problematic because $\mathrm{d}\tau = (1-|\vec{V}|^2)^{1/2}\,\mathrm{d}t \to 0$ in this limit. Proper time therefore cannot be used as a parameter along lightlike worldlines.

7.3.3.1 Affine parameter

In order to describe massless particles moving at the speed of light, we have to develop a new formalism independent of what we used for massive particles. We start with the fact that massless particles move along light rays, which we know how to draw in a Minkowski space-time diagram. Any such worldline can be parameterized as

$$\mathbf{x}(\lambda) = \mathbf{x}_0 + \lambda\mathbf{u} \tag{7.117}$$

with a real number λ and two constant 4-vectors, the initial event $\mathbf{x}_0$ and a lightlike 4-vector $\mathbf{u}$. The lightlike vector, $||\mathbf{u}||^2 = 0$, is the tangent vector to the worldline and has the same relationship with $\mathbf{x}(\lambda)$ as the 4-velocity:

$$\mathbf{u} = \frac{\mathrm{d}\mathbf{x}}{\mathrm{d}\lambda}\,. \tag{7.118}$$

For instance, if $\mathbf{u} = (1,1,0,0)^T$, we can solve (7.118) and obtain the equations $t(\lambda) = t_0 + \lambda$ and $(x(\lambda), y(\lambda), z(\lambda)) = (x_0 + \lambda, y_0, z_0)$ which describe a light ray in Minkowski space-time.

Whenever a parameter λ appears linearly in a parameterization of a light ray in Minkowski space-time, as in (7.117), we call it an **affine parameter**. An affine parameter is the replacement of proper time for a lightlike worldline, but it is not as unique: On a timelike worldline, we can redefine proper time by changing its initial value. The function $\tilde{\tau}(t) = \tau(t) + \tau_0$ obeys the same relationship $\mathrm{d}\tilde{\tau} = (1 - |\vec{V}|^2)^{1/2}\,\mathrm{d}t$ as $\tau(t)$. An affine parameter can be redefined in the same way, but we can also multiply it by a constant A: Using $A\lambda$ instead of λ in (7.117) gives us the same lightlike worldline because $A\mathbf{u}$ is a lightlike vector pointing in the same direction as $\mathbf{u}$.

In spite of the additional freedom in parameterizing lightlike worldlines, an affine parameter is distinguished by its linear appearance in (7.117). This behavior implies the equation of motion

$$\frac{\mathrm{d}^2\mathbf{x}}{\mathrm{d}\lambda^2} = \frac{\mathrm{d}\mathbf{u}}{\mathrm{d}\lambda} = 0\,, \tag{7.119}$$

consistent with the absence of any forces acting on light, other than gravity which we have not included yet. The equation of motion would be different if we had used a parameter μ in which $\mathbf{x}(\mu) = \mathbf{x}_0 + f(\mu)\mathbf{t}$, where $f(\mu)$ is some non-linear function and $\mathbf{t}$ a lightlike 4-vector. For such a parameterization, we have

$$\frac{\mathrm{d}\mathbf{x}}{\mathrm{d}\mu} = f'(\mu)\mathbf{t}\,, \tag{7.120}$$

such that $\mathbf{t}$ is proportional to the tangent vector, and

$$\frac{\mathrm{d}^2\mathbf{x}}{\mathrm{d}\mu^2} = f''(\mu)\mathbf{t} \neq 0\,. \tag{7.121}$$

The last equation implies that there would be apparent forces on light, but they are not observable because they are caused only by choosing a non-linear mathematical parameterization.

7.3.3.2 Wave 4-vector

The energy and spatial momentum of a massive particle are related to the velocity by $E = m\gamma_{\vec{V}}$ and $\vec{p} = m\gamma_{\vec{V}}\vec{V}$. Like the 4-velocity components, they do not have well-defined limits for $|\vec{V}| \to 1$, but they tell us that the mass has to approach zero in this limit for finite energy-momentum components to be possible.

However, the ratio $E/|\vec{p}| = 1/|\vec{V}|$ has a well-defined limit $E/|\vec{p}| \to 1$ for $|\vec{V}| \to 1$. The fact that this ratio should play an important role for light rays is also suggested by quantum mechanics, where we have the two relations

$$E = \hbar\omega \quad \text{and} \quad \vec{p} = \hbar\vec{k} \tag{7.122}$$

with Planck's constant $\hbar$, the frequency $\omega = 2\pi f$ of light (or a photon) and the wave vector $\vec{k}$. The wave vector has magnitude $|\vec{k}| = 2\pi/\lambda$ related to the wave length λ and points in the direction in which the light ray (or photon) moves. (Except for this place, we rarely use the wave length in the present context in order to avoid confusion with the affine parameter, for which the letter λ is customary too.) The frequency and wave length of light in vacuum are related to the speed of light by $\lambda f = c$ which implies the **dispersion relation**

$$\omega = |\vec{k}| \tag{7.123}$$

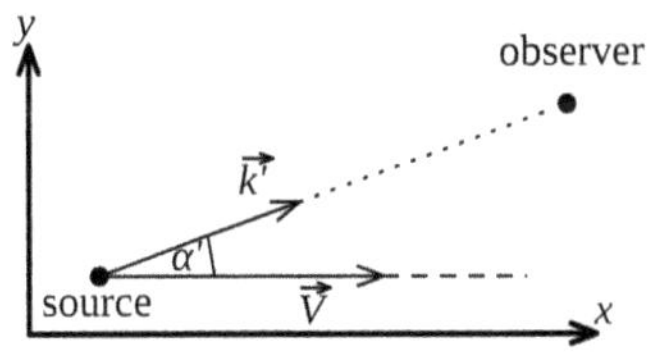

Figure 7.30 Light ray emitted by a moving source. The observer measures an angle α' and frequency $\omega' = |\vec{k}'|$

in terms of ω and $\vec{k}$ and with $c = 1$. If we combine (7.122) and (7.123), we obtain the ratio $E/|\vec{p}| = 1$, just as it is suggested by the limit of $E/|\vec{p}|$ of massive particles when $|\vec{V}| \to 1$.

We use the quantum-mechanics relationship between energy-momentum and frequency-wave number to derive a new 4-vector. We know that $\mathbf{p} = (E, \vec{p}^T)^T$ is a 4-vector, and $\hbar$, as a fundamental constant, is invariant according to the relativity principle. If we divide $\mathbf{p}$ by $\hbar$, we obtain a new 4-vector, the **wave 4-vector**

$$\mathbf{k} = \begin{pmatrix} \omega \\ \vec{k} \end{pmatrix} \tag{7.124}$$

of a light ray. The dispersion relation (7.123) implies that the wave 4-vector is always lightlike:

$$||\mathbf{k}||^2 = -\omega^2 + |\vec{k}|^2 = 0 . \tag{7.125}$$

7.3.3.3 Doppler shift and relativistic beaming

Transforming the wave 4-vector $\mathbf{k}$ to a new frame allows us to derive relationships between observations of light rays made by different observers. Consider a light source moving with velocity $\vec{V}$ along the x-axis in the inertial frame of an observer. If we know what physical mechanism is responsible for the emission of light, as for instance in light emitted by a star, we can use quantum mechanics to predict the possible frequencies ω in the spectrum. Or we can use lab experiments to measure the expected frequencies. For a given spectral line, we then know the time component of the wave vector $\mathbf{k} = (\omega, \vec{k}^T)^T$ in the rest frame of the source, and the magnitude of $\vec{k}$ is determined by the dispersion relation. The distant observer does not measure these components but rather those of the transformed $\mathbf{k}' = (\omega', \vec{k}'^T)^T$.

By assumption, the source is moving in the x-direction, but it may emit light in all possible directions. The velocity vector of the source and the spatial direction of an emitted light ray generically span a plane, which we choose as the x-y plane of the observer as shown in Figure 7.30. A single angle α' is sufficient to specify the direction of the light ray in the plane, given by the components

$$k^{x'} = |\vec{k}'| \cos(\alpha') = \omega' \cos(\alpha') \quad \text{and} \quad k^{y'} = |\vec{k}'| \sin(\alpha') = \omega' \sin(\alpha') \tag{7.126}$$

of the spatial wave vector $\vec{k}'$. Here, we use primes in the observer frame because these components are obtained in a frame different from the rest frame of the source. Even though there is no observer on the star, a different set of lab observations has been made in order to determine the expected spectrum of unprimed frequencies ω.

If we boost the observer's frame to speed $|\vec{V}|$ in the x-direction, it agrees with the rest frame of the source. A Lorentz boost therefore relates the frequencies ω' measured by the observer and ω measured in the rest frame of the source:

$$\omega = \gamma_{\vec{V}}(\omega' - |\vec{V}| k^{x'}) = \gamma_{\vec{V}}(1 - |\vec{V}| \cos(\alpha'))\omega' \tag{7.127}$$

which we can invert to obtain the equation

$$\omega' = \omega \frac{\sqrt{1-|\vec{V}|^2}}{1-|\vec{V}|\cos(\alpha')} \tag{7.128}$$

of the **relativistic Doppler shift**.

For small $|\vec{V}|$, we Taylor expand $(1-|\vec{V}|^2)^{1/2} \approx 1-\frac{1}{2}|\vec{V}|^2$ and $(1-|\vec{V}|\cos(\alpha'))^{-1} \approx 1+|\vec{V}|\cos(\alpha')$. In this limit, the result $\omega' \approx \omega(1+|\vec{V}|\cos(\alpha'))$ agrees with the acoustic Doppler shift. For any velocity, we have blue-shift, $\omega' > \omega$, for a source moving toward the observer ($\alpha' = 0$) and redshift, $\omega' < \omega$, for a source moving away from the observer ($\alpha' = \pi$): In these cases,

$$\omega' = \omega \frac{\sqrt{1-|\vec{V}|^2}}{1\mp|\vec{V}|} = \omega\sqrt{\frac{1\pm|\vec{V}|}{1\mp|\vec{V}|}} \qquad \begin{cases} >\omega & \text{if } +1 = \cos(\alpha') \\ <\omega & \text{if } -1 = \cos(\alpha') \end{cases} . \tag{7.129}$$

The **transverse Doppler shift**, $\omega' = \omega(1-|\vec{V}|^2)^{1/2} \neq \omega$ for $\alpha' = \pi/2$, is an effect of time dilation and does not occur in the acoustic version.

In addition to (7.127), a Lorentz boost of the wave 4-vector gives us the equation

$$k^x = \gamma_{\vec{V}}(k^{x'} - |\vec{V}|\omega') \tag{7.130}$$

for the x-components. Using (7.126) and (7.127), we write

$$\begin{aligned} k^x &= \gamma_{\vec{V}} k^{x'}\left(1-\frac{|\vec{V}|}{\cos(\alpha')}\right) \qquad (7.131)\\ &= \omega\cos(\alpha) = \omega'\cos(\alpha)\frac{1-|\vec{V}|\cos(\alpha')}{\sqrt{1-|\vec{V}|^2}} = \frac{k^{x'}}{\cos(\alpha')}\cos(\alpha)\,\gamma_{\vec{V}}(1-|\vec{V}|\cos(\alpha')). \end{aligned}$$

Comparing these two lines, we see that the angle changes according to

$$\cos(\alpha) = \frac{\cos(\alpha')-|\vec{V}|}{1-|\vec{V}|\cos(\alpha')} \tag{7.132}$$

with inverse

$$\cos(\alpha') = \frac{\cos(\alpha)+|\vec{V}|}{1+|\vec{V}|\cos(\alpha)} . \tag{7.133}$$

We determine implications of (7.133) by inserting a few specific angles. For $\alpha = 0$, or forward emission in the direction of the velocity, we obtain the same angle $\alpha' = 0$. For $\alpha = \frac{1}{2}\pi$, or transversal emission, we obtain $\cos(\alpha') = |\vec{V}|$. If $|\vec{V}| \approx 1$ is close to the speed of light, the angle $\alpha' \ll 1$ is very small, in contrast to the large initial angle $\alpha = \frac{1}{2}\pi$. By continuity, all light rays emitted into the forward hemisphere in the rest frame of the light source, defined by $|\alpha| \leq \frac{1}{2}\pi$, are emitted into a narrow cone in the forward direction in a frame in which the light source moves at a speed close to the speed of light. This effect, illustrated in Figure 7.31, is called **relativistic beaming**. Applications include synchrotron radiation, in which electromagnetic waves emitted by charged particles in accelerators are sent predominantly in the forward direction. The effect can also explain the large luminosity of some relativistic jets emitted by active astrophysical sources.

If we identify $\cos(\alpha)$ with the x-component of the velocity of an emitted light ray (with speed $c = 1$), we recognize the relativistic composition law of velocities in (7.133). The positions of primes

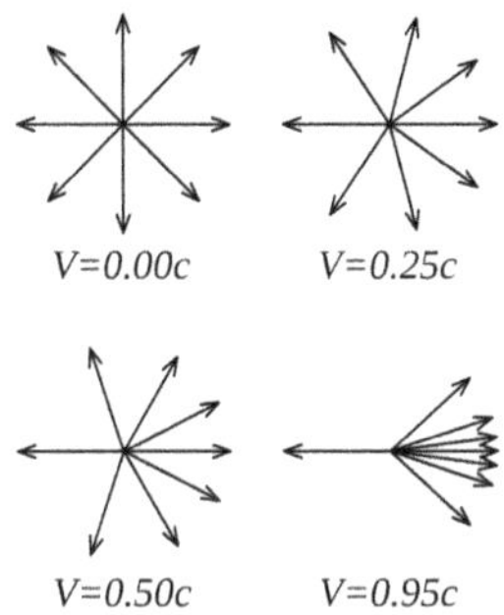

Figure 7.31 Relativistic beaming: If a source emits light isotropically in its rest frame ($V = 0$), it emits more light in the forward direction when it is moving, here shown to the right at three different speeds.

are reversed because in our derivation here we boosted the observer instead of the source. The y-components of the composition law correspond to the transformation of $\sin(\alpha)$,

$$\sin(\alpha') = \frac{\sin(\alpha)}{\gamma_{\vec{V}}(1+|\vec{V}|\cos(\alpha))}. \tag{7.134}$$

The factor of $1/\gamma_{\vec{V}} = (1-|\vec{V}|^2)^{1/2}$ further explains why most light rays are emitted in the forward direction if $|\vec{V}| \approx 1$: For relativistic velocities, $1/\gamma_{\vec{V}} \ll 1$ is very small and suppresses $\sin(\alpha')$ compared with $\sin(\alpha)$. Therefore, α' is much smaller than α for $|\alpha| \leq \frac{1}{2}\pi$.

The effect of relativistic beaming may come as a surprise because the angle, as a geometrical concept, is supposed to be preserved by frame transformations. However, α and α' are angles determined by Euclidean geometry on the plane spanned by the source velocity and the spatial direction of the light ray. Our transformations are Lorentz boosts, which preserve the Minkowski geometry of space-time but not Euclidean geometry. The space-time angle of a light ray, specified by the lightlike condition, is always preserved by Lorentz boosts. It is an infinite angle with any spacelike or timelike direction.

7.3.4 OBSERVABLES IN LOCAL INERTIAL FRAMES

All derivations so far referred to observers in inertial frames. We used **global coordinate systems** in which, in principle, all events in space-time have been marked by coordinates for space and time. Marking all events with coordinate values, at least in a region large enough to enclose all planned observations, requires much time and effort. In particular, an observer should synchronize clocks to make sure that time measurements at different places are correctly compared with one another. If the observer's frame changes during this process, previous synchronizations will be lost due to time dilation, and the frame can no longer be used. This is the physical reason why observations described by global coordinate systems must be based on inertial frames.

When we make observations, we are never in an inertial frame because we are rotating with Earth around the planet's axis, orbiting around the sun, and moving along with the solar system. Nevertheless, we expect that observations made during a short period of time can be described correctly without taking into account the motion of the astronomical bodies. This expectation is confirmed by the notion of local inertial frames.

Local inertial frames can describe (most) observations made by non-inertial observers. Instead of using a global coordinate system, it is often sufficient to stay close to the event when the observation is made, and to specify four independent directions for reference with a unit distance in each direction as shown in Figure 7.32. Measurements made in a small region around the observation

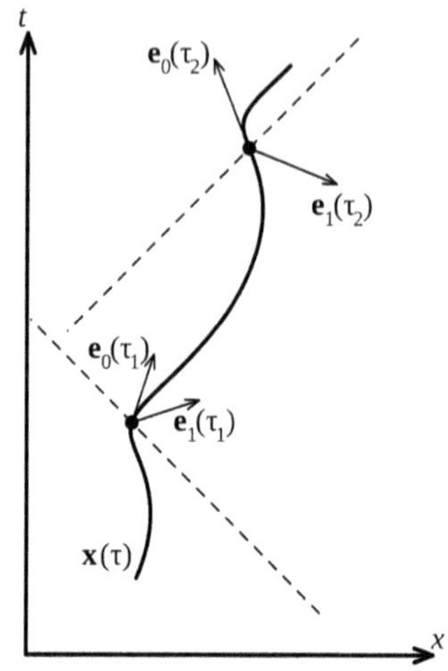

Figure 7.32 In a local inertial frame, we do not use global coordinate systems with axes spanning through all of space-time, but only reference directions around observation events. Whenever worldlines of observed particles or light rays (dashed lines) intersect the observer's worldline (solid), an orthonormal basis is chosen in order to determine 4-vector components. The basis vectors are adapted to the observer's motion by choosing the timelike basis vector to equal the observer's 4-velocity, $\mathbf{e}_0(\tau) = \mathbf{u}_{\text{obs}}(\tau)$, at each observation event. The timelike direction at any fixed time is therefore parallel to the time axis in an instantaneous rest frame of the observer. For a given $\mathbf{e}_0(\tau)$ in a 2-dimensional cross-section of space-time, the spacelike basis vector $\mathbf{e}_1(\tau)$ is determined by orthonormality and the standard orientation of frames, such that $\mathbf{e}_1(\tau)$ points rightward. If the observer is not inertial, the orthonormal basis of a local inertial frame is time-dependent: we have time-dependent vectors $\mathbf{e}_\alpha(\tau)$. Orthonormality is determined according to the Minkowski scalar product (and does not always look right-angled or equal-length).

event can then be described by projections onto these directions. Mathematically, we replace the coordinate system with a set of independent and, for convenience, orthogonal unit vectors at the observation event, or an orthonormal basis of four 4-vectors $\mathbf{e}_0$, $\mathbf{e}_1$, $\mathbf{e}_2$ and $\mathbf{e}_3$.

If the observer makes several successive observations but is accelerated between them, we adjust our basis vectors to account for rotations or boosts of the observer. We therefore provide a basis not just at one event but at all events on the observer's worldline. If the basis needs to be adjusted to account for acceleration, the basis vectors $\mathbf{e}_0(\tau)$, $\mathbf{e}_1(\tau)$, $\mathbf{e}_2(\tau)$ and $\mathbf{e}_3(\tau)$ depend on the observer's proper time τ. With this time dependence, we refer to the basis as a **local inertial frame**.

A local inertial frame is orthonormal at each time τ,

$$\mathbf{e}_\alpha(\tau) \cdot \mathbf{e}_\beta(\tau) = \eta_{\alpha\beta} \tag{7.135}$$

for all τ. Here, the subscript indices α and β on $\mathbf{e}$ label different basis vectors rather than their components. The Minkowski metric $\eta_{\alpha\beta}$ is used for a compact notation of all orthogonality and normalization conditions, including the timelike normalization of $||\mathbf{e}_0||^2 = -1$. Notice that $\eta_{\alpha\beta}$ also appears on the left-hand side of (7.135) if we write the scalar product in terms of components: A basis vector $\mathbf{e}_\alpha$ for a given α has components e_α^γ, which appear in the scalar product

$$\eta_{\gamma\delta} e_\alpha^\gamma e_\beta^\delta = \eta_{\alpha\beta} \,. \tag{7.136}$$

One of the basis vectors, $\mathbf{e}_0$, should be timelike. Since we normalize all basis vectors, a convenient choice for $\mathbf{e}_0 = \mathbf{u}_{\text{obs}}$ is the observer's 4-velocity $\mathbf{u}_{\text{obs}}$, or the tangent vector to the observer's worldline. If we include only the x-direction among the spatial ones, the vector $\mathbf{e}_1$ is determined by orthonormality if the basis has the usual orientation assumed for a frame. The vector $\mathbf{e}_1$ then points to the right if $\mathbf{e}_0$ points upwards in a diagram.

If all three spatial directions are included, specifying the spacelike basis vectors requires a choice of the usual three angles in space: Two spherical angles to determine the direction of a 3-axis, and an additional angle to determine the rotation of the 1–2 plane around the fixed 3-axis. The same three angles appeared in (4.242) on page 167. In the case of motion in the x-direction at speed V, a local inertial frame is given by

$$\mathbf{e}_0 = \begin{pmatrix} \gamma_V \\ \gamma_V V \\ 0 \\ 0 \end{pmatrix} \quad , \quad \mathbf{e}_1 = \begin{pmatrix} \gamma_V V \\ V \\ 0 \\ 0 \end{pmatrix} \quad , \quad \mathbf{e}_2 = \begin{pmatrix} 0 \\ 0 \\ 1 \\ 0 \end{pmatrix} \quad , \quad \mathbf{e}_3 = \begin{pmatrix} 0 \\ 0 \\ 0 \\ 1 \end{pmatrix} . \tag{7.137}$$

The first vector, $\mathbf{e}_0$, is the usual normalized 4-velocity for the given 3-velocity. The first spatial vector, $\mathbf{e}_1$, is obtained by swapping the first two components of $\mathbf{e}_0$. Thanks to the negative sign in the Minkowski scalar product, this step makes sure that $\mathbf{e}_1$ is orthogonal to $\mathbf{e}_0$, and it turns timelike normalization of $\mathbf{e}_0$ into spacelike normalization of $\mathbf{e}_1$.

Given a local inertial frame, we can, at any given time τ, decompose a 4-vector $\mathbf{p} = p^\alpha(\tau)e_\alpha(\tau)$ into components $p^\alpha(\tau)$. The original vector $\mathbf{p}$ may have been determined by specifying components in a global frame, for instance after solving equations of motion for a moving particle. These components, in general, do not correspond to values measured by a local observer, which rather require projections onto the observer's reference frame $\mathbf{e}_\alpha(\tau)$. The parameter τ refers to the time when the observation is made, for instance at the measurement event when a particle hits a detector. Geometrically, this event is the intersection of the particle's worldline with the observer's worldline. Using orthonormality of the local inertial frame, we compute the components of a vector by scalar products:

$$\mathbf{p} \cdot \mathbf{e}_\beta = (p^\alpha \mathbf{e}_\alpha) \cdot \mathbf{e}_\beta = p^\alpha (\mathbf{e}_\alpha \cdot \mathbf{e}_\beta) = p^\alpha \eta_{\alpha\beta} = p_\beta . \tag{7.138}$$

In the last step, we defined new components p_β of the vector $\mathbf{p}$ with a subscript index instead of the original superscript in p^α. We have to match index positions on both sides of the equation, but the expressions also show that p_β in general has values different from those of p^β. Nevertheless, subscript components are strictly related to superscript components because the matrix $(\eta_{\alpha\beta})$ is invertible. The time components differ, $p_0 = \eta_{0\beta} p^\beta = \eta_{00} p^0 = -p^0$ (only $\beta = 0$ contributes a non-zero term to the β-sum) while the spatial components are equal for both index positions, $p_1 = p^1$, $p_2 = p^2$ and $p_3 = p^3$ in Minkowski space-time. With this new ingredient in our notation, we may write the scalar product of two 4-vectors $\mathbf{a}$ and $\mathbf{b}$ as

$$\mathbf{a} \cdot \mathbf{b} = a^\alpha \eta_{\alpha\beta} b^\beta = a^\alpha b_\alpha = a_\beta b^\beta . \tag{7.139}$$

These definitions allow us to transfer our results for components of 4-vectors in global inertial frames to local inertial frames. We have identified the time component of the 4-momentum, $p^0 = E$ with a superscript as defined originally for all vectors, with the energy of a moving object. Instead of referring to global components, we now write this relationship as a projection on the local inertial frame:

$$E = p^0 = -p_0 = -\mathbf{p} \cdot \mathbf{e}_0 = -\mathbf{p} \cdot \mathbf{u}_{\text{obs}} . \tag{7.140}$$

As a scalar product of two 4-vectors, the energy is invariantly defined, such that the same value is obtained in any coordinate system, but it refers to the observer's state of motion by reference to the 4-velocity $\mathbf{u}_{\text{obs}}$.

As a specific example, consider a massive particle in its own rest frame, with 4-momentum $\mathbf{p} = (m,0,0,0)^T$ and a light ray with wave 4-vector $\mathbf{k} = (\omega_*, \omega_*, 0, 0)^T$ in the same frame. Both objects are measured by an observer moving in the x-direction with 4-velocity $\mathbf{u}_{\text{obs}} = (\gamma_V, \gamma_V V, 0, 0)^T$. This 4-vector has the same form as in a global inertial frame, but now $V(\tau)$ may be time-dependent for

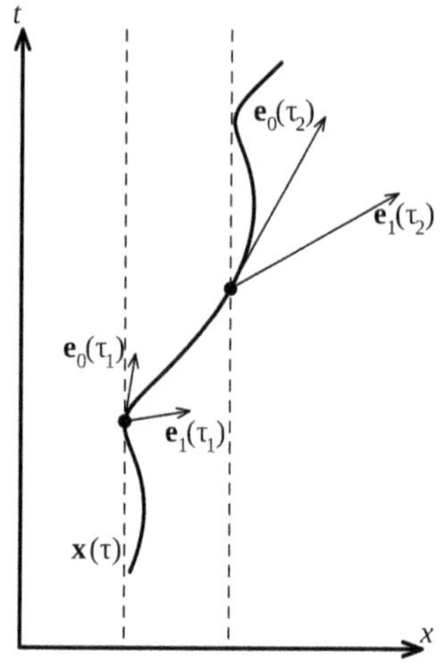

Figure 7.33 A local inertial frame cannot be used to describe direct length measurements, for instance to derive Lorentz contraction. If we need two or more events to specify a measurement, it is in general impossible to make them simultaneous if the observer is not inertial. In this diagram, the dashed curves are the worldlines of two ends of a rod at rest in the global frame. When they intersect the observer's worldline, they give rise to different local inertial frames, making it impossible for the observer to determine simultaneous events for the endpoints. Optical length measurements may be accessible in a local inertial frame.

an accelerated observer. The observer measures the energy $E = -\mathbf{p}\cdot\mathbf{u}_{\rm obs} = m\gamma_V$ of the particle and the frequency $\omega = -\mathbf{k}\cdot\mathbf{u}_{\rm obs} = \gamma_V\omega_*(1-V)$ of the light ray. Both equations, in particular the last one which equals our result for Doppler shift, are valid for accelerated observers moving in the x-direction.

Using the same massive particle and light ray, an observer moving along a circle, such as Earth's orbit, can be described by the 4-velocity $\mathbf{u} = (\gamma_V, \gamma_V V\cos(\Omega\tau), \gamma_V V\sin(\Omega\tau), 0)^T$. Again, the measured energy of the massive particle is $E = m\gamma_V$, and the observed light ray exhibits a periodic Doppler shift $\omega = \gamma_V\omega_*(1 - V\cos(\Omega\tau))$.

Local inertial frames allow us to describe measurements made by accelerated observers in small regions, but they cannot show how accelerated observers measure extended properties of space-time such as distances. For instance, a local inertial frame is not sufficient to determine simultaneous events relevant for a length measurement as required to derive Lorentz contraction. If the observer is accelerated during the length measurement, as in Figure 7.33, the procedure amounts to holding a ruler with shaking hands and does not give a reliable result. Mathematically, if there is acceleration, the geometry of space-time is not experienced in the same way at different events. A position-dependent metric $g_{\alpha\beta}$ replacing the constant $\eta_{\alpha\beta}$ is then required in the corresponding line element $ds^2 = g_{\alpha\beta}dx^\alpha dx^\beta$. The new geometrical notion related to such a position dependence is curvature.

An example that shows the appearance of position-dependent metrics, but not of curvature, for accelerated observers is obtained by transforming the Minkowski line element to a rotating coordinate system. Starting with coordinates (t,x,y,z), we transform them by non-linear functions to new coordinates $t' = t$, $x' = x\cos(\Omega t) + y\sin(\Omega t)$, $y' = -x\sin(\Omega t) + y\cos(\Omega t)$, $z' = z$ with a constant angular velocity Ω. If we use the Minkowski line element for the primed coordinates, it has the form

$$\begin{aligned} ds^2 &= -(dt')^2 + (dx')^2 + (dy')^2 + (dz')^2 \\ &= -\left(1 - \Omega^2(x^2+y^2)\right)dt^2 + 2\Omega(ydx - xdy)dt + dx^2 + dy^2 + dz^2. \end{aligned} \tag{7.141}$$

This line element is position-dependent, has cross terms (the corresponding metric $g_{\alpha\beta}$ is not diagonal), and lacks a dt^2-contribution at positions such that $\Omega\sqrt{x^2+y^2}=1$, where the rotating coordinate system moves at the speed of light. These properties are usually realized when we consider non-inertial coordinate systems, or space-times with curvature.

EXERCISES

1. Show that the wave equation

$$\left(-\frac{1}{c^2}\frac{\partial^2}{\partial t^2}+\frac{\partial^2}{\partial x^2}+\frac{\partial^2}{\partial y^2}+\frac{\partial^2}{\partial z^2}\right)u(t,x,y,z)=0$$

is invariant under a Lorentz transformation in the x-direction.

2. Show that the wave equation is not invariant under a Galilean transformation in the x-direction.
3. Show that Newton's second law is not invariant under a Lorentz transformation in the x-direction.
[Although *two* variables, t and x, are involved in the transformation, the *one*-parameter worldline $x(t)$ is mapped to a new parameterization $x'(t')$. Therefore, do not use the chain rule for a function of two variables; use the chain rule for the t-derivative only: $dx'/dt' = (dt/dt')\cdot(dx'(x,t)/dt)$.]
4. Define the event $A\colon (ct,x)=(1,1)$ and a worldline $\mathscr{L}$ as the set of events $\{(ct,x)=(3+\lambda,2)\}$, where λ takes all real values. Draw a space-time diagram with these objects.
For which values of λ is the separation between the corresponding event on $\mathscr{L}$ and A spacelike? For which values is it timelike or lightlike?
5. In the light clock used to derive the Minkowski line element, events A and C are separated by the displacement vector $(c\Delta t,\Delta x)=(2L,0)$ in the rest frame of the mirrors. In the moving frame, the displacement vector is $(c\Delta t',\Delta x')$ with

$$\Delta x' = -v\Delta t' \quad \text{and} \quad c\Delta t' = 2\sqrt{L^2+(\Delta x'/2)^2}. \tag{7.142}$$

Solve the two equations in (7.142) for $c\Delta t'$ and $\Delta x'$ in terms of L, v and c. Confirm that $(c\Delta t',\Delta x')$ is obtained from $(c\Delta t,\Delta x)$ by a Lorentz boost with speed v.
6. Following the example of rotations in the Euclidean line element, show that the Minkowski line element is invariant under a Lorentz boost.
7. An event A has coordinates $(ct_A,x_A)=(4,3)$.
   (i) Is the displacement vector from the origin to A spacelike, lightlike, or timelike? What time difference is measured by a clock that moves uniformly from the origin to A?
   (ii) Compute the displacement vector (ct'_A,x'_A) of A seen by an observer moving at half the speed of light in the x-direction. Is this vector spacelike, lightlike, or timelike?
   (iii) Repeat part (ii) with graphical methods: Draw event A in a space-time diagram with two frames (ct,x) and (ct',x') and read off the coordinates (ct'_A,x'_A) by measuring distances in your diagram.
   (iv) What is the speed of a second observer who experiences the origin and A at the same point in space? Compute the time difference that separates the two events in this frame.
8. In Frame 1 with coordinates (t,x), two events are given by $A\colon(0,0)$ and $B\colon(4,1)$. A second frame, Frame 2 with coordinates (t',x'), is moving at speed $v=\frac{1}{2}c$ in the x-direction.
   (i) Compute the components t'_B and x'_B of event B in Frame 2.
   (ii) Draw a diagram containing both frames and the two events.
   (iii) Compute the time difference between A and B measured by

(a) an observer in Frame 1,
(b) an observer in Frame 2, and
(c) a clock moving uniformly from A to B.

(iv) The clock introduced in part (iiic) has speed $w = x_B/t_B = \frac{1}{4}c$ in Frame 1, which is half the speed of Frame 2. Why do the values you obtained in (iiia) and (iiib) not agree with each other, even though $1/4$ is at the midpoint of the interval $[0,1/2]$?
(v) Use your results from part (i) to compute the speed w' of the clock in Frame 2.
(vi) Use relativistic composition of velocities to check your result from part (v).
(vii) If the clock is attached to one end of a rod with proper length $L_* = \sqrt{5}$, what is the length L measured in Frame 1? What is the length L' measured in Frame 2?

9. A Lorentz boost of (ct,x) with rapidity ρ can be written in matrix form as

$$\begin{pmatrix} ct' \\ x' \end{pmatrix} = \begin{pmatrix} \cosh\rho & -\sinh\rho \\ -\sinh\rho & \cosh\rho \end{pmatrix} \begin{pmatrix} ct \\ x \end{pmatrix}.$$

(i) Show that the composition of two Lorentz boosts — first from (ct,x) to (ct',x') with rapidity ρ_1, then from (ct',x') to (ct'',x'') with rapidity ρ_2 — is a Lorentz boost from (ct,x) to (ct'',x'') with rapidity $\rho = \rho_1 + \rho_2$.
(ii) Use this result to derive how the combined velocity v (given by ρ) depends on the two individual velocities, v_1 (given by ρ_1) and v_2 (given by ρ_2).

You may use the identity $\tanh(x+y) = (\tanh x + \tanh y)/(1+\tanh x \tanh y)$.

10. Two 4-vectors in Minkowski space-time are given by $\mathbf{a} = (2/\sqrt{3}, 1/\sqrt{3}, 0, 0)^T$ and $\mathbf{b} = (1,2,0,0)^T$. Compute

(i) $||\mathbf{a}||^2$,
(ii) $||\mathbf{b}||^2$, and
(iii) $\mathbf{a}\cdot\mathbf{b}$.

11. A particle moving in the x-direction has 4-velocity $\mathbf{u} = (\sqrt{5},2,0,0)^T$.

(i) Compute the norm squared of $\mathbf{u}$.
(ii) Derive the value of the speed $V = dx/dt$ of the particle in the frame in which the components of $\mathbf{u}$ have been specified.
(iii) What is the mass m of the particle if it has energy $E = 2m_0$, where m_0 is some mass unit?
(iv) Perform a Lorentz boost of the particle's 4-momentum to a frame moving at speed $v = -1/\sqrt{2}$ in the x-direction. What is the speed V' of the particle measured in this frame?

12. A particle has 4-velocity $(3/\sqrt{5}, 2/\sqrt{5}, 0, 0)^T$ in some frame.

(i) What is the speed of an observer who sees the particle at rest.
(ii) Boost the original 4-velocity to a frame moving at half the speed of light in the *negative* x-direction. Determine the speed V'' in this frame.
(iii) If the same particle has 4-momentum $(3,2,0,0)^T$, what is its mass? What is its energy?

13. A rod moves straight toward a camera at speed v in the direction of its axis. Using Lorentz contraction, the length L between simultaneous endpoints of the rod, measured in the rest frame of the camera, is shorter than the length L_* of a rod at rest.
In a single picture, the camera records all light rays that arrive at its lens at the same time, which therefore were reflected off the rod's two endpoints at different times. A (3D) picture taken by the camera thus shows a length different from both L and L_*.

(i) The length seen by the camera is the distance between the positions of the endpoints when light rays were reflected, such that they arrive at the camera at the same time: If it takes some time Δt for light to reach the near end from the far end of the moving rod, the length seen by the camera is $l = c\Delta t$. Use the diagram to derive a relation between l and L.

(ii) Does the length l appear shorter or longer than L_*?

14. Compute the hyperbolic angle subtended to the origin by a single branch of the unit hyperbola.
15. Which of the objects

$$\text{(i)} \quad \frac{dx^\alpha}{dt}$$

$$\text{(ii)} \quad \sqrt{\frac{d^2x^\alpha}{d\tau^2}}$$

$$\text{(iii)} \quad A\frac{dx^\alpha}{d\tau} + B\frac{d^2x^\alpha}{d\tau^2}$$

(with constants A and B that may have units) are components of 4-vectors? Why? As usual, the Greek letter α may take the values $0, 1, 2, 3$, labeling independent components.

16. A metric is determined by a matrix $g_{\alpha\beta}$ such that the scalar product of two vectors $\mathbf{a}$ and $\mathbf{b}$ with components a^α and b^β, respectively, is given by $\mathbf{a} \cdot \mathbf{b} = g_{\alpha\beta}a^\alpha b^\beta$. In general, the components of $g_{\alpha\beta}$ are not required to equal zero or ± 1 as in the Euclidean or Minkowski cases. The only condition in parts (i) and (ii) below is that $g_{\alpha\beta}$ as a matrix is symmetric and has non-zero determinant.

(i) Show that the scalar product can be written as $\mathbf{a}^T\mathbf{g}\mathbf{b}$ in matrix form.
(ii) The scalar product of two vectors should be invariant under transformations that change the vectors by $\mathbf{a}' = \Lambda\mathbf{a}$, with an invertible matrix Λ. Use this condition to derive the transformation law $\mathbf{g}' = (\Lambda^{-1})^T\mathbf{g}\Lambda^{-1}$ from the equation given in part (i).
(iii) Confirm that the Minkowski metric $g_{\alpha\beta} = \eta_{\alpha\beta}$ is invariant under a Lorentz boost, $\eta'_{\alpha\beta} = \eta_{\alpha\beta}$, by computing the matrix product of part (ii) with Λ the matrix of a Lorentz boost.

17. A particle with mass m moves along a circle in the x–y-plane: $x(t) = R\cos(\Omega t)$ and $y(t) = R\sin(\Omega t)$, where Ω is the constant orbiting frequency and R the radius.

(i) Determine the particle's proper time τ as a function of t.
(ii) Parameterize the position 4-vector along the particle's worldline as $\mathbf{x}(\tau)$.
(iii) Compute the 4-velocity of the particle and the 4-force $\mathbf{f}$.
(iv) What is the 3-force $\vec{F}$ experienced in this coordinate system?
(v) Transform to a frame that moves at speed v in the x-direction and derive the particle's 3-velocity and 3-force measured in this system.

18. A particle with mass m starts at rest and is subject to a constant force $\vec{F} = ma\vec{e}_x$ with a constant a and the unit vector $\vec{e}_x$ in the x-direction.

(i) Use the 3-force $\vec{F} = d\vec{p}/dt$ and integrate the equation of motion for $\vec{p}(t) = m\gamma_{\vec{V}(t)}\vec{V}(t)$.
(ii) Determine $\vec{V}(t)$ from the solution and show that it never reaches the speed of light.
(iii) Integrate your result for $\vec{V}(t)$ to find $\vec{x}(t)$ and show that the particle follows a hyperbola in space-time.
(iv) Compute the particle's proper time τ and parameterize the position as $\mathbf{x}(\tau)$.

19. The electromagnetic field can be arranged in matrix form with antisymmetric components $F_{\alpha\beta}$, as defined by the field-strength tensor

$$\mathbf{F} = \begin{pmatrix} 0 & E^x & E^y & E^z \\ -E^x & 0 & B^z & -B^y \\ -E^y & -B^z & 0 & B^x \\ -E^z & B^y & -B^x & 0 \end{pmatrix}.$$

As a matrix, $\mathbf{F}$ transforms like the metric $\mathbf{g}$ in a preceding exercise

(i) Apply a Lorentz boost Λ (along the x-axis, with velocity v) to $\mathbf{F}$ and read off the boosted electric and magnetic field vectors, $\vec{E}'$ and $\vec{B}'$.

(ii) Use your result to show that $\vec{E}\cdot\vec{B}$ and $|\vec{E}|^2 - |\vec{B}|^2$ are invariant.

20. The light clock used to derive the Minkowski line element has light rays bouncing back and forth between two mirrors. In the rest frame of the clock, the light rays move along a direction orthogonal to the mirror surfaces. In a moving frame, the angle between the light rays and mirror surfaces is different; see Figure 7.6. Show that its value is determined by relativistic beaming.

21. (i) Show that an orthonormal frame is defined by

$$\mathbf{e}_0 = \gamma \begin{pmatrix} 1 \\ -R\Omega\sin(\Omega\gamma\tau) \\ R\Omega\cos(\Omega\gamma\tau) \\ 0 \end{pmatrix}, \quad \mathbf{e}_1 = \gamma \begin{pmatrix} R\Omega \\ -\sin(\Omega\gamma\tau) \\ \cos(\Omega\gamma\tau) \\ 0 \end{pmatrix}$$
$$\mathbf{e}_2 = \begin{pmatrix} 0 \\ \cos(\Omega\gamma\tau) \\ \sin(\Omega\gamma\tau) \\ 0 \end{pmatrix}, \quad \mathbf{e}_3 = \begin{pmatrix} 0 \\ 0 \\ 0 \\ 1 \end{pmatrix} \tag{7.143}$$

with $\gamma = (1 - R^2\Omega^2)^{-1/2}$ and constants Ω and R.

(ii) The basis given in part (i) can be used as the local inertial frame of an observer orbiting with frequency Ω at radius R. What is the locally measured 4-velocity $\mathbf{u}'$ of a particle at rest in the global frame? What is the 3-velocity $\vec{V}'$?

(iii) What is the 3-force acting on the particle, as seen from the observer's frame?

22. A massive particle moving with velocity $\vec{V}$ has 4-momentum $\mathbf{p} = (m\gamma_{\vec{V}}, m\gamma_{\vec{V}}\vec{V}^T)^T$ in a global frame. Using a local inertial frame defined by the basis

$$\mathbf{e}_0 = \begin{pmatrix} \gamma \\ \gamma v \\ 0 \\ 0 \end{pmatrix}, \quad \mathbf{e}_1 = \begin{pmatrix} \gamma v \\ v \\ 0 \\ 0 \end{pmatrix}, \quad \mathbf{e}_2 = \begin{pmatrix} 0 \\ 0 \\ 1 \\ 0 \end{pmatrix}, \quad \mathbf{e}_3 = \begin{pmatrix} 0 \\ 0 \\ 0 \\ 1 \end{pmatrix},$$

compute the 4-momentum components $p^{0'} = -\mathbf{p}\cdot\mathbf{e}_0$, $p^{1'} = \mathbf{p}\cdot\mathbf{e}_1$, $p^{2'} = \mathbf{p}\cdot\mathbf{e}_2$ and $p^{3'} = \mathbf{p}\cdot\mathbf{e}_3$. Compute the 3-velocity $\vec{V}'$ corresponding to the momentum components $p^{\alpha'}$.

23. Consider a wave 4-vector $\mathbf{k} = (\omega, \omega\cos\alpha, \omega\sin\alpha, 0)^T$ and compute the components $k^{0'} = \omega' = -\mathbf{k}\cdot\mathbf{e}_0$, $k^{1'} = \mathbf{k}\cdot\mathbf{e}_1$, $k^{2'} = \mathbf{k}\cdot\mathbf{e}_2$ and $k^{3'} = \mathbf{k}\cdot\mathbf{e}_3$ using the local inertial frame from the preceding Exercise.

Compare your results with relativistic beaming.

Show that the components $k^{\alpha'}$ satisfy the dispersion relation of light.

8 Curved Spaces

Curvature is a central ingredient in the extension from special relativity to general relativity. We have an intuitive understanding of curvature based on pictures of curved lines and curved surfaces, but it turns out that these visible properties only form a small subset of features related to curvature. In fact, there are several independent kinds of curvature most of which are non-intuitive. Grasping this important notion therefore requires a systematic and abstract development.

8.1 CURVATURE

It is easy to determine whether the 1-dimensional graph of a function $f(x)$ is curved: If we do not trust our eyes, we compute $f''(x)$ and conclude that the graph is curved if $f''(x) \neq 0$. We also know how to tell whether a surface is curved by comparing it with a piece of a plane, but in mathematical terms the question of whether the object is curved is more difficult to answer by calculations for 2-dimensional surfaces, let alone 4-dimensional space-time.

There are different types of curvature. When we say that the graph of a function $f(x)$ is curved if $f''(x) \neq 0$, we are talking about **extrinsic curvature**: We have to see the graph drawn in a 2-dimensional diagram, or embedded in a higher-dimensional space. If we just know that an object is 1-dimensional, it is impossible to say whether it is curved: There are only a few mathematical laws in 1-dimensional geometry, given by adding or subtracting the lengths of intervals if they are adjacent or overlap. These laws are the same irrespective of how the 1-dimensional object is embedded in a plane or in space. Curvature exists for a 1-dimensional object only if we see how it is embedded in a surrounding space. The reference to an exterior space outside the curve implies that its curvature is extrinsic.

Extrinsic curvature is what we usually understand as curvature in an intuitive sense. If we walk on a curved line or on a surface in space, we say that it is curved because our orientation in the exterior space changes as we move along; see Figure 8.1. We can use this statement to glimpse, by analogy, an important implication of curvature in space-time: If we are surrounded by curved space-time, our space-time orientation varies as we change our position, even if we are not moving in space but just wait sometime. The space-time orientation is described by space-time angles, which include our velocity, or the rapidity. If space-time angles change as we move in a curved space-time, our velocity may change. A changing velocity implies acceleration, and acceleration is caused by a force. The force experienced in this way if space-time is curved is the gravitational force.

Before we can derive this result, we have to develop more space-time geometry in the presence of curvature. It is impossible to take an exterior view of space-time as we used it for our intuitive understanding of curved lines. It turns out that such an exterior view is not necessary because, for objects of two or more dimensions, there is also **intrinsic curvature** which can be determined just from geometry on the surface without reference to an exterior space. In two dimensions, examples of surfaces that we understand as being curved are the cylinder and the sphere. On both surfaces, our orientation in an embedding space changes as we move on the surface. Therefore, they both have extrinsic curvature. But only one of them has intrinsic curvature.

Intrinsic curvature implies that the surface has a geometry different from the flat geometry of the Euclidean plane. If we draw triangles or circles on a surface with intrinsic curvature, the relationships between side lengths and angles or the radius and circumference are different from what is implied by Euclid's laws. The cylinder has no intrinsic curvature because its geometry is locally the same as the geometry of a plane: We can construct a cylinder by rolling up a sheet of paper. If we draw sufficiently small triangles or circles on the sheet of paper before rolling it up, their side

DOI: 10.1201/9781003588702-10

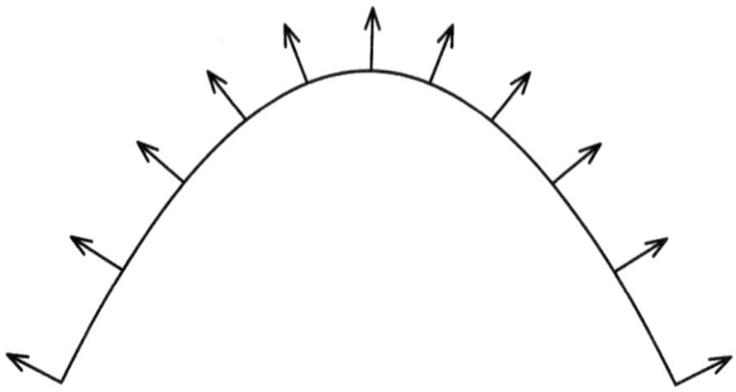

Figure 8.1 Extrinsic curvature of a line is determined by changing normal directions as we move along the curve.

lengths and angles are not modified by the procedure. Therefore, Euclid's laws remain valid. We have to make sure that a triangle or a circle is sufficiently small because it would cease to be a triangle or a circle if rolling up the sheet of paper causes some of its parts to overlap. In mathematical terms, rolling up a sheet of paper changes its topology, leading to overlap, but not its geometry.

The sphere cannot be constructed by rolling up a sheet of paper without wrinkling it or cutting and pasting. (Imagine sticking a Halloween template on the curved surface of a pumpkin.) If we wrinkle the sheet of paper or cut and paste, we change the side lengths of some triangles drawn on the sheet of paper, and the geometry differs from Euclid's laws. The sphere therefore has intrinsic curvature.

We describe the geometry on a surface by deriving a line element for length measurements on it. Starting with the Euclidean line element $\mathrm{d}s^2 = \mathrm{d}x^2 + \mathrm{d}y^2 + \mathrm{d}z^2$ of an embedding space, we restrict it to intervals that lie on the surface. For a cylinder with radius R, we choose coordinates ϕ and z to describe points on the surface, and embed them in 3-dimensional space by writing $x(z,\phi) = R\cos(\phi)$, $y(z,\phi) = R\sin(\phi)$ and $z(z,\phi) = z$. Using the chain rule in $\mathrm{d}x$, $\mathrm{d}y$ and $\mathrm{d}z$ for constant R and inserting the result in the line element gives

$$\mathrm{d}s^2 = \mathrm{d}z^2 + R^2\mathrm{d}\phi^2 \tag{8.1}$$

which differs from the Euclidean 2-dimensional line element only by a constant factor in the second term.

For a sphere of radius R, we use coordinates φ and ϑ on the surface and write the embedding in 3-dimensional space as $x(\vartheta,\varphi) = R\sin(\vartheta)\cos(\varphi)$, $y(\vartheta,\varphi) = R\sin(\vartheta)\sin(\varphi)$ and $z(\vartheta,\varphi) = R\cos(\vartheta)$. The chain rule in $\mathrm{d}x$, $\mathrm{d}y$ and $\mathrm{d}z$, again with constant R, then gives us the line element

$$\mathrm{d}s^2 = R^2(\mathrm{d}\vartheta^2 + \sin(\vartheta)^2\mathrm{d}\varphi^2). \tag{8.2}$$

Unlike the line element on the cylinder, there is a position-dependent coefficient $\sin(\vartheta)^2$ which changes the geometry.

The geometry on a surface that has intrinsic curvature is described by a line element with position-dependent coefficients. However, one of the complications with curvature is the fact that position-dependent coefficients in a line element can occur also for flat space. If we express the planar line element $\mathrm{d}s^2 = \mathrm{d}x^2 + \mathrm{d}y^2$ in polar coordinates, $x(r,\phi) = r\cos(\phi)$ and $y(r,\phi) = r\sin(\phi)$, we obtain $\mathrm{d}s^2 = \mathrm{d}r^2 + r^2\mathrm{d}\phi^2$. There is a position-dependent coefficient, given by the metric component r^2, but the geometry is the same as the geometry of the plane because we have just performed a coordinate transformation. How to disentangle coordinate effects from geometrical curvature effects, for instance deriving consequences of the coefficient $\sin(\vartheta)^2$ in the spherical line element (8.2) and r^2 in the planar line element expressed in polar coordinates, is one of the main questions related to a quantitative notion of curvature.

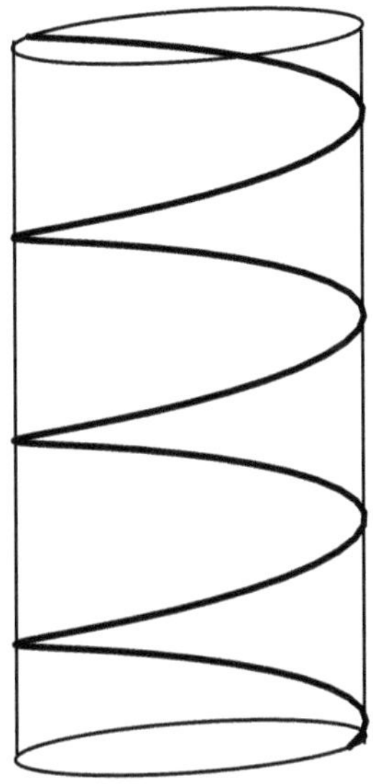

Figure 8.2 Geodesics on a cylinder: A helix, straight lines along the axis, and circle cross-sections.

We will work with position-dependent line elements and derive some implications of curvature. For now, we can see that the geometry on a sphere is non-Euclidean by looking at properties of triangles and circles. We first have to define the notion of a straight line, such as a side of a triangle, on a curved surface. In flat space, a straight line, understood as a line whose tangent vector always points in the same direction, gives us the shortest distance between two points. On a curved surface, it is more convenient to turn this statement around because the definition of geometry through a line element is based on measuring distances rather than directions. We define a **geodesic** between two points on a curved surface as the line with the shortest distance connecting the two points, generalizing the notion of a straight line. In Euclidean geometry, the notion of a parallel to a given straight line through a point is also important. It is the unique straight line through the given point not intersecting the given line. In Euclidean space, two different straight lines are either parallel or they intersect once.

We determine geodesics on a cylinder by using the construction of rolling up a sheet of paper. Straight lines on the sheet of paper determine geodesics on the cylinder because they minimize the distance between two given points, even though most of them are no longer straight as curves in the embedding space. A straight line on the sheet of paper is turned into one of three different types of geodesics shown in Figure 8.2: A line parallel to the axis of the cylinder, a circle transversal to the axis, or a helix. The Euclidean result that there is a unique parallel to any straight line through a given point remains correct. However, two different geodesics on the cylinder can intersect more than once as a result of rolling up the sheet of paper, a topological effect.

On a sphere, geodesics are the intersections of the sphere with planes through the center, like the equator or the meridians. (We will be able to demonstrate this fact in Section 8.6.6.) They are called **great circles**. There are no parallels because any two great circles have two intersection points. The geometry on the sphere is therefore non-Euclidean.

Triangles on a sphere have side lengths and angles that do not obey Euclid's laws. A triangle on a sphere consists of three pieces of great circles whose end points overlap. We construct a sample triangle by starting at the North pole and following two meridians with a right angle between them until they reach the equator. The enclosed piece of the equator is the third side of the triangle. Since meridians have right angles with the equator, all three angles in the triangle are right angles and

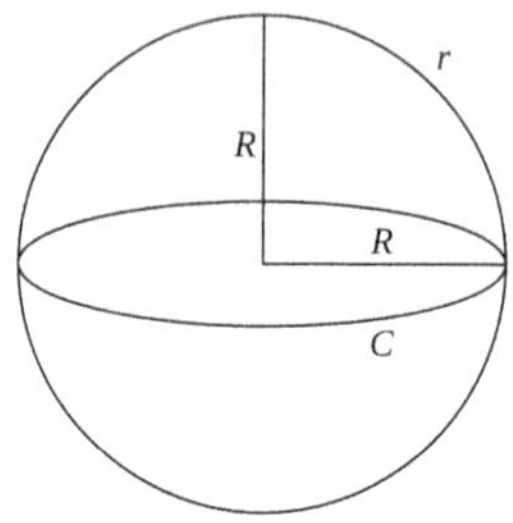

Figure 8.3 The equator as a circle around the North pole on a sphere. The radius of the circle in space is equal to the radius R of the sphere. The radius r of the circle on the sphere is the distance along a great circle from the North pole.

their sum is 270°, violating the Euclidean angle sum of 180°. We say that a surface has **positive curvature** if triangles on it have angle sums greater than 180°, and **negative curvature** if triangles on it have angle sums less than 180°.

As the last example, consider circles on a sphere, defined as sets of points that have equal distance, the radius r, from a center point. The distance is measured on the surface along geodesics, or pieces of great circles. In Euclidean space, the ratio of the circumference C, the length of the circle, over its radius is always equal to $C/r = 2\pi$. On a sphere, we may consider the equator as a circle around the North pole as shown in Figure 8.3. Its radius (again, measured on the sphere) is equal to a quarter of the circumference because it is a quarter of a meridian, and any meridian on a perfect sphere has a length equal to that of the equator. Therefore, $C/r = 4 < 2\pi$ in this case, and the Euclidean result is violated. We may even think of the South pole as a circle around the North pole. It is a single point, but it is also the set of all points with equal distance, half the length of the equator, from the North pole. As a single point, it has zero circumference but a non-zero radius, such that $C/r = 0$ in this case. As these examples show, the ratio depends on the circle, in contrast to the Euclidean behavior.

8.2 SPACE-TIME GEOMETRY AND GRAVITY

The gravitational force is related to space-time curvature. Curvature, a geometrical property, is described by the line element of space-time, which can no longer be of Minkowski form when the gravitational force is included. In particular, the constant components of the Minkowski metric must be replaced by suitable position-dependent functions. In order to derive these functions for a given setup of gravitational interactions, we need a new principle. Just as we referred to the relativity principle in our derivation of the Minkowski line element, we use the **equivalence principle** to derive the space-time metric in the presence of a constant gravitational force. The equivalence principle requires that the inertial mass of an object is always equal to its gravitational mass, an equivalence that is consistent with all measurements made so far but has no explanation in Newtonian mechanics.

More formally, the equivalence principle can be interpreted as stating that the laws of physics take the same form in a non-inertial frame with uniform acceleration, a constant $a = \dot{v}$, as they do in an inertial frame in which there is a constant gravitational force $g = a$. No force is therefore experienced in a system in free fall, provided it is sufficiently small such that the gravitational force can be considered constant within it: Gravity is then exactly compensated for by the acceleration of the system. In particular, the acceleration term in Newton's second law, $m'a$ with the inertial mass m', must be interchangeable with the gravitational term $mg = m'a$. The equality of these terms is equivalent to $m' = m$ if $g = a \neq 0$.

If all laws of physics are the same in a uniformly accelerated system and in a system with a constant gravitational force, the speed of light is the same in all such systems. We can use light

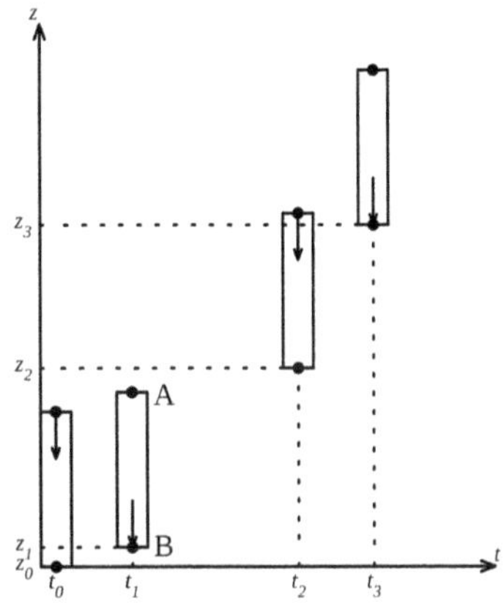

Figure 8.4 Light signals in an accelerated system. In the derivation of $\Delta\tau_B - \Delta\tau_A$, we use $t_2 = \Delta\tau_A$ and $t_3 = t_1 + \Delta\tau_B$, while t_1 is related to the height $h = z_A - z_B$ of the rocket.

signals to measure time, as in the light clock of special relativity, and compute their propagation in an accelerated system. Replacing the acceleration rate by a constant gravitational force then tells us how gravity affects measurements of proper time, which is one of the quantities determined by the space-time line element.

Figure 8.4 shows the spatial diagram of an accelerating system, a rocket moving upwards, and several instances where light rays are emitted or received. The emitter A is positioned at the top of the rocket of height h, and the receiver B at the bottom. We assume that A emits light rays in a periodic manner, separated by proper-time intervals $\Delta\tau_A$. They are periodically received at B, but with a different interval $\Delta\tau_B$ because they experience decreasing travel-time delays as the rocket speeds up. We expect $\Delta\tau_B < \Delta\tau_A$ in this set-up, but the precise relationship has to be derived from the motion of light rays.

The period $\Delta\tau_B$ shows how an observer at B sees the progress of time at A. Any difference in the period $\Delta\tau_A$ of the clock at A and the period $\Delta\tau_B$ of the same clock seen from B can be attributed to a time dilation effect caused by the positions of A and B in an accelerating system. Referring to the equivalence principle, the same effect applies to positions at different gravitational potentials $\Phi(z) = gz$ or at different altitudes z.

If synchronized clocks are placed at the bottom and the top, an observer can directly determine both time measurements, $\Delta\tau_A$ by looking at the synchronized clock at the bottom without any travel-time delays, and $\Delta\tau_B$ by looking toward the clock at the top based on signals received with varying travel-time delays as the rocket accelerates. Before the rocket was launched, it was inertial, making it possible to synchronize the two clocks.

In order to eliminate time-dilation effects from special relativity, we assume that the acceleration of the clock is sufficiently small so as to keep the rocket speed well below c. Therefore, $g\Delta\tau_A \ll c$ and $g\Delta\tau_B \ll c$. Here and in the following calculations, we explicitly write the speed of light because there are now various quantities with units of space over time that are not individual speed parameters.

If the height of the rocket is h and B is initially, at time $t = 0$, positioned at $z = 0$, the vertical positions of A and B are given by

$$z_A(t) = h + \frac{1}{2}gt^2 \quad \text{and} \quad z_B(t) = \frac{1}{2}gt^2 . \tag{8.3}$$

In these equations, t is the time coordinate of an inertial observer outside the rocket. For this coordinate, the clock period equals $\Delta t = \Delta\tau_A$ because the rocket clocks had been synchronized in this inertial frame before the rocket launched.

We use equation (8.3) to compute the distances travelled by two light signals sent from A to B, the first starting at time $t=0$ and the second one A-period later, at $t=\Delta\tau_A$. If the first signal arrives at B at time $t=t_1$, it travels a distance

$$z_A(0)-z_B(t_1)=h-\frac{1}{2}gt_1^2=ct_1 \tag{8.4}$$

where we use the constant speed of light in the second step. The second light ray travels a distance

$$z_A(\Delta\tau_A)-z_B(t_1+\Delta\tau_B)=h+\frac{1}{2}g\Delta\tau_A^2-\frac{1}{2}g(t_1+\Delta\tau_B)^2=c(t_1+\Delta\tau_B-\Delta\tau_A)\,. \tag{8.5}$$

Subtracting the two equations shows that the two clock periods have a non-zero difference because

$$c(\Delta\tau_A-\Delta\tau_B)=gt_1\Delta\tau_B-\frac{1}{2}g\Delta\tau_A^2+\frac{1}{2}g\Delta\tau_B^2 \tag{8.6}$$

requires $\Delta\tau_A\neq\Delta\tau_B$. The equation is quadratic for $\Delta\tau_B$ as a function of $\Delta\tau_A$, but it also depends on t_1. The relationship can be simplified if we use our approximation that g is small. The speed of the rocket during the first cycle of the clock is then small compared with the speed of light, such that the distance traveled by each light ray is close to the height h of the rocket, and $t_1\approx h/c$. Moreover, we use (8.6) to approximate the quadratic terms on the right as

$$\frac{1}{2}g(\Delta\tau_A^2-\Delta\tau_B^2)=g(\Delta\tau_A-\Delta\tau_B)\left(\Delta\tau_B+\frac{1}{2}(\Delta\tau_A-\Delta\tau_B)\right)\approx\frac{g^2t_1}{c}\left(1+\frac{1}{2}\frac{gt_1}{c}\right)\Delta\tau_B^2 \tag{8.7}$$

up to terms cubic in $\Delta\tau_A$ and $\Delta\tau_B$. The result is much smaller than the first term on the right of (8.6), $gt_1\Delta\tau_B$, because we assume $g\Delta\tau_B\ll c$. Keeping only terms linear in the periods, we conclude

$$\Delta\tau_B\approx\frac{\Delta\tau_A}{1+gh/c^2}\approx\Delta\tau_A\left(1-\frac{gh}{c^2}\right)\,. \tag{8.8}$$

Invoking the equivalence principle, we now equate gh with the difference of gravitational potentials at A and B, $gh=\Phi_A-\Phi_B$. The rate or frequency of signals emitted and received at different positions in a gravitational potential are then related by

$$\frac{1}{\Delta\tau_B}=\frac{1+gh/c^2}{\Delta\tau_A}=\frac{1}{\Delta\tau_A}\left(1+\frac{\Phi_A-\Phi_B}{c^2}\right)\,. \tag{8.9}$$

A higher frequency is observed if the receiver is positioned at a lower altitude compared with the emitter: A clock at higher altitude, seen from a lower altitude, seems to go fast. Commercial atomic clocks can detect this time difference if the altitude difference is just about ten meters, in which case $gh/c^2\sim 10^{-15}$, and experimental clocks can demonstrate the effect on the millimeter scale. The effect is important for the Global Positioning System (GPS), which computes the position on Earth by using travel times of signals from at least three satellites. For accurate travel times, the satellites need precisely synchronized clocks, for which clock rates must be compared in frames moving at different speeds and located at different altitudes.

In astrophysics, much larger time differences can be realized because the gravitational potential on the surface of a dense star with radius R and mass M, given by $\Phi_*=-GM/R$, is very different from the gravitational potential on Earth: Far away from the star, we have $\Phi_B\sim\Phi_\infty=0$ if we ignore the small gravitational potential on Earth, compared with the potential on the star. If a light ray is emitted with frequency ω_* on the star surface, the frequency ω_∞ measured on Earth is redshifted:

$$\omega_\infty=\left(1-\frac{GM}{Rc^2}\right)\omega_*<\omega_*\,, \tag{8.10}$$

assuming, as in our derivation, that $GM/Rc^2 \ll 1$ is small. For instance, for a white dwarf of about a solar mass and a radius of $R = 10^4$km, we have $(\omega_\infty - \omega_*)/\omega_* = -GM/(Rc^2) \sim -10^{-3}$. For a neutron star with about the same mass and $R = 10$km, $GM/(Rc^2)$ is not small and we need a full derivation from general relativity.

We are now ready to draw our main conclusion about space-time geometry. The presence of a gravitational force implies that clock rates at different positions differ from one another. Therefore, we need a position-dependent coefficient in the space-time line element in order to maintain the relationship between Δs^2 and proper time $\Delta\tau = \sqrt{-\Delta s^2/c^2}$. We derive the coefficient by starting with (8.9) and writing it in the form

$$\frac{1}{\Delta\tau_B} \approx \frac{1}{\Delta\tau_A}\frac{1+\Phi_A/c^2}{1+\Phi_B/c^2}\,. \tag{8.11}$$

The ratio

$$\frac{1+\Phi_x/c^2}{\Delta\tau_x} = \frac{1}{\Delta t}\,, \tag{8.12}$$

where x can be A or B or any other position, is therefore position-independent. We identify this value with the clock rate $1/\Delta t$ measured by an inertial observer far away from any mass, where $\Phi_\infty = 0$ and $1/\Delta\tau_\infty = 1/\Delta t$ corresponds to the time coordinate of an inertial frame. Using this result together with the definition of proper time, we write the line element of space-time in the presence of a gravitational potential as

$$\Delta s^2 = -c^2(\Delta\tau)^2 = -\left(1+\frac{\Phi(x)}{c^2}\right)^2 (c\Delta t)^2 + \cdots \approx -\left(1+2\frac{\Phi(x)}{c^2}\right)(c\Delta t)^2 + \cdots\,, \tag{8.13}$$

where we used the assumption $\Phi/c^2 \ll 1$ and indicated by dots that our derivation does not allow us to draw conclusions about spatial terms in the line element. At this point, we have confirmed that the gravitational force implies a position-dependent line element.

8.3 CURVED SPACE-TIME FROM NEWTON'S POTENTIAL

The equivalence principle implies gravitational time dilation in systems subject to a constant gravitational force. The line element of space-time then has a position-dependent coefficient in its time coefficient. For a weak gravitational field that is not spatially constant but time-independent, Newton's potential appears in the same form, and in the spatial components of the **weak-field line element**

$$\mathrm{d}s^2 = -\left(1+\frac{2\Phi(x^i)}{c^2}\right)(c\mathrm{d}t)^2 + \left(1-\frac{2\Phi(x^i)}{c^2}\right)(\mathrm{d}x^2+\mathrm{d}y^2+\mathrm{d}z^2)\,. \tag{8.14}$$

This line element and the underlying conditions will be derived in (9.90). The potential Φ is subject to Poisson's equation $\nabla^2\Phi = 4\pi G\mu$ with the mass density μ, and it obeys $\Phi(x^i)/c^2 \ll 1$ for a weak field.

The weak-field line element shows that the relativistic form of Newton's potential is not invariant, and it is not the component of a 4-vector or, like the electromagnetic field, of an antisymmetric field-strength tensor. It is rather a component of the space-time **metric** $g_{\alpha\beta}$ in $\mathrm{d}s^2 = g_{\alpha\beta}\mathrm{d}x^\alpha\mathrm{d}x^\beta$, a symmetric matrix because the components $\mathrm{d}x^\alpha$ and $\mathrm{d}x^\beta$ commute. If we change to a new frame by a coordinate transformation, the metric will take a different form, and Newton's potential may have to be adjusted. Unlike the electromagnetic field, moreover, the gravitational field appears directly in the metric and therefore affects distance measurements in space-time.

Position-dependent metric components imply varying clock rates if the clock's position in space-time is changing. A clock moving in space-time measures proper time along its worldline. We use the same invariant to define proper time as in special relativity, $\mathrm{d}\tau = \sqrt{-\mathrm{d}s^2/c^2}$, but now refer to the weak-field line element (8.14). Along a timelike curve between two events A and B, the proper-time difference is the integral $\tau_{AB} = \int_A^B \sqrt{-\mathrm{d}s^2/c^2}$, or

$$\tau_{AB} = \int_A^B \sqrt{\left(1+\frac{2\Phi}{c^2}\right)\mathrm{d}t^2 - \left(1-\frac{2\Phi}{c^2}\right)\frac{\mathrm{d}x^2+\mathrm{d}y^2+\mathrm{d}z^2}{c^2}} \approx \int_{t_A}^{t_B}\left(1-\frac{1}{2}\frac{|\vec{V}|^2}{c^2}+\frac{\Phi}{c^2}\right)\mathrm{d}t\,. \quad (8.15)$$

In the last step, we have ignored the term $\Phi|\vec{V}|^2/c^4$ which is much smaller than $|\vec{V}|^2/c^2$ and Φ/c^2 if the latter two are small. We have $\Phi/c^2 \ll 1$ because we assume the weak-field line element, and we require $|\vec{V}|^2/c^2 \ll 1$ for near-Newtonian dynamics.

The integral in (8.15) is familiar from classical mechanics. Except for the constant 1 added to the integrand and a constant factor of $-1/c^2$, it has the form of the time integral of kinetic energy minus potential energy, divided by the mass of the object. In classical mechanics, such an integral,

$$S(\vec{x},\mathrm{d}\vec{x}/\mathrm{d}t) = \int\left(\frac{1}{2}\left|\frac{\mathrm{d}\vec{x}}{\mathrm{d}t}\right|^2 - \Phi(\vec{x})\right)\mathrm{d}t\,, \quad (8.16)$$

is called the **action** (per mass), and the integrand

$$L(\vec{x},\mathrm{d}\vec{x}/\mathrm{d}t) = \frac{1}{2}\left|\frac{\mathrm{d}\vec{x}}{\mathrm{d}t}\right|^2 - \Phi(\vec{x}) \quad (8.17)$$

is the **Lagrange function** (per mass) of the system.

An object subject to the potential $\Phi(\vec{x})$ always moves such that the action takes on an extremum: a maximum, a minimum, or a saddle point. If we use this result in (8.15) it acquires a geometrical meaning: we learn that massive objects in a space-time with the weak-field line element always move such that their proper time between two events takes on an extremum among all possible worldlines connecting the two events. Such an extremal worldline in curved space-time is a **geodesic**. We saw that timelike straight lines in Minkowski space-time, in which there is no gravitational force, maximize the space-time distance between two events. In curved space-time with the weak-field line element, massive objects still move such that they maximize proper time, but the worldline differs from the straight line of special relativity.

We will later use extremization techniques, or variational methods, in order to derive the general equation for geodesics. It is therefore useful to recall how we show that motion in classical mechanics extremizes the action. We choose a family of curves $\vec{x}_\varepsilon(t)$ with a parameter ε in addition to time t, insert it in the action, and compute the extrema along the single variable ε. The choice is restricted by boundary conditions to make sure that all curves in the family connect the two given points, A and B: We require that $\vec{x}_\varepsilon(t_A) = \vec{x}_A$ and $\vec{x}_\varepsilon(t_B) = \vec{x}_B$ for all ε.

We compute the derivative of $S(\vec{x}_\varepsilon, \mathrm{d}\vec{x}_\varepsilon/\mathrm{d}t)$ by ε, treating $\vec{x}$ and $\dot{\vec{x}} = \mathrm{d}\vec{x}/\mathrm{d}t$ as independent parameters: At each point on the curve, we can independently vary the position and the tangent direction; see Figure 8.5. We obtain

$$\frac{\mathrm{d}S}{\mathrm{d}\varepsilon} = \int_{t_A}^{t_B}\left(\frac{\partial L}{\partial \dot{x}^i}\frac{\partial \dot{x}^i}{\partial \varepsilon} + \frac{\partial L}{\partial x^i}\frac{\partial x^i}{\partial \varepsilon}\right)\mathrm{d}t = \left.\left(\frac{\partial L}{\partial \dot{x}^i}\frac{\partial x^i}{\partial \varepsilon}\right)\right|_{t_A}^{t_B} + \int_{t_A}^{t_B}\left(-\frac{\mathrm{d}}{\mathrm{d}t}\frac{\partial L}{\partial \dot{x}^i} + \frac{\partial L}{\partial x^i}\right)\frac{\partial x^i}{\partial \varepsilon}\mathrm{d}t\,. \quad (8.18)$$

We applied the summation convention, summing over i in each product. Integration by parts is possible after writing $\partial \dot{x}^i/\partial\varepsilon = \mathrm{d}(\partial x^i/\partial\varepsilon)/\mathrm{d}t$ because derivatives by t, along the curves, and ε, transversal to the curves, commute.

Since the end points are fixed by our boundary conditions, we have $\partial x^i/\partial\varepsilon = 0$ at t_A and t_B. The boundary term is therefore zero. We have an extremum, $\mathrm{d}S/\mathrm{d}\varepsilon = 0$, if and only if the integrand vanishes because $\partial x^i/\partial\varepsilon$ is an arbitrary function if we consider all possible families of curves connecting points A and B. We obtain the **Euler–Lagrange equations**

$$-\frac{\mathrm{d}}{\mathrm{d}t}\frac{\partial L}{\partial \dot{x}^i} + \frac{\partial L}{\partial x^i} = 0\,, \quad (8.19)$$

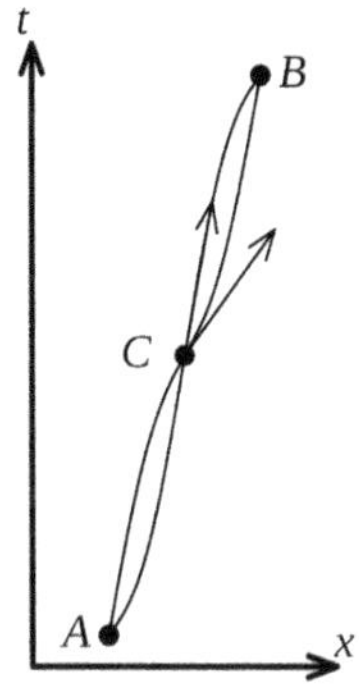

Figure 8.5 Variation of worldlines connecting two events, A and B. Since we are varying entire curves, at each t the position $x(t)$ and the slope $\dot{x}(t)$ are independent: At the event C, the two curves have the same position but different tangent directions.

valid for each value of the index i. Classical mechanics tells us that (8.19) is identical with the equations of motion in a system with Lagrange function L. In particular, for $L(\vec{x}, \mathrm{d}\vec{x}/\mathrm{d}t) = \frac{1}{2}|\mathrm{d}\vec{x}/\mathrm{d}t|^2 - \Phi(\vec{x})$ as in (8.15), the Euler-Lagrange equations imply $\mathrm{d}^2\vec{x}/\mathrm{d}t^2 = -\nabla\Phi$, or Newtonian motion in a gravitational potential Φ *if* the inertial mass equals the gravitational mass. By extremizing proper time between two events in space-time with the weak-field line element, we obtain the same motion as in Newtonian mechanics but with three important advantages:

1. No mass parameter appears in the extremization condition. We do not have to postulate two different parameters m and m' that just happen to be equal, as in Newtonian mechanics. The equivalence of inertial and gravitational mass is therefore explained by geodesic motion in curved space-time.
2. Motion in a gravitational field is interpreted geometrically as motion along a timelike geodesic in curved space-time.
3. Curved space-time and geodesic motion, unlike Newtonian gravity, can be generalized to strong gravitational fields. The required equations are provided by general relativity.

8.4 METRIC TENSOR

The geometry of space-time is described by a line element

$$\mathrm{d}s^2 = g_{\alpha\beta}\mathrm{d}x^\alpha \mathrm{d}x^\beta \tag{8.20}$$

with a symmetric metric tensor $g_{\alpha\beta} = g_{\beta\alpha}$. For flat space-time, $g_{\alpha\beta} = \eta_{\alpha\beta}$ is the Minkowski metric, but curvature or gravity leads to a different metric with position-dependent coefficients.

8.4.1 COORDINATE TRANSFORMATIONS

The equivalence principle tells us that a constant gravitational force has the same implications as uniform acceleration of an observer. When we apply this principle, we no longer require that the coordinate systems we use correspond to inertial frames. Coordinates can then be changed freely by non-linear functions $x^{\alpha'}(x^\beta)$ expressing the new coordinates $x^{\alpha'}$ as functions of the old ones, x^β. Here, index notation gives us a compact way of writing four functions $t'(t,x,y,z)$, $x'(t,x,y,z)$, $y'(t,x,y,z)$ and $z'(t,x,y,z)$. The only condition is that these functions be invertible, such that it is possible to transform back to the original coordinates $x^\alpha(x^{\beta'})$.

While the line element is always invariant under coordinate changes because it determines measurable distances according to the geometry, the metric coefficients depend on which coordinates we choose. We derive the transformation of $g_{\alpha\beta}$ by using the chain rule in (8.20), writing

$$\mathrm{d}x^\alpha = \frac{\partial x^\alpha}{\partial x^{0'}}\mathrm{d}x^{0'} + \frac{\partial x^\alpha}{\partial x^{1'}}\mathrm{d}x^{1'} + \frac{\partial x^\alpha}{\partial x^{2'}}\mathrm{d}x^{2'} + \frac{\partial x^\alpha}{\partial x^{3'}}\mathrm{d}x^{3'} = \frac{\partial x^\alpha}{\partial x^{\alpha'}}\mathrm{d}x^{\alpha'} . \tag{8.21}$$

In the last step, we extend our index notation by declaring that a superscript in the denominator, as in $\partial/\partial x^{\alpha'}$, counts as a subscript, and vice versa. The last expression in (8.21) then has a pair of indices α', once as a subscript and once as a superscript, and it is summed over. In general, there is no division defined for vectors. The only expressions where we have vector components in the denominator is therefore in derivatives.

If we insert (8.21), as well as a β-version of the same equation, in (8.20), we obtain

$$\mathrm{d}s^2 = g_{\alpha\beta}\mathrm{d}x^\alpha \mathrm{d}x^\beta = g_{\alpha\beta}\frac{\partial x^\alpha}{\partial x^{\alpha'}}\frac{\partial x^\beta}{\partial x^{\beta'}}\mathrm{d}x^{\alpha'}\mathrm{d}x^{\beta'} = g_{\alpha'\beta'}\mathrm{d}x^{\alpha'}\mathrm{d}x^{\beta'} . \tag{8.22}$$

By renaming αs into βs, we avoid using the same index more than twice in a product, an important consistency condition in index notation. In the last step, we have used the invariance of $\mathrm{d}s^2$, which must take the same value in any coordinate system. Comparing coefficients in (8.22), we obtain the transformation law

$$g_{\alpha'\beta'} = \frac{\partial x^\alpha}{\partial x^{\alpha'}}\frac{\partial x^\beta}{\partial x^{\beta'}} g_{\alpha\beta} \tag{8.23}$$

of the metric tensor. This transformation behavior is used to show that $g_{\alpha\beta}$ is a tensor, and not a mere matrix, just as we required a 4-vector in special relativity to not only have four components but also the correct transformation by a Lorentz boost. Any object with indices that transforms in the form of (8.23), that is with a factor of $\partial x^\alpha/\partial x^{\alpha'}$ for each subscript index and correspondingly a factor of $\partial x^{\alpha'}/\partial x^\alpha$ for each superscript index, is called a **tensor**.

As an example, in special relativity we consider linear coordinate transformations with linear functions $x^\alpha(x^{\alpha'})$ such that the partial derivatives $\partial x^\alpha/\partial x^{\alpha'} = \Lambda^\alpha{}_{\alpha'}$ are constant, and given by the boost matrix. The transformed metric can then be written in matrix notation as

$$g_{\alpha'\beta'} = \frac{\partial x^\alpha}{\partial x^{\alpha'}}\frac{\partial x^\beta}{\partial x^{\beta'}} g_{\alpha\beta} = \Lambda^\alpha{}_{\alpha'} g_{\alpha\beta}\Lambda^\beta{}_{\beta'} = (\Lambda^T \mathbf{g}\Lambda)_{\alpha'\beta'} . \tag{8.24}$$

We need a transpose of the first Λ to make sure that the summed-over α are next to each other. Whenever this is the case in a product of matrix components, we have the correct form of the matrix product pairing up rows of the first matrix with columns of the second matrix. One of the convenient features of abstract index notation is that we do not need to worry about such positioning because we simply write products of tensor components, with summation implied whenever a subscript is paired up with a superscript.

8.4.2 SINGULARITIES

The metric, like any symmetric matrix, can be diagonalized at a fixed space-time event and has a basis of eigenvectors. If one of the eigenvalues is zero, its eigenvector gives us a direction in which the geometrical distance does not change. We would no longer have three independent spatial directions and one time direction. We require that the metric is an invertible matrix in order to make sure that such a collapse of the usual four dimensions of space-time does not happen.

However, it is sometimes useful to choose coordinates in which the metric is not invertible somewhere, usually at isolated points. We then say that the metric has **singularities**. For instance, the line element on the 2-dimensional unit sphere in spherical coordinates is given by

$$\mathrm{d}s^2 = \mathrm{d}\vartheta^2 + \sin(\vartheta)^2\mathrm{d}\varphi^2 \tag{8.25}$$

with metric

$$\mathbf{g} = \begin{pmatrix} 1 & 0 \\ 0 & \sin(\vartheta)^2 \end{pmatrix}. \tag{8.26}$$

At the two poles, $\vartheta = 0$ or $\vartheta = \pi$, this matrix is not invertible. For a perfect sphere, the geometry at the poles is no different from the geometry at other points. Therefore, these singularities do not have any implications. They are merely a result of choosing coordinates such that curves of constant φ, the meridians, intersect at the poles. Changing φ at the poles does not change the point on the sphere, which is why $\mathrm{d}s^2$ has to be zero if $\vartheta = 0$ or $\vartheta = \pi$ is constant, even if $\mathrm{d}\varphi \neq 0$. Such a singularity is called a **coordinate singularity** because it is the result of choosing a specific set of coordinates. A coordinate singularity can be removed by transforming to a different coordinate system. An exercise on page 328 demonstrates how this can be done around a pole of the sphere.

Another example is given by the weak-field line element

$$\mathrm{d}s^2 = -\left(1 + \frac{2\Phi}{c^2}\right)(c\mathrm{d}t)^2 + \left(1 - \frac{2\Phi}{c^2}\right)(\mathrm{d}x^2 + \mathrm{d}y^2 + \mathrm{d}z^2). \tag{8.27}$$

Within the approximation we used so far, assuming $\Phi/c^2 \ll 1$, this metric is not singular because the eigenvalues are close to ± 1. However, a similar form of the metric can be used also for strong fields, including events where $2\Phi/c^2 = \pm 1$. At such points, the metric is not invertible because at least one of its eigenvalues is zero. As we will see later, one can transform to different coordinates in which the metric remains invertible where $2\Phi/c^2 = \pm 1$. These singularities are therefore coordinate singularities. The situation is different at $r = 0$ if we use Newton's potential $\Phi = -GM/r$, where $r = \sqrt{x^2 + y^2 + z^2}$ is the distance from the center of the spatial coordinate system. The metric then has infinite components at $r = 0$, which, as it turns out, cannot be transformed into a finite invertible metric by any coordinate change. Such a singularity is called a **physical singularity**. It is physical because it has implications for measurable quantities according to the geometry determined by $\mathrm{d}s^2$.

As in all our examples so far, we often use diagonal metrics in which there are no cross-terms in the line element

$$\mathrm{d}s^2 = g_{00}\mathrm{d}x^0\mathrm{d}x^0 + g_{11}\mathrm{d}x^1\mathrm{d}x^1 + g_{22}\mathrm{d}x^2\mathrm{d}x^2 + g_{33}\mathrm{d}x^3\mathrm{d}x^3 \tag{8.28}$$

where $g_{00} < 0$ if x^0 is the time coordinate. The diagonal form greatly simplifies calculations, for instance when we need the inverse matrix. However, coordinate transformations in general do not preserve the diagonal form. We will encounter an example when we try to eliminate the coordinate singularity of the extended weak-field metric at $2\Phi/c^2 = \pm 1$.

8.4.3 EQUIVALENCE PRINCIPLE

A geometrical version of the equivalence principle can be formulated in terms of the metric tensor. It is always possible to transform the metric at a given event P into the Minkowski metric:

$$g_{\alpha'\beta'}|_P = \eta_{\alpha'\beta'}\,. \tag{8.29}$$

At a fixed event, we can ignore the position-dependence of the metric components and diagonalize the symmetric matrix $g_{\alpha\beta}$. If the new matrix has eigenvalues λ_α different from ± 1, we absorb $\sqrt{|\lambda_\alpha|}$ in the coordinates: Defining new coordinates $x^{0'} = \sqrt{|\lambda_0|}x^0$ and so on, results in

$$\lambda_0(\mathrm{d}x^0)^2 + \lambda_1(\mathrm{d}x^1)^2 + \lambda_2(\mathrm{d}x^2)^2 + \lambda_3(\mathrm{d}x^3)^2 = -(\mathrm{d}x^{0'})^2 + (\mathrm{d}x^{1'})^2 + (\mathrm{d}x^{2'})^2 + (\mathrm{d}x^{3'})^2 \tag{8.30}$$

if λ_0 is the only negative eigenvalue and all λ_α can be assumed to be constant. This transformation, a basis change determined by diagonalization and absorbing the eigenvalues in coordinates, therefore leads to the Minkowski metric at P.

If we follow this procedure in a region around P, we usually have position-dependent eigenvalues which we cannot absorb completely in the coordinates. For instance, $\sqrt{\lambda_1}\mathrm{d}x^1 = \mathrm{d}(\sqrt{\lambda_1}x^1) - x^1\mathrm{d}\sqrt{\lambda_1} \neq \mathrm{d}(\sqrt{\lambda_1}x^1) = \mathrm{d}x^{1'}$ if λ_1 is not constant. Nevertheless, it is possible to show that the first derivatives of the metric can be transformed to zero at P,

$$\left.\frac{\partial g_{\alpha'\beta'}}{\partial x^{\gamma'}}\right|_P = 0\,, \tag{8.31}$$

but the second derivatives

$$\left.\frac{\partial^2 g_{\alpha'\beta'}}{\partial x^{\gamma'}\partial x^{\delta'}}\right|_P \neq 0 \tag{8.32}$$

cannot be transformed to zero if there is curvature. As we will see, the space-time curvature tensor is indeed related to second-order derivatives of the metric, and it has four indices just like α', β', γ' and δ' in (8.32). However, the expression in (8.32) is not a tensor: If we transform $g_{\alpha'\beta'}$ and $\partial x^{\gamma'}$ to new coordinates, we obtain factors with up to third-order derivatives of the new coordinates as functions of the old ones. Such an expression is not compatible with the tensor transformation law.

8.4.4 CAUSAL PROPERTIES

Given a line element $\mathrm{d}s^2 = g_{\alpha\beta}\mathrm{d}x^\alpha\mathrm{d}x^\beta$, we use the same classes of relationships between two events as defined in special relativity: We call two events **timelike separated** if the line element between them is negative, $\mathrm{d}s^2 < 0$, **lightlike separated** or **null** if $\mathrm{d}s^2 = 0$, and **spacelike separated** if $\mathrm{d}s^2 > 0$. Massive particles move along timelike worldlines, along which they experience proper time

$$\tau_{AB} = \int_A^B \sqrt{-\mathrm{d}s^2} = \int_A^B \sqrt{-g_{\alpha\beta}\mathrm{d}x^\alpha\mathrm{d}x^\beta} = \int_{t_A}^{t_B} \sqrt{-g_{\alpha\beta}\frac{\mathrm{d}x^\alpha}{\mathrm{d}t}\frac{\mathrm{d}x^\beta}{\mathrm{d}t}}\mathrm{d}t\,. \tag{8.33}$$

(We return to a choice of units such that $c = 1$.)

A **light ray** is a curve in space-time whose tangent vector is always lightlike. Therefore, the coordinates along a light ray obey the equation $\mathrm{d}s^2 = 0$. In a 2-dimensional space-time, for instance, with a diagonal metric in the line element $\mathrm{d}s^2 = g_{00}\mathrm{d}t^2 + g_{11}\mathrm{d}x^2$ with $g_{00} < 0$, the equation

$$\mathrm{d}s^2 = g_{00}\mathrm{d}t^2 + g_{11}\mathrm{d}x^2 = 0 \tag{8.34}$$

is equivalent to two differential equations,

$$\frac{\mathrm{d}t}{\mathrm{d}x} = \pm\sqrt{-\frac{g_{11}}{g_{00}}}\,. \tag{8.35}$$

We obtain two first-order differential equations from the possible sign choices when we take a square root to solve (8.34) for $\mathrm{d}t/\mathrm{d}x$. Their solutions correspond to two different kinds of light rays, one moving to the left (smaller x) and one moving to the right (larger x). In general, the right-hand side of (8.35) is not equal to ± 1, such that light rays are no longer given by diagonals in a generic coordinate system. In order to determine light rays or light cones in a curved space-time, we have to solve differential equations such as (8.35).

8.5 COSMOLOGY

As an extended example, we derive several properties of a curved space-time model which can be used to describe the expanding universe. We simplify the matter distribution in the universe by assuming that it is homogeneous, with the same density at all points in space at a given time, and isotropic, with the same pressure in all spatial directions. The geometry is then described by the **Friedmann–Robertson–Walker (FRW) line element**

$$\mathrm{d}s^2 = -N(t)^2\mathrm{d}t^2 + a(t)^2(\mathrm{d}x^2+\mathrm{d}y^2+\mathrm{d}z^2) \tag{8.36}$$

with two positive functions, $N(t) > 0$ and $a(t) > 0$, depending only on time. Here, we have also assumed that space has no intrinsic curvature, such that we can use the flat spatial line element $\mathrm{d}x^2+\mathrm{d}y^2+\mathrm{d}z^2$ rather than a line element as on the sphere or a version given by a hyperboloid.

8.5.1 SPACE-TIME ANALYSIS

If $\dot{a} > 0$, we interpret the line element (8.36) as the geometry of an expanding universe. We introduce a family of observers positioned at fixed values of the spatial coordinates, such that

$$\frac{\mathrm{d}x}{\mathrm{d}\tau} = \frac{\mathrm{d}y}{\mathrm{d}\tau} = \frac{\mathrm{d}z}{\mathrm{d}\tau} = 0 \tag{8.37}$$

along their timelike worldlines. The coordinate difference between two such observers, for instance Δx in the x-direction, is time-independent. However, the distance between them changes because, measured in the FRW geometry, it is equal to $\Delta s(t) = a(t)\Delta x$. Each such observer therefore sees the other observers in motion, away from the observer in an expanding universe. These observers are called **co-moving observers**: Although they do not actively move, they move along with the expansion of the universe.

Co-moving observers have timelike worldlines along which we compute the proper-time interval

$$\mathrm{d}\tau = \sqrt{-\mathrm{d}s^2} = N(t)\mathrm{d}t\,, \tag{8.38}$$

using (8.37). The factor $N(t)$ on the right-hand side of this equation depends only on t, and we integrate to obtain a function

$$\tau(t) = \int^t N(\tilde{t})\mathrm{d}\tilde{t}\,. \tag{8.39}$$

Since we assume $N(t) > 0$, $\tau(t)$ is monotonic and invertible. We introduce $\tau(t)$ as a new coordinate instead of t. Inserting (8.38) in (8.36) then gives us the simplified line element

$$\mathrm{d}s^2 = -\mathrm{d}\tau^2 + a(\tau)^2(\mathrm{d}x^2+\mathrm{d}y^2+\mathrm{d}z^2)\,. \tag{8.40}$$

The function $N(t)$ has been completely eliminated by a coordinate transformation. Therefore, it has no effect on the geometry of space-time or on cosmological measurements.

The function $a(\tau)$ cannot be eliminated in the same way because it depends on time but multiplies spatial differentials. In general, it is then impossible to find new coordinates x', y' and z' such that $a(\tau)^2(\mathrm{d}x^2+\mathrm{d}y^2+\mathrm{d}z^2) = (\mathrm{d}x')^2+(\mathrm{d}y')^2+(\mathrm{d}z')^2$. However, $a(\tau)$ is not completely observable either. We can change the spatial coordinates by multiplying all of them by a non-zero constant $\lambda \neq 0$: $x' = \lambda x$, $y' = \lambda y$ and $z' = \lambda z$. In these new coordinates, the line element, which is invariant and therefore equals (8.40), is

$$\mathrm{d}s^2 = -\mathrm{d}\tau^2 + \left(\frac{a(\tau)}{\lambda}\right)^2 \left((\mathrm{d}x')^2+(\mathrm{d}y')^2+(\mathrm{d}z')^2\right)\,. \tag{8.41}$$

It is impossible to use measurements or observations to distinguish between an FRW line element with a function $a(\tau)$, and one with a function $a(\tau)/\lambda$, where λ does not depend on time. (If λ depends on τ, (8.41) is not equal to (8.40) because there would be terms of the form $(\mathrm{d}\lambda/\mathrm{d}\tau)\mathrm{d}x\mathrm{d}\tau$ if $x' = \lambda(\tau)x$ is inserted in (8.41).) The possibility of rescaling the function $a(\tau)$ by a time-independent constant prevents it from being interpreted as a measure of the size of the universe, such as its radius. Instead, $a(\tau)$ is called the **scale factor** of the universe.

Something about $a(\tau)$ should be observable if this function describes the expansion of the universe. If we take two values $a(\tau_1)$ and $a(\tau_2)$ at two different times, the ratio $a(\tau_1)/a(\tau_2)$ is independent of the factor λ we choose in scaling spatial coordinates. It is easier to deal with differential quantities instead of ratios of values taken at two different times. In cosmology, we use the **Hubble parameter**

$$H = \frac{\dot{a}}{a} \tag{8.42}$$

where $\dot{a} = \mathrm{d}a/\mathrm{d}\tau$. This ratio is independent of the scaling parameter λ and can be measured by cosmological observations.

Edwin Hubble introduced this parameter as the constant of proportionality in his observed linear relationship between the escape velocity of distant stars, measured through redshift, and their distance from us. This relationship is explained by the FRW line element: Thanks to isotropy, we may always point our coordinate axes such that the star is on our x-axis. If the star is moving relative to us, its velocity equals $v_0 = \mathrm{d}\Delta x/\mathrm{d}\tau$ in terms of the coordinate distance Δx. The spatial distance of the star, according to FRW geometry, is equal to $R = \Delta s = a(\tau)\Delta x$. In an expanding space, there are two contributions to the velocity

$$V = \frac{\mathrm{d}R}{\mathrm{d}\tau} = \frac{\mathrm{d}a}{\mathrm{d}\tau}\Delta x + a\frac{\mathrm{d}\Delta x}{\mathrm{d}\tau} = HR + av_0 . \tag{8.43}$$

If the relative velocity v_0 is negligible, the linear **Hubble law** $V = HR$ is obtained. The escape velocity V is measured by the Doppler shift it implies on emitted light.

In addition to the Hubble parameter, one can define observable quantities depending on higher derivatives of $a(\tau)$. The next such expression is the **deceleration parameter**

$$q = -\frac{\ddot{a}a}{\dot{a}^2} . \tag{8.44}$$

The minus sign is included in the definition because one expects an attractive force such as gravity to slow down the expansion of any matter distribution in the universe. This parameter and those based on even higher derivatives are more difficult to measure than H.

Light rays need sometime to travel to us from distant sources. They contain information about the history of the universe or about the scale factor at earlier times. We can derive their curves in expanding space-time by solving the differential equation $\mathrm{d}s^2 = 0$ for the FRW line element, which implies

$$\frac{\mathrm{d}\tau}{\mathrm{d}x} = \pm a(\tau) \tag{8.45}$$

for a light ray moving in the x-direction. We may choose spatial coordinates such that $a(\tau_0) = 1$ at the time τ_0 when we are making a cosmological observation. Around this event, therefore, the light cone looks just like one in Minkowski space-time. However, if $a(\tau)$ is increasing, it was smaller in the past, and the slope (8.45) was smaller. Light rays in FRW space-time, therefore, are not diagonal lines but get steeper toward the future, as shown in Figure 8.6. Depending on how quickly $a(\tau)$ increased in the past, the size of a distant object as it appears to us in the sky may be smaller than it should actually be. If we have a theory about the formation of structures such as galaxies or galaxy clusters, we can compare the predicted average size with our measurements and derive information about $a(\tau)$. An important example for which this approach has been used is the **cosmic microwave background (CMB)** emitted in the hot early universe even before galaxies formed.

8.5.2 COSMOLOGICAL MODELS

Modeling the expansion of the universe requires an equation that tells us how $a(\tau)$ may be restricted. General relativity, as we will see later, gives us a differential equation

$$H^2 = \left(\frac{\dot{a}}{a}\right)^2 = \frac{8\pi G}{3}\rho , \tag{8.46}$$

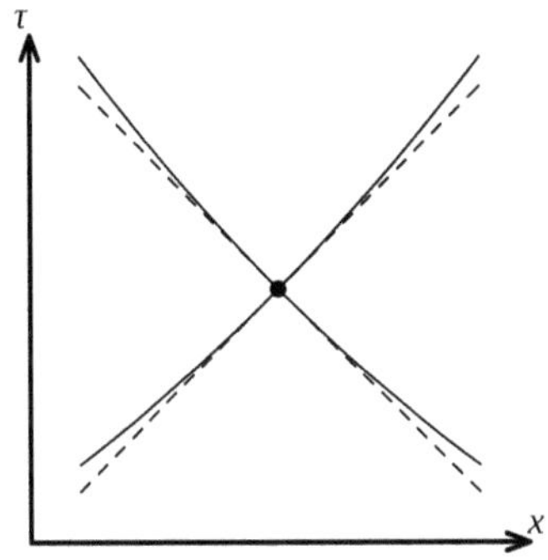

Figure 8.6 The light cone in an expanding FRW space-time, determined by the function $a(\tau)$ according to the given matter content in a cosmological era (here, pressureless dust). Observations in astronomy and cosmology are based on the past light cone, which is only a small fraction of our entire past. Dashed lines indicate the diagonals for comparison.

called the **Friedmann equation**. We continue to use coordinates such that $c = 1$, G is Newton's constant, and ρ the energy density of matter in the universe. In general, ρ is a function of τ because the density depends on the scale factor $a(\tau)$. For ordinary matter, $\rho(\tau)$ is proportional to $1/a(\tau)^3$ because its energy is diluted as the universe expands. For relativistic matter or electromagnetic radiation, which was dominant in the hot early universe, there is an additional redshift factor from the expansion such that $\rho(\tau)$ is proportional to $1/a(\tau)^4$. Once $\rho(\tau)$ has been specified in this way, depending on the matter content, we may be able to solve the first-order differential equation (8.46) for $a(\tau)$. It turns out that, for decreasing $\rho(a)$, there is always a time τ_∞ in the past when $a(\tau_\infty) = 0$. At this value, the FRW metric is not invertible, and we have a singularity. It is a physical singularity, called the **big-bang singularity**, because it corresponds to complete collapse of matter and therefore has physical consequences. In particular, if $a(\tau_\infty) = 0$, the energy density of radiation, which is proportional to $1/a(\tau)^4$, is infinite, and so is the temperature.

Although the large value of the energy density is in principle measurable, we cannot observe it directly. Hot matter is ionized, and free charges strongly scatter electromagnetic waves. The very early universe therefore was not transparent to electromagnetic waves. It does not give access to optical cosmological observations, just as we cannot see directly into the sun. Gravitational waves or neutrinos would not be subject to this limitation, but it is difficult to detect them reliably at such low intensities as expected for a cosmic background. The earliest time we can see using electromagnetic waves, called **recombination**, is when the expansion of the universe had diluted and cooled down matter so much that atoms formed from the previous ions and free electrons. From this time on, electromagnetic radiation interacts only weakly with matter in the universe. We see the remnants of this early radiation in the **cosmic microwave background**. At the time of recombination, the temperature of matter in the universe was close to 5000K, about the temperature on the surface of the sun. The expansion of the universe has by now cooled the average temperature down to about 2.7K, which for black-body radiation corresponds to the microwave range.

Detailed observations of the CMB in all possible directions have shown that it is nearly isotropic, that is the intensity does not depend much on the direction in which we point a microwave antenna. This observation is consistent with the expectation that galaxies or other structures had not yet formed at the time of recombination. Nevertheless, there are small density variations which we can measure using corresponding intensity fluctuations of the CMB over the sky. Measuring these structures gives us information about $a(\tau)$ at these early times.

The near isotropy of the CMB leads to a problem, called the **horizon problem**. If we look at two opposite directions in the sky, we receive microwave rays with almost the same intensity. In

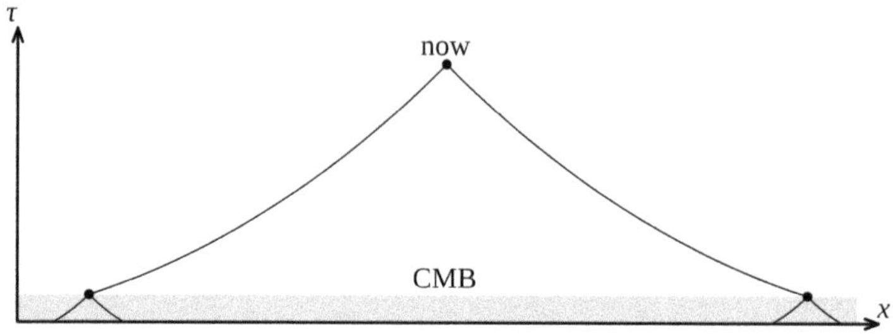

Figure 8.7 Illustration of the horizon problem: Electromagnetic background waves reaching us from opposite directions came from regions that did not have enough time to establish causal contact since the big-bang singularity.

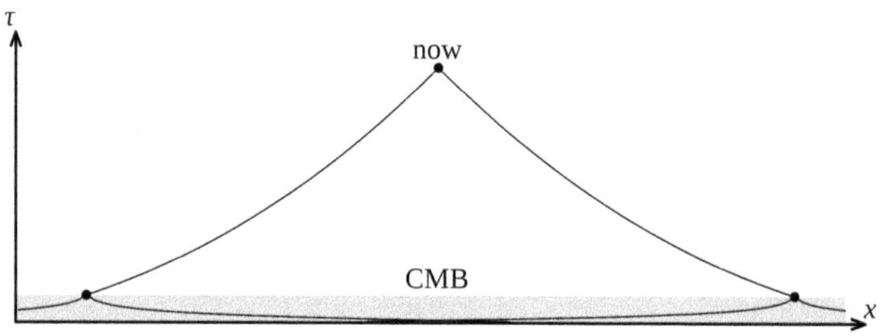

Figure 8.8 The effect of inflation on light cones: Accelerated early expansion implies that the past light cones spread out much more than in the non-accelerating case, and may be able to establish causal contact.

Figure 8.7, these two rays are shown as the two light rays moving in the positive and negative x-directions, arriving simultaneously at our observation time. The duration of the hot opaque phase of the universe, 380,000 years, was much smaller than the age of the universe, about 14 billion years. It is therefore difficult to see how the two events that we observe in opposite directions, which both happened at the time of recombination but far away from each other, could have been in causal contact: Their past light cones do not reach back far enough to have some overlap between recombination and the big-bang singularity. The maximum extension of the past light cones relative to a given observer defines the **cosmological horizon.** If the past light cones have no overlap, it is difficult to explain why the events sent out microwave rays of very similar intensities. The primary physical reason for a uniform intensity is given by matter and radiation in a thermodynamical equilibrium state. However, establishing an equilibrium state requires a period of interactions between the constituents of matter and radiation. Without causal contact, it is impossible to reach an equilibrium.

A possible resolution of the horizon problem is given by **cosmic inflation**, shown in Figure 8.8. If the scale factor increased very rapidly during the hot early phase of the universe, the past light cones spread out much more in this region of the space-time diagram. They could then partially overlap, as it indeed happens for instance when the scale factor increases exponentially. In particular, one can show that the acceleration $\ddot{a}$ must be positive for this behavior to be realized. However, we would expect that the expansion of the scale factor should slow down over time because gravity is always attractive and pulls matter distributions together. Nevertheless, some models of inflation are in good agreement with precise observations of the CMB.

Redshift observations of distant galaxies have shown that the expansion of the universe seems to have accelerated also in the more recent past, long after the universe became transparent. This recent phenomenon of acceleration is called **dark energy**, but is not known what could cause this

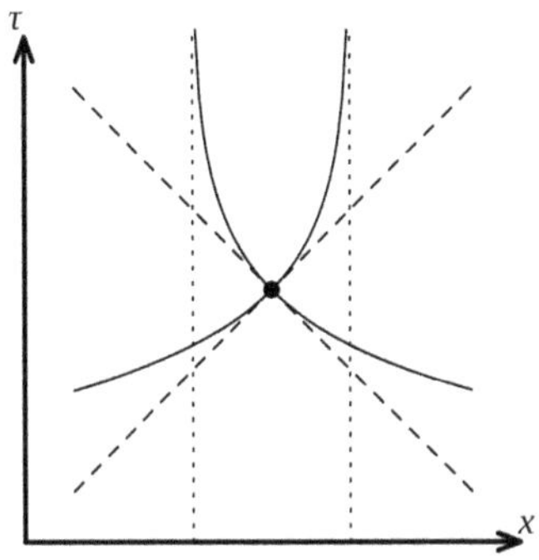

Figure 8.9 Future horizon: Accelerated expansion at late times, implied by dark energy, places an upper bound on the spatial coordinate distance Δx the light rays can travel during the duration of the universe. Regions beyond this distance (dotted lines) cannot be reached. Nevertheless, the light rays travel an infinite geometrical distance because $a(\tau)$ is unbounded in this case.

new form of energy that seems to have implications only on cosmological scales. Moreover, we will see that dark energy requires strong negative pressure throughout space at late times.

While accelerated expansion in the early universe may solve the horizon problem in the form of inflation, accelerated expansion implied by dark energy at late times leads to a different kind of **horizon** toward the future rather than the past: Because the expansion of space is accelerated, light rays cannot keep up with the expansion. They span only a finite spatial coordinate distance Δx even in an infinite amount of time, as can be seen in Figure 8.9. There are regions of the universe behind the horizon that we or our signals will never be able to reach. Nevertheless, the geometrical distance, given by the integral of $a(\tau)\mathrm{d}x$, is unbounded. While light rays travel an infinite geometrical distance during the life time of the universe, accelerated expansion implies that regions beyond a certain Δx are pulled away too rapidly and cannot be reached.

Another possible solution of the horizon problem is given by a universe model that does not have a big-bang singularity. Instead, the hot opaque phase might surround a region of very large but still finite energy density, caused perhaps by a previous collapse phase of the universe. If it is possible to avoid the singularity, light cones of events at the time of recombination could be extended further into the past than shown in Figure 8.7, until they have some overlap. However, turning collapse into expansion would require a repulsive force, and no such force is known that could overcome the attractive nature of gravity. The only possibility is that physics at extremely high densities is very different from the physics we know. Maybe even gravity could be repulsive at such densities. However, such a theory would have to be very different from general relativity because its equations make it possible to prove that a singularity appears in any relevant solution.

8.5.3 ISOTROPIC SINGULARITY THEOREM

In general relativity, any space-time line element that can describe the expansion of the universe or the gravitational collapse of matter must have a physical singularity. Such theorems were first proved by Stephen Hawking and Roger Penrose beginning in the 1960s. Before they became available, most researchers in gravitational physics believed that singularities such as the big bang in FRW solutions were merely a consequence of the special nature of these solutions with strong symmetries. If an initial matter distribution is homogeneous or spherically symmetric and starts collapsing, one expects that all its parts reach the same point at sometime, and the density becomes infinite. One also expects that deviations from an exactly symmetric distribution would lead to different particles missing one another in the collapse process, forming a large but still finite density at the collapse event.

Afterwards, the particles would just keep moving and disperse. However, this second expectation turns out to be incorrect, as shown by the singularity theorems of general relativity.

The full theorems are complicated because they consider space-time solutions in a general form, without any symmetries. Fortunately, however, the main ideas can be seen in an isotropic version that analyzes only isotropic (FRW) solutions but does not require us to specify the matter content. Importantly, we do not need to make assumptions about the dilution behavior of $\rho(\tau)$. The isotropic singularity theorem therefore shows that unknown physics at high densities, such as details of a quark-gluon plasma or other effects that might change the dilution behavior $\rho(\tau)\propto 1/a(\tau)^3$ of dust or $\rho(\tau)\propto 1/a(\tau)^4$ of radiation at very large densities, cannot prevent the big-bang singularity.

We combine the Friedmann equation

$$H^2=\left(\frac{\dot{a}}{a}\right)^2=\frac{8\pi G}{3}\rho \tag{8.47}$$

with the **continuity equation** for the dilution behavior of matter. This new equation can be derived from the first law of thermodynamics, which we expect to hold under all circumstances including very large densities and temperatures. We begin with the equation $\mathrm{d}E=-P\mathrm{d}V$: The change in energy of a matter system is given by the negative pressure P times its change in volume. We usually interpret the change as something caused by a thermodynamical process such as moving a piston, but we can also view it as a change caused by the expansion of the universe. We indicate this interpretation in our notation by dividing the first law by $\mathrm{d}\tau$:

$$\frac{\mathrm{d}E}{\mathrm{d}\tau}=-P\frac{\mathrm{d}V}{\mathrm{d}\tau}\,. \tag{8.48}$$

We do not know the volume of the universe at any given time. (Recall that $a(\tau)$ is the scale factor, not the radius or some other measure for the size of the universe.) However, since we assume that the matter distribution is homogeneous, we may choose any finite region and see how its volume and energy change in time. The volume of such a region is equal to $V(\tau)=V_0a(\tau)^3$, where V_0 is some unknown parameter that we could calculate by integrating $\mathrm{d}x\mathrm{d}y\mathrm{d}z$. However, we will not need this value because the energy in the same region, $E(\tau)=\rho(\tau)V(\tau)=\rho(\tau)V_0a(\tau)^3$, is also proportional to V_0. This parameter cancels out if we insert $E(\tau)$ and $V(\tau)$ in (8.48). We obtain

$$\frac{\mathrm{d}(\rho a^3)}{\mathrm{d}\tau}=-P\frac{\mathrm{d}a^3}{\mathrm{d}\tau} \tag{8.49}$$

or, after some simplifications,

$$\dot{\rho}+3H(\rho+P)=0\,. \tag{8.50}$$

This equation is the continuity equation of isotropic cosmology.

Taking a time derivative of (8.47), we can insert $\dot{\rho}$ as obtained from the continuity equation:

$$\begin{aligned} 2H\dot{H} &= \frac{\mathrm{d}H^2}{\mathrm{d}\tau}=\frac{8\pi G}{3}\dot{\rho}=-8\pi GH(\rho+P)\\ &= -\frac{8\pi G}{3}H(\rho+3P+2\rho)=-\frac{8\pi G}{3}H(\rho+3P)-2H^3\,. \end{aligned} \tag{8.51}$$

In the second line, we re-arranged the matter terms such that only the combination $\rho+3P$ appears because, as it turns out, all known matter satisfies

$$\rho+3P\geq 0\,. \tag{8.52}$$

This inequality is called the **strong energy condition**. For dust or any non-relativistic matter, we have $P\ll\rho$ because pressure is related to the momentum transfer of moving matter constituents, which for non-relativistic motion is much less than the rest energy contained in ρ. For electromagnetic radiation, $P = \frac{1}{3}\rho$ follows from inserting $\rho(a) \propto a^{-4}$ in the continuity equation. Both conditions satisfy $\rho + 3P \geq 0$.

An important exception to the strong energy condition is a constant positive energy density, given by a **cosmological constant** $\rho = \Lambda$. For $\dot{\rho} = 0$, the continuity equation implies $P = -\rho = -\Lambda$, such that $\rho + 3P = -2\Lambda < 0$. A constant energy density in the Friedmann equation implies exponential expansion, which is accelerated and could explain inflation or dark energy. However, no matter system is known that could produce the corresponding negative pressure throughout the whole universe.

Dividing (8.51) by $2H$ and using (8.52), we obtain

$$\dot{H} = -\frac{4\pi G}{3}(\rho + 3P) - H^2 \leq -H^2 . \tag{8.53}$$

This new inequality, $\dot{H} \leq -H^2$, is independent of matter terms and holds for any matter system that obeys the strong energy condition. It is the main ingredient of the isotropic singularity theorem. We will first use it to discuss why having accelerated expansion of the universe, as suggested by inflation or dark energy, is difficult to achieve.

1. We calculate

$$\dot{H} = \frac{\mathrm{d}}{\mathrm{d}\tau}\left(\frac{\dot{a}}{a}\right) = \frac{\ddot{a}}{a} - \left(\frac{\dot{a}}{a}\right)^2 . \tag{8.54}$$

   In terms of the scale factor, (8.53) is equivalent to the condition $\ddot{a} \leq 0$. We can have accelerated expansion only if we have enough matter that violates the strong energy condition. For instance, a common model of dark energy is the cosmological constant $\rho = \Lambda$, even though its physical nature remains unclear.
2. After dividing by $-H^2$, the inequality (8.53) has the equivalent form

$$-\frac{\dot{H}}{H^2} = \frac{\mathrm{d}}{\mathrm{d}\tau}\frac{1}{H} \geq 1 . \tag{8.55}$$

   Integrating both sides of this inequality from some past time $\tau < \tau_0$ to the present time τ_0, we obtain

$$\frac{1}{H(\tau_0)} - \frac{1}{H(\tau)} \geq \tau_0 - \tau \tag{8.56}$$

   where $H(\tau_0) > 0$ is the current value of the Hubble parameter. Consider another time in the past, defined by $\tau_1 = \tau_0 - 1/H(\tau_0) < \tau_0$. Using this value for τ in (8.56) implies

$$-\frac{1}{H(\tau_1)} \geq \tau_0 - \tau_1 - \frac{1}{H(\tau_0)} = 0 . \tag{8.57}$$

   Therefore, there was a time τ_1 in the past when $1/H(\tau_1) < 0$. If $H(\tau)$ is continuous and non-zero, $1/H(\tau)$ is continuous. The opposite signs of this function at τ_0 and τ_1 then imply a time τ_∞ between τ_1 and τ_0, where $1/H(\tau_\infty) = 0$ is zero; see Figure 8.10. But then, $H(\tau_\infty) = \infty$ and the Friedmann equation tells us that $\rho(\tau_\infty) = \infty$. We have a physical singularity.

General singularity theorems for cosmological solutions of general relativity are obtained without any assumptions about the form of the line element, except that there is one time, now, when it is consistent with an expanding universe. General relativity is therefore incomplete because it predicts that its solutions are valid only for a finite range in time. Nevertheless, the fact that general relativity is able to demonstrate its own incompleteness is impressive. Trying different matter ingredients does not help because the singularity theorems do not assume much about matter, except

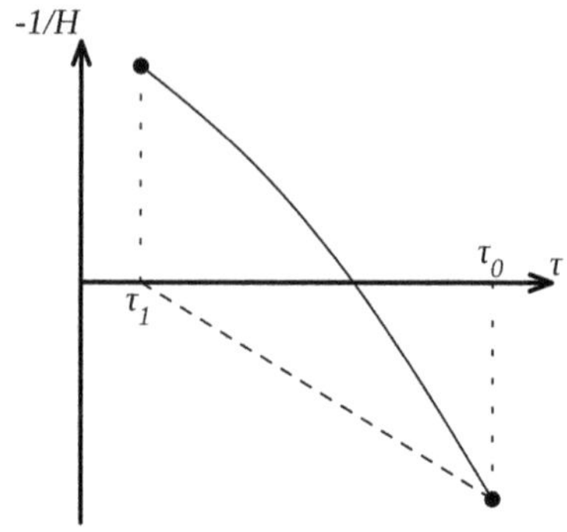

Figure 8.10 A sketch of the function $-1/H(\tau)$, bounded from below according to (8.56). The dashed line indicates the lower bound in this equation, evaluated at τ instead of τ_1.

for the strong energy condition. One possible way out may be to bring in quantum effects. For instance, the energy-time version of the uncertainty relation of quantum mechanics, $\sigma_E \Delta t \geq \hbar/2$ for the standard deviation σ_E of energy and Δt as discussed in Section 5.5.2, implies that the energy of matter might not always take a sharp value. In brief time intervals, there could be large energy fluctuations that allow ρ to become negative, violating the strong energy condition. However, large energy fluctuations are possible only during very brief time intervals, which do not seem relevant for cosmological evolution. Other quantum effects, not of matter but of space-time, might change the equations of general relativity at high temperatures close to the big bang, including the Friedmann equation. But no fully consistent quantum theory of gravity is known, as discussed in more detail in Chapter 13. For now, the singularity problem remains open.

8.5.4 QUANTUM MATTER IN THE EARLY UNIVERSE

The assumption that matter or energy in the early universe caused cosmic inflation can solve the horizon problem. More importantly, an analysis of how small density perturbations evolve during and after inflation produces results that are very close to what has been measured in tiny intensity variations of the cosmic microwave background. Even though we do not know what might cause inflation, the proposal is observationally testable.

8.5.4.1 Classical dynamics of field perturbations

For such an analysis, we have to extend the strictly homogeneous and isotropic FRW models by including a small degree of inhomogeneity. We do so by introducing a metric perturbation described by a function $u(\tau,x,y,z)$ on space-time. This field can be added to a metric component that is originally in FRW form, but in terms of evolution equations, it is more convenient to view it as a new field degree of freedom that evolves on an exact FRW space-time.

As an inhomogeneous evolving field, the function u should obey a covariant wave equation that can be derived from a field-theoretic Lagrange function. As a first approximation, we may assume a free field that does not self-interact or exchange energy with other fields. However, because its energy gets diluted as the universe expands, it must interact in some form with the underlying FRW space-time. These conditions are implemented by proposing a Lagrange function quadratic in u and its time and space derivatives, integrated over space with the volume element $a(\tau)^3 \mathrm{d}x\mathrm{d}y\mathrm{d}z$ implied by the FRW metric:

$$L = \frac{1}{2}\int\int\int\left(\left(\frac{\partial u}{\partial \tau}\right)^2 - \frac{1}{a(\tau)^2}\left(\left(\frac{\partial u}{\partial x}\right)^2 + \left(\frac{\partial u}{\partial y}\right)^2 + \left(\frac{\partial u}{\partial z}\right)^2\right)\right) a(\tau)^3 \mathrm{d}x\mathrm{d}y\mathrm{d}z. \tag{8.58}$$

The first term resembles the usual kinetic energy, while the remaining terms are determined by covariance and may be interpreted as a potential. We have used the FRW metric in order to take into account the correct geometry in spatial derivatives: A physically relevant change of the field in a given spatial coordinate direction x^i should be determined by a ratio of the form $\Delta u/\Delta s = \Delta u/(a\Delta x^i)$. The signs implied by the full FRW metric are consistent with a positive contribution from the kinetic energy and a negative contribution from the potential energy in a Lagrange function. The action implied by this Lagrange function is therefore covariant.

In this field theory with not only temporal but also spatial dependence of u, we have to extend **Euler–Lagrange equations** by treating all four derivatives like the time derivative in point-particle mechanics:

$$-\frac{\partial}{\partial x^\alpha}\frac{\partial L}{\partial(\partial u/\partial x^\alpha)}+\frac{\partial L}{\partial u}=0. \tag{8.59}$$

The α-index is summed over all four values because the partial derivative of L by $\partial u/\partial x^\alpha$ contains a superscript α in the denominator of its denominator. This index position counts as a superscript and combines correctly with the subscript α implied by $\partial/\partial x^\alpha$ according to the index notation.

Since $x^0 = \tau$, the dependence of $a(\tau)^3$ on τ produces additional terms compared with the usual wave equation in flat space-time. Computing all required derivatives implies

$$-\frac{\partial^2 u}{\partial \tau^2}+\frac{1}{a^2}\left(\frac{\partial^2 u}{\partial x^2}+\frac{\partial^2 u}{\partial y^2}+\frac{\partial^2 u}{\partial z^2}\right)-3H\frac{\partial u}{\partial \tau}=0 \tag{8.60}$$

with a friction term implied by the Hubble parameter, resembling a negative velocity-dependent force if $H > 0$.

Wave equations of this form can be solved by Fourier decomposition, writing $u(\tau,x,y,z)$ in a function basis of spatial plane waves:

$$u(\tau,x,y,z)=\frac{1}{(2\pi)^{3/2}}\int\int\int u_{\vec{k}}(\tau)\exp(i\vec{k}\cdot\vec{x})\mathrm{d}^3k. \tag{8.61}$$

We use a spatial basis instead of a space-time basis of functions because multiplication by τ-dependent factors from $a(\tau)$ would be transformed to complicated convolutions in the latter case. This step turns the wave equation into an ordinary differential equation for each $u_{\vec{k}}(\tau)$:

$$\frac{\mathrm{d}^2 u_{\vec{k}}}{\mathrm{d}\tau^2}+\frac{|\vec{k}|^2}{a^2}u_{\vec{k}}+3H\frac{\mathrm{d}u_{\vec{k}}}{\mathrm{d}\tau}=0. \tag{8.62}$$

We can further simplify the equation by transforming proper time τ to **conformal time** η defined such that $\mathrm{d}\tau = a\mathrm{d}\eta$. The resulting FRW line element is called **conformally flat** because it is proportional to the flat space-time line element. We also rescale the field modes to $v_{\vec{k}}(\eta) = a(\tau(\eta))u_{\vec{k}}(\tau(\eta))$, which then obey the **Mukhanov–Sasaki equation**

$$\frac{\mathrm{d}^2 v_{\vec{k}}}{\mathrm{d}\eta^2}+\left(|\vec{k}|^2-\frac{1}{a}\frac{\mathrm{d}^2 a}{\mathrm{d}\eta^2}\right)v_{\vec{k}}=0. \tag{8.63}$$

A positive cosmological constant Λ in the inflationary early universe implies a constant Hubble parameter $H = \sqrt{\Lambda}$. Solving the Friedmann equation in proper time, we therefore obtain $a(\tau) = a_0\exp(\sqrt{\Lambda}\tau)$ with an integration constant a_0 that has an arbitrary positive value because it can be rescaled by changing spatial coordinates. The transformation to conformal time is then determined by the differential equation $\mathrm{d}\eta/\mathrm{d}\tau = 1/a(\tau)$ such that $\eta(\tau) = -\Lambda^{-1/2}a(\tau)^{-1}$ or $\sqrt{\Lambda}a(\eta) = -1/\eta$. Note that η runs from $-\infty$ at the big bang to a finite negative value at a later stage. For the latest version of the wave equation, we need $a^{-1}\mathrm{d}^2a/\mathrm{d}\eta^2 = 2/\eta^2$. Inserting all this in (8.63), we obtain the differential equation

$$\frac{\mathrm{d}^2 v_{\vec{k}}}{\mathrm{d}\eta^2}+\left(|\vec{k}|^2-\frac{2}{\eta^2}\right)v_{\vec{k}}=0. \tag{8.64}$$

If the last term is small, as in the very early universe, we obtain standard plane-wave solutions. As the modes evolve toward larger negative η, the last term should be taken into account. Given the coefficients in the differential equation, a suitable functional form can be formulated as plane waves times a polynomial in $1/(|\vec{k}|\eta)$: A derivative of such a polynomial, when inserted in the differential equation, produces additional terms that may be able to cancel the $2/\eta^2$-term. It turns out that low-order polynomials are sufficient, as they appear in the complete solutions

$$v_{\vec{k}}(\eta) = A_{\vec{k}} \exp(-i|\vec{k}|\eta) \left(1 - \frac{i}{|\vec{k}|\eta}\right) + B_{\vec{k}} \exp(i|\vec{k}|\eta) \left(1 + \frac{i}{|\vec{k}|\eta}\right). \tag{8.65}$$

The $\vec{k}$-dependent coefficient functions $A_{\vec{k}}$ and $B_{\vec{k}}$ determine all contributions of plane waves at some initial conformal time.

8.5.4.2 Quantum physics of field perturbations

Quantum mechanics can be used to derive suitable values of the functions $A_{\vec{k}}$ and $B_{\vec{k}}$ by postulating that the initial state of our field in an empty universe was the ground state. Accelerated expansion of the universe can then function as a time-dependent force that excites the ground state to higher-energy states. After further cosmological evolution, we observe implications of the resulting spatial variations in the cosmic microwave background, or in the galaxy distribution at later times.

This is at least the general idea. In practice, it turns out that an exactly homogeneous initial state can develop inhomogeneous variations only if the corresponding Hamilton operator does not respect spatial translation symmetry. However, these symmetries are an important property of covariant field equations. This issue is similar to the measurement problem in quantum mechanics, but applied to the whole universe. In both cases, decoherence models can be used to provide better foundations to cosmological structure formation. Here, we will not present a discussion of such complicated models and instead use a common assumption made at a crucial point in inflationary cosmology that will produce the desired results.

The ground state in quantum mechanics is determined by the Hamilton operator. The classical analog can be derived by a Legendre transformation of the field-theory Lagrange function we used to derive the wave equation, also here with derivatives or momenta for all independent space-time coordinates. It is easier to derive the Hamilton function for a field theory in which we already used spatial Fourier transformation, such that we have independent modes for each spatial vector $\vec{k}$ and corresponding momenta $p_{\vec{k}} = \mathrm{d}v_{\vec{k}}/\mathrm{d}\eta$ of the modes. This expression takes the form

$$H = \frac{1}{2} \int\int\int \left(p_{\vec{k}}^2 + \left(|\vec{k}|^2 - \frac{1}{a}\frac{\mathrm{d}^2 a}{\mathrm{d}\eta^2}\right) v_{\vec{k}}^2\right) \mathrm{d}^3 k \tag{8.66}$$

Close to the big bang, $-\eta$ is very large because η approaches the value $-\infty$ at the big bang. We can therefore ignore the contribution $a^{-1}\mathrm{d}^2 a/\mathrm{d}\eta^2 = 2/\eta^2$. The remaining terms represent a harmonic oscillator for each mode with wave number $\vec{k}$. Comparing with the mechanics version of the harmonic oscillator, we identify the correct coefficients if we set $m = 1$ and use the frequency $\omega = |\vec{k}|$. The constant one does not have the correct mass units, but here we are looking for a formal relationship in order to find mathematical solutions.

For each $\vec{k}$, we obtain an independent oscillator because there are no coupling terms between the modes, which would require products of different mode functions in the Hamilton function. Quantum mechanics tells us that the expectation values of $\hat{v}_{\vec{k}}$ and $\hat{p}_{\vec{k}}$ are zero in the ground state. However, non-zero values are obtained for the variances $\sigma_{v_{\vec{k}}}^2 = \hbar/(2|\vec{k}|)$ for all field modes, inserting the formal values of $m = 1$ and $\omega = |\vec{k}|$ in the harmonic-oscillator result. The classically required time dependence in (8.65) is compatible with quantum-mechanics evolution $\exp(-i\eta E/\hbar) = \exp(-i|\vec{k}|\eta)$ if we use $E = \hbar\omega = \hbar|\vec{k}|$. Only the $A_{\vec{k}}$-term should therefore contribute, setting all $B_{\vec{k}} = 0$. The non-zero coefficients are then given by $A_{\vec{k}} = \sqrt{\hbar/(2|\vec{k}|)}$.

8.5.4.3 Power spectra

We now have a specific solution for $v_{\vec{k}}(\eta)$ or

$$u_{\vec{k}} = \frac{v_{\vec{k}}}{a} = \sqrt{\frac{\hbar}{2|\vec{k}|}} \frac{\exp(-i|\vec{k}|\eta)}{a} \left(1 - \frac{i}{|\vec{k}|\eta}\right) \tag{8.67}$$

for field modes evolving from their initial ground state. In order to determine implications for cosmological observations, we have to know the physical interpretation of the field u. The primary field for distant observations is the density distribution ρ of matter and energy because it directly affects the intensity of cosmic background emission in the hot early universe, or the visibility of objects in the galaxy distribution. The original field $u_{\vec{k}}$ plays a mathematical role as a perturbation of metric components, introduced in order to analyze deviations from exact spatial homogeneity in the FRW line element. According to (8.14), metric perturbations at weak gravitational fields are related to Newton's potential Φ. Newton's potential, in turn, is related to the matter density by Poisson's equation

$$\Delta\Phi = 4\pi G\rho \tag{8.68}$$

with the flat-space Laplacian Δ. In Newtonian gravity, ρ includes only the mass density, but the weak-field limit of general relativity shows that the full energy density ρ determines metric components via (8.68).

These observations allow us to identify $u_{\vec{k}}$ with the Fourier transformation of Newton's potential Φ, and subject it to the equation

$$-|\vec{k}|^2 u_{\vec{k}} = 4\pi G\rho_{\vec{k}}. \tag{8.69}$$

Thus,

$$\rho_{\vec{k}} = -\frac{|\vec{k}|^2}{4\pi G} u_{\vec{k}}. \tag{8.70}$$

Observational results in this case consider a large number of values obtained by aiming a telescope in all accessible directions within a given region, such that deviations from uniformity of the received radiation can be determined. Large data sets are evaluated by statistical methods, and a useful tool here is the classical correlation function

$$C_{\vec{k},\vec{k}'} = \langle \rho_{\vec{k}} \rho^*_{\vec{k}'} \rangle \tag{8.71}$$

computed from the ensemble of observational data. The reference system used to assign components to a spatial wave vector is arbitrary and may be changed by a rotation. We can therefore fix one of the vectors, $\vec{k}'$, or, in order to avoid this equally arbitrary choice, eliminate it by integration. This step results in the **matter power spectrum**

$$P_\rho(\vec{k}) = G^2 \int \langle \rho_{\vec{k}} \rho^*_{\vec{k}'} \rangle d^3k'. \tag{8.72}$$

Including the square of Newton's constant will allow us to express this value in a suitable power of length units. Near-isotropy implies that $P_\rho(\vec{k})$ only depends on the norm $|\vec{k}|$, which determines the wave length of a solution in the Fourier-transformed matter distribution. The power spectrum tells us how much a given wave length contributes to the image obtained by optical or microwave observations.

A power-law behavior is a useful parameterization for comparisons with observational data, defined by

$$P_\rho(\vec{k}) \propto |\vec{k}|^{n_s} \tag{8.73}$$

with the **scalar spectral index** n_s. (There is also a tensor spectral index for the distribution of primordial gravitational waves.)

Referring again to near uniformity of the matter distribution, we obtain a preferred value for n_s from symmetry arguments. Uniformity in the spatial distribution means that the observational data, in a first approximation, do not determine a preferred length scale. Therefore, the length units of $|\vec{k}|^{n_s}$ must agree with units implied by the general definition (8.72) of the power spectrum. The energy density times G has units of inverse length squared, as can be seen easily by looking at the Friedmann equation. Its Fourier transform is obtained by 3-dimensional spatial integration, which therefore has units of length. The wave number appears in the 3-dimensional integration of (8.72), in which each factor contributes units of inverse length. Combining units from all factors, $P_\rho(\vec{k})$ has units of inverse length, which is the same as the units of $|\vec{k}|$: We expect the scalar spectral index to be close to the value $n_s \approx 1$. A power spectrum with this spectral index is called **scale-free**.

8.5.4.4 Cosmic inflation and observations

A spectral index slightly different from one is usually obtained from actual observations because the distribution is not exactly uniform. Such a measurement can be compared with predictions obtained from the inflationary solution for $u_{\vec{k}}$. For a direct comparison, we transform the matter power spectrum to the power spectrum

$$P_u(\vec{k}) = \int \langle u_{\vec{k}} u^*_{\vec{k}'} \rangle \mathrm{d}^3k' \propto \frac{P_\rho(\vec{k})}{|\vec{k}|^4} \propto |\vec{k}|^{n_s-4}. \tag{8.74}$$

If the inflationary model is consistent with observations, the solution (8.67) should imply a power spectrum proportional to $|\vec{k}|^{-3}$. However, our solution for $u_{\vec{k}}$ so far depends on conformal time η, which we have to relate to $|\vec{k}|$ in a suitable way.

A relationship between these quantities is obtained if we look back at equation (8.62) and recall that our derivation of initial values from the quantum ground state was based on ignoring the last term in this equation, or the last term in the conformal-time version (8.63). During inflation, the coefficient H in this term is nearly constant, while the coefficient $1/a^2$ in the middle term decreases. Ignoring the last term is therefore justified at early times when we assumed the modes to be in their ground state, but it remains valid only for a finite amount of time in onward evolution.

Both terms are relevant during a certain transition period, but at later times, we obtain a new approximation by ignoring the middle term while keeping the last term. The resulting equation

$$\frac{\mathrm{d}^2 u_{\vec{k}}}{\mathrm{d}\tau^2} = -3H \frac{\mathrm{d}u_{\vec{k}}}{\mathrm{d}\tau}. \tag{8.75}$$

no longer has wave solutions. One solution of this ordinary differential equation is easy to see: a τ-independent $u_{\vec{k}}$. The second independent solution, for nearly-constant $H = \sqrt{\Lambda}$ during inflation, is given by $u_{\vec{k}}(\tau) \propto \exp(-3\sqrt{\Lambda}\tau)$. It decreases exponentially and can be ignored compared with the constant solution. Thus, when the third term in (8.62) or (8.63) becomes relevant, the evolved values of the modes $u_{\vec{k}}$ stop changing. The corresponding power spectrum is obtained by evaluating each mode at a time such that

$$|\vec{k}|^2 = \frac{1}{a}\frac{\mathrm{d}^2 a}{\mathrm{d}\eta^2} = \frac{2}{\eta^2} = 2H^2a^2. \tag{8.76}$$

For each mode, the conformal time η where the solution (8.67) should be evaluated, is now uniquely determined, through $a(\eta)$, by its wave number $|\vec{k}|$ and the constant Hubble parameter H during inflation. The Fourier mode is complex, but the norm squared

$$|u_{\vec{k}}|^2 = \frac{\hbar}{2|\vec{k}|}\frac{1}{a^2}\left(1 + \frac{1}{|\vec{k}|^2\eta^2}\right)\Bigg|_{2a^2=|\vec{k}|^2/H^2} = \frac{3}{4}\frac{\hbar H^2}{|\vec{k}|^3} \tag{8.77}$$

can be used to determine the $|\vec{k}|$-dependence of $P_u(\vec{k})$ defined in (8.74). It takes a simple form and is proportional to $|\vec{k}|^{-3}$, as required for a nearly scale-free spectrum. Small corrections to n_s are implied if H is not exactly constant, such that inflation can come to an end. In this way, specific models of cosmic inflation can be tested by observations.

In this derivation, we related classical initial conditions for field modes with quantum-mechanics results for standard deviations rather than expectation values. This assumption is a crucial ingredient that can be justified by decoherence models.

8.6 GEOMETRICAL PROPERTIES

Space-time geometries that significantly deviate from the simple FRW line element require non-trivial derivations of properties such as spatial volumes or areas, or of indications of curvature. In our cosmological discussion, we could simply use Euclidean flat space at a given value τ of proper time of co-moving observers and derive the expanding volume $V_0 a(\tau)^3$ of a region with coordinate size V_0. More advanced geometrical considerations are required in general because there are homogeneous and isotropic cosmological models that do not imply a flat spatial line element at a fixed value of proper time, and especially because homogeneity and isotropy are assumptions that are not suitable for important questions such as the space-time geometry around a compact star or a black hole.

The equivalence principle shows that gravity is related to space-time curvature. Complete curvature expressions are rather complicated, but a glimpse at some implications of curvature can be obtained from ratios of geometrical quantities, such as the circumference over the radius of a circle or the area over the radius of a sphere. These ingredients require formulas that allow us to compute the length, area, or volume of any region specified by certain coordinate relationships in a given space-time geometry. We consider some of these quantities in a simpler context by assuming that the metric tensor in a given coordinate system is diagonal, such that its line element is of the form

$$ds^2 = g_{00}(dx^0)^2 + g_{11}(dx^1)^2 + g_{22}(dx^2)^2 + g_{33}(dx^3)^2 . \tag{8.78}$$

All diagonal components are now allowed to be functions of the four coordinates x^α, such that they cannot be removed by integrations, as in $\sqrt{-g_{00}}dx^0 = d\tau$, used in FRW line elements, where $g_{00} = -N^2$ depended only on τ. We do assume that $g_{00} < 0$, distinguishing x^0 as the time coordinate.

In general, a diagonal form of the metric is not preserved by a coordinate transformation, and therefore saying that the metric is diagonal is not a covariant statement. Moreover, it is not guaranteed that a given metric can be brought to diagonal form by choosing suitable coordinates in all of space-time. It is always possible to diagonalize the line element at a given point, but the position dependence of metric components in general implies obstacles to the global diagonalization procedure. Nevertheless, this assumption in a given coordinate system, and applied to special geometries that obey the diagonalizability condition, simplifies calculations of various geometrical quantities or their ingredients. For instance, the inverse metric or its determinant are easy to obtain in the diagonal case.

8.6.1 COORDINATE LINES AND INDUCED METRICS

Another simplification implied by a diagonal metric is the computation of areas and volumes of regions bounded by **coordinate lines**, that is curves in space-time with constant values of all but one of the coordinates. The coordinate line ℓ^α belonging to a coordinate x^α for a fixed α is defined by the equations $x^\beta = c^\beta$ constant, for all $\beta \neq \alpha$. At any point where they intersect, coordinate lines of different coordinates are orthogonal if the metric is diagonal. The coordinate lines therefore determine an orthogonal grid, even if they are not straight lines as in Cartesian coordinates of flat space-time.

Using the metric, we compute the **proper length** along a coordinate line ℓ^α by integrating $\sqrt{\pm ds^2}$, the sign depending on whether the coordinate line is timelike ($\alpha = 0$) or spacelike ($\alpha \neq 0$). For instance, for $\alpha = 0$ we have the standard definition of proper time along the coordinate line, while $\alpha = 1$ gives

$$d\ell^1 = ds|_{dx^0=dx^2=dx^3=0} = \sqrt{g_{11}}\, dx^1 \,. \tag{8.79}$$

The orthogonal coordinate grid allows us to combine linear distances and compute areas and volumes. The **area element** is the area of a small, approximately square-shaped surface bounded by coordinate lines in two different directions. If we are interested in the area element at constant x^0 and constant x^3, we compute

$$dA^{12} = d\ell^1 d\ell^2 = \sqrt{g_{11}g_{22}}\, dx^1 dx^2 \,. \tag{8.80}$$

The spatial **3-volume element** is the volume of a small region at constant x^0, given by

$$dV = d\ell^1 d\ell^2 d\ell^3 = \sqrt{g_{11}g_{22}g_{33}}\, dx^1 dx^2 dx^3 \,. \tag{8.81}$$

It is also possible to combine the timelike coordinate line with two spatial coordinate lines and obtain a time-surface volume element such as

$$dV^{12} = d\ell^0 d\ell^1 d\ell^2 = \sqrt{-g_{00}g_{11}g_{22}}\, dx^0 dx^1 dx^2 \,. \tag{8.82}$$

This expression might appear if we integrate by parts (or apply the divergence theorem) in a 4-dimensional integral and obtain a combination of spacelike and timelike boundaries.

Such a 4-dimensional integral requires a space-time measure of its size. We do not have geometrical intuition for such an expression, but by analogy we define the volume of a 4-dimensional infinitesimal region by

$$dv = d\ell^0 d\ell^1 d\ell^2 d\ell^3 = \sqrt{-g_{00}g_{11}g_{22}g_{33}}\, dx^0 dx^1 dx^2 dx^3 = \sqrt{-\det g}\, dx^0 dx^1 dx^2 dx^3 \,, \tag{8.83}$$

called, for lack of a better word, the **4-volume element**. This quantity is rarely used in common geometrical expressions, but it appears in the action of general relativity or any field theory such as electrodynamics on relativistic space-time: The action is the time integral of the Lagrange function and therefore, for a theory of position-dependent fields, combines all space and time integrations. With hindsight, we recognize the FRW 4-volume element in the Lagrange function (8.58).

From a formal perspective, the 4-volume element is interesting because it is the only one of the above expressions that can be written in a simple form using the metric even if it is not diagonal: We always have

$$dv = \sqrt{-\det g}\, dx^0 dx^1 dx^2 dx^3 \,. \tag{8.84}$$

It is more complicated to compute the determinant of a non-diagonal metric, but we do not have to include additional terms in this formula. This is in contrast to the spatial area or volume element for non-diagonal metrics.

A general method for computing the latter quantities is to specify the surface or 3-dimensional region we are interested in by conditions on the spatial or space-time coordinates. We then derive the line element on the surface or in the 3-dimensional region by implementing the coordinate conditions in the space-time line element. The resulting line element may not be diagonal even if the original space-time line element is diagonal. However, the restricted line element or **induced line element** has the same dimension as the region we are interested in. In any such case, we simply integrate the determinant of the restricted metric over the region. This reasoning also applies to the length of a curve that is not a coordinate line.

As a simple example, consider Euclidean space with the diagonal line element $ds^2 = dx^2 + dy^2 + dz^2$. If we are interested in the area of a square on the plane defined by $y = z$, we cannot use coordinate lines in the way just introduced. Instead, we use the equation $y - z = 0$ to suggest a new coordinate $Z = y - z$ such that $\mathrm{d}Z = 0$ on the surface. The coordinate transformation from (x,y,z) to (x,y,Z) transforms the line element to the non-diagonal

$$\mathrm{d}s^2 = \mathrm{d}x^2 + 2\mathrm{d}y^2 - 2\mathrm{d}y\mathrm{d}Z + \mathrm{d}Z^2 . \tag{8.85}$$

We now have coordinate lines of x and y that lie on the surface.

The induced line element on the surface is obtained by inserting $\mathrm{d}Z = 0$: $\mathrm{d}s^2|_{\mathrm{d}Z=0} = \mathrm{d}x^2 + 2\mathrm{d}y^2$. The area of a square with side lengths Δx and Δy is given by

$$A = \int \sqrt{\det g}\mathrm{d}x\mathrm{d}y = \sqrt{2}\Delta x\Delta y . \tag{8.86}$$

Elementary Euclidean geometry shows that this result is correct because one side length of the square in the original 3-dimensional space is given by the diagonal in the $y = z$-plane, which has length $\sqrt{2}\Delta y$ if there is a y-difference Δy along this side. In this case, the induced metric is diagonal such that the area element is the product of the proper lengths $\mathrm{d}x$ and $\sqrt{2}\mathrm{d}y$.

8.6.2 SPHERES

As a familiar example, consider a 2-sphere in flat space. We use the Minkowski line element in polar coordinates,

$$\mathrm{d}s^2 = -\mathrm{d}t^2 + \mathrm{d}r^2 + r^2(\mathrm{d}\vartheta^2 + \sin(\vartheta)^2\mathrm{d}\varphi^2) , \tag{8.87}$$

and specify a 2-dimensional sphere by the conditions $r = R$ and t constant. Only ϑ and φ change on the sphere, where we have the area element

$$\mathrm{d}A = \sqrt{g_{\vartheta\vartheta}g_{\varphi\varphi}}\,\mathrm{d}\vartheta\mathrm{d}\varphi = r^2\sin(\vartheta)\mathrm{d}\vartheta\mathrm{d}\varphi . \tag{8.88}$$

The area of the sphere is obtained by integrating ϑ from 0 to π and φ from 0 to 2π: $A = \int \mathrm{d}A = 4\pi R^2$. The 3-volume element at constant t is

$$\mathrm{d}V = \sqrt{g_{rr}g_{\vartheta\vartheta}g_{\varphi\varphi}}\,\mathrm{d}r\mathrm{d}\vartheta\mathrm{d}\varphi = r^2\sin(\vartheta)\mathrm{d}r\mathrm{d}\vartheta\mathrm{d}\varphi . \tag{8.89}$$

The volume within the sphere equals $V = \frac{4}{3}\pi R^3$.

We can modify the line element in the previous example and compute properties of a 2-sphere in a **3-sphere**. A 3-sphere is *not* the 3-dimensional interior of a 2-sphere, but rather a 3-dimensional space with constant positive curvature. The interior of a 2-sphere, by contrast, has constant vanishing curvature. The 3-sphere generalizes the crucial properties of the 2-sphere: homogeneity (constant curvature), positive curvature, and the absence of a boundary. Only the first condition would be fulfilled by the interior of a 2-sphere. Because of the first and the last properties, a 3-sphere may be used as the spatial geometry of a universe model, replacing flat Euclidean space in (8.36).

The line element of the 3-sphere can be derived by generalizing the derivation of the line element of the 2-sphere as a surface in 3-dimensional flat space. We increase all dimensions by one, starting with 4-dimensional Euclidean space, choosing polar coordinates, and restricting the line element to a hypersurface on which the distance from the origin is constant. Starting with the 4-dimensional flat Euclidean line element

$$\mathrm{d}s^2 = \mathrm{d}w^2 + \mathrm{d}x^2 + \mathrm{d}y^2 + \mathrm{d}z^2 \tag{8.90}$$

with a new spatial coordinate w, the condition on the distance from the origin reads

$$w^2 + x^2 + y^2 + z^2 = \rho^2 \tag{8.91}$$

if ρ is the radius of the 3-sphere.

This condition is non-linear and involves all four coordinates. It is simplified by introducing a 4-dimensional version of spherical coordinates,

$$x = \rho\sin(\psi)\sin(\vartheta)\cos(\varphi) \quad (8.92)$$
$$y = \rho\sin(\psi)\sin(\vartheta)\sin(\varphi) \quad (8.93)$$
$$z = \rho\sin(\psi)\cos(\vartheta) \quad (8.94)$$
$$w = \rho\cos(\psi) \quad (8.95)$$

with a third angle ψ taking values from zero to π. Trigonometric identities can be used to confirm the distance condition. Moreover, in (x,y,z), we recognize the common form of 3-dimensional spherical coordinates on hypersurfaces with constant ψ, where we obtain a cross-section of the 3-sphere given by a 2-sphere with radius $\rho\sin(\psi)$. The 3-dimensional spherical coordinates are obtained by replacing the radius in all equations of the 2-dimensional version with $\rho\sin(\psi)$, and defining the fourth coordinate as $\rho\cos(\psi)$. This recipe can be extended to higher and higher dimensions, making sure that the distance condition is always satisfied. Also note that we can construct the 2-dimensional spherical coordinates in this way, starting with a single polar coordinate φ on a circle.

The distance condition is now linear and involves only one of the new coordinates, ρ. We define the 3-sphere line element as the 4-dimensional Euclidean line element restricted to constant ρ. After transforming from Cartesian coordinates (w,x,y,z) to spherical coordinates $(\rho,\psi,\vartheta,\varphi)$ and removing any $\mathrm{d}\rho$ because ρ is constant on the sphere, we obtain

$$\mathrm{d}s^2 = \rho^2\left(\mathrm{d}\psi^2 + \sin(\psi)^2\left(\mathrm{d}\vartheta^2 + \sin(\vartheta)^2\mathrm{d}\varphi^2\right)\right). \quad (8.96)$$

Also here, we recognize the 2-sphere line element on surfaces of constant ψ. This pattern again generalizes to higher-dimensional versions of the sphere, and also to **hyperboloids** if we start with Cartesian coordinates in Minkowski space-time rather than Euclidean space. In the latter case, some of the sines are replaced by hyperbolic versions.

It is sometimes convenient to apply a coordinate transformation, mapping ψ to a radial coordinate $r = \rho\sin(\psi)$. This transformation implies the line element

$$\mathrm{d}s^2 = -\mathrm{d}t^2 + \frac{\mathrm{d}r^2}{1 - r^2/\rho^2} + r^2(\mathrm{d}\vartheta^2 + \sin(\vartheta)^2\mathrm{d}\varphi^2) \quad (8.97)$$

if we also add the temporal contribution back in. Multiplying the spatial part by a squared scale factor, $a(\tau)^2$, would constitute a cosmological model with curved spatial hypersurfaces. The constant radius ρ of the 3-sphere determines the radial component of the line element, as well as the range $0 \leq r \leq \rho$ of the new r based on positivity of g_{rr}. In contrast to spherical coordinates in 3-dimensional Euclidean space, the range of r here is bounded, implying a compact 3-sphere.

The 3-sphere defines a new geometry about which standard Euclidean results do not tell us much. Nevertheless, we can compute some of its properties by considering combinations of coordinate lines. We are interested in a 2-sphere within the 3-sphere, defined by the conditions $r = R$ and constant t. We confirm that these conditions determine a 2-sphere because the line element (8.97) with $\mathrm{d}t = 0 = \mathrm{d}r$ is indeed the 2-sphere line element. The equator of such a 2-sphere is the coordinate line ℓ^φ with circumference

$$C = \int \mathrm{d}\ell^\varphi = \int_0^{2\pi}\sqrt{g_{\varphi\varphi}}\,\mathrm{d}\varphi = \int_0^{2\pi} R\mathrm{d}\varphi = 2\pi R. \quad (8.98)$$

Its radius S is the distance from the center ($r = 0$) to the 2-sphere ($r = R$) along the coordinate line ℓ^r:

$$S = \int \mathrm{d}\ell^r = \int_0^R \frac{1}{\sqrt{1 - r^2/\rho^2}}\mathrm{d}r = \rho\arcsin(R/\rho) \neq R. \quad (8.99)$$

The curved geometry of the 3-sphere implies that the radius S of the equator of a 2-sphere in it is not equal to the coordinate value $r = R$ where it is placed. The presence of curvature can be inferred from the fact that the ratio of circumference over radius of the equator,

$$\frac{C}{S} = 2\pi \frac{R/\rho}{\arcsin(R/\rho)}, \tag{8.100}$$

differs from 2π. It is close to the Euclidean value of 2π only for $R \ll \rho$.

8.6.3 VECTORS IN CURVED SPACE-TIME

Vectors represent directions in space or space-time. We usually picture a vector as a straight arrow, but the condition used here depends on a geometrical concept: straight lines in Euclidean geometry. If we are interested in specifying a direction on a curved surface embedded in 3-dimensional space, a 3-dimensional straight arrow does not completely lie on the surface.

A better interpretation of a direction is given by a short piece of a 1-dimensional curve. In flat space, this idea is equivalent to a straight arrow because we can simply use the tangent vector of the curve at a given point to represent the direction. However, we may just as well use a small or infinitesimal segment of the curve leaving the point. Such a segment is guaranteed to stay on the surface if the entire curve lies on the surface, and it does not require Euclidean properties such as straight lines.

For a given point on a curved surface, the directions of all possible curves through this point span the tangent plane at the point. If we represent tangent directions by straight arrows in the embedding 3-dimensional space, we can see another feature of curvature if we compare directions at different points. If the surface has extrinsic curvature in the embedding space, the representations of tangent planes by straight arrows, formulated at different points on the surface, are rotated with respect to each other. This rotation makes it difficult to compare directions or vectors at different points on a curved surface because a straight-arrow vector which is tangent to the surface at one point may not be tangent at another point when it is simply translated in the embedding space. And if we use infinitesimal segments of curves to represent directions directly in the surface, we cannot compare directions at different points because there is no unique curve between two points on a surface. Comparing directions on a curved surface, or in curved space or curved space-time, requires an additional mathematical ingredient, called **parallel transport** or a **connection**, that we will encounter later.

For now, it is sufficient to take into account the incomparability of directions at different points by forgoing the use of global coordinate axes, represented by a single basis of the Euclidean tangent plane as it is usually applied to compute components of a vector at any point in 3-dimensional space. For curved space-time, we do not use a set of just four vectors $\mathbf{e}_\alpha$, but instead specify a basis $\mathbf{e}_\alpha(\mathbf{x})$ at each event $\mathbf{x}$. This construction is similar to local inertial frames in special relativity, but now we do not refer to a fixed worldline of an observer. A collection of vectors at different positions can be expanded in our position-dependent basis,

$$\mathbf{a}(\mathbf{x}) = a^\alpha(\mathbf{x})\mathbf{e}_\alpha(\mathbf{x}), \tag{8.101}$$

and has position-dependent components $a^\alpha(\mathbf{x})$. If the vector depends on the position, we call it **vector field**.

We use the same calculational tools for vector fields that we already applied to vectors. For instance, we add two vector fields $\mathbf{a}(\mathbf{x})$ and $\mathbf{b}(\mathbf{x})$ by adding the components: $\mathbf{a}(\mathbf{x}) + \mathbf{b}(\mathbf{x})$ has components $a^\alpha(\mathbf{x}) + b^\alpha(\mathbf{x})$. However, we cannot add vectors at different positions in a curved space-time. The scalar product determines the function

$$\mathbf{a}(\mathbf{x}) \cdot \mathbf{b}(\mathbf{x}) = (\mathbf{e}_\alpha(\mathbf{x}) \cdot \mathbf{e}_\beta(\mathbf{x}))a^\alpha(\mathbf{x})b^\beta(\mathbf{x}) \tag{8.102}$$

with a position-dependent coefficient $\mathbf{e}_\alpha(\mathbf{x}) \cdot \mathbf{e}_\beta(\mathbf{x})$.

The metric components depend on the choice of our basis $\mathbf{e}_\alpha(\mathbf{x})$. Two options are most useful, depending on the calculation we intend to do.

1. A **coordinate basis** $\mathbf{e}_\alpha(\mathbf{x})$ is such that $\mathbf{e}_\alpha(\mathbf{x})\cdot\mathbf{e}_\beta(\mathbf{x}) = g_{\alpha\beta}(\mathbf{x})$ is equal to the metric in the line element of curved space-time for a given choice of coordinates. In this basis, the displacement vector has the coordinate differentials as its components: $\Delta\mathbf{x} = \Delta x^\alpha \mathbf{e}_\alpha(\mathbf{x})$, which implies the line element $||\Delta\mathbf{x}||^2 = g_{\alpha\beta}(\mathbf{x})\Delta x^\alpha \Delta x^\beta = \Delta s^2$.
   An elegant mathematical way of introducing vectors on curved spaces is given by the notion of a **directional derivative**. In flat space with a global basis $\mathbf{e}_\alpha$, the directional derivative defined by a vector $\mathbf{v} = v^\alpha \mathbf{e}_\alpha$ assigns the value
$$\mathbf{v}(f) = \mathbf{v}\cdot\nabla f = v^\alpha \frac{\partial f}{\partial x^\alpha} \tag{8.103}$$
   to a function f at a given point where the derivatives are evaluated. Functions can be defined on curved spaces by simply assigning a real or complex number to each point, expressed for instance in terms of coordinates. Comparing (8.103) with the basis decomposition of $\mathbf{v}$ suggests the common notation $\mathbf{e}_\alpha = \partial/\partial x^\alpha$ for the coordinate basis. The notion of a directional derivative is meaningful on a curved space because it only requires a map from functions to the value of the directional derivative. The mathematical requirements are that $\mathbf{v}(f)$ must be linear in f and obey the product rule $\mathbf{v}(fg) = f\mathbf{v}(g) + \mathbf{v}(f)g$ for two functions, f and g.
2. An **orthonormal basis** $\mathbf{e}_{\hat\alpha}(\mathbf{x})$ is such that $\mathbf{e}_{\hat\alpha}(\mathbf{x})\cdot\mathbf{e}_{\hat\beta}(\mathbf{x}) = \eta_{\hat\alpha\hat\beta}$, using hatted indices to indicate the basis choice. In this basis, we have $\mathbf{a}(\mathbf{x})\cdot\mathbf{b}(\mathbf{x}) = \eta_{\hat\alpha\hat\beta}a^{\hat\alpha}(\mathbf{x})b^{\hat\beta}(\mathbf{x})$ with the Minkowski metric $\eta_{\hat\alpha\hat\beta}$, even if the space-time is not Minkowski. The components of the displacement vector then cannot be the coordinate displacements. Therefore, the direction given by a vector that is part of an orthonormal basis in non-Minkowski space-time is not a simple coordinate translation.

We typically use a coordinate basis when we derive worldlines of moving objects because we need to solve differential equations for the coordinates. We use an orthonormal basis when we compute observable quantities, in the same way as we used local inertial frames in special relativity.

The coordinate basis and orthonormal basis are different from each other whenever the line element is not Minkowskian or Euclidean in Cartesian coordinates. For instance, the 2-dimensional plane in polar coordinates has the line element $\mathrm{d}s^2 = \mathrm{d}r^2 + r^2\mathrm{d}\varphi^2$. The coordinate basis $(\mathbf{e}_r, \mathbf{e}_\varphi)$ has vectors with scalar products
$$\mathbf{e}_r\cdot\mathbf{e}_r = 1 \quad , \quad \mathbf{e}_r\cdot\mathbf{e}_\varphi = 0 \quad , \quad \mathbf{e}_\varphi\cdot\mathbf{e}_\varphi = r^2 . \tag{8.104}$$
Only the last of these relations differs from an orthonormal basis because the vector $\mathbf{e}_\varphi$ in a coordinate basis increases in length as r increases. It is the length of a displacement by some small $\Delta\varphi$ in the angular direction, along a circle segment that gets longer as we move away from the origin of the polar system. The orthonormal basis, on the other hand, should have constant lengths for its vectors $(\mathbf{e}_{\hat r}, \mathbf{e}_{\hat\varphi})$. We can achieve the orthonormality conditions by defining $\mathbf{e}_{\hat r} = \mathbf{e}_r$ and $\mathbf{e}_{\hat\varphi} = r^{-1}\mathbf{e}_\varphi$.

It is always possible to find an orthonormal basis, even if the space-time is not Minkowskian. For instance, in a 2-dimensional cosmological model with line element $\mathrm{d}s^2 = -N(t)^2\mathrm{d}t^2 + a(t)^2\mathrm{d}x^2$ with $N(t) > 0$ and $a(t) > 0$, an orthonormal basis is given by $\mathbf{e}_{\hat t} = N(t)^{-1}\mathbf{e}_t$ and $\mathbf{e}_{\hat x} = a(t)^{-1}\mathbf{e}_x$ in terms of the coordinate basis. In this case, the orthonormal basis obeys $\mathbf{e}_{\hat\alpha}\cdot\mathbf{e}_{\hat\beta} = \eta_{\hat\alpha\hat\beta}$ even though there is no coordinate transformation that turns the metric into $\eta_{\alpha\beta}$ (unless $a(t)$ is constant).

8.6.4 ABSTRACT INDEX NOTATION

We are about to encounter our first multi-index objects. For their derivation it is useful to collect and summarize all the ingredients of **abstract index notation** introduced so far, and to provide a complete definition.

8.6.4.1 Rules

1. In any product, a given index can appear at most twice.
2. If an index value appears twice in a product, it must do so as a pair of one superscript and one subscript.
3. An index pair of a subscript and a superscript in a product is automatically summed over the allowed range $(0,1,2,3$ in space-time).
4. In a sum, each term must have matching single indices, while index pairs in two summands may or may not use the same index choices.
5. Both sides of an equation must have matching single indices, while index pairs on the two sides may or may not use the same index choices.
6. A free single index in an equation implies that the equation must hold for any value of this index. In space-time, such an equation represents $4N$ individual equations for its components, if N is the number of free single indices.

Examples of allowed terms are

$$a^{\alpha} \quad , \quad a_{\alpha} \quad , \quad a^{\alpha}+b^{\alpha} \quad , \quad a_{\alpha}b^{\alpha}=\sum_{\alpha=0}^{3}a_{\alpha}b^{\alpha} \quad , \quad a_{\alpha}b^{\alpha}c_{\beta}d^{\beta} \tag{8.105}$$

and $a_{\alpha}b^{\alpha}+c_{\alpha}d^{\alpha}=a_{\alpha}b^{\alpha}+c_{\beta}d^{\beta}$. An expression such as $a_{\alpha}b^{\alpha}c_{\alpha}d^{\alpha}$ is not allowed because it does not imply an unambiguous summation: Writing the summation symbol, it could be rendered in different forms, such as

$$\sum_{\alpha=0}^{3}(a_{\alpha}b^{\alpha}c_{\alpha}d^{\alpha}) \neq \left(\sum_{\alpha=0}^{3}a_{\alpha}b^{\alpha}\right)\left(\sum_{\alpha=0}^{3}c_{\alpha}d^{\alpha}\right) \neq \left(\sum_{\alpha=0}^{3}a_{\alpha}d^{\alpha}\right)\left(\sum_{\alpha=0}^{3}c_{\alpha}b^{\alpha}\right) \tag{8.106}$$

Using the same index pair in different terms in a sum is, however, allowed: If we use the same index pair twice in a sum, as in $a_{\alpha}b^{\alpha}+c_{\alpha}d^{\alpha}$, the two possible versions

$$\sum_{\alpha=0}^{3}(a_{\alpha}b^{\alpha}+c_{\alpha}d^{\alpha}) = \sum_{\alpha=0}^{3}a_{\alpha}b^{\alpha}+\sum_{\alpha=0}^{3}c_{\alpha}d^{\alpha} \tag{8.107}$$

are equal and there is no ambiguity.

8.6.4.2 Index positions

The first object we wrote with space-time indices was the position vector **x** in space-time. We chose a superscript index for its components x^{α}, $\alpha=0,1,2,3$. This choice was arbitrary but followed established customs in general relativity. If we had made the opposite choice as a subscript index, the following index positions would all have been reversed.

We then defined several vectorial and tensorial objects derived from the position vector:

1. The differential $\mathrm{d}x^{\alpha}$ must have a superscript just like x^{α}, and so do the 4-velocity $u^{\alpha}=\mathrm{d}x^{\alpha}/\mathrm{d}\tau$, the 4-momentum $p^{\alpha}=mu^{\alpha}$ and the 4-force $f^{\alpha}=\mathrm{d}p^{\alpha}/\mathrm{d}\tau$.
2. The chain rule applied to a coordinate transformation $x^{\alpha'}(x^{\beta})$ inserted in $\mathrm{d}x^{\alpha'}$ introduces a summation:

$$\mathrm{d}x^{\alpha'} = \frac{\partial x^{\alpha'}}{\partial x^{\beta}}\mathrm{d}x^{\beta}\,. \tag{8.108}$$

For the right-hand side to agree with our rules, there must be a subscript β. Therefore, a superscript index in the denominator is declared to count as a subscript. We usually do not divide by vector components. Therefore, the only case where we have an index in the denominator is a derivative such as (8.108).

3. Writing the line element $ds^2 = g_{\alpha\beta} dx^\alpha dx^\beta$ with generic coefficients given by a metric $g_{\alpha\beta}$ implies that the metric must have two subscript indices. The indices are then paired up correctly with the superscripts on dx^α and dx^β. The line element requires the displacement vector expressed in a coordinate basis $\mathbf{e}_\alpha$. The metric appears in the scalar products $g_{\alpha\beta} = \mathbf{e}_\alpha \cdot \mathbf{e}_\beta$. These scalar products are also available for an orthonormal basis, where we have $\mathbf{e}_{\hat{\alpha}} \cdot \mathbf{e}_{\hat{\beta}} = \eta_{\hat{\alpha}\hat{\beta}}$. This equation is compatible with the preceding one if we use $g_{\hat{\alpha}\hat{\beta}} = \eta_{\hat{\alpha}\hat{\beta}}$.

8.6.4.3 Lowering and raising indices

Our derivations in local inertial frames led to another definition: We often encounter expressions such as $g_{\alpha\beta} p^\beta$, for instance with the 4-momentum p^β. Since the metric is invertible, as long as there are no singularities, it is always possible to reconstruct the components p^β from the components $g_{\alpha\beta} p^\beta$. Therefore, both expressions contain the same information about the vector **p**. It is then convenient to denote them by the same symbol,

$$p_\alpha = g_{\alpha\beta} p^\beta , \tag{8.109}$$

where the difference between the two versions is indicated only by the index position. If we would like to emphasize the mathematical difference between a superscript object and a subscript object, we refer to the former as a vector and the latter as a **dual vector**. The metric allows us to lower a superscript index into a subscript, or to map a vector to a dual vector.

In order to reconstruct p^β from p_α, we need a transformation, or a matrix $t^{\beta\alpha}$ such that

$$p^\beta = t^{\beta\alpha} p_\alpha . \tag{8.110}$$

Inserting (8.109) in (8.110), we obtain

$$p^\beta = t^{\beta\alpha} (g_{\alpha\gamma} p^\gamma) = \left(t^{\beta\alpha} g_{\alpha\gamma} \right) p^\gamma . \tag{8.111}$$

We have to rename β when we insert (8.109) because we already have a free index β in (8.110). For (8.111) to be true for all p^β, we must have

$$t^{\beta\alpha} g_{\alpha\gamma} = \delta^\beta{}_\gamma \tag{8.112}$$

equal to the Kronecker delta, or the components of the identity matrix. This equation shows that the components $t^{\beta\alpha}$ are given by the inverse matrix of $g_{\alpha\beta}$:

$$t^{\beta\alpha} = (g^{-1})^{\beta\alpha} = g^{\beta\alpha} . \tag{8.113}$$

As indicated in the last step, taking the inverse matrix of $g_{\alpha\beta}$ is usually shown only by writing the same symbol, g, with superscript indices. Also here, invertibility implies that $g_{\alpha\beta}$ and $g^{\alpha\beta}$ contain the same information, such that we use the same symbol g and indicate the mathematical difference only by index positions. With this result, we raise indices by

$$p^\beta = g^{\beta\alpha} p_\alpha . \tag{8.114}$$

Equation (8.112) can now be re-interpreted as raising one index of the metric or lowering one index of the inverse metric:

$$g^{\beta\alpha} g_{\alpha\gamma} = \delta^\beta{}_\gamma . \tag{8.115}$$

In addition to $g_{\alpha\beta}$ and $g^{\alpha\beta}$, there is therefore a third related object in which the two indices have different positions. However, it does not contain any information about the geometry because it always equals the identity matrix. Note that the identity matrix appears with mixed index positions because, as the identity, it must preserve the index positions on vectors such as $\delta^\beta{}_\gamma p^\gamma = p^\beta$. Writing the subscript γ as the second index of $\delta^\beta{}_\gamma$ correctly pairs up rows and columns in a matrix product $\delta^\beta{}_\gamma p^\gamma = (\mathbb{I}\mathbf{p})^\beta$. In this equation, the first index must be a superscript in order to match the superscript on p^β, while the second index must be a subscript in order to be summed over with the superscript on p^γ. The metric or its inverse, by contrast, change index positions by lowering or raising operations.

We use the same relations in an orthonormal basis, where the equality $g_{\hat{\alpha}\hat{\beta}} = \mathbf{e}_{\hat{\alpha}} \cdot \mathbf{e}_{\hat{\beta}} = \eta_{\hat{\alpha}\hat{\beta}}$ simplifies conversions between superscript and subscript components: We always have $p_{\hat{0}} = -p^{\hat{0}}$ and $p_{\hat{i}} = p^{\hat{i}}$ for $i = 1,2,3$, even if space-time is curved.

8.6.4.4 Example: Lorentz 4-force

Given the 4-velocity u^β of a massive particle with charge q, we can combine it with the electromagnetic field-strength tensor $\mathscr{F}_{\alpha\beta}$ from (7.115) to obtain a vector $q\mathscr{F}_{\alpha\beta}u^\beta$ with a subscript index. The product combines components of the electric and magnetic fields correctly with the velocity components to produce the Lorentz force, but in $q\mathscr{F}_{\alpha\beta}u^\beta$ we do not have a free superscript index as required for a 4-force. We use the inverse metric to raise the index:

$$f^\gamma = -qg^{\gamma\alpha}\mathscr{F}_{\alpha\beta}u^\beta \,. \tag{8.116}$$

8.6.4.5 Tensor transformation law

Index positions on a tensor, generically $T^{\alpha_1\alpha_2\cdots\alpha_n}{}_{\beta_1\beta_2\cdots\beta_m}$, determine how the components should be transformed if we change coordinates from x^β to $x^{\alpha'}(x^\beta)$. We derived this transformation for the metric using the condition that the line element $\mathrm{d}s^2 = g_{\alpha\beta}\mathrm{d}x^\alpha\mathrm{d}x^\beta$ should be invariant, together with the chain rule (8.108):

$$g_{\alpha'\beta'} = \frac{\partial x^\alpha}{\partial x^{\alpha'}}\frac{\partial x^\beta}{\partial x^{\beta'}}g_{\alpha\beta} \,. \tag{8.117}$$

The index positions of α' and β' on the right correctly match those on the left because superscripts in the denominator count as subscripts.

Taking inverse matrices, we obtain the corresponding result for $g^{\alpha\beta}$:

$$g^{\alpha'\beta'} = \frac{\partial x^{\alpha'}}{\partial x^\alpha}\frac{\partial x^{\beta'}}{\partial x^\beta}g^{\alpha\beta} \,. \tag{8.118}$$

Recall that partial derivatives of an inverse function are obtained from partial derivatives of the original function by inverting the entire matrix $(\partial x^\alpha/\partial x^{\alpha'})$, rather than individual components; see also equation (8.121) below. For any other tensor, we determine the required factors of partial derivatives by making sure that all indices are paired up correctly, and positions of free indices match on both sides:

$$T^{\alpha_1'\alpha_2'\cdots\alpha_n'}{}_{\beta_1'\beta_2'\cdots\beta_m'} = \frac{\partial x^{\alpha_1'}}{\partial x^{\alpha_1}}\frac{\partial x^{\alpha_2'}}{\partial x^{\alpha_2}}\cdots\frac{\partial x^{\alpha_n'}}{\partial x^{\alpha_n}}\frac{\partial x^{\beta_1}}{\partial x^{\beta_1'}}\frac{\partial x^{\beta_2}}{\partial x^{\beta_2'}}\cdots\frac{\partial x^{\beta_m}}{\partial x^{\beta_m'}}T^{\alpha_1\alpha_2\cdots\alpha_n}{}_{\beta_1\beta_2\cdots\beta_m} \,. \tag{8.119}$$

The tensor transformation law implies that summed-over index pairs do not contribute to a transformation of the object they appear in. For instance, if we have a vector a^α and a dual vector b_α, the product $b_\alpha a^\alpha$ transforms according to

$$b_{\alpha'}a^{\alpha'} = \left(\frac{\partial x^\alpha}{\partial x^{\alpha'}}b_\alpha\right)\left(\frac{\partial x^{\alpha'}}{\partial x^\beta}a^\beta\right) = \left(\frac{\partial x^\alpha}{\partial x^{\alpha'}}\frac{\partial x^{\alpha'}}{\partial x^\beta}\right)b_\alpha a^\beta = \delta^\alpha{}_\beta b_\alpha a^\beta = b_\alpha a^\alpha \,, \tag{8.120}$$

such that $b_\alpha a^\alpha$ is invariant. In this calculation, we used the chain rule for multiple independent variables:

$$\frac{\partial x^\alpha}{\partial x^{\alpha'}}\frac{\partial x^{\alpha'}}{\partial x^\beta} = \frac{\partial x^\alpha}{\partial x^\beta} = \delta^\alpha{}_\beta\,. \tag{8.121}$$

Only free, single indices contribute to the transformation behavior. For instance, $b_\alpha g^{\alpha\beta}$ transforms according to

$$\begin{aligned} b_{\alpha'}g^{\alpha'\beta'} &= \left(\frac{\partial x^\alpha}{\partial x^{\alpha'}}b_\alpha\right)\left(\frac{\partial x^{\alpha'}}{\partial x^\beta}\frac{\partial x^{\beta'}}{\partial x^\gamma}g^{\beta\gamma}\right) = \left(\frac{\partial x^\alpha}{\partial x^{\alpha'}}\frac{\partial x^{\alpha'}}{\partial x^\beta}\right)\frac{\partial x^{\beta'}}{\partial x^\gamma}b_\alpha g^{\beta\gamma} \\ &= \delta^\alpha{}_\beta\frac{\partial x^{\beta'}}{\partial x^\gamma}b_\alpha g^{\beta\gamma} = \frac{\partial x^{\beta'}}{\partial x^\gamma}b_\beta g^{\beta\gamma}\,, \end{aligned} \tag{8.122}$$

in agreement with the transformation law of a tensor with a single subscript.

Equation (8.120) allows us to identify a mathematical interpretation of a single-subscript object that is independent of lowering the superscript of a vector field. Given an object b_α that obeys the tensor transformation law,

$$b_{\alpha'} = \frac{\partial x^\alpha}{\partial x^{\alpha'}}b_\alpha\,, \tag{8.123}$$

we can define a linear map from any given vector field a^α to the function $f = b_\alpha a^\alpha$. The index is summed-over and therefore does not appear on f. If we change coordinates, we map $f(x)$ to $f(x(x'))$ without an additional factor from partial derivatives because of (8.120). The expression $f = b_\alpha a^\alpha$ is indeed a function because it is uniquely defined by its evaluation at any given point. A single-subscript tensor or a **dual vector field**, also called a **co-vector field**, therefore defines a linear map from the space of all vector fields to the space of all functions on space-time. This linear map is compatible with matrix multiplication, $f = \mathbf{ba}$, if we write vectors as columns and dual vectors as rows. We usually follow this convention whenever we specify vector components, although it is not required for abstract index notation.

We have seen dual vector fields on a different space: In quantum mechanics, the reflected version $\langle\phi|$ of an abstract state $|\phi\rangle$ is a linear map from vectors to functions because it defines the complex number $\langle\phi|\psi\rangle$ for any other state $|\psi\rangle$. It is sufficient to consider vectors instead of vector fields in this case because quantum mechanics is based on a linear space. When we constructed the **density operator** $|\psi\rangle\langle\psi|$ of a pure state $|\psi\rangle$, we placed the vector and its dual in the opposite order. Picturing states $|\psi\rangle$ as column vectors and their duals $\langle\psi|$ as row vectors then implies that we no longer sum over matched rows and columns as in the inner product $\langle\psi|\psi\rangle$ but only multiply individual components. In abstract index notation, we drop the summation by choosing different labels for the two indices: Instead of $b_\alpha a^\alpha$ in the function defined by applying the dual vector field b_α to the vector field a^α, we obtain the **tensor product** $b_\beta a^\alpha$ as a 2-index object.

8.6.5 GEODESICS

A straight line in flat space is characterized by a tangent vector that points in the same direction everywhere on the line. In curved space, it is difficult to compare the directions of vectors at different points because the tangent planes are not the same everywhere, as seen from an embedding space. Instead of using the tangent vector, we derive analogs of straight lines in curved space(-time), or **geodesics**, by finding extrema of their proper distance. In Euclidean space, a straight line minimizes the distance between two given points. This condition can be evaluated without comparing directions at different points, if we apply the line element to compute the length of a curve. On curved space or space-time, it is therefore meaningful to use this result as a definition of a geodesic. We have to be careful and relax the minimization condition to extremization because the twin paradox taught us that straight timelike worldlines in Minkowski space-time maximize the proper distance between two events.

In anticipation of applications in general relativity, we consider the case of timelike geodesics in curved space-time. The defining property of these timelike worldlines is that they have an extremum of the proper-time duration when compared with all other curves connecting two given events A and B. We therefore consider the function

$$\tau_{AB} = \int_A^B \mathrm{d}\tau = \int_A^B \sqrt{-g_{\alpha\beta}\mathrm{d}x^\alpha \mathrm{d}x^\beta} = \int_A^B \sqrt{-g_{\alpha\beta}\frac{\mathrm{d}x^\alpha}{\mathrm{d}\sigma}\frac{\mathrm{d}x^\beta}{\mathrm{d}\sigma}}\mathrm{d}\sigma \tag{8.124}$$

which implicitly depends on the curve $\mathbf{x}(\sigma)$ along which we integrate. Here, σ is a real parameter which could be proper time, but we should keep it more general for now: If we use only τ as a curve parameter, equation (8.124) takes the form $\tau = \int \mathrm{d}\tau$ and does not contain information about the position-dependence of worldlines.

8.6.5.1 Euler–Lagrange equations

The proper-time difference τ_{AB} is extremized by a curve $\mathbf{x}(\sigma)$ for which the Euler–Lagrange equations

$$-\frac{\mathrm{d}}{\mathrm{d}\sigma}\frac{\partial L}{\partial(\mathrm{d}x^\alpha/\mathrm{d}\sigma)} + \frac{\partial L}{\partial x^\alpha} = 0 \tag{8.125}$$

are satisfied, using the Lagrange function

$$L(x^\alpha, \mathrm{d}x^\delta/\mathrm{d}\sigma) = \sqrt{-g_{\beta\gamma}(x^\epsilon)\frac{\mathrm{d}x^\beta}{\mathrm{d}\sigma}\frac{\mathrm{d}x^\gamma}{\mathrm{d}\sigma}} \tag{8.126}$$

from (8.124). We relabeled indices in order to avoid multiple occurrences of the index α used in (8.125). As in classical mechanics, page 287, we consider the space-time position x^α and the derivatives $\mathrm{d}x^\beta/\mathrm{d}\sigma$ as independent variables because we can independently change a point on the curve and the tangent vector. Here, the independent functions include those with an index $\alpha = 0$ or $\beta = 0$ because $x^0(\sigma)$ is a free function along a generic worldline.

Since coordinates and velocity components are independent, we have the partial derivatives

$$\frac{\partial x^\alpha}{\partial x^\beta} = \delta^\alpha{}_\beta \quad , \quad \frac{\partial(\mathrm{d}x^\alpha/\mathrm{d}\sigma)}{\partial(\mathrm{d}x^\beta/\mathrm{d}\sigma)} = \delta^\alpha{}_\beta \tag{8.127}$$

with the Kronecker delta $\delta^\alpha{}_\beta$: If $\alpha = \beta$, the numerator and denominator are equal in both derivatives, and the result is equal to one. If $\alpha \neq \beta$, we take the partial derivative of one variable by another independent one, and the result is zero. All mixed partial derivatives of the form

$$\frac{\partial(\mathrm{d}x^\alpha/\mathrm{d}\sigma)}{\partial x^\beta} = 0 \quad , \quad \frac{\partial x^\alpha}{\partial(\mathrm{d}x^\beta/\mathrm{d}\sigma)} = 0 \tag{8.128}$$

vanish because of the independence of position functions and their derivatives.

For our Lagrange function, we use the chain and product rules to obtain

$$\frac{\partial L}{\partial(\mathrm{d}x^\alpha/\mathrm{d}\sigma)} = -\frac{1}{2L}g_{\beta\gamma}\left(\delta^\beta{}_\alpha\frac{\mathrm{d}x^\gamma}{\mathrm{d}\sigma} + \frac{\mathrm{d}x^\beta}{\mathrm{d}\sigma}\delta^\gamma{}_\alpha\right) = -\frac{1}{L}g_{\alpha\beta}\frac{\mathrm{d}x^\beta}{\mathrm{d}\sigma}. \tag{8.129}$$

As seen before, we compute the sums over repeated indices β and γ, respectively, by noting that $\delta^\beta{}_\alpha$ is non-zero only if $\beta = \alpha$, and $\delta^\gamma{}_\alpha$ is non-zero only if $\gamma = \alpha$. Therefore, we simply drop the Kronecker deltas and set $\beta = \alpha$ in the first term and $\gamma = \alpha$ in the second. Both terms can then be combined by using the symmetry of $g_{\alpha\beta}$ and the fact that we are free to rename summed-over indices.

With this result, we obtain the Euler–Lagrange equation

$$\frac{\mathrm{d}}{\mathrm{d}\sigma}\left(\frac{1}{L}g_{\alpha\beta}\frac{\mathrm{d}x^\beta}{\mathrm{d}\sigma}\right)-\frac{1}{2L}\frac{\partial g_{\beta\gamma}}{\partial x^\alpha}\frac{\mathrm{d}x^\beta}{\mathrm{d}\sigma}\frac{\mathrm{d}x^\gamma}{\mathrm{d}\sigma}=0. \tag{8.130}$$

Solutions of this equation with suitable boundary conditions determine curves of extremal distance between two points. There is no unique solution because the generic parameterization by σ is not unique and can be changed without affecting the length of the curve. A specific parameterization is expressed by the function $L=\mathrm{d}\tau/\mathrm{d}\sigma$ along the curve, which equals the Lagrange function. Fixing the parameterization means choosing a function for $L(\sigma)$. We can exploit this freedom by simplifying equation (8.130), assuming a constant L, or $L=1$ since any constant value can be canceled out. Thus, we now assume a parameterization by proper time, without loss of generality because the position information is contained in the resulting equation

$$\frac{\mathrm{d}}{\mathrm{d}\tau}\left(g_{\alpha\beta}\frac{\mathrm{d}x^\beta}{\mathrm{d}\tau}\right)-\frac{1}{2}\frac{\partial g_{\beta\gamma}}{\partial x^\alpha}\frac{\mathrm{d}x^\beta}{\mathrm{d}\tau}\frac{\mathrm{d}x^\gamma}{\mathrm{d}\tau}=g_{\alpha\beta}\frac{\mathrm{d}^2x^\beta}{\mathrm{d}\tau^2}+\frac{\partial g_{\alpha\beta}}{\partial x^\gamma}\frac{\mathrm{d}x^\gamma}{\mathrm{d}\tau}\frac{\mathrm{d}x^\beta}{\mathrm{d}\tau}-\frac{1}{2}\frac{\partial g_{\beta\gamma}}{\partial x^\alpha}\frac{\mathrm{d}x^\beta}{\mathrm{d}\tau}\frac{\mathrm{d}x^\gamma}{\mathrm{d}\tau}=0. \tag{8.131}$$

We used the chain rule to compute the derivative $\mathrm{d}g_{\alpha\beta}/\mathrm{d}\tau=(\partial g_{\alpha\beta}/\partial x^\gamma)(\mathrm{d}x^\gamma/\mathrm{d}\tau)$.

In this equation, we recognize an acceleration term, given by second derivatives of the position, and velocity terms with position-dependent coefficients. We combine the velocity terms by factoring out $(\mathrm{d}x^\beta/\mathrm{d}\tau)(\mathrm{d}x^\gamma/\mathrm{d}\tau)$. We have to be careful, however, because this product is symmetric under exchange of β and γ due to commutativity of components, while the coefficients, $\partial g_{\alpha\beta}/\partial x^\gamma$ are not symmetric. Since a symmetric 4×4 matrix has ten independent components, only a certain subset of the 16 components of the non-symmetric coefficient matrix, or ten specific linear combinations of the 16 components, appear in the equation. We derive the relevant ones by writing

$$\frac{\partial g_{\alpha\beta}}{\partial x^\gamma}\frac{\mathrm{d}x^\gamma}{\mathrm{d}\tau}\frac{\mathrm{d}x^\beta}{\mathrm{d}\tau}=\frac{1}{2}\left(\frac{\partial g_{\alpha\beta}}{\partial x^\gamma}\frac{\mathrm{d}x^\gamma}{\mathrm{d}\tau}\frac{\mathrm{d}x^\beta}{\mathrm{d}\tau}+\frac{\partial g_{\alpha\beta}}{\partial x^\gamma}\frac{\mathrm{d}x^\beta}{\mathrm{d}\tau}\frac{\mathrm{d}x^\gamma}{\mathrm{d}\tau}\right)$$

$$=\frac{1}{2}\left(\frac{\partial g_{\alpha\gamma}}{\partial x^\beta}\frac{\mathrm{d}x^\beta}{\mathrm{d}\tau}\frac{\mathrm{d}x^\gamma}{\mathrm{d}\tau}+\frac{\partial g_{\alpha\beta}}{\partial x^\gamma}\frac{\mathrm{d}x^\beta}{\mathrm{d}\tau}\frac{\mathrm{d}x^\gamma}{\mathrm{d}\tau}\right)=\frac{1}{2}\left(\frac{\partial g_{\alpha\gamma}}{\partial x^\beta}+\frac{\partial g_{\alpha\beta}}{\partial x^\gamma}\right)\frac{\mathrm{d}x^\beta}{\mathrm{d}\tau}\frac{\mathrm{d}x^\gamma}{\mathrm{d}\tau}. \tag{8.132}$$

We first used symmetry of the component product, then relabeled $\beta\leftrightarrow\gamma$ in the second term, and finally factored out the velocity components. The new coefficient matrix is the symmetrization of the previous one, which has the same number of independent components as the symmetric product $(\mathrm{d}x^\beta/\mathrm{d}\tau)(\mathrm{d}x^\gamma/\mathrm{d}\tau)$.

Factoring out velocity components in all terms, we obtain the second-order differential equations

$$g_{\alpha\beta}\frac{\mathrm{d}^2x^\beta}{\mathrm{d}\tau^2}=-\frac{1}{2}\left(\frac{\partial g_{\alpha\beta}}{\partial x^\gamma}+\frac{\partial g_{\alpha\gamma}}{\partial x^\beta}-\frac{\partial g_{\beta\gamma}}{\partial x^\alpha}\right)\frac{\mathrm{d}x^\beta}{\mathrm{d}\tau}\frac{\mathrm{d}x^\gamma}{\mathrm{d}\tau}. \tag{8.133}$$

This result determines linear equations for the acceleration components $\mathrm{d}^2x^\beta/\mathrm{d}\tau^2$. We rewrite them directly for the components by multiplying with the inverse of the coefficient matrix, $g_{\alpha\beta}$, on the left. We introduced the notation $g^{\delta\alpha}$ for the inverse of the metric, such that $g^{\delta\alpha}g_{\alpha\beta}=\delta^\delta{}_\beta$. Therefore,

$$\frac{\mathrm{d}^2x^\delta}{\mathrm{d}\tau^2}=\delta^\delta{}_\beta\frac{\mathrm{d}^2x^\beta}{\mathrm{d}\tau^2}=g^{\delta\alpha}g_{\alpha\beta}\frac{\mathrm{d}^2x^\beta}{\mathrm{d}\tau^2}=-\frac{1}{2}g^{\delta\alpha}\left(\frac{\partial g_{\alpha\beta}}{\partial x^\gamma}+\frac{\partial g_{\alpha\gamma}}{\partial x^\beta}-\frac{\partial g_{\beta\gamma}}{\partial x^\alpha}\right)\frac{\mathrm{d}x^\beta}{\mathrm{d}\tau}\frac{\mathrm{d}x^\gamma}{\mathrm{d}\tau}. \tag{8.134}$$

8.6.5.2 Christoffel symbol

The position-dependent coefficients on the right-hand side of this equation,

$$\Gamma^\delta_{\beta\gamma}=\frac{1}{2}g^{\delta\alpha}\left(\frac{\partial g_{\alpha\beta}}{\partial x^\gamma}+\frac{\partial g_{\alpha\gamma}}{\partial x^\beta}-\frac{\partial g_{\beta\gamma}}{\partial x^\alpha}\right), \tag{8.135}$$

are referred to as the **Christoffel symbol**. With this definition, the **geodesic equation** takes the compact form

$$\frac{\mathrm{d}^2x^\delta}{\mathrm{d}\tau^2} = -\Gamma^\delta_{\beta\gamma}\frac{\mathrm{d}x^\beta}{\mathrm{d}\tau}\frac{\mathrm{d}x^\gamma}{\mathrm{d}\tau} \tag{8.136}$$

in the timelike case. We use the same form of the geodesic equation for spacelike worldlines after replacing proper time τ with proper length, $\mathrm{d}s = \sqrt{\mathrm{d}s^2}$.

For lightlike geodesics, our derivation cannot be used in the same way because $L = 0$ for all lightlike worldlines: There is no unique extremizing lightlike worldline between two given events as shown for Minkowski space-time in Figure 7.11. Moreover, lightlike worldlines have to obey another condition, $\mathrm{d}s^2 = 0$, which can be turned into a differential equation for the position vector if we divide by $\mathrm{d}\sigma^2$ for some curve parameter σ:

$$\frac{\mathrm{d}s^2}{\mathrm{d}\sigma^2} = g_{\alpha\beta}\frac{\mathrm{d}x^\alpha}{\mathrm{d}\sigma}\frac{\mathrm{d}x^\beta}{\mathrm{d}\sigma} = 0\,. \tag{8.137}$$

In general, this single equation does not imply unique solutions for light rays because we have to find four functions $x^\alpha(\sigma)$ with suitable initial values. In our 2-dimensional models, it was sufficient to use just this one equation because we only had to derive one function, such as $t(x)$, in order to specify a light ray.

If we combine (8.137) with all components of the geodesic equation, treating lightlike worldlines as limiting cases of timelike and spacelike ones, we obtain five differential equations for four functions. Such a set of equations could be over-determined, but it is possible to evade this problem by increasing the number of free functions: We state that we are looking for lightlike geodesics in a specific kind of parameterizations. Thus, in addition to the four position functions, we are also solving for a function $\lambda(\sigma)$ that determines a preferred class of parameters λ depending on the generic parameter σ. If

$$\frac{\mathrm{d}^2x^\delta}{\mathrm{d}\lambda^2} = -\Gamma^\delta_{\beta\gamma}\frac{\mathrm{d}x^\beta}{\mathrm{d}\lambda}\frac{\mathrm{d}x^\gamma}{\mathrm{d}\lambda} \tag{8.138}$$

holds in addition to $\mathrm{d}s^2/\mathrm{d}\lambda^2 = 0$, we refer to λ as an **affine parameter**. This definition generalizes the affine parameter introduced in special relativity, equation (7.119), where we did not have to solve for light rays because they were always along the diagonal. As in this case, the affine parameterization is not unique but can be changed by multiplying λ with a non-zero constant because both sides of (8.138) are quadratic in λ. For light rays, it is sometimes convenient to write

$$\frac{\mathrm{d}k^\delta}{\mathrm{d}\lambda} = -\Gamma^\delta_{\beta\gamma}k^\beta k^\gamma \tag{8.139}$$

if we are interested in changes of the wave 4-vector $\mathbf{k}$, which also obeys the dispersion relation $||\mathbf{k}||^2 = g_{\alpha\beta}k^\alpha k^\beta = 0$.

The coefficients $\Gamma^\delta_{\beta\gamma}$ do *not* form a tensor because a coordinate transformation, which changes $g_{\alpha\beta}$ to

$$g_{\alpha'\beta'} = \frac{\partial x^\alpha}{\partial x^{\alpha'}}\frac{\partial x^\beta}{\partial x^{\beta'}}g_{\alpha\beta}\,, \tag{8.140}$$

leads to factors with second-order derivatives of the coordinates when inserted in (8.135). Such factors are not appropriate for the transformation of a tensor with three indices, which should rather be of the form

$$T^{\delta'}_{\beta'\gamma'} = \frac{\partial x^\beta}{\partial x^{\beta'}}\frac{\partial x^\gamma}{\partial x^{\gamma'}}\frac{\partial x^{\delta'}}{\partial x^\delta}T^\delta_{\beta\gamma}\,. \tag{8.141}$$

This is the reason why we call $\Gamma^\delta_{\beta\gamma}$ the Christoffel *symbol*. In spite of their non-standard form, transformation properties of $\Gamma^\delta_{\beta\gamma}$ are important because they can be used as a guide toward the construction of curvature tensors, as we will see.

8.6.5.3 Examples

A simple example is given by Minkowski space-time in Cartesian coordinates, with $g_{\alpha\beta} = \eta_{\alpha\beta}$ constant. All partial derivatives of the metric are zero in this case, and therefore $\Gamma^{\gamma}_{\alpha\beta} = 0$. The geodesic equation is $\mathrm{d}^2x^{\alpha}/\mathrm{d}\tau^2 = 0$, solved by linear functions $x^{\alpha}(\tau)$.

An example with some non-zero components of the Christoffel symbol is the 2-plane in polar coordinates, with line element $\mathrm{d}s^2 = \mathrm{d}r^2 + r^2\mathrm{d}\varphi^2$. We derive the geodesic equation in two different ways, first by specializing the Euler–Lagrange equations to the specific metric, and then by a direct computation of the Christoffel symbol.

The Lagrange function for spacelike geodesics in the 2-plane with polar coordinates is

$$L(r,\varphi,\mathrm{d}r/\mathrm{d}\sigma,\mathrm{d}\varphi/\mathrm{d}\sigma) = \sqrt{\left(\frac{\mathrm{d}r}{\mathrm{d}\sigma}\right)^2 + r^2\left(\frac{\mathrm{d}\varphi}{\mathrm{d}\sigma}\right)^2}\,. \tag{8.142}$$

In this 2-dimensional space, we have two Euler–Lagrange equations, one for $x^{\alpha} = r$ and one for $x^{\alpha} = \varphi$. The first case gives

$$\frac{\mathrm{d}}{\mathrm{d}\sigma}\frac{\partial L}{\partial(\mathrm{d}r/\mathrm{d}\sigma)} = \frac{\mathrm{d}}{\mathrm{d}\sigma}\left(\frac{1}{L}\frac{\mathrm{d}r}{\mathrm{d}\sigma}\right) = \frac{\partial L}{\partial r} = \frac{r}{L}\left(\frac{\mathrm{d}\varphi}{\mathrm{d}\sigma}\right)^2\,, \tag{8.143}$$

and we have

$$\frac{\mathrm{d}}{\mathrm{d}\sigma}\frac{\partial L}{\partial(\mathrm{d}\varphi/\mathrm{d}\sigma)} = \frac{\mathrm{d}}{\mathrm{d}\sigma}\left(\frac{1}{L}r^2\frac{\mathrm{d}\varphi}{\mathrm{d}\sigma}\right) = \frac{\partial L}{\partial\varphi} = 0 \tag{8.144}$$

in the second case. We eliminate L, as in the general derivation, by using the proper length $\mathrm{d}s = L\mathrm{d}\sigma$ for our spacelike curves instead of a general curve parameter σ. The two components of the geodesic equation then are

$$\frac{\mathrm{d}^2r}{\mathrm{d}s^2} = r\left(\frac{\mathrm{d}\varphi}{\mathrm{d}s}\right)^2 \quad , \quad \frac{\mathrm{d}}{\mathrm{d}s}\left(r^2\frac{\mathrm{d}\varphi}{\mathrm{d}s}\right) = 0\,. \tag{8.145}$$

If we have motion in the 2-plane with the proper length s proportional to time, the first equation is Newton's equation of motion subject to a centrifugal force, and the second equation is conservation of angular momentum, both equations divided by the mass.

We now rederive these equations by a direct computation of the Christoffel symbol

$$\Gamma^{\delta}_{\beta\gamma} = \frac{1}{2}g^{\delta\alpha}\left(\frac{\partial g_{\alpha\beta}}{\partial x^{\gamma}} + \frac{\partial g_{\alpha\gamma}}{\partial x^{\beta}} - \frac{\partial g_{\beta\gamma}}{\partial x^{\alpha}}\right)\,. \tag{8.146}$$

In 2-dimensional space(-time), there are eight combinations of the three indices β, γ and δ, but not all eight components $\Gamma^{\delta}_{\beta\gamma}$ are independent: Some of them are related by the symmetry $\Gamma^{\delta}_{\beta\gamma} = \Gamma^{\delta}_{\gamma\beta}$ that follows directly from (8.146). In order to compute the Christoffel symbol, we have to go through all six remaining index combinations after taking into account the symmetry. For a specific combination, the index values of β, γ, and δ are fixed on the left-hand side of the equation and have to take the same values on the other side. The index α on the right remains unspecified and paired up in terms of subscripts and superscripts, which means that it is summed over all values it can take (two in this 2-dimensional example). But for a diagonal metric the inverse $g^{\delta\alpha}$ is non-zero only if $\alpha = \delta$. The sum over α then collapses to a single term.

We first write down the components of the metric and its inverse:

$$g_{rr} = 1 \quad , \quad g_{r\varphi} = 0 \quad , \quad g_{\varphi\varphi} = r^2 \tag{8.147}$$

and

$$g^{rr} = 1 \quad , \quad g^{r\varphi} = 0 \quad , \quad g^{\varphi\varphi} = \frac{1}{r^2}\,. \tag{8.148}$$

We then list the Christoffel components with specified indices α, β and γ, starting with

$$\Gamma^r_{\varphi\varphi} = \frac{1}{2}\sum_{\alpha=r,\varphi} g^{r\alpha}\left(\frac{\partial g_{\alpha\varphi}}{\partial\varphi} + \frac{\partial g_{\alpha\varphi}}{\partial\varphi} - \frac{\partial g_{\varphi\varphi}}{\partial x^\alpha}\right). \tag{8.149}$$

Here, the remaining sum is written explicitly and the index choices for β, γ and δ made on the left have been inserted on the right. If we write out the summed-over terms, we use the components of the metric and its inverse:

$$\Gamma^r_{\varphi\varphi} = \frac{1}{2}g^{rr}\left(2\frac{\partial g_{r\varphi}}{\partial\varphi} - \frac{\partial g_{\varphi\varphi}}{\partial r}\right) + \frac{1}{2}g^{r\varphi}\left(2\frac{g_{\varphi\varphi}}{\partial\varphi} - \frac{\partial g_{\varphi\varphi}}{\partial\varphi}\right) = -r. \tag{8.150}$$

Only one term is non-zero, given by $-\frac{1}{2}g^{rr}\partial g_{\varphi\varphi}/\partial r = -r$, and we have found our first non-zero Christoffel component.

The next index choice is

$$\Gamma^\varphi_{r\varphi} = \frac{1}{2}g^{\varphi\alpha}\left(\frac{\partial g_{\alpha\varphi}}{\partial r} + \frac{\partial g_{\alpha r}}{\partial\varphi} - \frac{\partial g_{\varphi r}}{\partial x^\alpha}\right) = \frac{1}{2}g^{\varphi\varphi}\left(\frac{\partial g_{\varphi\varphi}}{\partial r} + \frac{\partial g_{\varphi r}}{\partial\varphi} - \frac{\partial g_{\varphi r}}{\partial\varphi}\right) = \frac{1}{2}g^{\varphi\varphi}\frac{\partial g_{\varphi\varphi}}{\partial r} = \frac{1}{r}. \tag{8.151}$$

By symmetry, we obtain the third Christoffel symbol $\Gamma^\varphi_{\varphi r} = 1/r$. All remaining cases turn out to vanish. This outcome can easily be seen by a general argument, again using the fact that the metric is diagonal: For a non-zero component $\Gamma^\alpha_{\beta\gamma}$ with specific index values, at least one partial derivative obtained by a permutation of (α,β,γ) in $\partial g_{\alpha\beta}/\partial x^\gamma$ must be non-zero. For the planar line element, the only non-zero partial derivative of the metric is given by $\partial g_{\varphi\varphi}/\partial r$. Since the metric is diagonal, the set of index values $\{\varphi,\varphi,r\}$ in this expression must equal the set $\{\alpha,\beta,\gamma\}$ for a given non-zero component that we are looking for. The three non-zero components we considered already exhaust all permutations of (φ,φ,r), and therefore all other choices imply a vanishing Christoffel component. An argument of this form can be used whenever the metric is diagonal and has symmetry, such that the dependence of metric components on the coordinates is restricted. Such considerations are particularly useful in higher dimensions with a large initial number of Christoffel components.

Notice that there are some non-zero components of the Christoffel symbol in the planar example, even though the geometry is flat. Had we used Cartesian coordinates instead of polar coordinates, all Christoffel components would have been zero. This contrast is possible only because the Christoffel symbol does not form a tensor, for any tensor with components $T^\delta_{\beta\gamma} = 0$ all zero in one coordinate system would have all components equal to zero in any other coordinate system as a consequence of (8.141).

The geodesic equation is now determined by inserting the Christoffel symbol in

$$\frac{\mathrm{d}^2x^\delta}{\mathrm{d}s^2} = -\Gamma^\delta_{\beta\gamma}\frac{\mathrm{d}x^\beta}{\mathrm{d}s}\frac{\mathrm{d}x^\gamma}{\mathrm{d}s}. \tag{8.152}$$

Again, we list the two equations corresponding to the two possible choices for the free index δ, and in each case evaluate the sums over β and γ. For $x^\delta = r$, we have

$$\frac{\mathrm{d}^2r}{\mathrm{d}s^2} = -\Gamma^r_{\beta\gamma}\frac{\mathrm{d}x^\beta}{\mathrm{d}s}\frac{\mathrm{d}x^\gamma}{\mathrm{d}s} = -\left(\Gamma^r_{rr}\left(\frac{\mathrm{d}r}{\mathrm{d}s}\right)^2 + 2\Gamma^r_{r\varphi}\frac{\mathrm{d}r}{\mathrm{d}s}\frac{\mathrm{d}\varphi}{\mathrm{d}s} + \Gamma^r_{\varphi\varphi}\left(\frac{\mathrm{d}\varphi}{\mathrm{d}s}\right)^2\right) = r\left(\frac{\mathrm{d}\varphi}{\mathrm{d}s}\right)^2. \tag{8.153}$$

In the final step, we use our results from the computation of the Christoffel components, among which $\Gamma^r_{\varphi\varphi}$ was the only non-zero one with superscript r. The same results show that $\Gamma^\varphi_{r\varphi} = \Gamma^\varphi_{\varphi r}$ are the only non-zero Christoffel components with superscript φ. They appear in the second component of the geodesic equation, obtained for $x^\delta = \varphi$:

$$\frac{\mathrm{d}^2\varphi}{\mathrm{d}s^2} = -\Gamma^\varphi_{\beta\gamma}\frac{\mathrm{d}x^\beta}{\mathrm{d}s}\frac{\mathrm{d}x^\gamma}{\mathrm{d}s} = -2\Gamma^\varphi_{r\varphi}\frac{\mathrm{d}r}{\mathrm{d}s}\frac{\mathrm{d}\varphi}{\mathrm{d}s} = -\frac{2}{r}\frac{\mathrm{d}r}{\mathrm{d}s}\frac{\mathrm{d}\varphi}{\mathrm{d}s}. \tag{8.154}$$

This equation is equivalent to the conservation law in (8.145), although it appears here in a different form that does not make conservation obvious. For geodesics, there is an independent way of deriving conservation laws, based on symmetry of the geometry.

8.6.5.4 Solving the geodesic equation

The components of the geodesic equation are coupled non-linear second-order ordinary differential equations, as many as there are dimensions in space(-time). It is difficult to solve such equations, but in some cases conservation laws help. For instance, if we recognize that equation (8.154) can be written in the form (8.145), we can replace it with the first-order differential equation $r^2\mathrm{d}\varphi/\mathrm{d}s = \ell$, where the constant ℓ is interpreted as angular momentum per mass. A first-order equation is usually easier to solve than the original second-order differential equation. Whenever it is possible to replace a second-order differential equation with a first-order one, we call the latter a **first integral** of the former.

As the example suggests, first integrals are often implied by symmetry, in this case by the φ-independence of the line element, or rotational symmetry in the 2-plane. There is another first integral of the geodesic equation, which is not related to a symmetry of the line element. In the example, it is given by the identity

$$1 = \frac{\mathrm{d}s^2}{\mathrm{d}s^2} = g_{\alpha\beta}\frac{\mathrm{d}x^\alpha}{\mathrm{d}s}\frac{\mathrm{d}x^\beta}{\mathrm{d}s} = \left(\frac{\mathrm{d}r}{\mathrm{d}s}\right)^2 + r^2\left(\frac{\mathrm{d}\varphi}{\mathrm{d}s}\right)^2 . \tag{8.155}$$

It is a first integral because it depends only on first-order derivatives of r and φ as well as the function $r(s)$, and it holds for any solution of the geodesic equation.

We now have two first integrals in this 2-dimensional system, which allows us replace all second-order equations and to obtain a complete solution as we will see soon. For geodesics in space-time, the general first integral takes slightly different forms depending on whether we have timelike geodesics for which

$$-1 = \frac{\mathrm{d}s^2}{\mathrm{d}\tau^2} = g_{\alpha\beta}\frac{\mathrm{d}x^\alpha}{\mathrm{d}\tau}\frac{\mathrm{d}x^\beta}{\mathrm{d}\tau}, \tag{8.156}$$

lightlike geodesics for which

$$0 = \frac{\mathrm{d}s^2}{\mathrm{d}\lambda^2} = g_{\alpha\beta}\frac{\mathrm{d}x^\alpha}{\mathrm{d}\lambda}\frac{\mathrm{d}x^\beta}{\mathrm{d}\lambda}, \tag{8.157}$$

or spacelike geodesics for which we can use a 4-dimensional version of (8.155).

The relationship between symmetry and first integrals can be seen by returning to the Euler–Lagrange equations from which we obtained the geodesic equation. To be specific, we assume that the metric $g_{\alpha\beta}$ is independent of one of the coordinates, x^1, and therefore has a translation or rotation symmetry in this direction. The Lagrange function for geodesics, $L = \sqrt{-g_{\alpha\beta}(\mathrm{d}x^\alpha/\mathrm{d}\sigma)(\mathrm{d}x^\beta/\mathrm{d}\sigma)}$ in the timelike case, then obeys $\partial L/\partial x^1 = 0$, and the corresponding Euler–Lagrange equation

$$-\frac{\mathrm{d}}{\mathrm{d}\sigma}\frac{\partial L}{\partial(\mathrm{d}x^1/\mathrm{d}\sigma)} + \frac{\partial L}{\partial x^1} = 0 \tag{8.158}$$

immediately tells us that $\partial L/\partial(\mathrm{d}x^1/\mathrm{d}\sigma)$ is conserved along the geodesic.

Our general derivation of the geodesic equations on page 313 showed that

$$\frac{\partial L}{\partial(\mathrm{d}x^1/\mathrm{d}\sigma)} = -\frac{1}{L}g_{1\beta}\frac{\mathrm{d}x^\beta}{\mathrm{d}\sigma} = -g_{1\beta}u^\beta , \tag{8.159}$$

using the 4-velocity components $u^\beta = \mathrm{d}x^\beta/\mathrm{d}\tau = L^{-1}\mathrm{d}x^\beta/\mathrm{d}\sigma$. We can write this expression more compactly if we introduce the vector $\xi = (0,1,0,0)^T$ which points in the coordinate direction of the symmetry (x^1) and has constant components. The conserved quantity is then equal to the scalar product $\partial L/\partial(\mathrm{d}x^1/\mathrm{d}\sigma) = -\xi \cdot \mathbf{u}$. Such a vector ξ is called a **Killing vector**.

Back to our example of the 2-plane in polar coordinates, with line element $\mathrm{d}s^2 = \mathrm{d}r^2 + r^2\mathrm{d}\varphi^2$, we have the general first integral

$$\frac{\mathrm{d}s^2}{\mathrm{d}s^2} = g_{\alpha\beta}\frac{\mathrm{d}x^\alpha}{\mathrm{d}s}\frac{\mathrm{d}x^\beta}{\mathrm{d}s} = \left(\frac{\mathrm{d}r}{\mathrm{d}s}\right)^2 + r^2\left(\frac{\mathrm{d}\varphi}{\mathrm{d}s}\right)^2 = 1 \tag{8.160}$$

and, for $\xi = (\xi^r, \xi^\varphi)^T = (0,1)^T$ a Killing vector in the φ-direction, the conserved quantity

$$\xi \cdot \mathbf{u} = g_{\alpha\beta}\xi^\alpha u^\beta = g_{rr}\xi^r u^r + g_{r\varphi}(\xi^r u^\varphi + \xi^\varphi u^r) + g_{\varphi\varphi}\xi^\varphi u^\varphi = g_{\varphi\varphi}u^\varphi = r^2\frac{\mathrm{d}\varphi}{\mathrm{d}s}. \tag{8.161}$$

We have used the components $\xi^r = 0$ and $\xi^\varphi = 1$ of ξ, as well as the φ-component $u^\varphi = \mathrm{d}\varphi/\mathrm{d}s$ of the unit tangent vector $\mathbf{u} = (u^r, u^\varphi)^T = \mathrm{d}\mathbf{x}/\mathrm{d}s = (\mathrm{d}r/\mathrm{d}s, \mathrm{d}\varphi/\mathrm{d}s)^T$ in polar coordinates. Note that the Killing vector is expressed in the coordinate basis, and is not normalized according to the planar line element in polar coordinates.

The conserved quantity, as found before, is related to angular momentum (per mass). We set

$$r^2\frac{\mathrm{d}\varphi}{\mathrm{d}s} = \ell \tag{8.162}$$

equal to some constant, solve algebraically for $\mathrm{d}\varphi/\mathrm{d}s = \ell/r^2$, and insert this expression in (8.160):

$$1 = \left(\frac{\mathrm{d}r}{\mathrm{d}s}\right)^2 + r^2\left(\frac{\mathrm{d}\varphi}{\mathrm{d}s}\right)^2 = \left(\frac{\mathrm{d}r}{\mathrm{d}s}\right)^2 + \frac{\ell^2}{r^2} \tag{8.163}$$

which implies

$$\frac{\mathrm{d}r}{\mathrm{d}s} = \sqrt{1 - \frac{\ell^2}{r^2}}. \tag{8.164}$$

Combining the first integrals has allowed us to decouple the differential equation for $r(s)$ from $\varphi(s)$.

We could solve (8.164) by integrating the square root, but it turns out to be more convenient to derive an equation for $\varphi(r)$ by using both first integrals and the chain rule:

$$\frac{\mathrm{d}\varphi}{\mathrm{d}r} = \frac{\mathrm{d}\varphi/\mathrm{d}s}{\mathrm{d}r/\mathrm{d}s} = \frac{\ell/r^2}{\sqrt{1-\ell^2/r^2}}. \tag{8.165}$$

In this way, we eliminate the dependence on a parameterization. The fraction on the right-hand side can be integrated after using the substitution $z = \ell/\tilde{r}$:

$$\varphi(r) = \int^r \frac{\ell/\tilde{r}^2}{\sqrt{1-\ell^2/\tilde{r}^2}}\mathrm{d}\tilde{r} = -\int^{\ell/r}\frac{\mathrm{d}z}{\sqrt{1-z^2}} = \arccos(\ell/r) + \varphi_0. \tag{8.166}$$

This solution implies that $r\cos(\varphi - \varphi_0) = \ell$ along geodesics on the plane. We confirm that this is the correct expression of a straight line in polar coordinates by writing

$$r\cos(\varphi-\varphi_0) = r(\cos(\varphi)\cos(\varphi_0) + \sin(\varphi)\sin(\varphi_0)) = x\cos(\varphi_0) + y\sin(\varphi_0). \tag{8.167}$$

A linear combination of the Cartesian coordinates is therefore constant along a geodesic, which is the familiar description of a straight line. A specific solution given in terms of our parameters φ_0 and ℓ is normal to the unit vector $(\cos(\varphi_0), \sin(\varphi_0))^T$ and intersects the y-axis at $\ell/\sin(\varphi_0)$ unless $\varphi_0 = 0$ or $\varphi_0 = \pi$, in which case the line is parallel to the y-axis at $x = \ell/\cos(\varphi_0)$.

8.6.5.5 Cosmological redshift

As a space-time example, consider a lightlike geodesic in an expanding FRW universe, representing a light ray emitted from a distant star to us. In a local inertial frame, the emitted frequency equals $\omega_* = -\mathbf{k}\cdot\mathbf{u}_*$ with the 4-velocity $\mathbf{u}_*$ of an emitting atom on the star surface, and $\mathbf{k}$ the wave 4-vector along the lightlike geodesic. Once the light ray reaches our telescope, we measure the frequency $\omega = -\mathbf{k}\cdot\mathbf{u}$, with our 4-velocity $\mathbf{u}$. We ignore all relative velocities in order to focus on the expansion of space. Thus, both $\mathbf{u}_*$ and $\mathbf{u}$ are normalized vectors along the coordinate lines of proper time.

The FRW line element has Killing vectors $\mathbf{e}_x$, $\mathbf{e}_y$ and $\mathbf{e}_z$ in the coordinate basis. Without loss of generality, we may assume that the light ray is moving in the x-direction. The quantity $\mathbf{k}\cdot\mathbf{e}_x = k^x$, the x-component of the wave 4-vector, is therefore conserved. A lightlike vector moving in the x-direction has equal τ and x-components in an orthonormal local inertial frame: $-(k^{\hat{\tau}})^2 + (k^{\hat{x}})^2 = 0$ implies $|k^{\hat{\tau}}| = |k^{\hat{x}}|$. We have $k^{\hat{\tau}} = -\mathbf{k}\cdot\mathbf{u} = \omega$ because the 4-velocity is normalized. The spatial component in an orthonormal frame is related to the conserved $k^x = \mathbf{k}\cdot\mathbf{e}_x$ in the coordinate basis by

$$k^{\hat{x}} = \mathbf{k}\cdot\mathbf{e}_{\hat{x}} = \frac{\mathbf{k}\cdot\mathbf{e}_x}{a} = \frac{k^x}{a}. \tag{8.168}$$

These equations may be evaluated on the star and at our telescope. Therefore,

$$\omega_* = k_*^{\hat{\tau}} = |k_*^{\hat{x}}| = \frac{|k^x|}{a_*} = \frac{a|k^{\hat{x}}|}{a_*} = k^{\hat{\tau}}\frac{a}{a_*} = \frac{a}{a_*}\omega \tag{8.169}$$

with the scale factor a at the time of detection.

Redshift is measured in terms of wave lengths,

$$z = \frac{\lambda - \lambda_*}{\lambda_*} \tag{8.170}$$

as a shift relative to the wave lengths λ_* known from quantum mechanics or lab spectroscopy. Since wave lengths are inversely proportional to frequencies, we write

$$z = \frac{\omega_*}{\omega} - 1 = \frac{a}{a_*} - 1. \tag{8.171}$$

For astronomical distances much shorter than cosmological scales, the scale factor does not change much and we can use a Taylor expansion:

$$a \approx a_* + (\tau - \tau_*)\dot{a} \approx a + R\dot{a} \tag{8.172}$$

using the distance R and the speed of light, $c = 1$ in our units. Thus,

$$z \approx \frac{R\dot{a}}{a} = HR. \tag{8.173}$$

Comparing with Hubble's law (8.43), **cosmological redshift** is a direct measure of the escape velocity of distant stars.

8.6.6 SYMMETRY

Symmetry of a metric tensor is not always made obvious by the independence of its components on some of the coordinates. For instance, the 2-sphere should be invariant under rotations around any axis, but in standard spherical coordinates only rotations around the z-axis are an obvious symmetry, obtained from the independence of metric components on φ. Moreover, a symmetry is a coordinate-independent property of the metric because the geometry does not change in the corresponding direction. There should be a coordinate-independent definition of a symmetry, or of a Killing vector field.

Symmetry is formally related to coordinate transformations because we compare metric components at various position coordinates. One difference between a coordinate transformation and a symmetry transformation is that the former considers the metric tensor at a given point, where it may have different components depending on which coordinates we choose, while the latter maps points and compares the geometry around them. Formally, these notions are related because we may view the functions $x^{\alpha'}(x^\alpha)$ of a coordinate transformation at a given point as an actual mapping from the point x^α to a new point $x^{\alpha'}$, expressed in one and the same coordinate system. This observation may seem obvious, but it has an important implication: Comparing metrics in the context of symmetry transformations should make use of the tensor transformation law because of its prominence in coordinate transformations of tensors, such as the metric.

It is sufficient to consider an infinitesimal transformation from x^α to $y^\alpha = x^\alpha - v^\alpha$ with a vector field v^α that determines the direction and a short distance in which we move at each point. If $g_{\gamma\delta}(x)$ are the metric components at x^α, the metric components $h_{\alpha\beta}(y)$ at y^α are, in general, different. There are two modifications: First, we evaluate the component functions at a different point, which can be evaluated in the infinitesimal case by using a Taylor expansion. Secondly, because the same position change has a correspondence with a coordinate transformation, there are factors of $\partial x^\gamma/\partial y^\alpha$ from the tensor transformation law. Combining these terms, we write

$$\begin{aligned} h_{\alpha\beta}(y) &= \frac{\partial x^\gamma}{\partial y^\alpha}\frac{\partial x^\delta}{\partial y^\beta} g_{\gamma\delta}(y+v) \approx \left(\delta^\gamma{}_\alpha + \frac{\partial v^\gamma}{\partial x^\alpha}\right)\left(\delta^\delta{}_\beta + \frac{\partial v^\delta}{\partial x^\beta}\right)\left(g_{\gamma\delta}(y) + v^\epsilon \frac{\partial g_{\gamma\delta}}{\partial x^\epsilon}\right) \\ &\approx g_{\alpha\beta}(y) + g_{\beta\gamma}\frac{\partial v^\gamma}{\partial x^\alpha} + g_{\alpha\delta}\frac{\partial v^\delta}{\partial x^\beta} + v^\epsilon \frac{\partial g_{\alpha\beta}}{\partial x^\epsilon}\,. \end{aligned} \tag{8.174}$$

The difference $h_{\alpha\beta}(y) - g_{\alpha\beta}(y)$ determines the relationship between components of the metric $g_{\alpha\beta}$ at two different points because $h_{\alpha\beta}$ has been obtained by moving the original $g_{\alpha\beta}$ into a new position. It is called the **Lie derivative**

$$\mathcal{L}_v g_{\alpha\beta} = v^\epsilon \frac{\partial g_{\alpha\beta}}{\partial x^\epsilon} + g_{\beta\gamma}\frac{\partial v^\gamma}{\partial x^\alpha} + g_{\alpha\delta}\frac{\partial v^\delta}{\partial x^\beta} \tag{8.175}$$

of $g_{\alpha\beta}$ in the direction v^γ.

For later applications, we note that the Lie derivative can be applied to any tensor field by using the same combination of the tensor-transformation law with a Taylor expansion. For the Lie derivative of a vector field, this procedure results in the vector-field **commutator**

$$\mathcal{L}_v w^\alpha = v^\beta \frac{\partial w^\alpha}{\partial x^\beta} - w^\beta \frac{\partial v^\alpha}{\partial x^\beta} = [v,w]^\alpha\,. \tag{8.176}$$

Combining this basic expression with the Lie derivative of a function,

$$\mathcal{L}_v f = v^\beta \frac{\partial f}{\partial v^\beta}\,, \tag{8.177}$$

implies an efficient derivation of the Lie derivative of tensors from the product rule.

For a given metric $g_{\alpha\beta}$, a **Killing vector field** ξ^γ determines a direction in which the metric has a vanishing Lie derivative, $\mathcal{L}_\xi g_{\alpha\beta} = 0$ or

$$\xi^\epsilon \frac{\partial g_{\alpha\beta}}{\partial x^\epsilon} + g_{\beta\gamma}\frac{\partial \xi^\gamma}{\partial x^\alpha} + g_{\alpha\delta}\frac{\partial \xi^\delta}{\partial x^\beta} = 0\,. \tag{8.178}$$

The geometry is then the same at an initial point and at a point infinitesimally displaced in the direction of ξ^γ, and we obtain a symmetry. In this definition, we do not use properties of any specific coordinate system.

Consider the 2-sphere as an example. In this 2-dimensional case, the two symmetric indices in (8.175) imply three independent conditions on the two components of ξ^γ:

$$0 = \mathscr{L}_\xi g_{\vartheta\vartheta} = 2\frac{\partial \xi^\vartheta}{\partial x^\vartheta} \tag{8.179}$$

$$0 = \mathscr{L}_\xi g_{\varphi\varphi} = 2\sin(\vartheta)^2\left(\frac{\partial \xi^\varphi}{\partial x^\varphi} + \frac{\cos(\vartheta)}{\sin(\vartheta)}\xi^\vartheta\right) \tag{8.180}$$

$$0 = \mathscr{L}_\xi g_{\vartheta\varphi} = \sin(\vartheta)^2\frac{\partial \xi^\varphi}{\partial x^\vartheta} + \frac{\partial \xi^\vartheta}{\partial x^\varphi}\,. \tag{8.181}$$

According to the first equation, $\xi^\vartheta = f(\varphi)$ is a function depending only on φ. Using this result in the second equation,

$$\xi^\varphi = -\frac{\cos(\vartheta)}{\sin(\vartheta)}\int f(\varphi)\mathrm{d}\varphi + g(\vartheta) \tag{8.182}$$

with a free function $g(\vartheta)$ depending only on ϑ. Inserting these results in the third equation, we obtain

$$f'(\varphi) + \int f(\varphi)\mathrm{d}\varphi + g'(\vartheta) = 0\,. \tag{8.183}$$

This equation is separable and therefore implies the two ordinary differential equations

$$f'(\varphi) + \int f(\varphi)\mathrm{d}\varphi = a \quad , \quad g'(\vartheta) = -a \tag{8.184}$$

with a separation constant a. The meaning of this separation constant can be seen by observing that the first equation is of integro-differential form, in which the integral has unspecified limits. Choosing a is equivalent to specifying limits for the integral, such that we may set $a = 0$. The first one of these two equations implies $f''(\varphi) + f(\varphi) = 0$ upon taking an additional derivative, such that $f(\varphi) = b\sin(\varphi) + c\cos(\varphi)$. The function $g(\vartheta) = d$ must be constant. The components of a general Killing vector field on the 2-sphere are therefore given by

$$\xi^\vartheta = b\sin(\varphi) + c\cos(\varphi) \tag{8.185}$$

$$\xi^\varphi = \cot(\vartheta)(b\cos(\varphi) - c\sin(\varphi)) + d\,, \tag{8.186}$$

using $\int f(\varphi)\mathrm{d}\varphi = -f'(\varphi)$ after setting $a = 0$.

The three integration constants b, c and d determine three independent Killing vector fields. The constant d corresponds to the Killing vector field $\xi_z = \mathbf{e}_\varphi$ we already found, based on φ-independence of the metric or rotations around the z-axis. The remaining two Killing vector fields have the expressions

$$\xi_x = -\sin(\varphi)\mathbf{e}_\vartheta - \cot(\vartheta)\cos(\varphi)\mathbf{e}_\varphi \tag{8.187}$$

$$\xi_y = \cos(\varphi)\mathbf{e}_\vartheta - \cot(\vartheta)\sin(\varphi)\mathbf{e}_\varphi \tag{8.188}$$

in the coordinate basis, choosing $b = -1$ and $c = 1$, respectively. Writing the coordinate basis as $\mathbf{e}_\vartheta = \partial/\partial\vartheta$ and $\mathbf{e}_\varphi = \partial/\partial\varphi$, we recognize the angular-momentum operators (4.29) and (4.31) (without the factors of $\hbar/i$) that represent infinitesimal rotations in quantum mechanics.

The three independent Killing vector fields imply three conserved quantities along geodesics. Our derivation of the conservation law was based on the condition that the metric does not depend on one of the coordinates. Since then, we have generalized the notion of Killing vector fields, but the conservation law still applies because we may always choose coordinates, at least locally, such that one of them changes only in the direction of a given Killing vector field. The metric then does not depend on this coordinate. On the 2-sphere, therefore, we have three conserved quantities,

$$
\begin{aligned}
l_x &= \xi_x \cdot \mathbf{u} = -\sin(\varphi)u^\vartheta - \sin(\vartheta)\cos(\vartheta)\cos(\varphi)u^\varphi &\quad (8.189)\\
l_y &= \xi_y \cdot \mathbf{u} = \cos(\varphi)u^\vartheta - \sin(\vartheta)\cos(\vartheta)\sin(\varphi)u^\varphi &\quad (8.190)\\
l_z &= \xi_z \cdot \mathbf{u} = \sin(\vartheta)^2 u^\varphi . &\quad (8.191)
\end{aligned}
$$

These are three linear equations for two components, u^ϑ and u^φ. We eliminate u^ϑ by using the combination

$$l_x\cos(\varphi) + l_y\sin(\varphi) = -\sin(\vartheta)\cos(\vartheta)u^\varphi \quad (8.192)$$

of the first two equations. Together with the third equation, we obtain

$$l_x\cos(\varphi) + l_y\sin(\varphi) = -l_z\cot(\vartheta) \quad (8.193)$$

as a relationship between the coordinates φ and ϑ along a geodesic on the 2-sphere, determined by fixed values of l_x, l_y and l_z. Without doing any integrations, we have found an equation for all possible geodesics on the sphere because we had three conserved quantities in a 2-dimensional system.

Equation (8.193) indeed determines the well-known geodesics on the sphere because it takes the form

$$0 = l_x x(\vartheta,\varphi) + l_y y(\vartheta,\varphi) + l_z z(\vartheta,\varphi) = \vec{l}\cdot\vec{x}(\vartheta,\varphi) \quad (8.194)$$

upon multiplication by $\sin(\vartheta)$ and identifying Cartesian coordinates $\vec{x}(\vartheta,\varphi)$ of a point on the unit sphere. Because equation (8.194) is linear in Cartesian coordinates for a point on the sphere, it describes the intersection of the sphere with a plane. This plane goes through the origin because there is no constant shift in (8.194). The normal vector of the plane is equal to the vector $\vec{l}$, whose components are unrestricted. Therefore, the intersection of any plane through the origin with the sphere, or any **great circle**, is obtained as a solution of (8.193) with suitable values of l_x, l_y and l_z.

8.6.7 SINGULARITIES

In space-time, geodesics especially of the lightlike kind are a useful tool in the analysis of singularities. A singularity is obtained whenever the metric has a diverging component or zero determinant on some subset of events. In the diverging case, it is possible for a finite coordinate change to imply an infinite geometrical distance, and space-time seems to be ripping apart. In the zero-determinant case, the metric has a zero eigenvalue, and the corresponding eigenvector determines a direction in which there is no geometrical distance even if we change our coordinate values. Space-time then seems to collapse into a lower-dimensional object.

These outcomes are not necessarily physical conclusions because they may be implied by an unfortunate choice of coordinates. For instance, the determinant of the 2-sphere metric in spherical coordinates vanishes at the poles, but we obtain a lower-dimensional object only because the ϑ-coordinate lines intersect there. A non-zero change of φ then does not move us to a different point, and we indeed have a lower-dimensional subset at a fixed pole. It is possible to construct new coordinates that are non-singular at a pole, as in the exercise on page 328. In this case, the singularity is called a **coordinate singularity**. The big-bang singularity is an example of a **physical singularity** because it cannot be removed by a coordinate transformation. Its implication of infinite energy density would be a very real and observer-independent implication that is not merely a punishment for choosing bad coordinates.

It is not always possible to use simple shortcuts such as properties of the energy density in order to determine the physical nature of singularities. In the next chapter, we will see curvature invariants as a suitable alternative that can be used to provide purely geometrical indications of physical singularities: Such curvature invariants are coordinate independent and do not require assumptions about matter in extreme regimes. If we find one diverging curvature invariant, we can prove that a

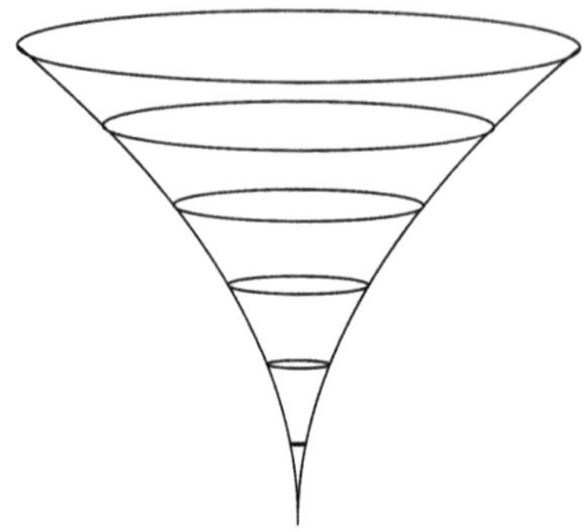

Figure 8.11 A sharp, singular tip on the surface of revolution defined by (8.195).

singularity is physical. An example of a physical singularity on a 2-dimensional surface is given by a sharp tip of a cone. Such a tip is realized on the surface of revolution with (ζ,φ), embedded in 3-dimensional flat space by

$$x=\zeta^2\cos(\varphi) \quad , \quad y=\zeta^2\sin(\varphi) \quad , \quad z=\zeta\,, \tag{8.195}$$

see Figure 8.11. The line element on the surface, $ds^2=\zeta^4 d\varphi^2+(1+4\zeta^2)d\zeta^2$, is singular at $\zeta=0$. The methods of an exercise on page 329 can be used to show that curvature diverges there, implying a physical singularity.

If a singularity is a mere coordinate singularity, we can prove this property by finding a coordinate transformation that implies a non-singular metric on the previously singular subset. The new coordinates might introduce singularities elsewhere, but in regions where the original coordinates were non-singular. In this case, there is a meaningful description of the geometry using multiple coordinate systems, or an **atlas** of several coordinate charts. Since we are now considering the huge set of non-linear coordinate transformations, it is, in general, not possible to guess a suitable one that may remove a coordinate singularity. In some cases, it is possible to do integrations such as $\int N(t)dt$ in the FRW line element and remove a vanishing metric coefficient. Another such example is the line element $-dt^2+x^2dx^2$: After integrating xdx, the new coordinate $z=\frac{1}{2}x^2$ implies a metric with nowhere vanishing determinant. However, this simple method is available only in special cases.

More generally, we can test whether a singularity is of coordinate form by aiming geodesics at it and checking whether their proper time, proper length, or affine parameter remains finite when we get there. This parameter, together with a second one that labels independent geodesics aimed from different starting points, may then serve as a new coordinate system in which one can explicitly compute the metric and see whether it still has a vanishing determinant. If it does not, one can extend the geometry beyond the coordinate singularity, which has thus been removed.

Consider the line element $ds^2=-X^2dT^2+dX^2$, which is singular at $X=0$. Light rays in this geometry are given by solutions of the equation $dT_\pm=\pm X^{-1}dX$, and thus obey $T_\pm(X)=V\pm\log|X|$ with an integration constant V. Changing the value of V translates the light ray in the T-direction. Starting at positive X and aiming light rays toward the singularity, we need left-moving ones determined by the negative sign in $T_\pm(X)$: $T_-(X)=V-\log|X|$.

We may use V as a new space-time coordinate, and complement it by a second coordinate that changes along light rays. Such a coordinate is given by the affine parameter λ, defined by requiring that light rays not only obey $ds^2=0$ but also the geodesic equation

$$\frac{d^2x^\delta}{d\lambda^2}=-\Gamma^\delta_{\beta\gamma}\frac{dx^\beta}{d\lambda}\frac{dx^\gamma}{d\lambda}\,. \tag{8.196}$$

Here, the non-zero Christoffel components are given by $\Gamma^X_{TT} = X$ and $\Gamma^T_{XT} = \Gamma^T_{TX} = 1/X$. Since the only non-zero partial derivative of metric components is $\partial g_{TT}/\partial X$ and the metric is diagonal, non-zero Christoffel components must have one X-index and two T-indices.

The two components of the geodesic equation are therefore given by

$$\frac{\mathrm{d}^2T}{\mathrm{d}\lambda^2} = -\frac{2}{X}\frac{\mathrm{d}T}{\mathrm{d}\lambda}\frac{\mathrm{d}X}{\mathrm{d}\lambda} \quad , \quad \frac{\mathrm{d}^2X}{\mathrm{d}\lambda^2} = -X\left(\frac{\mathrm{d}T}{\mathrm{d}\lambda}\right)^2 . \tag{8.197}$$

The lightlike condition implies $X^2(\mathrm{d}T/\mathrm{d}\lambda)^2 = (\mathrm{d}X/\mathrm{d}\lambda)^2$ which we can use to write the second geodesic equation as

$$0 = X\frac{\mathrm{d}^2X}{\mathrm{d}\lambda^2} + \left(\frac{\mathrm{d}X}{\mathrm{d}\lambda}\right)^2 = \frac{\mathrm{d}}{\mathrm{d}\lambda}\left(X\frac{\mathrm{d}X}{\mathrm{d}\lambda}\right), \tag{8.198}$$

solved by $X\mathrm{d}X/\mathrm{d}\lambda = A$, or $X(\lambda)^2 = 2A(\lambda-\lambda_0)$. Therefore, the affine parameter relative to its initial value is proportional to X^2. Using the affine parameter as a coordinate is equivalent to transforming from the original coordinate X to a new coordinate $Z = X^2$.

Including the parameter V that labels light rays, our 2-dimensional coordinate transformation reads $(T,X) \mapsto (V,Z) = (T+\log X, X^2)$. The new line element,

$$\mathrm{d}s^2 = -Z\mathrm{d}V^2 + \mathrm{d}V\mathrm{d}Z\,, \tag{8.199}$$

is no longer diagonal, but it is non-singular because its metric

$$\mathbf{g} = \begin{pmatrix} -Z & \frac{1}{2} \\ \frac{1}{2} & 0 \end{pmatrix} \tag{8.200}$$

has a non-vanishing determinant everywhere.

A space-time diagram illustrates the causal behavior around $Z = 0$. However, the new line element (8.199) does not have an obvious time coordinate. It has a negative contribution, but the corresponding $\mathrm{d}V$ is an increment in the parameter that labels light rays. Accordingly, if we set V constant, the line element completely vanishes. For a valid time coordinate, by contrast, we expect to obtain a spacelike hypersurface with a positive line element when we choose a constant value of time. We use this expectation for a formal definition of **time** in a line element: A time coordinate t is such that the line element $\mathrm{d}s^2|_{\mathrm{d}t=0} > 0$ is positive definite and may vanish only if there is no separation at all. The coordinate V does not obey this condition because it implies a zero line element even if $\mathrm{d}Z \neq 0$ while $\mathrm{d}V = 0$. The other coordinate, Z, is not a time coordinate either because $\mathrm{d}s^2|_{\mathrm{d}Z=0} = -Z\mathrm{d}V^2$ is not positive at $Z > 0$. Moving on to linear combinations of V and Z, we try $t = V - Z$:

$$\mathrm{d}s^2|_{\mathrm{d}V=\mathrm{d}Z} = (1-Z)\mathrm{d}Z^2 > 0 \quad \text{if } Z < 1\,. \tag{8.201}$$

The new t is not a time coordinate everywhere in space-time, but it may be used as time in a region around the former coordinate singularity at $Z = 0$.

We now derive light rays in the new line element and translate solutions to relationships between time t and space Z:

$$\mathrm{d}s^2 = \mathrm{d}V(-Z\mathrm{d}V + \mathrm{d}Z) = 0\,. \tag{8.202}$$

The product is zero if one of the two factors is zero, which gives rise to the following solutions:

1. $\mathrm{d}V = 0$ implies that V is constant, which should indeed be a solution because we derived V as a parameter that labels light rays. In terms of the new time coordinate, constant V implies the relationship

$$t(Z) = \text{const} - Z\,. \tag{8.203}$$

These solutions are straight lines in a (t,Z) space-time diagram, just like the downward diagonals in Minkowski space-time.

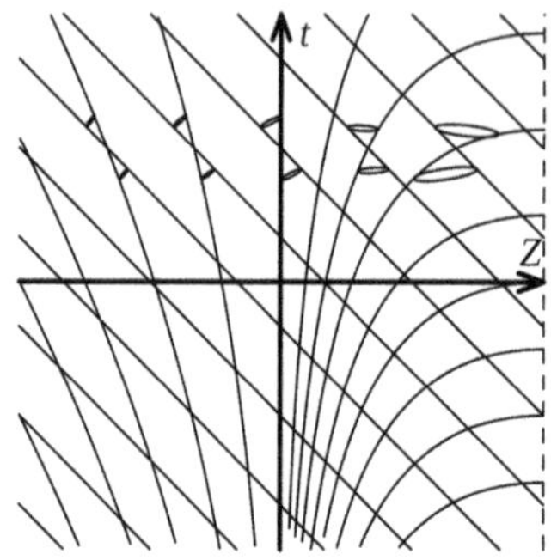

Figure 8.12 Space-time diagram of the line element (8.199), restricted to $Z < 1$. The coordinate transformation from the original line element has removed a coordinate singularity at $X = 0$, corresponding to $Z = 0$. At this value, the extended space-time diagram shows a horizon which cannot be crossed by light rays starting from values $Z < 0$.

2. $Z\mathrm{d}V = \mathrm{d}Z$ is solved by $V = \mathrm{const} + \log|Z|$. In terms of the new time coordinate,

$$t(Z) = \mathrm{const} + \log|Z| - Z. \tag{8.204}$$

   These solutions represent curves that reach a pole of $t \to -\infty$ at $Z = 0$. The function $t(Z)$ has zero derivative at $Z = 1$, and thus the light ray no longer moves toward larger t for $Z > 1$. This is why the coordinate t can be used as time only for $Z < 1$.
3. We now have two families of light rays, corresponding to left-moving and right-moving ones for $Z > 0$. There is a third solution that is not contained in these families, given by

$$Z = 0 \tag{8.205}$$

   which implies both $Z\mathrm{d}V = 0$ and $\mathrm{d}Z = 0$. This value corresponds to the coordinate singularity $X = 0$ of the original line element. We now see that the singularity did not represent a point in space or the worldline of an object at rest in space-time, but rather a light ray.

The space-time diagram in Figure 8.12 shows both families of light rays as well as a few sample light cones. For $Z > 0$, the light cones are slightly distorted compared with Minkowski light cones in Cartesian coordinates, but they have the usual future and past regions in which light rays can move to the left and to the right. In the extended region at $Z < 0$, however, all light rays move toward smaller values of Z. It is not possible for these light rays, or any messengers, to cross $Z = 0$ and reach positive values of Z.

The reason for this behavior can be seen by transforming to **double-null coordinates**, which consist of V as well as another parameter, U, that is constant along right-moving light rays in the original space-time. The previous derivation shows that $U = T - \log|X|$ in addition to $V = T + \log|X|$. The product $-\mathrm{d}U\mathrm{d}V = -\mathrm{d}T^2 + X^{-2}\mathrm{d}X^2$ is proportional to, or **conformally related** to, the original line element, such that

$$\mathrm{d}s^2 = -X(U,V)^2\mathrm{d}U\mathrm{d}V = -e^{V-U}\mathrm{d}U\mathrm{d}V. \tag{8.206}$$

The transformation does not remove the coordinate singularity at $X = 0$, but this set of events is no longer at finite coordinate values because it corresponds to $U \to \infty$ or $V \to -\infty$. We can map it back to finite values by integrating $e^V\mathrm{d}V = \mathrm{d}v$ and $-e^{-U}\mathrm{d}U = \mathrm{d}u$ in order to obtain new coordinates $v = e^V$ and $u = e^{-U}$ in which the former worldline $X = 0$ now consists of two worldlines, $u = 0$ or $v = 0$.

The last transformation results in the line element

$$ds^2 = du dv = -dt_M^2 + dx_M^2 \tag{8.207}$$

of 2-dimensional Minkowski space-time, where $t_M = \frac{1}{2}(v-u)$ and $x_M = \frac{1}{2}(v+u)$. Since there are no coordinate singularities in this line element, we may use the full range of real numbers for both t_M and x_M (or u and v). We have further extended our space-time description by deriving a well-defined geometry beyond the restricted ranges of u and v, and even beyond the region covered by V and Z which corresponds to $v > 0$.

As part of a general method, transforming from (U,V) to (u,v) amounts to expressing both families of light rays using double-null coordinates that are also affine parameters: As shown by (8.207), u is an affine parameter along constant-v light rays (where $t_M = -x_M + v$), and v is an affine parameter along constant-u light rays (where $t_M = x_M - u$). An affine parameter implies a constant rate of change along a light ray, generalizing the linear parameterization on flat space-time. Using light-ray parameters with a constant rate of change as coordinates avoids mapping parts of space-time to infinity, as happens when we use U and V instead of u and v.

The original range of coordinates (T,X) was incomplete because it covered only one quarter of Minkowski space-time, called the **Rindler wedge**; see Figure 8.13. The transformed coordinates (T,Z) that removed the coordinate singularity covered only half of Minkowski space-time because they extended the original region beyond one light ray, $u = 0$ called the **Rindler horizon**, but not beyond $v = 0$. In these coordinates, it is impossible to send messages from $Z < 0$ to $Z > 0$ simply because $Z = 0$ at finite V corresponds to the right-moving light ray $u = 0$ in the final coordinates. It is impossible to send a message to the right that catches up with a right-moving light ray. In spite of the unusual light-cone structure in the coordinates (T,Z), the worldline $Z = 0$ does not correspond to a barrier other than what is already implied by the speed of light as an upper limit.

Tracing back all the coordinate changes, our transformation from the initial coordinates (T,X) with $X > 0$ to the final coordinates (t_M, x_M) is given by

$$t_M = \frac{1}{2}(v-u) = \frac{1}{2}(e^V - e^{-U}) = \frac{1}{2}(Xe^T - Xe^{-T}) = X\sinh(T) \tag{8.208}$$

and

$$x_M = \frac{1}{2}(v+u) = \frac{1}{2}(e^V + e^{-U}) = \frac{1}{2}(Xe^T + Xe^{-T}) = X\cosh(T). \tag{8.209}$$

Comparing with (7.102), the original time coordinate T can be identified as the proper time of a non-inertial observer subject to constant acceleration. Varying X, we obtain a family of non-inertial observers whose worldlines are located in the right spacelike quadrant of Minkowski space-time, enclosed by the dashed lines of Figure 8.13. These observers do not receive light rays from the left spacelike quadrant because they keep accelerating: Even though they are always slower than light, their constant acceleration means that even light cannot catch up with them if it originated from an event sufficiently far to the left.

The coordinate Z, derived from left-moving light rays, extended the space-time geometry to the future timelike quadrant as seen from the perspective of the final Minkowski space-time. This transformation allowed us to determine that some light rays are unable to catch up with the observers. But this extension only included regions covered by the original worldlines with proper time T as well as left-moving light rays emitted from these worldlines. From the perspective of Minkowski space-time, these regions are the right spacelike quadrant and the future timelike quadrant. The final extension to complete Minkowski space-time required the past of right-moving light rays from these two regions. Since double-null coordinates include both left-moving and right-moving light rays, they were able to lead us to the final extension.

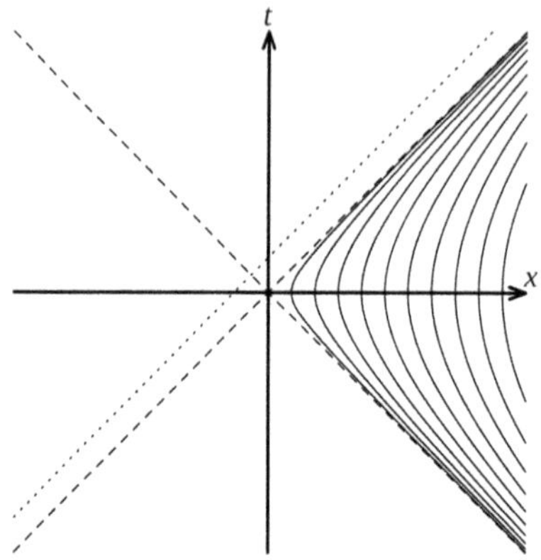

Figure 8.13 Minkowski space-time with the Rindler wedge, swept out by the worldlines of a family of uniformly accelerated observers. The dotted line indicates a right-moving light ray that is unable to catch up with the observers.

EXERCISES

1. The transformation

$$x = \tan(\tfrac{1}{2}\vartheta)\cos(\varphi) \quad , \quad y = \tan(\tfrac{1}{2}\vartheta)\sin(\varphi)$$

to **stereographic coordinates** can be used to show that the spherical line element

$$ds^2 = d\vartheta^2 + \sin(\vartheta)^2 d\varphi^2 = f(x,y)\,\left(dx^2 + dy^2\right)$$

equals the planar line element $dx^2 + dy^2$ multiplied by a position-dependent function $f(x,y)$. Determine $f(x,y)$ as a function of x and y.
Show that the North pole of the 2-sphere is a coordinate singularity.

2. Any real 1-dimensional curve has the same geometry because a generic line element $f(x)^2 dx^2$ in one variable can be written as $d\tilde{x}^2$ with $\tilde{x}(x) = \int f(x)dx$. A complex 1-dimensional curve with a complex parameter ζ, allows for inequivalent geometries.
Map the 2-dimensional real plane $\mathbb{R}^2$ to the complex line by defining $\zeta := x + iy \in \mathbb{C}$ and show that the line element transforms to $ds^2 = d\zeta d\zeta^*$. The next Exercise introduces an inequivalent geometry on the complex line.

3. Define the complex function $\zeta(\vartheta, \varphi) = e^{i\varphi}\cot(\vartheta/2)$.
   (i) Compute $d\zeta d\zeta^*$ in terms of the two angles ϑ and φ.
   (ii) Show that the equations

$$x = r\frac{\zeta + \zeta^*}{1 + \zeta\zeta^*} \quad , \quad y = -ir\frac{\zeta - \zeta^*}{1 + \zeta\zeta^*} \quad , \quad z = r\frac{\zeta\zeta^* - 1}{1 + \zeta\zeta^*} \tag{8.210}$$

   are equivalent to expressing Cartesian coordinates (x, y, z) in terms of spherical coordinates (r, ϑ, φ).
   (iii) Use (8.210) to transform the Euclidean line element $dx^2 + dy^2 + dz^2$ at constant $r^2 = x^2 + y^2 + z^2 = 1$ to an expression in terms of ζ and ζ^*.

4. A 3-dimensional hyperboloid in Minkowski space-time satisfies the condition

$$t^2 - x^2 - y^2 - z^2 = \rho^2$$

with a constant ρ.
Confirm that the hyperboloid can be parameterized by three angles ψ, ϑ, φ using

$$\begin{aligned} x &= \rho \sinh(\psi) \sin(\vartheta) \sin(\varphi) \\ y &= \rho \sinh(\psi) \sin(\vartheta) \cos(\varphi) \\ z &= \rho \sinh(\psi) \cos(\vartheta) \\ t &= \rho \cosh(\psi) . \end{aligned}$$

Derive the line element on the hyperboloid from the Minkowski line element in space-time.

5. Show that the line element you obtained in the preceding Exercise is equivalent to

$$\mathrm{d}s^2 = \frac{\mathrm{d}R^2}{1 - kR^2/\rho^2} + R^2(\mathrm{d}\vartheta^2 + \sin(\vartheta)^2 \mathrm{d}\varphi^2) \qquad (8.211)$$

for $k = -1$, while the 3-sphere line element (8.96) is equivalent to (8.211) with $k = 1$.
Show that (8.211) is equivalent to

$$\mathrm{d}s^2 = \frac{1}{(1 + \frac{1}{4}kr^2/\rho^2)^2}(\mathrm{d}r^2 + r^2(\mathrm{d}\vartheta^2 + \sin(\vartheta)^2 \mathrm{d}\varphi^2) . \qquad (8.212)$$

6. The curvature of a 2-dimensional surface at a given point can be obtained as follows: Assume that the surface is given as the set of all points (x, y, z) solving an equation $f(x, y) = z$, such that $f(0, 0) = 0$ and $\partial f / \partial x = 0 = \partial f / \partial y$ at $(x, y) = (0, 0)$. The determinant $G = \det K$ of the 2×2-matrix

$$K = \begin{pmatrix} \frac{\partial^2 f}{\partial x^2} & \frac{\partial^2 f}{\partial x \partial y} \\ \frac{\partial^2 f}{\partial y \partial x} & \frac{\partial^2 f}{\partial y^2} \end{pmatrix}$$

$(x, y) = (0, 0)$ happens to be independent of the chosen coordinates. It equals the **Gaussian curvature** of the surface at the point $(0, 0, 0)$.
Compute G for
   (i) a cylinder along the x-axis $[f(x, y) = 1 - \sqrt{1 - y^2}]$,
   (ii) a sphere $[f(x, y) = 1 - \sqrt{1 - x^2 - y^2}]$, and
   (iii) a hyperboloid $[f(x, y) = 1 - \sqrt{1 + x^2 - y^2}]$.

7. A **surface of revolution** in 3-dimensional Euclidean space is defined by the parameterization

$$x = f(\zeta) \cos(\varphi) \quad , \quad y = f(\zeta) \sin(\varphi) \quad , \quad z = \zeta$$

with two coordinates φ and ζ on the surface and a function $f(\zeta) \geq 0$ defining its shape.
   (i) Compute the line element of a general surface of revolution.
   (ii) Now choose $f(\zeta) = \sqrt{1 - \zeta^2}$ for $-1 \leq \zeta \leq 1$ and compute the line element. Show that the new coordinate ϑ defined by $\zeta = \cos(\vartheta)$ transforms the line element to that of the 2-sphere.
   (iii) The Gaussian curvature on a surface of revolution is given by the expression

$$G(\zeta) = \frac{-f''(\zeta)}{f(\zeta)(1 + f'(\zeta)^2)^2} \qquad (8.213)$$

   Compute G for the sphere in part (ii) and show that it is constant.

8. A 2-dimensional space-time region with coordinates (t,ξ) is obtained from 4-dimensional Minkowski space-time by imposing the equations

$$x = \sin(\xi)^2 \quad , \quad y = \cos(\xi)^2$$

and setting $z = 0$ constant.

   (i) Draw a diagram of the region in the $x-y$ plane swept out by ξ at constant t.
   (ii) Derive the line element of this 2-dimensional space-time, and its singularities.
   (iii) Compute the function $t(\xi)$ along light rays.
   (iv) Specialize your solutions to the two light rays through the event with coordinates $t = 1$ and $\xi = \frac{1}{8}\pi$. Draw a (t,ξ) space-time diagram of this light cone.

9. Assume a space-time with line element

$$ds^2 = -dt^2 + e^{2x}dx^2 + dy^2 + dz^2 .$$

   (i) Derive and solve the equation for light rays moving in the x-direction. Draw a space-time diagram with the light cone in the (t,x)-plane through the event $(t,x) = (1,1)$.
   (ii) Show that the 4-vectors

$$\mathbf{e}_0 = (1,0,0,0)^T \quad , \quad \mathbf{e}_1 = (0,e^{-x},0,0)^T \quad , \quad \mathbf{e}_2 = (0,0,1,0)^T \quad , \quad \mathbf{e}_3 = (0,0,0,1)^T$$

   form an orthonormal basis.
   (iii) In the local inertial frame defined by the basis from part (ii), compute the components $p^{0'} = -\mathbf{p}\cdot\mathbf{e}_0$ and $p^{1'} = \mathbf{p}\cdot\mathbf{e}_1$ of the 4-momentum $\mathbf{p} = (1, \frac{1}{2}e^{-x},0,0)^T$. What is the mass of a particle with this 4-momentum?
   (iv) Compute the 4-momentum $\mathbf{p}$ along a worldline given by

$$t(\tau) = \frac{1}{a}\sinh(a\tau) \quad , \quad x(\tau) = \ln(\cosh(a\tau)) - \ln(a)$$

   and $y(\tau) = 0 = z(\tau)$ with a constant a.
   (v) Again using basis given in part (ii), compute the components $p^{\alpha'}$ of the 4-momentum $\mathbf{p}(\tau)$ from part (iv). Derive the 4-force $\mathbf{f}'$.
   (vi) What is the 3-force $\vec{F}'$ that gives rise to motion along the worldline of part (iv)?

10. Write the following products in matrix notation:

   (i) $p_\alpha = \eta_{\alpha\beta}p^\beta$
   (ii) $\delta^\alpha{}_\beta = \eta^{\alpha\gamma}\eta_{\gamma\beta}$
   (iii) $\mathscr{F}^{\alpha\beta}\mathscr{F}_{\alpha\gamma} = \eta^{\alpha\delta}\eta^{\beta\varepsilon}\mathscr{F}_{\delta\varepsilon}\mathscr{F}_{\alpha\gamma}$

   Here, p^α are components of the 4-momentum $\mathbf{p}$, $\eta_{\alpha\beta}$ are components of the Minkowski metric η and $\eta^{\alpha\beta}$ components of its inverse, and $\mathscr{F}_{\alpha\beta}$ are components of the field-strength tensor $\mathscr{F}$.

11. Compute the Christoffel symbol of the 2-sphere with line element

$$ds^2 = d\vartheta^2 + \sin(\vartheta)^2 d\varphi^2 .$$

   Use your result to obtain both components of the geodesic equation on the 2-sphere. Confirm that all meridians and the equator correspond to solutions of this equation.

12. Compute the Christoffel symbol $\Gamma^\alpha_{\beta\gamma}$ of a 2-dimensional space-time with line element $ds^2 = -dt^2 + x^2dx^2$. Show that lightlike geodesics in this space-time are such that $t(\lambda)$ and $x(\lambda)^2$ are linear in the affine parameter λ.

13. Compute the Christoffel symbol $\Gamma^{\alpha}_{\beta\gamma}$ of the Friedmann–Robertson–Walker line element $\mathrm{d}s^2 = -\mathrm{d}t^2 + a(t)^2(\mathrm{d}x^2 + \mathrm{d}y^2 + \mathrm{d}z^2)$ with a time-dependent scale factor $a(t)$.
Show that the worldline $\mathbf{x}(\tau) = (\tau, 0, 0, 0)^T$ is a geodesic.
14. Derive (8.176) by following the procedure described on page 321.
15. Assume a 2-dimensional space-time with line element

$$\mathrm{d}s^2 = -\mathrm{d}t^2 + 2\sin(2\eta)^2\mathrm{d}\eta^2. \tag{8.214}$$

(i) Compute all non-zero components of the Christoffel symbol.
(ii) Write down the second-order equations for lightlike geodesics.
(iii) Show that the time coordinate t is linear in the affine parameter.
(iv) Show that left-moving light rays in the space-time (8.214) are determined by the relationship $t = v + \cos(2\eta)/\sqrt{2}$ with an integration constant v.
(v) Apply a coordinate transformation from (t, η) to $(t, v(t, \eta))$ and derive the new line element.
(vi) Derive the relationships between t and v for *all* light rays in the new coordinates.
(vii) Translate your results from part (vi) to relationships between time t and a new spatial coordinate defined by $\tilde{x} = v - t$. Draw a space-time diagram of these light rays.
(viii) Show that the vectors $\mathbf{e}_{\hat{0}} = (1, 0)^T$ and $\mathbf{e}_{\hat{1}} = (0, (\sqrt{2}\sin(2\eta))^{-1})^T$ define an orthonormal basis in the space-time given by (8.214).
(ix) A massive object moves along the geodesic given by $t(\tau) = \sqrt{2}\tau = -\cos(2\eta(\tau))$. Compute the relativistic velocity $\mathbf{u} = \mathrm{d}\mathbf{x}/\mathrm{d}\tau$ in the global frame determined by (t, η), as well as the components $u^{\hat{\alpha}}$ measured in the local inertial frame given in part (viii). What is the velocity component $V^{\hat{1}}$ in this frame?

9 General Relativity

Heuristic arguments seen in Section 8.1 suggest that space-time curvature implies the gravitational force. Curvature is related to changes of the geometry around events, and can be computed from derivatives of the metric $g_{\alpha\beta}$. However, individual partial derivatives of the metric, or the combinations of partial derivatives in the Christoffel symbol

$$\Gamma^{\alpha}_{\beta\gamma} = \frac{1}{2} g^{\alpha\delta} \left(\frac{\partial g_{\delta\beta}}{\partial x^{\gamma}} + \frac{\partial g_{\delta\gamma}}{\partial x^{\beta}} - \frac{\partial g_{\beta\gamma}}{\partial x^{\delta}} \right), \tag{9.1}$$

do not obey the tensor-transformation law because the partial derivative of a transformed metric depends on second or higher-order derivatives of the transformation functions $x^{\alpha'}(x^{\alpha})$. Nevertheless, the Christoffel symbol does contain information about curvature because it determines geodesics in curved space-time, and the behavior of geodesics can be used to infer specific deviations from Euclidean geometry. We extract this information by considering suitable covariant combinations of the Christoffel symbol with its own partial derivatives, such that the full expressions obey the tensor transformation law.

9.1 CURVATURE TENSORS

The Christoffel symbol is not a tensor because it does not obey the tensor transformation law. In order to see detailed violations of this law, we derive the transformation of $\Gamma^{\alpha}_{\beta\gamma}$ by inserting the tensor transformation of the metric and its inverse in (9.1):

$$\begin{aligned}
\Gamma^{\alpha}_{\beta\gamma} &= \frac{1}{2} \frac{\partial x^{\alpha}}{\partial x^{\alpha'}} \frac{\partial x^{\delta}}{\partial x^{\delta'}} g^{\alpha'\delta'} \\
&\quad \times \left(\frac{\partial}{\partial x^{\gamma}} \left(\frac{\partial x^{\varepsilon'}}{\partial x^{\delta}} \frac{\partial x^{\beta'}}{\partial x^{\beta}} g_{\varepsilon'\beta'} \right) + \frac{\partial}{\partial x^{\beta}} \left(\frac{\partial x^{\varepsilon'}}{\partial x^{\delta}} \frac{\partial x^{\gamma'}}{\partial x^{\gamma}} g_{\varepsilon'\gamma'} \right) - \frac{\partial}{\partial x^{\delta}} \left(\frac{\partial x^{\beta'}}{\partial x^{\beta}} \frac{\partial x^{\gamma'}}{\partial x^{\gamma}} g_{\beta'\gamma'} \right) \right) \\
&= \frac{\partial x^{\alpha}}{\partial x^{\alpha'}} \frac{\partial x^{\beta'}}{\partial x^{\beta}} \frac{\partial x^{\gamma'}}{\partial x^{\gamma}} \Gamma^{\alpha'}_{\beta'\gamma'} + \frac{1}{2} \frac{\partial x^{\alpha}}{\partial x^{\alpha'}} \frac{\partial x^{\delta}}{\partial x^{\delta'}} g^{\alpha'\delta'} \\
&\quad \times \left(\frac{\partial}{\partial x^{\gamma}} \left(\frac{\partial x^{\varepsilon'}}{\partial x^{\delta}} \frac{\partial x^{\beta'}}{\partial x^{\beta}} \right) g_{\varepsilon'\beta'} + \frac{\partial}{\partial x^{\beta}} \left(\frac{\partial x^{\varepsilon'}}{\partial x^{\delta}} \frac{\partial x^{\gamma'}}{\partial x^{\gamma}} \right) g_{\varepsilon'\gamma'} - \frac{\partial}{\partial x^{\delta}} \left(\frac{\partial x^{\beta'}}{\partial x^{\beta}} \frac{\partial x^{\gamma'}}{\partial x^{\gamma}} \right) g_{\beta'\gamma'} \right).
\end{aligned} \tag{9.2}$$

Taking partial derivatives only of the metric components produces the correct term for a tensor transformation law of a three-index object, written as the first term in the last equality. But second-order derivatives of the coordinate transformation, implied by the product rule, do not completely cancel out.

9.1.1 CANCELLATIONS

There are at least some cancellations after renaming suitable indices and using the fact that partial derivatives by coordinates in the same system commute with one another. But note that we should not commute partial derivatives by coordinates in different systems, such as $\partial^2/\partial x^{\alpha}\partial x^{\beta'}$. It is best to write all multiple derivatives by referring to coordinates in the same system.

Relabeling γ' and applying the product rule, we simplify the extra terms in (9.2) to

$$\frac{\partial}{\partial x^{\gamma}} \left(\frac{\partial x^{\varepsilon'}}{\partial x^{\delta}} \frac{\partial x^{\beta'}}{\partial x^{\beta}} \right) g_{\varepsilon'\beta'} + \frac{\partial}{\partial x^{\beta}} \left(\frac{\partial x^{\varepsilon'}}{\partial x^{\delta}} \frac{\partial x^{\beta'}}{\partial x^{\gamma}} \right) g_{\varepsilon'\beta'} - \frac{\partial}{\partial x^{\delta}} \left(\frac{\partial x^{\beta'}}{\partial x^{\beta}} \frac{\partial x^{\varepsilon'}}{\partial x^{\gamma}} \right) g_{\beta'\varepsilon'} = 2 g_{\varepsilon'\beta'} \frac{\partial x^{\varepsilon'}}{\partial x^{\delta}} \frac{\partial^2 x^{\beta'}}{\partial x^{\gamma} \partial x^{\beta}} \tag{9.3}$$

DOI: 10.1201/9781003588702-11

and obtain

$$\Gamma^{\alpha}_{\beta\gamma} = \frac{\partial x^{\alpha}}{\partial x^{\alpha'}}\frac{\partial x^{\beta'}}{\partial x^{\beta}}\frac{\partial x^{\gamma'}}{\partial x^{\gamma}}\Gamma^{\alpha'}_{\beta'\gamma'} + \frac{\partial x^{\alpha}}{\partial x^{\beta'}}\frac{\partial^2 x^{\beta'}}{\partial x^{\beta}\partial x^{\gamma}} \tag{9.4}$$

after using

$$\frac{\partial x^{\alpha}}{\partial x^{\alpha'}}g^{\alpha'\delta'}\frac{\partial x^{\delta}}{\partial x^{\delta'}}\frac{\partial x^{\varepsilon'}}{\partial x^{\delta}}g_{\varepsilon'\beta'} = \frac{\partial x^{\alpha}}{\partial x^{\alpha'}}g^{\alpha'\delta'}\delta^{\varepsilon'}{}_{\delta'}g_{\varepsilon'\beta'} = \frac{\partial x^{\alpha}}{\partial x^{\beta'}}\,. \tag{9.5}$$

One implication of curvature is focusing or defocusing of geodesics, such as converging meridians as they approach a pole. Since a single geodesic is determined by the Christoffel symbol, relationships between nearby geodesics should be related to derivatives of the Christoffel symbol. Even though taking additional partial derivatives initially complicates the transformation behavior, it also allows us to combine different expressions that will allow us to cancel out unwanted terms in the transformation behavior.

Continuing with (9.4), we obtain

$$\begin{aligned}\frac{\partial\Gamma^{\alpha}_{\beta\gamma}}{\partial x^{\delta}} &= \frac{\partial x^{\alpha}}{\partial x^{\alpha'}}\frac{\partial x^{\beta'}}{\partial x^{\beta}}\frac{\partial x^{\gamma'}}{\partial x^{\gamma}}\frac{\partial x^{\delta'}}{\partial x^{\delta}}\frac{\partial\Gamma^{\alpha'}_{\beta'\gamma'}}{\partial x^{\delta'}}\\ &\quad+\left(\frac{\partial^2 x^{\alpha}}{\partial x^{\alpha'}\partial x^{\delta'}}\frac{\partial x^{\beta'}}{\partial x^{\beta}}\frac{\partial x^{\gamma'}}{\partial x^{\gamma}}\frac{\partial x^{\delta'}}{\partial x^{\delta}}+\frac{\partial x^{\alpha}}{\partial x^{\alpha'}}\frac{\partial^2 x^{\beta'}}{\partial x^{\beta}\partial x^{\delta}}\frac{\partial x^{\gamma'}}{\partial x^{\gamma}}+\frac{\partial x^{\alpha}}{\partial x^{\alpha'}}\frac{\partial x^{\beta'}}{\partial x^{\beta}}\frac{\partial^2 x^{\gamma'}}{\partial x^{\gamma}\partial x^{\delta}}\right)\Gamma^{\alpha'}_{\beta'\gamma'}\\ &\quad+\frac{\partial^2 x^{\alpha}}{\partial x^{\delta'}\partial x^{\beta'}}\frac{\partial x^{\delta'}}{\partial x^{\delta}}\frac{\partial^2 x^{\beta'}}{\partial x^{\beta}\partial x^{\gamma}}+\frac{\partial x^{\alpha}}{\partial x^{\beta'}}\frac{\partial^3 x^{\beta'}}{\partial x^{\beta}\partial x^{\gamma}\partial x^{\delta}}\end{aligned} \tag{9.6}$$

We applied the product rule to all derivatives by x^{δ}, making sure that multiple derivatives are always taken in the same coordinate system. The last term is the only third-order partial derivative, and it is symmetric in β and δ. It therefore cancels out in the antisymmetric combination

$$\begin{aligned}\frac{\partial\Gamma^{\alpha}_{\delta\gamma}}{\partial x^{\beta}}-\frac{\partial\Gamma^{\alpha}_{\beta\gamma}}{\partial x^{\delta}} &= \frac{\partial x^{\alpha}}{\partial x^{\alpha'}}\frac{\partial x^{\beta'}}{\partial x^{\beta}}\frac{\partial x^{\gamma'}}{\partial x^{\gamma}}\frac{\partial x^{\delta'}}{\partial x^{\delta}}\left(\frac{\partial\Gamma^{\alpha'}_{\delta'\gamma'}}{\partial x^{\beta'}}-\frac{\partial\Gamma^{\alpha'}_{\beta'\gamma'}}{\partial x^{\delta'}}\right)\\ &\quad+\frac{\partial^2 x^{\alpha}}{\partial x^{\alpha'}\partial x^{\delta'}}\frac{\partial x^{\gamma'}}{\partial x^{\gamma}}\left(\frac{\partial x^{\beta'}}{\partial x^{\delta}}\frac{\partial x^{\delta'}}{\partial x^{\beta}}-\frac{\partial x^{\beta'}}{\partial x^{\beta}}\frac{\partial x^{\delta'}}{\partial x^{\delta}}\right)\Gamma^{\alpha'}_{\beta'\gamma'}\\ &\quad+\frac{\partial x^{\alpha}}{\partial x^{\alpha'}}\left(\frac{\partial x^{\beta'}}{\partial x^{\delta}}\frac{\partial^2 x^{\gamma'}}{\partial x^{\gamma}\partial x^{\beta}}-\frac{\partial x^{\beta'}}{\partial x^{\beta}}\frac{\partial^2 x^{\gamma'}}{\partial x^{\gamma}\partial x^{\delta}}\right)\Gamma^{\alpha'}_{\beta'\gamma'}\\ &\quad+\frac{\partial^2 x^{\alpha}}{\partial x^{\delta'}\partial x^{\beta'}}\left(\frac{\partial x^{\delta'}}{\partial x^{\beta}}\frac{\partial^2 x^{\beta'}}{\partial x^{\delta}\partial x^{\gamma}}-\frac{\partial x^{\delta'}}{\partial x^{\delta}}\frac{\partial^2 x^{\beta'}}{\partial x^{\beta}\partial x^{\gamma}}\right).\end{aligned} \tag{9.7}$$

In the last line, products of second-order partial derivatives of the coordinate transformation can also be obtained from products of transformed Christoffel symbols, which therefore may provide additional opportunities for cancellations. The product of two Christoffel symbols has six indices, but if two of them are summed over, four remain and could match the index structure of (9.7). In particular, the combination

$$\begin{aligned}\Gamma^{\alpha}_{\beta\varepsilon}\Gamma^{\varepsilon}_{\delta\gamma} &= \frac{\partial x^{\alpha}}{\partial x^{\alpha'}}\frac{\partial x^{\beta'}}{\partial x^{\beta}}\frac{\partial x^{\gamma'}}{\partial x^{\gamma}}\frac{\partial x^{\delta'}}{\partial x^{\delta}}\Gamma^{\alpha'}_{\beta'\varepsilon'}\Gamma^{\varepsilon'}_{\delta'\gamma'}+\frac{\partial x^{\alpha}}{\partial x^{\alpha'}}\frac{\partial x^{\beta'}}{\partial x^{\beta}}\frac{\partial^2\varepsilon'}{\partial x^{\gamma}\partial x^{\delta}}\Gamma^{\alpha'}_{\beta'\varepsilon'}\\ &\quad+\frac{\partial x^{\alpha}}{\partial x^{\beta'}}\frac{\partial^2 x^{\beta'}}{\partial x^{\beta}\partial x^{\varepsilon}}\frac{\partial x^{\varepsilon}}{\partial x^{\varepsilon'}}\frac{\partial x^{\delta'}}{\partial x^{\delta}}\frac{\partial x^{\gamma'}}{\partial x^{\gamma}}\Gamma^{\varepsilon'}_{\delta'\gamma'}+\frac{\partial x^{\alpha}}{\partial x^{\beta'}}\frac{\partial^2 x^{\beta'}}{\partial x^{\beta}\partial x^{\varepsilon}}\frac{\partial x^{\varepsilon}}{\partial x^{\varepsilon'}}\frac{\partial^2 x^{\varepsilon'}}{\partial x^{\delta}\partial x^{\gamma}},\end{aligned} \tag{9.8}$$

using again (9.4), matches half the extra terms from (9.7), while the remaining half appears in a version in which β and δ are exchanged. To see this, the identity

$$\frac{\partial x^{\alpha}}{\partial x^{\beta'}}\frac{\partial^2 x^{\beta'}}{\partial x^{\beta}\partial x^{\varepsilon}} = -\frac{\partial x^{\beta'}}{\partial x^{\beta}}\frac{\partial^2 x^{\alpha}}{\partial x^{\beta'}\partial x^{\varepsilon'}}\frac{\partial x^{\varepsilon'}}{\partial x^{\varepsilon}} \tag{9.9}$$

is useful, which follows after taking a partial derivative of

$$\frac{\partial x^{\alpha}}{\partial x^{\beta'}}\frac{\partial x^{\beta'}}{\partial x^{\beta}} = \frac{\partial x^{\alpha}}{\partial x^{\beta}} = \delta^{\alpha}{}_{\beta} \tag{9.10}$$

by x^{ε}.

Combining all four terms, we obtain a 4-index object that transforms like a tensor because all higher derivatives of transformation functions cancel out. The resulting **Riemann curvature tensor**

$$R_{\beta\delta\gamma}{}^{\alpha} = \frac{\partial\Gamma^{\alpha}_{\beta\gamma}}{\partial x^{\delta}} - \frac{\partial\Gamma^{\alpha}_{\delta\gamma}}{\partial x^{\beta}} + \Gamma^{\alpha}_{\delta\varepsilon}\Gamma^{\varepsilon}_{\beta\gamma} - \Gamma^{\alpha}_{\beta\varepsilon}\Gamma^{\varepsilon}_{\delta\gamma} \tag{9.11}$$

contains complete information on curvature of space or space-time. This tensor is antisymmetric in β and δ,

$$R_{\beta\delta\gamma}{}^{\alpha} = -R_{\delta\beta\gamma}{}^{\alpha}, \tag{9.12}$$

because we had to antisymmetrize in order to remove third-order partial derivatives of the coordinate transformation. Other symmetry properties follow from the detailed expression obtained by writing the tensor in terms of metric derivatives, some of which we will derive later in a more efficient way. Taking these into account, the Riemann tensor has $d^2(d^2-1)/12$ independent components in d dimensions, or 20 components in 4-dimensional space-time. In lower-dimensional examples, we see that there are (i) zero intrinsic curvature components in one dimension, (ii) one intrinsic curvature component on 2-dimensional surfaces (given by **Gaussian curvature**), and (iii) six intrinsic curvature components in three dimensions. Standard intuition allows us to grasp only the single function provided by Gaussian curvature.

Relevant information about curvature is often contained in simpler tensorial versions obtained from the Riemann tensor by summing over some indices. The **Ricci tensor** is defined by identifying α and δ in (9.11), thus summing over them:

$$R_{\beta\gamma} = R_{\beta\alpha\gamma}{}^{\alpha} = \frac{\partial\Gamma^{\alpha}_{\beta\gamma}}{\partial x^{\alpha}} - \frac{\partial\Gamma^{\alpha}_{\alpha\gamma}}{\partial x^{\beta}} + \Gamma^{\alpha}_{\alpha\varepsilon}\Gamma^{\varepsilon}_{\beta\gamma} - \Gamma^{\alpha}_{\beta\varepsilon}\Gamma^{\varepsilon}_{\alpha\gamma}. \tag{9.13}$$

The Ricci tensor is symmetric and therefore has ten independent components in 4-dimensional space-time. It has six independent components in three dimensions, which equals the number of independent components of the Riemann tensor. In three dimensions, all information about intrinsic curvature can be obtained from the Ricci tensor. In four dimensions, there are properties of curvature not determined by the Ricci tensor.

The remaining two indices can be summed over after raising one of them using the inverse metric, $g^{\beta\gamma}$. The result is the **Ricci scalar**

$$R = g^{\beta\gamma}R_{\beta\gamma}, \tag{9.14}$$

which is equal to the trace of the matrix given by $g^{\alpha\beta}R_{\beta\gamma} = (\mathbf{g}^{-1}\mathbf{R})^{\alpha}{}_{\gamma}$. On 2-dimensional surfaces, the Ricci scalar contains all the information about intrinsic curvature and is therefore related to Gaussian curvature.

We are now able to show that the 2-plane in polar coordinates really does not have curvature even though it has non-zero components (8.150) and (8.151) of the Christoffel symbol. These components, $\Gamma^{r}_{\varphi\varphi} = -r$ and $\Gamma^{\varphi}_{r\varphi} = \Gamma^{\varphi}_{\varphi r} = 1/r$ are functions of r, such that only r-derivatives contribute to components of the Ricci tensor. We compute the components one by one, just as we computed the Christoffel symbol: We make an index choice of β and γ for a given component on the left of (9.13), identify the same components on the right, and then check all values of the summed-over indices that produce non-zero terms. The identification of summed-over indices that produce non-zero contributions may be different in each term of a sum, such as α in (9.13), because the value of an individual product of components depends on how the factors are combined. For each of the four terms, we should make independent determinations of relevant values of α.

Starting with R_{rr}, we have

$$R_{rr} = \frac{\partial \Gamma^{\alpha}_{rr}}{\partial x^{\alpha}} - \frac{\partial \Gamma^{\alpha}_{\alpha r}}{\partial r} + \Gamma^{\alpha}_{\alpha\varepsilon}\Gamma^{\varepsilon}_{rr} - \Gamma^{\alpha}_{r\varepsilon}\Gamma^{\varepsilon}_{\alpha r} = -\frac{\partial \Gamma^{\varphi}_{\varphi r}}{\partial r} - (\Gamma^{\varphi}_{\varphi r})^2 = 0 . \tag{9.15}$$

Going through all possible terms of the first expression, none of them contribute because there is no non-zero Γ^{α}_{rr} with two r-subscripts. The second term requires $\alpha = \varphi$ for a non-zero contribution. Starting the analysis of the third term, $\Gamma^{\varepsilon}_{rr}$ requires $\varepsilon = \varphi$ for a non-zero factor, but then there is no non-zero $\Gamma^{\alpha}_{\alpha\varphi}$ with the same value of α in both positions. In the last term, a subscript r requires all other indices to be φ for a non-zero result. We therefore have two non-zero terms, but they cancel out when we evaluate them with the specific component $\Gamma^{\varphi}_{\varphi r}$.

The component $R_{\varphi\varphi}$ is slightly more involved. Using the previous procedure, we write

$$R_{\varphi\varphi} = \frac{\partial \Gamma^{\alpha}_{\varphi\varphi}}{\partial x^{\alpha}} - \frac{\partial \Gamma^{\alpha}_{\alpha\varphi}}{\partial \varphi} + \Gamma^{\alpha}_{\alpha\varepsilon}\Gamma^{\varepsilon}_{\varphi\varphi} - \Gamma^{\alpha}_{\varphi\varepsilon}\Gamma^{\varepsilon}_{\alpha\varphi} = \frac{\partial \Gamma^{r}_{\varphi\varphi}}{\partial r} + \Gamma^{\varphi}_{\varphi r}\Gamma^{r}_{\varphi\varphi} - \Gamma^{r}_{\varphi\varphi}\Gamma^{\varphi}_{\varphi r} - \Gamma^{\varphi}_{\varphi r}\Gamma^{r}_{\varphi\varphi} \tag{9.16}$$

The first three terms are similar to the previous case: The first one requires $\alpha = r$ for a non-zero derivative, the second one does not produce a non-zero result because it contains a derivative by φ, and the third term requires $\varepsilon = r$ which then requires $\alpha = \varphi$. In the fourth term, we can obtain a non-zero component of the form $\Gamma^{\alpha}_{\varepsilon\varphi}$ with two independent indices, α and ε: $\alpha = r$ and $\varepsilon = \varphi$, or $\alpha = \varphi$ and $\varepsilon = r$. Both index combinations may appear because α and ε are summed over independently. The double summation in the fourth term therefore produces two non-zero contributions, which is why we have three products of Christoffel components in (9.16). Two of them cancel out if we look at the specific index combinations, and the remaining terms cancel out after using our results for the Christoffel symbol. Thus, $R_{\varphi\varphi} = 0$ and

$$R = g^{\beta\gamma}R_{\beta\gamma} = R_{rr} + \frac{1}{r^2}R_{\varphi\varphi} = 0 . \tag{9.17}$$

We do not need $R_{r\varphi}$ because it does not contribute to the Ricci scalar if the metric is diagonal, but it can also be shown to vanish. The 2-plane in polar coordinates does not have curvature.

9.1.2 COVARIANT DERIVATIVE

If we have a tensor, its partial derivatives are not tensors. But suitable combinations of a partial derivative and products with the Christoffel symbol can be shown to be a tensor, based on similar cancellations that allowed us to construct curvature tensors. In this way, the Christoffel symbol gives rise to a **covariant derivative** ∇_{β} that preserves the tensorial character of objects it acts on. It also obeys the product rule, which characterizes it as a derivative.

Starting with a vector field v^{α}, we define its covariant derivative as the 2-tensor

$$\nabla_{\beta}v^{\alpha} = \partial_{\beta}v^{\alpha} + \Gamma^{\alpha}_{\beta\gamma}v^{\gamma} . \tag{9.18}$$

As in this expression, it is often convenient to shorten partial derivatives by defining $\partial_{\beta} = \partial/\partial x^{\beta}$. The geodesic equation then takes the compact form

$$u^{\beta}\nabla_{\beta}u^{\alpha} = 0 \tag{9.19}$$

where $\mathbf{u}$ is the tangent vector field. From this perspective, a geodesic is a straight line because its tangent vector is constant with respect to the covariant derivative. Using a common term from Euclidean geometry, we call the tangent vector field **parallel** because its direction at a given point of the geodesic is parallel to the direction at any other point on the geodesic, according to the covariant derivative implied by the geometry. Given a vector field v^α on a curve that need not be a geodesic, the direction at the endpoint of the curve is called the **parallel transport** of the direction at the starting point if $u^\beta \nabla_\beta v^\alpha = 0$ everywhere on the curve with the tangent vector field $\mathbf{u}$. In this way, the covariant derivative allows us to compare directions at different points on a curved space, provided they are connected by a curve. Using more than one curve between two points, the parallel transport of a vector at one point to the other point in general depends on the curve.

We demonstrate the tensorial nature of the covariant derivative by using (9.4) and $v^{\gamma'} = (\partial x^{\gamma'}/\partial x^\gamma) v^\gamma$, such that

$$\Gamma^\alpha_{\beta\gamma} v^\gamma = \frac{\partial x^\alpha}{\partial x^{\alpha'}} \frac{\partial x^{\beta'}}{\partial x^\beta} \Gamma^{\alpha'}_{\beta'\gamma} v^\gamma + \frac{\partial x^\alpha}{\partial x^{\beta'}} \frac{\partial x^\gamma}{\partial x^{\gamma'}} \frac{\partial^2 x^{\beta'}}{\partial x^\beta \partial x^\gamma} v^{\gamma'} . \tag{9.20}$$

The partial derivative transforms as

$$\begin{aligned} \frac{\partial v^{\alpha'}}{\partial x^{\beta'}} &= \frac{\partial x^\beta}{\partial x^{\beta'}} \frac{\partial}{\partial x^\beta} \left(\frac{\partial x^{\alpha'}}{\partial x^\alpha} v^\alpha \right) \\ &= \frac{\partial x^\beta}{\partial x^{\beta'}} \frac{\partial x^{\alpha'}}{\partial x^\alpha} \frac{\partial v^\alpha}{\partial x^\beta} + \frac{\partial x^\beta}{\partial x^{\beta'}} \frac{\partial^2 x^{\alpha'}}{\partial x^\beta \partial x^\alpha} \frac{\partial x^\alpha}{\partial x^{\gamma'}} v^{\gamma'} . \end{aligned} \tag{9.21}$$

Multiplying with inverse matrices of coordinate transformations and using (9.10) in various versions with renamed indices, we derive

$$\frac{\partial v^\alpha}{\partial x^\beta} = \frac{\partial x^{\beta'}}{\partial x^\beta} \frac{\partial x^\alpha}{\partial x^{\alpha'}} \frac{\partial v^{\alpha'}}{\partial x^{\beta'}} - \frac{\partial x^\alpha}{\partial x^{\alpha'}} \frac{\partial^2 x^{\alpha'}}{\partial x^\beta \partial x^\gamma} \frac{\partial x^\gamma}{\partial x^{\gamma'}} v^{\gamma'} . \tag{9.22}$$

After renaming α' as β' in the last term, the extra terms cancel out in the sum of (9.22) and (9.20). Thus, the covariant derivative of v^α transforms like a tensor.

Like any derivative, the covariant derivative should obey the product rule. This condition can be used to derive an expression for the covariant derivative of any tensor. We first observe that the partial derivative of a function, an object without any indices, does transform like a tensor: $\partial f/\partial x^\alpha = \nabla_\alpha f$. If the function is obtained by combining a vector field v^β with a dual vector field w_γ, $f = v^\beta w_\beta$, we can rewrite $\nabla_\alpha (v^\beta w_\beta)$ in two different ways:

$$\nabla_\alpha (v^\beta w_\beta) = \partial_\alpha (v^\beta w_\beta) = (\partial_\alpha v^\beta) w_\beta + v^\beta \partial_\alpha w_\beta \tag{9.23}$$

interpreting $v^\beta w_\beta$ as a function, and

$$\nabla_\alpha (v^\beta w_\beta) = (\nabla_\alpha v^\beta) w_\beta + v^\beta \nabla_\alpha w_\beta = (\partial_\alpha v^\beta + \Gamma^\beta_{\alpha\gamma} v^\gamma) w_\beta + v^\beta \nabla_\alpha w_\beta \tag{9.24}$$

by imposing the product rule on the covariant derivative. Equating these expressions and canceling terms, we obtain

$$v^\beta \nabla_\alpha w_\beta = v^\beta \partial_\alpha w_\beta - v^\gamma \Gamma^\beta_{\alpha\gamma} w_\beta \tag{9.25}$$

for any vector field v^β. If we relabel β and γ in the last term, we can factor out v^β and conclude that

$$\nabla_\alpha w_\beta = \partial_\alpha w_\beta - \Gamma^\gamma_{\alpha\beta} w_\gamma . \tag{9.26}$$

For a 2-index object such as the metric, we repeat the procedure and compute

$$\nabla_\alpha (g_{\beta\gamma} v^\beta w^\gamma) = (\partial_\alpha g_{\beta\gamma}) v^\beta w^\gamma + g_{\beta\gamma} (\partial_\alpha v^\beta) w^\gamma + g_{\beta\gamma} v^\beta (\partial_\alpha w^\gamma) \tag{9.27}$$

as well as

$$\nabla_\alpha (g_{\beta\gamma} v^\beta w^\gamma) = (\nabla_\alpha g_{\beta\gamma}) v^\beta w^\gamma + g_{\beta\gamma} (\partial_\alpha v^\beta + \Gamma^\beta_{\alpha\delta} v^\delta) w^\gamma + g_{\beta\gamma} v^\beta (\partial_\alpha w^\gamma + \Gamma^\gamma_{\alpha\delta} w^\delta) \tag{9.28}$$

Comparing the two results, all derivatives of the two vector fields cancel out, and we obtain

$$v^{\beta}w^{\gamma}\nabla_{\alpha}g_{\beta\gamma}=v^{\beta}w^{\gamma}\left(\partial_{\alpha}g_{\beta\gamma}-g_{\delta\gamma}\Gamma^{\delta}_{\alpha\beta}-g_{\beta\delta}\Gamma^{\delta}_{\alpha\gamma}\right). \tag{9.29}$$

The explicit form of the Christoffel symbol in terms of metric components,

$$g_{\gamma\delta}\Gamma^{\delta}_{\alpha\beta}=\frac{1}{2}(\partial_{\alpha}g_{\beta\gamma}+\partial_{\beta}g_{\gamma\alpha}-\partial_{\gamma}g_{\alpha\beta}), \tag{9.30}$$

shows that

$$\nabla_{\alpha}g_{\beta\gamma}=0. \tag{9.31}$$

The metric is therefore **covariantly constant**, even if it has position-dependent components in curved space-time or with non-Cartesian coordinates. An important consequence is an unambiguous interpretation of index positions because of identities such as

$$\nabla_{\alpha}v^{\beta}=\nabla_{\alpha}(g^{\beta\gamma}v_{\gamma})=g^{\beta\gamma}\nabla_{\alpha}v_{\gamma}. \tag{9.32}$$

We are free to raise and lower indices before or after taking covariant derivatives.

The product-rule method can be applied to the covariant derivative of any tensor. As a consequence, for each index on a tensor we either subtract or add a Christoffel term according to the equations for vectors and dual vectors. For instance, (9.29) shows that there are two Christoffel terms in the covariant derivative of a generic 2-subscript tensor $T_{\beta\gamma}$, one for β and one for γ. Both terms have a negative coefficient because they are obtained for subscript indices.

9.1.2.1 Riemann curvature tensor

We use (9.26) to show that covariant derivatives commute when they act on functions:

$$\nabla_{\alpha}(\nabla_{\beta}f)=\partial_{\alpha}(\nabla_{\beta}f)-\Gamma^{\gamma}_{\alpha\beta}\nabla_{\gamma}f=\partial_{\alpha}\partial_{\beta}f-\Gamma^{\gamma}_{\alpha\beta}\partial_{\gamma}f \tag{9.33}$$

is symmetric in α and β because partial derivatives commute and the Christoffel symbol is symmetric in its subscripts. However, covariant derivatives in general do not commute when they are applied to tensors. For instance,

$$\nabla_{\alpha}(\nabla_{\beta}w_{\gamma})=\partial_{\alpha}(\nabla_{\beta}w_{\gamma})-\Gamma^{\delta}_{\alpha\beta}\nabla_{\delta}w_{\gamma}-\Gamma^{\delta}_{\alpha\gamma}\nabla_{\beta}w_{\delta} \tag{9.34}$$

because, as a consequence of the product rule, each subscript index implies a subtraction of a Γ-term as in (9.26).

The middle term is symmetric in α and β thanks to $\Gamma^{\delta}_{\alpha\beta}$ and therefore does not contribute to any non-commutativity of covariant derivatives. Combining the remaining terms and expanding the covariant derivatives, we obtain

$$\begin{aligned}\nabla_{\alpha}(\nabla_{\beta}w_{\gamma})+\Gamma^{\delta}_{\alpha\beta}\nabla_{\delta}w_{\gamma} &= \partial_{\alpha}(\nabla_{\beta}w_{\gamma})-\Gamma^{\delta}_{\alpha\gamma}\nabla_{\beta}w_{\delta}\\ &= \partial_{\alpha}(\partial_{\beta}w_{\gamma}-\Gamma^{\delta}_{\beta\gamma}w_{\delta})-\Gamma^{\delta}_{\alpha\gamma}(\partial_{\beta}w_{\delta}-\Gamma^{\varepsilon}_{\beta\delta}w_{\varepsilon})\\ &= \partial_{\alpha}\partial_{\beta}w_{\gamma}-w_{\delta}\partial_{\alpha}\Gamma^{\delta}_{\beta\gamma}-\Gamma^{\delta}_{\beta\gamma}\partial_{\alpha}w_{\delta}-\Gamma^{\delta}_{\alpha\gamma}\partial_{\beta}w_{\delta}+\Gamma^{\delta}_{\alpha\gamma}\Gamma^{\varepsilon}_{\beta\delta}w_{\varepsilon}.\end{aligned} \tag{9.35}$$

The second-order partial derivative of w_{γ} as well as the combination of both first-order terms are again symmetric in α and β. Antisymmetrization therefore implies the commutator

$$(\nabla_{\alpha}\nabla_{\beta}-\nabla_{\beta}\nabla_{\alpha})w_{\gamma}=w_{\varepsilon}\left(\partial_{\beta}\Gamma^{\varepsilon}_{\alpha\gamma}-\partial_{\alpha}\Gamma^{\varepsilon}_{\beta\gamma}+\Gamma^{\delta}_{\alpha\gamma}\Gamma^{\varepsilon}_{\beta\delta}-\Gamma^{\delta}_{\beta\gamma}\Gamma^{\varepsilon}_{\alpha\delta}\right)=w_{\varepsilon}R_{\alpha\beta\gamma}{}^{\varepsilon}. \tag{9.36}$$

The Riemann curvature tensor determines non-commutativity of covariant derivatives.

Equation (9.36) may be used as a definition of the Riemann curvature tensor if the covariant derivative is already known, avoiding the lengthy cancellation arguments. It is also a convenient starting point for new derivations of properties related to curvature. An example is the **geodesic deviation equation** which has important implications such as the form of tidal forces or a fundamental phenomenon that constitutes the basis of gravitational-wave detectors. Mathematically, the geodesic deviation equation determines how the region occupied by a family of test masses in curved space-time is deformed as they evolve, each one following its own geodesic. In flat space-time, geodesics as straight lines imply that the region is simply translated as time changes, retaining its shape. In curved space-time, different geodesics may approach one another or depart from one another, such as the great circles on a sphere approaching and then leaving one of the poles. Different parts of an extended object may change their relative distance as the whole object moves, an effect attributed to **tidal forces** acting on the extended object.

We set up initial values by choosing the time coordinate t of an observer in a region that contains the object of interest, and defining spatial positions of all test masses on the spacelike hypersurface in the given region defined by $t = t_0$ constant. Each test mass then follows a geodesic if we ignore the small gravitational force between different test masses. As time changes, we obtain additional snapshots of the object at later spacelike hypersurfaces defined by $t = t_0 + \Delta t$. The position of each object at any time t is uniquely determined by the intersection of its geodesic with the spacelike hypersurface at time t. We assume that the time duration is short enough such that geodesics of test masses do not intersect one another. Each test mass then has a unique position. For an intrinsic description of the family of test masses, we transform from the observer's time t to proper time τ along each geodesic. This transformation may imply slight deformations of the original spacelike hypersurfaces from constant-t to constant-τ owing to time dilation and relative simultaneity. The tangent vector of a geodesic in the new coordinate basis is given by $\mathbf{t} = \mathbf{e}_0 = \partial_0$.

The relative distance between test masses at a fixed value of proper time is determined by differences of spatial coordinates x^i, $i = 1,2,3$. We may assume that these coordinates have been set up by the observer as a reference system. Considering a single pair of test masses, we can arrange the spatial coordinates such that the relative direction is given by the vector field $\mathbf{X} = \mathbf{e}_1 = \partial_1$, again in the coordinate basis. Since partial derivatives commute, it does not matter whether we first apply the directional derivative along $\mathbf{X}$ and then along $\mathbf{t}$, or the other way around: We have a vanishing **vector-field commutator** $[\mathbf{t},\mathbf{X}] = 0$, where

$$[\mathbf{t},\mathbf{X}]^\alpha = t^\beta \partial_\beta X^\alpha - X^\beta \partial_\beta t^\alpha = t^\beta \nabla_\beta X^\alpha - X^\beta \nabla_\beta t^\alpha \tag{9.37}$$

as already encountered in the Lie derivative (8.176). In this combination, Christoffel terms from the covariant derivatives cancel out. Therefore, the expression of the commutator is covariant and vanishes in any basis if it vanishes in the coordinate basis.

Relative forces or relative acceleration between test masses in geodesic motion are determined by the second-order time derivative

$$\frac{\mathrm{d}^2 X^\alpha}{\mathrm{d}\tau^2} = t^\beta \nabla_\beta (t^\gamma \nabla_\gamma X^\alpha) \tag{9.38}$$

where each τ-derivative is expressed by a covariant derivative, defining the relative acceleration in a covariant way. The expression amounts to a second-order partial derivative by τ in our adjusted coordinate basis. Notice that (9.38) does not describe a single geodesic because X^α is the relative distance between two test masses rather than the position along a geodesic.

The vanishing commutator $[\mathbf{t},\mathbf{X}] = 0$ tells us that we can write the final term in (9.38) as $t^\gamma \nabla_\gamma X^\alpha = X^\gamma \nabla_\gamma t^\alpha$. The product rule applied to the first derivative in (9.38) then implies

$$\frac{\mathrm{d}^2 X^\alpha}{\mathrm{d}\tau^2} = (t^\beta \nabla_\beta X^\gamma)(\nabla_\gamma t^\alpha) + t^\beta X^\gamma \nabla_\beta \nabla_\gamma t^\alpha = (X^\beta \nabla_\beta t^\gamma)(\nabla_\gamma t^\alpha) + t^\beta X^\gamma \nabla_\beta \nabla_\gamma t^\alpha \tag{9.39}$$

where we applied the zero commutator again in the first term. The new first term contains the combination $t^\gamma \nabla_\gamma t^\alpha = 0$, which vanishes by virtue of the geodesic equation (9.19), but with two parentheses standing in the way. We rewrite this term by a reverse application of the product rule:

$$\frac{\mathrm{d}^2 X^\alpha}{\mathrm{d}\tau^2} = X^\beta \nabla_\beta (t^\gamma \nabla_\gamma t^\alpha) - X^\beta t^\gamma \nabla_\beta \nabla_\gamma t^\alpha + t^\beta X^\gamma \nabla_\beta \nabla_\gamma t^\alpha . \tag{9.40}$$

Using the geodesic equation, the relative acceleration equals

$$\frac{\mathrm{d}^2 X^\alpha}{\mathrm{d}\tau^2} = t^\beta X^\gamma (\nabla_\beta \nabla_\gamma - \nabla_\gamma \nabla_\beta) t^\alpha \tag{9.41}$$

and can be expressed in terms of the Riemann curvature tensor (9.36):

$$\frac{\mathrm{d}^2 X^\alpha}{\mathrm{d}\tau^2} = t^\beta X^\gamma R_{\beta\gamma\delta}{}^{\varepsilon} t_\varepsilon g^{\alpha\delta} . \tag{9.42}$$

In general, the right-hand side of this **geodesic deviation equation** cannot be expressed solely in terms of the Ricci tensor. The Riemann tensor is therefore required for a complete description of the gravitational force on extended objects.

9.1.2.2 Einstein tensor

Another implication of (9.36) is the **Bianchi identity** which is an important consistency condition for the field equations of general relativity. We derive this identity by taking yet another derivative of the expressions from (9.36) in the form

$$\begin{aligned} &\nabla_\alpha (\nabla_\beta \nabla_\gamma - \nabla_\gamma \nabla_\beta) w_\delta + \nabla_\beta (\nabla_\gamma \nabla_\alpha - \nabla_\alpha \nabla_\gamma) w_\delta + \nabla_\gamma (\nabla_\alpha \nabla_\beta - \nabla_\beta \nabla_\alpha) w_\delta \\ = \; &(\nabla_\alpha \nabla_\beta - \nabla_\beta \nabla_\alpha) \nabla_\gamma w_\delta + (\nabla_\beta \nabla_\gamma - \nabla_\gamma \nabla_\beta) \nabla_\alpha w_\delta + (\nabla_\gamma \nabla_\alpha - \nabla_\alpha \nabla_\gamma) \nabla_\beta w_\delta . \end{aligned} \tag{9.43}$$

The entire combination is antisymmetric in the three indices α, β and γ, and each derivative term appears in the same form on both sides of the equation. However, commutators of derivatives are placed in different positions, such that we obtain a non-trivial relationship for the curvature tensor. The equality (9.43) therefore implies a curvature identity.

In the second line, the commutator acting on a 2-index object such as $\nabla_\gamma w_\delta$ can be computed by iterating the product rule, just as we derived the expression for a single covariant derivative of a 2-index object when we showed that the metric is covariantly constant. Here, the result is

$$(\nabla_\alpha \nabla_\beta - \nabla_\beta \nabla_\alpha) \nabla_\gamma w_\delta = R_{\alpha\beta\gamma}{}^{\varepsilon} \nabla_\varepsilon w_\delta + R_{\alpha\beta\delta}{}^{\varepsilon} \nabla_\gamma w_\varepsilon \tag{9.44}$$

with two curvature terms. The antisymmetrization in α, β and γ used in (9.43) contains the antisymmetrization of $R_{\alpha\beta\gamma}{}^{\varepsilon}$ in these indices, which happens to be zero: This combination of curvature components, according to (9.36), is obtained from the antisymmetrization of

$$\nabla_\alpha \nabla_\beta w_\gamma = \partial_\alpha (\partial_\beta w_\gamma - \Gamma^\delta_{\beta\gamma} w_\delta) - \Gamma^\delta_{\alpha\beta} (\partial_\delta w_\gamma - \Gamma^\varepsilon_{\delta\gamma} w_\varepsilon) - \Gamma^\delta_{\alpha\gamma} (\partial_\beta w_\delta - \Gamma^\varepsilon_{\beta\delta} w_\varepsilon) \tag{9.45}$$

if we use (9.35). Each term on the right is symmetric in one of the index pairs that can be obtained from (α, β, γ). Since complete antisymmetrization in all three indices includes antisymmetrization in each index pair, the result is zero:

$$R_{\alpha\beta\gamma}{}^{\delta} + R_{\beta\gamma\alpha}{}^{\delta} + R_{\gamma\alpha\beta}{}^{\delta} = 0 . \tag{9.46}$$

We need only three terms even though there are six permutations of three indices: Cyclic permutations are sufficient for complete antisymmetrization because we already know that the Riemann tensor is antisymmetric in its first two indices.

The first term in (9.44) does not contribute to the complete antisymmetrization of the first line from (9.43) in (α,β,γ). The second term does contribute, and it appears in a similar form if we evaluate terms in the first line of (9.43):

$$\nabla_\alpha(\nabla_\beta\nabla_\gamma-\nabla_\gamma\nabla_\beta)w_\delta=\nabla_\alpha(R_{\beta\gamma\delta}{}^\varepsilon w_\varepsilon)=(\nabla_\alpha R_{\beta\gamma\delta}{}^\varepsilon)w_\varepsilon+R_{\beta\gamma\delta}{}^\varepsilon\nabla_\alpha w_\varepsilon\,. \tag{9.47}$$

The last term here is the same as the last term in (9.44), up to a cyclic permutation of (α,β,γ) that does not change the complete antisymmetrization. The only remaining contribution from both sides in (9.43) is therefore the complete antisymmetrization of $\nabla_\alpha R_{\beta\gamma\delta}{}^\varepsilon$, which must then vanish:

$$\nabla_\alpha R_{\beta\gamma\delta}{}^\varepsilon+\nabla_\beta R_{\gamma\alpha\delta}{}^\varepsilon+\nabla_\gamma R_{\alpha\beta\delta}{}^\varepsilon=0\,. \tag{9.48}$$

This equation is the Bianchi identity, which has an important implication for the Ricci tensor if we identify, and therefore sum over, the indices α and ε, and multiply the equation by $g^{\beta\delta}$:

$$0=\nabla_\alpha(g^{\beta\delta}R_{\beta\gamma\delta}{}^\alpha)+g^{\beta\delta}\nabla_\beta R_{\gamma\alpha\delta}{}^\alpha+\nabla_\gamma(g^{\beta\delta}R_{\alpha\beta\delta}{}^\alpha)\,. \tag{9.49}$$

The last term is the negative covariant derivative of the Ricci scalar, negative because we have to flip α and β in order to apply the definition of the Ricci tensor in (9.13). The middle term contains the covariant derivative $\nabla_\beta R_{\gamma\delta}$, and so does the first term. The last statement is not obvious because the first term contains a sum over the first and third rather than second and fourth indices of the Riemann tensor. However, there is yet another identity that relates these two summations: If we consider $R_{\alpha\beta\gamma\delta}=g_{\delta\varepsilon}R_{\alpha\beta\gamma}{}^\varepsilon$, antisymmetry in the last two indices,

$$R_{\alpha\beta\gamma\delta}=-R_{\alpha\beta\delta\gamma}, \tag{9.50}$$

follows from metric compatibility:

$$0=(\nabla_\alpha\nabla_\beta-\nabla_\beta\nabla_\alpha)g_{\gamma\delta}=R_{\alpha\beta\gamma}{}^\varepsilon g_{\varepsilon\delta}+R_{\alpha\beta\delta}{}^\varepsilon g_{\gamma\varepsilon}=R_{\alpha\beta\gamma\delta}+R_{\alpha\beta\delta\gamma} \tag{9.51}$$

where we use a version of (9.44) for the covariant-derivative commutator acting on a 2-subscript tensor. Combined with antisymmetry in the first index pair, it follows that the Ricci tensor may be written as

$$R_{\alpha\beta}=R_{\alpha\gamma\beta}{}^\gamma=g^{\gamma\delta}R_{\alpha\gamma\beta\delta}=-g^{\gamma\delta}R_{\gamma\alpha\beta\delta}=g^{\gamma\delta}R_{\gamma\alpha\delta\beta}\,, \tag{9.52}$$

summing the Riemann tensor over its first and third indices.

We conclude that the first two terms in (9.49) are the same covariant derivative of the Ricci tensor, such that

$$2\nabla_\alpha R_\gamma{}^\alpha-\nabla_\gamma R=0\,. \tag{9.53}$$

Both terms can be combined by defining the **Einstein tensor**

$$G_{\alpha\beta}=R_{\alpha\beta}-\frac{1}{2}Rg_{\alpha\beta} \tag{9.54}$$

which obeys the important equation

$$g^{\alpha\beta}\nabla_\alpha G_{\beta\gamma}=0\,. \tag{9.55}$$

9.1.3 EINSTEIN'S EQUATION

Space-time curvature is related to the gravitational force, as confirmed in the weak-field case following the equivalence principle. For strong forces or large curvature, we expect that all metric components can independently differ from their flat-case values. A single gravitational potential is no longer sufficient to describe all aspects of curvature or of gravity. For a given matter distribution, the entire metric tensor should be determined by partial differential equations, generalizing the Poisson equation for Newton's potential.

The source of Newton's gravity is the mass density. In relativity, we should generalize this expression to the energy density because mass is the rest contribution to energy. For a point mass, we also know that the energy is not a scalar but the time component of the energy-momentum vector. The source term of a relativistic gravitational equation should therefore have at least one space-time index. A second index appears if we require that the formulation should not only include point masses but also energy distributions or fields. For such a distribution, an observer in a local inertial frame can measure not only the energy (density) but also directions in which the field is moving. The observed energy $E = -\mathbf{p}\cdot\mathbf{u}_{\rm obs} = -p_\alpha u^\alpha_{\rm obs}$ of a point mass with 4-momentum $\mathbf{p}$ is then replaced by the energy current

$$j_\beta = -T_{\alpha\beta}u^\alpha_{\rm obs} \tag{9.56}$$

with a 2-index object, the **energy-momentum tensor** $T_{\alpha\beta}$, representing the field's energy content.

9.1.3.1 Energy-momentum tensor

The time component of a relativistic current is the density, here the energy density

$$\rho = -j_\beta u^\beta_{\rm obs} = T_{\alpha\beta}u^\alpha_{\rm obs}u^\beta_{\rm obs}\,. \tag{9.57}$$

Since only the symmetric part of $T_{\alpha\beta}$ contributes to this expression, we require symmetry, $T_{\alpha\beta} = T_{\beta\alpha}$. This condition eliminates any contributions that would not contribute to the energy density in any reference frame.

The spatial components of the energy current are expressed as

$$j_K = j_\beta e^\beta_K = -T_{\alpha\beta}u^\alpha_{\rm obs}e^\beta_K \tag{9.58}$$

in a completion of $\mathbf{u}_{\rm obs}$ to an orthonormal basis or a local inertial frame with spatial reference directions $\mathbf{e}_K$, $K = 1,2,3$. If we boost the observer to a different speed, the spatial reference directions change along with $\mathbf{u}_{\rm obs}$ in order to preserve orthonormality. After the boost, the observer moves relative to the field, such that a previously stationary energy distribution may appear in flux. We also know from special relativity that boosting transforms energy into momentum. Accordingly, a non-zero spatial part of the energy current may be interpreted as energy flux or momentum density.

In addition to the energy current, which by symmetry determines the time-time and time-space components of $T_{\alpha\beta}$ in a given frame, the energy-momentum tensor has purely spatial components

$$T_{KL} = T_{\alpha\beta}e^\alpha_K e^\beta_L\,, \tag{9.59}$$

called the **spatial stress tensor**. Stress is closely related to energy density because it describes how a material reacts to a local force, minimizing its energy. Stress may be anisotropic if the material reacts in a direction different from the force. We therefore need a stress tensor, rather than a single stress coefficient. If there is no anisotropic stress, the stress tensor is determined by pressure P, such that $T_{KL} = P\delta_{KL}$.

An energy-momentum tensor in space-time is not an arbitrary set of components but must be consistent with energy conservation. Conservation of the energy current j_β means that the vector field $j^\alpha = g^{\alpha\beta}j_\beta$ must obey the continuity equation

$$\nabla_\alpha j^\alpha = 0\,. \tag{9.60}$$

This equation follows from the conditions that (i) it should reproduce the continuity equation (1.88) in flat space-time, where it indeed reads $\partial\rho/\partial t + \partial j^K/\partial x^K = 0$, and (ii) remain covariant in the presence of curvature. The latter condition requires us to replace partial derivatives with covariant derivatives.

Since j_β is derived from $T_{\alpha\beta}$, a more fundamental condition for **energy conservation** is

$$g^{\alpha\beta}\nabla_\alpha T_{\beta\gamma}=0\,. \tag{9.61}$$

This law of local energy conservation implies four partial differential equations for the components of the energy-momentum tensor. As an example, a **perfect fluid** is an energy distribution that is characterized only by the energy density ρ and pressure P in its own rest frame. If the fluid rest frame is specified by a 4-velocity vector field u^α in the space-time region of the fluid, the energy-momentum tensor takes the form

$$T_{\alpha\beta}=\rho u_\alpha u_\beta+P(g_{\alpha\beta}+u_\alpha u_\beta)\,. \tag{9.62}$$

The only 2-subscript tensors available in this case are $g_{\alpha\beta}$ and $u_\alpha u_\beta$, implying two independent coefficients. The specific appearance of ρ and P in $T_{\alpha\beta}$ is such that $\rho=T_{\alpha\beta}u^\alpha u^\beta$ and $P=T_{\alpha\beta}e^\alpha e^\beta$ for any normalized spacelike $\mathbf{e}$ with $\mathbf{u}\cdot\mathbf{e}=0$. A fluid rest frame in free-fall has a 4-velocity field u^α tangent to geodesics, such that

$$u^\alpha\nabla_\alpha u_\beta=0\,. \tag{9.63}$$

The **u**-component of the conservation law then reads

$$-u^\gamma g^{\alpha\beta}\nabla_\alpha T_{\beta\gamma}=u^\alpha\nabla_\alpha\rho+(\rho+P)g^{\alpha\beta}\nabla_\alpha u_\beta\,. \tag{9.64}$$

On an FRW space-time, this equation reproduces the cosmological continuity equation (8.50).

9.1.3.2 Field equations

The energy-momentum tensor is the source of curvature according to the equations of general relativity. As a symmetric tensor, it has the same number of independent components as the metric and can therefore appear in its field equations. The metric also obeys the local conservation law (9.61) because it is always covariantly constant. However, from Newton's gravity we know that second-order derivatives of the metric (just like Newton's potential) but not the metric itself should be proportional to energy-momentum.

Covariant expressions for second-order derivatives of the metric are given by curvature tensors. A 2-index curvature tensor that obeys the local conservation law is the Einstein tensor (9.54), $G_{\alpha\beta}=R_{\alpha\beta}-\frac{1}{2}g_{\alpha\beta}R$. It is therefore possible for the Einstein tensor to be proportional to the energy-momentum tensor. Choosing the coefficient such that Poisson's equation is implied in the weak-field limit, **Einstein's equation**

$$G_{\alpha\beta}=8\pi GT_{\alpha\beta} \tag{9.65}$$

follows. A given energy-momentum tensor implies second-order partial differential equations for the metric. The factor of G on the right is Newton's constant rather than the trace of the Einstein tensor, which we will not be using.

General covariance is a strong condition that determines the form of Einstein's equation almost uniquely as a set of partial differential equations for the metric tensor, assuming up to second-order derivatives. The only additional conserved 2-index tensor that could be included is the metric itself because it is covariantly constant. Such a term would not change the derivatives, but it may be viewed as a simple form of a new gravitational potential. Einstein's equation then takes the form

$$G_{\alpha\beta}+\Lambda g_{\alpha\beta}=8\pi GT_{\alpha\beta} \tag{9.66}$$

with a free coefficient Λ, the **cosmological constant**. From a mathematical perspective, the values of G and Λ are the only choices required to specify the equation.

We can see this uniqueness result more easily if we use an action principle instead of equations of motion. The action is a single function of the physical variables, in the case of gravity given by the space-time metric $g_{\alpha\beta}$. It should not have any free indices and must therefore be an invariant expression, rather than a covariant equation. Moreover, it should contain at most second-order derivatives of the metric, which means that we should not use products of the Ricci tensor to construct it: The integrated product of two second-order derivatives is, after integrating by parts, equivalent to the product of a third-order derivative and a first-order one. Moreover, the field-theory version of Euler–Lagrange equations as used in (8.59) would in this case contain a fourth-order derivative. (Euler-Lagrange equations include additional terms if the Lagrange function contains higher-order derivatives. But these terms do not remove this fourth-order derivative.)

The only possibility to eliminate the indices of the Ricci tensor while maintaining the required derivative order is to use the Ricci scalar $R = g^{\alpha\beta}R_{\alpha\beta}$ in linear form. An action should also be integrated over time, as well as space if we have fields like the metric which depend on spatial coordinates. The Ricci scalar can be integrated after multiplication by the 4-volume element $v = \sqrt{-\det g}\,\mathrm{d}^4x$ in order take into account the correct geometry of space-time. We arrive at the **Einstein–Hilbert action**

$$S[g] = \frac{1}{16\pi G}\int\int\int\int \sqrt{-\det g}\,(R(g) - 2\Lambda)\mathrm{d}^4x \tag{9.67}$$

of general relativity. Square brackets in $S[g]$ are used to indicate that this expression is a **functional**, a function of a function that in this case depends on the metric through an integration.

The field-theory version of Euler–Lagrange equations shows that the equations of motion implied by $S[g]$ are Einstein's equation in vacuum, with $T_{\alpha\beta} = 0$. The free constants G and Λ appear as the two coefficients of the integrand, a linear expression in the Ricci scalar $R(g)$. If we add a suitable matter action, such as the action of the electromagnetic field, the energy-momentum term is included in the resulting Euler–Lagrange equations with the correct coefficient of $8\pi G$, as determined by the choice of $(16\pi G)^{-1}$ in $S[g]$.

9.1.3.3 Gravitational waves

Small curvature effects can be analyzed by perturbation theory around Minkowski space-time. To this end, we write the metric as

$$g_{\alpha\beta} = \eta_{\alpha\beta} + \omega_{\alpha\beta} \tag{9.68}$$

with $\omega_{\alpha\beta} = \omega_{\beta\alpha}$ small in the sense that we ignore any expressions quadratic or higher order in this metric perturbation. Unlike a metric, $\omega_{\alpha\beta}$ is in general not invertible. If needed, we raise its indices by using the inverse space-time metric $g^{\alpha\beta}$ or, since quadratic or higher terms in $\omega_{\alpha\beta}$ are ignored, simply by using the inverse Minkowski metric $\eta^{\alpha\beta}$. We can then show that the inverse space-time metric in perturbed form is given by

$$g^{\alpha\beta} = \eta^{\alpha\beta} - \omega^{\alpha\beta} = \eta^{\alpha\beta} - \eta^{\alpha\gamma}\eta^{\beta\delta}\omega_{\gamma\delta} \tag{9.69}$$

because

$$g^{\alpha\beta}g_{\beta\gamma} = \left(\eta^{\alpha\beta} - \eta^{\alpha\gamma}\eta^{\beta\delta}\omega_{\gamma\delta}\right)(\eta_{\beta\gamma} + \omega_{\beta\gamma}) = \delta^{\alpha}{}_{\gamma} + \eta^{\alpha\beta}\omega_{\beta\gamma} - \eta^{\alpha\gamma}\delta^{\delta}{}_{\gamma}\omega_{\gamma\delta} = \delta^{\alpha}{}_{\gamma} \tag{9.70}$$

where we ignored a quadratic term in $\omega_{\alpha\beta}$. The sign change between (9.68) and (9.69) corresponds to the opposite signs in $1 + x$ and $(1 + x)^{-1} \approx 1 - x$ for $x \ll 1$.

Field equations for $\omega_{\alpha\beta}$ are determined by the Einstein tensor of (9.68). Proceeding in steps, we first note that the perturbed Christoffel symbol is strictly first-order in $\omega_{\alpha\beta}$, without a zeroth-order term because any partial derivative of $\eta_{\alpha\beta}$ vanishes:

$$\Gamma_{(1)}{}^{\gamma}{}_{\alpha\beta} = \frac{1}{2}\eta^{\gamma\delta}(\partial_\alpha\omega_{\beta\delta} + \partial_\beta\omega_{\alpha\delta} - \partial_\delta\omega_{\alpha\beta}). \tag{9.71}$$

In particular, ${\Gamma_{(1)}}^{\gamma}{}_{\gamma\beta} = \frac{1}{2}\partial_\beta \omega^\gamma_\gamma$. The first-order Ricci tensor then contains only the derivative terms

$$R^{(1)}_{\alpha\beta} = \partial_\gamma {\Gamma_{(1)}}^{\gamma}{}_{\alpha\beta} - \partial_\alpha {\Gamma_{(1)}}^{\gamma}{}_{\gamma\beta} = \frac{1}{2}\eta^{\gamma\delta}\partial_\delta(\partial_\beta\omega_{\alpha\gamma} + \partial_\alpha\omega_{\beta\gamma}) - \frac{1}{2}\eta^{\gamma\delta}\partial_\gamma\partial_\delta\omega_{\alpha\beta} - \frac{1}{2}\partial_\alpha\partial_\beta\omega^\gamma_\gamma. \quad (9.72)$$

The Ricci scalar equals

$$R^{(1)} = \eta^{\alpha\gamma}\eta^{\beta\delta}\partial_\alpha\partial_\beta\omega_{\gamma\delta} - \eta^{\alpha\beta}\partial_\alpha\partial_\beta\omega^\gamma_\gamma. \quad (9.73)$$

Before combining these expressions in the Einstein tensor, it is useful to note that all trace terms ω^γ_γ can be absorbed by working with the redefined perturbation

$$\bar{\omega}_{\alpha\beta} = \omega_{\alpha\beta} - \frac{1}{2}\eta_{\alpha\beta}\omega^\gamma_\gamma. \quad (9.74)$$

In terms of this expression, the Einstein tensor takes the form

$$G^{(1)}_{\alpha\beta} = -\frac{1}{2}\eta^{\gamma\delta}\partial_\gamma\partial_\delta\bar{\omega}_{\alpha\beta} + \frac{1}{2}\eta^{\gamma\delta}\partial_\gamma(\partial_\beta\bar{\omega}_{\alpha\delta} + \partial_\alpha\bar{\omega}_{\beta\delta}) - \frac{1}{2}\eta_{\alpha\beta}\eta^{\gamma\gamma'}\eta^{\delta\delta'}\partial_\gamma\partial_\delta\bar{\omega}_{\gamma'\delta'}. \quad (9.75)$$

The Einstein tensor vanishes in vacuum, presenting a set of linear partial differential equations for $\bar{\omega}_{\alpha\beta}$. In spite of the linear nature, the presence of various derivative terms complicates an analysis. The first term, $\eta^{\gamma\delta}\partial_\gamma\partial_\delta\bar{\omega}_{\alpha\beta}$, implies the standard wave equation (7.10) for which well-established solution procedures exist, but the remaining terms would be more complicated. A notable feature is that all of them contain a factor of $\eta^{\alpha\beta}\partial_\alpha\bar{\omega}_{\beta\gamma}$ after commuting some of the partial derivatives. This property is realized only after introducing $\bar{\omega}_{\alpha\beta}$.

The perturbed metric is not uniquely defined because we can make a small coordinate change which, if it is non-linear, changes the Minkowski metric. After writing the transformed metric in perturbed form, the new perturbed metric receives a contribution from the perturbed transformation of the Minkowski metric. A small change of the Minkowski metric is given by the Lie derivative (8.175) with $g_{\alpha\beta} = \eta_{\alpha\beta}$, $\mathscr{L}_\xi\eta_{\alpha\beta} = \partial_\alpha\xi_\beta + \partial_\beta\xi_\alpha$ with a vector field ξ^α. Therefore, a small coordinate change in the direction ξ^α implies that the original metric perturbation $\omega_{\alpha\beta}$ is transformed to

$$\omega'_{\alpha\beta} = \omega_{\alpha\beta} + \partial_\alpha\xi_\beta + \partial_\beta\xi_\alpha. \quad (9.76)$$

After such a transformation, the inconvenient term in the perturbed Einstein tensor takes the form

$$\eta^{\gamma\delta}\partial_\gamma\bar{\omega}'_{\delta\alpha} = \frac{1}{2}\eta^{\gamma\delta}\partial_\gamma(2\omega_{\delta\alpha} + 2\partial_\delta\xi_\alpha + 2\partial_\alpha\xi_\delta - \eta_{\delta\alpha}\eta^{\beta\varepsilon}(\omega_{\beta\varepsilon} + 2\partial_\beta\xi_\varepsilon)) = \eta^{\gamma\delta}(\partial_\gamma\bar{\omega}_{\delta\alpha} + \partial_\gamma\partial_\delta\xi_\alpha). \quad (9.77)$$

Notice a cancellation in the middle term because $\eta^{\gamma\delta}\partial_\gamma\partial_\alpha\xi_\delta = \eta^{\beta\varepsilon}\partial_\beta\partial_\alpha\xi_\varepsilon = \eta^{\gamma\delta}\eta_{\delta\alpha}\eta^{\beta\varepsilon}\partial_\beta\partial_\gamma\xi_\varepsilon$. We can make the result vanish if it is possible to find a vector field ξ_α such that

$$\eta^{\gamma\delta}\partial_\gamma\partial_\delta\xi_\alpha = -\eta^{\gamma\delta}\partial_\gamma\bar{\omega}_{\delta\alpha} \quad (9.78)$$

starting with a generic perturbation $\bar{\omega}_{\delta\alpha}$. Equation (9.78) is just the wave equation for ξ_α, with standard solutions on flat space-time.

A small vector field ξ^α as a solution of (9.78) determines an infinitesimal coordinate transformation. Therefore, it is always possible to find a coordinate system such that $\eta^{\gamma\delta}\partial_\gamma\bar{\omega}_{\delta\alpha} = 0$. We do not have to construct this system explicitly, but we now know that it is always possible to bring the perturbed Einstein's equation in vacuum to the form

$$\eta^{\gamma\delta}\partial_\gamma\partial_\delta\bar{\omega}_{\alpha\beta} = 0 \quad (9.79)$$

of the standard wave equation. In particular, there are solutions of **gravitational waves** travelling at the speed of light, a new kind of massless objects.

The perturbed metric has ten independent components, some of which are removed by the condition $\eta^{\gamma\delta}\partial_\gamma\bar{\omega}_{\delta\alpha} = 0$ on our special kind of coordinate system. However, such a system is not unique because (9.78) does not have a unique solution. A solution of the wave equation is unique if and only if we specify initial values for both ξ_α and its first time derivative on an initial spacelike hypersurface. Time derivatives of ξ_α appear in the coordinate change $\partial_\gamma\xi_\delta$ of the metric with $\gamma = 0$, which contribute to the perturbation components $\omega_{0\delta}$. We may therefore place conditions on these components for further restrictions of the coordinate system. A convenient choice is given by setting $\omega_{0i} = 0$ for $i = 1,2,3$, and choosing ω_{00} such that the trace $\eta^{\gamma\delta}\omega_{\gamma\delta}$ vanishes. These conditions imply $\omega_{\gamma\delta} = \bar{\omega}_{\gamma\delta}$. The original metric perturbation $\omega_{\gamma\delta} = \bar{\omega}_{\gamma\delta}$ then obeys the flat-space wave equation in vacuum.

After the last two choices, the condition $\eta^{\gamma\delta}\partial_\gamma\omega_{\delta\alpha} = 0$ evaluated for $\alpha = 0$ implies $-\partial_t\omega_{00} = 0$. The vacuum wave equation for ω_{00} is then reduced to the elliptic equation $\nabla^2\omega_{00} = 0$, which has a unique solution for given boundary values enclosing a spatial region. In particular, if the perturbation vanishes at large distances, far from any heavy masses, the boundary values are zero and the unique solution is $\omega_{00} = 0$. Together with the choice $\omega_{0i} = 0$ for $i = 1,2,3$, we see that all components $\omega_{0\alpha}$ of a vacuum solution can be eliminated by a suitable coordinate choice.

The spatial components ω_{ij} obey the wave equation $\eta^{\gamma\delta}\partial_\gamma\partial_\delta\omega_{ij} = 0$ in vacuum. In addition, we inherit the condition $\partial_i\omega^i{}_j = 0$ and a vanishing spatial trace $\omega^i{}_i = 0$ from our previous conditions, $\eta^{\gamma\delta}\partial_\gamma\bar{\omega}_{\delta\alpha} = 0$ and $\eta^{\gamma\delta}\omega_{\gamma\delta} = 0$, combined with $\omega_{0\alpha} = 0$. The former condition implies that gravitational waves are transversal: Using a plane-wave form $\omega_{jk} = A_{jk}\exp(ik_\alpha x^\alpha)$ in a sufficiently small region, with an amplitude matrix A_{jk} and a wave 4-vector $\mathbf{k}$, $\partial_i\omega^i{}_j = 0$ implies that the amplitudes are transversal to the spatial direction k^j of the wave: $k^jA_{jk} = 0$. A vanishing spatial trace of the amplitudes, $\delta^{jk}A_{jk} = 0$, is implied by the vanishing spatial trace of ω_{jk}.

Writing the amplitude matrix A_{jk} in a basis in which the wave vector k^j is the first basis vector and the remainder is a basis of the transversal plane, it takes the form

$$\mathbf{A} = \begin{pmatrix} 0 & 0 & 0 \\ 0 & A_{22} & A_{23} \\ 0 & A_{23} & -A_{22} \end{pmatrix} \tag{9.80}$$

using transversality, symmetry, and zero trace. The presence of two free functions implies that gravitational waves have two polarizations. If $A_{22} \neq 0$, the wave alternately stretches one transversal direction and contracts the other one. If $A_{23} \neq 0$, the wave stretches and contracts diagonal directions in the transversal plane, corresponding to the eigenvectors of the matrix (9.80) with $A_{22} = 0$.

Gravitational waves have two polarizations, just like electromagnetic waves. However, unlike the two orthogonal directions of a vector such as the electric field, which determines the independent linear polarizations of an electromagnetic wave, the two polarizations of the metric tensor, shown in Figure 9.1, are related by a rotation of 45°. In quantum mechanics, this difference implies that electromagnetic waves are described by photons with spin one, while gravitational waves are described by gravitons with spin two: A spin-2 representation of rotations contains factors of $e^{im\varphi} = e^{2i\varphi}$, such that an angle of 45° is sufficient to produce the other polarization.

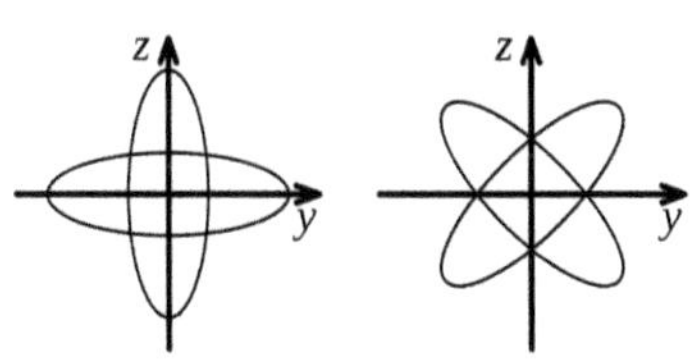

Figure 9.1 The two polarizations of a gravitational wave, each one indicated by alternating expansions and contractions in two orthogonal directions in a plane transversal to the wave vector.

A gravitational wave implies time-dependent curvature, and therefore time-dependent tidal forces on extended objects. Relevant effects can be estimated by applying the geodesic deviation equation (9.42) to a perturbed space-time metric and slowly-moving test masses. The latter condition implies that the geodesic tangent vector field t^α, corresponding to the 4-velocity $(\gamma_{\vec{V}}, \gamma_{\vec{V}}\vec{V})^T \approx (1,0,0,0)^T$ of non-relativistic motion, simply projects out time components of the Riemann curvature tensor:

$$\frac{\mathrm{d}^2 X^\alpha}{\mathrm{d}\tau^2} = t^\beta X^\gamma R_{\beta\gamma\delta}{}^\varepsilon t_\varepsilon g^{\alpha\delta} = -t^\beta X^\gamma t^\delta R_{\beta\gamma\delta}{}^\alpha \approx R_{\beta 00}{}^\alpha X^\beta\,, \tag{9.81}$$

where we used (9.50) in the second step. The perturbed Riemann components are given by

$$R_{\beta 00}{}^\alpha \approx \partial_0 \Gamma^\alpha_{\beta 0} - \partial_\beta \Gamma^\alpha_{00} \tag{9.82}$$

since we can ignore quadratic terms in $\omega_{\alpha\beta}$. We concluded that the metric perturbation of a gravitational wave in vacuum, which we may use because we are treating an extended object as a test mass, and with a suitable choice of coordinates, has only spatial components. Thus,

$$\Gamma^\alpha_{00} = 0 \quad \text{and} \quad \Gamma^\alpha_{\beta 0} = \frac{1}{2}\eta^{\alpha\gamma}\partial_0 \omega_{\beta\gamma} \tag{9.83}$$

and we obtain

$$R_{\beta 00}{}^\alpha \approx \frac{1}{2}\eta^{\alpha\gamma}\frac{\partial^2 \omega_{\beta\gamma}}{\partial t^2}\,. \tag{9.84}$$

A gravitational wave therefore implies a periodic tidal force on extended objects, proportional to the squared frequency of the wave. Such a force stretches and contracts the arms of an interferometer, which is how gravitational waves have been detected first by LIGO, the Laser Interferometer Gravitational-Wave Observatory, in 2015. Another and older proposal is given by bar detectors, large metal cylinders designed to be at resonance in an expected range of gravitational-wave frequencies. Pulsar-timing arrays, a third approach, do not make use of an extended test object but rather detect variations of precise pulsar frequencies implied by the changing proper lengths within the region of a gravitational wave.

9.1.3.4 Weak-field limit

Linear perturbations of the metric tensor can be used to determine a precise weak-field limit of general relativity, in which gravitational effects are small and motion is non-relativistic. The latter means that time derivatives of the metric can be ignored in the wave equation, and that the energy-momentum tensor is dominated by its time component: $T_{\alpha\beta} = \rho t_\alpha t_\beta$ with an absolute time direction t^α as in Newtonian gravity: The energy density contains a mass term, which in non-relativistic situations is much greater than the kinetic energy or any momentum transfer that would contribute to the pressure.

With these assumptions, the wave equation obtained from (9.75) splits into two conditions,

$$\nabla^2 \bar{\omega}_{00} = -16\pi G\rho \tag{9.85}$$

and

$$\nabla^2 \bar{\omega}_{\alpha\beta} = 0 \tag{9.86}$$

if $\alpha \neq 0$ or $\beta \neq 0$. In both equations, ∇^2 is the Laplace operator on flat space.

These equations are elliptic partial differential equations, which implies that they have unique solutions if we specify spatial boundary values for $\bar{\omega}_{\alpha\beta}$. Placing the boundary sufficiently far from any massive sources, we may assume that all boundary values vanish. The second equation then immediately implies the solution $\bar{\omega}_{\alpha\beta} = 0$ if $\alpha \neq 0$ or $\beta \neq 0$, while the solution for $\bar{\omega}_{00}$ depends on

ρ in the interior of our region. In the coordinate basis with t_α as the timelike vector field, we have $\bar{\omega}_{\alpha\beta} = \bar{\omega}_{00} t_\alpha t_\beta$.

Keeping $\bar{\omega}_{00}$ as an undetermined function, our partial solution is sufficient to transform $\bar{\omega}_{\alpha\beta}$ back to $\omega_{\alpha\beta}$. The trace of $\bar{\omega}_{\alpha\beta}$ is given by $\bar{\omega}^\alpha{}_\alpha = \eta^{00}\bar{\omega}_{00} = -\bar{\omega}_{00}$. Thus,

$$\omega_{00} = \bar{\omega}_{00} + \frac{1}{2}\eta_{00}\omega^\alpha{}_\alpha = \bar{\omega}_{00} - \frac{1}{2}\eta_{00}\bar{\omega}^\alpha{}_\alpha = \frac{1}{2}\bar{\omega}_{00}\,. \tag{9.87}$$

In the second step, we use the relationship $\bar{\omega}^\alpha{}_\alpha = -\omega^\alpha{}_\alpha$ between the traces of $\bar{\omega}_{\alpha\beta}$ and $\omega_{\alpha\beta}$, which follows from taking the trace of the definition (9.74).

Comparing (9.85) with Poisson's equation

$$\nabla^2\Phi = 4\pi G\rho \tag{9.88}$$

suggests to rewrite the perturbation equations in terms of $\Phi = -\frac{1}{4}\bar{\omega}_{00} = -\frac{1}{2}\omega_{00}$. The complete metric perturbation can then be written as

$$\omega_{\alpha\beta} = \bar{\omega}_{\alpha\beta} - \frac{1}{2}\eta_{\alpha\beta}\bar{\omega}^\gamma{}_\gamma = \left(t_\alpha t_\beta + \frac{1}{2}\eta_{\alpha\beta}\right)\bar{\omega}_{00} = -2(2t_\alpha t_\beta + \eta_{\alpha\beta})\Phi = -2\delta_{\alpha\beta}\Phi \tag{9.89}$$

Note that the terms $2t_\alpha t_\beta$ and $\eta_{\alpha\beta}$ combine to form a Kronecker delta in a space-time coordinate basis. In general, this expression is not a covariant space-time tensor, but it can appear here because the absolute Newtonian frame breaks covariance via the distinguished time direction t^α.

The space-time metric $g_{\alpha\beta} = \eta_{\alpha\beta} + \omega_{\alpha\beta}$ with (9.89) implies the line element

$$\mathrm{d}s^2 = -(1+2\Phi)\mathrm{d}t^2 + (1-2\Phi)(\mathrm{d}x^2 + \mathrm{d}y^2 + \mathrm{d}z^2)\,, \tag{9.90}$$

confirming (8.14).

9.2 GENERAL RELATIVITY IN HAMILTONIAN FORM

General covariance determines the form of Einstein's equation up to the choice of two constants. The Einstein tensor depends on up to second-order derivatives of the metric, and therefore we obtain second-order field equations. However, not all components of Einstein's equation are of second order: The local conservation law $0 = g^{\alpha\beta}\nabla_\alpha G_{\beta\gamma} = \nabla_\alpha G^\alpha{}_\gamma$ can be written as

$$\frac{\partial G^0{}_\gamma}{\partial t} = -\Gamma^0_{0\delta}G^\delta{}_\gamma + \Gamma^\delta_{0\gamma}G^0{}_\delta - \nabla_i G^i{}_\gamma \tag{9.91}$$

where $i = 1,2,3$ takes only spatial indices. Each term on the right is at most second order in time derivatives of the metric because we take only spatial derivatives in ∇_i, and the Christoffel components may introduce additional time derivatives only by multiplying with first-order time derivatives of the metric. If G^0_γ could contain terms of second order in time derivatives of the metric, the additional time derivative on the left would imply third-order time derivatives. Since such terms would have no match on the right, we conclude that $G^0{}_\gamma$ contains at most first-order time derivatives of the metric. These components may contain second-order spatial derivatives, to which our argument does not apply.

Although we may think of all components of $G_{\alpha\beta} = 8\pi G T_{\alpha\beta}$ as equal parts of the field equations because they are related to one another by general covariance, some of them play different roles from a mathematical perspective. The spatial part G_{ij} contains second-order time derivatives of the metric and therefore provides evolution equations in the standard sense. We usually try to find solutions of such equations by starting with values for the field, here the metric, and its first-order time derivatives at an initial time. The four components $G^0{}_\gamma$, however, depend on at most first-order time derivatives of the metric. These equations tell us that we are not free to choose arbitrary initial values for the evolution equations determined by G_{ij} because there are non-trivial relationships $G^0{}_\gamma = 8\pi G T^0{}_\gamma$ between the metric components and their first-order derivatives. Such additional conditions are required for general covariance in the form of the local conservation law of the Einstein tensor. They impose **constraints** on the initial values. Given these different types of equations, solving Einstein's equation is a non-standard problem, even if we set aside the non-linear nature of these partial differential equations.

9.2.1 METRIC COMPONENTS

The time derivative of the metric $g_{\alpha\beta}$, used in the preceding arguments, gives rise to another set of questions. The metric is defined on all of space-time and does not evolve in the usual way. There is a frame-dependent time direction in space-time, and the geometry may be changing in this direction. For instance, the spatial part of the FRW line element is rescaled as proper time of co-moving observers changes. However, only the spatial part in this example evolves in a meaningful way, but not the full metric. Moreover, the time direction and the spatial part of a metric are not unique in general.

9.2.1.1 Frames

We can define the spatial part of the metric if we use a specific choice of a global time coordinate t, such that its coordinate lines are indeed timelike. Setting this coordinate equal to some constant, $t = t_0$, then defines a spacelike hypersurface embedded in space-time. On this hypersurface, there is a unique induced metric as defined in Section 8.6.1, which we may identify as the **spatial metric** relative to the given time coordinate. An explicit derivation of the induced metric from the line element of the embedding space requires a parameterization of the embedded hypersurface. Alternatively, if we make use of the unit normal n^α on the hypersurface, defined such that $g_{\alpha\beta}n^\alpha n^\beta = -1$ and $g_{\alpha\beta}n^\alpha s^\beta$ for any s^β tangent to the hypersurface, we recognize that the tensor

$$h_{\alpha\beta} = g_{\alpha\beta} + n_\alpha n_\beta \tag{9.92}$$

determines properties of the induced metric because $h_{\alpha\beta}s^\beta = g_{\alpha\beta}s^\beta$ preserves geometrical properties for any spatial tangent direction, while $h_{\alpha\beta}n^\beta = 0$ shows that distances in the normal direction are removed. We already encountered the combination (9.92) in the energy-momentum tensor (9.62) of the perfect fluid. However, $h_{\alpha\beta}$ is still a space-time tensor because the indices may take all four values. We will see soon how a 3-dimensional spatial metric tensor can be derived from $h_{\alpha\beta}$.

A time coordinate defines spacelike hypersurfaces, but not a time direction. It only tells us that we evolve from a hypersurface at constant t_0 to some later hypersurface at constant $t_0 + \Delta t$, but it does not determine a unique point on the latter hypersurface that represents an older version of an initial point on the former hypersurface. A complete evolution picture requires a set of timelike worldlines. One option related to the coordinate t is given by its coordinate lines. However, time coordinate lines require a choice of spatial coordinates to be held constant on a given line. We have not introduced such coordinates in our construction, and working with evolving spatial geometries on hypersurfaces is more efficient if we do not have to use a fixed choice of spatial coordinates.

Instead of fixing spatial coordinates, we introduce a **time-evolution vector field** in order to specify a timelike direction everywhere in space-time. Such a vector field constitutes a minimal ingredient that tells us how a spatial point at time t_0 moves forward to a spatial point at $t_0 + \delta t$ an infinitesimal amount of time later, or some finite time later if we use the worldlines to which it is tangent. A time-evolution vector field is not unique and therefore requires additional choices. We capture its independent ingredients by formulating a suitable parameterization for a given family of spacelike hypersurfaces determined by a time coordinate.

Any family of spacelike hypersurfaces determines a timelike vector field, given by the unit normal n^α. This vector field may not always be the time direction we are interested in, but it can be used as part of a basis together with a spatial basis s_i^α of three vector fields tangent to the spatial hypersurfaces. Any time-evolution vector field can then be written as

$$t^\alpha = Nn^\alpha + M^i s_i^\alpha \tag{9.93}$$

with four functions N and M^i on space-time. We apply the summation convention to spatial indices i. The first function, N, is called the **lapse function** because it determines time lapses in the normal direction. The spatial components N^i form the **shift vector**. The time-evolution vector field need not be normalized, and therefore it does not equal the 4-velocity of an observer in a local inertial frame. If we normalize it, we obtain

$$\frac{t^\alpha}{\sqrt{-||t||^2}} = \frac{n^\alpha + s_i^\alpha M^i/N}{\sqrt{1-|\vec{M}|^2/N^2}}\,. \tag{9.94}$$

Relative to the normal frame n^α, the normalized time-evolution vector field takes the form of a 4-velocity with 3-velocity $\vec{V} = \vec{M}/N$.

Our new geometrical ingredients for a fixed time coordinate are given by $h_{\alpha\beta}$ together with N and M^i. It turns out that the set of all these functions is equivalent to the components of the space-time metric, but organized in a different way. Starting with the inverse-metric version of (9.92) and then inserting (9.93), we write

$$\begin{aligned} g^{\alpha\beta} &= h^{\alpha\beta} - n^\alpha n^\beta = h^{\alpha\beta} - \frac{1}{N^2}(t^\alpha - M^i s_i^\alpha)(t^\beta - M^j s_j^\beta) \\ &= -\frac{1}{N^2}t^\alpha t^\beta + \frac{M^i}{N^2}(t^\alpha s_i^\beta + t^\beta s_i^\alpha) + (h^{ij} - M^i M^j/N^2)s_i^\alpha s_j^\beta\,. \end{aligned} \tag{9.95}$$

Here, $h^{\alpha\beta}$ is obtained by raising the indices using the space-time inverse metric $g^{\alpha\beta}$. Even though $h_{\alpha\beta}$ is related to the induced metric, it is not invertible on space-time because it does not have components for directions normal to a hypersurface. We may now use the basis s_i^α on a spatial hypersurface and define the **spatial metric** $h_{ij} = h_{\alpha\beta}s_i^\alpha s_j^\beta$ on this hypersurface, which is invertible as a 3×3-matrix. We distinguish between h_{ij} and the space-time tensor $h_{\alpha\beta}$, to which it is closely related, by referring to the latter in the form $h^\alpha{}_\beta = g^{\alpha\gamma}h_{\gamma\beta}$ as the **spatial projector**: It preserves any vector tangent to the hypersurface, $h^\alpha{}_\beta s^\beta = s^\alpha$ if $s^\alpha n_\alpha = 0$, and maps a normal vector to the null vector, $h^\alpha{}_\beta n^\beta = 0$. The inverse spatial metric then appears in the spatial part of the inverse space-time metric (9.95), which has the time-time component $-1/N^2$ and time-space components M^i/N^2. An inverse space-time metric given in this form is called the **Arnowitt–Deser–Misner (ADM) decomposition**.

Just like the components of Einstein's equation, the metric components play different roles. The time-time and time-space components determine the frame of an observer corresponding to a set of constant-time hypersurfaces and a given time-evolution vector field. The spatial metric is then defined on each hypersurface and may be changing, or evolving. We expect evolution equations for h_{ij}, but the components N and M^i that determine the frame are not required to evolve in the usual way. Going through all derivatives in Einstein's equations indeed shows that second-order time derivatives appear for h_{ij}, but not for N and M^i.

9.2.1.2 Lagrange functional

Instead of analyzing the Einstein tensor directly, it is more convenient to decompose the Lagrange functional of general relativity into different components that may resemble kinetic or potential contributions, with corresponding dependencies on time derivatives as velocities. The new view on the space-time metric indeed allows us to define the velocity of the metric as

$$v_{ij} = \mathscr{L}_t h_{ij} \tag{9.96}$$

using the Lie derivative in the direction of the time-evolution vector field. Geometrically, this object is closely related to the **extrinsic curvature** of spacelike hypersurfaces in space-time. In general, the extrinsic-curvature tensor of a family of hypersurfaces with unit normal n^α is defined as

$$K_{\alpha\beta} = \frac{1}{2}\mathscr{L}_n h_{\alpha\beta}\,. \tag{9.97}$$

This tensor expresses how the induced metric changes if we expand the hypersurface slightly in the varying normal direction at each point. The bulges and crevices on a surface with extrinsic curvature mean that the induced metric gets stretched in this process, implying a non-zero Lie derivative. For a flat hyperplane, by contrast, we merely translate the plane if we move in the normal direction without changing its induced metric.

The extrinsic-curvature tensor is defined as a space-time tensor, but its non-zero contributions form a spatial tensor on each spacelike hypersurface because $n^\alpha K_{\alpha\beta} = 0 = n^\beta K_{\alpha\beta}$. The second equality follows from the first one because $K_{\alpha\beta} = K_{\beta\alpha}$ is a symmetric tensor. Applying the Lie derivative along n^γ to $0 = n^\alpha h_{\alpha\beta}$ and using the product rule,

$$0 = \mathscr{L}_n(n^\alpha h_{\alpha\beta}) = (\mathscr{L}_n n^\alpha)h_{\alpha\beta} + n^\alpha \mathscr{L}_n h_{\alpha\beta} = 2n^\alpha K_{\alpha\beta} \tag{9.98}$$

demonstrates the claimed property because $\mathscr{L}_n n^\alpha = [n,n]^\alpha = 0$ based on (8.176).

We evaluate the extrinsic-curvature tensor by using the expression (8.175) for the Lie derivative. Here,

$$\begin{aligned}\mathscr{L}_n h_{\alpha\beta} &= h_{\delta\beta}\frac{\partial n^\delta}{\partial x^\alpha} + h_{\alpha\delta}\frac{\partial n^\delta}{\partial x^\beta} + n^\varepsilon \frac{\partial h_{\alpha\beta}}{\partial x^\varepsilon} = \frac{1}{N}\mathscr{L}_{Nn} h_{\alpha\beta}\\ &= \frac{1}{N}\mathscr{L}_{t-M} h_{\alpha\beta} = \frac{1}{N}(\mathscr{L}_t h_{\alpha\beta} - \mathscr{L}_M h_{\alpha\beta})\,.\end{aligned} \tag{9.99}$$

In the first line, it is possible to move the lapse function N into the vector along which we take the derivative because, in spite of the product rule, any term that contains a derivative of the lapse function is multiplied by the vanishing $h_{\alpha\beta}n^\beta$. Since the resulting expression is linear in $\mathscr{L}_t h_{\alpha\beta}$, with non-zero components forming a spatial tensor just like $K_{\alpha\beta}$, we may use it instead of the metric velocity v_{ij}.

We now have an evolving field h_{ij} with its velocity, expressed in terms of the extrinsic-curvature tensor. Some of the relevant derivatives should appear in the Lagrange functional

$$L[g] = \frac{1}{16\pi G}\int\int\int \sqrt{-\det g}\,(R(g) - 2\Lambda)\mathrm{d}^3x \tag{9.100}$$

of general relativity via the space-time Ricci scalar R. Identifying the correct components is simplified by geometrical identities that relate the Ricci scalar R of a 4-dimensional embedding space to the Ricci scalar R_3 and the spatial extrinsic-curvature tensor $K_{ij} = s_i^\alpha s_j^\beta K_{\alpha\beta}$ of an embedded hypersurface. In particular, the Gauss–Codazzi relation, see equation (9.177) in Section 9.3, tells us that

$$R = R_3 + K_{ij}K^{ij} - (K^i{}_i)^2 \tag{9.101}$$

up to a covariant divergence that would only contribute a boundary term to the action integral. From a physics perspective, the right-hand side does take the form of a kinetic energy quadratic in extrinsic curvature, and a potential R_3 that only depends on the induced spatial metric and its spatial derivatives.

For the volume element, we have to compute the determinant of the space-time metric, expressed in terms of h_{ij}, N and M^i. Since we know the components of the inverse space-time metric (9.95), we can use a convenient shortcut provided by the co-matrix formula (3.62). Applying this equation to the time-time component of the inverse space-time metric, we write

$$g^{00} = \frac{M(0,0)}{\det g} = \frac{\det h}{\det g} \tag{9.102}$$

because removing the 0-components of the space-time metric, corresponding to the time direction t^α, results in the spatial metric. Equation (9.95) tells us that $g^{00} = -1/N^2$. Thus, we derive $\det g = -N^2 \det h$ or

$$\sqrt{-\det g} = N\sqrt{\det h}\,. \tag{9.103}$$

9.2.2 CONSTRAINTS

Combining the decomposition of metric and curvature components, we arrive at the Lagrange functional

$$\begin{aligned} L[h,N,M] &= \frac{1}{16\pi G}\int\int\int \sqrt{-\det g}\,(R(g)-2\Lambda)\mathrm{d}^3x \\ &= \frac{1}{16\pi G}\int\int\int N\sqrt{\det h}\,(R_3(h)+K_{ij}K^{ij}-(K^i{}_i)^2-2\Lambda)\mathrm{d}^3x\,. \end{aligned} \tag{9.104}$$

Formally, the spatial Ricci scalar R_3, together with the cosmological constant, serves as a potential of general relativity, while extrinsic curvature provides the kinetic contribution in terms of the metric velocity $\mathscr{L}_t h_{\alpha\beta}$. Extrinsic curvature also depends on N and M^i, but not through their time derivatives: The contribution to K_{ij} from the last term in (9.99),

$$s_i^\alpha s_j^\beta \mathscr{L}_M h_{\alpha\beta} = h_{j\gamma} s_i^\alpha \frac{\partial M^\gamma}{\partial x^\alpha} + h_{i\gamma} s_j^\beta \frac{\partial M^\gamma}{\partial x^\beta} + s_i^\alpha s_j^\beta M^k s_k^\gamma \frac{\partial h_{\alpha\beta}}{\partial x^\gamma}\,, \tag{9.105}$$

using (8.175), contains only derivatives of $M^\gamma = s_k^\gamma M^k$ in directions tangential to the hypersurface. The Lagrange functional then does not depend on velocity components of N and M^j, in which case Euler–Lagrange equations do not produce equations of motion. Instead, we obtain constraint equations that do not contain second-order time derivatives: Writing the integrand of $L[h,N,M]$ as $l(h,N,M)$ of which we can take partial derivatives, these Euler–Lagrange equations require

$$\frac{\partial l(h,N,M)}{\partial N} = 0 \tag{9.106}$$

as well as

$$\frac{\partial l(h,N,M)}{\partial M^i} - \partial_j\left(\frac{\partial l(h,N,M)}{\partial(\partial_j M^i)}\right) = 0\,. \tag{9.107}$$

The latter equation takes into account the fact that the Lagrange functional depends on $\partial_j M^i$, which can be turned into an M^i without a spatial derivative after integrating by parts. These four **constraints** correspond to the components $G^0{}_\gamma$ of the Einstein tensor which, as we have seen, do not depend on time derivatives of the metric.

Before we proceed, it is convenient to introduce a compact notation for derivatives of the form (9.107) that take into account spatial derivatives in an integrand. We rewrite the first equation, (9.106), as the **functional derivative** of the Lagrange functional, rather than its integrand:

$$\frac{\delta L[h,N,M]}{\delta N} = 0 \tag{9.108}$$

where the δ-derivative is defined by the basic expression

$$\frac{\delta N(x)}{\delta N(y)} = \delta^3(x-y) \tag{9.109}$$

together with the chain rule. On the right-hand side, $\delta^3(x-y)$ is Dirac's delta function $\delta^3(x-y) = \Pi_{i=1}^3 \delta(x^i - y^i)$ in three dimensions. The functional derivative can therefore be interpreted as varying a field by adding a delta peak at a given position. For non-linear functions of N, we apply the functional derivative by using the chain rule. For instance, for $L[h,N,M] = \int\int\int l(h,N,M)\mathrm{d}^3x$ with an integrand that does not depend on spatial derivatives of N, the chain rule implies

$$\frac{\delta L[h,N,M]}{\delta N(y)} = \int\int\int \frac{\partial l(h,N,M)}{\partial N(x)} \frac{\delta N(x)}{\delta N(y)} \mathrm{d}^3x = \frac{\partial l(h,N,M)}{\partial N(y)} \tag{9.110}$$

upon integrating out the delta function implied by (9.109). The appearance of a delta function shows that we should apply functional derivatives to integrated expressions such as $L[h,N,M]$ which depend on the complete independent functions, rather than the integrand $l(h,N,M)$ which at a given point depends only on evaluations of the independent functions and their partial derivatives.

From the perspective of a partial derivative by the value $N(x)$ at a given point, both $N(y)$ and $\partial_j N(y)$ are independent because we can change the value and the slope of a function independently at a given point; see our discussion of curve variations in the context of Figure 8.5. For the functional derivative, however, a field and its partial derivatives are not independent because we are varying the entire function, which determines its evaluation and its slope. Since a derivative by $N(y)$ and a spatial derivative by x^j are independent and therefore commuting operations, we have a non-trivial

$$\frac{\delta(\partial_j N(x))}{\delta N(y)} = \frac{\delta}{\delta N(y)} \frac{\partial N(x)}{\partial x^j} = \frac{\partial}{\partial x^j} \frac{\delta N(x)}{\delta N(y)} = \frac{\partial \delta^3(x-y)}{\partial x^j}. \tag{9.111}$$

Partial derivatives of the delta function are defined via integration by parts applied to any integral in which such derivatives may appear. Compare with our example in the context of the delta-function potential, as described in the caption of Figure 2.10. For instance, equation (9.107) is obtained as the functional derivative by M^i acting on a function that depends on both M^j and its first-order spatial derivatives:

$$\begin{aligned}
\frac{\delta L[h,N,M]}{\delta M^i(y)} &= \int\int\int \left(\frac{\partial l(h,N,M)}{\partial M^j(x)} \frac{\delta M^j(x)}{\delta M^i(y)} + \frac{\partial l(h,N,M)}{\partial(\partial_k M^j(x))} \frac{\delta(\partial_k M^j(x))}{\delta M^i(y)} \right) \mathrm{d}^3x \\
&= \int\int\int \left(\frac{\partial l(h,N,M)}{\partial M^j(x)} \delta^3(x-y)\delta_i^j + \frac{\partial l(h,N,M)}{\partial(\partial_k M^j(x))} \delta_i^j \partial_k \delta^3(x-y) \right) \mathrm{d}^3x \\
&= \int\int\int \left(\frac{\partial l(h,N,M)}{\partial M^i(x)} - \partial_k \frac{\partial l(h,N,M)}{\partial(\partial_k M^i(x))} \right) \delta^3(x-y)\mathrm{d}^3x \\
&= \frac{\partial l(h,N,M)}{\partial M^i(x)} - \partial_k \frac{\partial l(h,N,M)}{\partial(\partial_k M^i(x))}.
\end{aligned} \tag{9.112}$$

If there are higher-order spatial derivatives, all contributions to the functional derivative are produced by successive integrations by parts, matching terms in the Euler–Lagrange equations.

The spatial metric h_{ij} appears with its time derivative $\mathscr{L}_t h_{ij}$ in the Lagrange functional and is therefore subject to evolution rather than constraint equations. In the presence of constraints, it is convenient to derive these equations in Hamiltonian or canonical form by first replacing velocity components K_{ij} with momentum components

$$p^{ij}(x) = \frac{\delta L[h,N,M]}{\mathscr{L}_t h_{ij}(x)} = \frac{1}{2N} \frac{\delta L[h,N,M]}{\delta K_{ij}(x)} = \frac{\sqrt{\det h}}{16\pi G}(K^{ij} - K^k{}_k h^{ij}). \tag{9.113}$$

Momentum components are therefore obtained by a linear transformation of velocity components. Indices of K_{ij} and p^{ij} are raised or lowered by using the spatial metric h_{ij} and its inverse.

We invert this equation for K_{ij} in terms of p^{ij} by taking the trace on both sides of the equation, such that

$$p^i{}_i = \frac{\sqrt{\det h}}{16\pi G}(K^i{}_i - 3K^k{}_k) = -\frac{\sqrt{\det h}}{8\pi G}K^i{}_i, \tag{9.114}$$

and writing

$$K_{ij} = \frac{16\pi G}{\sqrt{\det h}}p_{ij} + K^k{}_k h_{ij} = \frac{16\pi G}{\sqrt{\det h}}\left(p_{ij} - \frac{1}{2}p^k{}_k h_{ij}\right). \tag{9.115}$$

Thus,

$$\mathcal{L}_t h_{ij} = \frac{16\pi GN}{\sqrt{\det h}}(2p_{ij} - p^k{}_k h_{ij}) + \mathcal{L}_M h_{ij} \tag{9.116}$$

using (9.97) and (9.99). We perform a Legendre transformation of our Lagrange functional $L[h,N,M]$ to the Hamilton functional

$$\begin{aligned} H[h,p,N,M] &= \int\!\!\int\!\!\int \left((\mathcal{L}_t h_{ij})p^{ij} - l(h,N,M)\right) \mathrm{d}^3x \qquad (9.117)\\ &= \int\!\!\int\!\!\int \left(\frac{16\pi GN}{\sqrt{\det h}}\left(p_{ij}p^{ij} - \frac{1}{2}(p^k{}_k)^2\right) + p^{ij}\mathcal{L}_M h_{ij} - \frac{N\sqrt{\det h}}{16\pi G}R_3\right)\mathrm{d}^3x. \end{aligned}$$

With (9.105), the middle term equals

$$\begin{aligned} \int\!\!\int\!\!\int p^{ij}\mathcal{L}_M h_{ij}\mathrm{d}^3x &= \int\!\!\int\!\!\int (2p^{ij}h_{jk}\partial_i M^k + p^{ij}M^k\partial_k h_{ij})\mathrm{d}^3x\\ &= \int\!\!\int\!\!\int (-2\partial_i(p^{ij}h_{jk}) + p^{ij}\partial_k h_{ij})M^k\mathrm{d}^3x \qquad (9.118) \end{aligned}$$

if we assume that p^{ij} or M^k approach zero at infinity, eliminating boundary terms in integrations by parts. But note that the spatial metric h_{ij} should always be invertible and therefore cannot vanish, even at infinity.

All functions N and M^k now appear as simple coefficients without derivatives, allowing us to derive explicit constraints: The **Hamiltonian constraint**

$$C_0 = \frac{\delta H}{\delta N} = \frac{16\pi G}{\sqrt{\det h}}\left(p_{ij}p^{ij} - \frac{1}{2}(p^k_k)^2\right) - \frac{\sqrt{\det h}}{16\pi G}R_3 = 0 \tag{9.119}$$

and the **diffeomorphism constraints**

$$C_k = \frac{\delta H}{\delta M^k} = -2\left(h_{jk}\partial_i p^{ij} + p^{ij}\partial_i h_{jk}\right) + p^{ij}\partial_k h_{ij} = 0. \tag{9.120}$$

In addition, the Hamilton function determines Hamilton's equations of motion

$$\mathcal{L}_t h_{ij} = \frac{\delta H[h,p,N,M]}{\delta p^{ij}} \quad \text{and} \quad \mathcal{L}_t p^{ij} = -\frac{\delta H[h,p,N,M]}{\delta h_{ij}} \tag{9.121}$$

which are too long to be shown here in complete form. We will evaluate them in simpler cases given by space-time solutions with symmetry. The combination of constraints and Hamilton's equations for the spatial metric and its momenta is equivalent to Einstein's equation for the space-time metric in a frame determined by N and M^j.

9.2.3 SYMMETRY

The Hamilton functional of general relativity is not unique because there are constraints C_0 and C_j that vanish for any admissible solution. These constraints appeared in the process of deriving momenta from velocities, before the Legendre transformation. The resulting Hamilton functional is ambiguous because we may add multiples of the constraints to its expression without changing its value on admissible solutions of the constraints. The general Hamilton functional is therefore given by

$$H_{\rm complete}[h,p,N,M,\lambda] = H[h,p,N,M] + \int\int\int(\lambda^0 C_0 + \lambda^j C_j)\mathrm{d}^3x \qquad (9.122)$$

where λ^0 and λ^j are arbitrary spatial functions.

9.2.3.1 Hamiltonian functional

If the Hamilton functional is ambiguous, the equations of motion it implies are not uniquely defined. Here, they depend on the arbitrary functions λ^0 and λ^j: Hamilton's equations are obtained from derivatives of the Hamilton functional by h_{ij} or p^{ij}, including derivatives of the constraints if we add multiples of them to the Hamilton functional. While the constraints vanish on admissible solutions, their derivatives by h_{ij} or p^{ij} need not vanish.

Upon closer inspection of the Hamilton functional (9.117) and the constraints, we notice that the ambiguity in (9.122) is of a rather simple form: The original Hamilton functional can be written as

$$H[h,p,N,M] = \int\int\int(NC_0 + M^j C_j)\mathrm{d}^3x \qquad (9.123)$$

because its expression is linear in N and M^j up to integration by parts. The complete Hamilton functional then simply replaces N and M^j with $N+\lambda^0$ and $M^j+\lambda^j$, resembling translation symmetry.

This observation tells us more about the reason behind the ambiguous evolution: From the perspective of general solutions, the functions N and M^j were ambiguous to begin with because they specified the frame of an observer, but general equations of motion do not come with a distinguished observer. In particular, the time direction and the time lapse are arbitrary in a relativistic theory. We should not expect unambiguous equations of motion because motion can be uniquely determined only if there is a specific notion of a time direction and progress. The constraints and the resulting ambiguities in the Hamiltonian formulation are closely related to these universal aspects of general covariance.

At this point, it is useful to discuss equations of motion on a general level by using the Poisson bracket, defined as

$$\{A,B\} = \int\int\int\left(\frac{\delta A}{\delta h_{ij}}\frac{\delta B}{\delta p^{ij}} - \frac{\delta A}{\delta p^{ij}}\frac{\delta B}{\delta h_{ij}}\right)\mathrm{d}^3x \qquad (9.124)$$

where we sum over i and j. This expression generalizes (3.201) to a Hamiltonian field theory, replacing partial derivatives with functional derivatives. Hamilton's equations then take the compact form

$$\mathcal{L}_t f[h,p] = \{f[h,p], H[h,p,N,M]\} \qquad (9.125)$$

for any functional $f[h,p]$, using the chain rule. If we add λ-terms to the Hamilton functional, we obtain equations of motion

$$\mathcal{L}_{t'} f[h,p] = \{f[h,p], H[h,p,N+\lambda^0, M+\lambda^j]\} \qquad (9.126)$$

with respect to a new time-evolution vector field $t^{\alpha'}$ defined by $N+\lambda^0$ and $M^j+\lambda^j$.

Since the Hamilton functional is linear in N and M^j, we can write the preceding equation as

$$\begin{aligned}\mathcal{L}_{t'} f[h,p] &= \{f[h,p], H[h,p,N,M] + C[\lambda^0,\lambda^j]\} \\ &= \{f[h,p], H[h,p,N,M]\} + \{f[h,p], C[\lambda^0,\lambda^j]\} \qquad (9.127)\end{aligned}$$

where we introduced the integrated constraint functionals

$$C[\lambda^0,\lambda^j] = \int\int\int (\lambda^0 C_0 + \lambda^j C_j)\mathrm{d}^3x. \tag{9.128}$$

Requiring $C[\lambda^0,\lambda^j] = 0$ for all spatial functions λ^0 and λ^j is equivalent to $C_0 = 0$ and $C_j = 0$.

The new time derivative then equals

$$\mathscr{L}_{t'} f[h,p] = \mathscr{L}_t f[h,p] + \delta_\lambda f[h,p] \tag{9.129}$$

with the **gauge transformation**

$$\delta_\lambda f[h,p] = \{f[h,p], C[\lambda^0,\lambda^j]\}. \tag{9.130}$$

In a covariant formulation, observer-independent expressions are **gauge-invariant**. They are invariant under gauge transformations that, in the present case, change the observer's frame. The presence of gauge transformations in general relativity tells us that any solution obtained in a fixed frame can be transformed to an equivalent solution in a different frame. The infinitesimal expression of this transformation is given by (9.130).

Gauge transformations are closely related to symmetry transformations. However, symmetry is a property that a given solution may or may not have, while gauge invariance is a requirement for any observer-independent observable. For instance, a Kepler ellipse as a solution of Newton's equations with a spherically symmetric potential is not symmetric. But an observer-independent observable such as the mass in $-m^2 = ||\mathbf{p}||^2$ must always be invariant under frame transformations. Symmetry, as used here, is a property of the potential or Hamilton function, or of a specific subset of solutions; gauge-invariance is a property of observables, provided the Hamilton function permits the corresponding gauge transformations.

9.2.3.2 Algebra

It should be possible to combine different gauge transformations, transforming from one observer to a second and then a third one. If the two individual transformations are $\delta_{\lambda_1} f[h,p] = \{f[h,p], C[\lambda_1^0,\lambda_1^j]\}$ and $\delta_{\lambda_2} f[h,p] = \{f[h,p], C[\lambda_2^0,\lambda_2^j]\}$, the combined transformation is

$$\delta_{\lambda_{12}} f[h,p] = \{\{f[h,p], C[\lambda_1^0,\lambda_1^j]\}, C[\lambda_2^0,\lambda_2^j]\} \tag{9.131}$$

if λ_1 is used first. Gauge transformations, just like symmetries, do not necessarily commute. We may then have a non-zero difference

$$\delta_{\lambda_{12}} f[h,p] - \delta_{\lambda_{21}} f[h,p]\} = \{\{f[h,p], C[\lambda_1^0,\lambda_1^j]\}, C[\lambda_2^0,\lambda_2^j]\} - \{\{f[h,p], C[\lambda_2^0,\lambda_2^j]\}, C[\lambda_1^0,\lambda_1^j]\}. \tag{9.132}$$

However, the difference should be another gauge transformation, in which case there are λ_3^0 and λ_3^j such that

$$\{f[h,p], C[\lambda_3^0,\lambda_3^j]\} = \{\{f[h,p], C[\lambda_1^0,\lambda_1^j]\}, C[\lambda_2^0,\lambda_2^j]\} - \{\{f[h,p], C[\lambda_2^0,\lambda_2^j]\}, C[\lambda_1^0,\lambda_1^j]\} \tag{9.133}$$

This expression can be evaluated by using the Jacobi identity (3.202), which holds for any Poisson bracket:

$$\begin{aligned} & \{\{f[h,p], C[\lambda_1^0,\lambda_1^j]\}, C[\lambda_2^0,\lambda_2^j]\} - \{\{f[h,p], C[\lambda_2^0,\lambda_2^j]\}, C[\lambda_1^0,\lambda_1^j]\} \\ = & \{f[h,p], \{C[\lambda_1^0,\lambda_1^j], C[\lambda_2^0,\lambda_2^j]\}\}. \end{aligned} \tag{9.134}$$

Comparing (9.133) and (9.134), we see that a consistent gauge theory requires the algebraic identity

$$\{C[\lambda_1^0,\lambda_1^j], C[\lambda_2^0,\lambda_2^j]\} = C[\lambda_3^0,\lambda_3^j] \tag{9.135}$$

with a suitable relationship between $(\lambda_3^0,\lambda_3^j)$ and the pair of $(\lambda_1^0,\lambda_1^j)$ and $(\lambda_2^0,\lambda_2^j)$. Such Poisson-bracket relations form the classical version of the Lie bracket in a Lie algebra, corresponding to commutator relationships such as (4.42).

In the present case, Poisson brackets of the constraints with at most one non-zero λ^0 can be evaluated rather easily and result in

$$\begin{aligned} \{C[0,\lambda_1^j],C[0,\lambda_2^j]\} &= C[0,\mathcal{L}_{\lambda_1}\lambda_2^j] &(9.136)\\ \{C[0,\lambda_1^j],C[\lambda_2^0,0]\} &= C[\mathcal{L}_{\lambda_1}\lambda_2^0]\,. &(9.137)\end{aligned}$$

They produce Lie derivatives on the right, where $\mathcal{L}_v f = v^i\partial_i f$ for a function f and $\mathcal{L}_v w^i = v^j\partial_j w^i - w^j\partial_j v^i$ for a vector field w^i; see (8.176). For two non-zero λ^0, a longer calculation with several integrations by parts shows that

$$\{C[\lambda_1^0,0],C[\lambda_2^0,0]\} = C[0,h^{ij}(\lambda_1^0\partial_j\lambda_2^0 - \lambda_2^0\partial_j\lambda_1^0)]\,. \tag{9.138}$$

These **hypersurface-deformation relations** describe infinitesimal properties of arbitrary local frame changes, which unlike Lorentz boosts are not required to use the same parameters everywhere in space. Therefore, instead of a finite number of rotation angles or the boost rapidity in global frame changes, we have free non-linear functions λ^0 and λ^j.

9.2.4 SPHERICALLY SYMMETRIC VACUUM SOLUTIONS

The terms in the Hamilton functional, just like the action of general relativity, are rather complicated expressions as functions of the spatial metric and the momentum. They can be evaluated more easily when specialized to classes of space-time solutions that obey a certain symmetry.

Here, we assume that all space-time metrics of interest in a given application are spherically symmetric, such that the line element can be expressed as

$$ds^2 = -N(t,x)^2dt^2 + L(t,x)^2\,(dx + M(t,x)dt)^2 + S(t,x)^2(d\vartheta^2 + \sin(\vartheta)^2d\varphi^2) \tag{9.139}$$

with the frame functions N and M (a single component of the radial shift vector field) as well as two functions, L and S, that specify the spatial metric components. We have two momentum functions, p_L and p_S, that depend on t and x.

As a tensor, the momenta p^{ij} may be written as

$$\mathbf{p} = \frac{p_L}{2L}\frac{\partial}{\partial x}\otimes\frac{\partial}{\partial x} + \frac{p_S}{4S}\left(\frac{\partial}{\partial\vartheta}\otimes\frac{\partial}{\partial\vartheta} + \frac{1}{\sin(\vartheta)^2}\frac{\partial}{\partial\varphi}\otimes\frac{\partial}{\partial\varphi}\right) \tag{9.140}$$

with tensor products in the coordinate basis. We have chosen coefficients in $\mathbf{p}$ such that $p^{ij}\mathcal{L}_t h_{ij} = \dot{L}p_L + \dot{S}p_S$ resembles the term $\dot{x}p$ in the Legendre transformation of a classical-mechanics system. This equation can be confirmed in matrix form, using

$$\mathbf{h} = \begin{pmatrix} L^2 & 0 & 0\\ 0 & S^2 & 0\\ 0 & 0 & S^2\end{pmatrix} \quad , \quad \mathbf{p} = \begin{pmatrix} p_L/(2L) & 0 & 0\\ 0 & p_S/(4S) & 0\\ 0 & 0 & p_S/(4S)\end{pmatrix} \tag{9.141}$$

If we assume vacuum solutions, the relevant contribution to the constraint C_0 is given by

$$h_{ik}h_{jl}p^{ij}p^{kl} - \frac{1}{2}(h_{ij}p^{ij})^2 = \mathrm{tr}(\mathbf{hphp}) - \frac{1}{2}(\mathrm{tr}(\mathbf{hp}))^2 = \frac{1}{8}(L^2p_L^2 - 2LSp_Lp_S)\,. \tag{9.142}$$

Therefore,

$$C_0 = 2\pi G\left(\frac{Lp_L^2}{S^2} - 2\frac{p_Lp_S}{S}\right) - \frac{LS^2}{16\pi G}R_3\,. \tag{9.143}$$

Including the spatial Ricci scalar of a general spherically symmetric metric, to be derived in Section 9.3, we obtain the Hamiltonian constraint

$$C_0 = 4\pi G\left(\frac{Lp_L^2}{2S^2} - \frac{p_L p_S}{S}\right) + \frac{1}{4\pi G}\left(\frac{1}{2}\frac{(S')^2}{L} + \frac{SS''}{L} - \frac{SS'L'}{L^2} - \frac{1}{2}L\right) \tag{9.144}$$

The diffeomorphism constraints have only the radial component

$$C_1 = -2\left(h_{jx}\partial_i p^{ij} + p^{ij}\partial_i h_{jx}\right) + p^{ij}\partial_x h_{ij} \tag{9.145}$$

where we specified the previous index on C as an x. Using the diagonal matrices (9.141), the first two terms contain only x-indices, while the last one is summed over the diagonal choices of (i,j):

$$\begin{aligned} C_1 &= -2(h_{xx}\partial_x p^{xx} + p^{xx}\partial_x h_{xx}) + (p^{xx}\partial_x h_{xx} + 2p^{\vartheta\vartheta}\partial_x h_{\vartheta\vartheta}) \\ &= -2h_{xx}\partial_x p^{xx} - p^{xx}\partial_x h_{xx} + 2p^{\vartheta\vartheta}\partial_x h_{\vartheta\vartheta} = -L^2\partial_x(p_L/L) - \frac{p_L}{2L}\partial_x L^2 + \frac{p_S}{2S}\partial_x S^2 \\ &= -Lp_L' + p_S S'. \end{aligned} \tag{9.146}$$

Space-time solutions are determined by functions L and S such that $C_0 = 0$ and $C_1 = 0$, in which p_L and p_S are related to $\dot{L}$ and $\dot{S}$ by Hamilton's equations $\dot{L} = \{L, C[N,M]\}$ and $\dot{S} = \{S, C[N,M]\}$ for suitable functions N and M in $C[N,M] = \int\int\int (NC_0 + MC_1)\mathrm{d}^3x$. Moreover, the time dependence of p_L and p_S must be such that $\dot{p}_L = \{p_L, C[N,M]\}$ and $\dot{p}_S = \{p_S, C[N,M]\}$. It is difficult to solve these equations for generic functions N and M that may depend on t and x. This challenging task would result in a space-time solution valid in any frame, which is rarely required. It is easier and sufficient to choose a suitable frame before solving equations, just as we usually derive geometrical properties in a given coordinate system.

A set of interesting solutions is obtained by imposing the physical condition that they be time-independent, or static. Such solutions could describe the space-time geometry around a central mass such as a star in its own rest frame. Time-independence implies that all time derivatives and momenta vanish. The constraint $C_1 = 0$ is then identically satisfied. Time-independence in the line element also means that the time-space component M should vanish. This condition implies that the time-evolution vector field points in the normal direction and does not introduce a spatial shift between different equal-time hypersurfaces.

At this point, we have the free functions N, L and S. We could try to choose N as part of specifying the frame, but we have seen that the time component of the metric describes gravitational redshift. Since we do not know yet how gravitational redshift changes as we approach a central mass with possibly strong gravitational field, we should keep $N(x)$ as a free function and find a way to solve for it. Like all the functions, it should be time-independent in the static case.

Another frame choice available to us is the spatial coordinate, which we generically called x because the metric (9.139) does not assign a specific geometrical meaning to it. We can freely transform the spatial coordinate. In the static case, we may transform from x to $r = S(x)$, where $S(x)$ appears in the angular component of (9.139), provided it is monotonic. Such a coordinate transformation is not available if S depends also on time, which is why we had to include a free S-function in the generic spherically symmetric line element. After the transformation, the static line element takes the form

$$\mathrm{d}s^2 = -N(r)^2\mathrm{d}t^2 + L(r)^2\mathrm{d}r^2 + r^2(\mathrm{d}\vartheta^2 + \sin(\vartheta)^2\mathrm{d}\varphi). \tag{9.147}$$

The coordinate r now determines the area $4\pi r^2$ of a sphere at constant r. However, the radius, measured from the center of the system, depends on the function $L(r)$.

With the choice of r as the coordinate such that $S(r) = r$, the constraint $C_0 = 0$ with vanishing momenta in (9.144) simplifies to the first-order differential equation

$$\frac{1}{2L} - \frac{rL'}{L^2} - \frac{1}{2}L = 0 \tag{9.148}$$

or

$$\frac{f'}{f(1-f)} = \frac{1}{r} \tag{9.149}$$

for $f = L^2$. Integrating on both sides, we solve for

$$L^2 = \frac{1}{1+C/r} \tag{9.150}$$

with an integration constant C.

The remaining metric component is N. We do not have a version of Hamilton's equation for N, and this function is not restricted by the constraints. However, it does appear in Hamilton's equations for $\dot{p}_L$ and $\dot{p}_S$. We set $p_L = 0$ and $p_S = 0$ when we solved the time-independent constraint, but we also have to make sure that these values stay zero as time changes. Therefore, requiring $\dot{p}_L = 0$ and $\dot{p}_S = 0$, in general, constitutes independent equations in addition to $p_L = 0$ and $p_S = 0$ imposed at an initial time. We can drop only one of these conditions here because we eliminated S by choosing the spatial coordinate. Since the S-function then disappears from the Hamilton function, Hamilton's equation $\dot{p}_S = -\delta C_0[N,0]/\delta S = 0$ with our choice of $M = 0$ is automatically satisfied.

We derive our final equation for $\dot{p}_L$ from Hamilton's equations:

$$\begin{aligned} \dot{p}_L &= -\frac{\delta C_0[N,0]}{\delta L} = -N\left(-\frac{1}{2L^2} + \frac{2rL'}{L^3} - \frac{1}{2}\right) - \frac{\mathrm{d}}{\mathrm{d}r}\left(\frac{rN}{L^2}\right) \\ &= -N\left(\frac{1}{2L^2} - \frac{1}{2}\right) - \frac{r}{L^2}N'. \end{aligned} \tag{9.151}$$

We had to integrate by parts in order to eliminate derivatives of delta functions. Setting $\dot{p}_L = 0$ and using (9.150), we obtain

$$\frac{2}{N}\frac{\mathrm{d}N}{\mathrm{d}r} = \frac{1}{r}(L^2-1) = -\frac{C}{r^2}\frac{1}{1+C/r} = \frac{1}{1+C/r}\frac{\mathrm{d}}{\mathrm{d}r}(1+C/r). \tag{9.152}$$

Logarithmic integration on both sides implies

$$N^2 = N_0^2(1+C/r) \tag{9.153}$$

with an integration constant N_0 that can be set equal to one by rescaling the time coordinate t.

We have derived the line element

$$\mathrm{d}s^2 = (1+C/r)\mathrm{d}t^2 + \frac{1}{1+C/r}\mathrm{d}r^2 + r^2(\mathrm{d}\vartheta^2 + \sin(\vartheta)^2\mathrm{d}\varphi^2). \tag{9.154}$$

Comparing the time component with the weak-field line element (8.14) at $r \gg C$ shows that $C = -2GM/c^2$ with the central mass M, or $C = -2GM$ in units such that $c = 1$. The space components do not resemble (8.14) because only the $\mathrm{d}r^2$-term is multiplied by a r-dependent function. However, compared with the time component, the space components are subdominant to leading relativistic order because they have an additional factor of velocity squared, such as $(\mathrm{d}r/\mathrm{d}t)^2$ or $r^2(\mathrm{d}\varphi/\mathrm{d}t)^2$. In the next chapter, we will see that the complete weak-field form can be achieved by a suitable coordinate transformation.

After identifying the mass parameter, we obtain the **Schwarzschild line element**

$$\mathrm{d}s^2 = \left(1-\frac{2GM}{r}\right)\mathrm{d}t^2 + \frac{1}{1-2GM/r}\mathrm{d}r^2 + r^2(\mathrm{d}\vartheta^2 + \sin(\vartheta)^2\mathrm{d}\varphi^2). \tag{9.155}$$

9.3 GAUSS–CODAZZI RELATIONS

It is often useful to relate the curvature tensor of an embedded hypersurface to its extrinsic curvature and the curvature tensor of the embedding space-time. Such identities were crucial in the derivation of the Hamiltonian formulation of general relativity, but they can also be used to break the derivation of curvature tensors into smaller building blocks by piecing together lower-dimensional contributions.

9.3.1 EXTRINSIC CURVATURE

We consider the case of a non-lightlike $(n-1)$-dimensional hypersurface in an n-dimensional space or space-time. The n-dimensional metric $g_{\alpha\beta}$ determines the unit normal vector field n^α, which may be timelike or spacelike such that $n^\alpha n_\alpha = \sigma = \pm 1$, and the induced metric from

$$h_{\alpha\beta} = g_{\alpha\beta} - \sigma n_\alpha n_\beta \,. \tag{9.156}$$

Both tensors are ingredients in the definition of the extrinsic-curvature tensor $K_{\alpha\beta}$. As a primary definition of the latter, we use

$$K_{\alpha\beta} = h_\alpha{}^\gamma h_\beta{}^\delta \nabla_\gamma n_\delta \tag{9.157}$$

with the covariant derivative determined by $g_{\alpha\beta}$. Compared with (9.97), which used the Lie derivative and was motivated by a relationship with changes in time of the spatial metric on a family of hypersurfaces, the new definition is closer to the heuristic introduction of extrinsic curvature on page 279, based on a changing normal direction as we move along the surface.

In (9.157), we use $h_{\alpha\beta}$ with a raised index,

$$h_\alpha{}^\gamma = g^{\gamma\delta} h_{\alpha\delta} = \delta_\alpha{}^\gamma - \sigma n_\alpha n^\gamma \,, \tag{9.158}$$

as a projection tensor that turns the covariant derivative $\nabla_\gamma n_\delta$ into a tensor on the family of hypersurfaces: $n^\alpha K_{\alpha\beta} = n^\beta K_{\alpha\beta} = 0$. In fact, a single projection is sufficient because

$$K_{\alpha\beta} = h_\beta{}^\delta h_\alpha{}^\gamma \nabla_\gamma n_\delta = (g_\beta{}^\delta - \sigma n_\beta n^\delta) h_\alpha{}^\gamma \nabla_\gamma n_\delta = h_\alpha{}^\gamma \nabla_\gamma n_\beta - \sigma n_\beta n^\delta h_\alpha{}^\gamma \nabla_\gamma n_\delta \tag{9.159}$$

using the fact that $g_\beta{}^\delta$ is covariantly constant. In the last term, $n^\delta \nabla_\gamma n_\delta = 0$ follows from $n^\delta n_\delta = \sigma$ constant. We conclude that

$$K_{\alpha\beta} = h_\alpha{}^\gamma \nabla_\gamma n_\beta \,. \tag{9.160}$$

This equation is equivalent to our original definition (9.97) of extrinsic curvature: We first write the Lie derivative in terms of the covariant derivative,

$$\mathcal{L}_n h_{\alpha\beta} = n^\gamma \nabla_\gamma h_{\alpha\beta} + h_{\gamma\beta} \nabla_\alpha n^\gamma + h_{\alpha\gamma} \nabla_\beta n^\gamma \,. \tag{9.161}$$

The Lie derivative of a tensor is always a tensor. In particular, we may replace partial derivatives in the general expressions for Lie derivatives with covariant derivatives without changing the result. We now use (9.156) in order to write all three occurrences of the projection tensor in terms of the space-time metric and the unit normal. Any covariant derivative of the space-time metric vanishes, and $n_\alpha \nabla_\beta n^\alpha = 0$ as used above. Thus,

$$\mathcal{L}_n h_{\alpha\beta} = -\sigma n^\gamma \nabla_\gamma (n_\alpha n_\beta) + \nabla_\alpha n_\beta + \nabla_\beta n_\alpha \,. \tag{9.162}$$

After applying the product rule in the first term, it can be combined with the remaining terms such that

$$2K_{\alpha\beta} = \mathcal{L}_n h_{\alpha\beta} = (g_\alpha{}^\gamma - \sigma n_\alpha n^\gamma) \nabla_\gamma n_\beta + (g_\beta{}^\gamma - \sigma n_\beta n^\gamma) \nabla_\gamma n_\alpha \,. \tag{9.163}$$

Recognizing the projection tensors in the two coefficients, which are equal up to renaming indices, proves that (9.160) or the equivalent (9.157) is indeed the same as (9.97). In particular, (9.160) is symmetric in α and β, even though this property is not obvious from the new definition. The vanishing antisymmetric part of (9.160) can be seen by writing the spatial projection as $s_i^\alpha s_j^\beta \nabla_\alpha n_\beta$ in a basis of commuting vector fields, $s_i^\alpha \nabla_\alpha s_j^\beta - s_j^\alpha \nabla_\alpha s_i^\beta = 0$, and using $s_j^\beta \nabla_\alpha n_\beta = -n_\beta \nabla_\alpha s_j^\beta$ from $s_j^\beta n_\beta = 0$.

9.3.2 PROJECTIONS

The projected covariant derivative of the normal vector field, used in (9.160), is an example of a general construction: For any tensor field $T^{\alpha\cdots}{}_{\beta\ldots}$, given a family of hypersurfaces with the corresponding projection tensor $h_{\alpha\beta}$, we define the induced covariant derivative as the projection

$$D_\gamma T^{\alpha\cdots}{}_{\beta\ldots} = h_\gamma{}^{\gamma'} h^\alpha{}_{\alpha'} \cdots h_\beta{}^{\beta'} \cdots \nabla_{\gamma'} T^{\alpha'\cdots}{}_{\beta'\ldots} \tag{9.164}$$

on all indices of the expression. The position of the subscript or the superscript as the first or second index of the projection tensor does not matter because $h_{\alpha\beta}$ is symmetric. We choose the positions such that summed-over indices, for instance α', on the projection tensor and the projected tensor appear next to each other.

This definition results in a tensor $D_\gamma T^{\alpha\cdots}{}_{\beta\ldots}$ that vanishes when summed over with the unit normal on any index. Applied to the projection tensor,

$$\begin{aligned} D_\alpha h_{\beta\gamma} &= h_\alpha{}^{\alpha'} h_\beta{}^{\beta'} h_\gamma{}^{\gamma'} \nabla_{\alpha'} h_{\beta'\gamma'} = h_\alpha{}^{\alpha'} h_\beta{}^{\beta'} h_\gamma{}^{\gamma'} \nabla_{\alpha'} (g_{\beta'\gamma'} - \sigma n_{\beta'} n_{\gamma'}) \\ &= -\sigma h_\alpha{}^{\alpha'} (h_\beta{}^{\beta'} n_{\beta'} h_\gamma{}^{\gamma'} \nabla_{\alpha'} n_{\gamma'} + h_\gamma{}^{\gamma'} n_{\gamma'} h_\beta{}^{\beta'} \nabla_{\alpha'} n_{\beta'}) = 0 \end{aligned} \tag{9.165}$$

because $h_\beta{}^{\beta'} n_{\beta'} = 0$. Therefore, D_α is the unique covariant derivative compatible with $h_{\beta\gamma}$ or the induced metric implied by its non-vanishing components, such that $h_{\beta\gamma}$ is covariantly constant with respect to D_α. In components, D_α is therefore identical with the covariant derivative derived from the Christoffel symbol of $h_{ij} = s_i^\alpha s_j^\beta h_{\alpha\beta}$ rather than $g_{\alpha\beta}$, with basis vectors s_i^α on our hypersurfaces. The relationship (9.164) between the induced and space-time covariant derivative now makes it possible to relate the corresponding curvature tensors.

We use (9.36) and introduce the induced Riemann tensor by

$$(D_\alpha D_\beta - D_\beta D_\alpha) w_\gamma = R^3_{\alpha\beta\gamma}{}^\delta w_\delta \tag{9.166}$$

where w_δ is tangent to hypersurfaces: $w_\delta n^\delta = 0$. In this equation,

$$\begin{aligned} D_\alpha D_\beta w_\gamma &= D_\alpha (h_\beta{}^\delta h_\gamma{}^\varepsilon \nabla_\delta w_\varepsilon) = h_\alpha{}^{\alpha'} h_\beta{}^{\beta'} h_\gamma{}^{\gamma'} \nabla_{\alpha'} (h_{\beta'}{}^\delta h_{\gamma'}{}^\varepsilon \nabla_\delta w_\varepsilon) \\ &= h_\alpha{}^{\alpha'} h_\beta{}^\delta h_\gamma{}^\varepsilon \nabla_{\alpha'} \nabla_\delta w_\varepsilon + h_\gamma{}^\varepsilon (h_\alpha{}^{\alpha'} h_\beta{}^{\beta'} \nabla_{\alpha'} h_{\beta'}{}^\delta) \nabla_\delta w_\varepsilon + h_\beta{}^\delta (h_\alpha{}^{\alpha'} h_\gamma{}^{\gamma'} \nabla_{\alpha'} h_{\gamma'}{}^\varepsilon) \nabla_\delta w_\varepsilon \end{aligned} \tag{9.167}$$

from the product rule. The first term, when antisymmetrized in α and β as it appears in (9.166), turns into a projection of the space-time Riemann tensor. In the second term, we write $\nabla_{\alpha'} h_{\beta'}{}^\delta = \nabla_{\alpha'} (g_{\beta'}{}^\delta - \sigma n_{\beta'} n^\delta)$, use the fact that the space-time metric is covariantly constant, and obtain extrinsic-curvature components from the projected covariant derivatives of the unit normal:

$$h_\alpha{}^{\alpha'} h_\beta{}^{\beta'} \nabla_{\alpha'} h_{\beta'}{}^\delta = -\sigma n^\delta h_\alpha{}^{\alpha'} h_\beta{}^{\beta'} \nabla_{\alpha'} n_{\beta'} = -\sigma n^\delta K_{\alpha\beta} \,. \tag{9.168}$$

This term is symmetric in α and β and therefore does not contribute to the antisymmetrization in (9.166).

In the last term of (9.167), we use a similar construction as well as $n^\varepsilon\nabla_\delta w_\varepsilon = -w_\varepsilon\nabla_\delta n^\varepsilon$ for hypersurface-tangent w_ε, derived from $w_\varepsilon n^\varepsilon = 0$, and conclude

$$h_\beta{}^\delta(h_\alpha{}^{\alpha'}h_\gamma{}^{\gamma'}\nabla_{\alpha'}h_{\gamma'}{}^\varepsilon)\nabla_\delta w_\varepsilon = -\sigma h_\beta{}^\delta K_{\alpha\gamma}n^\varepsilon\nabla_\delta w_\varepsilon = \sigma K_{\alpha\gamma}h_\beta{}^\delta w_\varepsilon\nabla_\delta n^\varepsilon = \sigma K_{\alpha\gamma}K_\beta^\varepsilon w_\varepsilon\,. \quad (9.169)$$

Antisymmetrizing and reading off the coefficients of w_ε implies the **Gauss equation**

$$h_\alpha{}^{\alpha'}h_\beta{}^{\beta'}h_\gamma{}^{\gamma'}h^\delta{}_{\delta'}R_{\alpha'\beta'\gamma'}{}^{\delta'} = R^3_{\alpha\beta\gamma}{}^\delta - \sigma K_{\alpha\gamma}K_\beta{}^\delta + \sigma K_{\beta\gamma}K_\alpha{}^\delta\,. \quad (9.170)$$

The projection of the Riemann tensor is not sufficient for an expression of the space-time Ricci scalar, which is the Riemann tensor contracted with the space-time metric. Writing the space-time metric as the induced metric plus normal terms, the missing contributions can be obtained from contractions of the Riemann tensor with a combination of normal vectors and the induced metric. Several such contributions vanish thanks to antisymmetry of some of the Riemann components, and we arrive at

$$R = (h^{\alpha\gamma}+\sigma n^\alpha n^\gamma)(h^{\beta\delta}+\sigma n^\beta n^\delta)R_{\alpha\beta\gamma\delta} = h^{\alpha\gamma}h^{\beta\delta}R_{\alpha\beta\gamma\delta} + 2\sigma R_{\alpha\beta}n^\alpha n^\beta\,. \quad (9.171)$$

All indices are hypersurface-tangent in the first term because each one appears on the projection tensor. This term is directly related to the projection that appears in the Gauss equation,

$$R = h^{\alpha\gamma}h^{\beta\delta}h_\alpha{}^{\alpha'}h_\beta{}^{\beta'}h_\gamma{}^{\gamma'}h_\delta{}^{\delta'}R_{\alpha'\beta'\gamma'\delta'} + 2\sigma R_{\alpha\beta}n^\alpha n^\beta\,, \quad (9.172)$$

after inserting additional projection tensors using $h^{\alpha\gamma}h_\alpha{}^{\alpha'} = h^{\alpha'\gamma}$.

The second term is given by $R_{\alpha\beta}n^\alpha n^\beta = R_{\alpha\gamma\beta}{}^\gamma n^\alpha n^\beta$. Antisymmetry in the third and fourth index of the Riemann tensor allows us to apply the definition of the Riemann tensor in terms of covariant derivatives:

$$\begin{aligned} R_{\alpha\beta}n^\alpha n^\beta &= -g^{\gamma\delta}n^\alpha R_{\alpha\gamma\delta}{}^\beta n_\beta = -g^{\gamma\delta}n^\alpha(\nabla_\alpha\nabla_\gamma - \nabla_\gamma\nabla_\alpha)n_\delta = -n^\alpha(\nabla_\alpha\nabla_\gamma-\nabla_\gamma\nabla_\alpha)n^\gamma \\ &= -\nabla_\alpha(n^\alpha\nabla_\gamma n^\gamma) + (\nabla_\alpha n^\alpha)(\nabla_\gamma n^\gamma) + \nabla_\gamma(n^\alpha\nabla_\alpha n^\gamma) - (\nabla_\gamma n^\alpha)(\nabla_\alpha n^\gamma)\,. \end{aligned} \quad (9.173)$$

We applied the reverse product rule in order to bring in the extrinsic-curvature tensor by using

$$\begin{aligned} (\nabla_\gamma n^\alpha)(\nabla_\alpha n^\gamma) &= g_\alpha{}^\delta g_\varepsilon{}^\gamma(\nabla_\gamma n^\alpha)(\nabla_\delta n^\varepsilon) \\ &= (h_\varepsilon{}^\gamma\nabla_\gamma n^\alpha)(h_\alpha{}^\delta\nabla_\delta n^\varepsilon) + \sigma h_\alpha{}^\delta n^\gamma n_\varepsilon(\nabla_\gamma n^\alpha)(\nabla_\delta n^\varepsilon) \\ &\quad + \sigma n_\alpha n^\delta h_\varepsilon{}^\gamma(\nabla_\gamma n^\alpha)(\nabla_\delta n^\varepsilon) + n_\alpha n^\delta n^\gamma n_\varepsilon(\nabla_\gamma n^\alpha)(\nabla_\delta n^\varepsilon) = K_\varepsilon{}^\alpha K_\alpha{}^\varepsilon \end{aligned} \quad (9.174)$$

and an analogous calculation for $(\nabla_\alpha n^\alpha)(\nabla_\gamma n^\gamma)$. We obtain

$$R_{\alpha\beta}n^\alpha n^\beta = (K_\alpha{}^\alpha)^2 - K_\alpha{}^\beta K_\beta{}^\alpha + \nabla_\alpha v^\alpha \quad (9.175)$$

with the vector field

$$v^\alpha = -n^\alpha\nabla_\gamma n^\gamma + n^\gamma\nabla_\gamma n^\alpha\,. \quad (9.176)$$

Inserting (9.170) and (9.175) in (9.172), we derive the **Gauss–Codazzi relation**

$$R = R_3 - \sigma K_{\alpha\beta}K^{\alpha\beta} + \sigma(K_\alpha{}^\alpha)^2 + 2\sigma\nabla_\alpha v^\alpha\,. \quad (9.177)$$

Because $K_{\alpha\beta}$ is a tensor on the family of hypersurfaces with vanishing normal components, we may write its appearance in terms of $K_{ij}K^{ij}$ and $K^i{}_i$, as used in (9.101): The inverse space-time metric $g^{\alpha\beta} = h^{\alpha\beta} - n^\alpha n^\beta$ implies

$$\begin{aligned} K_{\alpha\beta}K^{\alpha\beta} &= K_{\alpha\beta}K_{\gamma\delta}(h^{\alpha\gamma}-n^\alpha n^\gamma)(h^{\beta\delta}-n^\beta n^\delta) \\ &= K_{\alpha\beta}K_{\gamma\delta}(h^{ij}s_i^\alpha s_j^\gamma)(h^{kl}s_k^\beta s_l^\delta) = K_{ik}K_{jl}h^{ij}h^{kl} = K_{ik}K^{ik} \end{aligned} \quad (9.178)$$

with basis vector fields s_i^α tangent to the family of hypersurfaces. A similar projection implies $K^\alpha{}_\alpha = K^i{}_i$.

9.3.3 RICCI SCALAR OF SPHERICALLY SYMMETRIC SPACES

As an application, we obtain a short derivation of the Ricci scalar R of a general spherically symmetric spatial line element

$$\mathrm{d}s^2 = L(x)^2\mathrm{d}x^2 + S(x)^2(\mathrm{d}\vartheta^2 + \sin(\vartheta)^2\mathrm{d}\varphi^2) \tag{9.179}$$

as used in the constraint (9.144). The radial coordinate x is not required to equal the area radius r, such that $S(x)$ is a free function just like $L(x)$.

A 2-sphere at constant x is a spatial hypersurface in a 3-dimensional space with the line element (9.179). Thus, $\sigma = 1$ in the Gauss–Codazzi relation

$$R = R_2 - K_{\alpha\beta}K^{\alpha\beta} + (K_\alpha{}^\alpha)^2 + 2\nabla_\alpha v^\alpha\,. \tag{9.180}$$

The Ricci scalar of a 2-sphere with area $4\pi S^2$ is given by $R_2 = 2/S^2$.

The unit normal vector field $\mathbf{n} = L^{-1}\mathbf{e}_x$ together with the 2-sphere metric

$$\mathbf{h} = \begin{pmatrix} S^2 & 0 \\ 0 & \sin(\vartheta)^2 S^2 \end{pmatrix} \tag{9.181}$$

determines the extrinsic-curvature tensor

$$K_{\alpha\beta} = \frac{1}{2}\mathscr{L}_n h_{\alpha\beta} = \frac{1}{2}(n^\gamma\partial_\gamma h_{\alpha\beta} + h_{\gamma\beta}\partial_\alpha n^\gamma + h_{\alpha\gamma}\partial_\beta n^\gamma) \tag{9.182}$$

using the expression of the Lie derivative in terms of partial derivatives. Only the first term in (9.182) contributes to the angular components because any partial derivative of n^γ by ϑ or φ vanishes. Writing the surface tensor K_{ij} as a matrix, we compute

$$\mathbf{K} = \frac{1}{2}n^x\frac{\partial\mathbf{h}}{\partial x} = \frac{1}{L}\begin{pmatrix} SS' & 0 \\ 0 & SS'\sin(\vartheta)^2 \end{pmatrix} \tag{9.183}$$

and

$$-K_{\alpha\beta}K^{\alpha\beta} + (K_\alpha{}^\alpha)^2 = -\mathrm{tr}(\mathbf{K}\mathbf{h}^{-1}\mathbf{K}\mathbf{h}^{-1}) + \mathrm{tr}(\mathbf{K}\mathbf{h}^{-1})^2 = 2\left(\frac{S'}{SL}\right)^2\,. \tag{9.184}$$

When we used the Gauss–Codazzi relation in order to transform the action of general relativity to a Hamilton function, we were able to ignore the divergence $\nabla_\alpha v^\alpha$ because it integrates to a boundary term. Here, this term is relevant because the spatial Ricci scalar, when inserted in the Hamiltonian constraint C_0, is multiplied by the lapse function N before it is integrated. In this case, it does not imply a boundary term. The second term in $v^\alpha = -n^\alpha\nabla_\gamma n^\gamma + n^\gamma\nabla_\gamma n^\alpha$ is given by

$$n^\gamma\nabla_\gamma n^\alpha = n^\gamma(\partial_\gamma n^\alpha + \Gamma^\alpha_{\gamma\delta}n^\delta) = n^x\frac{\partial n^x}{\partial x} + n^x\Gamma^x_{xx}n^x = 0\,. \tag{9.185}$$

Here, only x-components contribute, and the remaining terms cancel out upon inserting $n^x = 1/L$ and $\Gamma^x_{xx} = \frac{1}{2}g^{xx}\partial g_{xx}/\partial x = L'/L$. For the first term, we obtain

$$\nabla_\gamma n^\gamma = \frac{\partial n^x}{\partial x} + \Gamma^\gamma_{\gamma x}n^x = 2\frac{S'}{SL}\,. \tag{9.186}$$

Notice that angular terms $\Gamma^\vartheta_{\vartheta x} = S'/S = \Gamma^\varphi_{\varphi x}$ contribute even though n^γ is non-zero only if $\gamma = x$: The intermediate sum introduced by expanding the covariant derivative in terms of the Christoffel symbol introduces additional contributions. With this result,

$$\begin{aligned} -\nabla_\alpha v^\alpha = \nabla_\alpha(n^\alpha\nabla_\gamma n^\gamma) &= 2\nabla_\alpha\left(\frac{S'}{SL^2}\mathbf{e}_x^\alpha\right) = 2\frac{\partial}{\partial x}\left(\frac{S'}{SL^2}\right) + 2\Gamma^\alpha_{\alpha x}\frac{S'}{SL^2} \\ &= 2\frac{S''}{SL^2} + 2\frac{(S')^2}{S^2L^2} - 2\frac{S'L'}{SL^3}\,. \end{aligned} \tag{9.187}$$

We conclude that the curvature contribution to the spherically symmetric constraint C_0 is given by

$$\begin{aligned} -\sqrt{\det h}R &= -\sqrt{\det h}\left(R_2 - K_{\alpha\beta}K^{\alpha\beta} + (K_\alpha{}^\alpha)^2 + 2\nabla_\alpha v^\alpha\right) \\ &= 4\frac{SS''}{L} - 4\frac{SS'L'}{L^2} + 2\frac{(S')^2}{L} - 2L. \end{aligned} \tag{9.188}$$

EXERCISES

1. Show that equation (8.175) for the Lie derivative of the metric can be written more compactly as
$$\nabla_\alpha v_\beta + \nabla_\beta v_\alpha = 0 \tag{9.189}$$
using the covariant derivative.
2. The covariant derivative can be used to define the covariant Laplacian
$$\nabla^2 f = g^{\alpha\beta}\nabla_\alpha\nabla_\beta f \tag{9.190}$$
of a function f on a curved space, or on flat space expressed in non-Cartesian coordinates. Derive equation (4.110) for the Laplacian on flat space in spherical coordinates.
3. Assume that a metric $\tilde{g}_{\alpha\beta}$ is defined by rescaling another metric, $g_{\alpha\beta}$, by a positive function: $\tilde{g}_{\alpha\beta} = \Omega^2 g_{\alpha\beta}$.
   (i) Express the Christoffel symbol $\tilde{\Gamma}^\alpha{}_{\beta\gamma}$ of the derivative operator compatible with $\tilde{g}_{\alpha\beta}$ in terms of Ω and the Christoffel symbol $\Gamma^\alpha{}_{\beta\gamma}$ of $g_{\alpha\beta}$.
   (ii) Use the result from part (i) to compute the Christoffel symbol of a metric obtained by rescaling the flat Euclidean metric $\delta_{\alpha\beta}$ by Ω^2.
   (iii) Apply this derivation in order to find all components of the Christoffel symbol on the unit 2-sphere in **stereographic coordinates** (x,y), introduced in an Exercise on page 328.
4. Define a closed curve C on the unit 2-sphere by starting at the North pole, moving along a meridian to an angle $0 < \vartheta_0 < \pi$, then moving along constant ϑ by an angle $0 < \varphi_0 < 2\pi$, and finally moving back to the North pole along a meridian.
   (i) Map the curve C to the plane by using stereographic coordinates.
   (ii) Compute the parallel transport of an arbitrary tangent vector v^α along each smooth part of C. Confirm that the vector norm remains constant.
   (iii) Derive the linear map acting on tangent vectors at the North pole, defined by parallel transport along the entire curve C. Compare with the area enclosed by the curve.
5. Compute the Ricci scalar of a 2-sphere with radius r, using spherical coordinates.
6. Compute the Riemann tensor and the Ricci scalar of a 2-sphere using stereographic coordinates.
7. Show that the Ricci scalar on a surface of revolution equals twice the Gaussian curvature (8.213).
8. Compute the Ricci tensor of the line element
$$\mathrm{d}s^2 = -X^2\mathrm{d}T^2 + \mathrm{d}X^2.$$
9. Derive the cosmological continuity equation from (9.64).
10. Derive the coordinate transformation of the amplitude matrix (9.80) corresponding to a rotation by an angle φ in the y-z plane. Show that this transformation corresponds to a rotation of the vector (H_{22}, H_{23}) by an angle of 2φ.

11. Compute the Ricci tensor for the Friedmann–Robertson–Walker line element.
Show that the tt-component of Einstein's equation for a Friedmann–Robertson–Walker universe is equivalent to the Friedmann equation

$$\left(\frac{\dot{a}}{a}\right)^2 = \frac{8\pi G}{3}\rho$$

where the energy density ρ is the tt-component of the stress-energy tensor.
How does a cosmological constant change the equation?

12. Derive the Friedmann equation and the cosmological continuity equation from the Hamiltonian formulation of general relativity.

10 Vacuum Solutions

If we assume that there is no matter in a space-time region that is spherically symmetric, the **Schwarzschild line element**

$$\mathrm{d}s^2 = -\left(1 - \frac{2GM}{c^2 r}\right)(c\mathrm{d}t)^2 + \frac{1}{1-2GM/(c^2 r)}\mathrm{d}r^2 + r^2(\mathrm{d}\vartheta^2 + \sin(\vartheta)^2\mathrm{d}\varphi^2) \tag{10.1}$$

with the constant mass $M \geq 0$ solves Einstein's equation, see equation (9.155). We confirmed the meaning of M as the central mass by comparing (9.155) with the time component of the weak-field line element, derived from the equivalence principle.

One could think that the absence of matter should imply the absence of curvature. Minkowski space-time would then be the only solution. However, it is possible to have purely gravitational binding energy, which like any other energy implies a gravitational force. This force is described by the curvature of (10.1). We can use the Schwarzschild line element also for the geometry outside a spherical star whose matter then provides the value of the mass M. However, we would have a different line element inside the star, where $T_{\alpha\beta} \neq 0$.

The line element (10.1) is an exact solution of Einstein's equation and can therefore be used even if $GM/(c^2 r)$ is not small, unlike the weak-field line element. We then have a singularity at the value of $r = R_S$, where

$$R_S = \frac{2GM}{c^2} \tag{10.2}$$

is called the **Schwarzschild radius**. This singularity, as we will see, is a coordinate singularity. In addition, there is a physical singularity at $r = 0$. For most astrophysical objects, the vacuum region around them does not include singularities because the Schwarzschild radius is usually much smaller than the radius of the star. For the sun, for instance, the Schwarzschild radius is about 3km, much smaller than the surface radius of $7 \cdot 10^5$km.

As shown by (10.2), masses can be strictly related to distances if we use the two fundamental constants c and G. We have already used c to relate time units to length units, setting $c = 1$. In general relativity, it is customary to set $G = 1$ as well, which allows us to eliminate mass units in favor of length units. In terms of length units, a mass unit M_{mass} takes the value

$$M_{\mathrm{length}} = \frac{GM_{\mathrm{mass}}}{c^2} \approx 0.742 \cdot 10^{-28}\frac{\mathrm{cm}}{\mathrm{g}} M_{\mathrm{mass}}\,. \tag{10.3}$$

For instance, the sun has a mass of $M = 1.47$km, and Earth's mass is 0.44cm. In these **geometrized units**, the Schwarzschild line element takes the form

$$\mathrm{d}s^2 = -\left(1 - \frac{2M}{r}\right)\mathrm{d}t^2 + \frac{1}{1-2M/r}\mathrm{d}r^2 + r^2(\mathrm{d}\vartheta^2 + \sin(\vartheta)^2\mathrm{d}\varphi^2)\,. \tag{10.4}$$

The Schwarzschild line element determines the geometry outside a spherical star. We will derive properties of geodesics using two Killing vectors: Since the metric is time independent, the first Killing vector is given by

$$\xi = \mathbf{e}_t = \frac{\partial}{\partial t} \leftrightarrow \begin{pmatrix} 1 \\ 0 \\ 0 \\ 0 \end{pmatrix}. \tag{10.5}$$

DOI: 10.1201/9781003588702-12

The second Killing vector follows from the independence of the metric on φ, and is given by

$$\eta = \mathbf{e}_\varphi = \frac{\partial}{\partial\varphi} \leftrightarrow \begin{pmatrix} 0 \\ 0 \\ 0 \\ 1 \end{pmatrix}. \tag{10.6}$$

Moreover, spherical symmetry allows us to describe the spatial coordinates (r, ϑ, φ) of any given geodesic within the equatorial plane, setting $\vartheta = \frac{1}{2}\pi$: As in classical mechanics, not only the magnitude but also the direction of angular momentum is conserved, given by the three constants l_i that appeared in our derivation of (8.194). Formally, we use the three spacelike Killing vector fields implied by spherical symmetry and obtain the same equation (8.193) from the conserved quantities as in flat space because the ϑ and φ components of the Schwarzschild line element are not modified by curvature. The same arguments as used for geodesics of the 2-sphere then show that the projection onto a constant-t hypersurface of any geodesic in the Schwarzschild space-time is contained in a single spatial plane. Spherical symmetry allows us to choose this plane as the equatorial plane.

We are therefore able to eliminate one of the free variables, ϑ, and we have three first integrals: two given by Killing vectors in addition to $\mathrm{d}s^2/\mathrm{d}\tau^2 = -1$ for timelike geodesics, or $\mathrm{d}s^2/\mathrm{d}\lambda^2 = 0$ for lightlike ones. These properties will allow us to solve the geodesic equation for motion around a central mass.

10.1 GEODESICS IN THE SCHWARZSCHILD SPACE-TIME

Timelike as well as lightlike geodesics in the Schwarzschild space-time imply observable deviations from Newtonian gravity in orbits of planets and light rays or other electromagnetic waves. These dynamical effects can be used to test general relativity by solar-systems observations, even though Newton's potential on these scales remains a good approximation of the static gravitational force.

10.1.1 RELATIVISTIC CORRECTIONS TO NEWTON'S POTENTIAL

The time component of the line element (10.1) is precisely of the weak-field form $-1-2\Phi(r)$ with Newton's potential $\Phi(r)$. However, the spatial part of the line element does not equal the required $(1-2\Phi(r))(\mathrm{d}x^2+\mathrm{d}y^2+\mathrm{d}z^2)$. Even if we use a Taylor expansion for small $\Phi(r)$, we obtain a potential-dependent term only in $\mathrm{d}r^2$ but not in the angular terms. We have to find a suitable coordinate transformation in order to determine weak-field implications beyond Newton's potential.

As derived in (9.90), the required form of the weak-field line element is

$$\mathrm{d}s^2 = -(1+2\Phi_{\rm GR}(R))\mathrm{d}t^2 + (1-2\Phi_{\rm GR}(R))(\mathrm{d}x^2+\mathrm{d}y^2+\mathrm{d}z^2) \tag{10.7}$$

with a potential $\Phi_{\rm GR}(R)$ as a function of $R=\sqrt{x^2+y^2+z^2}$, which is not necessarily the same as the Schwarzschild coordinate r. If we transform the spatial Cartesian coordinates (x,y,z) to spherical coordinates (R,ϑ,φ), we can compare coefficients with (10.1) and determine equations for $R(r)$. Since both versions of the line element are spherically symmetric, we may use the same angular coordinates in both cases.

The result is the equation

$$r = R\left(1+\frac{M}{2R}\right)^2 \tag{10.8}$$

which transforms the line element into its **isotropic** form

$$\mathrm{d}s^2 = -\left(\frac{R-M/2}{R+M/2}\right)^2\mathrm{d}t^2 + \left(1+\frac{M}{2R}\right)^4(\mathrm{d}R^2+R^2(\mathrm{d}\vartheta^2+\sin(\vartheta)^2\mathrm{d}\varphi^2)). \tag{10.9}$$

Comparing the time component with $-1-2\Phi_{GR}(R)$ and expanding to second order in M/R, we obtain Newton's potential with the leading relativistic correction,

$$\Phi_{GR}(R) = -\frac{M}{R} + \frac{M^2}{R^2} + \cdots \tag{10.10}$$

in the static case. The relativistic nature of the correction term can be seen by re-instating units in which c and G do not equal one. The ratio of the second term over the first term, M/R, should then be transformed into a unitless factor. Using (10.3), this is achieved by writing the ratio as $GM/(c^2R)$. In the non-relativistic limit, $c \to \infty$ and the correction is removed. We see that the correction is small in applications to solar-system physics by writing the ratio M/R as $\frac{1}{2}R_S/R$, where R_S is the Schwarzschild radius of the sun.

10.1.2 EFFECTIVE POTENTIAL

There are relevant relativistic corrections for orbiting objects if they are treated as non-static test masses or massless test light rays, moving along geodesics on a static Schwarzschild background. For this purpose, it is convenient to return to the line element in Schwarzschild coordinates, (t,r,ϑ,φ).

We first consider the case of timelike geodesics for massive objects. In space-time around a spherical mass, only two of the four polar space-time coordinates appear in the metric, while t and φ do not. This symmetry, using the Killing vectors (10.5) and (10.6), implies two conserved quantities, the energy (per mass) $e = -\mathbf{u}\cdot\xi$, and the angular momentum (per mass) $\ell = \mathbf{u}\cdot\eta$. Given the metric, we compute these quantities as

$$e = \left(1-\frac{2M}{r}\right)\frac{\mathrm{d}t}{\mathrm{d}\tau} \quad \text{and} \quad \ell = r^2\sin(\vartheta)^2\frac{\mathrm{d}\varphi}{\mathrm{d}\tau}\,. \tag{10.11}$$

If we are interested in a single geodesic, we can always place it in the equatorial plane where $\vartheta = \frac{1}{2}\pi$ and $\sin(\vartheta) = 1$. Moreover, $\mathrm{d}\vartheta/\mathrm{d}\tau = 0$ when ϑ is constant, and we have only three non-zero 4-velocity components, $\mathrm{d}t/\mathrm{d}\tau$, $\mathrm{d}r/\mathrm{d}\tau$ and $\mathrm{d}\varphi/\mathrm{d}\tau$. These components are related to one another by normalization of the 4-velocity, $||\mathbf{u}||^2 = -1$. With the Schwarzschild metric and our components of $\mathbf{u}$, the normalization equation is

$$-\left(1-\frac{2M}{r}\right)\left(\frac{\mathrm{d}t}{\mathrm{d}\tau}\right)^2 + \frac{1}{1-2M/r}\left(\frac{\mathrm{d}r}{\mathrm{d}\tau}\right)^2 + r^2\left(\frac{\mathrm{d}\varphi}{\mathrm{d}\tau}\right)^2 = -1\,. \tag{10.12}$$

We do not know the remaining 4-velocity components yet, but we can replace two of them with the conserved quantities e and ℓ. The normalization equation then reads

$$-\frac{e^2}{1-2M/r} + \frac{(\mathrm{d}r/\mathrm{d}\tau)^2}{1-2M/r} + \frac{\ell^2}{r^2} = -1\,. \tag{10.13}$$

The terms in this equation depend only on r and its rate of change with respect to τ. We have arrived at a differential equation for $r(\tau)$ which we can now analyze independently of the values of t and φ. First, however, it is useful to rewrite this equation as

$$\frac{e^2-1}{2} = \frac{1}{2}\left(\frac{\mathrm{d}r}{\mathrm{d}\tau}\right)^2 - \frac{M}{r} + \frac{\ell^2}{2r^2} - \frac{M\ell^2}{r^3}\,. \tag{10.14}$$

If we multiply the equations by a planet's mass m, the first three terms on the right look just like the radial kinetic energy, Newton's gravitational potential $-mM/r$, and the centrifugal energy $\frac{1}{2}m\ell^2/r^2$, well-known from classical mechanics with a central-mass potential. The last term, $-M\ell^2/r^3$, is new and implies relativistic corrections.

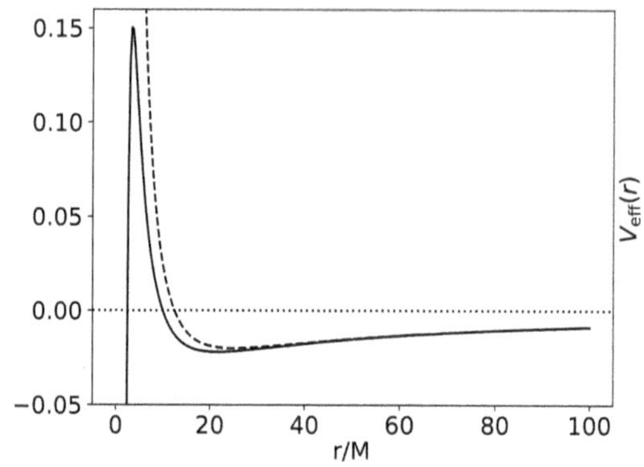

Figure 10.1 Effective potential for planetary orbits in general relativity (solid) and for Newtonian gravity (dashed). Deviations between the two curves are more pronounced at small r. Planetary orbits have energies near the local minimum of the potential, with r not changing much along a near-circular trajectory. The mass and angular momentum in this example are such that $\ell/M = 5$.

Given these familiar terms, we may interpret the equation as an energy-balance law. The constant on the left-hand side of (10.14), $\frac{1}{2}(e^2-1)$, is related to the non-relativistic energy if we approximate $e-1=\sqrt{1+(e^2-1)}-1 \approx \frac{1}{2}(e^2-1)$ for small e^2-1. The energy of a planet of mass m is then $E = me \approx m + \frac{1}{2}m(e^2-1)$, and therefore (10.14) is the energy minus the rest energy, divided by the mass. We should expect this quantity to equal kinetic plus potential energy, as realized by our formula. In compact form, the normalization or energy equation for a planet of mass m is

$$E - m = \frac{1}{2}m\left(\frac{\mathrm{d}r}{\mathrm{d}\tau}\right)^2 + mV_{\mathrm{eff}}(r) \tag{10.15}$$

with the **effective potential**

$$V_{\mathrm{eff}}(r) = -\frac{M}{r} + \frac{\ell^2}{2r^2} - \frac{M\ell^2}{r^3}. \tag{10.16}$$

We analyze the resulting motion using the technique of potential landscapes. It turns out that there are different qualitative forms of the effective potential, depending on the value of the ratio ℓ/M. For planets, we can estimate this ratio using Kepler's laws: The third law can be written as $\ell \approx \sqrt{R_S r}$ if we replace the sun's mass with its Schwarzschild radius. Thus, $\ell/M \approx \sqrt{r/R_S}$ is large. For such values, or any ratio $\ell/M > 4$, the effective potential has the form shown in Figure 10.1.

The potential has one local minimum, around which we have nearly-circular orbits. Since the potential energy cannot be greater than the conserved energy, there are barriers to smaller as well as larger r. The orbit has a maximum and a minimum distance to the sun, reached periodically along the orbit. The closer the energy is to the local minimum, the smaller the difference between maximum and minimum radius, and the more circular the orbit. These qualitative properties are identical in relativistic and Newtonian mechanics, but the precise orbiting behaviors differ because the potentials are not the same.

There is one characteristic difference between the relativistic and the Newtonian effective potentials: The relativistic potential has a maximum at a small value of r and drops down to negative infinity near $r = 0$, while the Newtonian one is unbounded from above and approaches positive infinity at $r = 0$. The Newtonian potential always has a barrier that keeps a planet from falling all the way to $r = 0$ (unless $\ell = 0$), even if its energy is large. In general relativity, this barrier is still present but is of finite height. It is therefore easier for an object to fall to small r. In Newtonian gravity, an object falling toward the sun accelerates more and more and reaches very high velocities

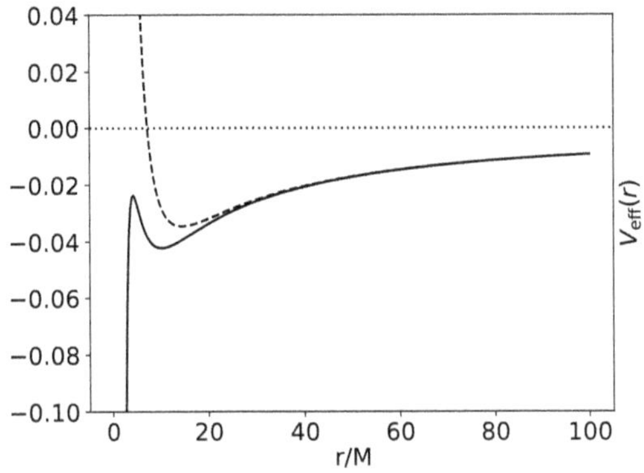

Figure 10.2 Effective potential for orbits of small angular momentum in general relativity (solid) and in Newtonian mechanics (dashed). The relativistic potential is negative for all values of r, while the form of the Newtonian potential is unchanged. Mass and angular momentum in this example are such that $\ell/M = 3.8$.

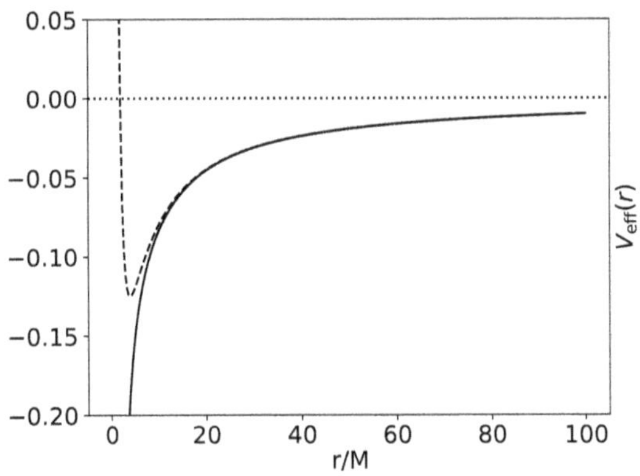

Figure 10.3 Effective potential for orbits of very small angular momentum in general relativity (solid) and in Newtonian mechanics (dashed). The relativistic potential is negative for all values of r, and does not have a local minimum, while the Newtonian potential is qualitatively unchanged. Mass and angular momentum in this example are such that $\ell/M = 2$.

that prevent it from falling further down. In relativity, the maximum speed is the speed of light; at some point, the angular velocity cannot be increased to prevent an object from falling all the way into the sun.

If the ratio ℓ/M gets smaller and reaches 4, the height of the potential barrier shrinks, the maximum reaching the value zero. An example for a ratio less than 4 is shown in Figure 10.2. If the ratio is decreased to $2\sqrt{3} \approx 3.4$, there is no local maximum at all and the effective potential is monotonic; see Figure 10.3. The local minimum disappears, and no bound orbit is possible for very small angular momenta. In Newtonian physics, on the other hand, bound orbits always exist as long as there is any angular momentum, $\ell \neq 0$.

For a large value of $\ell/M > 4$, the potential landscape as in Figure 10.1 leads to orbits of four types. If the energy equals the value of V_{eff} at the local minimum, there is no radial force on the planet. The radius stays constant and the planet moves on a circle around the sun. (There is another circular orbit at the local maximum, but it is unstable and plays no physical role.) If we raise the energy, but keep it negative, there is some kinetic energy left at the local minimum. The rate of

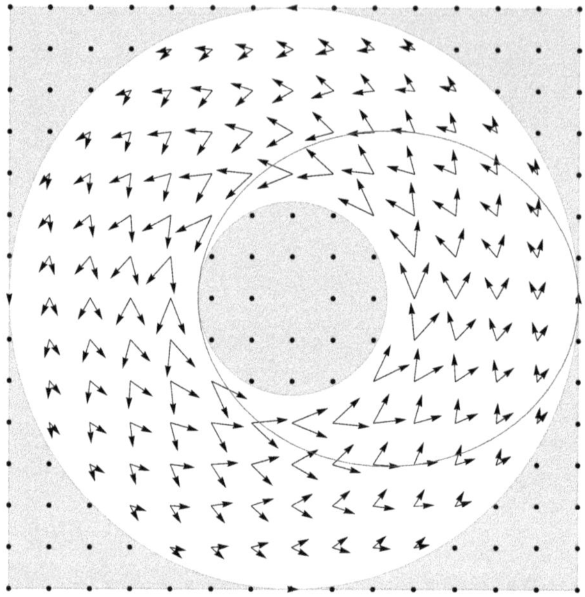

Figure 10.4 Vector field for the differential equation of Newton's gravitational force, with two arrows at each point for the positive and negative solutions of $\mathrm{d}r/\mathrm{d}t$ according to the quadratic energy equation. The conservation laws of energy and angular momentum determine the increments $\mathrm{d}r$ and $r\mathrm{d}\varphi$ during an interval $\mathrm{d}t$, which are then used to compute the vector field as the sum of $(\cos(\varphi)\mathrm{d}r, \sin(\varphi)\mathrm{d}r)$ and $(-r\sin(\varphi)\mathrm{d}\varphi, r\cos(\varphi)\mathrm{d}\varphi)$ at the position $(r\cos(\varphi), r\sin(\varphi))$. The energy is negative, such that solutions cannot enter the gray regions, bounded by the positions of turning points. One Kepler ellipse is shown.

change $\mathrm{d}r/\mathrm{d}\tau$ is non-zero, and the radius varies. It stops changing only when the potential walls are reached for the given energy, turning the planet's radius back toward the local minimum. The radius oscillates around the local minimum, implying an orbit of near-elliptic form. For the Newtonian potential, an exact ellipse results, as shown in Figure 10.4. The planet is bound to the sun and cannot escape to large values of r.

If the energy is raised so much that it becomes positive, the orbit is no longer bound. There may still be a barrier keeping the object away from small r, but no force prevents it from moving to larger and larger r. It approaches the sun at most once, and is then deflected back to a position far away. In Newtonian gravity, these orbits have the form of a parabola or a hyperbola. If the energy is increased further such that it becomes greater than the local maximum of the effective potential, there is no barrier whatsoever to stop the radial motion. If an object with such an energy falls toward the sun, it will not be held back until it smashes into the surface. Complete plunge orbits, with large radial energy, do not exist for the Newtonian potential, which has no finite maximum unless $\ell = 0$. A potential barrier always exists at small r, and a plunge is possible only if the energy is so high that the sun's radius is reached before the energy barrier.

10.1.3 ESCAPE VELOCITY

Consider a rocket launched straight up in radial motion. Its angular momentum, $\ell = 0$, vanishes. For the rocket to escape the gravitational force of Earth, its energy must be at least so large that the rocket comes to rest far from the center, in the limit where r is infinite. The 4-velocity at such a position is $\mathbf{u} = (1,0,0,0)^T$: It has zero spatial components and is normalized in the Schwarzschild metric at $r \to \infty$. Such a geodesic has the conserved energy per mass of $e = (1 - 2M/r)\mathrm{d}t/\mathrm{d}\tau = u^t = 1$, computed at $r \to \infty$. The radial velocity, obeying the normalization equation with $\ell = 0$, then satisfies $\frac{1}{2}(\mathrm{d}r/\mathrm{d}\tau)^2 - M/r = 0$, or $u^r = \mathrm{d}r/\mathrm{d}\tau = \sqrt{2M/r}$ for the radial component of the 4-velocity. We took the positive square root for outward radial motion.

Given unchanging angles, the angular components of the 4-velocity vanish, $u^\vartheta = 0 = u^\varphi$. The time component for all r follows from our conserved energy: $1 = e = (1-2M/r)u^t$ implies $u^t = 1/(1-2M/r)$. As a 4-vector,

$$\mathbf{u}(r) = \left(\frac{1}{1-2M/r}, \sqrt{\frac{2M}{r}}, 0, 0\right)^T. \tag{10.17}$$

The required energy as measured by an observer staying at the launch pad, at Earth's radius $r = R$, is $E_R = -m\mathbf{u}(R)\cdot\mathbf{u}_{\mathrm{obs}}$ with the rocket mass m. The observer is at rest, and therefore has a normalized 4-velocity $\mathbf{u}_{\mathrm{obs}} = ((1-2M/R)^{-1/2}, 0, 0, 0)^T$. The energy equals $E_R = -mg_{tt}u^t(R)u^t_{\mathrm{obs}} = m(1-2M/R)^{-1/2}$. Since the radius R is much greater than the Earth's Schwarzschild radius, the energy can be approximated as $E_R \approx m + mM/R$, which equals the rest energy plus the Newtonian potential energy.

A rocket with mass m needs an energy $E_R = m(1-2M/R)^{-1/2}$ to escape Earth's gravitational force, with M and R the planet's mass and radius. An observer at rest would attribute an energy of $E_R = m\gamma_V = m(1-V^2)^{-1/2}$ to the rocket moving at speed V. Direct comparison of the two versions of E_R shows that the **escape velocity** of a rocket is $V = \sqrt{2M/R} = \sqrt{R_S/R}$ with the Earth's Schwarzschild radius R_S. The ratio of the radii is small: $V \approx 0.00003$, implying an escape velocity much less than the speed of light.

For a denser stellar object, a much larger escape velocity close to the speed of light is required. If the surface radius is equal to the Schwarzschild radius, the escape velocity equals the speed of light. No massive object can be that fast, and no matter can escape from the Schwarzschild radius.

In addition, escaping signals are subject to **gravitational redshift**. The Schwarzschild line element allows us to extend our previous derivation of gravitational redshift based on the equivalence principle to stronger gravitational forces. In an astronomical situation, we are observers far away from a star, at a much smaller value of the gravitational potential than what is realized on the star's surface. Light emitted from the surface of the star is subject to different time dilation than we are, and the emitted frequency ω_* is redshifted to a smaller $\omega < \omega_*$ once light reaches our telescopes.

The line element determines gravitational redshift by the position dependence of its time component. Once emitted, light moves outward along a lightlike geodesic whose form follows from the position dependence of the line element. Lightlike geodesics obey $\mathrm{d}s^2 = 0$, and therefore a radial one with constant angles must satisfy the relationship $\mathrm{d}t = \pm\mathrm{d}r/(1-2M/r)$. For r much greater than the Schwarzschild radius, this equation is close to the one $\mathrm{d}t = \pm\mathrm{d}r$ for a straight diagonal. But at smaller r, near the star surface, the slope is greater. The worldlines of light near the surface, compared with flat-space light rays, initially stay closer to the worldline of the surface at constant r, as indicated in Figure 10.5, which is what we expect from gravitational attraction.

Another implication of the position-dependent line element is **time dilation** even for observers at rest. Proper time at vanishing velocity is related to coordinate time by $\mathrm{d}\tau = \sqrt{-\mathrm{d}s^2} = \sqrt{-g_{tt}}\,\mathrm{d}t$. For the Schwarzschild line element, $\mathrm{d}\tau = (1-2M/r)^{1/2}\mathrm{d}t$. On the surface of a star, proper time progresses more slowly than coordinate time of an outside observer. We can express time dilation as a modified time component of an observer's 4-velocity. The time component of $\mathbf{u}_{\mathrm{obs}} = \mathrm{d}\mathbf{x}/\mathrm{d}\tau$ is $u^t_{\mathrm{obs}} = \mathrm{d}t/\mathrm{d}\tau = (1-2M/r)^{-1/2}$ and depends on the displacement r from the center.

We use the 4-velocity of an observer in order to compute observed energies of objects or frequencies of light by means of a local inertial frame. For light, the frequency is $\omega = -\mathbf{k}\cdot\mathbf{u}_{\mathrm{obs}}$ with the wave 4-vector $\mathbf{k}$ on the lightlike geodesic. Both $\mathbf{k}$ and $\mathbf{u}_{\mathrm{obs}}$ change as we move along the geodesic. Also the frequency changes, but $\mathbf{k}\cdot\xi$ with $\xi = (1,0,0,0)^T$ is a conserved quantity, where ξ is our timelike Killing vector. Since $\mathbf{u}_{\mathrm{obs}}$, for an observer at rest, is proportional to ξ, we compute the shifted frequencies by relating $\mathbf{k}\cdot\mathbf{u}_{\mathrm{obs}}$ to $\mathbf{k}\cdot\xi$, writing $\omega_* = -\mathbf{k}\cdot\mathbf{u}_{\mathrm{obs}}(R) = -\mathbf{k}\cdot\xi(1-2M/R)^{-1/2}$.

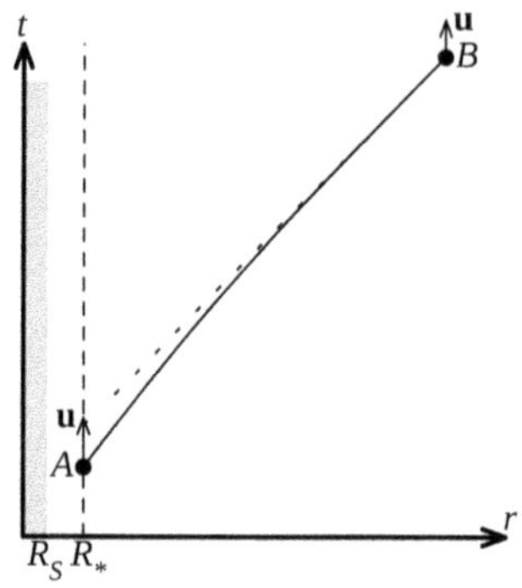

Figure 10.5 A space-time diagram with the radial displacement r from the center of a star. The star surface at $r = R_*$, outside of its Schwarzschild radius R_S, emits light outwards, seen by distant observers. Light moves along lightlike geodesics, obeying $ds^2 = 0$. In the coordinates r and t, these lines are not straight but bent according to the r-dependent metric coefficients. The r-dependence also implies that the 4-velocity $\mathbf{u}$ of an observer at constant r has a greater time-component closer to the surface.

We also have $\omega = -\mathbf{k}\cdot\mathbf{u}_{\text{obs}}(r) = -\mathbf{k}\cdot\xi$ for r much greater than the Schwarzschild radius, far away from the star's surface. Taken together, these equations imply

$$\omega = \sqrt{1-\frac{2M}{R}}\,\omega_* < \omega_*\,. \tag{10.18}$$

For small $2M/R$ we recognize our previous formula for gravitational redshift as a good approximation, but deviations occur for R closer to R_S.

10.1.4 KEPLER'S THIRD LAW

For the nearly circular orbit of a planet, the radius is close to the local minimum of the effective potential for given values of M (the solar mass) and ℓ (the planet's angular momentum). The local minimum can be found at a radius of

$$r_+ = \frac{\ell^2}{2M}\left(1+\sqrt{1-\frac{12M^2}{\ell^2}}\right)\,. \tag{10.19}$$

The smallest possible value, reached for $\ell/M = 2\sqrt{3}$ where the local minimum of the effective potential disappears, equals $r = 6M = 3R_S$ and is called the radius of the **innermost stable circular orbit** (or ISCO). Circular motion is impossible for $r < 3R_S$, a range that usually lies within the star's interior. But dense stars can have a surface radius less than $3R_S$, still greater than the Schwarzschild radius. Sometimes, the ISCO radius is observable via matter orbiting the compact object. For instance, an accretion disk of debris left over from a star's collapse, can be stable only in a range of radii greater than $3R_S$. Matter scattered to smaller radii does not orbit but rather falls all the way to the star's surface or the Schwarzschild radius. In practice, the ISCO radius also depends on the angular momentum of the central mass, which requires a non-spherical geometry.

For radii greater than $3R_S$, stable circular orbits are possible on which a planet may move around the star for a long time. Measured by an observer far away from the star, where the coordinate t is close to the observer's proper time, the angular velocity equals $\Omega = d\varphi/dt$. We relate Ω to our conserved quantities by writing $\Omega = (d\varphi/d\tau)/(dt/d\tau)$ and using the expressions $d\varphi/d\tau = \ell/r^2$

and $\mathrm{d}t/\mathrm{d}\tau = e/(1-2M/r)$. Thus,

$$\Omega = \frac{1-2M/r}{r^2}\frac{\ell}{e}. \tag{10.20}$$

The radius of a circular orbit does not change, $\mathrm{d}r/\mathrm{d}\tau = 0$, such that the normalization equation implies

$$e = \sqrt{\left(1-\frac{2M}{r}\right)\left(1+\frac{\ell^2}{r^2}\right)}. \tag{10.21}$$

With the orbit radius at the local minimum, $r = r_+$, we then have $\ell^2 = Mr/(1-3M/r)$ or

$$\frac{\ell}{e} = \frac{\sqrt{Mr}}{1-2M/r}. \tag{10.22}$$

Inserting this result in (10.20), we obtain

$$\Omega^2 = \frac{M}{r^3}. \tag{10.23}$$

The angular velocity $\Omega = \sqrt{M/r^3}$ implies a squared orbit period of $T^2 = 4\pi^2 r^3/M$, behaving like the cube of the orbit radius. Planets farther away from the sun take longer to complete one orbit. This law, derived here from general relativity, is the same as **Kepler's third law**. There are no correction terms compared to the Newtonian result, provided we determine the orbit period T by referring to the time coordinate t of an observer far from the sun. The orbit period measured on a planet such as Earth does not refer to t but rather to the planet's proper time τ. If the planet's velocity is small, it does not give rise to much time dilation; and if the orbit stays far away from the Schwarzschild radius of the sun, gravitational time dilation is not significant either. The formula of Kepler's third law is then an excellent approximation for all solar planets.

10.1.5 PERIHELION SHIFT

Mercury's orbit is an interesting test case for seeing whether Newton's law of gravity or Einstein's theory of general relativity provide a better fit to observations: For small r, the relativistic correction in the energy balance law for planetary motion is most significant. Mercury is close to the sun and not easy to observe, and its orbit had not been known well for a long time. But by 1855, observations had accumulated to a substantial period covering many orbits. Looking at the data, the astronomer Urbain Le Verrier noticed that Mercury does not obey Kepler's laws: it does not move along an exact ellipse. When observed over many of its periods, the would-be elliptical orbit rotates by shifting its point closest to the sun, the perihelion, by small amounts each time it goes around the sun; see Figure 10.6. In one century, the angular position of the perihelion rotates around the sun by just 1.5°.

Perihelion shift or its analogs for motion around objects other than the sun are not uncommon. The moon orbit has a shifting perigee, its closest encounter with Earth. This effect is much easier to observe and was already known to the ancient Greeks and Romans. In this case, the development of Fourier analysis made it possible to explain all deviations from Kepler's strict ellipse by the sun's influence and the Earth's non-spherical form. Even though the sun is far away, it is massive, and its influence on the Earth-moon system is well known from its contribution to the tides. With these additional effects, the moon orbit agrees with predictions of Newtonian mechanics, although Newton himself was unable to derive this result: Fourier analysis, required for complicated derivations of orbits in a 3-body problem combining the moon, Earth and the sun, was not available until the early 19th century.

In the case of Mercury's orbit, the sun rotates and is not exactly spherical, which accounts for some of the perihelion shift. Also, other planets, in particular Mercury's neighbor Venus, attract the

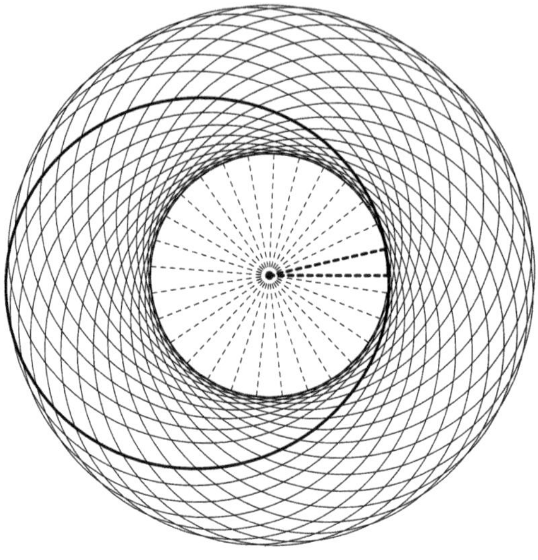

Figure 10.6 Perihelion shift: The Kepler ellipse of a planet rotates slowly, such that the point of closest encounter with the sun, the perihelion, shifts from orbit to orbit. The first cycle is drawn as a thick line. Dashed lines indicate the distance from the shifting perihelion to the focal point.

planet toward them and affect the orbit. Unlike the moon, however, Mercury holds on to a small perihelion shift of 0.012° per century, or $43''$ per century, even if all the known Newtonian effects are added up.

To account for the remaining shift, some astronomers suggested that there might be another planet even closer to the sun, which would be difficult to spot in the intense solar glare but could affect Mercury's orbit. (They even came up with a name for this planet: Vulcan.) General relativity provides a more elegant explanation without the need of new assumptions. Its effective potential modifies Newton's force, such that non-circular orbits are no longer closed ellipses.

The effective potential suggests viewing non-circular orbits as small oscillations around circular ones for a radius near the local minimum. The frequency of these oscillations depends on the force, which is given by the slope of the potential around the minimum. For Newton's force, the oscillation frequency happens to be identical to the orbital frequency ω. When a planet completes one orbit around the sun, the radial oscillation returns to its starting point. The correction from general relativity implies that the two frequencies are out of tune. The orbit does not close, and the perihelion shifts.

If we are interested in the behavior of small deviations from a circular orbit, the differential equation for $r(\tau)$ can be solved by approximation. For given M and ℓ, we call the circular radius $r_\circ$, and write the radial displacement along the non-circular orbit as $r = r_\circ + d$. Treating $d = r - r_\circ$ as small simplifies the derivation of the oscillation frequency.

We first consider the Newtonian effective potential $V_{\rm eff}(r) = -M/r + \frac{1}{2}\ell^2/r^2$, write it as a function of d and then use a Taylor expansion:

$$\begin{aligned} V_{\rm eff}(d) &= -\frac{M}{r_\circ + d} + \frac{\ell^2}{2(r_\circ + d)^2} = \frac{-2M(r_\circ + d) + \ell^2}{2r_\circ^2}\frac{1}{(1 + d/r_\circ)^2} \\ &= \frac{1}{2r_\circ^2}\left(\ell^2 - 2Mr_\circ - 2\frac{\ell^2 - Mr_\circ}{r_\circ}d + \frac{3\ell^2 - 2Mr_\circ}{r_\circ^2}d^2 + \cdots\right). \end{aligned} \quad (10.24)$$

The term of $V_{\rm eff}$ linear in d vanishes if $r_\circ$ is the radius at the local minimum, $r_\circ = \ell^2/M$. The coefficient of the quadratic term in $V_{\rm eff}$ is then $\omega_{\rm rad}^2 = \ell^2/r_\circ^4 = M/r_\circ^3$ with the radial frequency $\omega_{\rm rad}$

read off by identifying the quadratic term with a harmonic potential $\frac{1}{2}\omega_{\rm rad}^2 d^2$. A quadratic potential implies harmonic oscillations $d(t) \sim d_0 \sin(\omega_{\rm rad}\tau)$ around the circular radius. The frequency $\omega_{\rm rad}$ is the same as Kepler's frequency $\Omega = \sqrt{M/r_\circ^3}$ for one period of the orbit. The Newtonian orbit is therefore closed even if it is not circular.

The circular orbit in general relativity has a radius

$$r_\circ = r_+ = \frac{\ell^2}{2M}\left(1+\sqrt{1-\frac{12M^2}{\ell^2}}\right). \tag{10.25}$$

The radial frequency for oscillations around it is equal to $\omega_{\rm rad} = \ell^2(1-6M/r_\circ)/r_\circ^4$, computed in the same way as above by a quadratic expansion of the effective potential. The difference

$$\omega_{\rm prec} = \Omega - \omega_{\rm rad} = \frac{\ell}{r_\circ^2}\left(1-\sqrt{1-\frac{6M}{r_\circ}}\right) \approx \frac{3M\ell}{r_\circ^3} \tag{10.26}$$

is non-zero and gives us the angular rate of change of the perihelion, or the **precession frequency**. The angular velocity implies an angle difference of $\Delta\varphi = \omega_{\rm prec}/\Omega = 3M/r_\circ$ per orbit.

10.1.6 LIGHT

Lightlike geodesics in Schwarzschild space-time have the conserved quantities

$$e = -\mathbf{k}\cdot\xi = \left(1-\frac{2M}{r}\right)\frac{\mathrm{d}t}{\mathrm{d}\lambda} \quad \text{and} \quad \ell = \mathbf{k}\cdot\eta = r^2\sin(\vartheta)^2\frac{\mathrm{d}\varphi}{\mathrm{d}\lambda} \tag{10.27}$$

in terms of the wave 4-vector $\mathbf{k}$ and an affine parameter λ. We replace the normalization equation of timelike geodesics with the lightlike condition or dispersion relation

$$||\mathbf{k}||^2 = g_{\alpha\beta}\frac{\mathrm{d}x^\alpha}{\mathrm{d}\lambda}\frac{\mathrm{d}x^\beta}{\mathrm{d}\lambda} = 0. \tag{10.28}$$

Inserting the conserved quantities for the rates of change of t and φ in the dispersion relation and setting $\vartheta = \frac{1}{2}\pi$ constant for a light ray in the equatorial plane implies

$$-\frac{e^2}{1-2M/r} + \frac{1}{1-2M/r}\left(\frac{\mathrm{d}r}{\mathrm{d}\lambda}\right)^2 + \frac{\ell^2}{r^2} = 0. \tag{10.29}$$

The conserved quantities e and ℓ are useful for a mathematical separation of the components of the geodesic equation. However, unlike in the timelike case, they do not correspond to observable quantities because they are not invariant under rescalings of the affine parameter. Mimicking the effective-potential treatment of timelike geodesics but keeping in mind the scaling behavior, it is then useful to replace the previous energy constant e with a ratio such as $b = \ell/e$ that is scaling invariant.

Using this ratio, we write the lightlike condition as

$$\frac{1}{b^2} = \frac{1}{\ell^2}\left(\frac{\mathrm{d}r}{\mathrm{d}\lambda}\right)^2 + W_{\rm eff}(r) \tag{10.30}$$

with the effective potential

$$W_{\rm eff}(r) = \frac{1}{r^2}\left(1-\frac{2M}{r}\right). \tag{10.31}$$

The remaining constant ℓ in (10.30) is not scaling invariant, but it appears only in the invariant combination $\ell^2\mathrm{d}\lambda^2$.

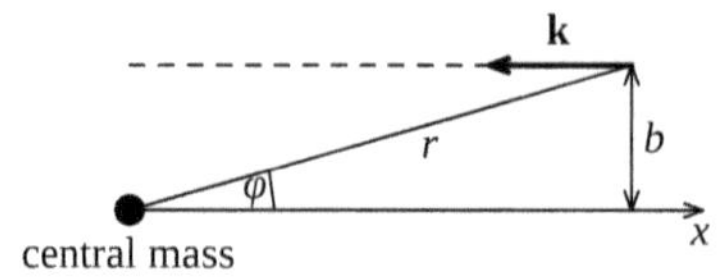

Figure 10.7 The impact parameter b by which a straight light ray would miss the central mass. For a straight line, the point of closest approach to the central mass is reached when $r = b$ equals the impact parameter. The turning-point condition $1/b^2 = 1/r^2 = W_{\text{eff}}(r)$ determines the classical effective potential defined by equation (10.30).

The ratio $b = \ell/e$ has units of length and determines the **impact parameter** of the light ray relative to the central mass. The geometry of straight lines implies that $|b|$ equals the distance by which a light ray would miss the central mass if it were to move without deflection, as illustrated in Figure 10.7. The sign of b determines whether the light ray moves around in a clockwise or counter-clockwise fashion.

In order to see these features, consider the case of $|b|$ much smaller than the distance r. A line connecting the central mass to a point on the light ray is then almost parallel to the light ray. Therefore, the change of r in time is almost equal to the speed of light, or the negative $\mathrm{d}r/\mathrm{d}t \approx -1$ when light is approaching the central mass. The connecting line is not exactly parallel to the light ray, however, and instead has a small angle $\varphi(r) \approx b/r$ with it. The angle changes as r changes, while b remains constant. We have

$$\frac{\mathrm{d}\varphi}{\mathrm{d}t} = \frac{\mathrm{d}\varphi}{\mathrm{d}r}\frac{\mathrm{d}r}{\mathrm{d}t} = -\frac{\mathrm{d}(b/r)}{\mathrm{d}r} = \frac{b}{r^2}. \tag{10.32}$$

We obtain the same result from the conserved quantities, again for large displacement r compared to b:

$$b = \frac{\ell}{e} \approx r^2\frac{\mathrm{d}\varphi/\mathrm{d}\lambda}{\mathrm{d}t/\mathrm{d}\lambda} = r^2\frac{\mathrm{d}\varphi}{\mathrm{d}t}. \tag{10.33}$$

This equivalence confirms the interpretation of b as the impact parameter.

If light moves along a straight line passing by the central mass, the position x along the ray changes by $\mathrm{d}x/\mathrm{d}t = -1$ on approach. In Euclidean geometry, the distance r from the central mass to a point on the light ray obeys $r^2 = x^2 + b^2$ according to Figure 10.7. It changes according to $2r\mathrm{d}r/\mathrm{d}t = 2x\mathrm{d}x/\mathrm{d}t$, or $\mathrm{d}r/\mathrm{d}t = (x/r)\mathrm{d}x/\mathrm{d}t = -(1-b^2/r^2)^{1/2}$. Writing this equation as

$$\frac{1}{b^2}\left(\frac{\mathrm{d}r}{\mathrm{d}t}\right)^2 + \frac{1}{r^2} = \frac{1}{b^2}, \tag{10.34}$$

we recognize the form of (10.30) with an effective potential $1/r^2$, and $\lambda = bt/\ell = t/e$. The turning point in r for a given light ray with impact parameter b is obtained when $1/b^2 = 1/r^2$ equals the effective potential, such that $r_{\text{min}} = b$. This derivation confirms that light rays in Euclidean geometry are straight lines that preserve the transversal distance b from the center of the polar coordinate system. According to general relativity, the central mass M changes the effective potential by adding $-2M/r^3$, implying an attractive force that deflects light rays as they pass close by.

For a light ray to pass by a star like the sun, the impact parameter b must be greater than the surface radius R, which in turn is much larger than the Schwarzschild radius $R_S = 2M$. With this condition, the minimum distance from the central mass is reduced from b to $r_{\text{min}} \approx b - M$ as the leading correction in an expansion by M/b, derived from $W_{\text{eff}}(r_{\text{min}}) = 1/b^2$. During the passing phase, the angle changes by an amount given by

$$\varphi = 2\int_{r_{\text{min}}}^{\infty}\frac{\mathrm{d}\varphi}{\mathrm{d}r}\mathrm{d}r = 2\int_{r_{\text{min}}}^{\infty}\frac{\mathrm{d}\varphi/\mathrm{d}\lambda}{\mathrm{d}r/\mathrm{d}\lambda}\mathrm{d}r = 2\int_{r_{\text{min}}}^{\infty}\frac{b/r^2}{\sqrt{1-b^2/r^2+2b^2M/r^3}}\mathrm{d}r. \tag{10.35}$$

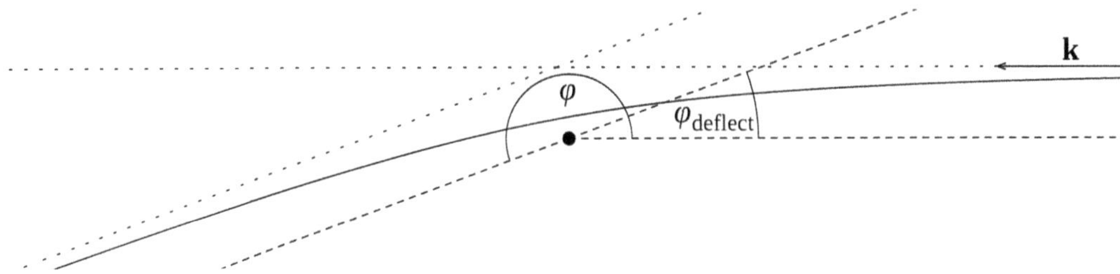

Figure 10.8 The deflection angle φ_{deflect} and the full angular difference $\varphi = \pi + \varphi_{\text{deflect}}$.

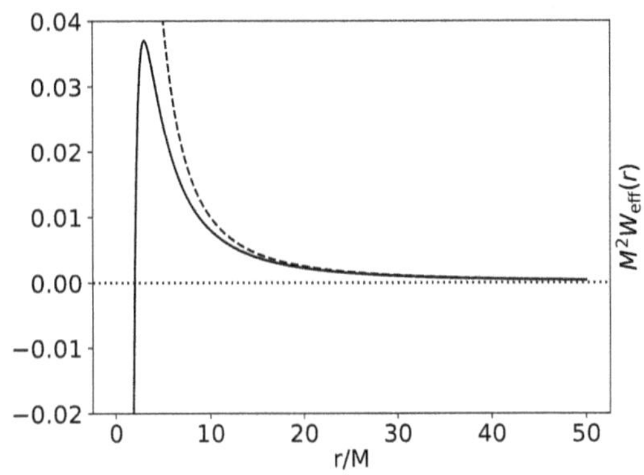

Figure 10.9 The effective potential for radial motion of light.

For $M = 0$, the integration can be performed using the inverse sine and results in the value $\varphi = \pi$ which, as illustrated in Figure 10.8, is expected for a straight line. If $M \neq 0$, the integral is elliptic and, even though b^2M/r^3 is much smaller than b^2/r^2, cannot easily be approximated because the dominant contributions are around the pole of the integrand. But because the minimum radius shifts by a relative amount of M/b, we expect the same qualitative behavior for the deflection angle $\varphi_{\text{deflect}} = \varphi - \pi$. For a light ray grazing the surface of the sun, $b = R$ is the solar radius, and φ_{deflect} equals approximately $0.0005°$. This tiny change of the direction of sight of stars was first measured during a complete solar eclipse in 1919, by a team led by Arthur Eddington.

In the potential landscape of Figure 10.9, any ray with small $1/b^2$ encounters a potential barrier. Arriving from larger r, the light ray is turned back to increasing r after reaching a minimum distance to the star. When b is smaller, however, $1/b^2$ may be greater than the potential at its peak, and no barrier limits r; this case is realized when b^2 increases above $27M^2$. The light ray then has a monotonically decreasing r while it spirals toward the star until it hits the surface at $r = R$.

If the impact parameter equals $b = \sqrt{27}M$, $1/b^2$ is at the maximum of the effective potential. Equation (10.30) can then be fulfilled with vanishing $\mathrm{d}r/\mathrm{d}\lambda$. A constant distance from the star means that such a light ray is in a circular orbit; light is captured and orbits around the star almost like a planet. The radius of the orbit is small, $r = 3M = 1.5R_S$, but greater than the Schwarzschild radius. For the sun, this place would be well inside its bulk where light cannot move without being scattered, but it may be outside of the surface of a dense star. The orbit is unstable because it corresponds to a local maximum of the effective potential: A slight displacement to smaller or larger r implies forces pulling the light ray away from the circular orbit. One would have to shine one's light in just the right direction, staying exactly at the circular radius, in order to see it come back after one orbit around the star.

The minimum impact parameter $b = \sqrt{27}M \approx 2.6R_S$ determines the transversal distance from the center along an asymptotic light ray that escapes outward from the unstable circular orbit. If a light ray escapes from the orbit and approaches our telescopes, we would see it at this distance from the central mass, rather than at the orbit radius of $1.5R_S$. If there are light sources around the central mass, such as an accretion disk, light can enter the circular orbit from multiple angles such that it is eventually sent out into our direction. Provided light can travel freely toward us, we expect to see a **photon ring** at the radius determined by b, or rather at a value that takes into account angular momentum of the central mass. Comparing the minimal impact parameter with the innermost stable circular orbit of massless objects at $3R_S$, the region in which we would see the photon ring should indeed by free of matter from the accretion disk.

10.2 BLACK HOLES

The Schwarzschild line element

$$ds^2 = -\left(1-\frac{2M}{r}\right)dt^2 + \frac{1}{1-2M/r}dr^2 + r^2(d\vartheta^2 + \sin(\vartheta)^2 d\varphi^2) \tag{10.36}$$

has vanishing or infinite components at $r = 0$ and $r = 2M$. If we extend the weak-field line element to stronger gravitational forces, starting at large $r \gg 2M$ and moving to smaller values, we first encounter a singularity at $r = 2M$, the Schwarzschild radius.

10.2.1 EDDINGTON–FINKELSTEIN COORDINATES

The Schwarzschild radius is a coordinate singularity. To show this, we use the methods of Section 8.6.7 and derive properties of lightlike geodesics approaching the Schwarzschild radius from greater values of r. Ingoing light rays in Schwarzschild space-time obey the equation

$$t = v - r - 2M\log\left|\frac{r}{2M} - 1\right| . \tag{10.37}$$

Moreover, using the lightlike condition, the radial component of the geodesic equation simplifies to $d^2r/d\lambda^2 = 0$, such that the affine parameter is proportional to r. It is not necessary to transform r, and we just transform t to $v(t,r)$ obtained by rewriting (10.37). The logarithm in this transformation shows that a finite value of v at $r = 2M$ implies an infinite t. While it is possible to extend v beyond this finite value, t is already infinite and does not show the correct behavior of space-time around the Schwarzschild radius.

The transformation (10.37) implies an invertible metric at $r = 2M$. We first compute dt for $t(v,r)$ as given in (10.37), and write the time contribution to the Schwarzschild line element as

$$\begin{aligned} -\left(1-\frac{2M}{r}\right)dt^2 &= -\left(1-\frac{2M}{r}\right)\left(dv - dr - \frac{1}{r/(2M)-1}dr\right)^2 \\ &= -\left(1-\frac{2M}{r}\right)dv^2 + 2dvdr - \frac{1}{1-2M/r}dr^2 . \end{aligned} \tag{10.38}$$

If we insert this result in the Schwarzschild line element, there is a cross-term $dvdr$, but the dr^2-terms cancel out: We obtain the **Schwarzschild line element in Eddington–Finkelstein coordinates** (v,r),

$$ds^2 = -\left(1-\frac{2M}{r}\right)dv^2 + 2dvdr + r^2(d\vartheta^2 + \sin(\vartheta)^2 d\varphi^2) . \tag{10.39}$$

The new metric retains a component $g_{vv} = -(1-2M/r)$ that vanishes at $r = 2M$, and the component $g_{rr} = 0$ is always zero in these coordinates. However, there are non-zero off-diagonal components $g_{vr} = g_{rv} = 1$, which implies that the determinant

$$\det(\mathbf{g})|_{r=2M} = \det \begin{pmatrix} 0 & 1 & 0 & 0 \\ 1 & 0 & 0 & 0 \\ 0 & 0 & 4M^2 & 0 \\ 0 & 0 & 0 & 4M^2\sin(\vartheta)^2 \end{pmatrix} \neq 0 \tag{10.40}$$

does not vanish at $r = 2M$, unless $\sin(\vartheta) = 0$ which corresponds to familiar coordinate singularities at the poles of a sphere. The new metric is invertible at the Schwarzschild radius, which therefore is a coordinate singularity of the original coordinates (t, r). The metric remains non-invertible at $r = 0$. At the end of this section, we will see that this is a physical singularity.

10.2.2 CAUSAL STRUCTURE

In order to understand the extended Schwarzschild geometry in Eddington–Finkelstein coordinates, we analyze its causal structure by solving the equation it implies for light rays. It is sufficient to use only radial light rays which have constant ϑ and φ and therefore $\mathrm{d}\vartheta = 0 = \mathrm{d}\varphi$. The equation for radial light rays,

$$0 = \mathrm{d}s^2 = -\left(1 - \frac{2M}{r}\right)\mathrm{d}v^2 + 2\mathrm{d}v\mathrm{d}r\,, \tag{10.41}$$

has three classes of solutions:

1. $\mathrm{d}v = 0$ implies $\mathrm{d}s^2 = 0$. Therefore, one class of light rays is given by $v = \text{const}$.
2. If $\mathrm{d}v \neq 0$, (10.41) implies $-(1-2M/r)\mathrm{d}v + 2\mathrm{d}r = 0$, or

$$\mathrm{d}v - \frac{2\mathrm{d}r}{1-2M/r} = 0\,. \tag{10.42}$$

   We integrate this equation to

$$v - 2\left(r + 2M\log\left|\frac{r}{2M} - 1\right|\right) = \text{const}\,. \tag{10.43}$$

3. The final solution is given by $r = 2M$, for which $1 - 2M/r = 0$ and $\mathrm{d}r = 0$. The worldline at constant r equal to the Schwarzschild radius is therefore not a spatial surface but a light ray. This result indicates that the geometry at small r may be rather counter-intuitive.

10.2.2.1 Space-time diagram

We use light rays to visualize the causal structure of our geometry by means of a space-time diagram. As seen in Section 8.6.7, we should be careful about the meaning of "time" in this context. A time coordinate $\tilde{t}$ in a region of space-time should be such that setting it to a constant value, $\mathrm{d}\tilde{t} = 0$, determines a spacelike hypersurface in space-time. In Minkowski space-time, this condition is satisfied for $\tilde{t} = t$, the usual time coordinate. But in Schwarzschild space-time, the coordinate we have called t does not satisfy the condition and is not a suitable time coordinate in a region that includes $r = 2M$: Setting $\mathrm{d}t = 0$ in the Schwarzschild line element gives us a 3-dimensional line element

$$\mathrm{d}s^2 = \frac{1}{1-2M/r}\mathrm{d}r^2 + r^2(\mathrm{d}\vartheta^2 + \sin(\vartheta)^2\mathrm{d}\varphi^2) \tag{10.44}$$

which is spacelike (positive) for $r > 2M$ but not for $r < 2M$. If $r < 2M$, the metric component g_{rr} is negative, which would suggest that r rather than t should be used as a time coordinate. In a region around $r = 2M$ that includes some values of r less than $2M$ and some greater than $2M$, neither t nor r is a suitable time coordinate.

We have already seen that transforming (t,r) to (v,r) in Eddington–Finkelstein coordinates is useful around $r = 2M$ because it eliminates the coordinate singularity. However, v is constant along light rays, such that setting $\mathrm{d}v = 0$ in the Schwarzschild line element with Eddington–Finkelstein coordinates does not give us a spacelike hypersurface but rather a collection of light rays. Therefore, v is not a good time coordinate either.

A time coordinate is given by $\tilde{t} = v - r$ in Eddington–Finkelstein coordinates: If we set $\mathrm{d}\tilde{t} = 0$, we have $\mathrm{d}v = \mathrm{d}r$ in (10.39), and

$$\mathrm{d}s^2|_{\mathrm{d}\tilde{t}=0} = \left(1 + \frac{2M}{r}\right)\mathrm{d}r^2 + r^2(\mathrm{d}\vartheta^2 + \sin(\vartheta)^2\mathrm{d}\varphi^2) \tag{10.45}$$

is positive. We will therefore draw space-time diagrams of the region around $r = 2M$ using r as a spatial coordinate and $\tilde{t} = v - r$ as time.

Our first family of light rays, with constant values of v, is given by downward diagonals in our space-time diagram, just as in Minkowski space-time: Written as a function $\tilde{t}(r)$, the condition $v = \text{const}$ takes the form $\tilde{t}(r) = \text{const} - r$. The other family of light rays, also written as a function $\tilde{t}(r)$, takes the form

$$\tilde{t}(r) = \text{const} + r + 4M\log\left|\frac{r}{2M} - 1\right|. \tag{10.46}$$

For $r \gg 2M$, the logarithm is subdominant and these light rays asymptote to the form familiar from the remaining light rays in Minkowski space-time, which are upward diagonals. For $r > 2M$ but close to $2M$, the logarithm is dominant and approaches $-\infty$. These light rays therefore asymptote to the light ray at $r = 2M$ in the distant past. For $r < 2M$, light rays have the same asymptotic behavior near $r = 2M$ and approach $r = 0$ at their other ends. These light rays, together with a few sample light cones, are shown in Figure 10.10.

For $r \gg 2M$, the light cones look just like slightly deformed versions of Minkowski light cones. For values of r closer to $2M$, the deviation of the second family of light rays from straight lines becomes important, and the light cones tip over slightly to the left. At $r = 2M$, this tipping-over is so strong that one of the light rays, which would usually move outwards to greater values of r, stays fixed at constant $r = 2M$. All other light rays in a light cone starting at $r = 2M$ move inwards, toward smaller values of r. Once we are in the region of $r < 2M$, all light rays move to smaller values of r, and so does any timelike worldline followed by a massive particle. Correspondingly, r may be used as a time coordinate in this region because $\mathrm{d}s^2|_{\mathrm{d}r=0} > 0$ if $r < 2M$: Always moving forward in time is then equivalent to always moving inward. All objects, even light, are trapped in the region of $r \leq 2M$. This geometry therefore describes a **black hole**.

10.2.2.2 Horizon

The hypersurface determined by $r = 2M$ separates the interior of the black hole, from which nothing can escape, from its exterior. This hypersurface consists of light rays and is not a spatial surface like the surface of a compact object, on which we could possibly land a satellite. Once we reach $r = 2M$, only light can stay fixed at this value. In order to distinguish $r = 2M$ from an ordinary surface, we call it the **horizon** of the black hole. If we approach the black hole, and cross $r = 2M$, we do not notice anything different about the geometry: In a small region around an event with $r \approx 2M$, we can just rotate our coordinate system such that all light cones point upwards in the new coordinates and resemble Minkowski space-time. The black-hole horizon therefore appears like the horizon on Earth: It looks like a boundary, but opens up on approach.

Even the fact that all light rays move to smaller r when $r < 2M$ cannot be noticed by a local observer in this region. Such an observer follows a timelike worldline which always points toward the interior of the future light cone, as shown in Figure 10.11. The local observer has the usual

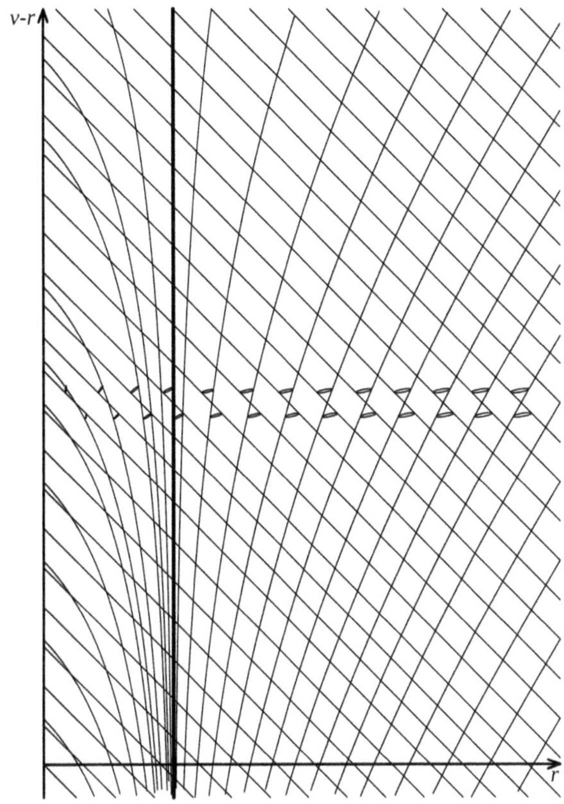

Figure 10.10 A diagram of Schwarzschild space-time in Eddington–Finkelstein coordinates. The horizon is drawn as a thick line.

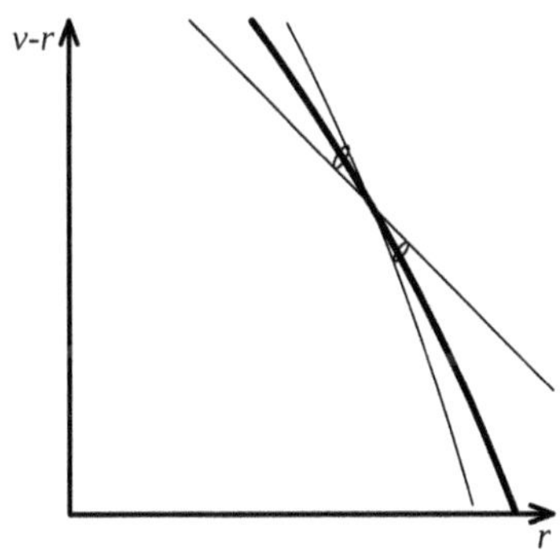

Figure 10.11 The worldline of an infalling observer (thick) in the region $r < 2M$, together with two light rays and the light cone through an event on the worldline. Even though all light rays are infalling in this region because $r(\lambda)$ decreases for both solutions, the observer sees one family of radial light rays moving in (to values of r less than the observer's $r(\tau)$) and one family of radial light rays moving out (to values of r greater than the observer's future $r(\tau)$). These properties characterize a trapped surface, given by the sphere at $r(\tau)$ parameterized by the polar angles ϑ and φ that are not shown in this plot.

perspective: There are two (ϑ, φ)-families of radial light rays, one of which are moving in toward the center and the other are moving out away from the center. From the global point of view, both families are infalling as described by their decreasing $r(\lambda)$, but one is falling in faster than the local observer, and one is falling in more slowly than the local observer. The former is seen as moving toward the center by the local observer, and the latter as moving away from the center. At any event of the worldline of an infalling local observer in the region $r < 2M$, the two families of radial light rays have the expected behavior with respect to the observer's $r(\tau)$ as a function of proper time. The globally infalling nature of all light rays cannot be noticed locally.

Such global properties are the crucial difference between the Schwarzschild geometry and the Rindler wedge, analyzed in Section 8.6.7. Here, the interior region does not open up but can be enclosed by a family of stationary observers. Moreover, each coordinate pair (v, r) corresponds to a 2-dimensional sphere. These properties differ from the example of the Rindler wedge. The Rindler horizon results from a coordinate singularity that can be resolved in a way similar to the Schwarzschild radius, but turns out to be a standard right-moving light ray in flat space-time. It presents a limit only to signals received by a specific family of accelerated observers; it is not a feature of space-time geometry. Unlike the Rindler horizon or cosmological horizons discussed in Section 8.5, the horizon of a black hole is an observer-independent concept because it is defined geometrically through its area, $4\pi R_{\mathrm{S}}^2$ for a spherical black hole.

The two cases of a Rindler horizon and a black-hole horizon are related by the equivalence principle. This principle turns the observer-dependent acceleration parameter into a physical parameter for the strength of gravity, such as the black-hole mass. However, since gravity is not homogeneous but spherical in the black-hole example, the equivalence principle turns this configuration into a family of observers accelerating away from a central point. Light emitted from points sufficiently close to the center cannot catch up with them. Unlike the right-moving observers in the Rindler wedge who do not have causal contact with an infinite spatial region to their left, the radially outgoing observers lack causal contact with a spatial region surrounded by them. The gravitational side of the equivalence principle represents the observers as stationary, such that the central region from which they do not receive light is surrounded by a constant-area surface at any given value of their time.

Physically, by including gravity through curvature, the interior is turned into a trapped region. If we include the two angles ϑ and φ, an event on the observer's worldline in the $(\tilde{t}, r)$-diagram of Figure 10.11 defines a sphere. We have a family of radially infalling observers set up on a sphere. As we move along the worldline, the sphere collapses to smaller areas. From a local perspective, the observers can send light rays to the interior of the sphere or to its exterior, but a global perspective shows that all these light rays are falling in together with the observers on their collapsing sphere. Such a sphere is an example of a **trapped surface**, the existence of which, according to singularity theorems, implies that there is a physical singularity in its future. This is the singularity at $r = 0$.

10.2.2.3 Gravitational collapse

The presence of a black-hole horizon can be noticed only by global properties, such as the behavior of light rays which connect this region with an outside observer at large $r \gg 2M$. These light rays would also be relevant for potential observations of the horizon, if it forms during the collapse of a heavy star. The interior of the star is not described by the Schwarzschild line element because it contains matter and does not fulfill the vacuum condition $T_{\alpha\beta} = 0$ under which Schwarzschild's solution is correct. But we can use the Schwarzschild line element to describe the exterior geometry of a collapsing star, provided the star is not rotating too rapidly. Figure 10.12 shows this situation, in which we see the light rays of the exterior geometry.

The surface radius of a collapsing star follows a timelike worldline crossing $r = 2M$, at which event the horizon forms. This worldline is not a geodesic because there is non-zero pressure in the

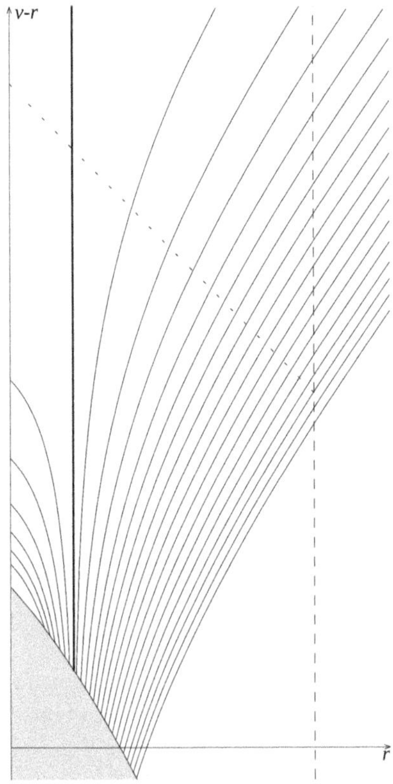

Figure 10.12 A diagram of a collapsing star (gray region), with a future geometry given by Schwarzschild space-time in Eddington–Finkelstein coordinates. To an outside observer (dashed line), the bending of light rays implies that the collapse seems to be delayed more and more as the star surface approaches the Schwarzschild radius. By the time the delays become noticeable, it is too late to send in a signal (dotted line) and call back an orbiting satellite.

collapsing star and the surface is not in free fall. Some of the outgoing light rays in the region $r > 2M$ will eventually reach an outside observer at $r_R \gg 2M$.

One implication of the formation of a horizon is strong gravitational redshift: A light ray emitted by the star with frequency ω_E is received by the observer with a different frequency, ω_R. If the star collapses sufficiently slowly, the surface is at almost constant $r = r_E$ close to $2M$ during the emission of a light ray. The frequency ω_E at such a place is determined with respect to proper time $d\tau_E = (-ds^2)^{1/2} = (1-2M/r_E)^{1/2}dt$. The observer at $r_R \gg 2M$ has proper time $d\tau_R = (1-2M/r_R)^{1/2}dt \approx dt$. The frequencies are inversely proportional to proper-time intervals, and therefore have a ratio of

$$\frac{\omega_R}{\omega_E} = \sqrt{\frac{1-2M/r_E}{1-2M/r_R}} \approx \sqrt{1-2M/r_E}\,, \qquad (10.47)$$

which is close to zero for $r_E \to 2M$. The approach to the horizon, as seen by an outside observer, is combined with a large amount of redshift. Most of the light emitted by a collapsing star will eventually be shifted out of the visible spectrum or even out of observationally accessible infrared ranges. The star appears to be dimming.

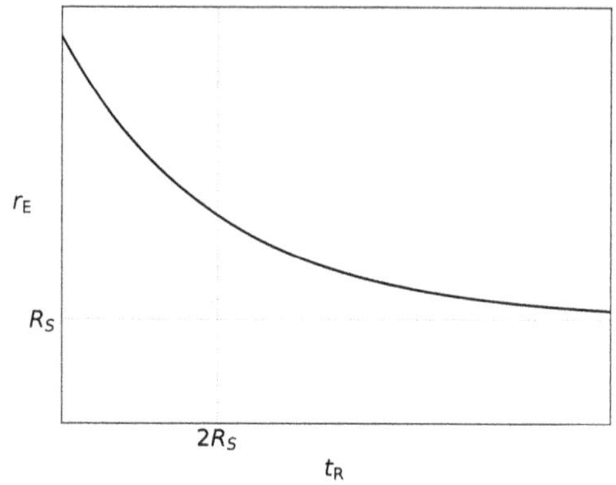

Figure 10.13 Collapse radius as a function of the observer's time.

Detailed properties of outgoing light rays allow us to derive the rate of dimming. An outgoing light ray in Eddington–Finkelstein coordinates obeys the equation

$$v - 2\left(r + 2M\log\left|\frac{r}{2M} - 1\right|\right) = C \tag{10.48}$$

where C is constant. We can compute the value of the constant using the observer's position $r_{\rm R}$ and the time $t_{\rm R}$ of a measurement. Since $r_{\rm R} \gg 2M$, the logarithm in (10.48) can be ignored and we have

$$C \approx v_{\rm R} - 2r_{\rm R} \approx t_{\rm R} - r_{\rm R}\,. \tag{10.49}$$

In the second step, we have used the relationship between v and

$$t = v - r - 2M\log\left|\frac{r}{2M} - 1\right| \tag{10.50}$$

and ignored the logarithm also here for $r = r_{\rm R} \gg 2M$.

The same light ray (same C) originated at a position $r_{\rm E} \approx 2M$. For such a value, the logarithm in (10.48) is dominant and we have

$$C \approx -4M\log\left|\frac{r_{\rm E}}{2M} - 1\right|\,. \tag{10.51}$$

Since this must be the same as the value of C computed at $r_{\rm R}$ for a single light ray, we have

$$-4M\log\left|\frac{r_{\rm E}}{2M} - 1\right| \approx t_{\rm R} - r_{\rm R}\,, \tag{10.52}$$

or

$$r_{\rm E}(t_{\rm R}) \approx 2M\left(1 + \exp\left(-\frac{t_{\rm R} - r_{\rm R}}{4M}\right)\right)\,. \tag{10.53}$$

The exponential function, plotted in Figure 10.13, describes how the surface radius $r_{\rm E}$ of the collapsing star appears to change when seen by the observer with time $t_{\rm R}$. The surface radius is always greater than the Schwarzschild radius because the outside observer does not see the horizon itself or the interior. Once the surface radius is close to $2M$, it approaches the Schwarzschild radius exponentially as seen by the external observer. The collapse appears to be very rapid, with a characteristic time scale of $4M$, given by the parameter in the denominator of the exponential function. For a solar-mass star, $4M$ equals $2 \cdot 10^{-5}$s, the time required for light to traverse twice the Schwarzschild radius of about 3km. Any direct view on the collapse of a star as large as the sun should take much longer, such that the rapid exponential dimming could be used as an observational signature of a star collapsing into a black hole. The rapid dimming is a consequence of outgoing light rays being delayed near the horizon, as shown by the space-time diagram (10.12).

10.2.2.4 Black-hole singularity

For a local observer falling in with the collapsing star, the surface does cross the Schwarzschild radius and then approaches $r = 0$. The radial coordinate r is the affine parameter of infalling radial lightlike geodesics, as used in our derivation of Eddington–Finkelstein coordinates. Zero radius is therefore reached in a finite amount of the affine parameter along lightlike geodesics. Similarly, for a massive object falling in on a timelike geodesic, $r = 0$ is reached after a finite amount of proper time.

Timelike geodesics are harder to solve for than lightlike geodesics, but one can estimate the proper-time difference by analyzing the timelike geodesic equation in the black-hole interior. Strictly in the interior, we can go back to Schwarzschild coordinates and enjoy the simplifications implied by a diagonal line element. However, in this space-time region, r plays the role of a time coordinate while t behaves like a spatial coordinate: If $r < 2M$, we have $\mathrm{d}s^2|_{\mathrm{d}r=0} = -(1-2M/r)\mathrm{d}t^2 > 0$ and $\mathrm{d}s^2|_{\mathrm{d}t=0} = \mathrm{d}r^2/(1-2M/r) < 0$. We indicate this new behavior by relabeling $r = 2M - T$ a time coordinate running from $T = 0$ at $r = 2M$ to $T = 2M$ at $r = 0$, and $t = R$ a spatial coordinate. In these coordinates, the interior Schwarzschild line element is given by

$$\mathrm{d}s^2 = -\frac{2M-T}{T}\mathrm{d}T^2 + \frac{T}{2M-T}\mathrm{d}R^2 . \tag{10.54}$$

We now have a spacelike Killing vector field $\xi = \mathbf{e}_R$, which implies the conserved quantity

$$\xi \cdot \mathbf{u} = \frac{T}{2M-T}\frac{\mathrm{d}R}{\mathrm{d}\tau} = C \tag{10.55}$$

along timelike geodesics. Together with timelike normalization of the 4-velocity $\mathbf{u}$, we obtain the differential equation

$$\frac{\mathrm{d}T}{\mathrm{d}\tau} = \sqrt{\frac{2MC^2+(1-C^2)T}{2M-T}} . \tag{10.56}$$

This equation is separable, but while the T-integral is available in closed form, it is rather clumsy. Instead of working with this expression, it is sufficient to estimate the integral over a finite T-range around any potential pole of the integrand. The presence of a pole depends on the value of C, but any possible divergence of the integrand is at most by an inverse square root, which integrates to a finite number. Therefore, it takes a finite amount of proper time to reach $T = 2M$ or $r = 0$. The singularity is relevant because it can in principle be reached by infalling objects.

However, $r = 0$ is a singularity in Schwarzschild and Eddington–Finkelstein coordinates, and we have already exploited our light-ray techniques to find alternative coordinates for an extension toward the future. It is possible to use double-null coordinates for an extension toward the past, following the procedure that lead to Figure 8.13. In the case of the Schwarzschild line element, this complete extension leads to the Kruskal space-time, but it does not resolve the singularity at $r = 0$. It turns out to be a physical singularity, which means that no coordinate transformation can remove it. This result is implied by the existence of curvature parameters that are invariant and become infinite at $r = 0$.

We have already seen the Ricci scalar $R = g^{\gamma\delta}R_{\gamma\delta}$ as an example of a curvature invariant. However, for vacuum solutions like the Schwarzschild metric, the Einstein tensor $G_{\alpha\beta} = R_{\alpha\beta} - \frac{1}{2}Rg_{\alpha\beta}$ is zero, which implies that $R_{\alpha\beta} = 0$ as well: If $G_{\alpha\beta} = 0$, we have $g^{\alpha\beta}G_{\alpha\beta} = g^{\alpha\beta}R_{\alpha\beta} - \frac{1}{2}Rg^{\alpha\beta}g_{\alpha\beta} = 0$. In this equation, we use $g^{\alpha\beta}R_{\alpha\beta} = R$ and $g^{\alpha\beta}g_{\alpha\beta} = 4$, the trace of the unit matrix $\mathbf{g}^{-1}\mathbf{g}$. Therefore, $g^{\alpha\beta}G_{\alpha\beta} = -R = 0$, and $G_{\alpha\beta} = 0$ implies $R_{\alpha\beta} = 0$. The Ricci tensor does not show any diverging curvature in vacuum solutions.

The Ricci tensor is not the most general curvature tensor. The most general such tensor is the **Riemann tensor** (9.11) from which the Ricci tensor is obtained as a summed-over subset of components, given by $R_{\alpha\beta} = R_{\alpha\varepsilon\beta}{}^{\varepsilon}$. While the Ricci tensor is zero for the Schwarzschild line element, the Riemann tensor has non-zero components. A direct, and long, calculation shows that

$$R_{\alpha\beta\gamma}{}^{\delta}R^{\alpha\beta\gamma}{}_{\delta}=\frac{48M^2}{r} \tag{10.57}$$

where the indices of the second Riemann tensor have to be raised and lowered using the metric or inverse metric:

$$R_{\alpha\beta\gamma}{}^{\delta}R^{\alpha\beta\gamma}{}_{\delta}=R_{\alpha\beta\gamma}{}^{\delta}R_{\mu\nu\rho}{}^{\sigma}g^{\alpha\mu}g^{\beta\nu}g^{\gamma\rho}g_{\delta\sigma}\,. \tag{10.58}$$

The combination of curvature components in (10.57) does not have any free indices and is therefore independent of coordinate choices. It is infinite at $r=0$, which shows that there is a physical singularity at this place which cannot be removed by coordinate transformations. The diverging curvature has physical implications in the form of strong tidal forces which pull apart even the smallest objects that come close. At the horizon, for comparison, the curvature invariant is finite and small: $R_{\alpha\beta\gamma}{}^{\delta}R^{\alpha\beta\gamma}{}_{\delta}=\frac{3}{4}M^{-4}=12/R_{\rm S}^4$. For a solar-mass black hole, the curvature radius $R_{\rm C}$, defined by $R_{\alpha\beta\gamma}{}^{\delta}R^{\alpha\beta\gamma}{}_{\delta}=1/R_{\rm C}^4$, is greater than 1km. Only objects of this size would experience noticeable tidal forces.

Tidal forces on an extended object are the result of forces pointing in different directions at various parts of the object. For instance, the tides on Earth are caused by a combination of the gravitational force of the moon and the centrifugal force on Earth, adding up to a varying force on Earth's surface and its oceans while the Earth–moon system orbits around its center of mass. The equilibrium configuration is therefore deformed compared with a planet that does not have a moon. Usually, only a large object like an entire planet can let us notice these forces because gravity is nearly uniform on the scale of the objects we usually deal with. However, if the gravitational force gets stronger, the corresponding space-time curvature increases. Curvature, as in (9.11), is related to second-order derivatives of the metric, and the gravitational force to first-order derivatives via the Christoffel symbols in geodesic equations. If curvature is large, the Christoffel symbol changes considerably on small scales, and different parts even of a small object experience noticeably different gravitational forces.

Formally, this result is expressed by the **geodesic deviation equation** (9.42) which holds for any collection of test masses even at strong curvature. If some components of the Riemann tensor are large, there is strong relative acceleration, implying tidal forces if we view the collection of test masses as an extended object. If curvature is large, even a small object experiences strong tidal forces. Any object that comes close to a physical singularity where the curvature is infinite will be torn apart by tidal forces, even if it is as small as an elementary particle. There is currently no theory that could tell us what then happens to matter and space-time.

10.2.3 ROTATING BLACK HOLES

Rotating black holes are described by the Kerr geometry, found by Roy Kerr in 1963. It has the non-diagonal line element

$$\begin{aligned}{\rm d}s^2 &= -\left(1-\frac{2Mr}{\rho^2}\right){\rm d}t^2-\frac{4Mar\sin(\vartheta)^2}{\rho^2}{\rm d}\varphi{\rm d}t+\frac{\rho^2}{\Delta}{\rm d}r^2+\rho^2{\rm d}\vartheta^2\\ &\quad+\left(r^2+a^2+\frac{2Mra^2\sin(\vartheta)^2}{\rho^2}\right)\sin(\vartheta)^2{\rm d}\varphi^2\end{aligned} \tag{10.59}$$

where

$$a=\frac{J}{M} \tag{10.60}$$

depends on the mass M and the angular momentum J of the black hole. Moreover, ρ and Δ are function given by

$$\rho(r,\vartheta)^2=r^2+a^2\cos(\vartheta)^2 \tag{10.61}$$

and

$$\Delta(r) = r^2 - 2Mr + a^2. \tag{10.62}$$

10.2.3.1 Singularities

The determinant of the metric equals

$$\det \mathbf{g} = -\rho^4 \sin(\vartheta)^2 \tag{10.63}$$

and vanishes at the following singularities:

1. $\vartheta = 0$ or $\vartheta = \pi$,
2. $\rho = 0$, which requires $r = 0$ and $\vartheta = \pi/2$,
3. $\Delta = 0$ such that $g_{rr} \to \infty$.

The first singularity determines the usual poles of spherical coordinates, which are coordinate singularities. The second one, $\rho = 0$, turns out to be a physical singularity. Geometrically, it takes the form of a ring rather than a point because $g_{\varphi\varphi}$ remains non-zero in the limit of $r \to 0$ and $\vartheta \to \pi/2$ at constant t. One spatial coordinate, φ, therefore retains non-zero distances at the singularity, unlike in the non-rotating case where $r^2(\mathrm{d}\vartheta^2 + \sin(\vartheta)^2\mathrm{d}\varphi^2)$ vanishes at the physical singularity.

The final case, $\Delta = 0$, is obtained at

$$r_\pm = M \pm \sqrt{M^2 - a^2}. \tag{10.64}$$

These solutions are real only if $a \leq M$, bounding the rotation rate of black holes. If this condition is violated, there is an exposed or naked singularity at $\rho = 0$ not covered by a horizon. In the non-rotating limit, $a \to 0$, we have $r_+ \to 2M$ and $r_- \to 0$, the two singularities of the Schwarzschild solution. For non-zero a, both r_+ and r_- are coordinate singularities, determining an outer horizon at r_+ and an inner horizon at r_-. In the Kerr geometry, it is possible to extend the original coordinate system through both coordinate singularities at r_+ and r_-, and to bypass the ring singularity. In the new coordinates, one then encounters another pair of an inner and an outer horizon, through which it is possible to exit into a nearly flat region extending to infinite radius that is not connected to the initial black-hole exterior. However, the existence of such an extension of space-time is not stable under the inclusion of matter.

10.2.3.2 Geodesics

For geodesics around a rotating central mass described by the Kerr geometry, we have two conserved quantities,

$$e = -\xi \cdot \mathbf{u} \quad \text{with } \xi = (1,0,0,0)^T \tag{10.65}$$

and

$$\ell = \eta \cdot \mathbf{u} \quad \text{with } \eta = (0,0,0,1)^T \tag{10.66}$$

in addition to normalization as a first integral. However, only the z-component ℓ of angular momentum is conserved, such that orbits are not guaranteed to be contained in a plane. We only have three immediate first integrals, which is not enough for replacing all four components of the second-order geodesic equation with first-order equations. Also the method of effective potentials then seems to be unavailable.

Fortunately, there is an additional conserved quantity that is not related to symmetry in an obvious way. It is called Carter's constant and takes the form

$$C = \rho^4 \left(\frac{\mathrm{d}\vartheta}{\mathrm{d}\tau}\right)^2 + \ell^2 \cot(\vartheta)^2 - a^2(e^2 - 1)\cos(\vartheta)^2. \tag{10.67}$$

This constant can be derived from a generalization of Killing vectors called Killing tensor: In the 2-index case of a Killing tensor, $\xi_{\alpha\beta} = \xi_{\beta\alpha}$ obeys the condition

$$\nabla_\alpha \xi_{\beta\gamma} + \nabla_\beta \xi_{\gamma\alpha} + \nabla_\gamma \xi_{\alpha\beta} = 0 \tag{10.68}$$

and implies the conserved quantity $\xi_{\alpha\beta} u^\alpha u^\beta$ along geodesics with tangent vector field u^α. Here, the conserved quantity C results in a differential equation for ϑ. Orbits in the equatorial plane correspond to the special case of $C = 0$.

Equatorial orbits can be described by effective potentials given by

$$V_{\rm eff}(r) = -\frac{M}{r} + \frac{\ell^2 - a^2(e^2-1)}{2r^2} - \frac{M(\ell - ae)^2}{r^3} \tag{10.69}$$

in the timelike case, and

$$W_{\rm eff}(r) = \frac{1}{r^2}\left(1 - \frac{a^2}{b^2}\right) - \frac{2M}{r^3}\left(1 - \sigma\frac{a}{|b|}\right)^2 \tag{10.70}$$

in the lightlike case. As used before, $b = \ell/e$ is the impact parameter, and we use an explicit sign factor $\sigma = \mathrm{sgn}(\ell)$ that distinguishes the co-rotating ($\sigma = 1$) from the counter-rotating case ($\sigma = -1$). For light passing by the black hole, we have $|b| > r_+ > M \geq a$, such that $a/|b| < 1$. All lightlike geodesics are therefore subject to a potential barrier proportional to $1/r^2$ with a positive coefficient, which descends to negative infinity at small r where the $1/r^3$-term is relevant. The barrier is smaller for $\sigma = -1$, such that more counter-rotating than co-rotating light is captured by the black hole.

EXERCISES

1. Complete the derivation of the Schwarzschild geometry in isotropic coordinates, (10.9):
   (i) Show that any static spherically symmetric metric can be written in isotropic coordinates, defined as coordinates in which the spatial part of the line element

   $$\mathrm{d}s^2 = -F(\tilde{r})\mathrm{d}t^2 + H(\tilde{r})(\mathrm{d}\tilde{r}^2 + \tilde{r}^2(\mathrm{d}\vartheta^2 + \sin(\vartheta)^2\mathrm{d}\varphi^2))$$

   is conformally flat.
   Demonstrate that the radial isotropic coordinate $\tilde{r}$, up to a constant factor, approaches the area radius r for $r \to \infty$ if the original metric approaches the Minkowski metric in this limit.
   (ii) Transform the Schwarzschild solution to isotropic coordinates.
   (iii) Use isotropic coordinates to compute the Newtonian limit of the Schwarzschild solution. Compare with (9.89).
2. Derive a coordinate system in which constant-time hypersurfaces in the Schwarzschild geometry are flat.
3. Radial motion of massive objects around a spherical central mass M in general relativity is determined by an effective potential $V_{\rm eff}(r)$. If an orbiting object is slightly disturbed from the stable circular orbit at radius r_+, its radial distance oscillates around the stable radius. Compute the corresponding oscillation frequency ω_r by Taylor expanding the potential $V_{\rm eff}(r) = V_{\rm eff}(r_+) + \frac{1}{2}\omega_r^2(r - r_+)^2 + \cdots$ around r_+.
   Expand the precession rate $\omega_{\rm p} := \Omega - \omega_r$, where $\Omega = \mathrm{d}\varphi/\mathrm{d}\tau$ for $r_+ \gg M$ to leading non-vanishing order.

4. Show that radially infalling lightlike worldlines in Schwarzschild space-time are given by

$$t = v - r - 2M\ln\left|\frac{r}{2M} - 1\right|$$

with a constant v.
Compute the t and r components of the lightlike geodesic equation for radial infall in Schwarzschild space-time. Show that r can be used as an affine parameter.
5. Compute the t and r components of the timelike geodesic equation for radial infall in the Schwarzschild space-time.
Use the normalization condition to write your result for $\mathrm{d}^2r/\mathrm{d}\tau^2$ as a function only of r and the mass M. Compare with Newton's gravitational force.
6. Compute the Ricci scalar for a 2-dimensional line element $\mathrm{d}s^2 = f(\vartheta)\mathrm{d}\vartheta^2 + g(\vartheta)\mathrm{d}\varphi^2$.
Check your result by applying it to the 2-sphere line element.
Use your result to derive the Ricci scalar on a 2-dimensional spacelike surface defined by constant t and r in the Kerr line element, assuming $r \gg M > a$ and keeping the leading order in an expansion by a/r and M/r that includes an M-dependent term.

Section III

Relativity and Quantum Physics

11 Quantum Mechanics with Newton's Potential

Newtonian gravity, at first sight, does not imply new examples for quantum mechanics because of its close similarity with static electric forces. A massive object moving in an approximately constant gravitational field near the surface of Earth is subject to a linear potential $V(z) = mgz$, while a 2-mass system is subject to Newton's potential $V(r) = -Gm_1m_2/r$ and therefore resembles the hydrogen model. Moreover, gravitational effects in lab systems are usually weak and would require sensitive measurements to be noticed.

As a first estimate, we start with the leading relativistic correction to Newton's potential, given by

$$V(r) = -\frac{GM}{r}\left(1 - \frac{GM}{c^2 r} + \cdots\right) \tag{11.1}$$

from (10.10). Relativistic corrections take the form of a power series in the dimensionless ratio $GM/(c^2 r)$. Similarly, we expect quantum corrections to take the form of a power series in a dimensionless ratio related to $\hbar$ and the distance r. Using $\hbar$, G as well as c, there is a unique combination with units of length, given by the Planck length $\ell_{\rm P} = \sqrt{G\hbar/c^3}$ which will be discussed in some more detail in Chapter 13.

Corrections linear in $\hbar$ would therefore be of the form

$$V(r) = -\frac{GM}{r}\left(1 - \frac{GM}{c^2 r} + \varepsilon \frac{G\hbar}{c^3 r^2} + \cdots\right) \tag{11.2}$$

with a numerical coefficient ε. The quantum contribution is much smaller than the leading relativistic correction: It is quadratic in $1/r$, and the Planck length $\ell_{\rm P} \sim 10^{-35}$m is much smaller than the Schwarzschild radius of, say, Earth. Certain states could imply quantum corrections of the order $\sqrt{\hbar}$ rather than $\hbar$, depending on how they are related to standard deviations, which are of the order $\sqrt{\hbar}$ if the state is close to saturation of the uncertainty relation. However, even though the relativistic and quantum corrections would then both be proportional to $1/r$, the tiny Planck length would still suppress quantum contributions.

11.1 NEWTONIAN PHASE SHIFTS

Testing quantum mechanics in the presence of gravitational forces constitutes a novel physical situation that opens up new challenges.[1] Because the gravitational force is much weaker than the electric force, microscopic gravitationally bound states are not stable under naturally occurring circumstances, and therefore there are no gravitational analogs of atoms. Systems in which quantum mechanics and gravity might co-exist in a measurable way must therefore be constructed in a dedicated manner.

Given the weak nature of the gravitational force, interferometry may be valuable in tests of its quantum nature. Light does not experience gravity in a Newtonian treatment, and using electrically charged particles in an interferometer would make any experiment sensitive to stray electric fields. However, neutron interferometry is not subject to these limitations and was indeed used in a first measurement of gravitational effects in quantum mechanics. Moreover, neutrons can be cooled to

[1] D. Carbey, P. C. E. Stamp, J. M. Taylor: Tabletop experiments for quantum gravity: a user's manual, Class. Quantum Grav. **36** (2019) 034001, arXiv:1807.11494

DOI: 10.1201/9781003588702-14

low temperatures, or moderated, such that non-relativistic quantum mechanics may be used. In 1975, Roberto Colella, Albert W. Overhauser and Samuel A. Werner were indeed able to measure phase shifts implied by a raised interferometer arm, confirming their agreement with a prediction based on Newton's potential.

The phase shift in an interferometer arm is given by kL, where k is the wave number and L the length of the arm: A plane wave has a phase according to the momentum eigenfunction $\exp(ikx)$. The wave number is determined by the dispersion relation, or by the energy equation. A free particle has the wave number $k=\sqrt{2mE}/\hbar$ as a consequence of

$$E=\frac{p^2}{2m}=\frac{\hbar^2k^2}{2m}. \tag{11.3}$$

If one of the interferometer arms is raised by a distance z, it implies an added constant potential compared with the free particle beam in the other arm.

The energy equation

$$E=\frac{p^2}{2m}+mgz=\frac{\hbar^2k^2}{2m}+mgz \tag{11.4}$$

in the presence of Newton's potential implies the wave number

$$k=\frac{\sqrt{2m(E-mgz)}}{\hbar}\approx\frac{\sqrt{2mE}}{\hbar}\left(1-\frac{mgz}{2E}\right) \tag{11.5}$$

after a Taylor expansion. The two beams therefore have wave numbers with a difference

$$\Delta\phi=L\Delta k=\frac{m^{3/2}gzL}{\sqrt{2E}\,\hbar}\approx\frac{m^2gzL\lambda}{2\pi\hbar^2} \tag{11.6}$$

with the wave length λ of the beam traded in for the energy. We recognize the phase shift as the product of the gravitational potential mgz times the time

$$\Delta t=\frac{L}{v}=\frac{mL}{p}=\frac{mL\lambda}{2\pi\hbar} \tag{11.7}$$

spent in the arm, corresponding to the time dependence $\exp(-iEt/\hbar)$ of a stationary state.

It is possible to use interferometry in order to construct specific quantum states. Once the initial beam splits up into two components, one in each arm, the full state is in a superposition of different position eigenstates. The precise position is not relevant provided we maintain information about which arm is occupied by a given contribution. We may therefore write the superposition state as

$$|\psi\rangle(0)=\frac{1}{\sqrt{2}}(|1\rangle+|2\rangle) \tag{11.8}$$

where "1" or "2" labels the interferometer arm. Including the mass and a gravitational potential, the two components of the superposition evolve with different phases, such that the state when the arms meet again at time T is given by

$$|\psi\rangle(T)=\frac{1}{\sqrt{2}}(e^{i\Delta\phi_1}|1\rangle+e^{i\Delta\phi_2}|2\rangle) \tag{11.9}$$

with a phase difference $\Delta\phi$ for each arm, depending on its altitude and the time T.

We may now combine two interferometers, A and B, such that we have four arms that may all be at different altitudes. The combined system evolves from the initial product state

$$\begin{aligned}|\psi\rangle_A\otimes|\psi\rangle_B(0) &= \frac{1}{2}(|1\rangle+|2\rangle)_A\otimes(|1\rangle+|2\rangle)_B \\ &= \frac{1}{2}(|1\rangle_A\otimes|1\rangle_B+|1\rangle_A\otimes|2\rangle_B+|2\rangle_A\otimes|1\rangle_B+|2\rangle_A\otimes|2\rangle_B)\end{aligned} \tag{11.10}$$

to a final state

$$|\psi\rangle_{AB}(T) = \frac{1}{2}(e^{i\Delta\phi_{11}}|1\rangle_A \otimes |1\rangle_B + e^{i\Delta\phi_{12}}|1\rangle_A \otimes |2\rangle_B + e^{i\Delta\phi_{21}}|2\rangle_A \otimes |1\rangle_B + e^{i\Delta\phi_{22}}|2\rangle_A \otimes |2\rangle_B) \quad (11.11)$$

that, in general, is not a product state. Therefore, a gravitational potential even in its classical linearized form may help to introduce quantum entanglement in evolving states, provided a suitable initial state can be prepared.

11.2 OPERATORS FOR NEWTON'S POTENTIAL

A similar conclusion can be reached if the linear gravitational potential is replaced by Newton's potential describing gravitational attraction between the massive particles moving through the arms. Alternatively, we may envision to test gravitational effects in quantum mechanics by analyzing two massive oscillators that interact through gravity. Since their positions are represented by operators, we obtain the potential operator

$$\hat{V} = -\frac{Gm_1m_2}{|d+\hat{x}_1-\hat{x}_2|} \quad (11.12)$$

if m_1 and m_2 are the two mass values and d is the distance between the rest points of the oscillators at displacements $\hat{x}_1$ and $\hat{x}_2$, respectively. In the classical analog, we can use a Taylor expansion if d is much bigger than the maximum of $x_1 - x_2$. We may then represent the same quantum system by the approximated potential operator

$$\hat{V} \approx -\frac{Gm_1m_2}{d} + \frac{Gm_1m_2}{d^2}(\hat{x}_1-\hat{x}_2) + \frac{Gm_1m_2}{d^3}(\hat{x}_1-\hat{x}_2)^2 + \cdots \quad (11.13)$$

To linear order, the two oscillators evolve like two separate systems in independent potentials depending on $\hat{x}_1$ and $\hat{x}_2$, respectively. They are connected only through the product m_1m_2 appearing in the coefficient, but there are no products of operators from both oscillators to this order. Since the kinetic energies also add up, the complete Hamilton operator $\hat{H} = \hat{H}_1 + \hat{H}_2$ is the sum of two commuting Hamilton operators for the subsystems, $[\hat{H}_1, \hat{H}_2] = 0$. In this case, the evolution operator $\hat{U}_{\hat{H}}(t) = \exp(-it\hat{H}/\hbar)$ can be factorized into $\exp(-it(\hat{H}_1+\hat{H}_2)/\hbar) = \exp(-it\hat{H}_1/\hbar)\exp(-it\hat{H}_2/\hbar)$. It follows that an initial product state always remains a product state, and no entanglement can be generated dynamically.

At quadratic order, we have an interaction term $\hat{x}_1\hat{x}_2$. The complete Hamilton operator is now of the form $\hat{H} = \hat{H}_1 + \hat{H}_2 + \hat{I}$ with an interaction Hamiltonian that does not commute with $\hat{H}_1$ or $\hat{H}_2$. The evolution operator no longer factorizes, and it may be possible to generate entanglement by dynamical evolution.

Since non-relativistic gravitational potentials are formally equivalent to static electric potentials, entanglement can be produced in both cases. For the electromagnetic field, we know that long-range static electric potentials are approximations or effective descriptions of underlying local quantum processes, given by the exchange of photons. An elementary process that produces entanglement is particle production by a photon, in which the spin eigenstate of the new particle pair, such as $|1,0\rangle$ since the photon has spin $\ell = 1$, may be entangled. In this case, a quantum process is required to produce entanglement because there is no classical analog of pair production. However, as the derivation of (11.11) shows, there are also classical processes, such as motion through gravitational potentials, that may produce entanglement of a quantum test object. Newtonian gravity does not provide an elementary exchange particle that could transmit the gravitational force. Instead, it defines non-local long-range interactions through potentials that may have an effect on the entanglement of an evolving state.

Newton's potential, even in the classical case, is not fundamental but rather obtained after performing several approximations in Einstein's equation as spelled out in our derivation of (9.90). It is therefore unclear whether we should directly quantize Newton's potential by inserting position operators, or start with an earlier quantization procedure performed at some level of general relativity and then derive a quantum weak-field limit. Such a procedure would require some knowledge of quantum gravity, which we will discuss in Chapter 13, but the number of different approximations, now related to both relativity and quantum physics, could result in potentials that are not simply obtained by inserting position operators. We still need positions in order to obtain the right units, but they could also be provided by a position expectation value, as in an effective potential of the form

$$V_{\text{eff}} = -\frac{Gm_1m_2}{|d + \langle \hat{x}_1 \rangle - \langle \hat{x}_2 \rangle|} \,. \tag{11.14}$$

Such a potential does not introduce quantum interaction terms because there are no products of operators in a Taylor expansion. Instead, the effective potential, when inserted in the Schrödinger equation, implies non-linear dynamics because the expectation values $\langle \hat{x}_1 \rangle$ and $\langle \hat{x}_2 \rangle$ depend on the same wave function that is multiplied by the potential in Schrödinger's equation. Solutions of a non-linear equation are no longer subject to the superposition principle, implying potentially testable deviations from standard quantum mechanics.

Another example of non-linear quantum mechanics is given by the Schrödinger–Newton equation, in which the potential is derived as a solution of Poisson's equation using a stationary wave function in order to define a static mass density $M|\Psi(t,x,y,z)|^2$. Treating the system in self-gravitating fashion, in which any small part of the wave function gravitationally interacts with all others, it experiences the time-independent gravitational potential

$$V_\Psi(\vec{x}) = -\int\int\int \frac{GM}{|\vec{x} - \vec{X}|} |\Psi(t,\vec{X})|^2 \mathrm{d}^3X \tag{11.15}$$

generated by its own wave function. The system is then subject to the non-linear Schrödinger equation

$$i\hbar\frac{\partial\Psi}{\partial t} = -\frac{\hbar^2}{2M}\nabla^2\Psi + V_\Psi(\vec{x})\Psi \,. \tag{11.16}$$

The treatment of the potential is similar to (11.12) in that both $\vec{x}$ and $\vec{X}$ play the role of operators, although in different ways: The operator $\vec{x}$ acts on the wave function in the Schrödinger equation, while $\vec{X}$ is treated as a multiplication operator in the definition (11.15). It can be written as $V_\Psi(\vec{x}) = -GM\langle|\vec{x} - \vec{X}|^{-1}\rangle_{\vec{X}}$ in which the expectation value is taken only for the $\vec{X}$-dependence at fixed $\vec{x}$.

The validity of the Schrödinger–Newton equation depends on whether it is meaningful to interpret the probability density of a state as an actual density of an extended object. It is therefore subject to conceptual questions surrounding the measurement problem of quantum mechanics.

12 Relativistic Quantum Mechanics

Special relativity introduces four independent coordinates t, x, y and z of Minkowski space-time, which in classical dynamics are independent functions of proper time τ for a massive object or the affine parameter λ for a massless object. Specifying motion therefore requires two times four initial values, given by the initial space-time position and velocity, subject to one condition that the 4-velocity be unit timelike or lightlike.

The number of independent initial values determines the phase space of a classical system. In non-relativistic classical mechanics, we have two times three initial values for the position $\vec{x}$ and velocity $\vec{V}$ or momentum $\vec{p} = m\vec{V}$. The non-relativistic phase space of a point particle is therefore 6-dimensional. The form of Hamilton's equations for the dynamics shows that this phase space has a Poisson bracket for position and momentum components, such that $\{x^j, p^k\} = \delta^{jk}$. Hamilton's equations are then expressed as

$$\dot{x}^j = \{x^j, H\} = \frac{\partial H}{\partial p^j} \quad , \quad \dot{p}^k = \{p^k, H\} = -\frac{\partial H}{\partial x^k} \tag{12.1}$$

for a given Hamilton function H. Quantum mechanics turns the Poisson bracket into the commutators $[\hat{x}^j, \hat{p}^k] = i\hbar\delta^{jk}\hat{\mathbb{I}}$ and Hamilton's equations into Heisenberg equations.

12.1 RELATIVISTIC PHASE SPACE AND ITS QUANTIZATION

The initial values for motion in special relativity cannot be treated in the same way. It is straightforward to include an additional pair of dimensions in a phase space with position coordinates x^α and momenta p^β. However, the correct phase space is odd-dimensional because the 4-velocity is normalized or lightlike, or alternatively because the 4-momentum has a fixed norm squared given by the mass in $||\mathbf{p}||^2 = -m^2$. An odd-dimensional phase space cannot be arranged into pairs of variables with standard canonical Poisson brackets or commutators.

12.1.1 CONSTRAINTS AND OBSERVABLES

We formulate the special-relativistic phase space by implementing the normalization condition in a second step, after introducing the fourth dimension for both position and momentum. The initial phase-space coordinates are given by x^α and p^β with Poisson brackets

$$\{x^\alpha, p^\beta\} = \eta^{\alpha\beta} . \tag{12.2}$$

The time components therefore have the opposite sign of their Poisson bracket compared with the spatial components, using $\eta^{00} = -1$. The use of the Minkowski metric rather than a Kronecker delta is required in order to make the equation covariant under Lorentz boosts: We need matching space-time indices on both sides of the equation. Alternatively, we have canonical Poisson brackets with all signs positive if we use the momentum p_β with a subscript index: $\{x^\alpha, p_\beta\} = \delta^\alpha{}_\beta$. The minus sign is then absorbed in $p_0 = \eta_{0\beta}p^\beta = -p^0$.

For a given mass m of a moving object, the eight variables x^α and p^β are subject to one **constraint**,

$$C = \eta_{\alpha\beta}p^\alpha p^\beta + m^2 = -(p^0)^2 + |\vec{p}|^2 + m^2 = 0 . \tag{12.3}$$

DOI: 10.1201/9781003588702-15

The presence of a constraint implies that it is impossible to define an unambiguous Hamilton function H because any multiple εC of C can be added to H without changing its value on admissible solutions, where ε may be a function of the phase-space variables. However, equations of motion depend on which version we use as a Hamilton function because partial derivatives of C by the phase-space variables are not necessarily zero on subsets of the phase space where $C = 0$. The constraint C then generates **gauge transformations**

$$\delta_\varepsilon x^\alpha = \{x^\alpha, \varepsilon C\} \quad , \quad \delta_\varepsilon p^\beta = \{p^\beta, \varepsilon C\} \tag{12.4}$$

through its contribution to Hamilton's equations implied by an ambiguous $H + \varepsilon C$.

Equations of motion in the presence of a constraint are unambiguous only if we consider combinations of x^α and p^β that have vanishing Poisson brackets with the constraint. Such gauge-invariant expressions are **Dirac observables**. The presence of a constraint means that the original phase-space variables are not observable in a frame-independent manner.

In the present example, (12.3), gauge transformations are given by

$$\delta_\varepsilon x^\alpha = \{x^\alpha, \varepsilon(\eta_{\beta\gamma} p^\beta p^\gamma + m^2)\} = \varepsilon \eta_{\beta\gamma}(\eta^{\alpha\beta} p^\gamma + \eta^{\alpha\gamma} p^\beta) + \{x^\alpha, \varepsilon\} C = 2\varepsilon p^\alpha \tag{12.5}$$

and $\delta_\varepsilon p^\beta = 0$. Therefore, all components p^β are gauge invariant, as well as the three differences $D^i = x^i - p^i x^0/p^0$ for $i = 1,2,3$. There are only three such variables because extending the definition of D^i to $i = 0$ would produce zero as a result. For a free particle, as described by the present constraint, the values of D^i are indeed constant along a given dynamical solution because we can write $D^i = x^i - V^i t$. The values of D^i are then the initial position of a trajectory, expressed in a way that is valid at all times.

Since we obtain the correct free-particle solutions directly from the constraint and its gauge transformations, we do not need a separate Hamilton function. This property is shared by all relativistic systems because the presence of an unconstrained Hamilton function would imply the existence of a preferred time variable that appears in Hamilton's equations. We have seen the same feature in the Hamiltonian formulation of general relativity. A constraint that also provides the dynamics is called a **Hamiltonian constraint**.

We conclude that there are only three independent position observables. There are initially four momentum components, but they are constrained by $C = 0$. Thus, we also have three independent momentum observables, which we can parameterize by the spatial components p^i such that the constraint implies $p^0 = \pm\sqrt{|\vec{p}|^2 + m^2}$. The relativistic phase space is therefore six-dimensional, just like the non-relativistic one. More precisely, we obtain two copies of a six-dimensional phase space, one for each sign choice in p^0. The same sign choice exists in non-relativistic mechanics because we may choose to solve equations of motion by evolving forward or backward in time. Relativity does not introduce new degrees of freedom, it only re-arranges the non-relativistic ones in order to make them Lorentz covariant.

Dirac observables determine dynamical properties of a system that are valid for any observer. In special relativity, we commonly do derivations in a frame-dependent way, computing quantities as they would be measured by an observer moving at a specific speed. Choosing an observer or a frame means that we are no longer able to perform gauge transformations. Mathematically, this outcome is achieved by specifying a second constraint, $F = 0$, that **fixes the gauge** by being zero only in one frame. Therefore, F should not be gauge invariant, and we require $\{F, C\} \neq 0$. The combination of two constraints, $F = 0$ and $C = 0$, reduces the initial eight-dimensional phase space to an even six-dimensional one, as in non-relativistic physics.

For instance, we may be interested in a frame in which x^0 is our time coordinate t. This statement is expressed by the constraint $F = x^0 - t = 0$ which indeed has a non-zero Poisson bracket $\{F, C\} = 2p^0 \neq 0$: The energy p^0 should always be non-zero for a dynamical object. The six-dimensional phase space can be parameterized by the remaining components that are free after the constraints are imposed: x^i and p^i, while $x^0 = t$ must equal the time value and p^0 is determined as before by $C = 0$.

A choice of frame implies specific dynamics with respect to the new time variable t, and therefore an unambiguous Hamilton function. Any Hamilton function in this system is of the form NC with a phase-space function N: The previous multiplier ε is usually renamed and referred to as the **lapse function** N when applied to the gauge-fixed dynamics. Dynamics compatible with the gauge fixing $F = 0$ is determined by a Hamilton function $H = NC$ that preserves the gauge fixing in the sense that $\{F,H\} + \partial F/\partial t$=0. Since the gauge fixing brings in a new variable, t, we include a partial derivative by t in Hamilton's equation, keeping track of its changes. The new condition implies the differential equation

$$\{F,NC\} = -\frac{\partial(NC)}{\partial p^0} = -\frac{\partial N}{\partial p^0}(-(p^0)^2 + |\vec{p}|^2 + m^2) + 2p^0 N = 1\,. \tag{12.6}$$

An obvious solution is $N \propto 1/C$, but this choice would not result in non-trivial dynamics. Instead, we may choose one of the two solutions

$$N_{\pm} = \frac{1}{p^0 \pm \sqrt{|\vec{p}|^2 + m^2}} \tag{12.7}$$

for which the differential equation in (12.6) can be checked directly.

Solving partial differential equations with non-constant coefficients, such as (12.6), is not easy, but there is a convenient shortcut in the present case: For any F linear in a configuration variable such as x^0, solutions such as $N_{\pm}$ can be found by factorizing the constraint C into terms linear in the momentum p^0 of x^0, here

$$C = -\left(p^0 - \sqrt{|\vec{p}|^2 + m^2}\right)\left(p^0 + \sqrt{|\vec{p}|^2 + m^2}\right)\,. \tag{12.8}$$

Choosing N^{-1} as the product of all but one of the factors, we make sure that the remaining term is linear in p^0 and obeys the required condition thanks to $\{x^0, -p^0\} = 1$.

For $N = N_+$, the compatible Hamilton function

$$H = N_+ C = -p^0 + \sqrt{|\vec{p}|^2 + m^2} \tag{12.9}$$

implies Hamilton's equations

$$\dot{x}^0 = \{x^0, H\} = \{x^0, -p^0\} = 1 = \frac{\mathrm{d}t}{\mathrm{d}t} \tag{12.10}$$

and

$$\dot{x}^i = \{x^i, H\} = \{x^i, \sqrt{|\vec{p}|^2 + m^2}\} = \frac{p^i}{\sqrt{|\vec{p}|^2 + m^2}} = \frac{u^i}{\sqrt{1 + |\vec{u}|^2}} = V^i = \frac{\mathrm{d}x^i}{\mathrm{d}t} \tag{12.11}$$

using the relativistic momentum in $\vec{p}/m = \vec{u} = \gamma_{\vec{V}} \vec{V}$. The equations of motion therefore implement the correct choice of coordinate time t.

A frame transformation is obtained by choosing a different gauge fixing, such as $F' = x^0 - \alpha x^1 - t = 0$. Variables A in this frame must be compatible with $F' = 0$ in the sense that $\{A, F'\} = 0$. Three independent choices are given by $x^1 - \alpha x^0$, x^2 and x^3. Together with $t = x^0 - \alpha x^1$ as the solution of $F' = 0$, we obtain an infinitesimal Lorentz boost compared with the original gauge fixing. With this observation, we can simplify the procedure of finding a compatible Hamilton function $H' = N'C$ by writing the invariant $\eta_{\alpha\beta} p^\alpha p^\beta$ in

$$C = -(p^0)^2 + |\vec{p}|^2 + m^2 = -(p^{0'})^2 + |\vec{p}'|^2 + m^2 \tag{12.12}$$

in terms of $\mathbf{p}'$ obtained from $\mathbf{p}$ by the same infinitesimal Lorentz boost that transforms the position coordinates:

$$N' = \frac{1}{p^{0'} + \sqrt{|\vec{p}'|^2 + m^2}} = \frac{1}{p^0 - \alpha p^1 + \sqrt{(p^1 - \alpha p^0)^2 + (p^2)^2 + (p^3)^2 + m^2}}. \tag{12.13}$$

Thus,

$$H' = N'C = -p^0 + \alpha p^1 + \sqrt{(p^1 - \alpha p^0)^2 + (p^2)^2 + (p^3)^2 + m^2} \tag{12.14}$$

implies the correct evolution equations in the new frame.

In any frame, we have the option of choosing N_+ or N_- as the lapse function. Our detailed calculations confirmed the correct dynamics for $N = N_+$. Choosing the other solution for the lapse function, $N = N_-$, implies the opposite sign of $\sqrt{|\vec{p}|^2 + m^2}$ in the Hamilton function. This choice therefore describes time-reversed solutions.

12.1.2 HILBERT SPACES

The quantum version of a classical system can be defined by representing basic canonical variables such as x and p as operators on a Hilbert space and expressing the Hamilton function or other observables in terms of these operators. For a relativistic system, we have four pairs of canonical variables, x^α and p^β, which are turned into operators $\hat{x}^\alpha$ and $\hat{p}^\alpha$. Choosing the position representation, we have wave functions $\psi(t,x,y,z)$ with $\hat{x}^\alpha$ as multiplication operators, and $\hat{p}^0 = i\hbar\partial/\partial x^0$ as well as $\hat{p}^i = -i\hbar\partial/\partial x^i$ as derivative operators. The sign choices imply the required commutators

$$[\hat{x}^\alpha, \hat{p}^\beta] = i\hbar\eta^{\alpha\beta}\hat{\mathbb{I}}. \tag{12.15}$$

The spatial components $\hat{x}^i$ and $\hat{p}^j$ are just as their corresponding versions in non-relativistic 3-dimensional quantum mechanics. Now, we have a new pair of operators given by the time operator $\hat{x}^0 = \hat{t}$ and the energy operator $\hat{E} = \hat{p}^0 = i\hbar\partial/\partial t$. Note that the time operator may be self-adjoint without contradicting the results of Section 5.5: At this stage, we do not have any restrictions on the range of energies. Such restrictions appear once we impose the constraint which, as we will see, changes the behavior of the time operator.

Instead of a Hamilton function, relativistic systems require a Hamiltonian constraint. We should therefore use our basic position and momentum operators and turn the constraint C into an operator $\hat{C}$. Since the constraint vanishes on admissible classical configurations, the constraint operator has to vanish when applied to a quantum state: $\hat{C}|\psi\rangle = 0$. This condition ensures that the expectation value $\langle\hat{C}\rangle$, the variance σ_C^2, and all higher moments of $\hat{C}$ vanish in an admissible state. Since the classical constraint is real, we require that $\hat{C} = \hat{C}^\dagger$ is self-adjoint. It then generates a unitary family of quantum **gauge transformations**,

$$\hat{U}_{\hat{C}}(\delta) = \exp(-i\delta\hat{C}/\hbar). \tag{12.16}$$

Every admissible state is gauge invariant because

$$\hat{U}_{\hat{C}}(\delta)|\psi\rangle = \sum_{n=0}^{\infty} \frac{(i\delta)^n}{n!}\hat{C}^n|\psi\rangle = |\psi\rangle \tag{12.17}$$

if $\hat{C}|\psi\rangle = 0$.

The constraint equation $\hat{C}|\psi\rangle = 0$ means that we are looking for eigenstates of $\hat{C}$ with eigenvalue zero. If zero is in the continuous part of the spectrum of $\hat{C}$, the eigenfunctions are not normalizable and are therefore not contained in the original Hilbert space. This is the case for instance if we use a constraint as in the preceding classical discussion because the energy operator $\hat{E}$, just like a spatial momentum operator, has a continuous spectrum. It is therefore impossible to implement the constraint on the original Hilbert space.

A general formal proof of this impossibility only assumes that there is an operator $\hat{t}$, such as the time operator, with a canonical commutation relation $[\hat{C},\hat{t}] = i\hbar$. Assuming that there are normalizable solutions $|\psi\rangle$ of the constraint equation then implies the contradiction

$$i\hbar = \langle\psi|[\hat{C},\hat{t}]|\psi\rangle = \langle\psi|\hat{C}\hat{t} - \hat{t}\hat{C}|\psi\rangle = 0\,. \tag{12.18}$$

In the last step, we used the self-adjointness of $\hat{C}$, such that

$$\langle\psi|\hat{C} = (\hat{C}^\dagger|\psi\rangle)^\dagger = (\hat{C}|\psi\rangle)^\dagger = 0\,. \tag{12.19}$$

Our previous constraints are subject to this general result because they obey the condition $[\hat{N}\hat{C},\hat{t}] = -i\hbar$ after including the lapse function.

In our specific example, the constraint equation in the position representation implies the **Klein–Gordon equation**

$$-\frac{\partial^2\psi}{\partial t^2} + \nabla^2\psi = m^2\psi\,, \tag{12.20}$$

a second-order partial differential equation that requires an initial wave function and its first time derivative for a unique solution. We can find solutions by using the factorization

$$\hat{C} = -\hat{E}^2 + |\hat{\vec{p}}|^2 + m^2 = -\left(-\hat{E} - \sqrt{|\hat{\vec{p}}|^2+m^2}\right)\left(-\hat{E} + \sqrt{|\hat{\vec{p}}|^2+m^2}\right) = -\hat{C}_-\hat{C}_+\,. \tag{12.21}$$

The square root is defined in the momentum representation:

$$\sqrt{|\hat{\vec{p}}|^2+m^2}\,\phi(\vec{k}) = \sqrt{\hbar^2|\vec{k}|^2+m^2}\,\phi(\vec{k}) \tag{12.22}$$

including only spatial components $\vec{k}$.

The two factors $\hat{C}_\pm$ in (12.21) commute, and therefore $\hat{C}|\phi\rangle = 0$ if and only if one of the factors annihilates the state: $\hat{C}_+|\phi\rangle = 0$ or $\hat{C}_-|\phi\rangle = 0$. Using the energy operator, the equations take the form

$$\hat{E}|\phi\rangle = i\hbar\frac{\partial|\phi\rangle}{\partial t} = \pm\sqrt{|\hat{\vec{p}}|^2+m^2}\,|\phi\rangle \tag{12.23}$$

and resemble the Schrödinger equation with Hamilton operator

$$\hat{H}_\pm = \pm\sqrt{|\hat{\vec{p}}|^2+m^2}\,. \tag{12.24}$$

Solutions are therefore given by $|\phi\rangle_\pm(t) = \exp(-it\hat{H}_\pm/\hbar)|\phi\rangle_\pm(0)$, or

$$\phi_\pm(t,\vec{k}) = c_\pm(\vec{k})\exp\left(\mp i\sqrt{k^2+k_{\rm C}^2}\,t\right) \tag{12.25}$$

in the momentum representation, using the reduced Compton wave number $k_{\rm C} = 2\pi/\lambda_{\rm C}$ with $\lambda_{\rm C} = m/(2\pi\hbar)$. The two functions $c_\pm(\vec{k})$ correspond to the required freedom of initial values for a second-order partial differential equation.

These solutions, as eigenfunctions with eigenvalue zero, are not normalizable. The original Hilbert space included wave functions of four coordinates, $\psi(t,x,y,z)$ in the position representation, or $\phi(t,p^x,p^y,p^z)$ in the spatial momentum representation. Such wave functions are normalized by the inner product

$$\langle\phi|\phi\rangle = \int\int\int\int|\phi(t,p^x,p^y,p^z)|^2 {\rm d}t{\rm d}^3p\,. \tag{12.26}$$

The time integration is necessary because we included $\hat{t}$ as one of the basic operators, with an energy operator $\hat{E}$ that is self-adjoint only if the t-integration is included. However, any $\phi(t,p^x,p^y,p^z) = \phi_\pm(t,\vec{p}/\hbar)$ has a complex norm independent of t, such that the t-integration results in an infinite norm.

Our explicit solutions indicate a simple way around the normalization problem: If we just omit the t-integration, we can make sure that the wave functions are normalized, provided we choose normalized coefficient functions $c_\pm(\vec{k})$. If t no longer appears in the normalization integral, we replace the 4-variable wave function $\phi(t,p^x,p^y,p^z)$ with the time-dependent 3-variable wave function $\Phi(t,\vec{k})$ as used in non-relativistic quantum mechanics. Formally, this step implies that we have to replace the original or **kinematical Hilbert space** of 4-variable wave functions with the **physical Hilbert space** of t-parameterized 3-variable wave functions. Along with the time integration in the inner product, the self-adjoint kinematical time operator $\hat{t}$ disappears once we introduce the physical Hilbert space.

Such a transition from a kinematical to a physical Hilbert space is always required if there is a constraint, but it may be more complicated than just omitting one of the integrations. Even in the present case, the construction of the physical Hilbert space turns out to be subtle because of superselection rules.

12.1.3 SUPERSELECTION AND ANTIMATTER

So far, we have only considered separate solutions $\Phi_\pm(t,\vec{k})$ of the factorized constraint equations $\hat{C}_\pm|\Phi\rangle_\pm = 0$. We should expect that the general solution of the original constraint equation $\hat{C}|\Phi\rangle = 0$ is a superposition $\Phi(t,\vec{k}) = \Phi_+(t,\vec{k}) + \Phi_-(t,\vec{k})$. However, while each separate solution $\Phi_+(t,\vec{k})$ or $\Phi_-(t,\vec{k})$ has a conserved norm in time, where

$$\langle\Phi_\pm|\Phi_\pm\rangle = \int\int\int |c_\pm(\vec{k})|^2 \mathrm{d}^3k \tag{12.27}$$

is time-independent, the superposition does not:

$$\langle\Phi|\Phi\rangle = \int\int\int \left(|c_+(\vec{k})|^2 + |c_-(\vec{k})|^2 + 2\mathrm{Re}\left(c_+(\vec{k})^* c_-(\vec{k}) e^{2i\sqrt{k^2+k_C^2}\,t}\right)\right) \mathrm{d}^3x. \tag{12.28}$$

A superposition $\Phi_+(t,\vec{k}) + \Phi_-(t,\vec{k})$, therefore, is not subject to unitary evolution

Thus, while we may take arbitrary superpositions of wave functions Φ_+ with different coefficient functions c_+, or of wave functions Φ_- with different coefficient functions c_-, we are not allowed to consider superpositions of a Φ_+ with a Φ_- as a valid wave function. Such restrictions of the superposition principle are called **superselection rules**. In the present case, the superselection rule determines the physical Hilbert space as

$$\mathscr{H}_{\mathrm{phys}} = \mathscr{H}_+ \oplus \mathscr{H}_- \tag{12.29}$$

where $\mathscr{H}_\pm$ is the Hilbert space of functions that can be written as a superposition of the form $\Phi_\pm$. The sign $\oplus$ indicates that a wave function in $\mathscr{H}_+$ is defined as orthogonal to any state in $\mathscr{H}_-$. This declaration modifies the kinematical inner product (12.26) by eliminating the interference term in (12.28). For two wave functions in the same $\mathscr{H}_\pm$, the physical inner product is defined by (12.27).

The relativistic constraint equation implies two orthogonal quantum systems, one with wave functions Φ_+ and one with wave functions Φ_-. The former describe a particle with mass m, while the latter describe an **antiparticle** with the same mass. Formally, an antiparticle appears as a particle moving backward in time because the sign of its Hamilton operator is reversed. However, this sign difference affects only the phase factor, turning $\exp(-itE/\hbar)$ into $\exp(+itE/\hbar)$, while phase differences between contributions to a superposition state Φ_- may be the same as in a corresponding superposition Φ_+. Therefore, the formal backward direction of time is not an observable feature.

An observable difference between particles and antiparticles is obtained if they have electric charge. An electric potential can be introduced systematically in the Hamilton operator by replacing E with $E - q\phi$ in the constraint. In non-relativistic quantum mechanics, this substitution simply adds the potential $q\phi$. In the relativistic constraint, (12.21), it adds $q\phi$ to the Hamilton operator implied by C_+, and it adds $-q\phi$ to the Hamilton operator implied by C_-. Therefore, particles and antiparticles have charges of opposite signs.

12.2 QUANTUM REFERENCE FRAMES

The commutators (12.15) are covariant under Lorentz transformations: Classically, both x^α and p^β transform like 4-vectors, $x^{\alpha'} = \Lambda^{\alpha'}{}_\alpha x^\alpha$ with a matrix $\Lambda^{\alpha'}{}_\alpha$ that has constant coefficients. The basic operators then obey the same transformation law without requiring decisions about how to order a quantized $\Lambda^{\alpha'}{}_\alpha$ and the operators $\hat{x}^\alpha$ and $\hat{p}^\beta$. For the same reason, we can move the transformation matrix into or out of the commutator, such that

$$[\hat{x}^{\alpha'}, \hat{p}^{\beta'}] = [\Lambda^{\alpha'}{}_\alpha \hat{x}^\alpha, \Lambda^{\beta'}{}_\beta \hat{p}^\beta] = \Lambda^{\alpha'}{}_\alpha [\hat{x}^\alpha, \hat{p}^\beta] \Lambda^{\beta'}{}_\beta = i\hbar \left(\Lambda \eta \Lambda^T\right)^{\alpha'\beta'} \hat{\mathbb{I}} = i\hbar \eta^{\alpha'\beta'} \hat{\mathbb{I}} \tag{12.30}$$

using Lorentz invariance of the Minkowski metric. Since the basic commutators are covariant, we can use the same kinematical Hilbert space for all reference frames.

12.2.1 FRAMES FOR RELATIVISTIC PARTICLES

We implement a specific frame by choosing an operator $\hat{F}$ such that $[\hat{F}, \hat{C}] = i\hbar\hat{\mathbb{I}}$. Our corresponding classical expressions for F were linear in x^α and can directly be turned into operators such as $\hat{F} = \hat{x}^0 - t\hat{\mathbb{I}}$ with the correct commutator required for $[\hat{F}, \hat{C}]$. The corresponding Hamilton function, classically, was given by $H = NC$ with a non-linear function N depending on momentum components. The classical case gave us two choices, $N_\pm$ in (12.7), which can directly be turned into operators

$$\hat{N}_\pm = \frac{1}{\hat{p}^0 \pm \sqrt{|\hat{\vec{p}}|^2 + m^2}} \tag{12.31}$$

in the momentum representation. These operators, as well as the corresponding Hamilton operators $\hat{H}_\pm = \hat{N}_\pm \hat{C}$, act on the kinematical Hilbert space, but they are well-defined also on the physical Hilbert space (12.29) because they preserve the orthogonal subspaces $\mathscr{H}_\pm$. The Hamilton operators $\hat{H}_\pm$ as well operate on the physical Hilbert space.

For a given frame, $\hat{F}$, the physical Hilbert space is obtained by evaluating the kinematical wave function $\psi(t, \vec{p})$ at a fixed value of $x^0 = t$, and using the standard inner product from 3-dimensional quantum mechanics for the $\vec{p}$-integrations, with a subtlety implied by the superselection rule. If we choose a different frame, $\hat{F}' = \hat{x}^{0'} - t$, we take a different cross-section of the kinematical Hilbert space by evaluating wave functions at constant $x^{0'}$, and then integrating over transformed momentum components $\vec{p}\,'$ that include a dependence on p^0 when compared with the original frame. As a consequence, there is no immediate relationship between the physical Hilbert spaces corresponding to two different frame choices within the same system.

On the kinematical Hilbert space, we have a unitary operator that maps $\psi(t,x,y,z)$ to $\psi(t',x',y',z')$, where unitarity follows from the invariance of the kinematical integration measure $dtdxdydz$ thanks to $\det(\partial x^{\alpha'}/\partial x^\alpha) = \det(\Lambda^{\alpha'}{}_\alpha) = 1$. However, this argument is not available on a given physical Hilbert space, where we fix t or t'. The remaining 3-dimensional integration is not guaranteed to have a Lorentz-invariant measure.

Two different physical Hilbert spaces obtained by a frame transformation within the same system are related in some way because they make use of the same type of functions, depending on the same number of variables. If we choose an orthonormal basis in each Hilbert space, we can easily define a unitary transformation between them by mapping a basis vector to the corresponding one in the other Hilbert space. However, such a definition is basis-dependent and does not constitute an intrinsic relationship between the Hilbert spaces. Compared with spatial symmetry, space-time symmetry in relativistic quantum mechanics is more complicated and cannot be represented by a single unitary transformation on a given physical Hilbert space.

The free relativistic particle requires several specific constructions for a well-defined quantum description, including new ingredients compared with non-relativistic quantum mechanics. If the particle is not free, the quantum theory gets even more complicated. For instance, any position-dependent contributions to the constraint, which could be implied by an electric field acting on a charged particle, or by a time-dependent mass, eliminates some of the initial steps performed for the free particle.

In the example of a time-dependent mass, $\hat{C} = -\hat{E}^2 + |\hat{\vec{p}}|^2 + m(\hat{t})^2$ (a simpler version of a covariant position-dependent expression such as $m(||\hat{\mathbf{x}}||^2)$), the non-zero commutator $[\hat{E}, \hat{t}] = i\hbar\hat{\mathbb{I}}$ implies that we can no longer factorize the quadratic constraints: We have

$$-\left(\hat{E} + \sqrt{|\hat{\vec{p}}|^2 + m(\hat{t})^2}\right)\left(\hat{E} - \sqrt{|\hat{\vec{p}}|^2 + m(\hat{t})^2}\right) = -\hat{E}^2 + |\hat{\vec{p}}|^2 + m(\hat{t})^2 - \left[\hat{E}, \sqrt{|\hat{\vec{p}}|^2 + m(\hat{t})^2}\right] \tag{12.32}$$

with a non-zero

$$\left[\hat{E}, \sqrt{|\hat{\vec{p}}|^2 + m(\hat{t})^2}\right] = \frac{i\hbar m(\hat{t})\dot{m}(\hat{t})}{\sqrt{|\hat{\vec{p}}|^2 + m(\hat{t})^2}}. \tag{12.33}$$

Without a factorization of $\hat{C}$ into terms linear in $\hat{E}$, it is difficult to find suitable lapse operators $\hat{N}$ that could determine a valid Hamilton operator $\hat{H} = \hat{N}\hat{C}$ for a given frame choice $\hat{F}$. Moreover, if we manage to find a suitable $\hat{N}$, it is likely to inherit some of the position dependence of $\hat{C}$, in such a way that it no longer commutes with $\hat{C}$. The operator $\hat{N}\hat{C}$ is then non-Hermitian and does not generate norm-preserving evolution. The symmetric ordering $\frac{1}{2}(\hat{N}\hat{C} + \hat{C}\hat{N})$ would be Hermitian, but if the commutator $[\hat{N}, \hat{C}] \neq 0$ is non-zero, it would not imply the same cancellations that make $\hat{N}\hat{C}$ compatible with the frame choice $\hat{F}$. For these reasons, quantum reference frames in relativistic systems remain poorly understood.

12.2.2 QUANTUM COSMOLOGICAL MODELS

The requirements of general relativity are even more challenging. General relativity provides a large number of examples that correspond to position-dependent coefficients in a relativistic constraint. After all, position-dependent metric components are a hallmark of curvature or non-Cartesian coordinates, and the metric features prominently in relativistic constraints. Moreover, the complete equations of general relativity in the Hamiltonian formulation consist of several constraints with spatial dependence of their lapse functions. A quantum treatment then requires complicated consistency conditions between the independent gauge transformations of all the constraints.

A tractable class of examples is given by quantum cosmology, which applies methods from quantum mechanics to the dynamics of Friedmann–Robertson–Walker models. The setting is provided by general relativity in Hamiltonian form, restricted to isotropic line elements with the spatial metric $h_{ij} = a(t)^2\sigma_{ij}$ in terms of the scale factor $a(t)$ as a function of some time coordinate t. Here, $\sigma_{ij} = \delta_{ij}$ is the Kronecker delta for flat constant-time hypersurfaces and is determined by (8.96) for constant-time hypersurfaces of constant positive curvature, or by a hyperbolic version of this line element for constant negative curvature.

Extrinsic curvature (9.97) of constant-t hypersurfaces is given by $K_{ij} = a\dot{a}N^{-1}\sigma_{ij}$ using the unit normal $n^\alpha = N^{-1}(\partial/\partial t)^\alpha$. The trace of this tensor equals $K^i{}_i = K_{ij}h^{ij} = 3\dot{a}/(Na)$. We obtain the momentum tensor

$$p^{ij} = \frac{a^3}{16\pi G}\left(\frac{\dot{a}}{Na^3} - \frac{K^i{}_i}{a^2}\right)\sigma^{ij} = -\frac{1}{8\pi G}\frac{\dot{a}}{N}\sigma^{ij} \tag{12.34}$$

from (9.113).

Since there is no spatial dependence in these models, the constraint C_k, defined in (9.120), vanishes identically, while the constraint C_0 in (9.119) is reduced to

$$C = \frac{16\pi G}{a^3}\left(\frac{3}{64\pi^2 G^2}\frac{\dot{a}^2 a^4}{N^2} - \frac{1}{2}\left(\frac{3}{8\pi G}\frac{\dot{a}a^2}{N}\right)^2\right) - \frac{a^3}{16\pi G}R_3 = -\frac{3}{8\pi G}\frac{\dot{a}^2 a}{N^2} - \frac{a^3}{16\pi G}R_3 \quad (12.35)$$

Factors such as $a^{-3}\sqrt{\det h} = \sqrt{\det\sigma}$, which may be coordinate dependent if space has intrinsic curvature or we use non-Cartesian coordinates in flat space, can be removed by integrating the original C_0 over a finite spatial region. Homogeneity tells us that the choice of this region does not change the constraint, up to a factor of $V_0 = a^{-3}\int\int\int\sqrt{\det h}d^3x$ given by the coordinate volume. We already saw such a parameter in Section 8.5.3. Without loss of generality, we may assume $V_0 = 1$.

The spatial Ricci scalar can be derived from (9.188), after recognizing that all spatially isotropic geometries are of the spherically symmetric form (9.179) in a coordinate system (ψ,ϑ,φ) as used in (8.96) with $L(\psi) = a$ constant while $S(\psi) = a\psi$ for spatially flat constant-time hypersurfaces, $S(\psi) = a\sin(\psi)$ for constant-time 3-sphere hypersurfaces with constant positive curvature, (8.96), and $S(\psi) = a\sinh(\psi)$ for constant-time hypersurfaces with constant negative curvature in the hyperbolic version of the 3-sphere. In these cases, the spatial Ricci scalar takes the simple form $R_3 = 6k/a^2$ with k the sign of constant spatial curvature.

We obtain the constraint

$$C = -\frac{3a^3}{8\pi G}\left(\frac{\dot{a}^2}{N^2 a^2} + \frac{k}{a^2}\right). \quad (12.36)$$

In the presence of matter, in which case the full constraint C_0 is replaced by $C_0 + \sqrt{\det h}\,\rho$ with the energy density ρ, we derive the correct Friedmann equation

$$\frac{\dot{a}^2}{N^2 a^2} + \frac{k}{a^2} = \frac{8\pi G}{3}\rho \quad (12.37)$$

in terms of the Hubble parameter $\dot{a}/(Na)$ in a generic time coordinate t.

The momentum tensor (12.34) appears in the Poisson bracket

$$\{h_{ij}, p^{kl}\} = \frac{1}{2}(\delta_i{}^k\delta_j{}^l + \delta_j{}^k\delta_i{}^l) \quad (12.38)$$

where we symmetrized on the right because both h_{ij} and p^{kl} are symmetric tensors. Multiplying with σ_{kl} and the inverse σ^{ij}, this bracket implies the relationship

$$\{h^j{}_j, p^k{}_k\} = \sigma^{ij}\sigma_{kl}\{h_{ij}, p^{kl}\} = \sigma^{ij}\sigma_{ij}\delta^j{}_j = 3. \quad (12.39)$$

Inserting the traces of the spatial metric $h_{ij} = a^2\sigma_{ij}$ and of the momentum tensor (12.34),

$$3 = \left\{3a^2, -\frac{3\dot{a}}{8\pi GN}\right\} \quad (12.40)$$

shows that the momentum of a^2, defined by $\{a^2, p_{a^2}\} = 1$, is given by $p_{a^2} = -3\dot{a}/(8\pi GN)$. Since

$$1 = \left\{a^2, -\frac{3\dot{a}}{8\pi GN}\right\} = -\frac{3a}{4\pi G}\{a, \dot{a}/N\} \quad (12.41)$$

based on the definition (3.201) of a Poisson bracket in terms of partial derivatives, the momentum of the scale factor a with $\{a, p_a\} = 1$ is

$$p_a = -\frac{3}{4\pi G}\frac{a\dot{a}}{N}. \quad (12.42)$$

In terms of this momentum, the constraint in the presence of matter equals

$$C = -\frac{2\pi G}{3}\frac{p_a^2}{a} - \frac{3ka}{8\pi G} + a^3\rho\,. \tag{12.43}$$

Matter may be described by a function $\rho(a)$, such as $\rho(a) \propto 1/a^3$ for non-relativistic matter, or in terms of a dynamical variable ϕ, a scalar field, with its own momentum p_ϕ and a Hamilton function, such as

$$H_\phi = \frac{1}{2}p_\phi^2 + V(\phi) \tag{12.44}$$

in terms of a kinetic energy and a potential. In relativistic theories, the mass is not contained in the kinetic energy but rather appears in a quadratic contribution $\frac{1}{2}m^2\phi^2$ to the potential.

The constraint equation $C = 0$, when turned into an operator, implies a second-order differential equation for wave functions in the "position" representation, depending on a and ϕ. This equation resembles the Klein–Gordon equation (12.20), but it does not have constant coefficients. Quantum cosmology therefore has to face all the complications related to a non-factorizable quantum constraint $\hat{C}$.

Additional issues are presented by the general notion of arbitrary non-linear coordinate changes in a general relativistic setting, in particular if we try to relax the strong assumption of strict spatial homogeneity. For non-linear coordinate changes, the transformation matrix in (12.30) is not constant, such that the basic commutator $[\hat{x}^\alpha, \hat{p}^\beta]$ for a point particle cannot be maintained. Applied to the gravitational field, basic Poisson brackets are determined by the Hamiltonian formulation of general relativity, such that (12.38) has to be turned into a meaningful commutator relationship. As used in models of quantum cosmology, this step can be performed with standard methods of quantum mechanics if the geometry is spatially homogeneous and therefore presents a finite number of degrees of freedom. But it is more complicated in the general case of position-dependent gravitational fields that require methods from quantum field theory. In gravity, an additional challenge is the requirement to maintain covariance conditions, which are classically represented by Poisson brackets (9.138) and need to be turned into well-defined commutator relationships for operators. All these questions remain open and have given rise to a variety of approaches that aim to address the main problems of quantum gravity.

13 Quantum Gravity

General relativity has been successful in observational tests, but it predicts its own limitations in the form of space-time singularities found in many relevant solutions. Black-hole singularities can in principle be reached in a finite amount of time, and the big-bang singularity may affect initial values of the universe that determine properties of the cosmic background waves. A complete description of the universe that tells us what happens at the physical singularities of general relativity requires an extended theory.

One expectation is that quantum effects may be essential for an understanding of what is described by general relativity as singularities. General relativity can be combined with the quantum theory of matter. Such a combined theory of quantum physics on curved space-time is necessary for instance in the behavior of matter at high density in the very early universe, as seen in Section 8.5.4, or of nuclear matter in neutron stars. However, the question of how to describe the quantum physics of space-time itself, or of the gravitational force, remains to be resolved.

13.1 DIMENSIONAL ARGUMENTS AND EFFECTIVE FIELD THEORY

Qualitative arguments indicate that an application of quantum physics to space-time, or a theory of quantum gravity, might on extreme scales be a better description of the universe than general relativity. If we combine quantum physics with gravitational physics, there are three relevant fundamental constants: Planck's constant $\hbar$ in addition to Newton's constant G and the speed of light c. It turns out that these three parameters can be combined in a unique way to a quantity with the units of length, called the **Planck length**

$$\ell_{\mathrm{P}} = \sqrt{\frac{G\hbar}{c^3}} \sim 10^{-35}\mathrm{m}. \tag{13.1}$$

Its extremely small value shows that *direct* measurements of quantum-gravity effects are unlikely to be made in the near future.

The smallest length scales currently probed using particle colliders are about 10^{-20}m, but even these observations do not constitute direct length measurements. The primary measurement in particle colliders is of angle-dependent scattering probabilities, or scattering cross-sections, of all the particles created by the energy of a collision event. In a known potential, scattering angles are related to impact parameters with units of length. It is therefore possible to express scattering measurements as indirect length measurements in tests of particle-physics interactions. Measuring on the scale of the tiny Planck length either requires much higher energies, or dedicated indirect searches based on a detailed understanding of quantum-gravitational interactions.

While geometrized units make it possible to express all physical quantities in units of length, other units are sometimes useful if they can suggest alternative measurement procedures. Using the speed of light, the Planck length can be converted into the **Planck time** $t_{\mathrm{P}} = 5.4\cdot 10^{-44}$s. The collider length scale corresponds to a time of $10^{-29}\mathrm{s} = 10\mathrm{qs}$ expressed in quectoseconds, using the smallest named unit prefix at the time of this writing. The shortest time duration from a direct measurement is the photon traversal time of $24.7\cdot 10^{-20}\mathrm{s} = 247\mathrm{zs}$ (zeptoseconds) through a hydrogen molecule.[1]

Similarly, one can find a unique combination of the fundamental constants with the units of a density, called the **Planck density**. Its value amounts to more than one trillion solar masses in a proton-sized region. These are extreme scales, but if we approach something described by general relativity as a physical singularity, where curvature or the density of matter may become infinite,

[1] S. Grundmann et al., Zeptosecond birth time delay in molecular photoionization, *Science* **370** (2020) 339–341.

DOI: 10.1201/9781003588702-16

they will eventually be reached. We therefore have to understand quantum gravity before we can say what happens at the big bang or inside black holes. Perhaps, such a theory could modify the attractive nature of classical gravity, in such a way that gravitational collapse can be stopped and a new kind of dense and compact but non-singular object remains.

Dimensional arguments may be used to analyze the possible form of force laws or interaction terms expressed by an action principle. In Newtonian gravity, Newton's constant provides the correct units to turn the product of two masses divided by the distance into an energy, resulting in the potential as one term in an action from which the force can be derived. Mathematically, we may interpret the resulting $1/r$-behavior as the leading term in a Taylor expansion by $1/r$, such that higher-order contributions would only be relevant on tiny distances on which the gravitational force has not been measured in detail. However, because r has units of length, each term in the expansion would require a new dimensionful coefficient. A single dimensionful coupling constant is the most efficient way of specifying a force or an interaction. This statement does not rule out higher-order terms in the expansion, but it suggests that we should interpret each of them as a new physical phenomenon or a new mass-dependent force, rather than an extended description of gravity.

The behavior is different in special relativity because we have a new fundamental constant, the finite value of the speed of light, which allows us to turn a mass parameter M into a length GM/c^2, as seen in the Schwarzschild radius. Thus, a Taylor expansion in $GM/(c^2r)$ does not require a new dimensionful coefficient at each order. Instead, there is now an infinite number of dimensionless coefficients that determine a generic mass-dependent potential as a Taylor expansion in $1/r$. The two terms shown in (11.1) may then be extended to arbitrary orders, and the short-distance behavior remains undetermined from the perspective of special relativity. However, all these contributions are expressed in terms of a single fundamental constant, G, with specific numerical coefficients determined by the condition of **general covariance** as shown by the derivation of (11.1) from the Newtonian limit of general relativity. Symmetry is therefore important for removing ambiguities in consistent force laws or interactions.

The property of general covariance applies to solutions derived from general relativity as well as the theory itself, expressed by the Einstein–Hilbert action (9.67). If the integrand of a gravitational action is required to be invariant and have the correct units of an action, only the cosmological constant Λ and Newton's constant G are free. The latter also makes it possible to obtain units of an action from the basic field given by the metric. As suggested in the example of a Taylor-expanded Newtonian force, the two allowed terms are interpreted as two different phenomena, the gravitational interaction related to G and dark energy related to Λ. No other terms are possible in dynamical laws described by a metric if we require general covariance and the appearance of only the classical fundamental constants.

Quantum gravity, at first sight, seems to make the gravitational interaction more ambiguous because it must implement a new fundamental constant, $\hbar$. The Planck length can then be used in mass-independent higher-order terms in a Taylor expansion of the gravitational potential by $1/r$, as seen in (11.2). Moreover, the action is no longer determined by a few parameters because the combination $\ell_{\rm P}^2 R$ with the Ricci scalar R is unitless. Based on dimensional arguments, the classical integrand of the action can then be extended to a Taylor series of the form

$$f(R) = -2\Lambda + R + R(c_1\ell_{\rm P}^2 R + c_2\ell_{\rm P}^4 R^2 + \cdots) \tag{13.2}$$

with an infinite number of numerical coefficients c_j. The result is an infinite number of inequivalent $f(R)$-theories of gravity. Gravity is even more ambiguous because combining higher-order R-terms with other curvature invariants, such as $R_{\alpha\beta}R^{\alpha\beta}$ and $R_{\alpha\beta\gamma\delta}R^{\alpha\beta\gamma\delta}$ or fully contracted products of covariant derivatives of the Riemann tensor, generalizes the action beyond $f(R)$-gravity.

Such theories are called **effective field theories** of a given interaction. An effective field theory is defined by the condition that it obey a given symmetry, here general covariance, and depend on a specific set of fields, here the space-time metric, that model a physical phenomenon. In general, these conditions can be fulfilled with an infinite number of free parameters, but they may still be restrictive in situations in which only one or a few leading terms in an expansion are relevant. The general framework of an effective field theory is therefore testable by observations. However, an effective field theory is usually not considered fundamental because of its high degree of ambiguity.

If a gravitational theory has a large degree of ambiguity, general covariance is no longer able to predict specific numerical coefficients of an $\hbar$-dependent potential such as (11.2). In addition, it remains unknown how general covariance can be implemented consistently at the quantum level. If this symmetry property can be achieved, it is likely to require further consistency conditions that may help to restrict the free parameters in a general action principle or in an expansion of the gravitational potential. Some approaches to quantum gravity have revealed at least the presence of additional consistency conditions, even though they have not yet been implemented in complete detail.

13.2 APPROACHES

Various complications found in attempts to apply quantum physics to the gravitational force have been analyzed in different ways, giving rise to a number of approaches with distinct conceptual underpinnings as well as mathematical and numerical methods. The summary given in this section focuses on properties and results that are closely related to material covered in the preceding parts of this book, while avoiding technical discussions of quantum field theory or numerical methods. The varying amounts of details provided for the approaches and their placing in the following list do not reflect a ranking of their importance or relevance.

13.2.1 PERTURBATIVE QUANTUM GRAVITY

Newton's potential was the first description of classical gravity, but it is not fundamental because it is only realized as a specific behavior of metric components in the weak-field limit of general relativity. Quantum theories based on Newton's potential are therefore effective descriptions that include some but not all relevant features of gravity. In a given situation, it may be difficult to determine which features are relevant if we do not know the full theory that is supposed to be approximated by an effective description.

Instead of using only Newton's potential, weak gravitational fields can be described more completely by assuming a perturbative metric $g_{\alpha\beta} = \eta_{\alpha\beta} + \omega_{\alpha\beta}$ close to Minkowski space-time, which is called the **background space-time** in this context. In this approach, we are not asking how space-time itself can appear as a quantum object, and instead analyze quantum features of small perturbations around it, such as gravitational waves. Such an approach is called **background dependent** because its results are likely to depend on which classical background metric is chosen.

In contrast to the linearized treatment of gravitational waves in Section 9.1.3, perturbative quantum gravity takes into account non-quadratic metric terms in the gravitational interaction Hamiltonian, or in the gravitational action principle. Methods of quantum field theory then make it possible to derive quantities such as scattering probabilities of quantized gravitational waves, or of gravitons.

In this setting, the gravitational force is described by an exchange of gravitons, just as photons describe the quantized transmission of energy by the electromagnetic field. However, the gravitational interaction is more complicated because the gravitational field carries energy and couples to energy or mass. Therefore, gravitons have complicated self-interactions, while photons are electrically neutral and do not interact with one another. Analyzing graviton interactions shows that perturbative quantum gravity is not renormalizable: The series of all possible self-interaction terms of gravitons implies an effective field theory with an infinite number of undetermined coefficients. Low orders of this series may imply meaningful results about some aspects of graviton scattering, but the theory is necessarily an approximation and cannot be fundamental.[2]

[2] C. P. Burgess, Quantum Gravity in Everyday Life: General Relativity as an Effective Field Theory, *Living Rev. Relativity* **7** (2004) 5.

13.2.2 ASYMPTOTIC SAFETY

In perturbative quantum gravity, the main parameter that determines graviton scattering is Newton's constant G. Focusing on this constant resembles a Newtonian picture on the gravitational force, even though space-time metrics are used in this approach.

A broader perspective may include the cosmological constant as another parameter that determines some aspects of space-time physics. Newton's constant and the cosmological constant are the only free parameters in the equations of motion or the action of general relativity, but quantum physics may be more general in particular if it is applied at large curvature. We may then add higher-order curvature invariants to the classical action, such as R^2 or $R_{\alpha\beta\gamma\delta}R^{\alpha\beta\gamma\delta}$ or higher powers in a higher-curvature action of effective field theory. Each of these terms may have an independent numerical coefficient in the action, vastly increasing the number of free parameters. If suitable terms are included in graviton scattering, it might be possible to achieve cancellations that produce less ambiguous results.

One problem of effective field theory is the introduction of an infinite number of free parameters, which can never be determined by a finite number of measurements. The approach of asymptotically safe gravity tries to demonstrate that all dynamical properties of gravitons may be described by a finite number of parameters. Specific studies are usually based on a proposed tractable subclass of higher-curvature actions, and they are largely numerical. There is some evidence that perturbative quantum gravity may be asymptotically safe.[3]

13.2.3 CAUSAL DYNAMICAL TRIANGULATIONS

Like asymptotic safety scenarios, causal dynamical triangulations are evaluated mainly by numerical means. But instead of working with a metric tensor, they are based on a discretization of space-time that is expected to give rise to continuous Riemannian geometry only in a large-scale approximation.

The discretization is based on a classical approximation technique in general relativity called Regge calculus. In this formalism, curvature is expressed through deficit angles, which generalize the basic idea that one can construct surfaces such as a cone by cutting a triangle out of a planar sheet of paper and gluing the new edges onto each other. The angle at the cut-out corner of the triangle is the deficit angle and represents the cone's curvature. We can see its extrinsic curvature in an embedding in 3-dimensional space, and intrinsic curvature is implied by cutting out pieces of the original flat plane. Regge calculus is based on higher-dimensional versions.

Causal dynamical triangulations discretize space-time by representing it as a set of glued tetrahedra or their 4-dimensional versions, called 4-simplices. Depending on how many 4-simplices meet at a given corner, there may be a positive or negative deficit angle that introduces a discrete quantum of curvature. One of the dimensions is interpreted as timelike, such that the corresponding deficit angles are hyperbolic, representing deficit rapidities. By summing over all corners, one can compute the integrated Ricci scalar, or the value of the gravitational action for a given discretization. A classical solution corresponds to an extremum of the action. Quantum physics of these discretizations can be expressed by using them as a statistical ensemble with a probability distribution determined by the action. In this way, statistical properties of a quantum version of space-time can be computed numerically.[4]

13.2.4 CAUSAL SETS

Causal sets present a different discretization of space-time as sets of individual events with causality relations between them. Instead of light cones in continuous space-time, as causal set assigns to each

[3] M. Niedermaier and M. Reuter, The Asymptotic Safety Scenario in Quantum Gravity, *Living Rev. Relativity* **9** (2006) 5.

[4] J. Ambjrn, A. Görlich, J. Jurkiewicz and R. Loll, Nonperturbative quantum gravity, *Phys. Rept.* **519** (2012) 127–210, arXiv:1203.3591

event the sets of other events that lie in its past or future. Here, the analysis is mainly mathematical rather than numerical. It classifies different types of causal sets, some of which can be viewed as approximation of continuous space-time, while others appear collapsed and do not offer a large-scale approximation.[5]

13.2.5 NON-COMMUTATIVE AND NON-ASSOCIATIVE GEOMETRIES

We usually represent space or space-time as sets of points with certain mathematical structures such as a metric that determines distances. We often use functions on a given space or space-time in order to perform specific calculations or to include physical fields such as matter. It turns out that this picture can be reversed: We can characterize a space or space-time by properties of the set of functions that can be supported by it. Formally, this reversal is obtained by interpreting the evaluation $f(x)$ of a function at a point as $x(f) = f(x)$, defining x as a mapping from a function f to a real or complex number with a value given by $f(x)$. This mapping is an algebra homomorphism: $x(f_1+f_2) = f_1(x)+f_2(x) = x(f_1)+f(f_2)$ and $x(f_1f_2) = f_1(x)f_2(x) = x(f_1)f(f_2)$. One can therefore define points in a space as the set of algebra homomorphisms from functions on this space to real or complex numbers.

In geometry, we usually work with functions whose product is commutative and associative. If these conditions are relaxed, new algebras of functions are obtained that may also have homomorphisms to the real or complex numbers. These homomorphisms then define new spaces or space-times with a non-commutative or non-associative geometry.

Tensor calculus may be defined by using the Riemannian definitions but paying due attention to the ordering or bracketing of factors in a product. For instance, for a vector field v^α we have the directional derivative $v(f) = v^\alpha \partial_\alpha f$ that obeys the product rule when it is applied to the product of functions. In the non-commutative context, the product rule should be written as $v(f_1f_2) = f_1v(f_2)+v(f_1)f_2$, always keeping f_1 to the left and f_2 to the right. If there are several factors, as in the contributions to the Riemann tensor, much care is needed to make sure that consistent relationships are obtained, especially in the non-associative case.[6]

13.2.6 STRING THEORY

String theory starts with a formulation of particles as 1-dimensional extended objects, or strings. If a string has internal tension and moves in a background space-time with metric $g_{\alpha\beta}$, its dynamics depends on the curvature of space-time because of tidal forces. We can formalize the dynamics by introducing a reference frame on the string. Unlike the reference frame of a point particle, which only requires the 4-velocity $\mathbf{u}$ along the 1-dimensional worldline, the 2-dimensional worldsheet swept out by a string requires a timelike and a spacelike direction. Following the Hamiltonian formulation of general relativity while applying it to the 2-dimensional worldsheet, we specify the timelike direction $t^a = Nn^a + Ms^a$ in terms of the lapse function N and the 1-dimensional shift vector M, relative to the unit normal n^a and an orthogonal unit spatial direction s^a. (We use indices $a,b,\ldots$ in two dimensions.) For any 2-dimensional metric m_{ab}, the sum $m_{ab}+n_an_b$ defines the 1-dimensional induced spatial metric on worldlines normal to n^a because $n^am_{ab} = 0 = m_{ab}n^b$. Writing a general 1-dimensional metric as qs_as_b with a positive function q, we obtain the inverse frame metric

$$m^{ab} \quad = \quad -n^an^b + \frac{1}{q}s^as^b = -\frac{1}{N^2}(t^a - Ms^a)(t^b - Ms^b) + \frac{1}{q}s^as^b$$

[5] S. Surya, The causal set approach to quantum gravity, arXiv:1903.11544

[6] S. Doplicher, K. Fredenhagen and J. E. Roberts, The quantum structure of spacetime at the Planck scale and quantum fields, *Commun. Math. Phys.* **172** (1995) 187–220, hep-th/0303037; P. Aschieri, M. Dimitrijevic, F. Meyer and J. Wess, Noncommutative geometry and gravity, *Class. Quant. Grav.* **23** (2006) 1883, hep-th/0510059; R. Blumenhagen and M. Fuchs, Towards a Theory of Nonassociative Gravity *JHEP* **07** (2016) 019, arXiv:1604.03253

$$= -\frac{1}{N^2}t^at^b + \frac{M}{N^2}(t^as^b + t^bs^a) + \left(\frac{1}{q} - \frac{M^2}{N^2}\right)s^as^b\,. \tag{13.3}$$

The **Polyakov action** is defined by combining the frame metric m^{ab} with the induced metric h_{ab} defined by the embedding of the worldsheet in a space-time with metric $g_{\alpha\beta}$. Using worldsheet coordinates $y^a = (\tau, \sigma)$, the induced metric

$$h_{ab} = \frac{\partial x^\alpha}{\partial y^a}\frac{\partial x^\beta}{\partial y^b}g_{\alpha\beta} \tag{13.4}$$

is combined with the frame metric in the action

$$S[x^\alpha] = -\frac{1}{2\alpha'}\int_{-\infty}^{\infty}\int_0^1 \sqrt{-\det m}\, m^{ab}\frac{\partial x^\alpha}{\partial y^a}\frac{\partial x^\beta}{\partial y^b}g_{\alpha\beta}\mathrm{d}\tau\mathrm{d}\sigma \tag{13.5}$$

if the string is assumed to be closed with coordinate range $\sigma \in [0,1)$. (The frame length $\int_0^1 \sqrt{q}\mathrm{d}\sigma$ remains free and may change as the string gets stretched.) The coefficient $1/\alpha'$ determines the **string tension**. Based on units, the string tension may replace Newton's constant in the definition of the string length $\ell_s = \sqrt{\hbar\alpha'}$ as an analog of the Planck length. The value of the action depends on the embedding functions $x^\alpha(y^a)$ of the string worldsheet.

For a Hamiltonian formulation, we write the action in the explicit form

$$\begin{aligned} S[x^\alpha] &= \frac{1}{2\alpha'}\int_{-\infty}^{\infty}\int_0^1 N\sqrt{q} \\ &\quad \times \left(\frac{1}{N^2}\frac{\partial x^\alpha}{\partial \tau}\frac{\partial x^\beta}{\partial \tau} - 2\frac{M}{N^2}\frac{\partial x^\alpha}{\partial \tau}\frac{\partial x^\beta}{\partial \sigma} - \left(\frac{1}{q} - \frac{M^2}{N^2}\right)\frac{\partial x^\alpha}{\partial \sigma}\frac{\partial x^\beta}{\partial \sigma}\right)g_{\alpha\beta}\mathrm{d}\tau\mathrm{d}\sigma \end{aligned} \tag{13.6}$$

based on (13.3), and derive the momenta

$$p_\alpha = \frac{\delta S}{\delta(\partial x^\alpha/\partial\tau)} = \frac{\sqrt{q}}{\alpha' N}g_{\alpha\beta}\left(\frac{\partial x^\beta}{\partial\tau} - M\frac{\partial x^\beta}{\partial\sigma}\right). \tag{13.7}$$

The Hamilton functional H of this 2-dimensional field theory is a spatial integral, such that $\int_{-\infty}^{\infty}\mathrm{d}\tau$ equals the Legendre transformation of the action:

$$\begin{aligned} \int_{-\infty}^{\infty} H\mathrm{d}\tau &= \int_{-\infty}^{\infty}\int_0^1 p_\alpha\frac{\partial x^\alpha}{\partial\tau}\mathrm{d}\tau\mathrm{d}\sigma - S[x^\alpha] \\ &= \int_{-\infty}^{\infty}\int_0^1 \left(\frac{N}{2\sqrt{q}}\left(\alpha' p_\alpha p_\beta g^{\alpha\beta} + \frac{1}{\alpha'}\frac{\partial x^\alpha}{\partial\sigma}\frac{\partial x^\beta}{\partial\sigma}g_{\alpha\beta}\right) + Mp_\alpha\frac{\partial x^\alpha}{\partial\sigma}\right)\mathrm{d}\tau\mathrm{d}\sigma\,. \end{aligned} \tag{13.8}$$

This expression resembles the gravitational Hamilton functional (9.117) in that all terms are linear in N or M. Also here, these functions determine the frame, such that observer-independent physics should not depend on their values. This covariance condition is implemented by requiring that the functional derivatives $\delta H/\delta N$ and $\delta H/\delta M$ vanish. Here, for simplicity, we may combine N and q in the form $N/\sqrt{q}$ because the Hamilton functional depends only on this combination. We obtain the constraints

$$C = \frac{1}{2}\left(\alpha' p_\alpha p_\beta g^{\alpha\beta} + \frac{1}{\alpha'}\frac{\partial x^\alpha}{\partial\sigma}\frac{\partial x^\beta}{\partial\sigma}g_{\alpha\beta}\right) = 0 \tag{13.9}$$

and

$$D = p_\alpha\frac{\partial x^\alpha}{\partial\sigma} = 0\,. \tag{13.10}$$

As in the Hamiltonian formulation of general relativity, the constraints determine equations of motion as well as gauge transformations. The mathematical form of gauge transformations is given by Poisson brackets of the constraints, which in the present case are

$$\{C[N_1/\sqrt{q}],C[N_2/\sqrt{q}]\} = D[q^{-1}(N_1\partial N_2/\partial\sigma - N_2\partial N_1/\partial\sigma)] \quad (13.11)$$
$$\{C[N/\sqrt{q}],D[M]\} = C[(N/\sqrt{q})\partial M/\partial\sigma - M\partial(N/\sqrt{q})/\partial\sigma] \quad (13.12)$$
$$\{D[M_1],D[M_2]\} = D[M_1\partial M_2/\partial\sigma - M_2\partial M_1/\partial\sigma]\,. \quad (13.13)$$

These equations closely resemble the hypersurface-deformation relations (9.138) combined with (9.136) and (9.137). However, they are simpler in the present case because q only appears through the combination $N/\sqrt{q}$, and can therefore be eliminated by redefining the frame functions as $\lambda = N/\sqrt{q}$, together with the original M. The linear combinations

$$V^{\pm}[\lambda] = C[\lambda] \pm D[\lambda] \quad (13.14)$$

then obey

$$\{V^{\pm}[\lambda_1],V^{\mp}[\lambda_2]\} = 0 \quad (13.15)$$

and

$$\{V^{\pm}[\lambda_1],V^{\pm}[\lambda_2]\} = \pm 2V^{\pm}[\lambda_1\partial\lambda_2/\partial\sigma - \lambda_2\partial\lambda_1/\partial\sigma]\,. \quad (13.16)$$

Up to a constant coefficient, this relation is identical with two versions of (13.13), one for V^+ and one for V_-. Hypersurface deformations in two dimensions can therefore be rewritten as two copies of spatial deformations in one dimension. The latter symmetries, in infinitesimal form, are represented by the relations (13.13) and called the **Virasoro algebra**.

Given the simplified form of the constraints, it is possible to solve them or turn them into operators. If the background space-time is Minkowskian, $g_{\alpha\beta} = \eta_{\alpha\beta}$, the Hamilton functional is quadratic in the fields $x^\alpha(\sigma)$ and their momenta. After expressing these functions on the bounded range of σ as a Fourier series, the Hamilton functional is turned into a series of infinitely many copies of the harmonic-oscillator Hamilton function, one for each mode x_n^α. Quantization proceeds by introducing an infinite number of ladder operators, labeled by the mode number n and the Minkowski index α. As a major implication, the tensor product of two ladder operators of the form $\hat{a}_{n_1}^\alpha \otimes \hat{a}_{n_2}^\beta$ can be used to define a 2-index tensorial excitation of quantum states. If additional conditions are fulfilled, these excitations may include a metric tensor, or a quantized perturbation $\omega_{\alpha\beta}$ representing a graviton on the background metric $\eta_{\alpha\beta}$.

String theory may therefore be used as a background-dependent quantum theory of gravity. Compared with full general relativity, the worldsheet formulation simplified the constraints and their gauge transformations, representing hypersurface deformations. The constraint

$$C[1] = \frac{1}{2}\int_0^1 \left(\alpha' p_\alpha p_\beta g^{\alpha\beta} + \frac{1}{\alpha'}\frac{\partial x^\alpha}{\partial\sigma}\frac{\partial x^\beta}{\partial\sigma}g_{\alpha\beta}\right)\frac{\mathrm{d}\sigma}{\sqrt{q}} = 0\,, \quad (13.17)$$

evaluated for a constant lapse function $N = 1$, is of particular interest because it can be used to derive the mass of a string through $||\mathbf{p}||^2 = -m^2$. The analysis is greatly simplified if we assume a background Minkowski space-time, $g_{\alpha\beta} = \eta_{\alpha\beta}$, such that the constraint is quadratic in $\partial x^\alpha/\partial\sigma$ as well as p_α.

However, p_α at a given time τ is not a single 4-vector but rather a σ-dependent function, describing dynamical deformations of the string. The overall mass is related to a suitable average of p_α over the string, which can be identified with the zero-mode in a Fourier expansion. To this end, we first define a convenient angle coordinate on the string by

$$\phi(\sigma) = 2\pi\frac{\int_0^\sigma \sqrt{q}\mathrm{d}\tilde{\sigma}}{\int_0^1 \sqrt{q}\mathrm{d}\tilde{\sigma}} \quad (13.18)$$

and write an orthonormal function basis on the string as

$$e_n(\phi) = \frac{1}{\sqrt{2\pi}} \exp(in\phi) . \tag{13.19}$$

The position and momentum components along the string then have Fourier modes given by

$$x^\alpha(\phi) = \sum_{n=-\infty}^{\infty} x_n^\alpha e_n(\phi) \tag{13.20}$$

and

$$p_\alpha(\phi) = \sum_{n=-\infty}^{\infty} p_{n\alpha} e_{-n}(\phi) . \tag{13.21}$$

Reality of $x^\alpha(\phi)$ and $p_\alpha(\phi)$ implies $x_{-n}^\alpha = (x_n^\alpha)^*$ and $p_{-n\alpha} = p_{n\alpha}^*$.

The zero-modes x_0^α and $p_{0\alpha}$ are real. The latter are suitable candidates for the average momentum, from which we derive the mass spectrum. Starting with

$$\begin{aligned} C[1] &= \frac{1}{2}\alpha' \sum_{n=-\infty}^{\infty} \left(p_{n\alpha} p_{-n\beta} \eta^{\alpha\beta} + \frac{4\pi^2 n^2}{\alpha'^2} x_n^\alpha x_{-n}^\beta \eta_{\alpha\beta} \right) \\ &= \frac{1}{2}\alpha' \left(p_{0\alpha} p_{0\beta} \eta^{\alpha\beta} + 2\sum_{n=1}^{\infty} \left(p_{n\alpha} p_{n\beta}^* \eta^{\alpha\beta} + \frac{4\pi^2 n^2}{\alpha'^2} x_n^\alpha (x_n^\beta)^* \eta_{\alpha\beta} \right) \right) = 0 , \end{aligned} \tag{13.22}$$

we obtain the mass

$$m^2 = -||\mathbf{p}_0||^2 = 2\sum_{n=1}^{\infty} \left(p_{n\alpha} p_{n\beta}^* \eta^{\alpha\beta} + \frac{4\pi^2 n^2}{\alpha'^2} x_n^\alpha (x_n^\beta)^* \eta_{\alpha\beta} \right) . \tag{13.23}$$

The average position x_0^α does not contribute to the mass because the x_n^α-terms are multiplied by n^2. The other position modes contribute to the mass because they describe how the string gets stretched, which changes the energy depending on the string tension α'.

The expression (13.23) contains terms that are not gauge-invariant because the components $p_{n\alpha}$ and x_n^α are subject to a large set of gauge transformations generated by the constraints with non-constant lapse function N and shift vector M. Since there are two independent constraints, we can reduce the mass to its relevant contributions in a suitable frame by imposing gauge-fixing conditions on two sets of the mode functions, such as those for $\alpha = 0$ and $\alpha = 1$. If we choose the conditions $p_{n0} = -2\pi n \alpha'^{-1} g_{0\beta} x_n^\beta$ and $p_{n1} = -2\pi n \alpha'^{-1} g_{1\beta} x_n^\beta$, the time components and first spatial components are eliminated from the mass. The first condition also removes a potentially destabilizing negative contribution to the mass, which would be implied by $\eta_{00} < 0$ if time components were present.

The mass now takes the manifestly positive form

$$m^2 = 2\sum_{n=1}^{\infty} \left(|p_{n2}|^2 + |p_{n3}|^2 + \frac{4\pi^2 n^2}{\alpha'^2} (|x_n^2|^2 + |x_n^3|^2) \right) \tag{13.24}$$

including only 2-modes and 3-modes. The real and imaginary parts of each mode contribute independent degrees of freedom, called left-moving and right-moving string excitations. On a Minkowski background, as assumed in the derivation, each of them has a quadratic expression for the mass, which we quantize to a harmonic-oscillator Hamilton operator with a formal identification of the frequency $\omega = 2\pi n/\alpha'$ and unit mass. Inserting harmonic-oscillator energy eigenvalues for the 2-modes and the 3-modes, we obtain the quantized masses

$$m_j^2 = 4\pi m_s^2 \sum_{n=1}^{\infty} (j_{n2} + j_{n3} + 1) n \tag{13.25}$$

with the string mass $m_s = \sqrt{\hbar/\alpha'}$. The quantum numbers j_{n2} and j_{n3} are non-negative integers for each $n > 0$.

The mass spectrum appears to be discrete, but so far we just obtain an infinite value from (13.25) for any choice of quantum numbers because of the infinite $\sum_{n=1}^{\infty} n$. This contribution results from the non-zero ground-state energy of the harmonic oscillator, but now added up infinitely many times because we have quantized an infinite number of modes. Such infinities are a general consequence of quantum field theory, the quantum theory of position-dependent degrees of freedom. The appearance of infinities requires a procedure of **regularization**. We have taken the formal analogy with the harmonic oscillator in quantum mechanics too far by assuming that each mode receives the same ground-state energy as a single harmonic oscillator. We need an additional construction that tells us how each non-excited string mode contributes to the energy or the mass. We could simply decide that each mode should contribute a vanishing ground-state energy, removing the infinite term. This proposal amounts to changing the harmonic-oscillator potential by adding the constant $-\frac{1}{2}\hbar\omega$. Adding a constant term to the potential does not change the classical dynamics, but the $\hbar$-dependence of the constant term is conceptually problematic: It means that we have to change the classical potential in anticipation of quantization problems.

A more elegant regularization that takes into account the specific form of the infinite term $\sum_{n=1}^{\infty} n$ starts with the observation that this formal sum is the value of the **zeta function** $\zeta(z) = \sum_{n=1}^{\infty} n^{-z}$ at $z = -1$. This function turns out to have a well-defined expression for any complex number z, but it is not always represented by the series form. The complete function, discussed in complex analysis, results in the finite value $\zeta(-1) = -1/12$. The regularized mass spectrum is then given by

$$m_j^2 = 4\pi m_s^2 \left(\sum_{n=1}^{\infty} (j_{n2} + j_{n3})n - \frac{1}{12} \right) . \tag{13.26}$$

The mass values are now finite provided only a finite number of j_{n2} and j_{n3} are non-zero.

However, the lowest mode, or the ground state obtained for $j_{n2} = 0$ and $j_{n3} = 0$, has a negative m^2. Moreover, since $1/12$ is non-integer, there are no massless excitations that could be used to construct a graviton. It is easier to solve the second problem, at least in terms of formal constructions. The value of $-1/12$ in (13.26) is multiplied by $\frac{1}{2}(4-2) = 1$ with a factor of $1/2$ from the ground-state energy of a single harmonic oscillator, and a factor of the number $4-2$ of remaining string functions after fixing the gauge. If the string moves in a d-dimensional space-time, this factor is replaced by $\frac{1}{2}(d-2)$, such that the constant contribution to the regularized mass spectrum equals $(d-2)/24$. If $d = 26$, we obtain massless modes for either $j_{12} = 1$ or $j_{23} = 1$. This solution requires a higher-dimensional space-time, which may be consistent with observations if all but four of them are circles with tiny radii. An interesting consistency result is that the required number of dimensions is indeed an integer.

The mode with $m^2 < 0$, called a **tachyon**, can be eliminated if the string excitations include additional fermionic fields, defined as fields with mode operators that have canonical **anti-commutators** $[\hat{A}, \hat{B}]_+ = \hat{A}\hat{B} + \hat{B}\hat{A}$ instead of canonical commutators. This solution also requires a specific symmetry transformation between the original string modes and the new fermionic ones, such that the negative contribution can be canceled out. A symmetry that relates fields to fermionic fields is called a **supersymmetry**. The value of the contribution $-1/12$ then changes, which means that the higher-dimensional argument no longer applies in the given form. However, detailed constructions show that the desired number of massless modes remains valid provided the dimension is reduced to $d = 10$, still an integer and closer to the observed four dimensions.[7]

[7] D. Lüst and S. Theisen, Lectures on String Theory, Springer (1989); J. Polchinski, String Theory I and II, Cambridge University Press (1998).

13.2.7 CANONICAL QUANTUM GRAVITY

Canonical quantum gravity is the most direct attempt to quantize gravity. Starting with the Hamiltonian formulation of general relativity, also known as the **canonical formulation** of general relativity, and following the methods of Section 12.2, the constraints C_0 in (9.119) and C_k in (9.120) should be turned into operators, and the resulting constraint equations should be solved in a suitable way including perhaps a choice of a quantum reference frame. Such an approach to quantum gravity would be background independent because it quantizes the full metric in the form of the spatial part h_{ij}, while the time-space components N and M^i are left to the choice of frame.

While appealing, this approach immediately faces severe challenges, even before general questions about quantum reference frames appear, as discussed in Section 12.2.1. On a more basic level, if there is no background space-time, we are unable to use Fourier transformation of metric components in order to turn the constraint for the gravitational field into known operators such as the Hamilton operator of the harmonic oscillator: Fourier transformation in three or four dimensions requires a scalar product for a decomposition of a function into plane waves $\exp(i\mathbf{k}\cdot\mathbf{x})$ or $\exp(i\vec{k}\cdot\vec{x})$, which depends on the existence of a metric that is not an operator.

The underlying mathematical problem can be seen by starting with the Poisson bracket $\{\phi(x), p_\phi(y)\}$ for a scalar field. Since field values at different positions x are independent, the Poisson bracket evaluated at $x \neq y$ must vanish, just like $\{x_j, p_k\} = 0$ for $j \neq k$ in two or more dimensions. Moreover, the integral

$$\int_{-\infty}^{\infty} \{\phi(x), p_\phi(y)\} \mathrm{d}x = \left\{ \int_{-\infty}^{\infty} \phi(x) \mathrm{d}x, p(y) \right\} \neq 0 \tag{13.27}$$

should be non-zero because the infinite integration range of x on the right includes the value y, whatever it might be. A non-vanishing integral of a function $\{\phi(x), p_\phi(y)\}$ that depends on x and y and equals zero whenever $x \neq y$ implies that the function must be proportional to the delta function. Setting the proportionality constant equal to one then implies the canonical Poisson bracket

$$\{\phi(x), p_\phi(y)\} = \delta(x-y). \tag{13.28}$$

Delta functions such as (13.28) are meaningful in classical expressions, as long as we make sure that we never obtain a product of delta functions with poles at the same point. However, trying to find operators for $\phi(x)$ and $p_\phi(y)$ such that their commutator equals (13.28) times $i\hbar$ is not meaningful because the delta function is not a finite coefficient: If a single delta function would be a meaningful operator, its square should also be a meaningful operator in spite of the fact that the product of delta functions with poles at the same point is not well-defined. (A delta function may result in a well-defined operator if it is added to another operator, as in the case of the Hamilton operator of the delta-function potential analyzed in Section 2.5.1. In this case, the kinetic-energy operator makes it possible to derive eigenfunctions and eigenvalues of the Hamilton operator via (dis)continuity conditions without considering products of the delta function. The resulting energy spectrum implies a meaningful definition of the Hamilton operator.)

We can eliminate delta functions by integration. In particular, if we replace the set of values $\phi(x)$ with the set of all $\phi_{\mathscr{I}} = \int_{\mathscr{I}} \phi(x)\mathrm{d}x$ for finite open intervals $\mathscr{I} \subset \mathbb{R}$, we obtain the finite result

$$\{\phi_{\mathscr{I}}, p_\phi(y)\} = \int_{\mathscr{I}} \delta(x-y)\mathrm{d}x = \begin{cases} 0 & \text{if } y \notin \mathscr{I} \\ 1 & \text{if } y \in \mathscr{I} \end{cases}. \tag{13.29}$$

Moreover, if we know sufficiently many $\phi_{\mathscr{I}}$, we can approximate any individual value $\phi(x)$ in a limiting process.

It is more convenient to work with "smeared" fields $\phi_f = \int_{-\infty}^{\infty} f(x)\phi(x)\mathrm{d}x$ for a suitable set of functions $f(x)$, such as those of compact support. In this case,

$$\{\phi_f, p_\phi(y)\} = f(y). \tag{13.30}$$

Formal operators representing this relation with the correct commutator act on functionals $\psi[\phi]$ that assign a complex number to any function ϕ. The integrals ϕ_f are then turned into multiplication operators, while $\hat{p}_\phi(y) = -i\hbar\delta/\delta\phi(y)$ are functional derivatives.

If we apply such a construction to a field on curved space, we must make sure that the correct volume element is used, including a factor of $\sqrt{\det h}$ with the spatial metric h_{ij}. This ingredient is problematic if we are trying to quantize the metric itself because it would result in a non-linear expression that is not compatible with canonical commutation relations. Fortunately, the factor of $\sqrt{\det h}$ is not necessary if we smear the momentum instead of the field: In a field theory on curved space-time, the Lagrange functional contains the term

$$L[\phi] = \frac{1}{2}\int d^3xN\sqrt{\det h}\dot{\phi}^2 + \cdots \tag{13.31}$$

where $\sqrt{\det h}$ is the required volume element implied by $\sqrt{-\det g} = N\sqrt{\det h}$, and $\dot{\phi}^2$ is the time-derivative contribution from a covariant $-g^{\alpha\beta}(\nabla_\alpha\phi)(\nabla_\beta\phi)$. Therefore, the momentum

$$p_\phi(y) = \frac{\delta L}{\delta\dot{\phi}(y)} = N\sqrt{\det h}\,\dot{\phi}(y) \tag{13.32}$$

contains the correct volume element. The smearing $p_{\phi,g} = \int\int\int g(y)p_\phi(y)d^3y$ is then invariant under spatial coordinate changes, and it is linear in the field momentum. We obtain a formal representation in the momentum representation, with $\hat{\phi}(x)$ a functional derivative.

For canonical quantum gravity, we have to extend the smearing procedure to tensorial objects such as h_{ij} and p^{jk}. We can only integrate index-free functions because the coefficients $\partial x^{j'}/\partial x^j$ in the tensor transformation law are in general position-dependent and cannot be moved out of an integral. Therefore, the integral of a transformed tensor does not equal the transformed integral: Integrals are not covariant unless they are applied to scalar functions without indices.

Initial constructions of canonical quantum gravity, referred to as **Wheeler–DeWitt quantum gravity**, did not introduce complete smearing procedures. Instead, momentum operators were formally defined as functional derivatives

$$\hat{p}^{ij} = \frac{\hbar}{i}\frac{\delta}{\delta h_{ij}} \tag{13.33}$$

with indices, acting on wave functions that depend on the metric tensor.

These constructions are formal because it is difficult to understand the specific function space and to equip it with a well-defined inner product. Moreover, the Hamiltonian constraint is turned into functional differential equations for which it is hard to develop general solution procedures. It also remained unclear how constraint operators can be found that preserve the required relationships (9.138) for general covariance in Hamiltonian form, in particular because the right-hand side of these equations requires constraints evaluated for metric-dependent shift vectors. If the metric is quantized, the definition of such an expression requires much care related to how to order the metric operator and the constraint operator. These questions remain unresolved.

13.2.8 LOOP QUANTUM GRAVITY

The approach of loop quantum gravity builds on canonical quantum gravity by reconsidering the choice of smearings and basic operators. The spatial metric and its momentum turn out to be rather inconvenient for a smearing procedure. Moreover, positivity of the metric in the sense that $\det h > 0$ implies additional conditions that could complicate a quantum representation.[8]

[8] J. Klauder: The affine quantum gravity program, Class. Quantum Grav. **19** (2002) 817; gr-qc/0110098

Turning to the second problem first, we can avoid positivity restrictions if we replace the metric with a **triad**, defined as a set of three vector fields e^i_I, $I = 1,2,3$, that form an orthonormal basis:

$$h_{ij}e^i_I e^j_J = \delta_{IJ} . \tag{13.34}$$

A basis in three dimensions may be right-handed or left-handed, such that $\det(e^i_I)$ may take either sign. As long as $\det(e^i_I) \neq 0$, we can invert the matrix with components e^i_I and obtain the inverse e^I_i. As with the metric and its inverse, the inverse relationship is indicated only by the position of indices. Multiplying (13.34) with two inverse triads,

$$\delta_{IJ}e^I_k e^J_l = h_{ij}e^i_I e^j_J e^I_k e^J_l = h_{ij}\delta^i{}_k\delta^j{}_l = h_{kl} , \tag{13.35}$$

we see that the metric can be expressed as a unique function of the inverse triad. Conversely, however, the triad is not uniquely defined by the metric because it may be rotated without changing the metric. This freedom implies a new gauge transformation in addition to coordinate changes.

We include a volume element in the triad by defining the **densitized triad**

$$E^i_I = \sqrt{\det h}\, e^i_I = |\det(e^J_j)|\, e^i_I . \tag{13.36}$$

The index I labels basis vectors and does not transform under a coordinate change, but it transforms under triad rotations. The remaining index i can be eliminated in a surface integral if we combine it with the index of the **co-normal** n_i of the surface. Unlike the normal n^i which requires raising the index, the co-normal n_i, defined as a dual vector field, does not depend on the metric because it can be expressed as $n_i = \partial_i s$ if the surface S is defined as the constant-level set $s(x^j) = \text{const}$: This definition implies that $v^i n_i = v^i \partial_i s = 0$ for any vector field v^i that is tangent to the surface and therefore represents a direction in which s is constant. The densitized triad smeared over a 2-dimensional surface,

$$F_{f,S} = \int_S f^I E^i_I n_i \mathrm{d}^2 y \tag{13.37}$$

with three smearing functions f^I, is analogous to the flux of the electric field.

Two-dimensional integration does not completely remove 3-dimensional delta functions in a Poisson bracket of gravitational fields of Hamiltonian general relativity. For well-defined expressions suitable for a commutator, we have to combine fluxes with 1-dimensional integrations of the gravitational momenta. A 1-dimensional integration over a curve presents us with the tangent vector t^i as an intrinsic 1-index object, analogous to the co-normal n_i on a surface. In a smeared object, it should be combined with a version of the momenta p^{kl} that have a single subscript index. Since the momenta are related to extrinsic curvature, we can start with K_{ij} and use the triad to define $K^I_i = K_{ij}e^j_J \delta^{IJ}$ with a single spatial index i. The canonical pair (K^I_i, E^j_J) was first used by Julian Schwinger.

There is another object with the same index structure as K^I_i, which may therefore be considered as an additional contribution to curve-integrable expressions. This object appears when we introduce a covariant derivative of the triad e^i_I. Corresponding to the two different types of indices, I and i, there are now two meanings of covariance: Spatial coordinate transformations that appear in the tensor transformation law $e^{i'}_I = (\partial x^{i'}/\partial x^i)e^i_I$, and triad rotations $e^i_{I'} = R_{I'}{}^I e^i_I$ with 3-dimensional rotation matrices $R_{I'}{}^I$. The latter may be position dependent (in a smooth way) because we can rotate triads by different angles at different positions. If this is the case, the original covariant derivative is not rotation-covariant because

$$\nabla_j(R_{I'}{}^I e^i_I) = R_{I'}{}^I \nabla_j e^i_I + e^i_I \partial_j R_{I'}{}^I \tag{13.38}$$

does not obey the same transformation law as e^i_I even though $\nabla_j e^i_I$ has the same index structure regarding rotations, given by a single subscript I.

If we intend to include rotations in covariance laws, we have to add an additional term to the original covariant derivative such that the second term in (13.38) cancels out, just as the Christoffel term makes it possible to cancel out derivatives of coordinate transformations implied by acting with a partial derivative on a tensor. We define the new covariant derivative as

$$\mathscr{D}_j e^i_I = \nabla_j e^i_I - \mathscr{C}_{jI}{}^J e^i_J = \partial_j e^i_I + \Gamma^i_{jk} e^k_I - \mathscr{C}_{jI}{}^J e^i_J \tag{13.39}$$

with the **spin connection** $\mathscr{C}_{jI}{}^J$. We will derive the spin connection by requiring two conditions: (i) It implies a covariant transformation law $\mathscr{D}_j e^i_{I'} = R_{I'}{}^I \mathscr{D}_j e^i_I$, where $e^i_{I'} = R_{I'}{}^I e^i_I$. The entire object $\mathscr{D}_j e^i_I$ then behaves like a 1-subscript tensor under triad rotations. (ii) The triad is **covariantly constant**: $\mathscr{D}_j e^i_I = 0$. As we saw in the case of the Christoffel symbol, a non-tensorial object may be non-zero in some coordinate systems or frames even if it vanishes in one system or frame. Condition (i) is therefore a prerequisite for condition (ii).

A **connection** is the general mathematical object that turns a partial derivative into a covariant derivative for any symmetry or gauge transformation. The Christoffel symbol is another example of a connection. The spin connection has to transform under a rotation such that (13.39) is covariant, while it must transform as a tensor under coordinate changes in order to maintain the covariance of ∇_j under such transformations. Combining (13.38) and (13.39), the first condition implies

$$\begin{aligned} \mathscr{D}_j e^i_{I'} &= \nabla_j e^i_{I'} - \mathscr{C}_{jI'}{}^{J'} e^i_{J'} = R_{I'}{}^I \nabla_j e^i_I + e^i_I \partial_j R_{I'}{}^I - \mathscr{C}_{jI'}{}^{J'} e^i_{J'} \\ &= R_{I'}{}^I \mathscr{D}_j e^i_I = R_{I'}{}^I \nabla_j e^i_I - R_{I'}{}^I \mathscr{C}_{jI}{}^J e^i_J \end{aligned} \tag{13.40}$$

and therefore

$$\mathscr{C}_{jI'}{}^{J'} = R_{I'}{}^I R^{J'}{}_J \mathscr{C}_{jI}{}^J + R^{J'}{}_I \partial_j R_{I'}{}^I \tag{13.41}$$

with the inverse rotation matrix $R^{J'}{}_J = e^{J'}_i e^i_J$. The derivative term $\partial_j R_{I'}{}^I$ implies that the spin connection is not a tensor, but like the Christoffel symbol it is used to turn a non-tensorial derivative into a covariant derivative. Under coordinate changes, $C_{iJ}{}^I$ does transform like a tensor, corresponding to the single subscript i.

The spin connection is determined if we require that the triad be **covariantly constant**, just as the metric under ∇_α:

$$0 = \mathscr{D}_j e^i_I = \partial_j e^i_I + \Gamma^i_{jk} e^k_I - \mathscr{C}_{jI}{}^J e^i_J \tag{13.42}$$

implies

$$\mathscr{C}_{jI}{}^J = e^J_i \partial_j e^i_I + e^J_i \Gamma^i_{jk} e^k_I . \tag{13.43}$$

Since this expression depends only on the triad, it can be combined with extrinsic curvature without changing Poisson brackets with the triad. While K^I_i has one index I, $\Gamma_{jI}{}^J$ has two indices I and J. Nevertheless, these two objects can be combined by using the completely antisymmetric tensor ε_{IJK} in three dimensions, normalized by $\varepsilon_{123} = 1$. The result is the Ashtekar–Barbero connection $A^I_i = \Gamma_{iJ}{}^K \varepsilon^{IJL} \delta_{KL} + \gamma K^I_i$ with a free constant γ, called the Barbero–Immirzi parameter.

The Ashtekar–Barbero connection can be integrated over a curve if we multiply it by the curve's tangent vector. Integration may be interpreted as solving a differential equation. A connection implies a natural differential equation for vector fields, given by **parallel transport**: If t^i is the tangent vector field on a spatial curve, another vector field v^j on the curve is parallel-transported if $t^i \nabla_i v^j = 0$, or $t^i \partial_i v^j = -t^i \Gamma^j_{ik} v^k$. This definition can be applied to any tensor, including tensors with a rotation index I if we replace ∇_i with $\mathscr{D}_i$ or D_i, where the latter is defined using $A^K_i \varepsilon_{KIL} \delta^{LJ}$ instead of $\mathscr{C}^J_{jI}$.

Parallel transport with a non-zero connection means that moving along the curve requires an infinitesimal rotation of the transported vector at every step. Integration along the curve results in a position-dependent rotation matrix $h(s)^I{}_J$, such that the vector field w^I is parallel transported if $w^I(s) = h(s)^I{}_J w^J(0)$ along the curve, starting with $w^I(0)$ at one end. Since $w^I(0)$ is constant for a given vector field, the rotational parallel transport can be expressed in matrix form through $h(s)^I{}_J$, without specifying any particular vector field. Inserting $w^I(s)$ in the condition

$$0 = t^i D_i w^J = t^i \partial_i w^J + t^i A_i^K \varepsilon_{KJL} \delta^{LI} w^J , \tag{13.44}$$

we obtain the differential equation

$$\frac{\mathrm{d}h(s)^I{}_J}{\mathrm{d}s} = t^i \partial_i h(s)^I{}_J = -t^i A_i^K \varepsilon_{KML} \delta^{LI} h(s)^M{}_J = t^i A_i^K \varepsilon_K{}^I{}_M h(s)^M{}_J , \tag{13.45}$$

using δ^{LI} to raise an index in the last step.

The term $t^i A_i^K \varepsilon_K{}^I{}_M$ represents an infinitesimal rotation with position-dependent angles $t^i A_i^K$ because the matrices L_K, $K = 1,2,3$, with components $(L_K)^I{}_M = \varepsilon_K{}^I{}_M$ exponentiate to rotation matrices around the three Cartesian axes. For instance,

$$\exp(\varphi L_3) = \exp \begin{pmatrix} 0 & \varphi & 0 \\ -\varphi & 0 & 0 \\ 0 & 0 & 0 \end{pmatrix} = \begin{pmatrix} \cos(\varphi) & \sin(\varphi) & 0 \\ -\sin(\varphi) & \cos(\varphi) & 0 \\ 0 & 0 & 0 \end{pmatrix} \tag{13.46}$$

can be derived by the same methods that led us to (4.236).

It is easier to do calculations with 2×2 matrices rather than 3×3 ones. Here, this transition can be achieved by using spin-1/2 matrices instead of the L_K, which correspond to spin one in its 3-dimensional representation. Infinitesimal rotations in the spin-1/2 representation are given by the Pauli matrices σ_K. We therefore replace $\varepsilon_K{}^I{}_M$ in (13.45) with $(\sigma_K)^I{}_M$ and use 2×2 matrices $h(s)$ along the curve. Parallel transport is then expressed by the differential equation.

$$\frac{\mathrm{d}h(s)}{\mathrm{d}s} = t^i A_i^I \sigma_I h(s) . \tag{13.47}$$

For given s, the expression $h(s)$ represents a 3-dimensional rotation in the form (4.242). It is subject to the gauge transformation $h(1) \to R(0)^{-1} h(1) R(1)$ if we rotate the triad by $R_{I'}{}^I$ such that $R(0)$ and $R(1)$ are the corresponding rotations at the endpoints of the curve, also expressed in the form (4.242). If the curve is closed, the trace $\mathrm{tr}(h(1))$, called the **holonomy** of A_i^I, is invariant under rotations because $R(0) = R(1)$ in this case.

The differential equation has a unique solution if we use the initial condition that $h(0) = \mathbb{I}$ is the identity matrix. The form of the equation suggests that $h(s)$ is related to the matrix exponential of the curve-integrated $A_i^I \sigma_I$, but this expression must be defined in a specific ordering because the integrand has non-commuting expressions at different points. The differential equation amounts to an implicit definition of the ordering.

Using parallel transport or holonomies to represent gravitational variables is distinguished by covariance under rotations: While K_i^I may be integrated over a curve in a coordinate-invariant manner, the integral $\int K_i^I t^i \mathrm{d}s$ is not covariant under position-dependent rotations. It is possible to make this expression fully covariant by introducing smearing functions f_I as used for fluxes, but the existence of holonomies implies a covariant integration without any additional input, other than a curve.

The combined $2+1$ integrations of a flux and a parallel transport or holonomy remove the 3-dimensional delta function from the basic Poisson bracket. The resulting Poisson bracket is non-zero only if the curve and the surface intersect. This observation suggests a discrete picture upon quantization because intersection numbers take integer values. A specific representation that demonstrates this result has been constructed in loop quantum gravity, based on a complete construction of a kinematical Hilbert space with a well-defined inner product for wave functions.

In a representation of wave functions depending on the connection, the expressions for parallel transport are multiplication operators that change the dependence along a discrete structure, or a network, determined by all the curves used in multiplication operations in order to construct a specific state. In this representation, parallel transport plays the role of ladder operators, while the ground state is given by a constant function on the infinite-dimensional space of all connections. Making mathematical sense of this infinite-dimensional space has been a success of loop quantum gravity.[9]

A specific orthonormal basis of the Hilbert space constructed in loop quantum gravity is given by **spin networks**, in which a spin-s representation is used on a spatial network link in order to determine the connection dependence through parallel transport. Basis states are therefore represented by networks in 3-dimensional space, labeled by half-integer quantum numbers on the links as well as additional information on vertices that connect intersecting links in a manner that is gauge invariant under triad rotations. The discreteness of spatial networks and of spin quantum numbers suggests a discretization of space in this approach. Related methods that work with generalizations of the spin-network basis are given by spin-foam models and group-field theory.[10] Note, however, that spin networks do not have direct physical meaning since they merely define a convenient choice of basis states. Also the appearance of spin is not related to physical rotations, but rather to triad rotations as gauge transformations that led us to Pauli matrices in expressions for parallel transport.

By applying fluxes and holonomies, loop quantum gravity has been able to construct a well-defined kinematical representation for canonical quantum gravity, including a kinematical Hilbert space with a well-defined inner product. The major remaining problem is finding a valid representation of the constraints (9.119) and (9.120) that respects the algebraic conditions (9.138). While the classical constraints are polynomial in extrinsic curvature or connection components, the basic kinematical operators of loop quantum gravity require exponentials of the connection. Thus, the constraints can be represented only if they are modified, or regularized, so as to depend on the connection in exponential form.

At the same time, the constraints must closely approximate their classical analogs when the arguments of exponential functions for parallel transport are small. This condition is related to low curvature, but in a non-covariant form that separates extrinsic and intrinsic curvature as determined by the connection. Consistent modifications are expected to be highly restricted by the condition that the resulting theory must respect formal covariance conditions, even while one does not expect space-time to retain its classical form after quantization. Formal covariance conditions are implied by a consistent implementation of gauge transformations, given in the present case by the hypersurface-deformation relation (9.138). The operators resulting from a regularized quantization of the constraints must obey a suitable commutator version of this classical relation.

Such a representation remains out of reach, and it is not known how one would solve the resulting operator equations if they could be found. It is not even clear whether a consistent representation of the constraints with holonomy modifications can exist in general. There are consistent examples in model systems, such as spherically symmetric space-times and some special cases of gravitational waves, but they have to obey a large number of non-trivial conditions. In the meantime, potential implications are being explored mainly by analyzing how the substitution of exponentials of the connection in the classical canonical equations may change properties of general relativity. One example is the boundedness of holonomies as functions of the connection, which resemble the bounded matrix coefficients of (4.242) and could be related to bounded curvature in space-time.

13.3 OUTLOOK

In spite of their different physical implications, relativity and quantum physics share many formal features. Both theories are geometrical, which is easy to see in the case of special and general relativity but more hidden in the case of quantum mechanics. Moreover, the geometry underlying

[9]A. Ashtekar, J. Lewandowski, D. Marolf, J. Mourao and T. Thiemann, Quantization of Diffeomorphism Invariant Theories of Connections with Local Degrees of Freedom, *J. Math. Phys.* **36** (1995) 6456, gr-qc/9504018

[10]A. Perez, Introduction to Loop Quantum Gravity and Spin Foams, gr-qc/0409061

quantum mechanics is less familiar and less intuitive than Riemannian geometry used in general relativity.

From a mathematical perspective, geometry can be defined in various ways. In a broad sense, geometry provides means to characterize curves and surfaces beyond a basic definition as subsets of a space or space-time. Such characterizations need not be length measures as provided by a metric, and may instead include the specification of distinguished classes of curves, such as geodesics. On a more formal level, a humble but important ingredient in constructions of Riemannian geometry is the ability to raise and lower indices. Without this operation, we would not be able to turn partial derivatives of functions or tensor components, where $\partial/\partial x^\alpha$ counts as a subscript, into directions that could play the role of a tangent vector field of a distinguished curve. The existence of a tensor calculus in which it is possible to raise and lower indices is a basic requirement of geometry, and it can be obtained not only from a metric tensor.

Mathematically, the operation of raising and lowering indices is a map from the tangent space V at a given point to its **dual vector space** V^*, defined as the set of linear maps from V to real or complex numbers: If we lower the index of a vector in V, mapping t^α to t_β, the result defines a covariant linear map $t(v) = t_\beta v^\beta$ from vectors to numbers. Therefore, t^α represents an element of V, while t_β represents an element of V^*. In Riemannian geometry, we do not distinguish between the letter t assigned to both t^α and t_β because the metric determines a vector-space isomorphism from V to V^*. But from a mathematical perspective, V and V^* have independent definitions, and they are unrelated if we do not have a metric tensor or an alternative geometrical object.

A one-to-one map from t^α to t_β requires an invertible matrix $m_{\alpha\beta}$ such that $t_\beta = m_{\beta\alpha}t^\alpha$, and therefore $t^\alpha = m^{\alpha\beta}t_\beta$ with $m^{\alpha\beta}m_{\beta\gamma} = \delta^\alpha{}_\gamma$. The metric is a symmetric version of $m_{\alpha\beta}$, but symmetry is not required for invertibility. Since any matrix $m_{\alpha\beta}$ can be written as the sum of a symmetric and an antisymmetric matrix,

$$m_{\alpha\beta} = \frac{1}{2}(m_{\alpha\beta} + m_{\beta\alpha}) + \frac{1}{2}(m_{\alpha\beta} - m_{\beta\alpha}), \tag{13.48}$$

the alternative to a symmetric matrix $g_{\alpha\beta}$ as a means to raising and lowering indices is an antisymmetric invertible matric $\Omega_{\alpha\beta}$, such that $\Omega_{\alpha\beta} = -\Omega_{\beta\alpha}$.

The geometry defined by an antisymmetric invertible 2-tensor $\Omega_{\alpha\beta}$ that also obeys the differential equation

$$\partial_\alpha\Omega_{\beta\gamma} + \partial_\beta\Omega_{\gamma\alpha} + \partial_\gamma\Omega_{\alpha\beta} = 0 \tag{13.49}$$

is called **symplectic geometry**, and $\Omega_{\alpha\beta}$ is its **symplectic form**. The two conditions characterize $\Omega_{\alpha\beta}$ as a closed 2-form in the language of differential forms. Antisymmetry of an invertible real matrix implies that all eigenvalues are imaginary and come in pairs $\lambda_\pm = \pm i\mu$ with real numbers $\mu \neq 0$. The determinant of $\Omega_{\alpha\beta}$ is therefore a positive real number. The pairwise occurrence of eigenvalues implies that a space equipped with symplectic geometry must have an even dimension. In symplectic geometry, all vectors are "lightlike" because $\Omega_{\alpha\beta}v^\alpha v^\beta = 0$ due to antisymmetry. Every vector is therefore orthogonal to itself, where two vectors v^α and w^β are (symplectically) orthogonal if $\Omega_{\alpha\beta}v^\alpha w^\beta = 0$. Since $\Omega_{\alpha\beta}$ is required to be invertible as a matrix, it has a non-zero determinant $\det(\Omega)$ that defines the invariant volume element $\sqrt{\det(\Omega)}\,\mathrm{d}^n x$ with coordinates x^α, just like $\sqrt{|\det(g)|}\,\mathrm{d}^n x$ in Riemannian geometry. Integrating the volume element over a region determines its size.

Given a hypersurface with coordinates y^i, embedded in a space with symplectic geometry by functions $x^\alpha(y^i)$, we define the **presymplectic form**

$$\pi_{ij} = \frac{\partial x^\alpha}{\partial y^i}\frac{\partial x^\beta}{\partial y^j}\Omega_{\alpha\beta} \tag{13.50}$$

on the hypersurface. Like $\Omega_{\alpha\beta}$, π_{ij} is antisymmetric. However, it is not guaranteed to be invertible and therefore does not necessarily define a symplectic form. If π_{ij} is invertible for a given hypersurface, the integral of its volume form $\sqrt{\det(\pi)}\,\mathrm{d}^m y$ defines the area.

A function H on a symplectic space, also called a **phase space**, determines a dual vector field $\partial_\alpha H$. Using the inverse $\Omega^{\beta\alpha}$ to raise the index, we obtain the **Hamiltonian vector field**

$$X_H^\beta = \Omega^{\beta\alpha} \frac{\partial H}{\partial x^\alpha} \tag{13.51}$$

of H. The inverse $\Pi^{\alpha\beta} = \Omega^{\alpha\beta}$ is called the **Poisson tensor** of the symplectic space and defines a **Poisson bracket**,

$$\{f,g\} = \Pi^{\alpha\beta} \frac{\partial f}{\partial x^\alpha} \frac{\partial g}{\partial x^\beta} \tag{13.52}$$

for functions f and g on the symplectic space. The directional derivative of a function f along the Hamiltonian vector field equals

$$X_H^\beta \partial_\beta f = \left(\Pi^{\beta\alpha} \frac{\partial H}{\partial x^\alpha}\right) \frac{\partial f}{\partial x^\beta} = \{f,H\}. \tag{13.53}$$

If H is the Hamilton function of a classical-mechanics system, the Hamiltonian vector field represents time evolution and we obtain Hamilton's equations from the Poisson bracket.

Using $\Pi^{\alpha\beta}\Omega_{\beta\gamma} = \delta^\alpha{}_\gamma$, the condition (13.49) can be shown to imply

$$\Pi^{\alpha\delta}\partial_\delta \Pi^{\beta\gamma} + \Pi^{\beta\delta}\partial_\delta \Pi^{\gamma\alpha} + \Pi^{\gamma\delta}\partial_\delta \Pi^{\alpha\beta} = 0 \tag{13.54}$$

which is equivalent to the **Jacobi identity**

$$\{f,\{g,h\}\} + \{g,\{h,f\}\} + \{h,\{f,g\}\} = 0 \tag{13.55}$$

for the Poisson bracket. With this identity, the Poisson bracket has a close relationship with the vector-field commutator (9.37) because

$$X_{\{f,g\}}^\alpha = -[X_f, X_g]^\alpha. \tag{13.56}$$

The map from functions f to Hamiltonian vector fields X_f can be interpreted as a **representation** of the Poisson bracket. In quantum mechanics, we are looking for alternative representations of the **basic Poisson bracket** defined for a restricted set of basic degrees of freedom, such as x and p in 1-dimensional quantum mechanics. The Poisson bracket $\{x,p\} = 1$ on the 3-dimensional function space given by linear combinations of 1, x and p is represented by operators $\hat{\mathbb{I}}$, $\hat{x}$ and $\hat{p}$ on a Hilbert space. The resulting algebra is the Heisenberg algebra with commutators (3.232). In contrast to the representation by Hamiltonian vector fields, there is a theorem attributed to Hilbrand Groenewold and Leon van Hove which shows that it is impossible to represent all functions f of x and p as operators such that $[\hat{f},\hat{g}] = \hat{h}$ whenever $h = \{f,g\}$. The Groenewold–van Hove theorem implies that the Poisson bracket, or the Poisson tensor, receives $\hbar$-dependent quantum corrections, just like a Hamilton operator.

An extended or **quantum Poisson bracket** for expectation values $\langle\hat{A}\rangle$, interpreted as functions on a space of operators, is defined by

$$\{\langle\hat{A}\rangle, \langle\hat{B}\rangle\} = \frac{\langle[\hat{A},\hat{B}]\rangle}{i\hbar}. \tag{13.57}$$

Using the product rule, this expression can be generalized to polynomials in expectation values, such as variances or higher moments. Since the commutator is antisymmetric and always obeys the Jacobi identity (4.40), (13.57) fulfills all the requirements on a Poisson bracket. Schrödinger evolution, via Heisenberg's equations (3.230), is equivalent to Hamilton's equations

$$\frac{\mathrm{d}\langle\hat{A}\rangle}{\mathrm{d}t} = \{\langle\hat{A}\rangle, \langle\hat{H}\rangle\}\,. \tag{13.58}$$

The existence of this quantum Poisson bracket shows that quantum mechanics is geometrical, with fundamental structures defined by a Poisson tensor.

For basic operators $\hat{x}$ and $\hat{p}$, we reproduce the classical Poisson bracket

$$\{\langle\hat{x}\rangle, \langle\hat{p}\rangle\} = \frac{\langle[\hat{x},\hat{p}]\rangle}{i\hbar} = 1 \tag{13.59}$$

However, polynomials in x and p are in general subject to ordering ambiguities: For instance, we may quantize the real function x^2p^2 as the symmetric operators $\frac{1}{2}(\hat{x}^2\hat{p}^2+\hat{p}^2\hat{x}^2)$, $\hat{x}\hat{p}^2\hat{x}$, $\hat{p}\hat{x}^2\hat{p}$, or linear combinations of these choices. These expressions differ by $\hbar$-dependent terms, as in

$$\frac{1}{2}(\hat{x}^2\hat{p}^2+\hat{p}^2\hat{x}^2) = \frac{1}{2}(\hat{x}\hat{p}^2\hat{x}+\hat{x}[\hat{x},\hat{p}^2]+\hat{x}\hat{p}^2\hat{x}+[\hat{p}^2,\hat{x}]\hat{x}) = \hat{x}\hat{p}^2\hat{x}+i\hbar[\hat{x},\hat{p}] = \hat{x}\hat{p}^2\hat{x}-\hbar^2\,. \tag{13.60}$$

Depending on how we choose to quantize the function $h(x,p)=9x^2p^2$ which equals $h=\{f,g\}$ with $f(x)=x^3$ and $g(p)=p^3$, the quantum Poisson bracket

$$\{\langle\hat{x}^3\rangle, \langle\hat{p}^3\rangle\} = \frac{\langle[\hat{x}^3,\hat{p}^3]\rangle}{i\hbar} = 3\langle\hat{x}^2\hat{p}^2+\hat{x}\hat{p}^2\hat{x}+\hat{p}^2\hat{x}^2\rangle \tag{13.61}$$

receives different corrections proportional to $\hbar^2$. For instance, if $\hat{h}=\frac{9}{2}(\hat{x}^2\hat{p}^2+\hat{p}^2\hat{x}^2)$ we obtain $\{\langle\hat{f}\rangle,\langle\hat{g}\rangle\}=\langle\hat{h}\rangle+3\hbar^2$, while $\{\langle\hat{f}\rangle,\langle\hat{g}\rangle\}=\langle\hat{h}\rangle-6\hbar^2$ if $\hat{h}=9\hat{x}\hat{p}^2\hat{x}$.

In general terms, quantum mechanics is a representation problem in which we look for Hilbert-space versions of the basic Poisson bracket of a classical system. In many cases, such representations are not unique even for a given choice of basic degrees of freedom, and the choice of basic expressions is far from unique too. In classical mechanics, we can make arbitrary **canonical transformations** on phase space, mapping (x,p) to $(x'(x,p),p'(x,p))$ such that

$$\{x'(x,p),p'(x,p)\} = \frac{\partial x'}{\partial x}\frac{\partial p'}{\partial p} - \frac{\partial x'}{\partial p}\frac{\partial p'}{\partial x} = 1\,. \tag{13.62}$$

Hamilton's equations in the new coordinates are then equivalent to Hamilton's equations in the old coordinates. But if the transformation is non-linear, such as $x'=\frac{1}{2}x^2$ and $p'=p/x$, Hilbert-space representations of the old and the new coordinates are in general inequivalent, for instance (again) owing to ordering ambiguities. The choice of representation then implies physical consequences for predicted measurement results, such that a valid representation can be established by using observations.

In quantum gravity, we have to combine the geometrical structures of quantum mechanics and general relativity. Just as we turn the classical Poisson bracket into a quantum Poisson bracket or the commutator of operators, we should find a suitable operator version of the symmetric bracket, or the scalar product $(v,w)=g_{\alpha\beta}v^\alpha w^\beta$, defined by Riemannian geometry. We then need a feasible solution procedure for the resulting constraint operators in terms of a quantum metric, as well as geometrical or physical ways of interpreting states and expectation values.

This requirement presents a challenging problem of observables because the familiar physical interpretations of space-time solutions in classical general relativity are based on expressions derived in specific coordinate systems. Predictions for observables are coordinate invariant, even if they are related to the frame of a specific observer, but their derivations would be tedious if they were performed using coordinate-independent expressions in every single step. Using coordinates in quantum gravity would require a generalization of the classical scalar product (v, w) to a quantum scalar product on an extended space-time, just as we generalized the classical Poisson bracket to a quantum Poisson bracket. However, even the quantum Poisson bracket depends on the choice of basic degrees of freedom, which is likely the case also for the more complicated construction of a quantum scalar product. If this is the case, we cannot expect to maintain full covariance under non-linear coordinate transformations.

At this point, the various approaches to quantum gravity make different assumptions about what might replace the classical version of coordinate-independent observables. Background-dependent approaches such as perturbative quantum gravity, asymptotically safe quantum gravity, or string theory use a background space-time, often in a preferred coordinate system, in order to derive physical results for propagation of quantum space-time excitations. Discrete approaches such as causal dynamical triangulations, causal sets, or loop quantum gravity propose alternative formulations of space-time and potential observables. None of these approaches can be considered complete in their description of covariant quantum space-times independent of assumed auxiliary structures, or in the reliability of their physical predictions. Compared with quantum mechanics, the representation problem is much more challenging, and there are, for now, no clear observational results that could help us distinguish between different choices. Observations are needed to guide theoretical developments, but at the same time, theoretical predictions are required in order to know where we should look for observable implications. At the present stage of developments, it is therefore important to develop all the approaches as much as possible until reliable suggestions can be made for observational tests.

Index

For Product Safety Concerns and Information please contact our EU
representative GPSR@taylorandfrancis.com
Taylor & Francis Verlag GmbH, Kaufingerstraße 24, 80331 München, Germany

www.ingramcontent.com/pod-product-compliance
Ingram Content Group UK Ltd.
Pitfield, Milton Keynes, MK11 3LW, UK
UKHW052218190926
475702UK00006B/1072
* 9 7 8 1 0 3 2 9 6 2 3 4 4 *